HIGH INTENSITY AND HIGH BRIGHTNESS HADRON BEAMS

Related Titles from AIP Conference Proceedings and the Subseries on Accelerators, Beams, and Instrumentation

648 Beam Instrumentation Workshop 2002: Tenth Workshop
Edited by Gary Smith and Thomas Russo, December 2002, 0-7354-0103-9

647 Advanced Accelerator Concepts: Tenth Workshop
Edited by Christopher Clayton and Patrick Muggli, December 2002, 0-7354-0102-0

639 Production and Neutralization of Negative Ions and Beams: Ninth International Symposium on the Production and Neutralization of Negative Ions and Beams
Edited by Martin P. Stockli, November 2002, 0-7354-0094-6

592 High Quality Beams: Joint US-CERN-JAPAN-RUSSIA Accelerator School
Edited by S. I. Kurokawa, S. Y. Lee, J. Miles, E. A. Perevedentsev, November 2001, 0-7354-0034-2

581 Physics of, and Science with, the X-Ray Free-Electron Laser: 19th Advanced ICFA Beam Dynamics Workshop
Edited by C. Pellegini, S. Chattopadhyay, M. Cornacchia, and I. Lindau, August 2001, 0-7354-0022-9

569 Advanced Accelerator Concepts: Ninth Workshop
Edited by Patrick L. Colestock and Sandra Kelley, June 2001, CD-ROM included, 0-7354-0005-9

546 Beam Instrumentation Workshop 2000: Ninth Workshop
Edited by Kenneth D. Jacobs and R. Coles Sibley III, December 2000, 1-56396-975-0

536 Instrumentation in Elementary Particle Physics: VIII ICFA School
Edited by Sehban Kartal, September 2000, 1-56396-960-2

496 Workshop on Instabilities of High Intensity Hadron Beams in Rings
Edited by T. Roser and S. Y. Zhang, December 1999, 1-56396-910-6

480 Space Charge Dominated Beam Physics for Heavy Ion Fusion
Edited by Yuri K. Batygin, June 1999, 1-56396-860-6

468 Nonlinear and Collective Phenomena in Beam Physics — 1998 Workshop: International Committee on Future Accelerators
Edited by Swapan Chattopadhyay, Max Cornacchia, and Claudio Pellegrini, April 1999, 1-56396-862-2

422 Instrumentation in Elementary Particle Physics: The VII ICFA School
Edited by G. Herrera Corral and M. Sosa Aquino, March 1998, 1-56396-763-4

To learn more about these titles, or the AIP Conference Proceedings Series, please visit the webpage **http://proceedings.aip.org/proceedings**

HIGH INTENSITY AND HIGH BRIGHTNESS HADRON BEAMS

20th ICFA International Beam Dynamics Workshop on High Intensity and High Brightness Hadron Beams
ICFA-HB2002

Batavia, Illinois *8-12 April 2002*

Sponsored by:

Fermilab

ICFA

KEK

EDITORS

Weiren Chou
Fermi National Accelerator Laboratory
Batavia, Illinois

Yoshiharu Mori
High Energy Accelerator Research Organization (KEK)
Ibaraki, Japan

David Neuffer
Jean-François Ostiguy
Fermi National Accelerator Laboratory
Batavia, Illinois

CD-ROM INCLUDED

AMERICAN INSTITUTE OF PHYSICS

Melville, New York, 2002
AIP CONFERENCE PROCEEDINGS ■ VOLUME 642

Editors:

Weiren Chou
Fermi National Accelerator Laboratory
P.O. Box 500
Batavia, IL 60510-0500
USA
E-mail: chou@fnal.gov

Yoshiharu Mori
High Energy Accelerator Research Organization (KEK)
Oho 1-1
Tsukaba, Ibaraki 305-0801
JAPAN
E-mail: yoshiharu.mori@kek.jp

David Neuffer
Jean-François Ostiguy
Fermi National Accelerator Laboratory
P.O. Box 500
Batavia, IL 60510-0500
USA
E-mail: neuffer@fnal.gov
ostiguy@fnal.gov

L.C. Catalog Card No. 2002113720
ISBN 0-7354-0097-0
ISSN 0094-243X

Printed in the United States of America

CONTENTS

PLENARY SESSIONS: PHYSICS AND MACHINES OVERVIEW

WORKING GROUP I: CIRCULAR ACCELERATORS

WORKING GROUP II: LINEAR ACCELERATORS

WORKING GROUP III: BEAMLINES AND TARGETS

SESSION A: LATTICES

SESSION B: BEAM LOSS, COLLIMATION, AND SHIELDING

JOINT SESSION C AND G: FFAGs AND CYCLOTRONS

SESSION D: NEW IDEAS

SESSION E: SPACE CHARGE SIMULATIONS

SESSION F: REMOTE HANDLING

SESSION H: ION SOURCES

SESSION I: SPACE CHARGE EXPERIMENTS

SESSION J: HADRON BEAM COOLING

SESSION K: H^- STRIPPING

SESSION L: ELECTRON CLOUD EFFECTS

Preface

The 20th ICFA Advanced Beam Dynamics Workshop took place at Fermilab (Batavia, Illinois, USA) on April 8-12, 2002. Fermilab and KEK co-sponsored the Workshop. The theme of the Workshop was "High-Intensity and High-Brightness Hadron Beams." This was the first ICFA workshop fully devoted to this subject. A broad range of topics was discussed, including reviews of the performance of existing high-intensity hadron machines, overviews of planned high-intensity hadron sources and projects, accelerator physics issues, technical systems designs, and applications of these beams in high-energy and nuclear physics as well as in fusion studies, medicine, industrial uses, and other fields.

The Workshop was held at an opportune time. Several hadron accelerators recently have been completed, several are under construction, and many more are planned or projected to meet future needs. Recently completed accelerators include the AGS Booster and the Fermilab Main Injector. Both of these are being developed to obtain high-intensity high-brightness hadron beams. The Japan Hadron Facility (JHF) at JAERI/KEK and the Spallation Neutron Source (SNS) at Oak Ridge National Laboratory are under construction and will increase medium-energy proton beams and neutron source intensities by an order of magnitude. The Large Hadron Collider (LHC) at CERN will use high-brightness proton beams provided by the CERN PS and SPS to explore the energy frontier. Future projects such as the Proton Driver, CERN SPL, AHF, ESS, Radioactive Ion Beams, etc. are also being developed.

The paramount advance in recent high-energy physics research is the convincing evidence for neutrino oscillations and neutrino mass. High-intensity hadron beams are needed to obtain the high-intensity and high-quality neutrino beams required to determine the parameters in the multi-neutrino system. Existing hadron beams will be used at maximum intensities in new beam lines (NuMI, MiniBooNE, K2K, and CNGS experiments) to explore neutrino properties. Higher intensities using new facilities (JHF, SPL, Proton Driver, upgraded Main Injector, neutrino factories, etc.) will also be needed. It was emphasized at the workshop that the U.S. high-energy physics community must prepare a proposal for a mid-size accelerator project (a proton driver) in addition to the large-size linear collider project in order to guarantee a future in the particle physics frontiers for this community.

The Workshop had about 160 participants and included about 140 presentations on various aspects of high-intensity and high-brightness hadron beams. Reports on the great majority of these are included in these proceedings. There were 4 plenary sessions, 3 working groups and 11 parallel sessions. A guided tour at Fermilab was also provided on the last day. A complete program can be found at the end of the proceedings.

The first two plenary sessions covered topics of general interest, including the physics based on high-intensity secondary particles obtained from intense proton sources (muon, neutrino, kaon, neutron, pion, etc.), and an overview of machines under construction (JHF and SNS) or being proposed (synchrotron-based proton drivers, linac-based proton drivers, and an 8-GeV superconducting proton linac). The working groups discussed various aspects concerning circular accelerators, linear accelerators, beam lines and targets. Each of the parallel sessions focused on a specific topic. The three working groups and eleven parallel sessions and their conveners are listed below:

Working Groups	Conveners
WG I - Circular Accelerators	L. Teng (ANL), R. Cappi (CERN)
WG II - Linear Accelerators	J-M. Lagniel (CEA-Saclay), R. Garoby (CERN)
WG III - Beamlines and Targets	C. Moore (Fermilab), K. McDonald (Princeton U.)

Parallel Sessions	Conveners
A. Lattices	A. Thiessen (LANL)
B. Beam Loss, Collimation, and Shielding	N. Mokhov (Fermilab)
C and G (Joint) FFAGs and Cyclotrons	R. Baartman (Triumf), Y. Mori (KEK),

	P. Schmelzbach (PSI)
D. New Ideas	K. Takayama (KEK), C. Ankenbrandt (Fermilab)
E. Space Charge Simulations	I. Hofmann (GSI), R. Ryne (LBL)
F. Remote Handling	D. Pushka (Fermilab)
H. Ion Sources	H. Klein (U. of Frankfurt), K-N. Leung (LBL)
I. Space Charge Experiments	S. Machida (KEK)
J. Hadron Beam Cooling	S. Nagaitsev (Fermilab)
K. H^- Stripping	I. Yamane (KEK)
L. Electron Cloud Effects	R. Macek (LANL)

At the Workshop were comprehensive discussions on critical features and problems of high-intensity and high-brightness hadron beams, including those of past, present, and future facilities. The final two plenary sessions included reports from the neutrino physics study groups at Fermilab and BNL. Both of these studies were initiated by the lab directors. There was also a report on high brightness ion beams for medicine, industry and fusion from LBL. The conveners of each working group and parallel session gave summary reports. S. Holmes, Associate Director of Fermilab for Accelerators, presented closing remarks.

The Workshop was a highly productive event for the organizers and the participants. It strengthened the communication and collaboration between accelerator physicists and high-energy physicists as well as among accelerator physicists working in the same field but at different institutions. We take this opportunity to thank the Local Organizing Committee, the Program Committee, the International Advisory Committee and the working group organizers and session conveners for their help with the Workshop. We would especially like to thank Patti Poole, Cynthia Sazama and Suzanne Weber for their untiring efforts in organizing the meeting, and to Ernie Malamud for preparing the workshop web site (http://www-bd.fnal.gov/HB2002). Viewgraphs of most of the presentations can be found on this web site, including the very few that are not represented by proceedings papers.

In view of the worldwide strong interest in high-intensity high-brightness hadron beams, it was decided to have another ICFA workshop on this subject around 2004-2005 in Europe. The organizers will be Ingo Hofmann (GSI) and Jean-Michel Lagniel (CEA-Saclay).

Weiren Chou, Yoshiharu Mori, David Neuffer, and Jean-François Ostiguy
Editors

Workshop Co-Chairmen

Weiren Chou	Fermilab
Yoshiharu Mori	KEK

International Advisory Committee

Yanglai Cho	ANL
Shouxian Fang	IHEP/Beijing
Ingo Hofmann	GSI
Steve Holmes	Fermilab
John Jowett	CERN
Thomas Kirk	BNL
Paul Lisowski	LANL
Shoji Nagamiya	KEK
Satoshi Ozaki	BNL
Robert Palmer	BNL
Vasily Parkhomchuk	BINP
Ewan Paterson	SLAC
Grahame Rees	RAL
Andrew Sessler	LBL
Erich Steiner	PSI
Lee Teng	ANL
Maury Tigner	Cornell U.
Albrecht Wagner	DESY
Carlo Wyss	CERN
Yoshishige Yamazaki	KEK
ICFA Beam Dynamics Panel	

Program Committee

Roberto Cappi	CERN
Byung-Ho Choi	KAERI
Ronald Davidson	PPPL
Gerardo Dutto	TRIUMF
John Galambos	ORNL
Roland Garoby	CERN
Xialing Guan	IHEP/Beijing
Jean-Michel Lagniel	CEA-Saclay
S.Y. Lee	Indiana U.
Ka-Ngo Leung	LBL
Trevor Linnecar	CERN
Robert Macek	LANL
Shinji Machida	KEK
Ernie Malamud	Fermilab
John Marriner	Fermilab
Steve Milton	ANL
Michiko Minty	DESY
Chris Prior	RAL
Thomas Roser	BNL
Gennady Stupakov	SLAC
Mike Syphers	Fermilab
Ken Takayama	KEK
Arch Thiessen	LANL
Jie Wei	BNL
Hideaki Yokomizo	JAERI

Local Organizing Committee

Jeffrey A. Appel
Weiren Chou (Chair)
Kiyomi Koba
David Neuffer
Jean-François Ostiguy
Patti Poole
Zubao Qian
Cynthia Sazama
Suzanne Weber

PLENARY SESSIONS

PHYSICS AND MACHINES OVERVIEW

Introductory Remarks: Reflections on the Reasons for Building Accelerators

Andrew M. Sessler

Lawrence Berkeley Laboratory Berkeley, CA 94720

INTRODUCTION

This is a good occasion, just when Fermilab is considering whether or not to build a new machine, and just when KEK/JAERI has initiated a new machine, to look back in history and review the reasons given – even believed by the proponents – for constructing a new accelerator. And then, of course, look at what really happened. Now there is a fair amount of history, more than 80 years of it, as accelerators go back to the 20's. Perhaps it is useful to categorize accelerators into groups. The first might be Energy Frontier Machines. Now that, at first sight, might not be relevant, but it is. For a driver can be used to make neutrino beams of high energy and, of course, become the driver for a neutrino factory, and some day, perhaps, for a muon collider. Plenty of frontier things. The second category might be Special Purpose Machines, such as high intensity machines. Many of these are not in high energy physics, such as spallation sources, cyclotrons for nuclear physics, and medical accelerators. The relevance to this Workshop is obvious. A third category might be Machines for Advancing the Accelerator Art, where, once again, the relevance is obvious.

There are machines built for political reasons. This is not a disparagement at all, for some of our most productive accelerators were built primarily for political reasons, such as building — or even maintaining — infra-structure in a country or a laboratory. I will mention the political/sociological motives as we go along through the three categories that are inclusive of all accelerators.

Now, there have been a great many accelerators, even in high energy physics, so I can't be exhaustive here (I may be in some other venue). So I shall just pick a few to give you what I hope will be a balanced historical view.

ACCELERATORS ON THE ENERGY FRONTIER

Consider, first, Accelerators on the Energy Frontier. The very earliest accelerators were, of course, in this category. I am thinking of the electrostatic machines that were constructed for, and did achieve, artificial nuclear reactions. Then, of course, there were, in the 30's, the cyclotrons. (And that was it, electrostatic machines and cyclotrons, until after WWII !) The cyclotrons were built with no deeper thought, as far as I can see, except to get to ever-higher energy. But the intuition was right: no one dreamed of the wealth of physics they would be opened up: nuclear isotopes, the discovery of tritium, and then all the transuranics (including plutonium and all that followed in reactors and bombs), nuclear medicine, and, then, just after WWII, the production of pions.

The Cosmotron was the first accelerator built at Brookhaven, simply to get to even higher energies than previously achieved. It rather quickly led TD Lee and CN Yang to interpret the data on weak decays as violations of parity; a result no one envisioned when the device was initiated (Nobel Prize).

The Bevatron was a political deal (the proposal was one page long) as was the AGS. Something for Berkeley on the West Coast and something for Brookhaven on the East Coast; in those days the AEC didn't see the middle of the country. The Bevatron was built to discover, and it did, the anti-proton, but its real contribution was the discovery of "resonances", which weren't even dreamed of when it was proposed and constructed (Two Nobel Prizes). The AGS was used to discover two neutrinos, time reversal non-invariance, and the J/Psi particle, three Nobel Prizes and three phenomena not even dreamed of when the AGS was constructed.

Electron linacs were developed after WWII and electron scattering off nuclei allowed the first precision determination of the nuclear radius and charge distribution. Subsequently, and in good measure because of this success, SLAC was constructed, and the most important discovery on that machine was the disclosure of the parton structure of the nucleon, a completely unexpected result (Nobel Prize). Electron storage rings were built to study QED at ever-higher energies, i.e., at ever-smaller distances. But no one expected the J/Psi particle and the November revolution (Nobel Prize). I could tell you about some more modern machines, but I believe you can see, already, that the really interesting frontier ma-

CP642, *High Intensity and High Brightness Hadron Beams: 20th ICFA Advanced Beam Dynamics Workshop on High Intensity and High Brightness Hadron Beams,* edited by W. Chou, Y. Mori, D. Neuffer, and J.-F. Ostiguy
2002 American Institute of Physics 0-7354-0097-0

chines were machines with which unexpected phenomena were disclosed. Of course it didn't always turn out that frontier machines produced unexpected phenomena. I can think of PEP, PETRA, Tristan, all built in the enthusiasm following the November revolution and all producing minimal physics. Some machines were built for strong political reasons, such as the PS to bring Europe into high-energy physics, or the Beijing ring for China, or the proton accelerator at KEK for Japan. Essentially all served the designated purpose, and, as such, were very important to the sociology of high-energy physics; some also produced very important physics.

I believe one can be rather confident that a driver at Fermilab, or the driver underway at KEK/JAERI will be employed to study frontier physics. Whether, or not, significant discoveries will be made with these drivers is, of course, beyond me, but with the first signs of physics beyond the Standard model, now in hand, it would take a very pessimistic person to believe that little will come of these machines.

SPECIAL PURPOSE MACHINES

Turning our attention to Special Purpose Machines, the largest number of these is, of course, those produced for medical purposes; both the production of isotopes for diagnostic purposes and the therapy of cancer. One third of the patients in hospitals have a procedure involving nuclear isotopes, which, often, are made by accelerators. In regard to therapy, 1.75 M patients a year in the US receive accelerator treatments. Counting just electron machines, there are more than 3000, which at about 1 M$ apiece, is an investment of 3 B$. This is the major spin-off of the accelerator art to the economy. Besides these two accelerator spin-offs of high-energy physics, there are the large superconducting magnets used in MRI. These three items are, perhaps, the major spin-offs of high-energy physics. Then, of course, if we go up in energy, there are the electron synchrotrons. The last I looked there were some 70 around the world. The big ones are the ESRF in Europe, the APS in the US, and Spring 8 in Japan. Major investments; the APS must be about 800 M$. The impact on society is usually indirect - not quite so direct as treating a cancer patient - but very real in the long term. Then there are the spallation neutron sources, the small ones that already operate, and the very large ones under construction in the US and in Japan (with Europe probably to follow). The SNS is, to set the scale, a 1.3 B$ device. Still going up in energy there are the B-Factories. Certainly built with a very clear, and well-defined, purpose in mind. These devices, both the ones at KEK and at SLAC, have performed even better than promised and are well on the way to making significant discoveries about time reversal non-invariance. In sum, one sees that the machines built for a special purpose have, in all cases that I am aware of, worked as expected. I believe that also will be the case for the JHF and the proposed booster here.

MACHINES FOR ADVANCING THE ACCELERATOR ART

All new concepts, when first developed experimentally went through this phase. In the old days it was easy. The first cyclotron was 5 inches across. The first test of the phase focusing concept was on a 300 MeV electron synchrotron, the first linacs were small devices, the first test of AG was on a Cornell electron synchrotron.

Three rather large tests come to mind and illustrate the need – or the perceived need at the time – for extensive (and expensive) testing. The first was the concept of colliding proton beams. The theory was developed, primarily here in the Midwest, various aspects (such as rf stacking) were tested in the Midwest, and then the concept was taken seriously by CERN and the long-term stability was tested on a very low energy electron ring. But complete test required the ISR, which was probably, the most expensive accelerator development device ever made. Very little physics was done on the ISR (in part because the interesting events were at 90 degrees, not forward, and only in the last experiments were physicists even looking at such events). The second was the concept of stochastic cooling, which didn't require a high-energy machine for testing. The testing was done on ICE, which was, in itself, non-trivial in size and cost, and very complex indeed. The resulting physics, the discovery of the Z and the W, were great events. The third example is linear colliders of electrons. Here a very large device, the SLC, was built and I remember how it was greeted by skepticism on the part of the high-energy community (as was also true for the other two examples). In fact the SLC was originally only an accelerator project, but later Z-physics was done on the device (as well as, of course, on LEP).

The accelerator art has only been able to move forward, in very significant steps, with major tests. Sure, many advances were made in bench studies, or relatively minor studies (such as the pretzel scheme studies at Cornell, or RFQs). For muons we are going to require rather major tests, such as the targetry work at Brookhaven, MUCOOL here, and the International Cooling Experiment, MICE. Eventually, however, as with the ISR, it all has to be put together and having a driver at Fermilab, or using he driver at KEK/JAERI, will be the necessary front end for a neutrino factory, and eventually, a muon collider. I am confident that the accelerator art will continue to take bold steps; for if it doesn't that will mean the end of high-energy physics, at least as we have known it for the last 80 years, and I don't think any of us are envisioning that. But we need more than "vision"; we need action to prevent the demise of high-energy physics.

REPRIEVE

In this session, and really for most of this week, we shall look in detail, and explore, the accelerator aspects of high intensity and high brightness hadron beams. The technical issues are many-fold and quite fascinating. I hope the result of all these discussions of multitudinous effects at a variety of machines will be renewed enthusiasm for moving ahead here at Fermilab. For there are good reasons for moving ahead; at least as good as the reasons given in the past for constructing new machines. And, think, without that past optimism we would be in a very different place today.

JHF Physics

J. Imazato

Institute of Particle and Nuclear Studies, High Energy Accelerator Research Organization (KEK)
Oho 1-1, Tskuba-shi, Ibaraki-ken, 305-0801 Japan

Abstract. The construction of a high-intensity proton accelerator complex has started in Japan. It will provide a variety of beams such as neutrons, muons, kaons, and neutrinos *etc.* Diverse physics and applied physics topics will be addressed here, with the first beams scheduled for the year 2007. In this talk, the most important physics objectives and a few typical examples of experiments will be presented.

INTRODUCTION

A high-intensity proton accelerator complex is now being built in Japan to extend the horizons of hadron beam sciences in the 21st century [1]. The accelerator complex (so far called JHF [1]) consists of a 600-MeV linac, a 3-GeV rapid cycling synchrotron (RCS), a 50-GeV synchrotron (PS), and several experimental facilities. It is a joint project of Japan Atomic Energy Research Institute (JAERI) and High Energy Accelerator Research Organization (KEK). The construction has started last year and it is expected to be completed by the end of Japanese fiscal year 2006 and to provide first beam in 2007. The configuration of the accelerator complex is shown in Fig.1. The facility is being built in the Tokai site of JAERI on the Pacific Ocean coast in the neighborhood of already existing research reactors. The present activities of the 12-GeV PS and the neutron spallation source at KEK will move there.

The JHF is designed to be an intensity frontier facility and a variety of secondary beams will be produced there. The RCS with 1 MW of power will be used for a spallation neutron source and muon beams for materials and life sciences, by means of neutron diffraction/scattering and μSR techniques, respectively. The 50-GeV PS beams with a 15 μA current, corresponding to 0.76 MW of power, will lend high fluxes to secondary meson beams such as kaons and pions for particle and nuclear physics. A Neutrino beam will be also one of the most important beams. The present world's proton machines have typically 0.1 MW level of power both for low-energy and high-energy machines (see Fig.3) with one exception of the PSI cyclotron. Our RCS and 50-GeV PS will attain ten times higher level of 1MW together with the SNS machine under construction. Details of the accelerator are presented elsewhere in this workshop [3].

[1] An offi cial name of the accelerator facility will be given soon.

PARTICLE PHYSICS

Experimental facility

The physics potential of high-intensity machines has been discussed previously [2]. There will be two extracted beams from the 50-GeV PS: one is the slow extraction beam to be used for counter experiments and the other is the fast extraction beam used for neutrino beam production. Fig.2 shows the present design of an experimental hall for the slow beam with 3 primary beam lines and 3 or 4 target stations from which secondary lines are extracted. The expected beam intensity for a few typical secondary lines are compared in Fig.3 with the present machines. We will reap the benefits of beams of about 100 times more intense than at KEK 12-GeV PS and about 3-4 times as those of BNL-AGS.

Kaon decays

We may characterize the particle physics at JHF as the high-precision frontier physics using high-intensity beams. One of the major fields of interest is the rare kaon decay. Here, one tests the standard model (SM) and look for new physics as a deviation from it. One may stress the stand point of symmetry-violation physics of CP, T and CPT. This physics is a nice complement to B meson decay experiments and also to collider energy-frontier experiments.

A topic of current interest is the determination of CKM matrix and confirmation of the unitary triangle, or the determination of the ρ and η Wolfenstein parameters. The most useful decays are the flavor changing modes of $K_L \to \pi^0 \nu\bar{\nu}$ and $K^+ \to \pi^+ \nu\bar{\nu}$ (Fig.4). The former is a direct CP-violating process dominated by short distance contributions. The decay amplitude in SM is proportional to $\mathrm{Im}(V_{td})$ or to the parameter η, namely to the height of the triangle. The current estimate of the branching ratio is 3×10^{-11}. A pilot experiment (E391a at KEK-PS), aiming at a sensitivity of $10^{-9\sim-10}$, is currently in prepa-

CP642, *High Intensity and High Brightness Hadron Beams: 20th ICFA Advanced Beam Dynamics Workshop on High Intensity and High Brightness Hadron Beams*, edited by W. Chou, Y. Mori, D. Neuffer, and J.-F. Ostiguy

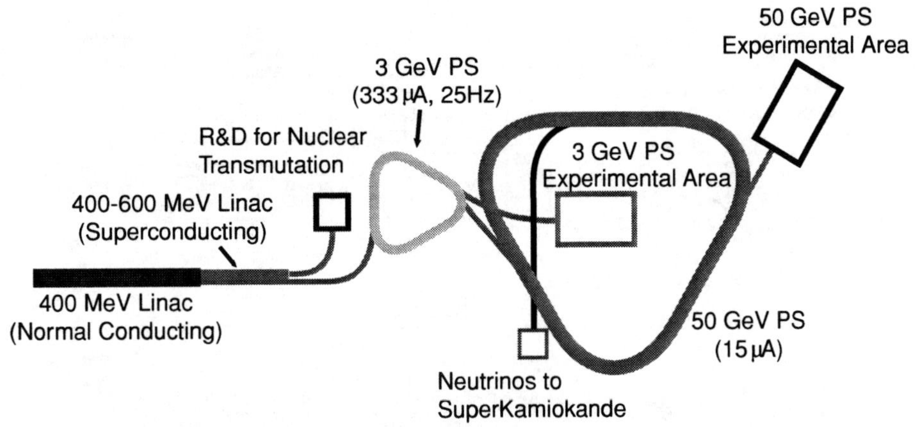

FIGURE 1. Configuration of the JHF accelerator complex

ration. BNL-AGS is also preparing an experiment (KOPIO) aiming at 10^{-12} sensitivity. The E391a setup, when moved to the 50-GeV PS, will achieve a sensitivity of 10^{-14} in a runtime of 3×10^7 sec accumulating 1000 events. The η parameter can be determined to a precision better than 10%, a result inaccessible at current-day facilities.

The $K^+ \rightarrow \pi^+ \nu\bar{\nu}$ decay is complementary to K_L decay and has a SM amplitude proportional to $\sqrt{(\rho-\rho_0)^2+\eta^2}$ which is rather sensitive to ρ, the horizontal coordinate of the triangle peak. The present SM prediction of the branching ratio is $(0.75 \pm 0.29) \times 10^{-10}$. This decay mode has long history of measurements of increasing sensitivities, and two events have recently been found in the E787 experiment at BNL-AGS. From the SM estimate, one expects to obsereve 100 events at JHF. One of the interests in investigating the unitary triangle lies in observing differences between kaon decays and B meson decays, where the contribution from new physics might appear as a several percent effect.

Time reversal invariance violation (T violation) is also important physics complementary to CP violation. One may test CP violation models assuming the CPT invariance. T violation can be searched for in the $K^+ \rightarrow \pi^0 \mu^+ \nu$ decay by looking at muon transverse polarization (P_T) which is the normal component of polarization to the decay plane. Because it is a T-odd correlation and the final state interaction is very small (10^{-6}), non-vanishing P_T is a signature of T violation. The interesting feature here is the non-existence of a SM contribution. Thus, P_T is sensitive to some of new physics, such as the multi-Higgs models, leptoquark models and a class of supersymmetric models. Now the KEK E246 experiment is improving its limit at the $P_T \sim 10^{-3}$ order. More than order of magnitude improvement (to the level less than $P_T \sim 10^{-4}$) can be expected at JHF.

Neutrino oscillation

Neutrino oscillation physics is becoming more and more important after the evidence for oscillation was found in atmospheric neutrino observation by Super-Kamiokande (SK) as well as in solar neutrinos at SNO. The K2K experiment, running at KEK 12-GeV PS, also confirms the oscillations. At JHF, one can take advantage of the high fluxes of ν_μ beams (100 times more intense than at K2K) to vastly improve these measurements. The distance to SK is 295 km and just the best condition for the relevant mass difference region and neutrino energy region. The current interest is the determination of mass difference, Δm^2, relevant to the muon neutrino disappearance and the mixing angle $\sin^2 2\theta_{23}$ of the MNS mixing matrix, and determination of $\sin^2 2\theta_{\mu e}$ (or $\sin^2 2\theta_{13}$) from an electron neutrino appearance measurement. The present constraint from SK is $1.6 \times 10^{-3} < \Delta m^2 < 3.9 \times 10^{-3}$ eV^2, and $\sin^2 2\theta_{23} > 0.97$ (90% C.L.). The recent result from K2K is also showing a similar region of $1.5 \times 10^{-3} < \Delta m^2 < 3.9 \times 10^{-3}$ eV^2 at $\sin^2 2\theta = 1$ (90% C.L.) with an expectation of corresponding improvement after the planned 10^{20} protons on target. The sensitivity at JHF will be $\delta(\sin^2 2\theta_{23}) \sim 0.01$ and $\delta(\Delta m^2) \leq 1 \times 10^{-4}$ eV^2 after a 5 year run.

As for the electron neutrino appearance experiment, one may push the limit of $\sin^2 2\theta_{\mu e}$ down to 0.3% level in 5 years improving the present limit from a reactor experiment by a factor 20 for the relevant mass difference region.

A neutrino beamline is now being designed based on the off-axis beam scheme with a neutrino beam in the direction at a finite angle from the pion beam axis. One can obtain a narrow energy spectrum with relatively high peak intensity. In the present case the off-axis angle of 2 degrees looks most promising. After a power upgrade, CP violation studies will be also within the scope.

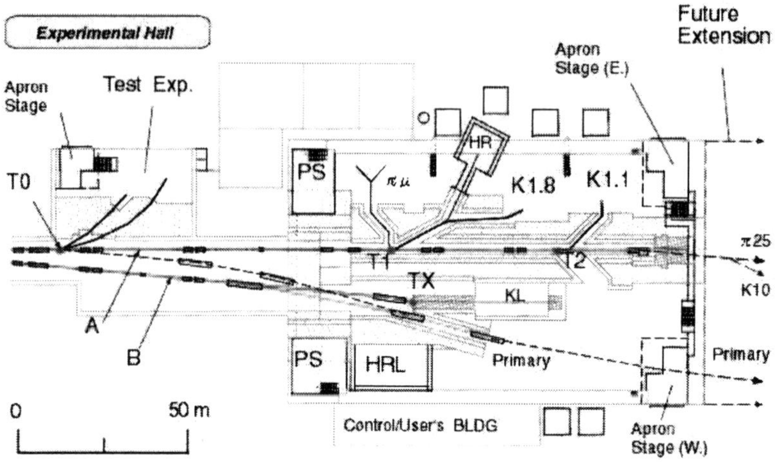

FIGURE 2. Experimental hall for slow extracted beams at the 50-GeV PS

Muon decays

Lepton flavor physics will be studied using high-intensity muon beams. Lepton number violating rare decays such as $\mu \to e\gamma$, $\mu \to 3e$, $\mu - e$ conversion and $Mu - \bar{M}u$ conversion have been studied since long improving their limits. These decays can be pursued at the 50 GeV PS to result in unprecedented precision. One of the most interesting decays is the $\mu - e$ conversion in which a μ^- in the muonic atom ground state is absorbed by the nucleus emitting an electron. This process has a very high sensitivity to SUSY-GUT, one of the recent theories beyond the SM. Now the MECO experiment at BNL-AGS is being prepared aiming for a sensitivity of 10^{-16}. An experiment at JHF may exceed this limit.

In order to achieve such a high sensitivity, the PRISM collaboration is considering to make a high-intensity muon source by means of phase rotation technique using a fixed field alternating gradient (FFAG) synchrotron in conjunction with a pion capture and decay sections with superconducting magnets.

NUCLEAR PHYSICS

One of the main features of nuclear physics at JHF is the use of strong kaon beams which enable so-called strangeness nuclear physics by introducing s quarks into nuclei. The conventional S=-1 systems of Λ and Σ hypernuclei have been studied at the present accelerator laboratories, KEK and BNL, and they will be pursued further. Hyperon states deep inside the nuclear medium can be studied as well as the hyperon nucleon interactions. By means of γ spectroscopy one may also open up a new field of high-resolution hypernuclear spectroscopy.

The high flux of kaon beams enables us to explore the S=-2 system in terms of the (K^-,K^+) reaction leading to Ξ nuclei and double Λ hypernuclei. The knowledge of S=-2 nuclei is up to now very poor. Only four events were found for double Λ nuclei and the Ξ nucleus is yet to be found. As for the H dibaryon only a lower bound of mass has been given. There is also a theoretical prediction about the existence of a mixed states of Ξ, 2Λ and H.

In order to perform the spectroscopy of the S=-2 system with (K^-,K^+) reactions of very small cross sections, we need a K^- beam of high quality and a spectrometer of high performance. The SKS superconducting spectrometer currently in use at KEK will be employed at JHF. A beamline with the beam momentum of 1.8 GeV/c is now being designed.

The hyperon-nucleon interaction can be also studied in scattering. The cross section data are still very poor in comparison with the nucleon- nucleon scattering. The interests will be the understanding of the baryon-baryon interaction in terms of $SU(3)$ scheme together with NN interaction. It is also desired to determine the sign (attractive or repulsive) and magnitude of the potential core and the strength of spin dependent forces.

There are of course many interesting fields in addition to the strangeness physics. Hadrons embedded in nu-

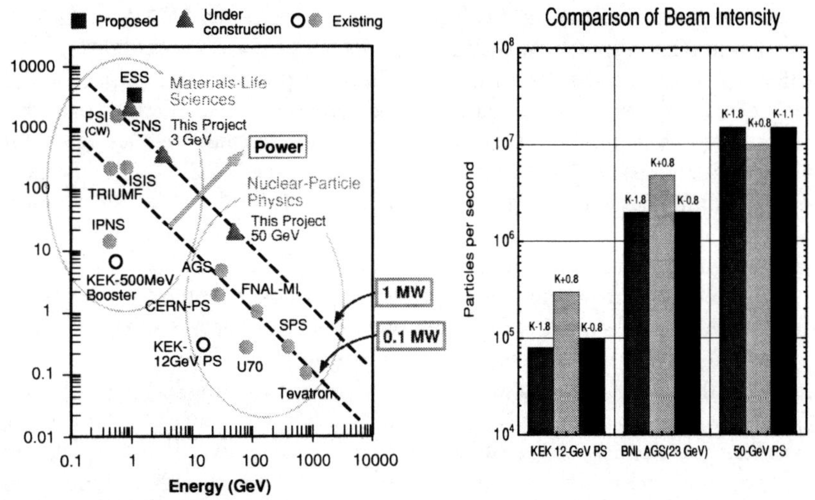

FIGURE 3. Beam power of world's proton machines and expected kaon beam intensities at JHF

clear matter may provide a hint to understand the origin of hadron mass, which can be argued in QCD theories. There are calculations of the vector meson mass in nuclear matter and experimental studies have been started. The spectroscopy of hadron itself will be one of the major fields, too. Current interest is in topics such as search (or identification) for exotic states with gluonic degrees of freedom such as glueballs and hybrids.

MATERIALS AND LIFE SCIENCES

Pulsed high flux neutron beams

Neutron diffraction has a wide range of applications in many sciences. In the planned experimental hall, there will be nearly 25 beamlines equipped with unique diffractometers and spectrometers of high performance, and materials and life sciences will be performed. The 25 Hz pulsed beam enables high-precision time-of-flight measurements to determine the scattering function $S(Q,\omega)$ over a wide range of Q and ω in a high resolution setting. Although the average flux is smaller than the most powerful reactor beam by a factor of four, the peak intensity will be 200 times higher.

High flux beam will enable the studies of:

- materials with large unit structure such as biomolecules and soft matters (with smaller samples),
- weak signals from new type of weak or exotic excitations in condensed matter, and
- dynamical properties in materials by means of the time dependent measurements.

The polarized neutron beam will be proportionally strong enough for many applications. The industrial applications of radiography will become more and more important, too.

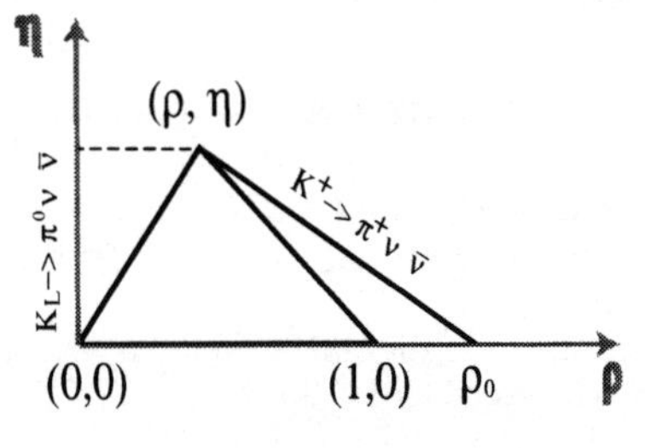

FIGURE 4. Kaon rare decays and the unitary triangle

Sciences with neutrons

In this respect one of the most important fields in which prominent development is expected is structural biology. In conjunction with the nature of neutron scattering that the sensitivity is rather uniform over atomic number Z in contrast to X ray which has the sensitivity only to heavy elements, the positions of hydrogen and hydration in proteins, DNA and physiologically important materials will be routinely determined. Physiological functions will be studied in inelastic scattering through excitations.

A new field will be opened in the soft-matter science such as polymer and liquid crystal. These materials form very beautiful higher-order complex structure in a large spatial scale. To realize the maximum potential of this probe, it is essential that we cover a wide Q-ω range with high flux neutron beams. The density fluctuation is known to play an important role in the structural formation, which might be investigated in time evolution of $S(Q,\omega)$. The response of soft matter to external fields is also an important subject.

The recent observation of quantum effects in the spin excitation in $CuGeO_3$ suggests another major field at our facility. In the scattering function, weak strength of magnetic scattering from electrons in the one-dimensional system was observed outside the classical excitation region. Such a weak signal provides means to study spin dynamics in magnetic substances. Similar studies of lattice dynamics, electron dynamics and orbital dynamics will be possible. One of the interests there will be the understanding of high T_c superconductivity.

High sensitivity for light elements opens up several applications of dynamic process measurements. Recently, an observation of the drifting of Li ions during the charging and discharging of a lithium battery has been reported. The high neutron fluxes at JHF will open up many applications of this kind.

Industrial applications of neutrons is another facet of JHF. The high flux beams make the radiography of the interiors of bulk materials feasible. One can exploit this technique to characterize new materials from the images created under various conditions of multi-parameter space such as concentration, temperature, pressure *etc.* The scattering measurements under extreme conditions of pressure and temperature will become possible, too.

The use of polarized neutrons is beneficial not only for the study of magnetic materials but also for the study of soft matter and biological samples to reduce incoherent backgrounds originated by hydrogen nuclei, or to apply a spin contrast variation technique.

Sciences with muons

Muon beams are also powerful tools in the material science. The muon spin rotation (μSR) is an established technique today at meson factories. Another very promising application will be the production of ultra-slow muon which can be accelerated to a beam with very small emittance. The muonium dissociation from material into vacuum with thermal energy and the subsequent laser ionization has been tested at KEK. One of the most interesting applications of negative muons is the muon catalyzed fusion. A negative muon can catalyze many times of D-T fusion in cycles through the resonant formation of a D-T molecular state. There is long history of this research, and JHF will contribute to further advancements in this field.

CONSTRUCTION STATUS

The first phase of our project was approved with a total budget of 133.5 BYen. It includes the linac up to 400 MeV with normal conducting cavities, the 3-GeV RCS and the 50-GeV PS. Initially the beam energy will probably be 30-40 GeV. As for experimental facilities, the neutron/muon hall and a half size of the nuclear and particle physics hall will be built. We have to wait for a subsequent budget for the completion of the facility. A separate budget request is submitted already this summer for a neutrino beam line.

The construction has already started last year and the accelerator components have been ordered. The ground breaking ceremony is planned for this fall. The project team is now busy with designing experimental facilities and beamlines aiming for the completion at the end of FY2006 and we expect first beam in 2007. We anticipate that the rated full beams will be delivered soon after the accelerator commissioning.

SUMMARY

JHF will be a unique high-intensity proton machine facility in the world with the 3-GeV RCS for neutron and muon sciences and the 50 GeV PS for particle and nuclear physics. In the future an R&D facility for nuclear transmutation (ADS) is also planned at the 600-MeV extension of the linac. This accelerator complex was realized as the joint project of two laboratories with activities in some common and distinct research fields. JHF aims for an international research center which is open to world-wide users. We also seek international cooperation at the facility construction phase. For example, we are recycling the retired magnets *etc.* from facilities all over the world, for their next active life at JHF.

ACKNOWLEDGMENTS

I would like to thank all members of the project team of KEK and JAERI for daily discussions and providing me with materials for this talk.

REFERENCES

1. The Joint Project for High-Intensity Proton Accelerators; KEK Report 99-4, JAERI-Tech 99-056, JHF-99-3; July 1999.
2. Proceedings of the International Workshop on JHF Science; edited by J.Chiba *et al.*, KEK Proceedings 98-5, JHF-98-2, 1998.
3. Y. Mori, "JHF machines", Talk in this workshop.

Neutrino Oscillations at Proton Accelerators

Douglas Michael

California Institute of Technology
Pasadena, CA 91125 USA

Abstract. Data from many different experiments have started to build a first glimpse of the phenomenology associated with neutrino oscillations. Results on atmospheric and solar neutrinos are particularly clear while a third result from LSND suggests a possibly very complex oscillation phenomenology. As impressive as the results from current experiments are, it is clear that we are just getting started on a long-term experimental program to understand neutrino masses, mixings and the physics which produce them. A number of exciting fundamental physics possibilities exist, including that neutrino oscillations could demonstrate CP or CPT violation and could be tied to exotic high-energy phenomena including strings and extra dimensions. A complete exploration of oscillation phenomena demands many experiments, including those possible using neutrino beams produced at high energy proton accelerators. Most existing neutrino experiments are statistics limited even though they use gigantic detectors. High intensity proton beams are essential for producing the intense neutrino beams which we need for next generation neutrino oscillation experiments.

OSCILLATION PHENOMENOLOGY

Neutrino oscillations can arise if there is a non-exact alignment between flavor and mass eigenstates. Assuming three flavors and mass eigenstates, the most general mixing can be described by the following unitary transformation:

$$\begin{pmatrix} \nu_e \\ \nu_\mu \\ \nu_\tau \end{pmatrix} = \begin{pmatrix} U_{e1} & U_{e2} & U_{e3} \\ U_{\mu 1} & U_{\mu 2} & U_{\mu 3} \\ U_{\tau 1} & U_{\tau 2} & U_{\tau 3} \end{pmatrix} \begin{pmatrix} \nu_1 \\ \nu_2 \\ \nu_3 \end{pmatrix}$$

The oscillation probability is then given by

$$P(\alpha \to \beta) =$$
$$\delta_{\alpha\beta} - 4\Sigma\Sigma U_{\alpha i} U_{\beta i} U_{\alpha j} U_{\beta j} \sin^2((m_i^2 - m_j^2)L/4E)$$

where m_i are the mass eigenstates, L is the distance that the neutrino has propagated and E is the neutrino energy.

One can keep expanding this matrix to any arbitrary number of neutrino flavors and mass eigenstates to account for any number of possible sterile neutrinos or very massive neutrinos. Additional light "active" neutrinos (those that have normal couplings to the Z^0) are ruled out by results from LEP and SLC [1]. We also note that there is no a priori requirement that the neutrino and anti-neutrino matrices or masses must be identical. However, if they are different it requires CPT violation. It has been suggested that such violation is possible due to non-local string theories and/or that while CPT is exactly conserved in higher dimensions that in 4-dimensions the conservation is only approximate [2].

A specific form of the mixing matrix for the three known active flavors has been developed by Maki, Nakagawa and Sakata and is known as the "MNS matrix" [3]:

$$U = \begin{pmatrix} c_{13}c_{12} & c_{13}s_{12} & s_{13}e^{-i\delta} \\ -c_{23}s_{12} - s_{13}s_{23}c_{12}e^{i\delta} & c_{23}c_{12} - s_{13}s_{23}s_{12}e^{i\delta} & c_{13}s_{23} \\ s_{23}s_{12} - s_{13}c_{23}c_{12}e^{i\delta} & -s_{23}c_{12} - s_{13}c_{23}s_{12}e^{i\delta} & c_{13}c_{23} \end{pmatrix}$$

Where $c_{ij} = \cos\theta_{ij}$ and $s_{ij} = \sin\theta_{ij}$. This is completely analogous to the CKM matrix for the quark sector. Of importance, we note that a CP-violating phase, δ is possible. Several authors have suggested that this phase may be measurable in future oscillation experiments [4, 5, 6, 7, 8].

A final important feature of neutrino oscillation phenomenology are matter effects. When neutrinos propagate through matter the oscillations can change depending on the type of neutrinos involved. Space requires limiting discussion on this interesting point other than to note that matter effects may be important in considering future long-baseline neutrino oscillation experiments.

REVIEW OF EXISTING OSCILLATION DATA

Experimental evidence exists for three, apparently different, neutrino oscillation effects. The complete set of results now favors solar neutrino parameters with large (but not maximal) mixing and Δm^2 in the range of several $\times 10^{-5}$ eV^2. The atmospheric oscillation parameters show apparently maximal mixing and with $\Delta m^2 \sim 3 \times 10^{-3}$ eV^2. The LSND oscillation parameters have very small mixing (less than $\sim 1\%$ probability) and with Δm^2 in the range 0.1-1 eV^2. We note that if all of the experimental measurements and their errors are correct that just three active neutrino flavors provide only two Δm^2 values and hence cannot explain all of the experimental effects. However, all of the effects can be explained

CP642, *High Intensity and High Brightness Hadron Beams: 20th ICFA Advanced Beam Dynamics Workshop on High Intensity and High Brightness Hadron Beams,* edited by W. Chou, Y. Mori, D. Neuffer, and J.-F. Ostiguy

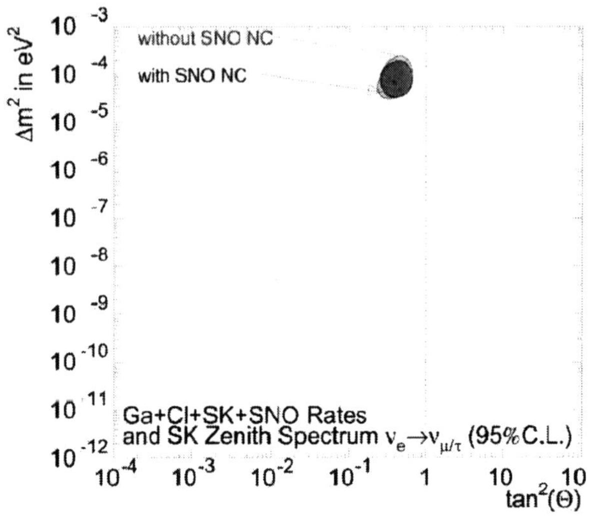

FIGURE 1. Compilation of allowed solar neutrino oscillation parameters. Only one region of parameter space remains at 90% CL. Adapted from reference [9].

if there are sterile neutrinos involved and/or if neutrinos and anti-neutrinos oscillate differently. A final observation is that neutrino masses suggested by the LSND Δm^2 can result in a significant (but not dominant) hot-dark-matter contribution by neutrinos in the universe.

Solar Neutrinos

In the last few years, data from Super-Kamiokande and SNO have significantly constrained the allowed solar neutrino parameters and phenomenology. The most important data for constraining the oscillation parameters comes from Super-Kamiokande. Super-Kamiokande observes no anomalous seasonal variation of the flux of neutrinos from the sun, observes an average flux of 0.45 that expected with no energy dependence and with no dependence on whether neutrinos have passed through the earth on their way from the sun to the detector or passed only through a small amount of overburden above the experiment (i.e. there is no "day/night" effect) [9]. The lack of anomalous seasonal variation rules out various "just-so" vacuum oscillation solutions. The lack of any modulation in the observed neutrino energy and the lack of any day/night effect constrains the oscillation parameters to relatively high mixing and Δm^2 in the vicinity of several $\times 10^{-5}$ eV^2. Figure 1 shows the summary of the allowed solar oscillation parameters with data combined from both Super-Kamiokande and SNO along with older experiments.

The SNO data offers additional information compared to Super-Kamiokande due to the heavy water in SNO. Both Super-K and SNO measure some events resulting from elastic scattering of neutrinos off electrons in

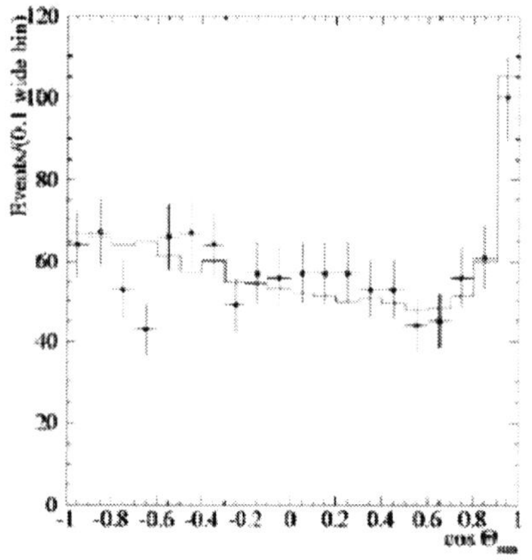

FIGURE 2. The observed rate of neutrino interactions in SNO vs the cosine of the angle from the sun along with a fi t to the different types of neutrino interactions and background. Adapted from reference [10].

the water. This has both a charged and neutral current contribution. In SNO, there are two additional interactions: Charged-current dissociation of the the deuterium into two protons and an electron and the corresponding neutral current dissociation into a proton and a neutron. Figure 2 shows the observed rate of neutrino interactions in SNO vs the cosine of the angle from the sun. A contribution from elastic scattering of electrons is seen near $\cos\theta = 1.0$, an overall normalization is visible resulting from a combination of NC deuterium dissociation and background and an overall slope, consistent with a $1 - 1/3\cos\theta$ distribution is visible which results from the CC dissociation of deuterium. When combined with the results from Super-K on elastic scatters, one can determine with good precision both the total flux of neutrinos from the sun and the remaining flux of ν_e at the earth. The observed flux of all neutrino types of $5.44 \pm 0.99 \times 10^6$ $cm^{-2}s^{-1}$ [10] is in impressive agreement with the standard solar model [11] value $5.44 + 1.01 - 0.81 \times 10^6\,cm^{-2}s^{-1}$. This results in a measured combined flux of $\nu_{\mu,\tau}$ of $3.7 \pm 1.1 \times 10^6\,cm^{-2}s^{-1}$.

The next important step in measurement of the "solar" neutrinos will come from the Kamland experiment, now taking data in Japan [12] Kamland is measuring the flux of $\overline{\nu_e}$ coming from several reactors in the vicinity of Kamioka with typical baseline distances between 100-200 km. Kamland should see a clear oscillation signature in these neutrinos if the suggested solar oscillation parameters are correct and if neutrinos and anti-neutrinos oscillate in the same way. The results from Kamland will be very interesting regardless of the outcome, but may imply different future experimental programs. Should Kamland not observe oscillations consistent with those suggested from the solar measurements then one possibility is that there is CPT violation in the neutrino sector. This would mean that LSND could be the result of a higher Δm^2 for $\overline{\nu_e}$ than for ν_e. It could also mean that there could be differences in ν_μ and $\overline{\nu_\mu}$ oscillations with near atmospheric parameters and differences in the rate at which these may oscillate to ν_e. On the other hand, it could mean that the solar oscillations are more subtle than currently supposed, or that some existing experimental results are incorrect. However, should Kamland see the expected oscillation signature at relatively high Δm^2 and large mixing then it is possible that future long-baseline oscillation experiments will be able to observe CP violation. We eagerly await the results of this experiment.

Atmospheric Neutrinos

The atmospheric neutrino data provides some of the strongest evidence that a classic oscillation effect is occurring in the neutrino sector (as opposed to any other means of neutrino flavor conversion). The data from Super-Kamiokande [13] are by far the most important for these measurements but there are also (pre)confirming data from MACRO, Soudan 2, Kamiokande and IMB [14].

The atmospheric oscillation effect is consistent with pure $\nu_\mu - \nu_\tau$ mixing and with that mixing being consistent between neutrinos and anti-neutrinos. However, the possibility of admixtures or differences in the oscillations still exist at the tens of percent level. Figure 3 shows the observed and predicted distributions of atmospheric neutrino events in Super-Kamiokande for ν_μ and ν_e. Several different observed energy regions are shown. The observed electron-like events are in good agreement with expectations with no oscillation while the observed muon-like events are in very poor agreement with a no oscillation hypothesis. However, the agreement with a simple $\nu_\mu - \nu_\tau$ oscillation model with $\Delta m^2 = 0.0028\,\mathrm{eV}^2$ and $\sin^2 2\theta = 1.0$ is excellent.

Accelerator-based long-baseline neutrino oscillation experiments are designed to study these oscillation effects with higher precision than the existing atmospheric data.

LSND Oscillation Effect

LSND (Los Alamos Scintillator Neutrino Detector) is the first accelerator experiment to report a positive observation of an apparent neutrino oscillation effect [15]. LSND observed the appearance of $\overline{\nu_e}$ in a $\overline{\nu_\mu}$ beam. Figure 4 shows the observed number of apparent $\overline{\nu_e}$ events compared to the expectations (from backgrounds) as a function of observed energy. The neutrino beam is produced from pions decaying at rest. This result suggests oscillation parameters as shown in figure 5. Since the solar neutrino Δm^2 is apparently much smaller, the parameters suggested from LSND must mean either that neutrinos and anti-neutrinos can have different Δm^2 or that there are sterile neutrinos involved (with large Δm^2) or that some experiments are wrong.

NEAR-TERM OSCILLATION EXPERIMENTS AT PROTON ACCELERATORS

Several experiments are either already running or in construction using accelerator neutrino beams to make further oscillation measurements. These experiments all already demonstrate the keen demand which neutrino oscillation experiments place on the intensity of high energy proton sources.

Addressing the LSND Effect: Mini-BooNE

The Mini-Boone (Mini-Booster Neutrino Experiment) [16] has just recently begun operation at Fermilab. It uses a neutrino beam generated from 8 GeV protons from the Fermilab Booster. The nominal number of protons expected on target for the complete neutrino exposure is about 6×10^{20}. The recent discussion that neutrinos and anti-neutrinos may oscillate differently has also generated much interest in running with an anti-neutrino beam as well as the neutrino beam. This reduces the number of events by about a factor of three and hence for equal statistics requires roughly 18×10^{20} protons on target.

The Mini-BooNE experiment uses a 1 KT mineral oil scintillation/Cerenkov detector to observe neutrino events in the energy range from 200-2000 MeV. The baseline is about 300 m. Figure 5 shows the allowed parameter space from the LSND and other measurements and the expected sensitivity for the Mini-Boone experiment in two years of running (for neutrinos). It is becoming fairly clear that a conclusive result may not be determined until the experiment has been able to achieve this exposure for both neutrinos and anti-neutrinos. With the nominally planned Booster intensity this could mean as much as 8 years of running. Clearly, a higher intensity of protons from the Booster is very desirable for this experiment.

The first long baseline experiment: K2K

The first "long-baseline" experiment, K2K, started operation in 1999. It utilizes the Super-Kamiokande detector and a beamline built for the 12 GeV PS at KEK to send a neutrino beam 270 km through the earth to that detector. Although at 40 kT the Super-K detector is easily the most massive available today for this type of measurement, the experiment is limited by statistical error

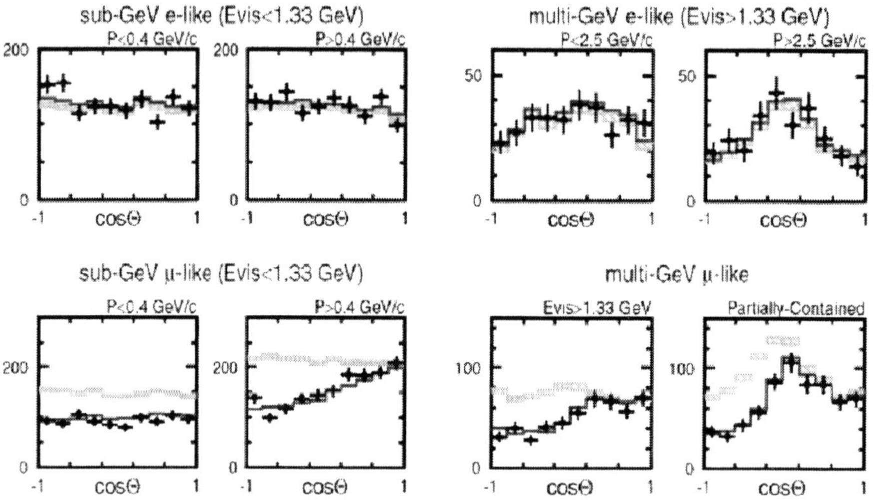

FIGURE 3. Observed and predicted atmospheric neutrino events in Super-Kamiokande. Adapted from [13].

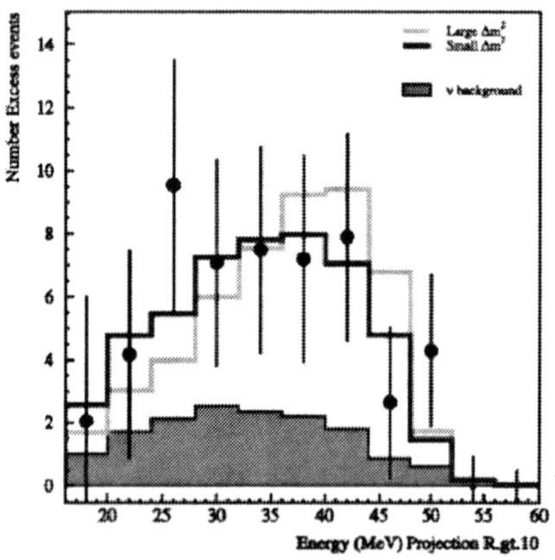

FIGURE 4. The observed number of $\bar{\nu}_e$ events in LSND vs neutrino energy compared to the expectation from backgrounds. Adapted from reference [15].

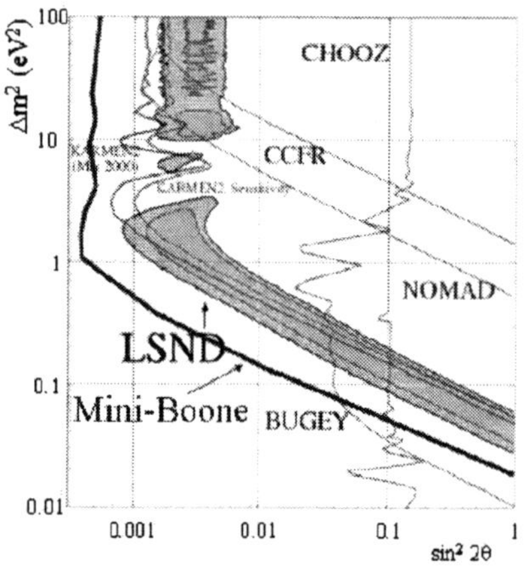

FIGURE 5. Allowed oscillation parameters for $\bar{\nu}_\mu$ to $\bar{\nu}_e$ based on LSND and other experiments. Mini-BooNE will clearly cover the allowed range of parameters.

due to the relatively low intensity of the existing 12 GeV machine. This may illustrate better than any existing experiment the relative importance of high-intensity proton accelerators as being a key element in the ability to make precise oscillation measurements.

K2K uses a "standard" neutrino beam configuration. The 12 GeV protons are aimed in the direction (on axis) of Super-Kamiokande and then impinge on a target to produce pions. The pions are focussed using a double-horn system and then directed down a decay pipe which is $\sim$ 150 m long. The peak neutrino energy is about 1.3 GeV. To date, a total of 5.6×10^{19} protons have been delivered to the K2K neutrino beam target. This has resulted in a total of 29 single-ring muon-like events observed at Super-Kamiokande with the expected number being 42 ± 6 events with no oscillations. An oscillation analysis on these events shows that the resulting oscillation parameters are in good agreement with those measured from the atmospheric neutrinos [18].

MINOS and the NuMI beamline at Fermilab

The next step in precision in long-baseline experiments will come from the MINOS (Main Injector Neutrino Oscillation Search) experiment using the NuMI (Neutrinos at the Main Injector) beamline at Fermilab. MINOS will make use of the intense neutrino beam afforded by the Main Injector in order to make precise measurements on neutrino oscillations associated with the "atmospheric" neutrino oscillations. By 2005, the

Main Injector will afford a very intense proton beam:

- 120 GeV protons
- 1.9 s cycle time
- 4×10^{13} protons per cycle
- 0.4 MW!
- 3.6×10^{20} protons per year.

The MINOS experiment uses a 735 km baseline and a peak neutrino energy of 3 GeV. Although the MINOS experiment is smaller than Super-Kamiokande (about a factor of 6 in fiducial mass) and the baseline is longer, the relatively large neutrino flux from the NuMI beamline more than compensates so the event rate in the MINOS far detector will be about 100 times that in Super-Kamiokande with K2K. This increase in neutrino flux permits better statistical precision and also qualitatively different types of measurements.

The NuMI beamline is very similar to that of the K2K beamline only somewhat larger in all dimensions. A graphite target is used due to the very high proton power. The energy of the neutrino beam can be tuned. A "low energy" beam is available with peak energy of 3 GeV.

The MINOS experiment utilizes a near and far detector of very similar construction in order to permit precision measurement of oscillation parameters, clear demonstration of the oscillation signature and precise determination of the flavor participation of all neutrino types involved in the oscillations. The MINOS detectors are sampling calorimeters with 2.54 cm thick iron absorbers interleaved with 1 cm thick plastic scintillator strips which are 4 cm wide. The iron in the detectors is magnetized with typical field of 1.5 T.

The MINOS far detector has a total mass of 5.4 kT. Roughly 5000 ν_μ charged-current events are expected, in the absence of oscillations, in the far detector for two years of running. These statistics are sufficiently high that an unambiguous oscillation signature should be visible. Figure 6 shows the expected energy distribution of ν_μ CC events measured in the MINOS far detector with and without oscillations for several different Δm^2 and $\sin^2 2\theta = 0.9$.

In addition to measurement of the oscillation signature and parameters, MINOS will provide measurements of the participation of all types of neutrinos in the oscillations. By looking for appearance of ν_e CC events in the far detector (or lack thereof) MINOS can determine the oscillation probability between ν_μ and ν_e at the 1-2% level. By comparing the number of NC events in the near and far detector MINOS can determine the level of oscillation of ν_μ to ν_{sterile} at the several percent level. Neutrinos which disappear but do not oscillate to ν_e or ν_{sterile} must oscillate to ν_τ so these are measured indirectly at the 10% level from the other measurements.

MINOS and the NuMI beamline are currently under construction. The first protons are scheduled on target in December 2004. The MINOS far detector is now half complete and the first half is being commissioned now for running for atmospheric neutrino measurements. The second half of MINOS is scheduled for completion by June 2003. Because of the magnetic field of MINOS, it will provide unique measurements on atmospheric neutrinos while awaiting completion of the NuMI beamline.

OPERA, ICARUS and the CNGS beamline

The final long-baseline project which is already in construction is the CERN to Gran Sasso (CNGS) beamline and experiments [17]. The CNGS beamline will use 400 GeV protons from the SPS to create a high energy neutrino beam ($E_{\text{avg}} = 17$ GeV) aimed at the Gran Sasso. The initial primary goal of this beamline and the experiments at the Gran Sasso will be to demonstrate direct appearance of ν_τ CC events in the nominally ν_μ beam. This will be possible in these experiments due to the combination of the high energy beam and large mass, very fine-grained detectors which are capable of identifying the ν_τ events. This beam will commissioned in 2006.

The two detectors which will initially utilize this beam at the Gran Sasso are Opera and Icarus. Opera will use plates of emulsion interleaved with thin lead plates to look for CC ν_τ events where the τ is produced in the lead and then subsequently decays in a gap downstream where a kinked track will be observed in the emulsion. Opera plans a total detector mass of about 2 kT, dominated by the mass of the lead. Icarus is a large liquid argon TPC which will use techniques similar to those used by the NOMAD short-baseline oscillation experiment to identify CC ν_τ events on a statistical basis through the transverse momentum imbalance which results when a τ decays to a muon or electron. Icarus is currently building and commissioning a 600 T detector and plans for a 3000 T detector by the time the CERN beam is commissioned. These detectors will have similar sensitivity to τ appearance with about 10 events expected in 5 years of operation with < 1 event background.

NEXT GENERATION OSCILLATION EXPERIMENTS AT PROTON ACCELERATORS

Even as the first round of very challenging long-baseline experiments is still under construction, already plans are developing for next-generation experiments. The current plans are primarily driven by the desire to search for a very small fraction of ν_μ to ν_e oscillation (perhaps at the 1% level?). Once such an oscillation is identified the goal will be to search for whether there is a difference be-

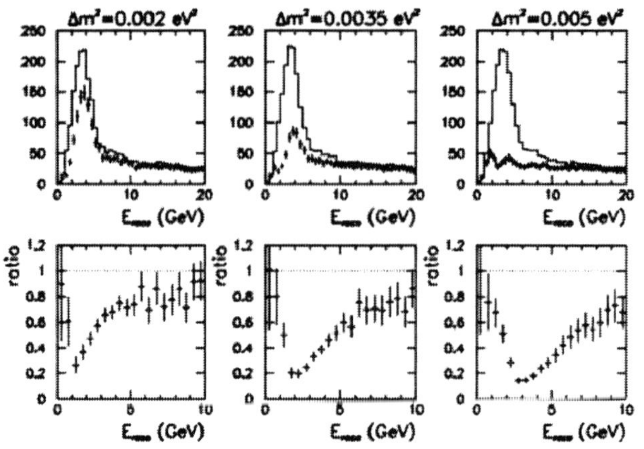

FIGURE 6. The top row of plots shows the expected energy distribution of ν_μ CC events measured in the MINOS far detector with and without oscillations for several different Δm^2 and $\sin^2 2\theta = 0.9$. The bottom row of plots shows the ratio of the spectra with and without oscillations. A clear dip in the ratio of neutrino energies with oscillations compared to the unoscillated spectrum will be observed within the expected range of Δm^2.

tween neutrinos and anti-neutrinos in these oscillations, resulting from CP or CPT violation or some combination of the two. Several theory papers in the last year have suggested that in fact this kind of subtlety in neutrino oscillations could exist [4, 5, 6, 7, 8].

In the previous section, we have seen that with existing detectors, accelerators and neutrino beam facilities that we are already pushing hard on technology to start to make precise measurements on oscillations. Taking the next step will require both larger, more-sensitive detectors and higher fluxes of neutrino beams, driven by more intense beams of protons.

Superbeams: The next step?

A number of ideas have been discussed for future high-intensity neutrino beams. However, it appears that a consensus has developed over the last couple of years that the next step in long-baseline oscillation physics will be built around "Superbeams". The term "Superbeam" is somewhat poorly defined, and yet most people have some idea what it means. The term generally refers to a neutrino beam of standard, or nearly standard construction but with a very intense proton source and where the beam may be used in somewhat novel ways compared to existing experiments. Most of these plans call for use of "off-axis" neutrino beams (discussed further below).

For future experiments, total proton beam power in the range of $\sim 1-4$ MW are generally discussed. This is based in part on the capabilities of proton accelerators that are either being designed or planned. It is also based in part on a minimization of the total cost for future experiments, regardless of any explicit proton accelerator design. Generally, it is realized that once one will be spending on order \$100M or more for gigantic detectors that equivalent investment in proton acceleration capabilities is likely to offer comparable value to the ultimate neutrino measurement capabilities. In general, both very large detector mass and the highest possible proton power will be necessary for the best sensitivity.

New high-intensity long-baseline experiments have been proposed for CERN, JHF, Brookhaven and Fermilab. Here, some of the possibilities are illustrated with the JHF to Super-Kamiokande proposal and the proposal for an off-axis experiment using an enhanced NuMI beam.

The JHF neutrino superbeam to Super-K/Hyper-K

A new 50 GeV proton synchrotron is under construction at Tokai-mura, Japan. The new facility is generally referred to (in neutrino circles at least) as the "Japanese Hadron Facility", JHF. The first-phase plans for this machine call for a cycle rate of 0.29 Hz and for a total of 3.3×10^{14} protons to be accelerated to 50 GeV every cycle, corresponding to a total power of 0.77 MW. In addition, the plans call for an off-axis beam with a novel means of tuning the beam energy given that the far detector location is already determined by the location of the existing Super-Kamiokande detector. As with other "superbeam" experiments currently being planned, the primary goal for this experiment is to search for ν_e appearance, and if observed to compare with $\bar{\nu}_e$ appearance to search for CP and/or CPT violation. Due to shortness of space and because this experiment is described in more detail elsewhere in these proceedings [19], we curtail further description here.

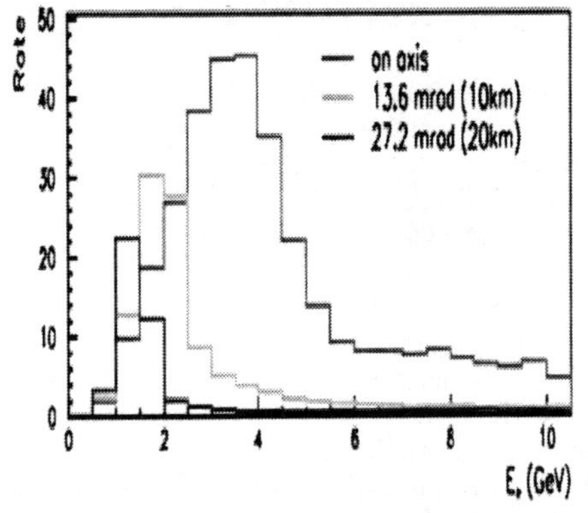

FIGURE 7. The rate of neutrino events vs neutrino energy for the NuMI on-axis and two off-axis beams.

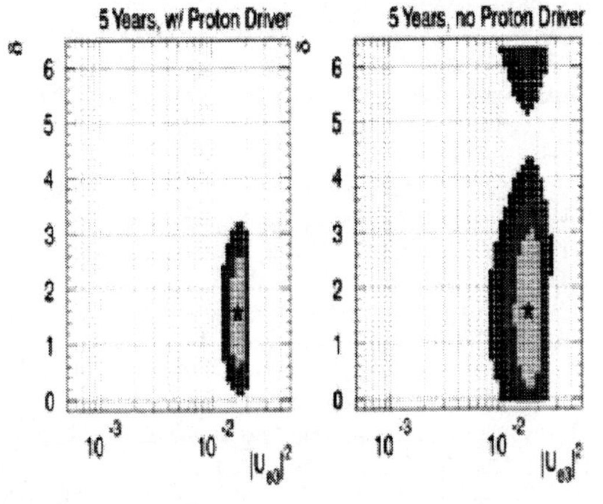

FIGURE 8. Comparison of possible CP violation measurements with a 20 kT off-axis experiment using the NuMI beamline with and without a new proton driver. δ is the CP violating phase and U_{e3}^2 is the relevant mixing angle for ν_e appearance. The measurement contours for 1,2 and 3 σ are shown for fi ve years of running (sharing between ν and $\bar{\nu}$ running) with a new proton driver on the left and without a new proton driver on the right. Although the oscillation parameters are unlikely to be exactly as shown here, the relative capabilities for precise measurements is similar for a wide range of parameters. Adapted from [20, 21].

Bulking up the NuMI beamline: From Intense to Super and off-axis

As noted above, the NuMI beamline will already have 0.4 MW of proton power starting in 2005. If not already a "super-beam", then it certainly is already well on its way. Recently new studies have been made for possible upgrades to the NuMI beamline and a new off-axis experiment. Part of the planning is to consider upgrades to the proton intensity through the Main Injector and upgrades to the NuMI beamline.

Figure 7 illustrates why off axis beams are of particular interest in general for measurement of ν_e appearance and specifically shows the case for the NuMI off-axis beam [21]. The rate of neutrino events vs neutrino energy per kT-year is shown at the distance of 735 km from Fermilab for the on-axis and two off-axis beams, one at 13.6 mR and one at 27.2 mR. One observes that although the total rate of neutrino events is lower for the off-axis locations that the rate at certain low neutrino energies is higher than for the on-axis location. Of additional importance for ν_e appearance is that the off-axis beams have relatively less high-energy tail which can contribute background, primarily from NC single π^0 events.

Several experimental concepts have been considered for a new off-axis experiment in the NuMI beamline [21, 22, 23]. Some of the technologies which are being considered include sampling calorimeters with a variety of absorbers and active elements and liquid argon. The sampling calorimeters generally attempt to reduce the average Z of the detector components from those in MINOS (dominated by the 1 inch thick iron plates) and provide finer longitudinal and transverse sampling. It is expected that these improvements will reduce the background from NC π^0 events to well below that from the intrinsic ν_e's in the beam. Is is expected that detectors with mass around 20 kT will provide good initial measurement capabilities.

Ultimately, the goal is not to simply measure that there is some small admixture of ν_μ to ν_e oscillation but to measure whether there is CP violation in the neutrino sector. Since this requires running with anti-neutrinos as well as neutrinos and looking for relatively subtle differences between the two on an already small signal... it is clear that one will wish to combine both the largest affordable detector and the most intense beam possible.

There are two, partially overlapping paths to providing higher intensity beams for the NuMI beamline. First, investments can be made in the existing accelerator complex, particularly in the Booster and Main Injector to increase the number of protons which can be delivered. Some modest improvements are already underway to meet the needs for the Collider Run II B and for MINOS and Mini-BooNE. Additional investment possibilities in the existing complex have been identified which could result in as much as a factor of two increase in intensity compared to the nominal NuMI design [24]. To make even greater increases in the intensity, Fermilab has also been studying the possibility of replacing the existing Linac and Booster with a new proton source, either a new 8 GeV synchrotron or a new 8 GeV LINAC. Although more expensive than simply investing in the existing accelerator complex, these machines, coupled with investments in the Main Injector (which overlap in any of the cases) could bring the total proton beam power for the NuMI beamline to 2-4 MW [25]. Such intense beams would likely require additional shielding upgrades in the target region of the current NuMI beamline.

Should a new proton driver become available at Fermilab, the potential for very precise neutrino oscillation measurements exists. Figure 8 shows the relative precision of CP violation measurement that might be possible using a new proton driver compared to the measurement possible simply using the nominal design flux for the NuMI beam. It is clear that impressive precision can be reached with appropriate upgrades to the "existing" experimental complex.

THE LONG-TERM FUTURE

It is generally expected that in the long-term that neutrino beams generated from muon storage rings will provide the best oscillation measurements. Such beams will have unique properties:

- Very intense beams, on order tens to hundreds of times more intense than planned superbeams.
- Relatively clean narrow-band beams, though measurement of charge in detectors is important since the starting beams will simultaneously consist equal numbers of ν_μ and $\bar{\nu}_e$ or $\bar{\nu_\mu}$ and ν_e.
- Intense, high energy ν_e and $\bar{\nu}_e$ beams! At the same time as it leads to specific detector requirements, these beams also add very significant measurement capabilities achievable in no other way.

Because there are a number of technical issues to be addressed, some of which are further developed for the super-beams, it is now generally realized that experiments based on these muon storage ring beams will begin sometime well into the next decade at the earliest.

CONCLUSIONS

Neutrino oscillations present an ever more compelling avenue towards deeper understanding of fundamental physics beyond the standard model. The oscillation phenomenology can be complex and very interesting. What we have already learned about oscillations has required a wide array of different types of experiments and detectors, including accelerator based experiments. Our future understanding of oscillations will continue to depend on a wide array of experiments, and even more on measurements in long-baseline accelerator experiments. We are just getting started on a series of precision measurements which high-energy accelerator experiments can offer. There are many exciting prospects and specific projects being defined. One of the major themes which unites all of these projects is the need for very intense proton beams for production of the intense neutrino beams which are essential to the precision future experiments. The future of neutrino physics rests on our ability to generate, control and utilize these beams.

REFERENCES

1. Many measurements... See in example M. Akrawy etal, Phys.Lett.B231:530,1989.
2. G. Barenboim, J. Beacom, L. Borrisov, B. Kayser, Phys.Lett.B537:227,2002.
3. Z. Maki, M. Hakagawa, S. Sakata, Prog.Theor.Phys.28:870,1962.
4. Probing CPT Violation with Atmospheric Neutrinos, S. Skadhauge, hep-ph0112189.
5. Neutrinos as the Messenger of CPT Violation, G. Barenboim, L. Borissov, J. Lykken, A. Smirnov, Submitted to Phys.Rev.Let. hep-ph/0108199.
6. CPT Violation and the Nature of Neutrinos, G. Barenboim, J. Beacom, L. Borissov, B. Kayser, hep-ph/0203261.
7. Neutrinos that violate CPT and the Experiments That Love Them, G. Barenboim, L. Borissov, J. Lykken, hep-ph/0201080.
8. Interpreting the LSND Anomaly: Sterile Neutrinos of CPT Violation or...?, A. Strumia, hep-ph/0201134.
9. Super-Kamiokande Collaboration, presented by M. Smy at Neutrino 2002, 'Super-Kamiokande's solar neutrino results", Munich, May 2002.
10. The SNO Collaboration, Presented by A. McDonald at TAUP 2001, Gran Sasso, Italy, Sep. 2001.
11. J. Bahcall, M. Pinsonneault, S. Basu, Astophys.J.555:990, 2001.
12. J. Shirai for the Kamland Collaboration, Presented at Neutrino 2002, Munich, Germany, May 2002.
13. Most recent results presented by M. Shiozawa at Neutrino 2002, Munich, Germany, May 2002.
14. Most recent results presented by M. Goodman at Neutrino 2002, Munich, Germany, May 2002.
15. C. Athanassopoulos etal., Phys.Rev.Lett.81:1774,1998.
16. 'The Mini-BooNE Experiment", Presented by R. Tayloe at Neutrino 2002, Munich, Germany, May 2002.
17. For recent status see talk by S. Katsenevas presented at Neutrino 2002, Munich, Germany, May 2002.
18. K. Nishikawa for the K2K Collaboration, 'Results from K2K", presented at Neutrino 2002, Munich, May 2002.
19. J. Imazato, 'JHF Physics", Presented at ICFA Workshop on high intensity hadron beams, Fermilab, April 2002, Published in these proceedings.
20. G. Barenboim, A. De Gouvea, M. Szleper and M. Velasco, 'Neutrino oscillations with a proton driver upgrade and an off axis detector: A case study", hep-ph/0204208 Apr. 2002. Also, M. Velasco, these proceedings.
21. G. Barenboim, etal., 'Physics potential at FNAL with stronger proton sources", hep-ph/0206025, Jun. 2002.
22. Also see various presentations from the 'Workshop on new initiatives for the NuMI Neutrino Beam", Fermilab, May 2002, http://www-numi.fnal.gov/fnal_minos/new_initiatives/new_initiatives.html.
23. 'Letter of Intent for an Off-Axis Detector for the NuMI Beamline", June 2002, http://www-numi.fnal.gov/fnal_minos/new_initiatives/loi.html.
24. Report of the NuMI Proton Intensity Working Group, July 2002, http://hep.caltech.edu/ michael/numipiwg/.
25. Studies on a new proton driver for Fermilab. W. Chou and W. Foster editors. Reports in preparation.

Progress in Bright Ion Beams for Industry, Medicine and Fusion at LBNL

Joe W. Kwan

Lawrence Berkeley National Laboratory
1 Cyclotron Road, Berkeley, CA 94546 USA

Abstract. Recent progresses at LBNL in developing ion beams for industry, radiation therapy and inertial fusion applications were discussed. The highlights include ion beam lithography, boron neutron capture therapy (BNCT), and heavy ion fusion (HIF) drivers using multiple linacs.

INTRODUCTION

Accelerator research at LBNL, in the Accelerator and Fusion Research Division, has a portfolio of diverse applications including areas outside high energy and nuclear physics such as semi-conductor industry, radiation therapy and fusion. This paper briefly reports the recent progress in the development of ion beam lithography, neutron capture radiation therapy, and heavy ion driven inertial fusion.

SEMI-CONDUCTOR INDUSTRY

It is well known that ion beams are useful for ion implantation in the semi-conductor industry. For this purpose, LBNL has developed ion sources such as the multi-cusp gas plasma sources and the Metal Vapor Vacuum Arc (MEVVA) sources. Another application that is under development is the ion projection lithography system.

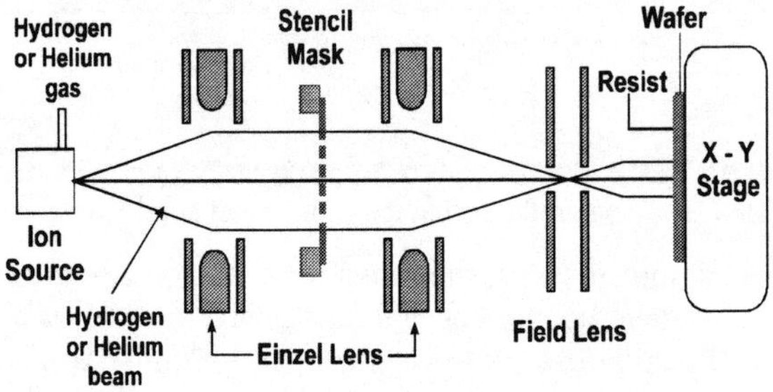

FIGURE 1. Ion projection lithography using a stencil mask.

In a conventional lithography system, a photon beam is projected through a stencil mask to produce images on a wafer that is covered by a layer of photo-resist. As the industry advances, the dimension of components is now approaching the photon diffraction limit of ~ 100 nm. Ion beams can potentially extend the limit down to ~ 5 nm. Figure 1 shows a schematic of an ion projection lithography system by replacing the photon beam with an ion beam.

Nevertheless, constant ion bombardment on the stencil mask limits its lifetime, thus affecting both the replacement cost and downtime. A maskless micro-beam reduction lithography system, as shown in Fig. 2, can improve the situation[1]. Here the exposure pattern is due to the aperture plate located at the ion source. The goal is to produce an array of micro-beams with the ability to switch on each micro-beam individually.

Figure 3 shows the concept of constructing micro-beam channels using layers of insulator. By applying

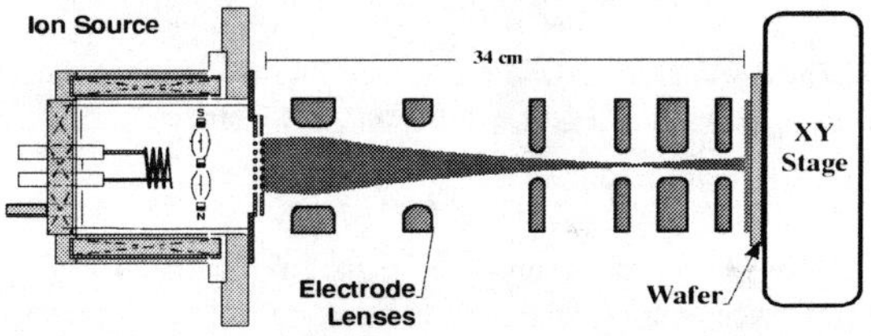

FIGURE 2. Maskless micro-beam reduction lithography.

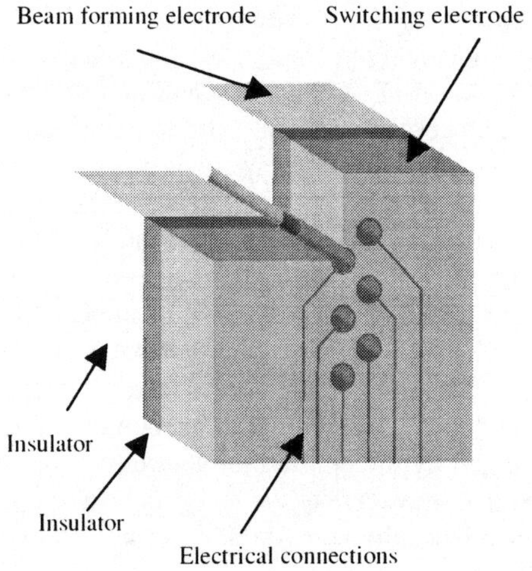

FIGURE 3. A micro-beam channel made of insulator layers.

CP642, *High Intensity and High Brightness Hadron Beams: 20th ICFA Advanced Beam Dynamics Workshop on High Intensity and High Brightness Hadron Beams,* edited by W. Chou, Y. Mori, D. Neuffer, and J.-F. Ostiguy

the appropriate voltages, a micro-beam can be gated either on or off. Separate layers of electrodes can be used to control rows of micro-beams in x and y directions for multiplexing.

RADIATION THERAPY

LBNL is a pioneer in radiation therapy since the early days of Bevatron. Nowadays proton and hadron therapies are common medical facilities worldwide. The trend is to develop facilities that are cost-effective, compact, and has good beam and dose controls. Figure 4 shows the size comparison of various types of machines. The new proposal is to build a 6-m diameter super-conducting cyclotron for 250 MeV/u (q/m=1/2) light ion therapy.

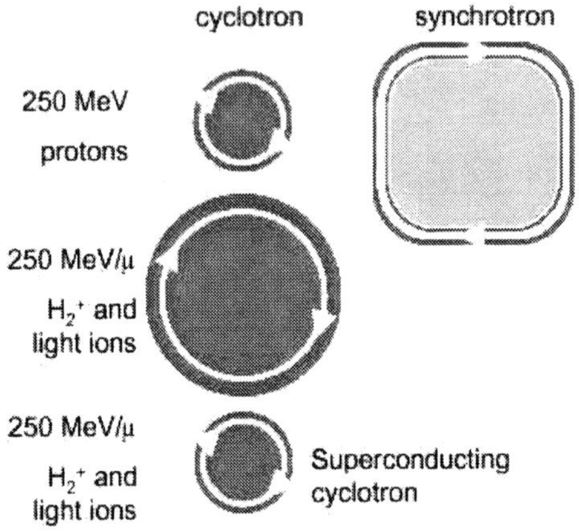

FIGURE 4. Comparing the sizes of various types of medical accelerators.

Another new challenge is to produce an accelerator-based neutron source for boron neutron capture therapy (BNCT). The concept of BNCT has been known for many years [2], but the practice lacks an effective boron-absorbing drug and neutron source of the correct energy spectrum. The basic principle of BNCT is to administer the patient with a tumor-seeking chemical that contains boron. After a high concentration of boron is gathered at the tumor cells, the patient receives neutron radiation dose. Since the ^{11}B formed by capturing a neutron is unstable and will decay into ^{7}Li + ^{4}He with the emission of alpha radiation. The short-range alpha can destroy the tumor's DNA while leaving the surrounding normal cells unharmed by the epithermal neutrons. This type of treatment is most effective for glioblastoma brain tumors that are usually untreatable by other means.

Recently there is significant progress in developing tumor-seeking drugs with boron content e.g. BOPP. Hence a successful BNCT clinical trail is expected if we can provide the treatment with an appropriate neutron source. Figure 5 illustrates the difference in reactor-based and accelerator-based neutron energy spectrum. The most useful energy range for BNCT is between 1 keV to 30 keV. These neutrons can be obtained by bombarding 2.5 MeV protons on a Li target with a Al/AlF$_3$ moderator. The necessary dosage, to be delivered within a reasonable time (~ a few hours), requires a 20 mA dc proton beam.

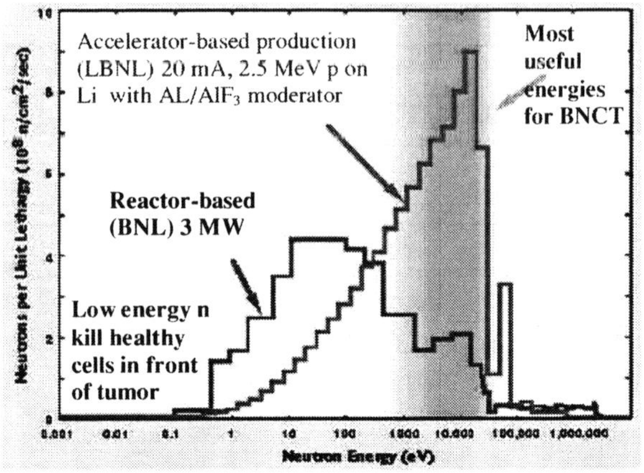

FIGURE 5. Accelerator-based neutron sources are superior to reactors for neutron therapy.

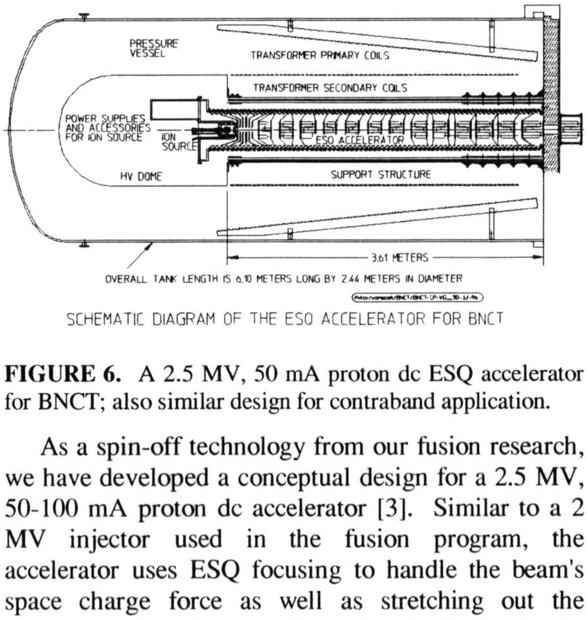

FIGURE 6. A 2.5 MV, 50 mA proton dc ESQ accelerator for BNCT; also similar design for contraband application.

As a spin-off technology from our fusion research, we have developed a conceptual design for a 2.5 MV, 50-100 mA proton dc accelerator [3]. Similar to a 2 MV injector used in the fusion program, the accelerator uses ESQ focusing to handle the beam's space charge force as well as stretching out the accelerator length to reduce longitudinal electric gradient (for preventing breakdowns). A schematic diagram of this accelerator is shown in Fig. 6.

The same accelerator design can be used for contraband applications. In that case a 1.76 MeV, 10 mA proton beam is directed onto a carbon target. The reaction $^{13}C(p,\gamma)^{14}N$ releases the 9.17 MeV γ radiation and subsequent resonant absorption by ^{14}N will detect explosives and drugs [4].

INERTIAL FUSION

Ion beams are essential to progress in fusion research. There are two main approaches to fusion: magnetic confinement and inertial confinement. The most advanced magnetic confinement concept are the tokamak, a device that has a toroidal shape with magnetic field generated by field coils in combination with the self-field generated by the plasma current. Significant progresses have been made by injecting energetic deuterium beams to heat the plasma, drive current and control the density profile. These "neutral beam injectors" (developed by LBNL) typically have ~ 100 keV energy; the beam ions are converted into neutral particles before entering the tokamak. The new generation of tokamak is larger, thus requiring beams of up to 1 MeV energy to penetrate the plasma.

While ion beams are auxiliary components in magnetic fusion devices, they are the main components in an inertial fusion machine using ion beam drivers. In comparison, ion beams can be more advantageous than laser beams because of the higher power efficiency (~30%), higher duty rate (~ 10 Hz) and more robust against radiation damages at the target chamber windows.

The goal of a heavy ion fusion driver is to deliver enough beam power to ignite a fusion target about 5 mm in diameter (see Fig.7). To do so, the required beam energy is ~ 5 MJ with a pulse length ~ 10 ns in order to achieve a peak power of ~ 10^{15} W/cm^2. Furthermore, the energy must be deposited within a short penetration range, e.g. 0.02 to 0.2 g/cm^2 of the target material. For heavy ions with atomic masses ~ 200, the allowable kinetic energy is < 10 GeV.

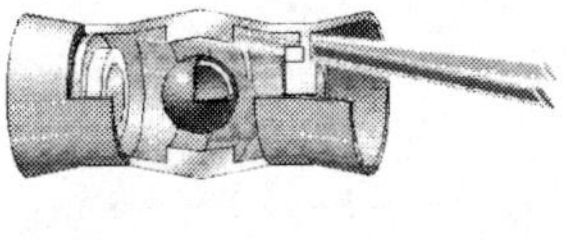

FIGURE 7. A typical IFE indirect-drive (hohlraum) target. Beam power is converted to x-ray radiation, compressing the target material to reach high density and ultimately high temperature for ignition.

Figure 8 is the block diagram of a typical heavy ion fusion driver using induction linacs [8]. Starting from a 2 MV injector (that may contain up to 100 beams at ~ 1 A per beam), the ions will be accelerated to ~ 100 MV using electrostatic quadrupole (ESQ) focusing. At higher ion velocity, magnetic quadrupole focusing becomes more effective. Combining beams at this point may be preferred in order to optimize the overall cost and to rematch beams into the magnetic quadrupole lattices.

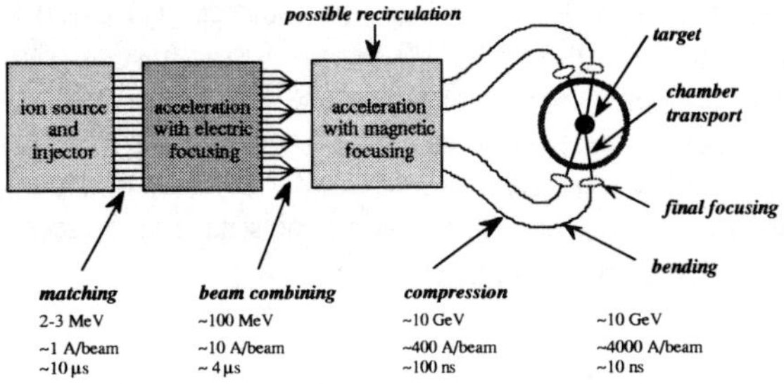

FIGURE 8. Block diagram of a typical Heavy Ion Beam Driver for IFE.

Present HIF Experiments

At present the HIF program is doing proof-of-principle experiments in order to learn how to build high current multiple beam injectors, how to transport space charge dominated beams through quadrupoles, and how to neutralized the space charge during final focusing into the target chamber. Figure 9 is a photograph of the High Current Experiment (HCX) at LBNL [6]. The goal is to transport 0.5A of K^+ at 1.6 MeV energy through 30-40 ESQ modules and 4 or more magnetic quadrupoles in order to study the dynamic aperture, steering, scraping, emittance growth, and electron cloud effects.

FIGURE 9. High Current Experiment (HCX) to study heavy ion beam transport at 0.25µC/m line charge density.

To be cost effective, a fusion power plant driver must consist of multiple beams that share a common set of induction cores. Since the cost of the induction cores is a major cost factor, it is necessary to minimize the core size by building arrays of compact super-conducting quadrupoles. Figure 10 is an example of how to pack multiple quadrupole channels.

The traditional HIF injector, as the one used in HCX shown in Fig. 9, is based on using a large diameter surface ionization source. This type of injector is considered too large (and too costly) when it comes to assembling ~ 100 beams for a driver. The large diameter beams also require a long matching-and-steering section that further complicates matter. Therefore we are developing a new type of compact injector that is based on using plasma ion sources coupled with mini-beamlet channels [7].

In the new design, more than a hundred of high intensity mini-beamlets are extracted and accelerated to > 1 MeV inside individual channels before they are merged together to form a single beam of 0.5 -1.0 A beam current (K^+ equivalent). By controlling the aiming of beamlets, a merged beam can be quickly matched into the ESQ channel. Computer simulation (using a 3D PIC code) showed that the emittance of the merged beam was acceptable and the required brightness could be achieved if the mini-beamlets have current density > 100 mA/cm^2. Figure 11 gives a comparison of the size reduction between the traditional and the advanced designs.

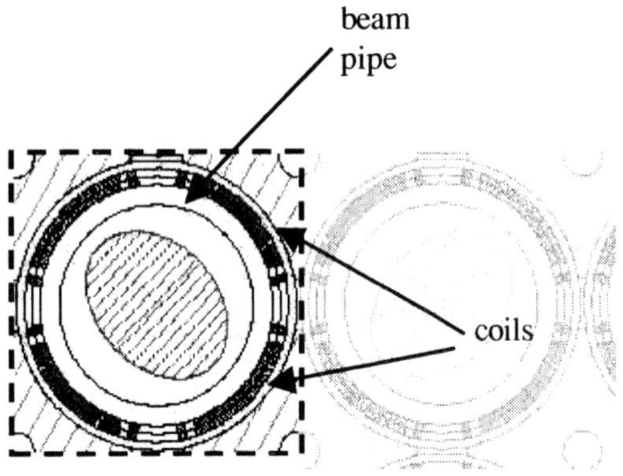

FIGURE 10. A HIF driver consists of an array of linacs using compact super-conducting quadrupoles.

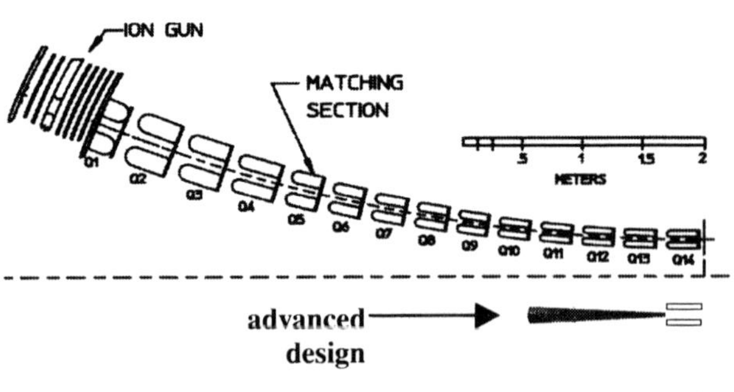

FIGURE 11. The HIF program is developing a new type of compact injector to replace the conventional type.

Future HIF Projects and Plans

The HIF program is proposing to build the next facility called the Integrated Beam Experiment (IBX), at about $50M total project cost. The experiment will be driver-scale in beam current but only short pulse (~1 μs) and limited in the number of parallel beams (1 upgradable to 4). This experiment will demonstrate integrated beam models for injection, acceleration, drift compression, final focus and chamber transport.

In our development path, on the way to eventually reaching a commercial IFE power plant, we expect another $300M facility containing prototype multiple beams (>4) with full pulse length and beam energy at ~ few 100 MeV. Successful demonstration of experiments on this facility will lead into the realization of a $1B class DEMO at beam energies of a few GeV.

ACKNOWLEDGMENTS

I would like to thank Dr. William Barletta for recomending me to give this invited talk, as well as his contribution to the viewgraphs. I am also thankful for the graphic materials provided by Dr. Ka-Ngo Leung and Dr. Bill Chiu. This work is supported by the Office of Fusion Energy, US DOE under contract No. DE-AC03-76SF00098.

REFERENCES

1. Y. Lee, et al, “Maskless Ion Beam Projection Lithography System”, in Proc. the International Conf. Micro- and Nano-Engineering, Septemer 22-24 1998, Leuven, Belgium.
2. Advances in Neutron Capture Therapy, Vol.I, Medicine and Physics, , Eds; B. Larsson, J. Crawford and R. Weinreich,(Elsevier Science B. V.), Amsterdam (1997).
3. J.W. Kwan et al., CP392, Application of Accelerators in Research and Industry, Eds: J. L. Duggan and I. L. Morgan, AIP Press, New York (1997) 1313.
4. S.T. Melnychuk, E. Kamykowski, J. Sredniawski, T. Debiak, R. Ruegg, B. Milton, in Proc. of the Particle Accelerator Conference, New York, 1999, p587.
5. R.O. Bangerter, in Proc. of Inter. Symp. on Heavy Ion Inertial Fusion, Frascati, Italy, 25-28 May, 1993, Nuovo Cimento, vol. 106A, N.11, p1445, (1993).
6. P. Seidl, et al, in Proc. of Inter. Symp. on Heavy Ion Inertial Fusion, Moscow, Russia, May 27-31, 2002, to be published in *Laser and Particle Beams*.
7. J.W. Kwan, et al, Nucl. Instrum. and Methods (A), 464, p379, (2001)

High Intensity Proton Accelerator Facility in Japan

Yoshiharu Mori

KEK, High Energy Accelerator Research Organization

Abstract. The accelerator complex of the joint project in Japan is described.

INTRODUCTION

One of the frontiers of accelerator science is the path toward the highest beam power. Many new accelerators are based on this philosophy. In proton accelerators, the high beam power allows production of a variety of intense secondary particle beams such as kaons, neutrons , muons, neutrinos. antitprotons , and short-lived radioactive nuclear beams. In nuclear and particle physics, an example using these secondary beams is to measure rare processes such as neutrino oscillations and CP violations. In addition, sciences and technologies other than particle and nuclear physics can be carried out by using these secondary beams. These sciences and technologies include a) material and life sciences with neutron and muon beam, b) accelerator -driven nuclear transmutation of long-lived nuclides in nuclear waste, and others related to the various secondary particles. In order to realize these requirements, the new high intensity proton facility is under construction in Japan as a joint project of the Japan Atomic Energy Research Institute (JAERI) and the High Energy Accelerator Research Organization (KEK). The location of the facility is the JAERI/Tokai site. The project has evolved from the Neutron Science Project (NSP) of JAERI and the Japan Hadron Facility (JHF) project of KEK. In Fig. 1, expected beam powers in the present facility are compared with those available in the world.

FACILITY OUTLINE

The facility comprises a 400 (600)-MeV linac, a 3GeV raid-cycling synchrotron (RCS) and a 50-GeV synchrotron (MR) as shown in Fig. 2. A half of the linac are injected to the RCS , while the other half are further accelerated up to 600 MeV by the superconducting (SC) linac. The RCS provides a beam power of 1MW to the pulsed spallation neutron source with the repetition rate of 25Hz, while the 50-GeV MR provides a beam current of 15 microA with a period of about 3-3.6 s to either the particle and nuclear physics or the neutrino production target. In case of the neutrino physics, the total beams are

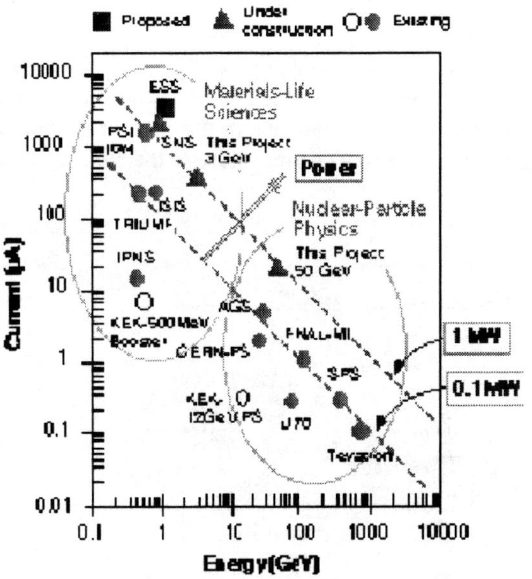

FIGURE 1. Beam power to be achieved by the facility.

extracted with the one-turn extraction. The time structure of the MR cycle and bunch configurations in the MR are shown in Figs. 3 and 4.

The phase I of the project was approved for the construction starting from April, 2001. The phase I will be completed by March 2007. In the phase I the linac will be constructed only for the RCS injection (400MeV). The 50-GeV MR will be operated with an energy of 30 GeV. Although the neutrino production target is not included in phase I, the effort will be immediately started for the approval of the phase II in order to start the experiments for the long-base line neutrino experiments and the ADS.

In Table 1, the main beam parameters of the accelerator complex are listed. The H- beams, which are produced with the negative ion source are accelerated to 400 MeV by the linac. The beams are chopped with a chopping rate of 56 %. The two buckets in the RCS are waiting for the beam injection. The injection continues for 500 microsec, while the magnet system of the RCS is operated with sinusoidal waves of a frequency of 25 Hz. After the beams are accelerated to 3 GeV, the beams are fast extracted for most of times to the muon production target and neutron production target located in a series in the

CP642, *High Intensity and High Brightness Hadron Beams: 20th ICFA Advanced Beam Dynamics Workshop on High Intensity and High Brightness Hadron Beams*, edited by W. Chou, Y. Mori, D. Neuffer, and J.-F. Ostiguy

Material and Life Science Experimental Hall. Every three second, however, the beams are extracted to the MR. The two buckets among the nine buckets in the MR accept the two bunches from the RCS at a time. This is repeated four time as shown in Fig. 4. After the last two bunches are injected , the ramping is immediately started, and the beams are slowly extracted for 0.7 s. to the Nuclear and Particle Physics Experimental Area in one case after the beams are accelerated up to the energy of 50 GeV. The beams, in other case, are extracted with one-turn extraction kicker and septum magnets to the neutrino production target. The total main ring cycle in the slow extraction mode is about 3.6 sec.

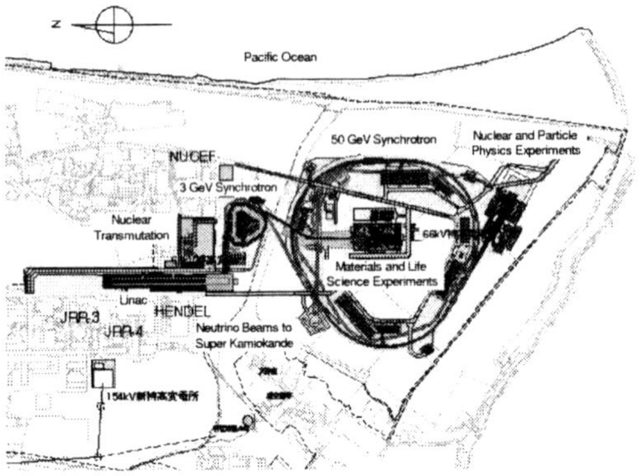

FIGURE 2. Layout of the facility.

TABLE 1 Main Beam Parameters

Linac	
Energy for RCS injection	400 MeV
Energy for ADS	600 MeV
Peak Current	50 mA
Beam Pulse Length	500 ms
Repetition Rate	50 Hz
RCS	
Extraction Beam Energy	3 GeV
Repetition	25 Hz
Average Beam Current	333 mA
Extraction Scheme	Fast
MR	
Extraction Beam Energy	50 GeV
Average Beam Current	15 mA
Repetition	0.3 Hz
Extraction Scheme	Fast, and Slow

In addition to the spallation neutron source and the nuclear and particle physics experimental areas, the accelerators provide the beams to the experimental areas for the muon science and the accelerator-driven nuclear waste transmutation system (ADS). The muon production target is located in a series in front of the neutron production target within the Materials and Life Science Experimental Area as mentioned above. Thus, the requirement for the accelerator complex from the muon science is only the beam pulse length (100 ns) in this case. Fortunately, this requirement is consistent with the present accelerator scheme.

For the ADS, the continuous wave (CW) beams are ultimately required. The linac to be operated in both the pulsed mode and the CW mode has been once proposed for the NSP, since the CW proton linac can be operated in the pulsed mode. However, this kind of linac is very expensive. If the peak beam current in the CW mode is lower, by a factor of three or more, than that in the pulsed mode, the linac may be more expensive than two linacs, that is, CW one and pulsed one. For this reason, the ADS experiments, which are costly compatible with the pulsed neutron source, are limited to those which use the pulsed beams. Since the pulse length the 3-GeV beam from the RCS is too short for the ADS experiment, the 600-MeV linac is necessary for the ADS experiment.

Accelerator Complex

a) Linac

The linac comprises a volume-production type of H- ion source, a 50-keV low-energy beam transport (LEBT), a 3-MeV, 324-MHz Radio-Frequency Quadrupole (RFQ) linac, a 50-MeV, 324-MHz Drift-Tube Linac (DTL), a 200-MeV, 324-MHz Separated DTL (SDTL), and a 400-MeV, 972-MHz high-energy linac as shown in Fig. 5. Table 2 summarizes the main parameters. The linac will be operated at a repetition rate of 50Hz. The 400-MeV beam is transported and injected to the RCS at a repetition rate of 25 Hz. At the other half of the repetition, the 400-MeV beam is further accelerated to 600MeV and is used for the ADS.

TABLE 2 Main parameters of the 600-MeV linac.

Energy	600 MeV
Repetition	50 Hz
Beam Pulse Length	500 μs
Chopping Rate	56 %
RFQ, DTL, SDTL Frequency	324 MHz
ACS, SCC Frequency	648 MHz

Peak Current	50 mA
Linac Average Current	1.25 mA
Average Current after chopping	700 μA
Total Length	350 m
	(393 m with a debuncher)

We have developed the smallest-possible electro quadrupole magnets. The electromagnet coils are produced by fully using the electroforming method and the wire cutting. In this way, it becomes possible to use a frequency of 324 MHz for the DTL starting from 3 MeV. Definitely, the klystrons can be used for this frequency. However, the huge power feeding system is necessary for exciting these electromagnets. Another problem arising from the high accelerating frequency is that the accelerating energy of the RFQ linac is quite limited (2 ~ 2.5 MeV for ~400MHz), since the four-vane type of the RFQ cannot exceed four times as long as the free-space wave length. This problem is solved by the invention of the p-mode stabilizing loop (PISL), which is also used for the SNS. The PISLs eliminate any effect of the deflecting field, resulting in the high quality of the accelerating and focusing fields.

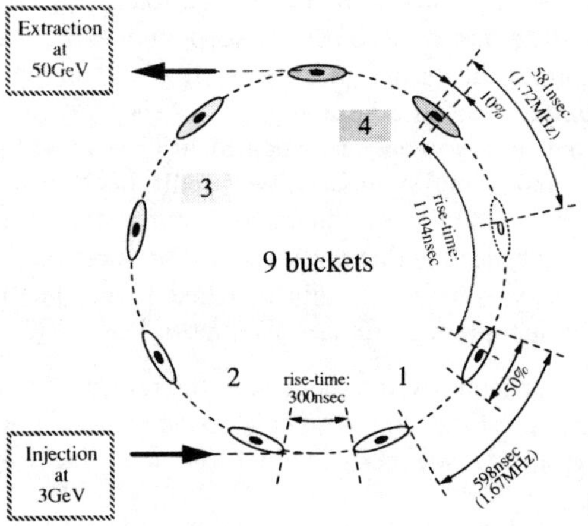

FIGURE 3. Bunch configurations in the 50-GeV MR.

Another feature of the linac design is that the longitudinal transition (200 MeV from SDTL to ACS) is separated from the transverse transition (50 MeV from DTL to SDTL). It is well known that the beam loss and beam quality degradation arise at the transitions. The separation of the two transitions give us more flexibility in order to avoid the mismatching at the transition, which gives rise to the halo formation. It should be emphasized that the linac is used as an injector to the RCS. The most stringent re alignment of 0.05 ~ 0.1 mm is necessary for the quadrupole magnets. In this context, the axial symmetry is perhaps important. This is one of the reasons for developing the Annular-Ring Coupled Structure (ACS) for the high-energy linac structure. The axial symmetry also imply the easy manufacturing and the mechanical stability of the structure.

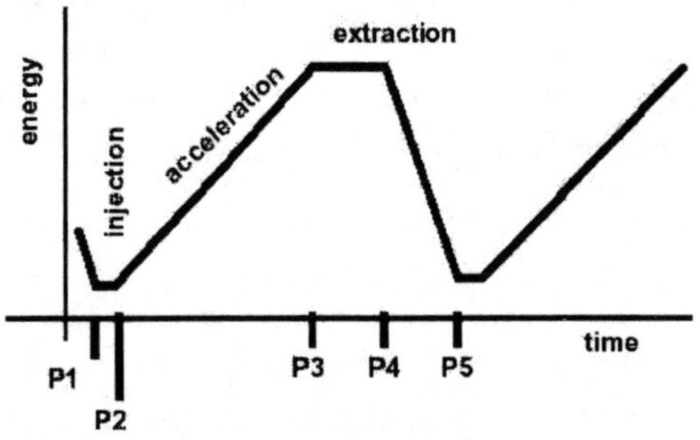

P1 - P2(injection)	0.17 s
P2 - P3(acceleration)	0.13+1.7+0.13 s
P3 - P4(extraction)	0.7 s
P4 - P5	0.7 s
total	3.53 s
slow extraction of 50GeV	
duty factor	0.20
average current	15μA

FIGURE 4. Time sturucture of the Main Ring cycle.

The medium-energy beam transport (MEBT) is also an important component in the proton linac, in particular for the injector linac. First of all, the beam from the RFQ should be matched to the DTL both longitudinally and transversely. Second, this is the place where one can chop the beam, which the ring RF separatrix cannot accept for its phase. The chopping is very difficult to do, since the chopping field should rise and fall, respectively, in between the two bunches. Otherwise, the beams partly deflected by the chopper would be accelerated, eventually giving rise to the high-energy beam loss. The RF chopper has been devised, and developed for this. Another difficulty in the chopper is that any scraper or stopper cannot stand the beam loss of all the chopped beams. The beams will be partly chopped before entering the RFQ linac, by decelerating the beam below the energy acceptance of the RFQ.

We are developing the ion sources both with and without cesium. At first we attempted the ion source without cesium, that is, purely volume production, since we prefer cesium-free ion source in order to avoid the possible decrease in the discharge limit of the following RFQ. However, the peak beam current of the cesium-free ion source is limited to 16 mA so far. Further improvement of the cesium-free ion source is under way. On the other hand, the cesium-seeded ion source being developed as a back up (of course, useless, if the RFQ cannot allow the use of the cesium) has already produced sufficient peak beam current. After the arc discharge power supply is upgraded, the peak beam current was increased in proportion to the arc power up to 70 mA (above the required value) with an aperture size of 8 mmϕ. The emittance measured is small enough. At present the effort is concentrated on the increase in its lifetime, which is one half of the required value.

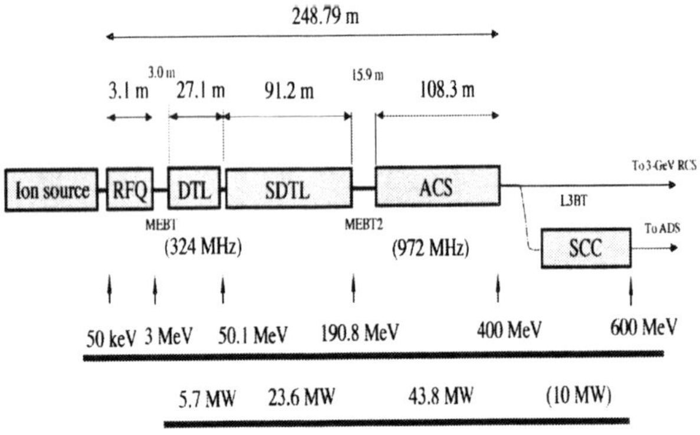

FIGURE.5 Layout of the linac

The Lorentz detuning which becomes dynamic under the pulse operation should be accurately compensated. The SCC has been recently power-tested with the same pulse mode as required. The detuning is periodic from pulse to pulse. The amount of the static detuning was in agreement with the simulation [14] within a few percent. This detuning will be accurately compensated, if one uses a system of one SC cavity per one klystron. However, a system of two SC cavities per one klystron is only competitive in cost with the normal conducting (NC) system, if one uses the SCL for the acceleration from 200 MeV to 400 MeV. Therefore, the feasibility of the 400-MeV SCL as an RCS injector is dependent upon how similar the detuning of the two cavities are to each other. It is recently realized that the high field gradient imposes further severe phase-amplitude control for the same deviation of the beam energy. For the same reason as the larger acceptance, the random kick or walk and the synchrotron oscillation during the course of the acceleration through the higher field gradient cavities becomes larger in the direction of the Dp/p in the longitudinal phase space. Under the presence of the Lorentz detuning the field control of the SCL is obviously much harder than the NC linac. For this reason, we have finally decided to use the NC linac up to 400. The commissioning of the 3-MeV RFQ linac has been started last March. The beam transmission through the RFQ was in agreement with the designed value.

b) Rapid Circulating Synchrotron

We have chosen the lattice with three-folding symmetry. We need three long straight sections. One is dedicated to the long RF acceleration section, another to the injection and collimation, and the other to the extraction. The latter two sections will suffer from a lot of radioactivity, in particular, the injection/collimation section. It is preferable to keep the RF section apart from these radioactive sections, since the maintenance of the RF components are usually required more frequently than other components. The circumference of the RCS is limited by two factors. One is the beam pulse length of less than 1 ms for the neutron production, and the other is the circumference of the MR. As seen from Fig. 6, the present circumference for the MR is perhaps the maximum, if one attempts to keep the MR within the campus. If one increases the circumference of the RCS, the number of the beam transfer from the RCS to the MR must be decreased, resulting in the decrease in the beam current of the MR. Once the circumference of the RCS is thus limited, the three-folding symmetry should be taken in order to keep one long straight section for the sufficient RF acceleration. Although the advantages and the disadvantages of the three-folding symmetry have been investigated in comparison with the four-folding symmetry, we finally decided to use the three-folding symmetry partly for this reason. Another reason is that the lattice with the three-folding symmetry is geometrically matched to the landform rather than the four-folding symmetry.

Other features incorporated in the lattice design are as follows. First, the straight sections are made of no-dispersive lattice in order to avoid the synchro-beta coupling. Second, the transition energy is chosen far above the operation energy. The disadvantages of the RCS have been discussed in Sec. 1.2 in comparison with the AR scheme. The RCS design should solve these problems. First of all, the space charge limit on the beam current should be increased as much as possible. For this purpose, we attempt to increase the beam emittance as large as possible, keeping the gap of the bending magnets fixed. Perhaps, this is most cost-effective method. Specifically, we set the gap of the bending magnets as 210 mm. Then, we attempted to make the beta function there as small as possible. As a result, we could keep the physical aperture of 486 π mm mrad. The collimator acceptance was chosen two third as wide as the physical aperture in order to restrict the beam loss at the collimator. Finally, the painting emittance of 216 π mm mrad was chosen two third as wide as the collimator acceptance. This implies that the emittance growth is allows up to 1.5 times after the injection.

TABLE 3. Main parameters of the 3-GeV synchrotron.

Energy	3 GeV
Beam Intensity	8.3 x 1013 ppp
Repetition	25 Hz
Average Beam Current	333 mA
Beam Power	1.0 MW
Circumference	348.3 m
Magnetic Rigidity	3.18 ~ 12.76 Tm
Lattice Cell Structure	(3-Cell FODO x 2module arc + 3-Cell Straight) x 3
Typical Tune	(6.72, 6.35)
Momentum Compaction Factor	0.012 (no transition below 3 GeV)
Transition g	9.17
Total Number of Cells	27
The Number of Bend. Magnets	24
Magnetic Field	0.27 ~ 1.1 T
The Number of Quadrupoles	60
Maximum Field Gradient	4.6 T/m
Harmonic Number	2
RF Frequency	1.36 ~ 1.86 MHz
Ave. Circulating Beam Current	9 ~ 12.4 A
RF Voltage	467 kV
RF Voltage per Cavity	42 kV (14 kV/gap)
The Number of RF Cavities	11 (+1)
Painting Emittance at Injection	216 π mm.mrad
Collimator Acceptance	324 π mm.mrad
Physical Aperture	486 π mm.mrad
Beam Emittance at Extraction	81 π mm.mrad
Bunch. Factor with 2nd harmonic	0.41
Incoherent Tune Shift	0.16
Bunch. Factor without 2nd harm.	0.27
Incoherent Tune Shift	0.24

The measure of the space charge effect is represented by the incoherent Laslett tune shift (spread). The value of the tune shift for the beam power of 1 MW is 0.24 with a bunching factor of 0.27, while it will come down to 0.16, if the bunching factor is improved to 0.41 by introducing the second harmonics into the RF accelerating field. Although the emittance growth should be carefully estimated on the basis of the beam simulation, the tune shift of 0.16 looks reasonable for keeping the emittance growth within 1.5 times. Taking all of these features into the lattice design, we have eleven families of magnet power supplies. The precise tracking of each of families is necessary. It is really a technical challenge how to precisely track this large number of families of the magnets.

One of the most difficult problems inherent to the high-energy RCS was solved by the innovative development of the accelerating cavity loaded with magnetic alloy(MA) [16-18], one of which is FINEMET. This cavity can generate the field gradient of over 50 kV/m (potentially over 100 kV/m) which is several times as high as conventional ferrite-loaded cavities. For this reason the RF system becomes a reasonable size even for the high-energy RCS. Further power test and beam test of the MA-loaded cavities are being continued after several successful experiments. As an injector the RCS has to match its beam longitudinally for the injection to the MR. For this reason the transition gamma should be much higher than 3 GeV, although the ring circumference becomes longer than the low transition gamma lattice. In addition the beam should be elongated in order to avoid a fast blow up just after the injection.

c) Main Ring

The striking feature of the MR lattice is the choice of the imaginary transition gamma. The imaginary transition gamma is realized by the missing bend method, in which the beta modulation is relatively small. The missing bend structure generates the negative dispersion at bending magnets, resulting in the imaginary transition gamma. Similarly to the RCS, we make the dispersionless straight section in order to avoid the synchro-betatron coupling. The beam emittance at the injection is chosen as 54 π mm mrad, since it corresponds to a Laslett tune shift of $\lesssim$ 0.14 with a bunching factor of 0.27 and a form factor of 1.7. The sizes of the magnets are quite reasonable by this choice. One serious problem is that this injection emittance is based upon the following assumption. The painting emittance at the 400-MeV injection to the RCS is 144 π mm mrad, which grows by a factor of 1.5, being adiabatically damped to 54 p mm mrad. Since the RCS collimator acceptance is 324 π mm mrad, the extracted beams from the RCS can extend beyond this value of emittance. The beams located between 54 π mm mrad and 324 π mm mrad will be lost at the collimator located at the beam transport from the RCS to the MR. If the emittance growth is more than estimated, the beam loss at the collimator will limit the beam current. The emittances related to this matter are listed in Table. 5.

It is also noted that the painting emittance for the MR injection cycle is by a factor of 1.5 less than that for the neutron production cycle. In other words, we have to change the painting emittance depending upon different cycles. The Laslett tune shift for the MR injection cycle is larger (- 0.22) than that for the neutron production cycle (- 0.16).

TABLE 4. Main parameters of the 50 GeV Main Ring.

Energy	50 GeV
Beam Intensity	3.3 x 1014 ppp
Repetition	0.3 Hz
Average Beam Current	15 mA
Beam Power	0.75 MW
Circumference	1567.5 m
Magnetic Rigidity	12.8 ~ 170 Tm
Lattice Cell Structure	Arc(3-Cell DOFO x 8 module) + Insertion(2-matching cell + 3-Straight Cell + 2-matching cell)
Typical Horizontal Tune	22.3
Typical Vertical Tune	17.3-22.3
Momentum Compaction Factor	-0.001 (imaginary gT)
Number of Bending Magnets	96 (5.85 m)
Magnetic Field	0.143 ~ 1.9 T
Total Number of Quads.	216 (0.86,1.26, 1.46, 1.56, 1.66, 1.76, 1.86 m)
Number of Quadrupole Family	11
Maximum Field Gradient	18 T/m
Harmonic Number	9
RF Frequency	1.67 ~ 1.72 MHz
Beam Current (fundamental)	19 ~ 25 A
RF Voltage	280 kV
RF Voltage per Cavity	47 kV (16 kV/gap)
Number of RF Cavities	6
Beam Emittance at Injection	54 π mm mrad
Beam Emittance at Ext.(30 GeV)	10 π mm mrad
Beam Emittance at Ext. (50 GeV)	6.1 π mm mrad

TABLE 5. Emittance and acceptance (π mm mrad) in the MR cycle

	Unnorm.	Norm.	Collimator	Physical
RCS				
Injection	144	146	324	486
Extraction	54	220		
BT from RCS to MR				
	54	220	54	120
MR				
injection	54	220	54~81	81
extraction (30 GeV)				
	10	330		81
extraction (50GeV)				
	6.1	330		81

The slow extraction scheme is most difficult issue to solve for this kind of high-intensity, high-energy proton synchrotron. Only the one percent beam loss is allowed during the slow extraction process. An electrostatic septum (80 mmf tungsten wires with rhenium) is being developed for this purpose. The voltage of 230 kV, which is higher than the necessary value of 170 kV, has been already supplied to the electrodes. Although the beam simulation results satisfy the above requirement, the further improvement in the beam loss simulation will be necessary for increasing the margin, which is needed for this kind of the beam loss/radioactivity elimination.

The RF system of the MR will also use cavities loaded with the same MA as that of the 3-GeV ring. However, the Q value will be optimized for the MR. The tunabliity of the Q value by cutting the MA core, which is also developed for this project, is another important advantage of the MA-loaded cavity.

Summary

The accelerator scheme for the high-intensity proton accelerator facility project in Japan is described. This facility has several unique featres. First, the RCS scheme is chosen for the MW proton machine producing the pulsed spallation neutrons. Second, the MR aims to be a MW proton machine also for the several-10 GeV region.

REFERENCES

[1] The Joint Project for High-Intensity Proton Accelerators, KEK Report 99-4, JHF-99-3 and JAERI-Tech 99-056 (1999).

Synchrotron Based Proton Drivers

Weiren Chou

Fermi National Accelerator Laboratory
P.O. Box 500, Batavia, IL 60510, USA

Abstract. Proton drivers are the proton sources that produce intense short proton bunches. They have a wide range of applications. This paper discusses the proton drivers based on high-intensity proton synchrotrons. It gives a review of the high-intensity proton sources over the world and a brief report on recent developments in this field in the U.S. high-energy physics (HEP) community. The Fermilab Proton Driver is used as a case study for a number of challenging technical design issues.

INTRODUCTION

Intense proton sources have been around for years. At this time, the highest beam power from a synchrotron is 160 kW at the ISIS (0.8 GeV) at the Rutherford Appleton Laboratory (RAL) in England, from an accumulator is 64 kW at the PSR (0.8 GeV) at the Los Alamos National Laboratory (LANL) in the U.S., and from a cyclotron is 1 MW at the PSI (0.59 GeV) in Switzerland. A proton driver differs from these sources in the following aspects: (1) The beam energy E is higher; (2) The bunch length σ is shorter; (3) The beam power P is larger. These differences come from the requirements of physics experiments, in particular, the neutrino oscillation experiments. Typical parameters of a proton driver are: $E \geq 4$ GeV, $\sigma \leq 3$ ns (rms), $P \geq 1$ MW. When the proton beam energy is below 4 GeV, π^-/μ^- yield from a carbon target would be too low to be useful. When the proton bunch length is longer than 3 ns, the production rate (i.e., number of π/μ particles per unit proton beam power) and π^+/π^- or μ^+/μ^- polarization ratio would be uneconomical. Because the physics case is strong and the capital cost is modest (less than 1/10 of the cost of a linear collider), proton drivers have attracted worldwide attention. A recently issued U.S. HEPAP Sub-Panel Report identified such a facility as a possible candidate for a construction project in the U.S. starting in the middle of this decade.[1]

There are two types of proton drivers: one is synchrotron-based, like the ISIS; another is linac-based (a linac plus an accumulator), like the PSR. Each type has its pros and cons. Compared with a linac-based system, for given beam power, a synchrotron has the advantage of lower cost, higher beam energy and lower beam current. Its injection beam power is lower. Hence, the stripping foil is less demanding and larger injection loss could be tolerated. On the other hand, however, a synchrotron is more difficult to design, build and operate than an accumulator. The hardware is more challenging and the reliability is not as high.

There are many similarities between the two types of proton drivers, in particular in the linac and linac front-end part. However, the design of a synchrotron and an accumulator is quite different. This paper will focus on the synchrotron-based proton drivers. A paper discussing linac-based ones can be found in Ref. 2.

OVERVIEW OF HIGH INTENSITY PROTON SOURCES

Table 1 is a survey carried out during the Snowmass 2001 Workshop. It gives the major parameters of high intensity proton sources over the world, including machines existing, under construction and proposed. In addition to the ISIS and PSR, several other existing machines also provide considerable beam power: AGS, IPNS, Fermilab Booster, Main Injector and SPS. There are two big accelerator projects currently under construction. One is the Spallation Neutron Source (SNS) at the Oak Ridge National Laboratory (ORNL) in the U.S. It consists of a 1 GeV superconducting linac and an accumulator. The beam power is 1.4 MW. Another is the JHF at the JAERI/KEK in Japan. It has a 400 MeV linac, a rapid cycling 3 GeV synchrotron at 1 MW, and a slow ramping 50 GeV synchrotron at 0.75 MW. There are a number of proton driver proposals from several labs, including Fermilab and BNL in the U.S., and CERN and RAL in Europe. There are also various proposals of high intensity proton sources for applications other than a proton driver, e.g., nuclear waste transmutation and plutonium production (AAA), spallation neutron sources (ESS, KOMAC), proton radiography (AHF), and accelerator driven system for multiple purposes (CONCERT, IHEP/China).

CP642, *High Intensity and High Brightness Hadron Beams: 20th ICFA Advanced Beam Dynamics Workshop on High Intensity and High Brightness Hadron Beams*, edited by W. Chou, Y. Mori, D. Neuffer, and J.-F. Ostiguy

TABLE 1. High Intensity Proton Sources

Machine	Flux (10^{13}/pulse)	Rep Rate (Hz)	Flux† (10^{20}/year)	Energy (GeV)	Power (MW)
Existing:					
RAL ISIS	2.5	50	125	0.8	0.16
BNL AGS	7	0.5	3.5	24	0.13
LANL PSR	2.5	20	50	0.8	0.064
ANL IPNS	0.3	30	9	0.45	0.0065
Fermilab Booster (*)	0.5	7.5	3.8	8	0.05
Fermilab Main Injector	3	0.54	1.6	120	0.3
CERN SPS	4.8	0.17	0.8	400	0.5
Under construction:					
ORNL SNS	14	60	840	1	1.4
JHF 50 GeV	32	0.3	10	50	0.75
JHF 3 GeV	8	25	200	3	1
Proton Driver proposals:					
Fermilab Study I	3	15	45	16	1.2
Fermilab Study II	2.5	15	37.5	8	0.5
Fermilab Study II Upgrade	10	15	150	8	2
Fermilab MI Upgrade	15	0.65	9.8	120	1.9
BNL Phase I	10	2.5	25	24	1
BNL Phase II	20	5	100	24	4
CERN SPL	23	50	1100	2.2	4
RAL 15 GeV (**)	6.6	25	165	15	4
RAL 5 GeV (**)	10	50	500	5	4
Other proposals:					
Europe ESS (**)	46.8	50	2340	1.334	5
Europe CONCERT	234	50	12000	1.334	25
LANL AAA	-	CW	62500	1	100
LANL AHF	3	0.04	0.03	50	0.003
KOMAC/Korea	-	CW	12500	1	20
IHEP/China	1.6	25	40	1.6	0.1

† 1 year = 1×10^7 seconds.
(*) Including planned improvements.
(**) Based on 2-ring design.

RECENT DEVELOPMENT IN THE U.S. HEP COMMUNITY

In July 2001, about 1,200 physicists gathered at Snowmass, Colorado, U.S.A. for a 3-week workshop. The topic was the future of the high-energy physics program in the U.S. One of the working groups was focused on high-intensity proton sources. After a 3-week intensive study, this group issued a 33-page report.[3] This report emphasized that the U.S. high-energy physics program needs an intense proton source, a 1-4 MW Proton Driver by the end of this decade. It also identified areas of accelerator R&D needed to achieve the required performance of a Proton Driver, i.e., a comprehensive and prioritized 26-point plan. This plan serves as the basis for future research and development of high-intensity proton machines including both linacs and synchrotrons.

In the meantime, based on the U.S. HEPAP Sub-Panel recommendation, the directors of Fermilab and BNL had, respectively, initiated proton driver design studies at the two labs. The reports of these studies have either been published or will be released soon.[4,5]

TECHNICAL CHALLENGES

The design of a high-intensity proton synchrotron involves a number of technical challenges. It requires a careful balance on the performance of various technical systems. It also calls for trade-offs between performances and costs. In the following we will list the major design issues and highlight the critical ones. The Fermilab 8-GeV Proton Driver, which is shown in Figure 1, will be used as an example in these discussions.

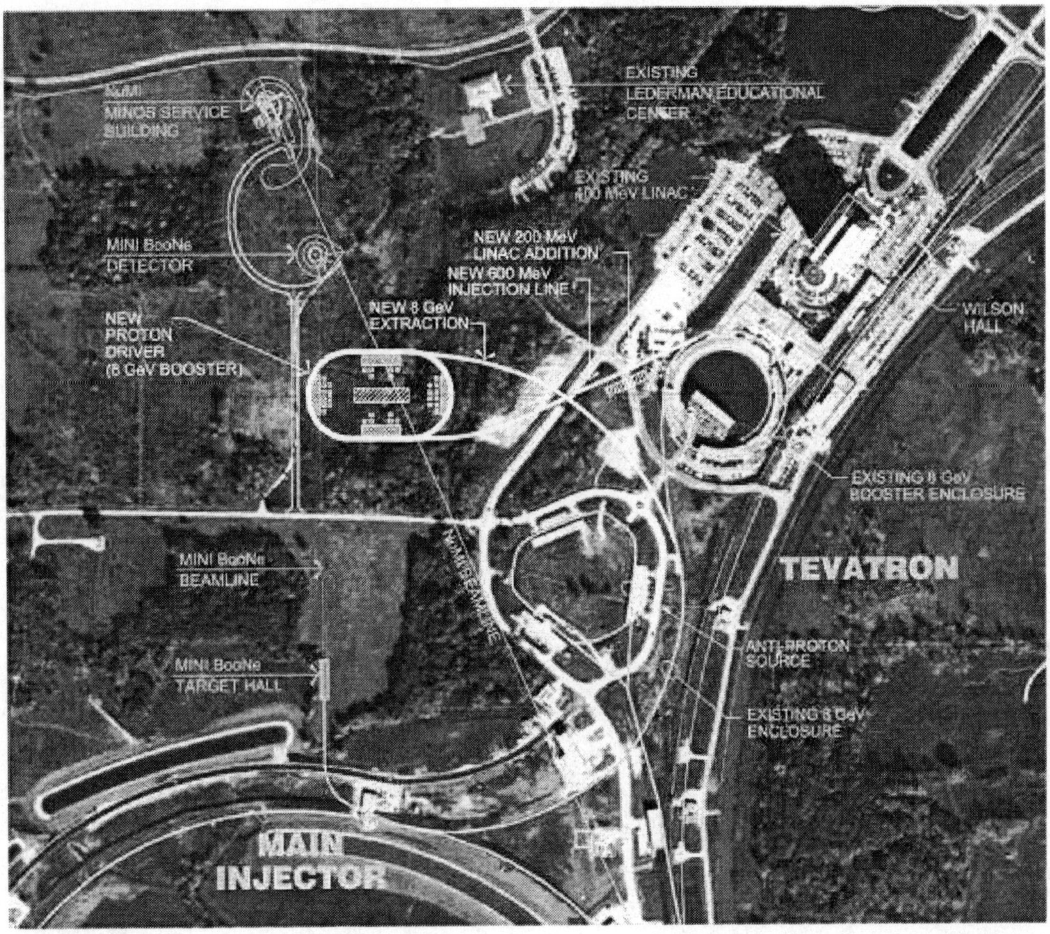

FIGURE 1. The layout of the Fermilab 8-GeV Proton Driver (the ring of the racetrack shape).

1. Lattice Design

Lattice is the foundation of a synchrotron. It is worth every effort to design the best lattice as one can. It would be a mistake to pick a lattice in hurry due to other factors (e.g., pressed by the project schedule, which did happen in the past). A poorly chosen lattice will have adverse effects for the life of the machine. A proton driver has two basic requirements on the lattice: (1) transition free, (2) zero-dispersion in the rf straight sections. The former is to avoid particle loss and emittance dilution during transition crossing; the latter to avoid the synchro-betatron coupling resonance.

For a medium-energy synchrotron (above 6 GeV), the regular FODO lattice (in which $\gamma_t \propto \sqrt{R}$, R the machine radius) is ruled out because it would use too many bending magnets in order to achieve (1). There are several lattices that have been investigated to obtain either a high or an imaginary γ_t. For example, (a) a flexible momentum compaction (FMC) lattice, which is basically a singlet 3-cell modular structure with a missing (or short) dipole in the mid-cell.[6,7,8] (b) a doublet 3-cell modular structure with a missing (or short) dipole in the mid-cell.[9] Figure 2 is an example of (b), which is designed for the Fermilab 8-GeV Proton Driver.

The choice of phase advance per module is of critical importance in this type of lattice. There are two reasons. (i) The chromaticity sextupoles are placed in the mid-cell, where the beta-function peaks and available space exists. In order to cancel the higher order effects of these sextupoles, they need to be paired properly. (ii) The phase advance per arc in the horizontal plane must be multiple of 2π in order to get zero dispersion in the straights without using dispersion suppressors (which are space consuming).

Other requirements in the lattice design include: ample space for correctors (steering magnets, trim quadrupoles, chromaticity and harmonic sextupoles, etc.), ample space for diagnostics, low beta and dispersion functions (to make the beam size small), large dynamic aperture (to accommodate beam halo), and large momentum acceptance (to allow for bunch compression). Table 2 lists the lattice parameters of the Fermilab 8-GeV Proton Driver.

TABLE 2. Lattice Parameters of the Fermilab 8-GeV Proton Driver

Circumference (m)	474.2
Super-periodicity	2
Number of straight sections	2
Length of each arc (m)	161.66
Length of each straight section (m)	75.44
Injection kinetic energy (MeV)	600
Extraction kinetic energy (GeV)	8
Injection dipole field (T)	0.2
Peak dipole field (T)	1.5
Bending radius (m)	19.77
Peak quadrupole gradient (T/m)	10
Good field region	4" × 6"
Max β_x, β_y (m)	15.14, 20.33
Min β_x, β_y (m)	4.105, 4.57
Max D_x in the arcs (m)	2.52
Dispersion in the straight sections	0
Transition γ_t	13.8
Horizontal, vertical tune ν_x, ν_y	11.747, 8.684
Natural chromaticity ξ_x, ξ_y	-13.6, -11.9
Momentum acceptance $\Delta p/p$	±1%
Dynamic aperture	> 120 π

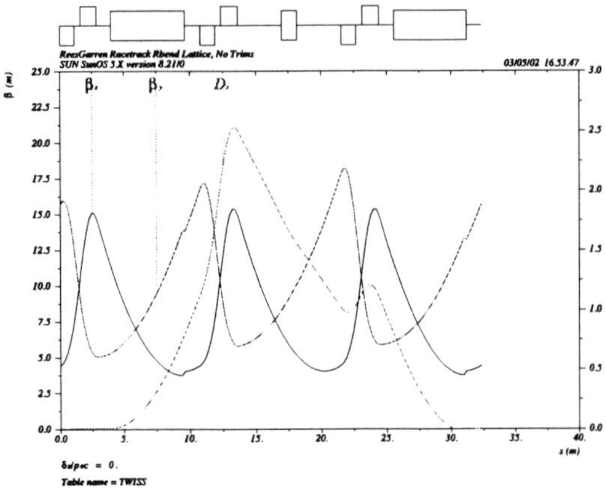

FIGURE 2. Arc module and lattice functions of the Fermilab 8-GeV Proton Driver. Each module has three doublet cells. The dipole in the mid-cell is short. The phase advance per module is 0.8 and 0.6 in the h- and v-plane, respectively. There are five modules in each arc.

2. Space Charge

Amongst numerous beam physics issues, the space charge is a major concern. It is usually the bottleneck limiting the beam intensity in an intense proton source. A useful scaling factor is the Laslett tune shift $\Delta\nu \propto (N/\varepsilon_N) \times (1/\beta\gamma^2)$, in which N is number of particles per bunch, ε_N the normalized transverse emittance, β and γ the relativistic factors. It shows the space charge effect is most severe at injection because the beam energy is low. The situation becomes worse for high-intensity machines not only because the intensity is high but also because the injection time is long. Numerical simulation is the main tool to study this effect. A number of 1D, 2D and 3D codes have been or are being written at many institutions. An example is shown in Figure 3. These codes are particularly useful to the design of the injection kicker current waveform for achieving uniform particle distribution in the beam, reducing emittance dilution and minimizing average number of hits per particle on the stripping foil during the phase space painting process. Several other measures, e.g., tune ramp, inductive inserts, quadrupole mode damper and electron beam compensation are under investigation for possible cures of the space charge effects. This is an active research field.

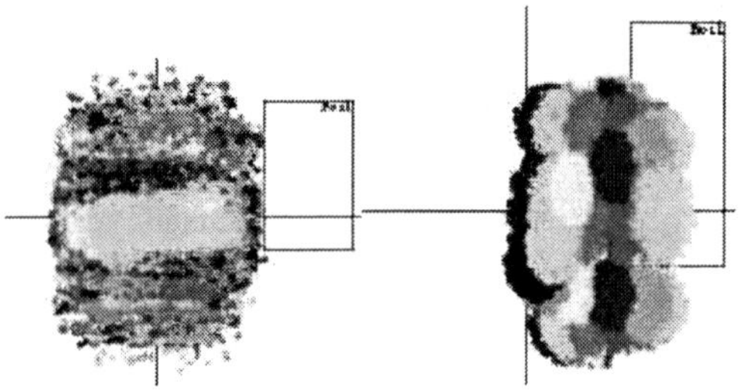

FIGURE 3. Space charge simulation using Track2D (by C. Prior). It shows the particle distribution after 45 turns injection in the Fermilab Proton Driver with (left) and without (right) the space charge effects.

3. Electron Cloud Effect

The electron cloud effect (ECE) has been the No. 1 problem limiting the PSR beam intensity for many years. Recent observations and successful cures of this effect on the CERN SPS, PEP-II and KEK B-factory have stimulated worldwide interest. At this moment, there are six proton machines that have reported observations of ECE. They are: ISR, CERN PS, SPS with LHC beams, SPS with fixed target beams, PSR and RHIC. A key parameter for the ECE seems to be the volume density of particles. It is interesting to notice that, despite enormous differences among these machines in beam energy, number of particles per bunch and bunch size, the volume density takes a remarkably similar value (about $0.2 \pm 0.1 \times 10^8$ /mm^3) when reaching the ECE threshold. [10] This is called the critical mass phenomenon.

It is believed that the ECE is mainly due to secondary electron yield from the wall. Reducing primary electrons (which come from beam loss and stripping foil in proton machines) does not seem to be helpful. It should be pointed out that, by far all reported ECE are either in DC machines (accumulators and storage rings) or AC machines in DC operation (i.e., on flat top or flat bottom). No ECE has been seen in AC machines during ramping. This implies that AC machines could be immune to ECE.

However, all these claims are based on empirical observations or numerical simulations. Lack of a reliable theory for understanding and analyzing the ECE is a loophole that urgently needs to be filled.

4. Other Beam Dynamics Issues

In addition to the space charge and ECE, there are several other beam dynamics issues important to the proton driver design.

- Microwave instability of bunched beam below transition. Because the machine will always operate below transition, the negative mass instability due to space charge would not occur. Would then this machine be immune to the microwave instability?

- Bunch rotation with path length dependence on momentum spread $\Delta p/p$ and space charge tune shift $\Delta\nu$. This is a new problem for proton drivers. Bunch rotation is necessary for obtaining short bunch length (a basic feature of a proton driver). However, due to large momentum spread (a few percent) and large tune shift (a few tenth), the dependence of the path length ΔL on $\Delta p/p$ and $\Delta\nu$ can no longer be ignored. In other words, the momentum compaction factor $\alpha = (\Delta L/L) / (\Delta p/p)$ cannot be treated as a constant during bunch rotation. It is dependent upon the momentum and amplitude of each particle. This results in a longer bunch after rotation.

- A split between the horizontal and vertical tunes is required in order to avoid the strong resonance $2\nu_x - 2\nu_y = 0$ that could be excited by the space charge. However, it is not clear how big the split needs to be. Does it have to be an integer? Or would a half-integer suffice?

5. Beam Loss, Collimation and Remote Handling

The rule of thumb for allowable uncontrolled beam loss in an accelerator enclosure is 1 W/m so that hands-on maintenance can be performed. This number is based on the operation experiences of many machines in many years as well as on numerical simulations at many labs. It is now widely accepted as a design criterion for high-intensity machines. For a 1-MW, 100-m machine, this would mean the loss had to be below 10^{-4}, a mission impossible! To solve this problem, collimators are introduced to localize the beam loss. A 2-stage (i.e., primary + secondary) collimator system can absorb more than 99% lost particles and leave most of the enclosure below 1 W/m. A well-designed collimator system not only has high efficiency, but also is not susceptible to parameter changes (tunes, closed orbit, different stages during the cycle, etc.).

The beam power deposited onto the components near the collimator area can reach as high as ~kW/m. It is a difficult but also critical problem how to handle these components in case they need to be repaired or replaced. Invaluable experiences can be learned from LANSCE (LANL) and PSI. [11] These machines have been handling MW beams for years and have designed several remote-handling systems that work reliably.

6. Negative Ion Sources

Modern high-intensity circular proton machines almost universally adopt the charge exchange injection. The main requirements of the negative ion sources are high intensity (~100 mA), high brightness (rms normalized emittance < 0.2π mm-mrad), high duty factor (several percent) and long lifetime (> 2 months). Low noise surface plasma sources with Cesium catalysis and volume sources are widely used to achieve these goals.

7. Chopper

In order to reduce the injection loss during rf capture, chopping the beam at low energy is crucial. The function of the chopper is to create a macro-structure in the linac beam so that it can fit properly into the rf buckets in the ring. The requirements of a chopper are: fast rise- and fall-time (tens of nsec), short physical length (to minimize the space charge effect, which is dominant at low energies), and flat top and bottom in the current waveform (to minimize the energy jittering in a beam). The ideal place to chop the beam is at the ion source, because the beam energy is the lowest. But the rise- and fall-time would be long (hundreds of nsec) due to the slow response of the plasma. The next best place is in the LEBT (low energy beam transfer) between the ion source and the RFQ. There are two designs. One is the LBL design for the SNS, which places the chopper (made of split electrodes) right after the Einzel lenses. [12] Another is the Fermilab-KEK design, which places the chopper in front of the RFQ. [13] The latter is a pulsed beam transformer made of Finemet cores and does the chopping by using the narrow energy window of the RFQ. It is now installed on the HIMAC linac in Chiba, Japan. A schematic drawing of this chopper is shown in Figure 4. There are also choppers made of traveling wave deflectors placed after the RFQ, which have been in use at the LANL and BNL.

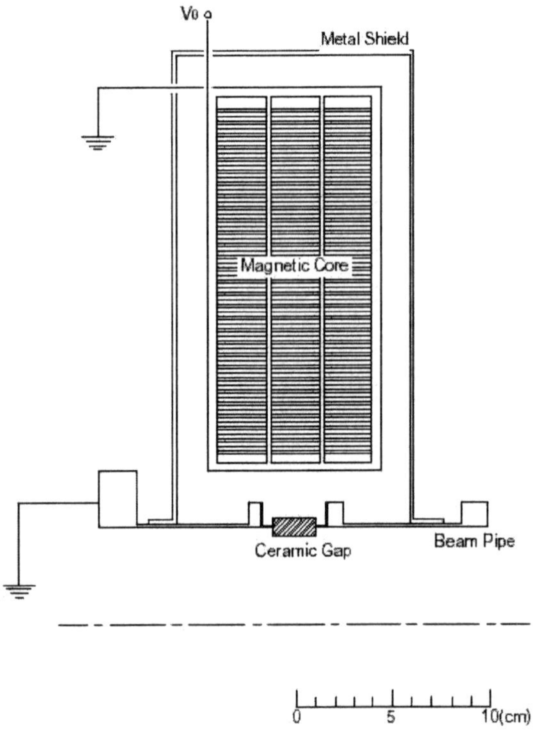

FIGURE 4. A schematic drawing of an rf chopper made of three Finemet cores. It is installed on the HIMAC linac.

8. H^- Injection

This is a complicated part in the proton driver design and has many technical issues involved. Most of particles losses in a synchrotron usually occur at this stage as well as during the rf capture immediately following it.

H^- particles are injected into the ring via a charge exchange process, in which the electrons are stripped by a foil and dumped, and the H^+ (proton) particles stay in the ring. This process takes hundreds or even thousands turns. The stripping foil must be able to stand high temperatures and large shock waves, and must have high efficiency and reasonable lifetime. The unstripped H^-, H^0 and electrons should be collected. Lorenz stripping of H^- ions in a magnetic field must be avoided. Phase space painting in the transverse and longitudinal planes needs to be employed in order to obtain a uniform distribution of particles in a bunch so that the spaces charge and transient beam loading effects can be reduced. Painting also helps minimize the average number of hits per particle on the foil. The emittance dilution due to Coulomb scattering needs to be controlled. Carbon foil is widely used. R&D on diamond foil and laser stripping is being pursued.

9. Slow Extraction

Although the efficiency of one-turn fast extraction can exceed 99%, it is much lower for multi-turn slow extractions. At high-intensity operation, the beam loss in existing machines during slow extraction is usually around 4-5%. This is not acceptable for the next generation of high-intensity machines, in which the beam power will be 1 MW or higher and one percent loss would mean 10 kW or higher. This is a serious problem when physics programs require slow extractions (which is the case for KAMI and CKM at the Fermilab Main Injector, and for kaon and nuclear physics at the JHF). Workshops and beam experiments are planned for tackling this problem.

10. Hardware

Although a number of proton synchrotrons have been built in the past half-century, hardware for MW machines presents particular challenges.

10.1. Magnets

Magnets are one of the most expensive technical systems of a synchrotron. A critical parameter in the magnet design is the vertical aperture of the main bending magnets. The magnet cost is essentially proportional to the aperture. It should be large enough to accommodate a full size beam and its halo. The following criterion was adopted in the Fermilab Proton Driver design:

$$A = \{3\ \varepsilon_N \times \beta_{max} / \beta\gamma\}^{1/2} + D_{max} \times \Delta p/p + \text{c.o.d.}$$

in which A is the half aperture, ε_N the normalized 100% beam emittance, β_{max} the maximum beta-function, D_{max} the maximum dispersion, c.o.d. the closed orbit distortion. The parameter 3 is the estimated size of the beam halo relative to the beam size.

Because this is an AC machine, field tracking between the dipoles and quadrupoles at high field is an important issue. Trim quads or trim coils are needed. The peak dipole field should not exceed 1.5 Tesla. The peak quadrupole gradient is limited by the saturation at the pole root (not pole tip).

The choice of the coil turn number per pole is a tradeoff between the coil AC loss and voltage-to-ground. The former requires the use of many small size coils, whereas the latter requires the opposite, namely, small number of turns. There are two ways to compromise. One is to employ stranded conductor coils, as shown in Figure 5, which was adopted in the JHF 3-GeV ring design. Another is to connect several coils in parallel at the magnet ends, as done in the ISIS. The ratio of the AC *vs.* DC coil loss should be kept around 2-3. The voltage-to-ground should not exceed a few kV. [14]

The aperture and good field region should include a rectangular area (instead of an elliptical area). This is because there will be a significant number of particles residing in the corners of the rectangle. The Fermilab 8-GeV Proton Driver design chose a 4 in × 6 in rectangle.

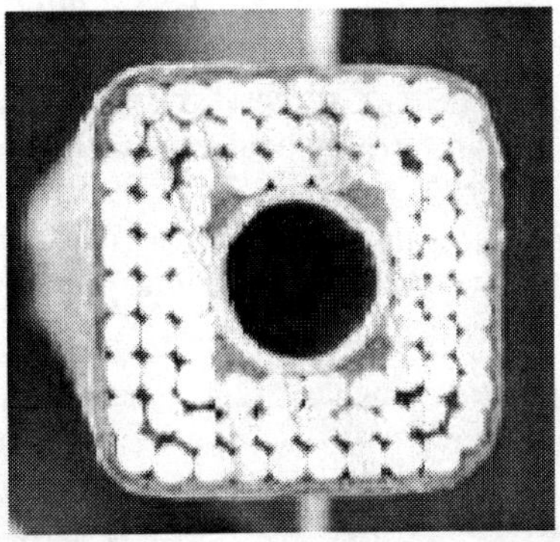

FIGURE 5. Stranded conductor coils for reducing coil AC losses.

10.2. Power Supplies

This is another expensive technical system. There are several choices for the power supplies in a rapid cycling machine. (1) A single harmonic resonant system, e.g., the Fermilab Booster which resonates at 15 Hz. (2) A dual-harmonic resonant system, e.g., the Fermilab Proton Driver design which uses a 15 Hz component plus a 12.5% 30 Hz component as shown below: [15]

$$I(t) = I_0 - I\,cos(2\pi ft) + 0.125\,I\,sin(4\pi ft)$$

in which f = 15 Hz, I_0 and I are two constants determined by the injection and peak current. The advantage of this system is that the peak value of dB/dt is decreased by 25%, as shown in Figure 6, which leads to a saving of the peak rf power by the same amount. (3) A programmable ramp system, e.g., the AGS Booster and AGS. Although this is a most versatile system (e.g., allowing for a front porch and a flat top), it is also most expensive.

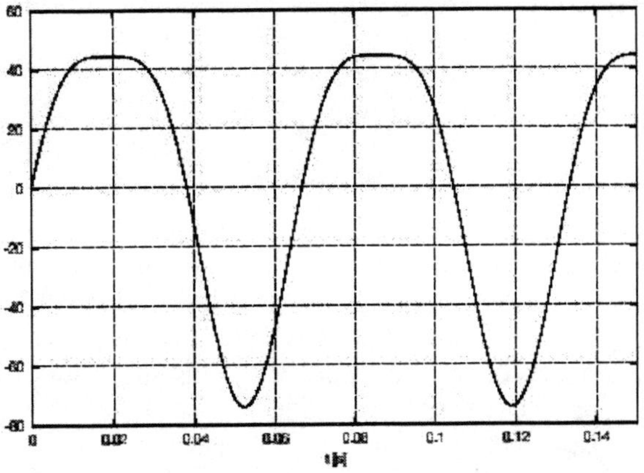

FIGURE 6. Waveform of the time derivative of the B-field (dB/dt in T/s) generated by a dual-harmonic power supply system. Compared to a single harmonic system, the peak value during up-ramp is decreased by 25%.

FIGURE 7. A 7.5 MHz Finemet rf cavity installed in the Fermilab Main Injector.

10.3. RF

The rf system is demanding, because it must deliver a large amount of power to the beam in a short period. In addition, it must be tunable, because the particle revolution frequency increases during acceleration. Cavities with ferrite tuners have been in use for decades. Recently the development of the Finemet cavities at the KEK has aroused strong interest at many laboratories. Thanks to a US-Japan collaboration, Fermilab has built a 7.5 MHz, 15 kV Finemet cavity and installed it in the Main Injector for bunch coalescing, as shown in Figure 7. [16] The main advantages of the Finemet cores are high accelerating gradient and wide bandwidth. The former is especially important for high-intensity small size rings, in which space is precious. The main concern, however, is its high power consumption. For example, the Fermilab Finemet cavity needs a 200 kW power amplifier to drive it. New types of magnetic alloys are under investigation for performance improvement.

10.4. Vacuum

Vacuum pipe for a rapid cycling machine is probably one of the most challenging items. Ceramic pipe with a metallic cage inside has been successfully employed at the ISIS. However, this is a costly solution, because it occupies a significant portion of the magnet aperture. Assuming the ceramic wall and the cage need a 1-in vertical space, the magnet aperture would have to be increased from 4-in to 5-in in the Fermilab Proton Driver, a 25% increase. This will directly be translated to a 25% increase in the magnet and power supply costs, equivalent to tens of millions dollars. Therefore, it was rejected by the Fermilab Proton Driver design.

Thin metallic pipe is an alternative. However, it must be very thin (several mils) in order to minimize the eddy current effects (pipe heating and induced magnetic field). Such a thin pipe is mechanically unstable under vacuum. Several methods have been tried to enhance its stability, including ceramic shields, metallic ribs and spiral lining. The first two do not look promising. The third one is under investigation.

In the Fermilab Proton Driver design, a different approach was adopted. The magnets employ external vacuum skins like those in the Fermilab Booster. Perforated metallic liners are used in the magnet gap to provide a low-impedance environment for the beam as shown in Figure 8.[17]

10.5. Diagnostics

A system that can diagnose beam parameters during multi-turn injection is highly desirable. The method for fast, accurate non-invasive tune measurement is being developed. A circulating beam profile monitor covering a large dynamic range with turn-by-turn speed will be crucial for studying beam halo. (A similar instrument has been developed for the linac beam halo experiment at LANL.[2])

11. New Ideas

In the past several years, there are a number of new or revitalized ideas proposed to the proton driver study. For example:

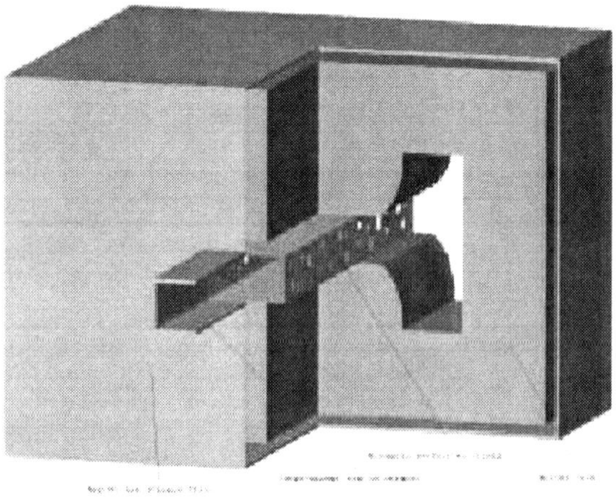

FIGURE 8. Corner section of a canned dipole with a perforated metallic liner.

• Inductive inserts: They are made of ferrite rings and also can have bias current for impedance tuning. Their inductive impedance would fully or partially compensate the space charge impedance, which is capacitive. The first successful experiment was at the LANL.[18] Two ferrite modules made by Fermilab have been installed in the PSR. They help increase the e-p instability threshold, which is a major bottleneck of that machine. Another experiment at the Fermilab Booster is being planned (Figure 9).

FIGURE 9. Inductive inserts in the Fermilab Booster.

• Induction synchrotron: This is a longitudinally separated function machine. In other words, the longitudinal focusing and acceleration are carried out by two separate rf systems. The former uses barrier rf buckets, the latter a constant rf voltage curve. One useful feature of this type of machine is tunable bunch lengths. So the so-called superbunch acceleration could be possible.[19]

• Barrier rf stacking: The application of Finemet and other magnetic alloys makes it possible to build broadband barrier rf cavities with high voltage (~10 kV or higher). They can be used to stack beams in the longitudinal phase space. This is particularly useful when the beam intensity of a synchrotron is limited by its injector (e.g., the intensity of the Fermilab Main Injector is limited by its Booster). Compared to the slip stacking, an advantage of barrier rf stacking is the greatly reduced beam loading effects due to a much lower peak beam current.[20]

• Fixed field alternating gradient (FFAG) accelerator: Although MURA first proposed this idea about 40 years ago, it was almost forgotten. Only the recent activities at the KEK brought it back to the world's attention. KEK has successfully built a 1 MeV proton FFAG and is building a large 150 MeV one.[21] FFAG is an ideal machine for high intensity beams. Its repetition rate can be much higher than a rapid cycling synchrotron (in the range of kHz). One problem of the FFAG, though, is that it is difficult (if not impossible) to fit it into an existing accelerator complex, which usually consists of a linac and a cascade of synchrotrons.

• Repetition rate increase in existing synchrotrons: This is a brute force approach but can be appealing because it is straightforward. For example, in the BNL Proton Driver design, one proposal is to increase the AGS repetition rate from 0.5 Hz to 2.5 Hz.[22] The Fermilab Main Injector upgrade also includes a rep rate increase (from 0.53 Hz to 0.65 Hz).[5]

CONCLUSIONS

Proton drivers are a hot topic in today's accelerator community. Because of their versatile applications, modest costs and great potentials to serve future big projects (a neutrino factory, a muon collider or a VLHC), the designs are being pursued in numerous laboratories over the world. There are many technical challenges. But there are no showstoppers towards the construction of such a facility. The world needs more than one proton driver. International collaborations on a number of R&D items have been formed. Steady progress and fresh ideas can be expected in the coming years in this dynamic field.

ACKNOWLEDGMENTS

A large portion of this paper is based on the Fermilab Proton Driver Study conducted by a group of physicists and engineers from Fermilab and several other institutions.

This work is sponsored by Universities Research Association Inc. under Contract No. DE-AC02-76CH03000 with the United States Department of Energy.

REFERENCES

1. U.S. HEPAP Sub-Panel Report on Long Range Planning for U.S. High Energy Physics, January 2002, http://doe-hep.hep.net/HEPAP/lrp_report0102.pdf
2. Wangler, T.P., "Linac Based Proton Drivers," these proceedings.
3. Chou, W. and Wei J., editors, "Report of the Snowmass M6 Working Group on High Intensity Proton Sources," FERMILAB-Conf-01/396, BNL-52639, August 10, 2001.
4. "The Proton Driver Design Study," FERMILAB-TM-2136, December 2000.
5. "Proton Driver Study II - Part 1," FERMILAB-TM-2169, May 2002.
6. Lee, S.Y. et al., *Phys. Rev.* **E48**, 3040 (1993).
7. Wienands, U. et al., Proc. 1992 HEACC (Hamburg, Germany), p. 1070.
8. See Chapter 3 of Ref. 4.
9. See Chapter 3 of Ref. 5.
10. Chou, W., "Summary Report of Session VI," Proc. ECLOUD'02 Workshop, CERN, Geneva, April 15-18, 2002, CERN Yellow Report 2002-001.
11. Wagner, E., "Remote Handling and Shielding at PSI," these proceedings.
12. Staples, J. et al., Proc. 1999 PAC (New York, USA), p.1961.
13. Chou, W. et al., "Design and Measurements of a Pulsed Beam Transformer as a Chopper," KEK Report 98-10 (September 1998); Chou, W. et al., Proc. 1999 PAC (New York, USA), p. 565; Shirakabe, Y. et al., Proc 2000 EPAC (Vienna, Austria), p. 2468.
14. Ostiguy, J.-F. and Mills, F., Proc. 2001 PAC (Chicago, USA), p. 3248.
15. Jach, C. and Wolff, D., Proc. 2001 PAC (Chicago, USA), p. 3248.
16. Wildman, D. et al., Proc. 2001 PAC (Chicago, USA), p. 882.
17. See Chapter 8 of Ref. 4.
18. Ng, K-Y. et al., Proc. 2001 PAC (Chicago, USA), p. 2890.
19. Yakayama, K. et al., Proc. 2002 EPAC (Paris, France), p. 998.
20. Ng, K-Y., "Doubling MI Beam Intensity using RF Barriers," these proceedigns.
21. Yoshimoto, M. et al., Proc. 2001 PAC (Chicago, USA), p. 51.
22. Weng, W.T. and Roser, T., these proceedings.

The Spallation Neutron Source Project – Physical and Technical Challenges[1]

Jie Wei[2]

for the Spallation Neutron Source Collaboration, USA

Abstract. The Spallation Neutron Source (SNS) is designed to reach an average proton beam power of 1.4 MW for pulsed neutron production. This paper summarizes design aspects and physical challenges to the project.

INTRODUCTION

The SNS project, designed to reach an average beam power above 1.4 MW for pulsed neutron production, is presently in the fourth year of a seven-year construction cycle at ORNL (Fig. 1) [1]. The accelerator system operates at a repetition rate of 60 Hz and an average current of 1.6 mA. It consists of an H^- RF volume source of 48 mA peak current at 6% duty, a low-energy beam transport (LEBT) housing a first-stage beam chopper with ± 20 ns rise/fall time; a 402.5 MHz, 4-vane radio-frequency-quadrupole (RFQ); a medium-energy beam transport (MEBT) housing a second-stage chopper ($< \pm 10$ ns rise/fall), an adjustable beam-halo scraper, diagnostics devices, and matching quadrupoles; a 402.5 MHz, 6-tank drift-tube-linac (DTL) with permanent-magnet quadrupoles; a 805 MHz, 4-module coupled-cavity-linac (CCL); a 805 MHz, superconducting RF (SRF) linac of medium- and high-β cavities accelerating the beam to the full energy; a high-energy beam transport (HEBT) for diagnostics, transverse and longitudinal collimation, matching, energy correction and painting; and an accumulator ring compressing the 1 GeV, 1 ms pulse to 650 ns for delivery onto the target through a ring-target beam transport (RTBT).

Table 1 lists major parameters. The energy acceptance of the ring is about ± 50 MeV, mainly due to conditions for a tolerable H^- and H^0 stripping loss. The back-up scenario corresponds to the case if the surface field of the SRF cavity is lower than expected (37.5 MV/m). Extra tunnel space (71 m) is reserved to extend the linac length for a higher output energy. Table 2 shows evolution of beam parameters during the cycle including expected energy, horizontal (H), vertical (V), and longitudinal (L) acceptances and emittances, and controlled and uncontrolled beam losses.

FIGURE 1. Layout of the Spallation Neutron Source.

TABLE 1. Spallation Neutron Source primary parameters.

	Baseline	Back-up
Kinetic energy, E_k [MeV]	1000	975
Uncertainty, ΔE_k (95%) [MeV]	± 15	± 15
SRF cryo-module number	$11+12$	$11+15$
SRF cavity number	$33+48$	$33+60$
Peak field E_p ($\beta = 0.61$) [MV/m]	27.5	27.5
ΔE_p ($\beta = 0.61$) [MV/m]	± 2.5	± 2.5
Peak field E_p ($\beta = 0.81$) [MV/m]	35	27.5
ΔE_p ($\beta = 0.81$) [MV/m]	$+2.5/-7.5$	± 2.5
Beam power on target, P_{max} [MW]	1.4	1.7
Pulse length on target [ns]	695	699
Chopper beam-on duty factor [%]	68	68
Linac macro pulse duty factor [%]	6.0	6.0
Ave. macropulse H^- current [mA]	26	32
Linac ave. beam current [mA]	1.6	1.9
Ring rf frequency [MHz]	1.058	1.054
Ring injection time [ms]	1.0	1.0
Ring bunch intensity [10^{14}]	1.6	1.9
Ring space-charge tune spread	0.15	0.20

DESIGN PHILOSOPHY

The primary concern is that radio-activation caused by excessive uncontrolled beam loss can limit the machine's availability and maintainability. Based on operational experiences, hands-on maintenance demands that the average uncontrolled beam loss does not exceed 1 W beam power per tunnel-meter [2]. Uncontrolled losses are usu-

[1] SNS is managed by UT-Battelle, LLC, under contract DE-AC05-00OR22725 for the U.S. Department of Energy. SNS is a partnership of six national laboratories: Argonne, Brookhaven, Jefferson, Lawrence Berkeley, Los Alamos, and Oak Ridge.

[2] Brookhaven National Laboratory, and on a joint appointment with the Oak Ridge National Laboratory for the SNS Project.

CP642, *High Intensity and High Brightness Hadron Beams: 20th ICFA Advanced Beam Dynamics Workshop on High Intensity and High Brightness Hadron Beams,* edited by W. Chou, Y. Mori, D. Neuffer, and J.-F. Ostiguy

TABLE 2. Beam parameter evolution across the SNS accelerators. The aperture and acceptance do not include scrapers and collimators. Notes are: a) corresponding to 27% chopped beam; b) corresponding to 5% chopped beam; c) beam loss on the transverse and momentum collimators; d) including total 4% of beam escaping foil and 0.2and f) corresponding to 20% beam loss averaged over RFQ length.

	LEBT	RFQ	MEBT	DTL	CCL	SRF 1	SRF 2	HEBT	Ring	RTBT	Unit
E_k (out)	0.065	2.5	2.5	86.8	185.6	387	1000	1000	1000	1000	MeV
Length	0.12	3.8	3.6	36.6	55.1	64.2	94.7	169.5	248.0	150.8	m
Peak current	47	38	38	38	38	38	38	38	9×10^4	9×10^4	mA
Min. trans. aperture		7	32	25	30	80	80	50	200		mm
Min. H acceptance			250	38	19	57	50	26	480	480	μm
H emit. out, $\varepsilon_{un,rms}$	17	2.9	3.7	0.75	0.59	0.41	0.23	0.26	24	24	μm
Min. V acceptance			51	42	18	55	39	26	480	400	μm
V emit. out, $\varepsilon_{un,rms}$	17	2.9	3.7	0.75	0.59	0.41	0.23	0.26	24	24	μm
Min. L acceptance			4.7	2.4	7.4	7.2	18		$19\times10^5/\pi$		$10^{-5}\pi$eVs
L emit. out, rms		7.6	10	12	14	17	23		$2\times10^7/\pi$		$10^{-7}\pi$eVs
Loss (control)	0.05^a	N/A	0.2^b	N/A	N/A	N/A	N/A	5^c	62^d	58^e	kW
Loss (uncontroll)	70	100^f	2	1	1	0.2	0.2	<1	1	<1	W/m
H emit. out, $\varepsilon_{N,rms}$	0.2	0.21	0.27	0.33	0.39	0.41	0.41	0.46	44	44	μm
V emit. out, $\varepsilon_{N,rms}$	0.2	0.21	0.27	0.33	0.39	0.41	0.41	0.46	44	44	μm

ally attributed to 1) mismatch upon change of linac structure, lattice, and frequency; 2) space-charge effects including envelope and parametric resonances and non-equipartition in the linac, and resonance crossing and instability enhancement in ring; 3) limited physical and momentum acceptance; 4) premature H^- and H^0 stripping and ring injection foil scattering; 5) magnetic errors, fringe fields, and misalignments; 6) instabilities (resistive impedances due to e.g. extraction kicker, and electron cloud); and 7) accidental loss due to system malfunction (ion source and linac, ring extraction kickers).

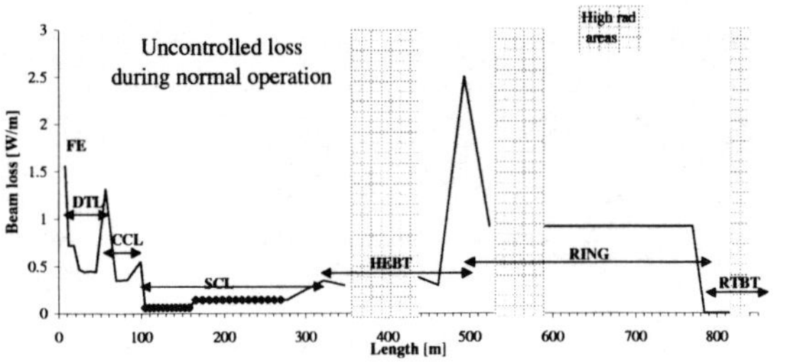

Source: Data from N. Catalan-Lasheras, et al

FIGURE 2. Expected beam loss across the SNS accelerator complex. The uncontrolled beam loss is at 1 W/m level.

SNS addresses the above seven issues by adopting a low-loss design philosophy [3]. Above all, foreseen losses are localized to shielded areas using 1) adjustable scrapers in the MEBT; 2) transverse and momentum collimators in the HEBT prior ring injection; 3) two-stage transverse collimation and momentum cleaning with beam-in-gap (BIG) kicker in the ring; 4) collimator protection in the RTBT, and 5) beam-gap cleaning with LEBT and MEBT choppers and ring BIG kicker (Fig. 2).

Emphasis is also put on machine's flexibility and reliability. The SRF linac allows operation with one failed cavity/klystron; the ring accepts $\pm5\%$ variation in linac output energy; a wide ring tuning range avoids resonances; a robust injection allows independent horizontal, vertical, and longitudinal painting; adjustable collimation systems accommodate variable beam size; and design reserve and redundancy ensure a high availability (e.g., spare cryo-module for a quick replacement, power supplies compatible with 1.3 GeV energy, multi-foil exchange, spare kicker power supply (PFN), and aperture clearance for one-kicker failure).

Finally, the facility is designed with the potential to reach a beam energy up to 1.3 GeV and a beam power higher than 2 MW, capable of supplying a second neutron target. The higher energy can be reached by upgrading the superconducting RF cavity gradient and klystron power supplies, and by filling the presently unoccupied linac tunnel spaces with up to 9 additional cryo-modules. The ring is capable of accommodating the energy and power increase without extensive hardware change – space is reserved for two additional extraction kickers, and for the replacement of 2 injection-chicane dipoles to satisfy H^0 stripping conditions [4].

ACCELERATOR DESIGN CHOICES

Superconducting vs. Warm Linac

The SRF linac operating at 805 MHz frequency accelerates the H^- beam from 186 MeV to top energy (Fig. 3). Comparing with the original normal-conducting (warm) CCL linac, the SRF linac provides a high accelerating gradient (11 - 16 MV/m) capable of reach-

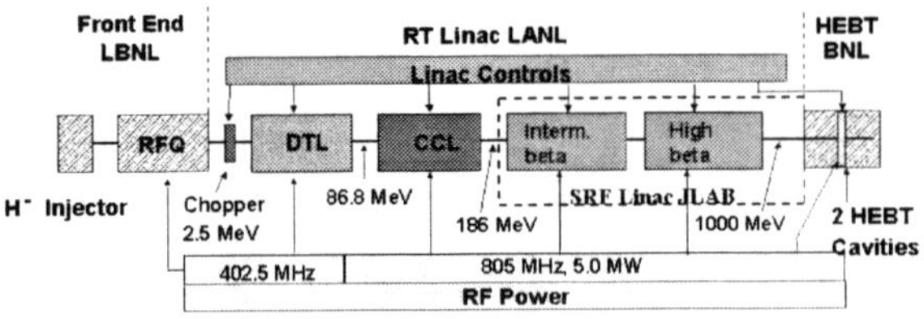

FIGURE 3. SNS front-end and linac stucture.

ing a higher beam energy, encounters less beam loss and halo scraping due to its larger bore radius, is immune to one cavity/klystron failure, operates at a better vacuum, and is expected to have higher reliability and availability. The selection of two types of SRF cavities allows for economic savings and future energy upgrades. On the other hand, the relatively large phase slip requires detailed error-sensitivity analysis. The choice of cavity geometric β value is based on a smooth transition from the warm section linac, a maximized final output energy, and a comfortable transition from medium- to high-β section with tolerance to one cavity failure. We also choose constant-gradient, continuous focusing to maximize the accelerating field strength [5].

Considering the tight construction schedule, a moderate peak surface field of 27.5 MV/m is chosen for the medium-β cavity. Benefiting from electro-polishing, a higher peak field of 35 MV/m is assumed for the high-β cavity. In order to reduce uncertainties in RF controls of an ion ($\beta < 1$) beam under Lorentz detuning, microphonics, beam transients and injecting energy offset, we decide to drive each cavity with its own klystron using independent amplitude and phase control.

Accumulator Ring vs. RCS

During the first year of construction, a study was performed comparing the present structure of full-energy linac plus accumulator ring to a rapid-cycling-synchrotron (RCS) design: a 60 Hz, 400 MeV linac feeds two, vertically stacked RCSs accelerating the proton beam to 2 GeV energy. The biggest challenge to the RCS design is from the stringent (1 W/m) beam-loss criterion: although relaxed by a factor of 5, still only 0.4% uncontrolled loss is allowed for a 2 MW beam power assuming 90% collimation efficiency. On the other hand, among existing rings the lowest loss of about 0.3% is achieved at LANL's PSR, a 800 MeV accumulator, as opposed to typical losses of a few to tens of percent in RCSs (e.g. ISIS, FNAL and AGS Boosters).

As opposed to the accumulator, the RCSs operating at 30 Hz require a high RF voltage (about 400 kV per ring at 1.4 - 1.9 MHz) for fast acceleration, a large magnet aperture to accommodate the space charge at a lower energy, ceramic vacuum pipes with detailed RF shielding, and high-performance power supplies. Minimization of

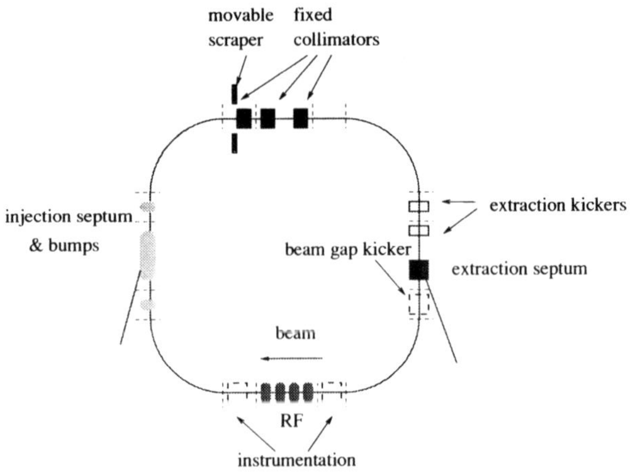

FIGURE 4. SNS accumulator ring layout.

magnetic errors due to eddy current, ramping, saturation, and power-supply tracking is non-trivial. The study concluded that the required RCS design is technically more demanding and less cost effective [4].

Permanent magnets were considered as an option for the accumulator ring magnets. Electromagnetic magnets were chosen instead, given the uncertainty in the linac energy. This choice is especially appropriate to accommodate later-adopted SRF linac.

Ring's FODO-doublet Lattice

The four-fold symmetric ring lattice contains four dispersion-free straights, each housing injection, collimation, RF, and extraction, as shown in Fig. 4. Each achromatic arc consists of 4 FODO cells with 90° horizontal phase advance.

After optimization, the ring lattice has doublet straights [3]. The lattice combines the FODO structure's simplicity and ease of correction with the doublet structure's flexibility for injection and collimation. Injection at a dispersion-free region allows independently adjustable painting in the transverse (with orbit bumps in the ring) and longitudinal (with an energy-spreading phase-modulated RF cavity in the HEBT) directions for a robust operation. The 12.5 m-long uninterrupted straight section with a flexible phase advance further improves collimation efficiency. Comparing with the original all-FODO lattice, matching between the arcs and the straights increases the arc acceptance by 50% with the same magnet aperture (Fig. 5).

CHALLENGES & LESSONS LEARNED

Front End & Warm Linac

Tight optical focusing used for chopping and antichopping in a long MEBT is a source of beam-halo generation. Studies show that even without the antichopper, partially deflected particles are still mostly contained by

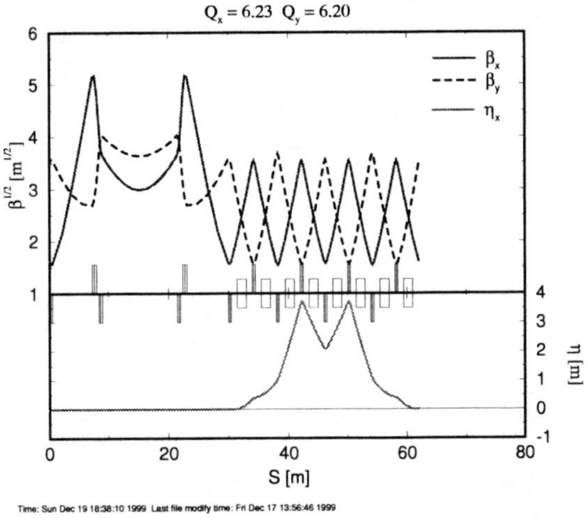

FIGURE 5. SNS ring lattice super-period of FODO/doublet structure. The lattice periodicity is 4.

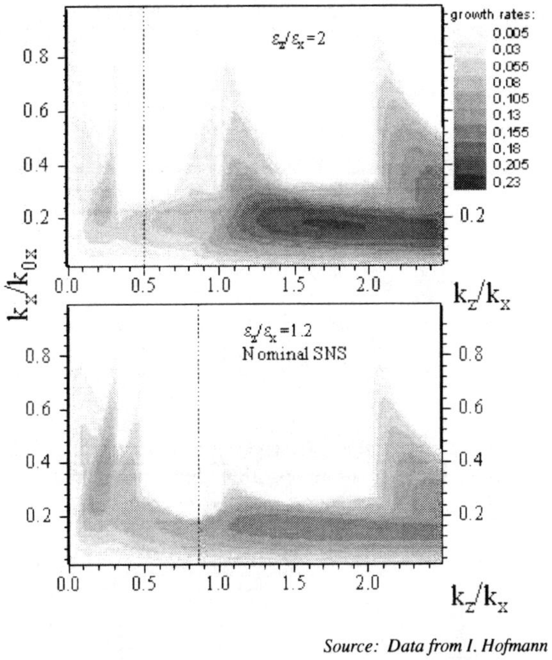

FIGURE 6. Analytical resonance chart showing Instability growth rate due to space-charge coupling resonance. The effects become important only when the transverse and longitudinal tunes are on resonance, and when the emittances differ signifi cantly. The dash lines indicate equipartition.

the envelope of the nominal unchopped beam [6]. The MEBT quadrupoles are thus made independently adjustable so that alternative optics can be realized, avoiding tight focusing at the antichopper or MEBT chopper.

Permanent-magnet quadrupoles are used in the DTL due to the tight geometry (402.5 MHz starting at 2.5 MeV), although electromagnetic quadrupoles could be used at DTL tank 3 and beyond. During 1999, the aperture of CCL was reduced from 4 to 3 cm for cost savings. Later when SRF linac is adopted, simulated beam loss often occurs near the end of CCL as the focusing strength is reduced to match the SRF optics.

A key challenge in linac performance is to minimize beam emittance growth and centroid jitter in both transverse and longitudinal directions upon ring injection, reducing foil traversal, scattering and radio-activation. The warm DTL operating at 402.5 MHz is expected to be less sensitive to vibrational noises than most existing linacs operating at 200 MHz. A tight RF control (<0.5% amplitude and 0.5° phase error) warrants tolerable energy variation before the beam enters energy correction and spreading cavities in the HEBT [7].

Superconducting RF Linac

Using only two types of cavity β for over 800 MeV of acceleration compromises the equipartition law. Space-charge coupling can cause transverse and longitudinal emittance exchange when the emittance ratio meets resonance conditions (Fig. 6) [8]. In addition, depending on the level of initial mismatch, space-charge parametric halo may develop in the linac. Efforts were made to reserve an economically affordable large aperture, and to reserve tunability in the MEBT, CCL and SRF linac.

Effects of higher-order modes (HOM) on the cavities is another issue. Overlapping of beam and HOM spectrum is possible because of the pulsed time structure of the beam and the fact that the beam frequency shifts with variable ring energy and repetition rate (e.g. for some two-target operation scenarios). Fortunately, transverse and longitudinal (beam break-up) instabilities are minor issues for an ion beam in the presence of a cavity-to-cavity frequency spread [9]. HOM dampers are implemented only for the purpose of power dissipation [10].

The SRF linac performance is limited by the available klystrons power (550 kW). Up to 40% RF-power is reserved for compensation of cavity errors (Lorntz detuning, microphonics, coupling loss, frequency setting), klystron loss, and missing-cavity tuning. To reduce such overhead, each SRF cavity is equipped with a piezo crystal driven fast tuner to compensate for the Lorentz force.

Ring and Transport

Solid-steel, as opposed to laminated-steel, was selected for most ring and transport magnet cores for cost savings. Individually, good field quality ($<10^{-4}$ relative error at full acceptance) is achieved. However, excessive (up to 0.25%) magnet-to-magnet variation is found in the dipole transfer function and its current dependence, as shown in Fig. 7 [11]. These dipoles are shimmed to achieve below 10^{-4} variation for 1 GeV operation, and sorted according to 1.3 GeV measurement data to minimize orbit corrector strength.

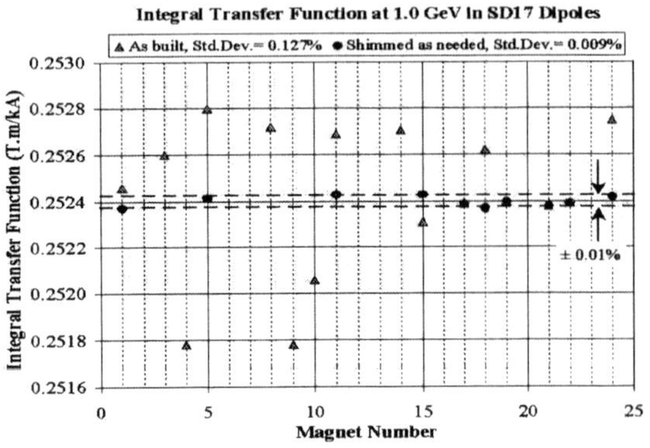

Source: Data from P. Wanderer, A. Jain, et al

FIGURE 7. Variation of integral transfer function of SNS ring dipole magnets before and after shimming. The measurement current corresponds to 1 GeV beam energy.

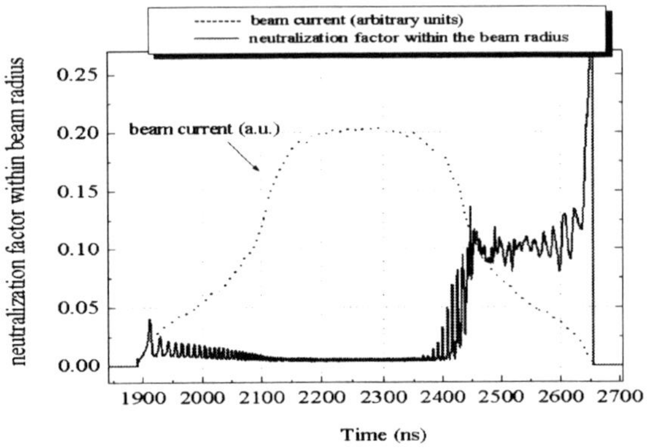

Source: Data from M. Pivi, M. Furman

FIGURE 8. Simulation of single-bunch electron multipacting with a peak secondary-emission yield of 2.

Main ring challenges include meeting the target requirements on the peak current density, minimizing uncontrolled beam loss, and controlling collective effects (space charge, instabilities, electron cloud (Fig. 8)) [4]. Efforts are made to minimize leading sources of beam-coupling impedance (Fig. 9 [12]), and to enhance Landau damping [4].

High-performance beam diagnostics is needed to accommodate the large variation of beam parameters, and for machine protection across the entire facility. Laser-wire monitors are under test for possible implementation in the SRF linac for a clean operation, and luminescence profile monitors are under test to reduce space-charge and electron-cloud complications in the ring.

SUMMARY

By adopting superconducting RF technology for the linac and by fully optimizing the accumulator ring design, the Spallation Neutron Source project, half way towards its completion, is meeting the challenge to be a next-generation, high-power accelerator facility.

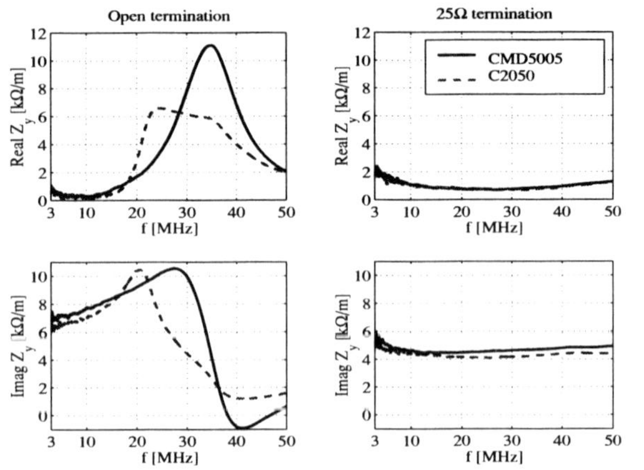

Source: Data from D. Davino, H. Hahn

FIGURE 9. Comparison of bench-measured coupling impedance for open and 25 Ω PFN termination, and high (1600) and medium (100) permeability ferrite of the ring extraction-kicker assembly.

I am indebted to my colleagues, especially those participating in SNS accelerator-physics discussions.

REFERENCES

1. N. Holtkamp, EPAC (2002)
2. N. Catalan-Lasheras et al, SNS Notes SNS/AP/7 (2001)
3. J. Wei et al, Phys. Rev. ST-AB **3**, 080101 (2000)
4. J. Wei et al, EPAC'00, 981; PAC (2001) 319; EPAC (2002)
5. J. Stovall et al, LINAC 2000, p.605
6. D. Jeon et al, SNS 104050000-TD0010-R01 (2002)
7. D. Raparia et al, Workshop HB2002, FNAL (2002)
8. I. Hofmann, et al, PAC (2001) 2902
9. R. Sundelin et al PAC'01 1984; D. Jeon et al, PAC'01 2063
10. S. Kim, et al, PAC (2001) 1128
11. P. Wanderer et al, EPAC (2002)
12. D. Davino et al, EPAC (2002)

Linac Based Proton Drivers

Thomas P. Wangler

Los Alamos National Laboratory
Los Alamos, NM 87544

Abstract. An overview is presented of modern high-power proton linacs, including a survey of worldwide applications that are presently underway. The recent trend towards RF superconducting linacs is also discussed. We present some preliminary results from the beam-halo experiment recently carried out at Los Alamos.

INTRODUCTION

Proton linacs are capable of producing high-energy, high-intensity beams with low emittance. The high-intensity performance is possible because strong focusing can be easily provided, and because the beam traverses the accelerator in a single pass, avoiding repetitive error conditions that lead to destructive beam resonances as in high-current circular machines. The linac can operate at any duty factor, all the way to 100% duty or a continuous wave (CW), which results in acceleration of beams with high average current. At present, there are two major R&D topics in this field: 1) linac applications of RF superconducting (SC) technology, and 2) beam halo formation in high-intensity beams. The latter topic is motivated by the desire to control beam losses and radioactivation of the accelerator, and the need for improved reliability and availability.

This paper is divided into two parts. First, we present an overview of the architecture of modern high-power proton linacs, including a survey of worldwide applications that are presently underway, and we discuss the recent trend toward RF SC linacs. Then, we present some preliminary results from the beam-halo experiment recently carried out at Los Alamos.

HIGH POWER PROTON LINAC OVERVIEW

Figure 1 shows a block diagram of a modern proton linac. The first linac accelerating structure is the radiofrequency quadrupole or RFQ, which bunches, electrically focuses, and accelerates the beam from the DC injector to a final velocity in the range of about β= 0.06 to 0.12. The RFQ is followed by a magnetically-focused section for acceleration of medium-velocity particles consisting of either a normal-conducting drift-tube linac (DTL), or "intermediate-velocity" SC cavities. The third section, beginning at a beam velocity of near β=0.5, consists of either a normal-conducting coupled-cavity linac (CCL), or SC elliptical cavities for high-velocity particles. Magnetic focusing is used throughout the high-velocity section.

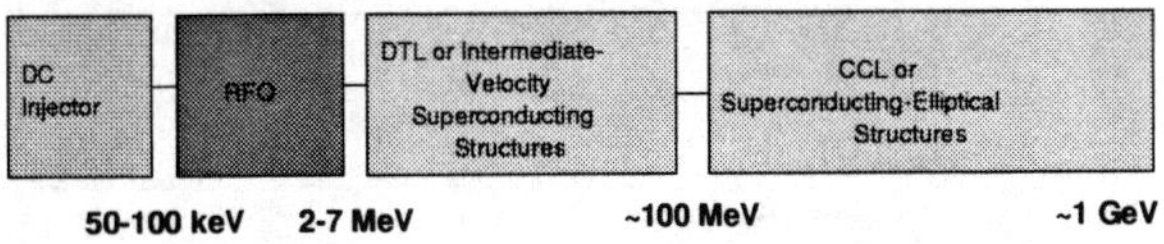

FIGURE 1. Block diagram of a modern high-power proton linac.

Table I shows the parameters for high-power linac projects being designed and/or being constructed. Included are linacs for both H^+ and H^- beams, as well as the IFMIF project that accelerates a deuteron beam. More details about most of these linacs are presented in other papers at this workshop. Most of these linacs would deliver final beams in the megawatt average beam-power range, which qualifies them as high-power accelerators. Applications include linacs for nuclear-waste transmutation (KEK/JAERI, KOMAC, TRASCO), neutrino factory (CERN SPL, FNAL 8-GeV), fusion-materials studies (IFMIF), injectors for spallation-neutron sources (LANSCE, SNS, ESS), and proton radiography (LANSCE). The LANSCE linac is the only high-power proton linac that has operated; first beam was 30 years ago. It should be noted that the LANSCE linac is the only one that does not have an RFQ. The SNS, KEK/JAERI, and KOMAC linacs are in various stages of construction, and will become the first examples of the modern proton linac architecture that was just described.

A new trend during the past decade has been the interest in SC proton linacs. All the linacs in Table I except for LANSCE and IFMIF have superconducting sections either in the baseline design or as an option. Technology progress during the past decade, including a large increase in accelerating gradients for SC cavities, higher power input couplers, and pulsed beam SC electron-linac experience at the TESLA Test Facility, has made RF SC linacs more attractive for proton-linac applications.

CP642, *High Intensity and High Brightness Hadron Beams: 20th ICFA Advanced Beam Dynamics Workshop on High Intensity and High Brightness Hadron Beams*, edited by W. Chou, Y. Mori, D. Neuffer, and J.-F. Ostiguy
2002 American Institute of Physics 0-7354-0097-0

There are several significant advantages of SC linacs that account for this trend. First, operating costs are reduced because of the reduced power dissipation. Second are the advantages of a larger affordable bore radius, which relaxes alignment, steering, and matching tolerances, reduces beam losses and activation, eases commissioning, and improves beam availability. Third, the SC-linac design approach with independently-phased short cavities provides flexibility that allows for continuing operation even with the failure of an accelerating module. Fourth, a worldwide industrial capability exists for fabrication of SC cavities and cryomodules. And finally, SC cavity performance is still improving with higher accelerating gradients and higher cavity quality factors. The SNS linac is scheduled in 2006 to become the first SC proton linac, and its operation will provide even more experience for pulsed SC linac operation.

TABLE 1. High Power Linac Survey

Name	Ion	Pulse length (msec)	Rep rate (Hz)	Duty factor (%)	I_{bunch} (mA)	$I_{Average}$ (mA)	Energy (GeV)	$P_{Average}$ (MW)	Start date
LANSCE	H^+/H^-	0.625	100/20	6.2/1.2	16/9.1	1.0/0.1	0.8	0.8/0.08	ON
SNS	H^-	1.0	60	6.0	38	1.4	1.0	1.4	2006
CERN SPL	H^-	2.8	50	14.0	22	1.8	2.2	4.0	?
ESS Short Pulse	H^-	1.2	50	6.0	114	3.75	1.33	5+5	2010
ESS Long Pulse	H^- or H^+	2/2.5	16.67	4.2	114/90				
FNAL 8 GeV	$H^+/H^-/e^-$	1.0	10	1.0	25	0.25	8.0	2.0	?
KEK/JAERI 400 MeV	H^-	0.5	50/25	2.5	50	0.7	0.4	0.28/0.14	2006
KEK/JAERI 600 MeV			25	1.25		0.35	0.6	0.21	?
TRASCO	H^+	CW	CW	100	30	30	>1.0	>30	?
IFMIF	D^+	CW	CW	100	2X125	2X125	0.040	10.0	2010
KOMAC(KAERI)	H^+/H^-	CW	CW	100	20	20	0.1(1.0)	2.0(20)	2011(?)

LOS ALAMOS BEAM HALO EXPERIMENT

Theoretical research during the past decade has led to a picture of space-charge forces in mismatched beams as a major source of beam halo in high-current proton beams. The nonlinear space-charge force experienced by individual particles, while the beam is undergoing coherent rms-size oscillations, slowly drives some particles to larger amplitudes to form a halo. The mechanism has been described by a particle-core model,[1,2,3] and identified as a parametric resonance.[2]

An experiment has been carried out at Los Alamos to test these ideas. A 52-quadrupole FODO beam-transport channel was installed at the end of the Low-Energy Demonstration Accelerator (LEDA) 6.7-MeV, CW, 350-MHz RFQ to carry out a first experimental study of beam-halo formation in a proton beam.[4,5,6] The channel was long enough for the development of about 10 mismatch oscillations, enough to observe the initial stages of emittance growth and halo formation caused by mismatch. A large compliment of beam diagnostics was provided.[7] The key diagnostic component was a unique transverse beam profile scanner[8] consisting of a thin, 33 μ diameter, carbon wire for measurement of the dense beam core, and a pair of 1.5-mm thick scraper plates to provide greater sensitivity for measurement of the outer halo regions. Each wire and its associated pair of scraper plates were mounted on a common movable frame. As the protons passed through the wire a signal was induced from secondary electron emission. The scraper plates were water-cooled and were constructed of graphite brazed onto copper. A signal was induced by the beam protons that stopped in the plates. The data from the wires and the plates were combined using computer software to produce a single distribution with a dynamic intensity range of about 10^5:1. Nine measurement stations, shown in Fig. 2, were located midway between pairs of quadrupoles; at each location both the horizontal and vertical projected distributions were measured. The first station was located after quadrupole 4. The next four stations were located after the beam had debunched at quadrupoles 20, 22, 24, and 26. The last four stations were located after quadrupoles 45, 47, 49, and 51.

FIGURE 2. Block diagram of the 52 quadrupole transport channel showing the locations of the 9 profile scanners.

Beam matching was done by adjusting the first four quadrupoles to produce equal rms sizes in both x and y. We estimated a measurement uncertainty for the rms beam sizes about 50μ, caused mostly by transverse beam jitter. A least-squares-fitting procedure was used based on measurements of derivatives of rms sizes with respect to the four matching quadrupole gradients. The Courant-Snyder parameters of the matched beam can be calculated from the beam dynamics using the measured values of the matched rms beam sizes. Pure mode mismatches were calculated by scaling the Courant-Snyder parameters to the desired values and finding the settings of the matching quadrupoles that produced these values. For example, multiplying each of the four transverse Courant-Snyder ellipse parameters α_x, β_x, α_y, and β_y by the same scaling factor produced a pure breathing-mode mismatch. The mismatch strength was measured by a parameter μ, which equals the ratio of the rms size of the initial mismatched beam to that of the matched beam.

The beam from the ion source was pulsed at 1 Hz to allow the use of interceptive beam diagnostics. The measurement cycle consisted of the following steps. The CW RFQ was de-energized with an RF blanking pulse, when the ion source was turned on. While the 75-keV dc-injector beam current increased, the beam was injected into the unpowered RFQ. After about 2 msec, the RF blanking pulse was removed; the RFQ fields approached a steady state after another 5 μsec. The ion source and RF fields remained on for another 30 μs for data acquisition. The beam-profile diagnostics were all in a fixed position during a measurement, and only one wire or scraper was in the beam at a time; all other wires and scrapers were out of the beam-pipe aperture. The wire or scraper in the beam accumulated beam-induced charge over about 30 μsec. Only the charge collected over the last 10 μsec, when the beam had best approximated a steady state, was selected for the recorded data. Then the DC injector was turned off, and the beam profile scanners

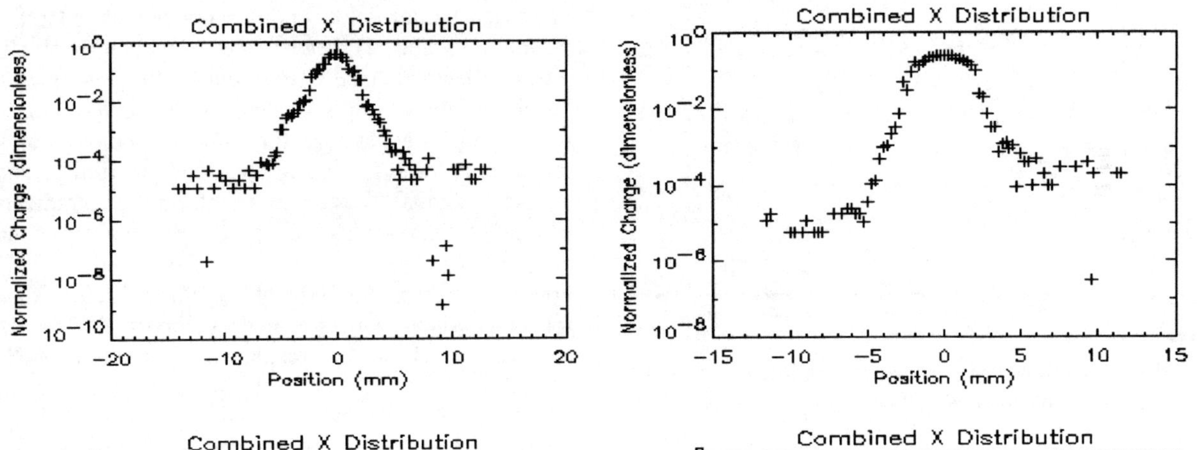

FIGURE 3. Measured beam profiles in x at 75 mA: (upper left) scanner 22, μ=1.0, matched,;(upper right) scanner 22, μ=1.5 mismatched,;(lower left) scanner 51, μ=1.0 matched,;(lower right) scanner 51, μ=1.5 mismatched.

were repositioned for the next cycle.

Among the methods used to characterize the beam were: 1) the rms emittance, and 2) the beam widths at different fractions of the peak intensity. Rms emittances at scanners 20 and 45 were calculated from a least-squares procedure using rms measurements from the profiles at the four scanners in each cluster.

PRELIMINARY RESULTS FOR 75 MA

Measurements were made at 16, 50, 75, and 100 mA and for a range of mismatch parameters, μ. In this paper we report preliminary results at 75 mA. The measured transverse profiles for the matched beams have Gaussian-like profiles in the linear plots (not shown). The semilog plots for the matched beams (see Fig. 3) show evidence of an extended low-density halo that may be present in the input beam. The measured transverse profiles for the mismatched beams show (see Fig. 3) structure in the form of shoulders in the semilog plots, and some asymmetric profiles at 10^{-3} level. Figures 4 and 5 show the x and y rms sizes at the different scanners for a matched and mismatched case, respectively.

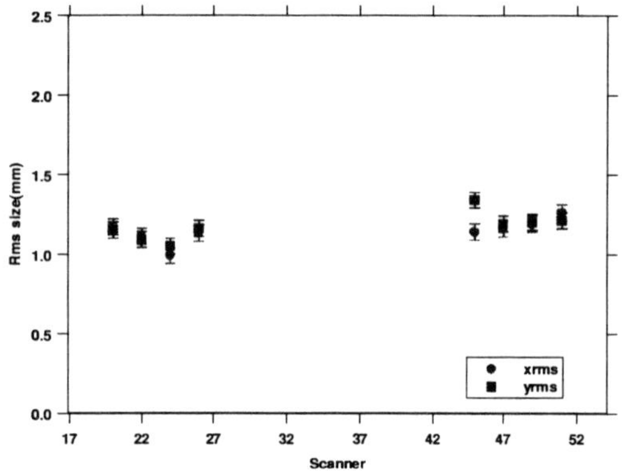

FIGURE 4. Rms displacements at 8 profile scanners for matched beam case (μ=1.0) at 75 mA.

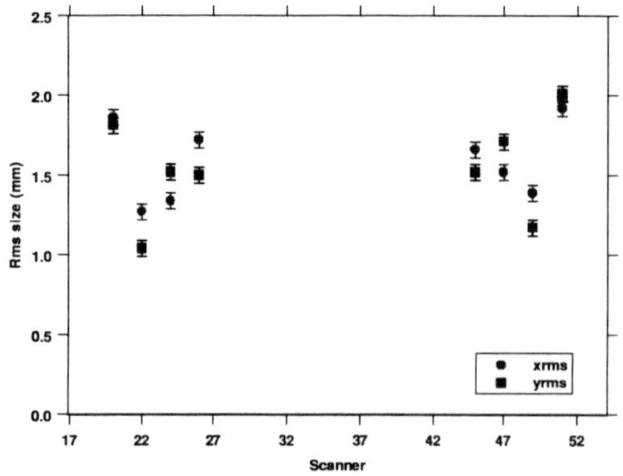

FIGURE 5. Rms displacements at 8 profile scanners for the breathing mode mismatch case (μ=1.5) at 75 mA.

Analysis of the rms-emittance data for comparison with free-energy theory is now in progress. Preliminary results for breathing-mode mismatches at 75 mA suggest a rapid rate of emittance growth that is consistent with a large fractional transfer of free energy to thermal energy within only a few mismatch oscillations.

Initial multiparticle simulation work has resulted in smaller beam widths and emittance growth rates for the mismatched beams than are experimentally observed. However, the input 6D phase space distribution was not measured in the experiment. Simulations of the experiment using different possible input distributions show that the input distributions with more extended tails generate more emittance growth. Good agreement between experiment and simulation may require detailed knowledge of the initial transverse phase-space distributions, especially for the tails.

SUMMARY

Proton linacs are capable of producing high-energy, high-intensity beams with low emittance. At present, there are two major R&D topics in the proton linear accelerator field: 1) linac applications of RF SC technology, and 2) beam halo formation in high-intensity beams. A new trend during the past decade has been the growing interest in SC proton linacs, resulting from advantages for reduced operating costs, larger bore radius, greater operating flexibility, and reduced capital costs as accelerating gradients continue to increase. A beam halo experiment has been carried out at Los Alamos, and analysis of the data is now in progress. Analysis of the rms-emittance data for comparison with free-energy theory is now in progress. Preliminary results for breathing-mode mismatches at 75 mA suggest a rapid rate of emittance growth that is consistent with a large fractional transfer of free energy to thermal energy within only few mismatch oscillations.

ACKNOWLEDGMENTS

Thanks to R. Garoby, J. M. Lagniel, R. Ferdinand, I. Masanori, C. Pagani, J. Wei, W. Park, and others who provided input or help for the high-power linac survey presented in Table 1. I wish to thank Ingo Hofmann for sending us a draft of his paper on anisotropy effects in mismatched beams, which is very helpful for interpreting our experimental results. I also want to acknowledge the contributions of my colleagues on the beam-halo experiment team, particularly C. K. Allen, K. C. D. Chan, P. Colestock,

K. R. Crandall, R. W. Garnett, J. D. Gilpatrick, W. Lysenko, J. Qiang, J. D. Schneider, M. E. Schulze, R. Sheffield, and H. V. Smith for the work that was described in this paper.

REFERENCES

1. J.S. O'Connell, T.P.Wangler, R.S.Mills, and K.R.Crandall, Proc. of 1993 Part. Accel. Conf., IEEE Catalog No. CH3279-7, 3657-3659.

2. R.L. Gluckstern, Phys. Rev.Lett. 73, 1247 (1994).

3. For an overview, see T.P.Wangler, K.R.Crandall, R.Ryne, and T.S.Wang, Phys.Rev. ST Accel. Beams 1 (084201)1998.

4. P.L.Colestock, et.al, "Measurements of Halo Generation for a Proton Beam in a FODO Channel," Proc. 2001 Part. Accel. Conf., IEEE Catalog No. 01CH37268, 170-172.

5. Martin Schulze, et al, "Characterization of the Proton Beam from the 6.7 MeV LEDA RFQ," Proc. 2001 Part. Accel., IEEE Catalog No. 01CH37268, 591-593.

6. T.P.Wangler, et al., Experimental Study of Proton-Beam Halo Induced by Beam Mismatch in LEDA," Proc. 2001 Part. Accel., IEEE Catalog No. 01CH37268, 2923-2925.

7. J.D.Gilpatrick, et al., "Experience with the Low Energy Demonstration Accelerator (LEDA) Halo Experiment Beam Instrumentation," Proc. 2001 Part. Accel., IEEE Catalog No. 01CH37268, 2311-2313.

8. J.D.Gilpatrick, et al., "Beam-Profile Instrumentation for Beam Halo Experiment: Overall Description and Operation," Proc. 2001 Part. Accel., IEEE Catalog No. 01CH37268, 525-527.

WORKING GROUP I

CIRCULAR ACCELERATORS

Working Group I: Circular Accelerators

Summary Report

L. Teng*

Argonne National Laboratory, Argonne, IL

Working Group I was very well attended by some 60 people at the peak. The two half-day sessions were fully loaded with 16 talks. Since it did not appear possible to have scheduled time dedicated to structured "discussion" or "work," we tried to allow as much time as possible after each talk for comments and discussions. The 15 talks (one withdrawn) are widespread in subject matter and content. They can roughly be grouped into the following categories.

Projects	
3-GeV Ring at the JHF	F. Noda (JAERI)
AGS High Power Upgrade Plan	W.T. Weng (BNL)
AHF Project	A. Thiessen (LANL)
Beam Studies and Simulations	
High Density and High Intensity Beams at CERN-PS	R. Cappi (CERN)
Impedance Reduction in the SPS	E. Shaposhnikova (CERN)
Emittance Dilution in HERA-p	R. Wanzenberg (DESY)
Investigations of Beam Tune and Stability in the IPNS-RCS	J. Dooling (ANL)
Multiparticle Longitudinal Dynamics Tracking Code ESME	J. MacLachlan (FNAL)
UAL-Development and First Application	N. Malitsky (BNL)
Hardware (JHF and PD)	
Longitudinal Dynamics and RF Hardware in JHF	M. Yoshii (KEK)
JHF Beam Injection and Extraction	I. Sakai (KEK)
High Intensity Proton Beam Profile Monitor	T. Toyama (KEK)
Proton Driver Magnet Design	F. Ostiguy (FNAL)
Dual-Harmonic Resonant Power Supply	C. Jach (FNAL)
Proton Driver Vacuum System	T. Anderson (FNAL)

Since full-length papers from the speakers will follow, in this summary I will only review some interesting features that struck my own fancy and point out some connections and similarities between the various subjects discussed.

Of the three projects reported, the Japan Hadron Facility (JHF) is the only one funded and under construction. The parameters and general features of JHF have already been reported in other sessions at this workshop. Other than aiming for high intensity, the unique feature of the project is that the utilities of the beam at each stage of acceleration (600 MeV, 3 GeV, 50 GeV) have been investigated and planned from the beginning. Thus, it will likely prove to be a very productive facility.

The planned high-power upgrades of the AGS are quite ambitious but straightforward in concept. The two stages of beam power upgrade to 0.47 MW and to 2 MW should be attainable; however, much effort and care are required.

The Advanced Hydro Facility (AHF) is actually an elaborate proton radiography facility. It consists of a slow-cycling synchrotron that provides very high per-pulse intensity. The 6×10^{13} protons/pulse in the ring are fast extracted into 12 beamlines that converge onto the target. Each beam consists of 29 bunches with 1×10^{11} protons/bunch (allowing for losses), which is adequate for one "frame" of the radiograph. As expected, this interesting machine requires several innovative design features.

A great deal of effort has been devoted to the study of high-intensity beam behaviors (emittance growths, coherent instabilities, transition crossing losses, etc.) in the CERN-PS and SPS. These are difficult measurements. The dynamic behaviors of the beam are difficult to measure, and the beam parameters are hard to control. Nevertheless, persistent and meticulous efforts have paid off. They have brought agreement between measurement and computation,

* Work supported by the U.S. Department of Energy, Office of Basic Energy Sciences, under Contract No. W-31-109-ENG-38.

CP642, *High Intensity and High Brightness Hadron Beams: 20th ICFA Advanced Beam Dynamics Workshop on High Intensity and High Brightness Hadron Beams,* edited by W. Chou, Y. Mori, D. Neuffer, and J.-F. Ostiguy

thereby establishing a reliable understanding of the phenomena.

In the HERA-p ring there is an observed luminosity decay due to intrabeam scattering and a factor of 6 growth in longitudinal emittance due to coupled bunch instability. On the other hand, in the Argonne Intense Pulsed Neutron Source — Rapid Cycling Synchrotron (IPNS-RCS) the longitudinal emittance is intentionally blown up using a phase shaker (scrambler) to avoid instability.

Two computing developments were reported. The long available longitudinal dynamics program ESME is now equipped with multiparticle capabilities to simulate high intensity beam motions. Both the space-charge and the beam-environment impedances can be input directly. Two interesting movies were shown, which exhibit bucket formation in beams interacting with empty rf cavities. An ambitious Unified Accelerator Library (UAL) composed of some ten programs has been compiled, debugged, and benchmarked at BNL. An initial application on the SNS ring has been successfully run.

Several hardware innovations were advanced for both the JHF and the Proton Driver (PD) projects. In the JHF 50-GeV ring we have the following:

1. Split ferrite rings are used in the rf cavities to control the Q.
2. Bipolar kickers and septa were designed for fast beam extraction and abort towards both the inside and the outside of the ring.
3. A clever gas sheet blasted across the beam is developed to use as the screen for displaying the profile of a very intense beam.

For the 15-Hz PD ring:

1. A dual-harmonic resonant power supply is designed to give a slower rise time and faster fall time.
2. To minimize the ring magnet gap, hence the size and cost of the magnet, the magnet-in-vacuum design is again contemplated. A low-temperature (150°C) bake will hopefully keep outgassing under control. The impedance of the exposed magnet laminations would be reduced by a liner or a cage structure. Thin silicon-steel laminations and multi-small-wire cables with radiation-resistant ceramic insulation and water cooling will be used for the magnets.

It s clear that for high intensity machines, beam loss is a major concern. Therefore, reliable and precise evaluation of the beam behavior based on high intensity beam dynamics is of paramount importance.

3 GeV RCS at the JKJ

Fumiaki Noda and JKJ Accelerator group

Japan Atomic Energy Research Institute, Tokai, Naka, Ibaraki 319-1195, Japan

Abstract. 3GeV RCS at the JAERI-KEK joint project (JKJ) is a rapid cycling synchrotron designed for high intensity proton beam. The designed output power is 1MW with a repetition rate of 25 Hz. In this paper, the outline of 3GeV RCS, key issues to achieve the goal, R&D status and time schedule of construction are reported.

INTRODUCTION

Japan Atomic Energy Research Institute (JAERI) and High Energy Accelerator Research Organization (KEK) are proposed the joint project for a high intensity proton accelerator facility [1-2]. The accelerator complex consists of a 0.4GeV linac, a 3GeV rapid-cycling synchrotron (RCS) and a 50GeV synchrotron (MR) at this facility. The major requirement for the 3GeV RCS is 1MW output beam power. It is very important for 3GeV RCS to control the beam loss and man/machine protection strategy from a radiation.

OUTLINE OF 3GEV RCS

The 3GeV Rapid-Cycling Synchrotron (RCS) at the JKJ will have a threefold symmetric lattice. Figure 1 shows the overview of 3GeV RSC. Each super-period consists of two 3-DOFO modules with missing bends in arc and 3-DOFO in insertion. The arc module has a missing bend cell for chromaticity correction magnets and longitudinal primary collimator. The insertion straights are used for the injection /collimation, the extraction and the RF acceleration. These insertion straights are dispersion free. See Reference [3] for more details of 3GeV RCS.

The main parameters are shown in table 1. The injection beam energy is 0.4GeV. The H- beam is injected by charge exchange multi-turn injection method and accelerated to 3GeV. The output beam power is 1 MW with a repetition rate of 25 Hz. The output beam is led to the spallation neutron and muon facility and the 50GeV MR.

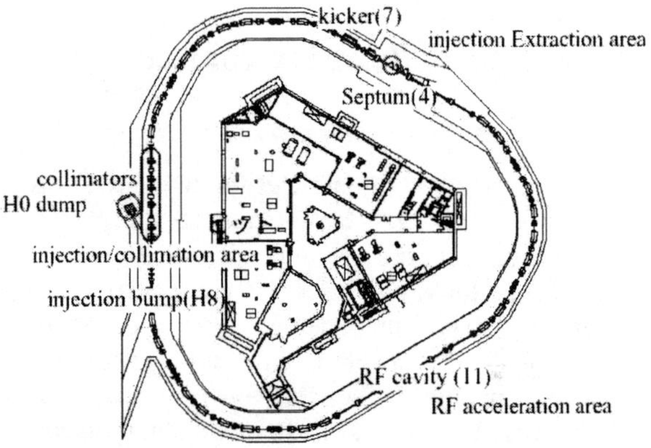

FIGURE 1. Schematic view of 3GeV RCS

TABLE 1. Main parameters of 3GeV RCS

Parameter	value
Circumference	348.333 m
Injection Energy	0.4GeV
Extraction Energy	3.0 GeV
Extracted beam power	1MW
Particle Per Pulse	8.3 x 10^{13}
Revolution Period	
at Injection/ Extraction	1.629 /1.196 µs
Repetition Rate	25 Hz
Ramping Pattern	Sinusoid
Injection period	~500□s
Circulating Current	
at Injection/ Extraction	8.2 /11.1 A
Bunching Factor	
at Injection	
Fundamental only	0.3
with 2nd harmonics	0.4
at Extraction	
for 3GeV users	0.2
for 50GeV ring	0.3

KEY ISSUES OF 3GEV RCS

In the design of 3GeV RCS, it is very important to decrease a beam loss. For that purpose, beam density

CP642, *High Intensity and High Brightness Hadron Beams: 20th ICFA Advanced Beam Dynamics Workshop on High Intensity and High Brightness Hadron Beams,* edited by W. Chou, Y. Mori, D. Neuffer, and J.-F. Ostiguy

should be controlled by a painting injection in transverse and in longitudinal. These are effective methods to defusing a space charge force. Simultaneously, it is need to control a beam loss (localization) and decrease an un-controllable beam loss. An adequate aperture/collimator ratio is a matter of importance for localization of beam loss. It is necessary to keep the large acceptance for extraction line, because the beam power at extraction energy reaches to 1MW.

Beam loss control

Space-charge force reduction

The number of particles in the 3 GeV RCS is 8.3×10^{13} ppp for 1MW output beam power. It is necessary to do painting injection with charge exchange multi-turn injection for reduction of space-charge force. The beam emittance is intentionally enlarged to 216 π mm-mrad at the injection by phase space painting. Simultaneously, dual harmonic acceleration system and longitudinal painting is important issue for reduction of space-charge force. It makes a bunching factor larger than one of simple RF system. We rely on that bunching factor is 0.4.

The incoherent space charge tune shift will be about -0.15 by above treatments.

See Reference [4-5] for more details of the design of painting injection system in transverse and longitudinal phase space.

Collimation system

Transverse and longitudinal collimation systems are set in the 3GeV RCS. Collimator system consists of primary collimator to scatter halo beam and secondary collimator to collect the halo beam. The aim of this collimator system is to localize the halo beam. It enable to keep 1W/m uncontrolled beam around the ring. Transverse primary collimator and secondary collimators set a dispersion free area. In addition, longitudinal primary collimator sets the missing bend area in arc module. This point has a large dispersion and small beta function for transverse. Longitudinal primary collimator is thinner than transverse one. So scattered halo beam at the longitudinal primary collimator passes through the arc section without beam loss. The passed hallo beam is scatted by transverse primary collimator again and is collimated by transverse secondary collimators.

Large acceptance for transverse and longitudinal space

The beam emittancce and apertures are shown in Table 2. The beam emittance is intentionally enlarged to 216π mm-mrad at the injection by phase space painting. On the other hand, collimator acceptance and physical aperture are 324 and 486π mm-mrad, respectively. The ring must also accept a maximum beam momentum spread of ±1 percentage.

The acceptance of extraction line is 324π mm-mrad. This value is equal to ring collimator aperture.

TABLE 2. Emittance and Apertures

Parameter	value
Transverse	
Injection beam	4 π.mm-mrad
Painting emittance	216 π.mm-mrad
Collimator aperture	324 π.mm-mrad
Physical aperture	486 π.mm-mrad
Extraction aperture	324 π.mm-mrad
Longitudinal (dp/p)	
Injection	0.1%
Collimator	1%

Other key issues

A damage of foil cause directly beam loss. Therefore, a long life and reliable charge exchange foil and an auto-changer mechanism are developing now. A prototype auto-changer was constructed already.

The large aperture bending magnet and quadrupole magnet were constructed already and performed the many studies. In these magnets, the coil conductor is aluminum stranded conductor in order to reduce an eddy current loss. See Reference [6] for more details. For another magnets, correction magnet and kicker magnets were constructed.

3GeV RCS is rapid cycling synchrotron, so it is necessary to high gradient RF cavity. In 3GeV RCS, the beam loading is very heavy. A leak from RF bucket causes a beam loss at acceleration and extraction. See Reference [7] for more details.

For vacuum chamber of rapid cycling synchrotron, it is necessary to avoid any harmful effect of the eddy current. On the other hand, vacuum chamber should be treated the RF shield. Now we have developed long and large aperture ceramic chamber with RF shield and TiN coating (inner surface) for bending magnet. See Reference [8] for more details

Monitor system is also very important for the beam diagnostic. A prototype beam position monitor was constructed and performed the many studies.

BEAM LOSS ESTIMATION

Beam loss at the injection

It is important for H- injection to care the Lorentz stripping loss and excited H0 beam loss. The magnetic field of injection line is limited 0.55T. The Lorentz stripping loss rate is about 10-6/m and total beam loss is less than a few watts. Charge exchange efficiency is 99.8%. The beam power of excited H^0 is about 0.4kW. For the magnet passed through an excited H^0 beam, the magnetic field set 0.2T between n=5 ad 6. Here n is the excitation states level. Therefore, the uncontrollable beam loss cause by excited H^0 is about 6W($n \geq 6$). The rest of excited H^0 ($n \leq 5$) is led to H^0 beam dump, which designed 1kW capacity.

Beam loss during a acceleration

The beam tracking with apace-charge force using Accsim, PATRASH and Simpsons codes are performed. Preliminary estimations show the 3% (4kW at injection) of beam are over the collimation aperture (324 π.mm-mrad). The detail tracking are doing now. See Reference [9] for more details of beam tracking. In addition, beam tracking with fringe field are doing now. See Reference [10] for more details

The beam collimator arrangement and the collimation efficiency are calculated by the STRUCT code. The collimation efficiency (≡beam loss at the collimator region/beam loss (4kW) is about 98%. The rest of beam loss (about 80W) is scattered around the ring. See Reference [11] for more details of beam collimation system.

TABLE 3. Summary of beam loss estimation

Parameter	value
Injection line	
Lorentz stripping loss	<1W (un-control)/1kW*
Injection area	
Excited H^0	6W (un-control)/1kW*
H^0 beam dump	0.4kW (control)/1kW*
Collimator area	4kW (control)/4kW*
Extraction area	~0 (un-control) /1kW*
Another area	total about 80W (un-control) <1W/m

*Upper value is a beam loss at normal operation.
Lower value is design value of building and components

SUMMARY AND SCHEDULE OF 3GEV RCS

Beam loss limits the output beam power of accelerator. Therefore, we should reduce the beam loss and control the beam loss. In this paper, key issues are described from the point of view of the beam loss control. The design of 3GeV RCS is in final stage. Now fine-tuning and feedback from R&D are performed. For more details of 3GeV RCS, see the Technical Design Report for High-Intensity Proton Accelerator Facility Project (Reference [12]).

The construction of facility and production of instruments will start in October, 2002 arraignment will start in April 2005. Moreover, the first beam of 3GeV RCS will be in October 2006.

REFERENCES

1. Imazato, J., "JHF Physics", in these proceedings.
2. Mori, Y., "Accelerator Complex of Joint Project in Japan", in these proceedings
3. Shigaki, K. *et al.*, "The JKJ Lattice", in these proceedings.
4. Sakai, I., et al, "H$^-$ Painting Injection System for the JKJ 3-GeV High-Intensity Proton Synchrotron ", Proceedings of EPAC2002
5. Yoshii, M., "Longitudinal dynamics and rf hardware", in these proceedings
6. Tani, N., et al, "Design of a Dipole Magnet for The 3-GeV Proton Synchrotron of The JAERI/KEK Joint Project", Proceedings of EPAC2002
7. Yamamoto. M., et al, "RF system for the 3-GeV Proton Synchrotron of The JAERI/KEK Joint Project", Proceedings of EPAC2002
8. Kinsho, M., et al, "Vacuum system design for the 3GeV-proton synchrotron of JAERI-KEK joint project", J. Vac. Sci. Technol. A20(3), May/Jun 2002, p829-832
9. Shimozaki, Y. *et al.*, "HALO-FORMATION AND BEAM LOSS IN THE 3GEV RING OF THE JOINT PROJECT", Proceedings of EPAC2002
10. Molodojentsev, A., "Tracking Studies for the JKJ Lattice", in these proceedings.
11. Yamamoto, K., *et al.*, "Beam Collimator Design for The 3GeV Synchrotron of The JAERI-KEK Joint Project ", Proceedings of PAC '01 (2001).
12. Accelerator Technical Design Report for High-Intensity Proton Accelerator Facility Project, edited by Y.Yamazaki *et al.* January 2001.

The AGS High Power Upgrade Plan*

W.T. Weng and T. Roser

Brookhaven National Laboratory
Upton, NY 11973, USA

Abstract. BNL could provide a Megawatt class neutrino beam from the AGS for very long baseline neutrino experiments. We have studied two possible approaches to upgrade the AGS to 1.0 MW beam power. The first is the linac option, comprising a new superconducting linac injector of 1.2 GeV, accelerating 9×10^{13} proton per pulse in the AGS to 28 GeV at 2.5 Hz. The second option is to extend the existing 200 MeV linac to 400 MeV, ramp the Booster to 2.5 GeV at 6 Hz, add a new 2.5 GeV accumulator ring in the AGS tunnel, and finally ramp the AGS to 28 GeV at 2.5 Hz. Due to the simplicity of the linac approach and minimum interference with the on going research program, the linac option is the preferred one.

INTRODUCTION

We have examined [1] possible upgrades to the AGS complex that would meet the requirements for the proton beam for a 1 MW neutrino super beam facility. Those requirements are summarized in Table 1 and a layout of the upgraded AGS is shown in Figure 1. Since the present number of protons per fill is already close to the required number, the upgrade will focus on increasing the repetition rate and on reducing beam losses to avoid excessive shielding requirements and to maintain the machine components serviceable by hand. It is also important to maintain all the present capabilities of the AGS, in particular its role as injector to RHIC.

We are proposing here to build a superconducting upgrade to the existing 200 MeV Linac to an energy of 1.2 GeV for direct H^- injection into the AGS. This will be discussed in the next section. The minimum ramp time to full energy is presently 0.5 s, which will have to be upgraded to reach the required repetition rate of 2.5 Hz. The required upgrade of the AGS power supply will be described in third section. Finally, the increase ramp rate requires a substantial upgrade to the AGS rf system.

TABLE 1. AGS Proton Driver Parameters.

Total beam power	1 MW
Beam energy	28 GeV
Average beam current	42 μA
Cycle time	400 ms
Number of protons per fill	9×10^{13}
Number of bunches per fill	6
Protons per bunch	1.6×10^{13}
Injection turns	360
Repetition rate	2.5 Hz
Pulse length	1.08 ms
Chopping rate	0.65
Linac average/peak current	20/30 mA

The front end consists of a high-intensity negative-ion source, followed by a 750 keV RFQ, and the first 5 tanks of the existing room temperature Drift-Tube Linac (DTL) to accelerate protons to 116 MeV. The SCL is made of three sections, each with its own energy range, and different cavity-cryostat arrangement.

The proposed new injector for the AGS is a 1.2 GeV superconducting linac upgrade with an average output beam power of about 50 kW. The injection energy is still low enough to control beam losses due to stripping of the negative ions that are used for multi-turn injection into the AGS. The duty cycle is about 0.5%. The injection into the AGS is modeled after the SNS [2]. However, the repetition rate and consequently the average beam power is much lower. The larger circumference of the AGS also reduces the number of foil traversals. Beam losses at the injection into AGS are estimated to be about 3% of controlled losses [1]. The AGS injection parameters are also summarized in Table 1.

SUPERCONDUCTING LINAC

The superconducting linacs accelerate the proton beam fro, 116 MeV to 1.2 GeV. The presented configuration follows a similar design described in detail [3]. All three linacs are built up from a sequence of identical periods.

The major parameters of the three sections of the SCL are given in Table 3. The low energy section operates at 805 MHz and accelerates from 116 to 400 MeV.

* Work performed under the auspices of the U.S. Department of Energy.

CP642, *High Intensity and High Brightness Hadron Beams: 20th ICFA Advanced Beam Dynamics Workshop on High Intensity and High Brightness Hadron Beams*, edited by W. Chou, Y. Mori, D. Neuffer, and J.-F. Ostiguy

The following two sections, accelerating to 800 MeV and 1.2 GeV respectively, operate at 1.61 GHz. A higher frequency is desirable for obtaining a larger accelerating gradient with a more compact structure and reduced cost.

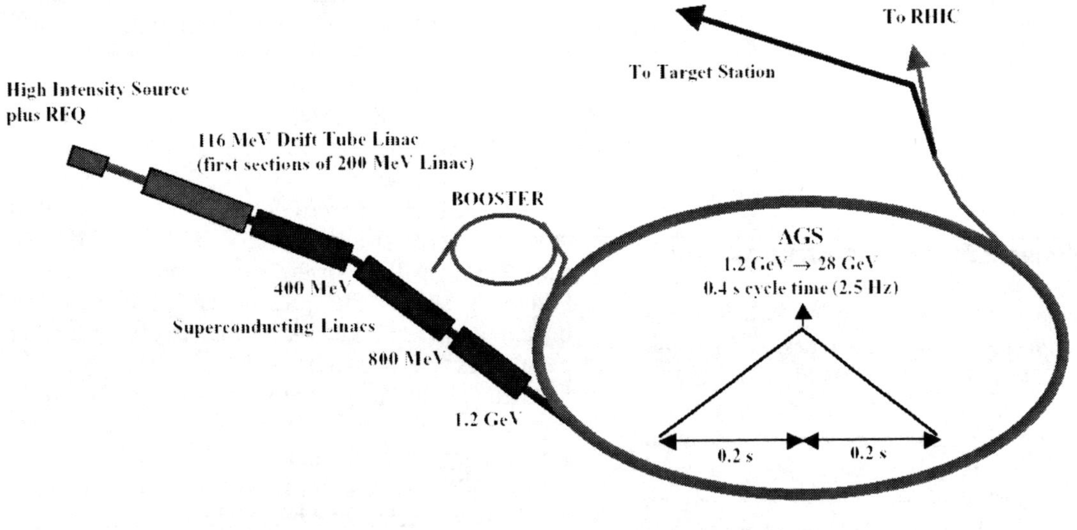

FIGURE 1. AGS Proton Driver Layout.

TABLE 2. Parameters of the Superconducting Linac.

	Low	Medium	High
Beam Power, kW	16	32	48
Kinetic Energy, MeV	116-400	400-800	800-1200
Velocity, cβ	0.4560	0.7131	0.8418
	0.7131	0.8418	0.8986
Frequency, MHz	805	1610	1610
Protons/μBunch, 10^8	9.32	9.32	9.32
Temperature, °K	2.0	2.0	2.0
Cells/Cavity	4	8	8
Cavities/Cryo-Module	4	4	4
Cell Length, cm	9.68	6.98	8.05
Cell reference velocity, $c\beta_0$	0.520	0.750	0.865
Cavity internal diameter	10	5	5
Cavity Separation, cm	32	16	16
Cold-to-Warm transition	30	30	30
Accelerator Gradient, MeV/m	11.9	22.0	21.5
Cavities/Klystron	4	4	4
No. of Klystrons	18	10	9
Klystron Power, kW	720	1920	2160
Energy Gain/Period, MeV	16.0	42.7	48.0
Length of a period, m	4.2	4.4	4.7
Total length, m	75.4	43.9	42.6

AGS MAIN POWER SUPPLY UPGRADE

The AGS ring consists of 240 magnets hooked up in series with a total resistance R = 0.27 Ω and a total inductance L = 0.75 H. Two stations of power supplies are each capable of delivering up to 4500 V and 6000 A. The two stations are connected in series and the magnet coils are arranged to have a total resistance R/2 and a total inductance of L/2. The grounding of the power supply is done only in one place, in the middle of station 1 or 2 through a resistive network. With this grounding configuration, the maximum voltage to ground in the magnets will not exceed 2500 Volt. The magnets are hi-potted to 3000 Volt to ground, prior of each starting of the AGS MMPS after long maintenance periods.

To cycle the AGS ring to 24 GeV at 2.5 pulses per second and with ramp time of 250 ms, the magnet peak current is 4300 Amp and the peak voltage is 25 kV. The total average power dissipated in the AGS magnets has been estimated to be 3.7 MW. To limit the AGS coil voltage to ground to 2.5 kV the AGS magnets will need to be divided into six identical sections, each powered similarly to present half AGS power system, except that now the magnet loads is 1/6 of the total resistance and inductance.

The peak power required is approximately 110 MW exceeding the 50 MW rating of the existing motor generator. The new motor-generator should also operate with 6 or 12 phases to limit or even eliminate phase-shifting transformers so that every power supply system generates 24 pulses.

AGS RF SYSTEM UPGRADE

At 2.5 Hz the peak acceleration rate is three times the present value for the AGS. With 10 accelerating stations each station will need to supply 270 kW peak power to the beam. The present power amplifier

TABLE 3. AGS Beam Power With Accumulator

	Now	AGS at 1 Hz	AGS at 2.5 Hz
Linac Energy (MeV)	200	400	400
Booster Intensity (ppp)	1.5×10^{13}	2.0×10^{13}	2.0×10^{13}
Booster energy (GeV)	1.8	2.5	2.5
Booster Cycles	4	6	6
AGS energy (GeV)	24	28	28
AGS Intensity	36	120	300
AGS Rep Rate (Hz)	0.6	1.0	2.5
AGS Current (μA)	5.75	19.2	48
AGS Intensity (ppp)	6×10^{13}	12×10^{13}	12×10^{13}
AGS power (kw)	138	538	1344

design, employing a 300 kW power tetrode will be suitable to drive the cavities and supply power to the beam. The number of power amplifiers will be double so that two amplifiers of the present design drive each station. This follows from the necessity to supply 2.5 times the rf voltage. An AGS rf station comprises four acceleration gaps surrounded by 0.35 m of ferrite stacks. The maximum voltage capability of a gap is limited by the ability of the ferrite to supply the magnetic induction. When the AGS operates at 0.5 Hz the gap voltage is 10 kV. At 2.5 Hz we will need up to 25 kV per gap.

The next concern is the power dissipation in ferrite and the thermal stress that is created by differential heating due to rf losses in the bulk of the material. We know from experience that below 300 mW/cm^3 ferrites can be adequately cooled.

ACCUMULATOR RING OPTION

There is another possible approach to provide 1 MW beam power by the AGS, this is to add an accumulator ring in the AGS [4]. In the first phase the linac will be improved to inject protons to the booster at 400 MeV (at present it is 200 MeV), and the booster energy increased to 2.5 GeV from 1.8 GeV. The addition of a fixed field accumulator storage ring in the AGS main tunnel will increase the AGS input beam from the present 4 booster pulses per AGS acceleration to 6 booster pulses per AGS acceleration and, at the same time, increase the AGS frequency from 0.6 Hz to 1.0 Hz. The AGS power increase would be from 0.14 to 0.53 MW. The new accumulator will be in the same tunnel as the AGS. Figure 2 shows the present and proposed AGS injection modes. The AGS intensity upgrades and parameters are shown in Table 3. The fixed field magnets of the accumulator ring will be essentially copies of the permanent magnets produced for a similar purpose at Fermilab [5]. In a second phase of the upgrades the AGS repetition rate will be increased to 2.5 Hz to reach a total beam power of 1.3 MW.

In this way, the injection time of 0.6 sec from the booster to the AGS is reduced to almost zero by pre-storing 6 batches of booster beam in the accumulator ring as shown in Figure 2.

The linac option is chosen as the baseline for its simplicity, only involves the linac and the AGS, and the clear upgrade path to 4.0 MW in the future. The accumulator option requires addition/modification of four accelerators, the linac, the booster, the accumulator ring and the AGS. In addition, the installation of the accumulator ring in the AGS tunnel will, inevitably, interference with existing RHIC program.

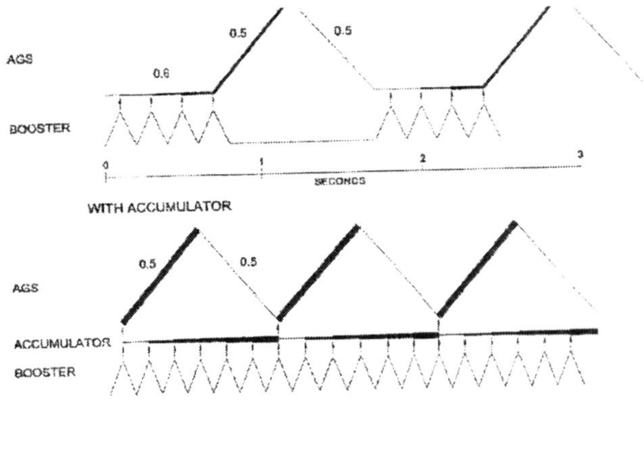

FIGURE 2. Time sequence of injecting pulses into the AGS. Top picture shows that at the moment 4 booster pulses are injected into the AGS during the period when the AGS magnets are at low field. In the new proposed configuration in the bottom picture the booster will inject 6 pulses into the accumulator which will store the beam until the AGS is at low field and then transfer the beam into the AGS.

REFERENCES

1. Neutrino Factory Feasibility Study II, http://server.cap.bnl.gov/mumu/studyii/
2. Wei, J., et. al., Low-loss design for the high-intensity accumulator ring of the Spallation Neutron Source, *Phys. Rev. ST Accel. Beams* **3**, 080101 (2000).
3. Ruggiero, A.G., A Superconducting Linac as a New Injector to the BNL AGS, BNL Internal Report, C-A/AP/40, February 2001.
4. Neutrino Oscillation Experiments, Letter of Intent to BNL PAC, D. Beavis, M. Diwan, et. al., April 2002.
5. Bill Foster, FNAL, private communication.

High-Density and High-Intensity Beams at the CERN PS

R. Cappi, M. Giovannozzi, M. Martini, E. Métral, G. Métral, A.-S. Muller, R. Steerenberg

CERN PS Division, 1211 Geneva 23, Switzerland

Abstract. The production of high-density and high-intensity beams has been a constant activity at the CERN Proton Synchrotron (PS) since many years. This paper is a status report of the present situation, including also a discussion of ongoing studies and future activities.

INTRODUCTION

Two different types of proton beams will be considered in this paper, namely high-density and high-intensity ones. The first set is characterized by having a large ratio $N_b / \varepsilon^*_{x,y}$, where N_b is the number of protons per bunch and $\varepsilon^*_{x,y}$ the normalized transverse (horizontal/vertical) rms emittance. As an example, the nominal LHC beam generated by the PS [1] falls in this category, the high-density being a necessary condition for achieving the high-luminosity required by the LHC.

In the second set, only the total number of protons N_t needs to be maximised. The PS normally delivers such beams either for neutrino experiments[1], as is the case of the future CERN Neutrino to Gran Sasso (CNGS) beam [2], or for the neutron Time-of-Flight (nTOF) Facility [3].

Although the longitudinal emittance ε_l is also an issue, its value being dictated by the total number of bunches, their shape or length, collective effect considerations and also hardware limitations, this paper focus mainly on transverse beam dynamics.

HIGH-DENSITY BEAMS

The Nominal LHC Beam

The beam parameters of the nominal LHC beam are [1]: $N_t \approx 8\times10^{12}$ ppp[2], $\varepsilon^*_{x,y} \leq 3$ μm. This beam density exceeds typical PS performance by more that a factor of two. The nominal figures can be achieved only by applying some special manipulations: i) inject two, instead of one, PS-Booster (PSB) pulses into the PS (i.e. doubling the final value for N_t while keeping $\varepsilon^*_{x,y}$ unchanged); ii) avoid any emittance dilution effect. The double-batch injection process is shown in Fig. 1.

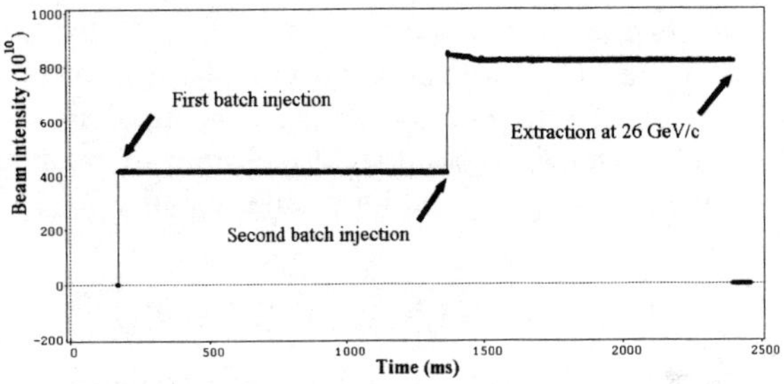

FIGURE 1. PS intensity versus time. The double-batch injection is clearly seen as a step increase.

The two injections at 1.4 GeV kinetic energy are clearly visible, spaced by 1.2 s (the repetition rate of PSB). The total transmission is practically 100%; neither losses nor emittance blow-up occur during the long injection flat-bottom.

The evolution of $\varepsilon^*_{x,y}$ along the PS cycle, from PSB extraction to PS extraction and also in the transfer line between the PS and the Super Proton Synchrotron (SPS) are shown in Fig. 2.

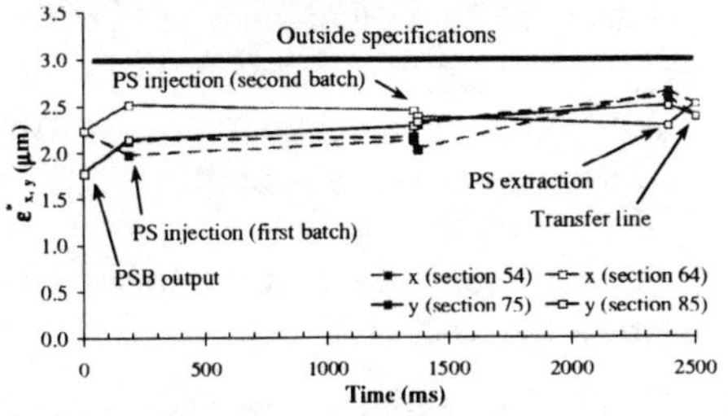

FIGURE 2. Evolution of $\varepsilon^*_{x,y}$ along the PS cycle. The emittances are measured at PSB output, in the PS after the 1st batch injection, before and after the 2nd batch injection, before extraction, and in the transfer line towards the SPS.

Different instruments are used to measure $\varepsilon^*_{x,y}$. At PSB extraction, beam emittances are obtained by means of SEM-grids located in a dedicated measurement line. In the PS ring, two pairs of wire scanners (each performing H/V plane measurements) are located in different machine sections. Finally, SEM-wires in the transfer line between the PS and the SPS measure the extracted beam emittance. The

[1] The relevant parameter is the particle flux, N_t f with f the repetition rate. For simplicity, a fixed machine cycle length will be considered.
[2] For the so-called *ultimate* LHC beam the intensity increases to $N_t \approx 1.3\times10^{13}$ ppp keeping the same transverse emittances.

CP642, *High Intensity and High Brightness Hadron Beams: 20th ICFA Advanced Beam Dynamics Workshop on High Intensity and High Brightness Hadron Beams,* edited by W. Chou, Y. Mori, D. Neuffer, and J.-F. Ostiguy

average emittance value is well below the 3 μm limit in both planes, with an accuracy of ± 10 %.

A number of modifications in the PSB/PS chain, together with fine adjustments of beam parameters, had to be applied to achieve the nominal characteristics of the LHC beam [4]. In particular, to avoid emittance blow-up during the 1.2 s long flat-bottom, three main approaches were used: i) increase of the PSB extraction energy from 1 to 1.4 GeV to reduce the space-charge incoherent tune shift from -3.5 to -2.3; ii) transverse higher-order mode head-tail instabilities cured in the PS by x-y coupling generated with skew quadrupoles [5]; iii) fine adjustment of both tune and chromaticity all along the PS cycle. This includes the fast change of sign at transition crossing to avoid head-tail instability.

HIGH-INTENSITY BEAMS

The PS routinely produces two flavours of high-intensity beam, namely single- and multi-bunch.

Single-Bunch Beams

A high-intensity, short, single-bunch beam is delivered to the nTOF facility. The experiments require a beam of the highest possible intensity, together with rms bunch length σ_b shorter than 7 ns, to allow for high precision time-of-flight measurements. In practice, typical beam parameters are: $N_b \approx 7\times10^{12}$ p/b and σ_b about 6 ns at extraction (occurring at 20 GeV/*c*) [6].

The transverse beam emittances delivered by the PSB are quite large for these beams. Usually, at injection, the 2σ physical transverse emittances are $\varepsilon_x(2\sigma)$ = 23 μm and $\varepsilon_y(2\sigma)$ = 12 μm, slightly beyond the limit defined by the machine acceptance of 60 and 20 μm. This explains why beam losses cannot be reduced below the level of 10-20 %.

The longitudinal emittance cannot exceed 2.5 eVs, otherwise it would not be possible to meet the requirement on the final bunch length. Moreover, its lower limit is dictated by the presence of a vertical Transverse Mode Coupling (TMC) instability at transition energy [7,8] (see Fig. 3). To cure the instability, ε_l has to be increased from 1.5 eVs, as delivered by the PSB, to about 2.3 eVs (see Refs. 7, 8 for more details), by means of controlled longitudinal blow-up. A non-adiabatic RF gymnastics generates the final short bunch: the RF phase is flipped by 180º and the bunch placed on the unstable phase. Then, the phase is set back to its original value, after sufficient time has elapsed for the bunch to be lengthened. The compression ratio is about 2.5 (see Fig. 4).

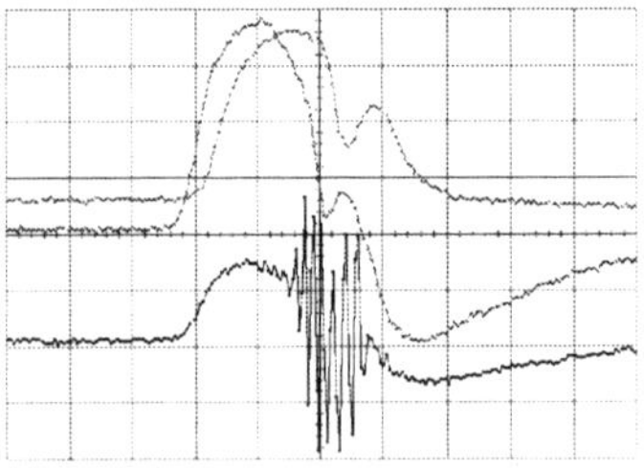

FIGURE 3. Vertical TMC instability at transition for high-intensity, single-bunch beam as observed by a pick-up (upper trace Σ, centre Δx, lower Δy, hor. scale is 10 ns/div).

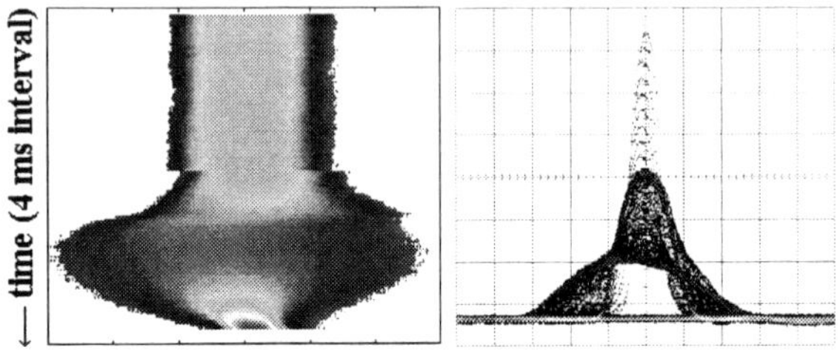

FIGURE 4. Bunch compression by RF phase jump. The beam distribution vs. time and bunch length is shown on the left. The superposition of the beam-profiles at different times is shown on the right (hor. scale is 20 ns/div).

Multi-Bunch Beams

A typical high-intensity multi-bunch beam is the one required by neutrino experiments. Presently, the PS is capable of delivering $N_t \approx 3.3\times10^{13}$ ppp in 8 bunches, with $\varepsilon_{x,y}(2\sigma)$ and ε_l similar to nTOF beam values. For the CNGS beam, a study is under way [9] aiming at approaching $N_t \approx 5\times10^{13}$ ppp. Improvements are foreseen in the PS both at low and high energy (tests for doubling PSB repetition rate are also in progress [10]).

Improvements At Low Energy

By injecting two PSB batches, similarly to what is done for the LHC beam, it is possible to obtain 8 bunches circulating in the PS with $N_b = 6.2\times10^{12}$ p/b, total bunch length of about 190 ns, $\varepsilon_x(2\sigma)$ = 22 μm, and $\varepsilon_y(2\sigma)$ = 9 μm. The emittance values just fit the PS acceptance with only few percent of losses. The generation of flat bunches in the PSB [11] would allow to decrease the space charge tune-shift to acceptable values, such as –0.15 and –0.27 in the horizontal and vertical plane respectively.

Improvements At High Energy

Presently, the so-called Continuous Transfer (CT) [12] is used to extract the multi-bunch high-intensity beam. This is a five-turn extraction mode, where the beam is split by means of an electrostatic septum. Hence, losses (10-20 %) are intrinsic and unavoidable. To increase the overall intensity and minimize radiation damage, a new approach has been proposed [13-14].

Using sextupoles and octupoles to create stable islands, and varying the linear tune, it is possible to capture the beam adiabatically into the islands of a fourth-order resonance. Once the beam is split, the extraction to the SPS can take place with fewer losses. Figure 5 reports the results of numerical simulations.

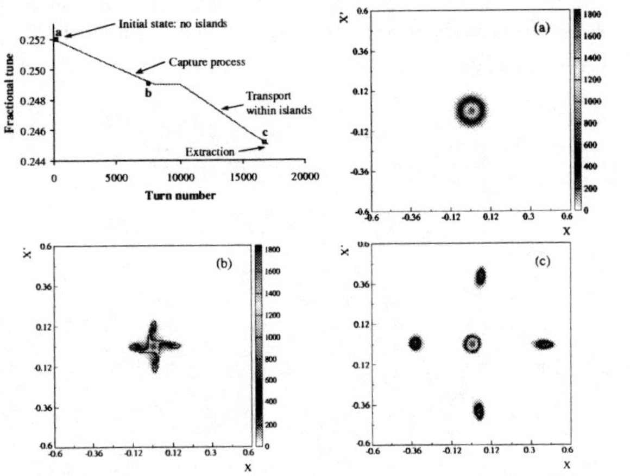

FIGURE 5. Results of numerical simulations of beam splitting obtained by means of a fourth-order resonance.

Preliminary analysis of phase space topology by means of multi-turn beam position measurements [15] confirms the possibility of generating four stable islands (see Fig. 6).

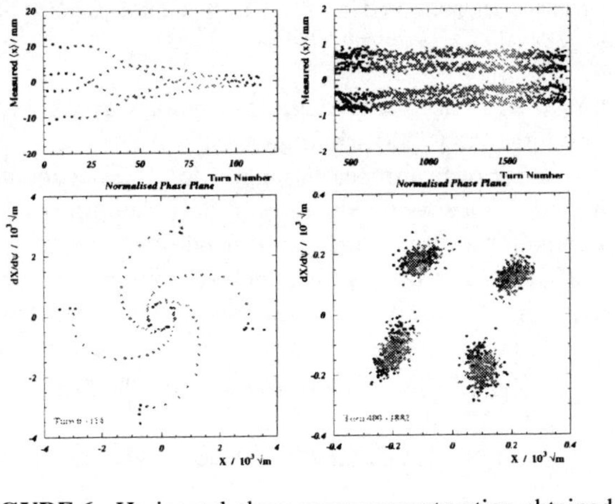

FIGURE 6. Horizontal phase space reconstruction obtained with multi-turn measurements of kicked beam in the PS.

CONCLUSION

CERN PS activities related with high-density beams will continue, with special emphasis on production of the ultimate, and of the single-bunch pilot LHC beams. As far as the high-intensity beams (CNGS) are concerned, studies will be devoted mainly to the intensity increase of the multi-bunch beam and to the tests of the extraction based on adiabatic trapping in islands of transverse phase space, aiming at reducing losses and increasing machine reliability.

REFERENCES

1. Benedikt, M., et al., "Performance of the LHC Pre-Injectors", CERN-PS-2001-011-DR, 2001.
2. Elsener, K., et al., "The CERN neutrino beam to Gran Sasso (NGS): conceptual technical design", CERN-98-02, 1998.
3. Andriamonje, S. A., et al., "Neutron TOF facility (PS213): technical design report", CERN-INTC-2000-004, 2000.
4. Benedikt, M. (Ed.), et al., "The PS complex as proton pre-injector for the LHC: design and implementation report", CERN-2000-003, 2000.
5. Cappi, R., Métral, E., and Garoby, R., "Collective Effects in the CERN-PS beam for LHC" in *Workshop on Instabilities of High Intensity Hadron Beams in Rings*, edited by T. Roser et al., AIP Conference Proceedings 496, New York: American Institute of Physics, 1999, pp. 116-30.
6. Cappi, R., Giovannozzi, M., and Métral, G., "The Proton Beams for the New Time-of-Flight Neutron Facility at the CERN-PS" in *7th EPAC Conference*, edited by J. Poole et al., Geneva: European Phys. Soc., 2000, pp. 1295-7.
7. Cappi, R., Métral, E., and Métral, G., "Beam Break-up Instability'' in *7th EPAC Conference*, edited by J. Poole et al., Geneva: European Phys. Soc., 2000, pp. 1152-4.
8. Métral, E., "Stability Criteria for High-Intensity Single-Bunch Beams in Synchrotrons", CERN-PS-2002-022-AE, 2002.
9. Cappi, R., et al., "Increasing the Proton Intensity of PS and SPS", CERN-PS-2001-041-AE, 2001.
10. Benedikt, M., Carli, C., and Chanel, M., private communication.
11. Carli, C., and Chanel, M., "New Methods to Create Hollow Bunches", these Proceedings.
12. Bovet, C., et al., "The Fast-shaving ejection for beam transfer from the CPS to the CERN 300 GeV machine" in *1973 PAC Conference*, edited by D. W. Dupen, New York: IEEE, 1973, pp. 438-40.
13. Cappi, R., and Giovannozzi, M., *Phys. Rev. Lett.* **88**, 104801 (2002).
14. Cappi, R., and Giovannozzi, M., "Adiabatic Capture of Charged Particles in Islands of Phase Space: a New Method for Multi-Turn Extraction", CERN-PS-2002-014-AE, 2002.
15. Angoletta, M. E., et al., "Analysis of Multi-Turn Beam Position Measurements in the CERN PS", CERN-PS-2002-018-AE, 2002.

Results from the Impedance Reduction in the CERN SPS

E. Shaposhnikova, T. Bohl and T. Linnecar

CERN, Geneva, Switzerland

Abstract. The future use of the SPS as injector for the LHC requires both a high intensity and a high quality beam. For many years the longitudinal single bunch instability was one of the most serious limitations to obtaining the nominal beam parameters. The main sources of impedance causing this instability were found a few years ago from measurements of the spectrum of unstable single bunches. Recently the major part of a programme to reduce this impedance was completed. Reference beam measurements with RF on and RF off have allowed the changes in bunch stability as well as in the coupling impedance to be seen.

IMPEDANCE IDENTIFICATION

The impedance reduction programme started in the SPS a few years ago to prepare the old machine for its new role as injector of LHC. Earlier estimations, based on measurements of microwave instability threshold, suggested [1] that the LHC beam with nominal bunch parameters (intensity $N = 1.1 \times 10^{11}$ and emittance $\varepsilon = 0.35$ eVs) will be unstable already on the 26 GeV injection plateau. This forecast was confirmed in 1999, when LHC beam was injected for the first time into the SPS. Prior to this [2], measurements with single long bunches and RF off had allowed the dominant resonant impedances with high R/Q to be seen as peaks in unstable beam spectra, Fig. 1.

Most of the impedance sources were identified very quickly. They were

- RF cavities

f MHz	type	R/Q kOhm	Q	operation mode
100	SW	1.4	150	leptons
200	SW	3.87	80	leptons
200	TW	25	130	protons, ions
352	SC	0.93	500	leptons
800	TW	6.5	150	protons

- Pumping ports ($\sim$1000) with total R/Q = (25 - 45) kOhm and Q$\sim$ 50 at 1.5, 1.9, 2.4 GHz etc.

However the source of impedance at 400 MHz which can be seen as a dominant peak in Fig. 1 was not obvious. The high signal amplitude suggested some significant impedance, comparable to the impedance of the 200 MHz RF system, also seen well in this spectrum.

With the end of LEP running, all lepton equipment was removed from the SPS ring including the 3 RF systems. The two RF systems with the highest R/Q remain in use.

The pumping port impedance was proved, both by measurements and by simulations, to be the main source of microwave instability. The installation of complex shields with sliding contacts involved displacing 400 dipole magnets in the ring. This work was completed during the 2000/2001 shutdown [3].

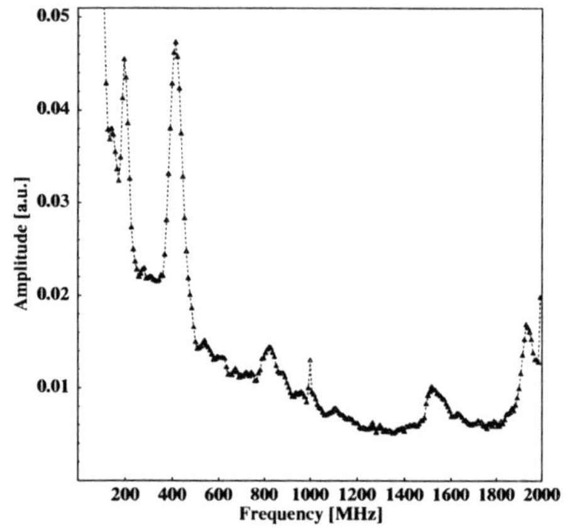

FIGURE 1. Averaged (over 15 acquisitions) projection of bunch spectra measured in 1999. Bunch parameters: $N = 6.0 \times 10^{10}$, $\varepsilon = 0.22$ eVs, bunch length $\tau = 25$ ns.

A series of beam measurements done before and after the impedance reduction allowed both the improvement in bunch stability and changes at particular frequencies to be seen. The results for bunch stability are presented in [4] and summarised at the end of this paper. Below we will present results devoted mainly to the search and elimination of the 400 MHz impedance source.

INSTABILITY AT 400 MHZ

The instability at 400 MHz was first observed in 1996 in the measurements with long single bunches injected into the SPS with RF off at 26 GeV. It was seen on a spectrum analyzer connected to the wideband pick-up as a strong signal growing at frequencies around 400 MHz. The instability was studied both below and above transition.

A signal at 400 MHz will always be seen in such measurements as a second harmonic generated by the nonlinearity of the instability developing due to the large

CP642, *High Intensity and High Brightness Hadron Beams: 20th ICFA Advanced Beam Dynamics Workshop on High Intensity and High Brightness Hadron Beams*, edited by W. Chou, Y. Mori, D. Neuffer, and J.-F. Ostiguy

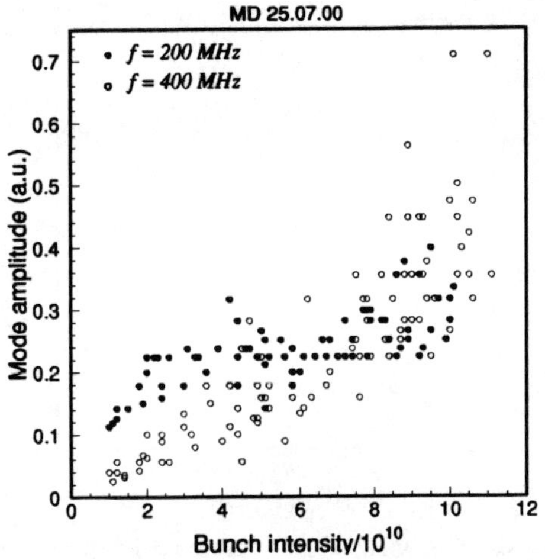

FIGURE 2. Mode amplitude at 200 MHz and 400 MHz as a function of bunch intensity measured at 26 GeV in 2000 ($\tau = 21$ ns, $\varepsilon = 0.24$ eVs).

impedance of the 200 MHz TW RF system. However particle simulations show [5] that the ratio of maximum amplitudes of the 400 and 200 MHz signals generated by the TW cavity impedance, $A_{400}/A_{200} \leq 0.5$ and the results of the measurements shown in Figs. 1,2 can be explained only by the presence of significant impedance at 400 MHz. This impedance could also be a cause of the single bunch instability observed at 26 GeV with RF on.

Estimations show that for a cavity-like object to have its lowest resonant frequency, TM_{010}, at 400 MHz, a radius around 30 cm is required. The list of such objects found in 1996 and studied later is presented in Table 1.

TABLE 1. Possible 400 MHz band sources in the SPS in 1996.

n	Element	No.
1	400 MHz, LHC prototype cavity	1
	MSL, lepton injection	2
	MKLE electron extraction	1
	MKLP positron extraction	1
	MKA antiproton injection	2
2	MSE, extraction septum	10
	MST, extraction septum, thin	6
	ZS electrostatic septum	10
3	MKE extraction	3
	MKP proton injection	3

The first group in this Table contains elements which were removed from the ring in 1998 (400 MHz cavity) and during 2000/2001 shutdown (lepton equipment). Unlike the second group in Table 1, they were never seriously suspected of being a source of single bunch instability due to their small R/Q. The second group of elements was carefully studied due to their large quantity.

Calculations of a simplified model of the MSE and MST septa did not give significant impedance, however bench measurements showed a clear peak at the frequencies of interest. These cavity-like objects were shielded during 1998-1999, but the instability was still there. Measurements of A_{200} (reference) and A_{400} following these changes indicated a 50% increase in intensity at which $A_{400} \sim A_{200}$ in 2000, compared to similar measurements done in 1999 although with longer bunches (26 ns). Later measurements showed a strong dependence of A_{400}/A_{200} on bunch length which could also have contributed to this result [5].

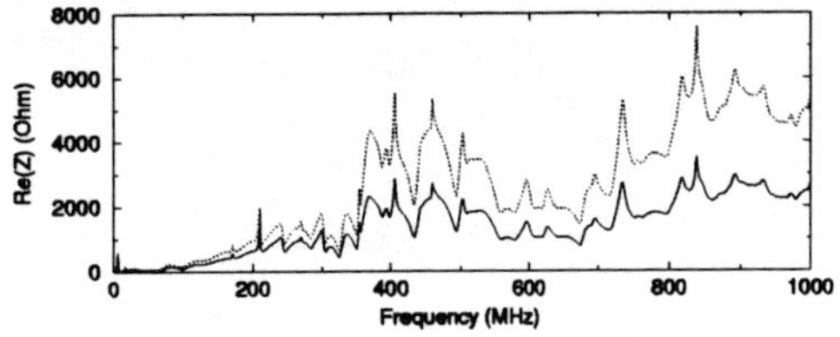

FIGURE 3. Real part of MKE impedance measured in the laboratory with wire method. Solid and dotted lines: standard and improved log formula. Data taken from [6].

Another possible candidate for the 400 MHz impedance was the electrostatic septum ZS. There are 10 of them in the ring and these devices are very difficult to shield. Fortunately bench measurements showed small impedance at 400 MHz.

Bench measurement done on the MKE and MKP kickers suggested large impedance around 400 MHz. Results of measurements for the real part of the MKE kicker impedance are presented in Fig. 3. During the 2000/2001 shutdown some kickers were removed from the ring and screens were installed in others. Measurements done in 2001 showed no instability. They are presented below.

RESULTS

Measurements of mode amplitude at 200 MHz and 400 MHz as a function of intensity are presented in Fig. 4. Compared to Fig. 2, the amplitude of the 400 MHz signal stays well below the 200 MHz signal.

Measurements of growth rates of the 400 MHz instability do not need any calibration, but they are difficult due to the impure exponential character of signal growth. Growth rates of the signal at 200 MHz and 400 MHz, measured in 2000 and 2001 as a function of intensity assuming exponential growth are shown in Fig. 6.

The spectrum measured in 2001 with bunch parameters similar to those in Fig. 1 is presented in Fig. 7. The 400 MHz signal is now significantly lower and appears as the second harmonic of the signal at 200 MHz. Peaks at 1.5 and 1.9 GHz have disappeared as well due to the pumping ports shielding.

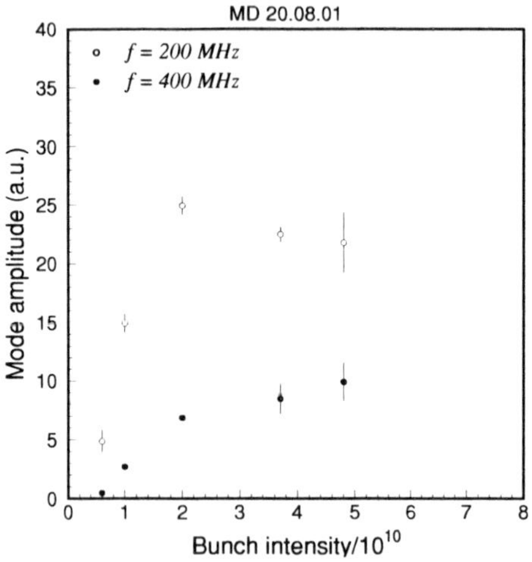

FIGURE 4. Mode amplitude at 200 MHz and 400 MHz as a function of bunch intensity measured at 26 GeV in 2001 ($\tau = 21$ ns, $\varepsilon = 0.24$ eVs).

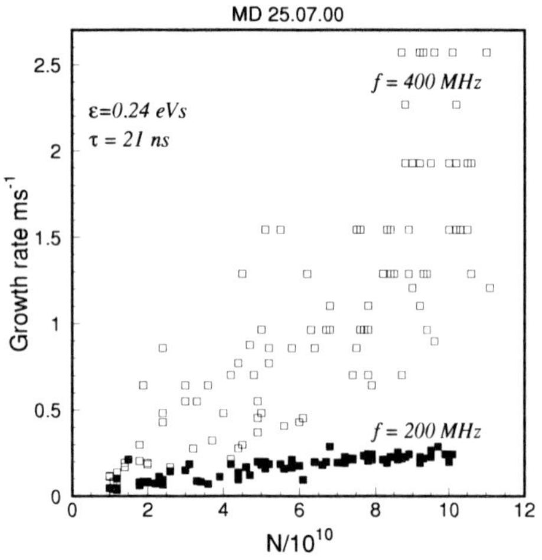

FIGURE 5. 200 MHz and 400 MHz mode growth rate as a function of bunch intensity in 2000 ($\tau = 21$ ns, $\varepsilon = 0.24$ eVs).

Comparison of the measurements of quadrupole frequency shift from 1999 and 2001 gives a factor 2.5 reduction in effective inductive impedance $\mathrm{Im}Z/n$. Measurements of bunch length as a function of intensity demonstrate a factor 7 decrease in the slope of the bunch lengthening curve and a more than 10 times increase in microwave instability threshold for the range of bunch parameters used in the experiment.

In 2002 an LHC beam with bunch intensity up to 1.4×10^{11} was stable at 26 GeV.

Results for transverse impedance reduction are presented in [7].

We are grateful to PS and SL operation groups for providing us with various beams during our studies.

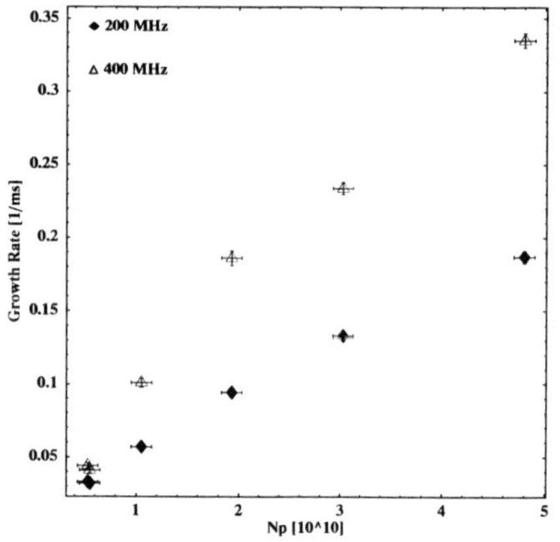

FIGURE 6. 200 MHz and 400 MHz mode growth rate as a function of bunch intensity in 2001 ($\tau = 22$ ns, $\varepsilon = 0.2$ eVs).

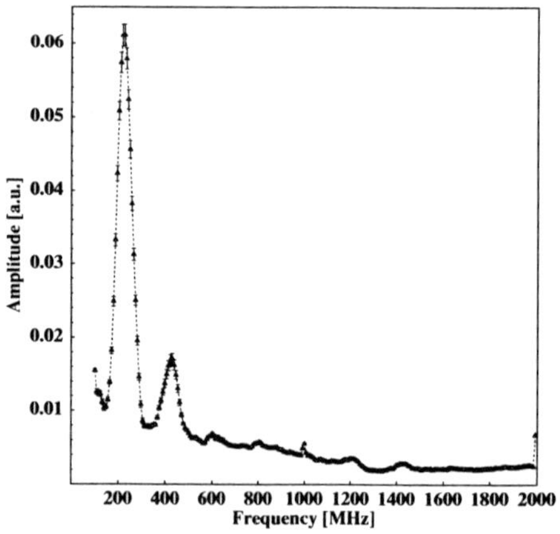

FIGURE 7. Bunch spectra in 2001. Bunch parameters: $N = 6.0 \times 10^{10}$, $\varepsilon = 0.22$ eVs, $\tau = 25$ ns.

REFERENCES

1. T. Linnecar, E. Shaposhnikova, Requirements for beam parameters in the SPS when used as LHC injector, CERN SL Note/94-87 (RF), 1994.
2. T. Bohl, T. Linnecar, E. Shaposhnikova, Measuring the resonance structure of accelerator impedance with single bunches, Phys. Rev. Lett., 3109, 1997.
3. P. Collier et al, Reducing the SPS impedance, Proc. EPAC 2002.
4. T. Bohl, T. Linnecar, E. Shaposhnikova, Impedance reduction in the CERN SPS as seen from longitudinal beam measurements, Proc. EPAC 2002.
5. E. Shaposhnikova, 400 MHz impedance - where are we? Proc. LHC Workshop 2001, Chamonix, CERN-SL-2001-003 DI.
6. F. Caspers et al, Impedance measurement of the SPS MKE kicker by means of the coaxial wire method, PS/RF/Note 2000-004, 2000.
7. H. Burkhardt, G. Rumolo, F. Zimmermann, Coherent beam oscillations and transverse impedance in the SPS, Proc. EPAC 2002.

Emittance Dilution In HERA-p: Lessons and Overview

R. Wanzenberg

DESY, Notkestr. 85, 22603 Hamburg, Germany

Abstract. Protons of an energy of 920 GeV collide with positrons or electrons (energy 27.5 GeV) in the HERA accelerators at DESY. To obtain a high luminosity any emittance dilution must be minimized. Different sources of potential emittance growth are discussed. Effects arising from a coherent injection oscillation or a mismatch of the beta-functions are the main sources of transverse emittance dilution at HERA. In addition a longitudinal emittance growth during the acceleration of the protons has been observed in HERA. Finally the emittance growth due to intra-beam scattering is briefly discussed.

INTRODUCTION

HERA, with a circumference of about 6.3 km, is the largest accelerator facility at DESY in Hamburg. A 920 GeV proton beam and a 27.5 GeV polarized positron or electron beam are provided for four high energy physics experiments. H1 and ZEUS are e/p collider experiments, while two fixed target experiments make use of the polarized e^+ (or e^-)-beam (HERMES) or the halo of the p-beam (HERA-B). The two accelerator rings for the p and e-beam are shown in Fig. 1. The basic beam parameters which were used until the year 2000 are listed in Table 1. During a typical run, a peak luminosity of 1.8 10^{31} $cm^{-2}s^{-1}$ and a specific luminosity of 0.8 10^{30} $cm^{-2}s^{-1}mA^{-2}$ have been achieved (see Fig. 6). In 2001 the interaction regions were upgraded for an increase in the specific luminosity by a factor of 2.76. Ref. [1] presents further details on these HERA parameters and future plans.

FIGURE 1. Photograph of the HERA tunnel. The superconducting magnets of the proton ring are installed above the (normal conducting) magnets of the electron/ positron ring.

SOURCES OF EMITTANCE DILUTION

An emittance dilution is caused by filamentation of a coherent oscillation. Possible sources are an injection oscillation, a mismatch of the linear beam optics (beta-function or dispersion), nonlinear fields or coherent beam instabilities. The different effects are estimated in the following sections for the HERA-p ring based on the data and formulas from Ref. [2].

TABLE 1. HERA parameters (pre upgrade).

Parameter	HERA-p	HERA-e
E / GeV	920	27.5
I / mA	100	50
# bunches	180	189
$N / 10^{10}$	7.3	3.5
ε_x / nm	5.0	41
$\varepsilon_y/\varepsilon_x$	1	0.1
σ_z / cm	19	1.1

Scattering of particles, such as residual gas scattering or intra-beam beam scattering also results in an emittance growth. Intra-beam scattering (IBS) is a multiple Coulomb scattering process of the particles in the bunch, which cause an emittance growth of the beam in all three beam dimensions. The longitudinal emittance growth due to IBS is a good candidate for the observed generation of coasting beam, which may result into background problems for HERA-B [3].

TRANSVERSE EMITTANCE DILUTION

The emittance dilution in the transverse planes is mainly caused from injection errors at HERA. After each magnet cycle a careful adjustment of the tunes, energy and injection trajectory is necessary since the magnetic fields in the superconducting magnets differ from cycle-to-cycle due to the induced persistent currents. The reproducibility of the injection trajectory, the precision of the beam position monitors (BPMs) of about 0.5 mm (resolution 0.1 mm) and the kicker ripple of about 1 % results into a total injection trajectory error of about 1 mm. This is achieved after pilot injection of a few proton bunches which are used to adjust the trajectory of the injected beam to the closed orbit. The maximum beta-function at the

CP642, *High Intensity and High Brightness Hadron Beams: 20th ICFA Advanced Beam Dynamics Workshop on High Intensity and High Brightness Hadron Beams,* edited by W. Chou, Y. Mori, D. Neuffer, and J.-F. Ostiguy

injection energy of 40 GeV is about 90 m, while the emittance of the injected proton beam is 0.12 µm. The emittance dilution due to injection errors is about 8 %.

Other sources of emittance dilution are of less significance. The emittance dilution due to a mismatch of the beta function is about 3 % assuming a beta-beat of $\Delta\beta / \beta = 10\ \%$. The emittance dilution due a dispersion mismatch, coupling in the transfer line and nonlinear fields in the transfer line is not larger than about 1 %. The total emittance dilution during injection amounts to about 10 % - 15 % (Ref [2]).

LONGITUDINAL EMITTANCE DILUTION

Two rf-systems with frequencies of 52 MHz (harmonic number h = 1100) and 208 MHz (h = 4400) are installed in the HERA-p ring. The proton bunch is essentially injected into a bucket of the 52 MHz rf-systems (total voltage 140 kV) with an additional small bucket deformation from to the 208 MHz rf-system (about 10 kV) included to increase Landau damping. During the ramp from 40 GeV to 920 GeV the voltage of the 208 MHz rf-systems is increased (to about 600 kV) to compress the bunch longitudinally. The bunch length and the energy of a proton bunch during the ramp of the HERA-p ring is shown in Fig. 2.

.

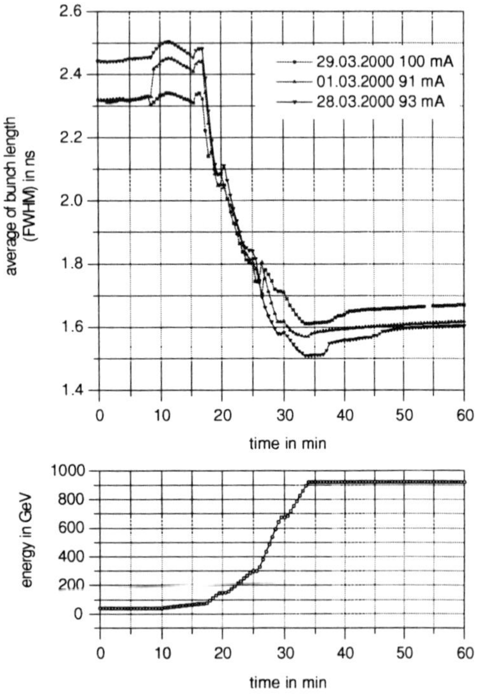

FIGURE 2. Bunch length and energy of a proton bunch during the ramp of the HERA-p ring (Adapted from [4].) .

The bunch length is observed to decreased from about 2.4 ns to 1.6 ns (FWHM). These measurements are based on signals from a resistive gap monitor. One would expect an even smaller bunch length if the longitudinal emittance would be not diluted during the ramp. The longitudinal emittance is increased by a factor of 6 during the ramp, which is shown in Fig. 3.

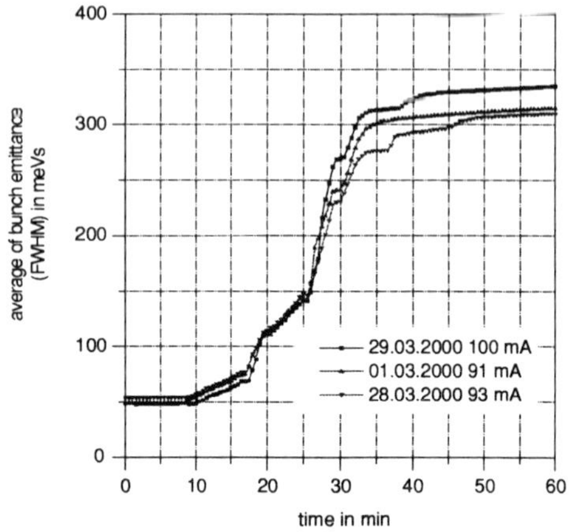

FIGURE 3. Longitudinal emittance of a proton bunch during the ramp of the HERAp ring (Adapted from [4].)

An analysis of multi-bunch phase oscillation of the proton bunches (see Ref. [4]) indicates that a possible cause of the emittance dilution is a coupled bunch instability. The modal spectrum of multi-bunch phase oscillations is shown in Fig. 4 . The modes l=1, which has a phase shift of 2π along the bunch train, and l=163 (phase shift about 3π/4) are the dominant modes in the spectrum.

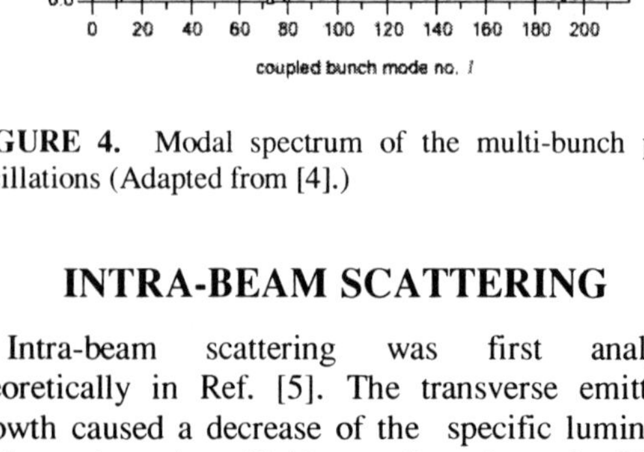

FIGURE 4. Modal spectrum of the multi-bunch phase oscillations (Adapted from [4].)

INTRA-BEAM SCATTERING

Intra-beam scattering was first analyzed theoretically in Ref. [5]. The transverse emittance growth caused a decrease of the specific luminosity during a long (say 10 h) run. Data from the HERA archive are shown in Fig. 6. The average specific luminosity decrease during that run was about 10 %. The archive data show also the influence of orbit drifts on the specific luminosity, which could be remedied by resteering of the beam in the interaction region.

During the same luminosity run the bunch length of proton bunch number 144 was recorded. The increase of the full width half maximum (FWHM) of the longitudinal proton bunch profile from initially of about 2.25 ns to about 2.5 ns during the run is show in Fig. 5 (see also Ref. [6]). A fit to recorded data shows that the initial increase in bunch length is 2 % per hour. This in good agreement with intra-beam scattering theory of Ref. [7]. An application of the formula:

$$\frac{1}{\tau_s} = \left\langle A \tfrac{\sigma_h}{\sigma_p} f(a,b,q) \right\rangle$$

(the function f and the parameters a, b and q are defined in Ref. [7])

to the HERA lattice predicted an initial growth of the bunch length of 1.8 %. An alternative analysis of the bunch length data, which is based on rf-noise, can be found in Ref. [8].

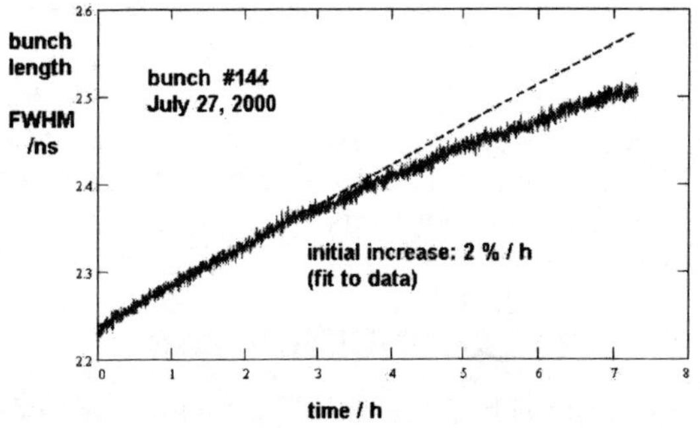

FIGURE 5. Bunch length (FWHM) of bunch number 144 during the luminosity run on July 27, 2000.

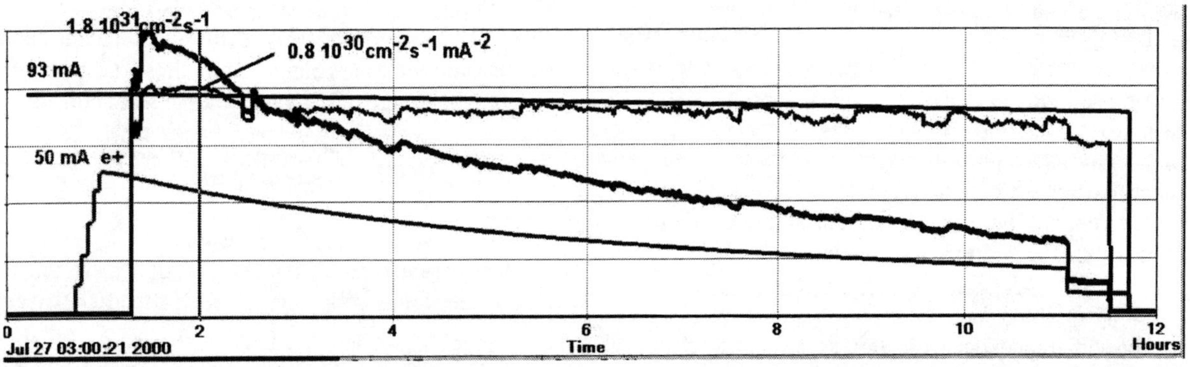

FIGURE 6. The proton current (93 mA, black line), the positon current (initially 50 mA, red line), the luminosity (initially 1.8 10^{31} $cm^{-2}s^{-1}$, brown line) and the specific luminosity (about 0.8 10^{30} $cm^{-2}s^{-1}mA^{-2}$, blue line) are shown versus time during a luminosity run at July 27, 2000.

ACKNOWLEDGMENTS

I would like to thank G. Hoffstaetter, E. Vogel and F. Willeke, for helpful discussions.

REFERENCES

1. Hoffstaetter, G.H., "Future possibilities for HERA" in *7th European Particle Accelerator Conference (EPAC 2000)*, Vienna, Austria, 26-30 Jun 2000
2. Willeke, F. "Emittance Dilution in Proton Rings" in *High Quality Beams, Joint US-CERN-JAPAN-RUSSIA Acc. School*, edited by S. I. Kurokawa et al., AIP Conference Proceedings 592, New York: American Institute of Physics, 2001, pp. 85-117
3. HERA-B Target Group, Ehret, K. et al., Nucl. Instrum. Meth. **A** 456 , 206-216 (2001).
4. Vogel, E., *Fast Longitudinal Diagnostics for the HERA Proton Ring*, Thesis, University of Hamburg, Hamburg, 2002, DESY-THESIS-2002-010.
5. Piwinski, A, Proc. 9th Int. Conf. on High Energy Acc., SLAC, Stanford, 1974, p. 405.
6. Zorzano, M.P., Wanzenberg R., *Intrabeam Scattering and the Coasting Beam in the HERA Proton Ring,* DESY HERA 00-06, Oct 2000
7. Piwinski, A, "Intra-Beam Scattering" in *CERN Acc. School -1991*, edited by S. C. Turmer ,CERN 92-01, Geneva,1992, pp. 226-238.
8. Ivanov, S., Lebedev, O., *Coasting Beam in HERA-p Ring* DESY HERA 01-03, Oct. 2001

Multiparticle Dynamics in the E-ϕ Tracking Code ESME

James A. MacLachlan

Fermi National Accelerator Laboratory[1]

Abstract. ESME has developed over a twenty year period from its origins as a program for modeling rf gymnastics to a rather general facility for that fraction of beam dynamics of synchrotrons and storage rings which can be properly treated in the two dimensional longitudinal phase space. The features of this program which serve particularly for multiparticle calculations are described, some uderlying principles are noted, and illustrative results are given.

INTRODUCING ESME

Multiparticle tracking has established utility for modeling evolution of longitudinal phase space distributions of particles in accelerators as they respond to the rf in acceleration or bunch manipulation. ESME has been primarily developed for design studies of machine cycles and rf gymnastics using single particle equations of motion.

One goes from single particle to multiparticle dynamics by calculating the beam current every time step and including its effect on the single particle motion. However, the number of macroparticles needed and bandwidth required for quantities in the frequency domain need careful attention. It is very easy to generate spectacular spurious instabilities by long time steps or too few macroparticles. Considerable attention has been given in recent years to these considerations and to development of facilities for multiparticle dynamics.

Space Charge Model

The self-impedance from the direct interparticle forces is derived from an electrostatic calculation in the beam rest frame transformed to the lab frame.[1] The force at the n-th time step arises from the gradient of the azimuthal charge distribution $\Lambda_n(\Theta)$, which can be evaluated from its Fourier series for the frequency domain or directly for time domain. The impedance representing the self force is

$$\frac{Z_m^{sc}}{m} = -i\frac{Z_0 g}{2\beta\gamma^2}, \tag{1}$$

where $Z_0 = \sqrt{\mu_0/\varepsilon_0} = 377\Omega$ and g is a factor containing the dependence on beam and vacuum chamber transverse dimensions. For a uniform cylindrical beam of radius a centered in a smooth beampipe of radius b

$$g = 1 + 2\log(\frac{b}{a}). \tag{2}$$

In ESME this factor is scaled with beam momentum to account for the change in beam radius and rolled off at very high frequency to approximate the exact solution at high mode numbers.

Frequency Domain Facilities

ESME is fundamentally time domain, but beam current and image currents can be Fourier analyzed at harmonics of synchronous circulation frequency $\Omega_{s,n}$. At the end of the n-th time step the beam has an azimuthal charge distribution $\Lambda_n(\Theta)$. Assuming that the distribution is practically unchanged in a single time step (usually one beam turn), it gives rise to a beam current

$$I_{b,n}(\Omega_{s,n}t) = \frac{e\Omega_{s,n}}{2\pi}\sum_m \Lambda_{m,n}e^{i(-m\Omega_{s,n}t+\psi_{m,n})}, \tag{3}$$

here expressed as a sum of phasors.

The current induces voltage through the total longitudinal coupling impedance $Z_{||}(\omega)$; this quantity evaluated at $m\Omega_{s,n}$ is denoted by the phasor $Z_{m,n}e^{i\chi_{m,n}}$. The beam-induced voltage is applied to each particle at time t_n when the synchronous particle is at the gap; that voltage depends on the relative phase between the particle and the current. The synchronous particle is defined to have phase 0. Thus, the energy increment for the i-th particle on the n-th turn resulting from the beam current is

$$eV_{i,n}^b = -\frac{Ne^2\Omega_{s,n}}{2\pi}\sum_m \Lambda_{m,n}Z_{m,n}e^{i(m\Theta_{i,n}+\psi_{m,n}+\chi_{m,n})}. \tag{4}$$

The apparent absence of frequencies other than circulation harmonics is not a defect in this formulation. Synchrotron sidebands are generated in the tracking the same way they are generated in a synchrotron — by the phase modulation.

Time Domain Facilities

The gradient of Λ is calculated at each particle location by cubic spline interpolation over three bins of a charge histogram. For simple resonances, a Green's function solution is used which permits turn-to-turn accumulation

[1] Work supported by the U. S. Department of Energy under contract number DE-AC02-76CH03000.

CP642, *High Intensity and High Brightness Hadron Beams: 20th ICFA Advanced Beam Dynamics Workshop on High Intensity and High Brightness Hadron Beams,* edited by W. Chou, Y. Mori, D. Neuffer, and J.-F. Ostiguy

of the wakefield. The wakefield for arbitrary distribution and arbitrary $Z_{\|}$ is calculated using the response to triangular unit current pulse calculated by TCBI or the like convolved with the current distribution. One may freely inter-mix potentials expressed in time domain and frequency domain. The calculation of bunch disruption occurring with heavy transient beam loading corrected by feed forward and feedback around the cavities plus global phase and voltage feedback loops, for example, is an application which would draw on nearly all of these features.

Scaling Concept

By inspection of the single particle equations of motion, it appears that the phase space motion can be accelerated by scaling the phase slip factor η and the potential by the same factor, hereafter λ.[2] Both the single particle and collective potential are scaled the same. Indeed, particle distributions have been practically identical when time t in an un-scaled calculation is compared to time t/λ in a scaled calculation for a variety of test calculations. The obvious gain is a factor λ^{-1} in computing time by speeding up the clock in the scaled calculation. However, more importantly, scaling of the time means that frequencies like the rf frequency and resonance frequencies in $Z_{\|}$ are also scaled. The frequencies of Fourier components are scaled, and, for $\lambda > 1$, fewer harmonics are needed to span the range of a given $Z_{\|}(\omega)$. For a particularly interesting example consider the space charge equivalent impedance; it is nearly linear in frequency so that just a few Fourier harmonics would appear necessary to characterize it. The evaluation is not compromised by widely spaced frequency sampling. Because the number of macroparticles can be scaled by λ^{-3} when the number of bins is reduced by λ^{-1} with same level of numerical noise, there is a potential gain from scaling of λ^4 in many cases. This cubic rule for macroparticle number has been shown rigorously for space charge in time domain[3] and heuristically in frequency domain for any smooth $Z_{\|}(\omega)$[4]

DEMONSTRATION APPLICATIONS

The two examples following are chosen for their suggestivity rather than as exhibits of results of special importance in themselves. Nonetheless, the first example, self bunching of a coasting beam, may surprise even some experienced beam physicists.

Coasting Beam Self-Bunching

Excitation of a passive resonator by coasting beam and beam response to the resonant voltage is an archetypal multiparticle dynamics problem. One expects that beam will self-bunch above threshold, with the bunches decelerated from the initial beam energy. This example illustrates that bunching is dependent on the relation of resonant frequency to nearest harmonic of beam circulation frequency. Animations using parameters of the Los Alamos PSR with addition of an h=3 resonator ($R_{\text{shunt}} = 300$ Ohm, $Q = 100$) show a distinctly different qualitative behavior depending whether the resonant frequency is one percent above or below three times the beam circulation frequency. The distributions after 160 ms are shown in Figs. 1 and 2; animated GIF's may be viewed at `www-ap.fnal.gov/~jmaclach` as psr3hi.gif and psr3lo.gif. In either case, mean energy is decelerated equally, in buckets when the resonance is above the beam circulation harmonic or by phase displacement with empty buckets when it is below. The analogy to Robinson instability (h=1) is suggestive.

Negative Mass Instability

Attempts to compare the Hardt[5] analysis of negative mass instability (NMI) to numerical tracking have run up against particle statistics problems deriving from computer speed and memory limitations. The relaxation of these constraints because of computer developments invites another look. Furthermore, time scaling should ameliorate both time and storage problems. In addition, numerically quiet distributions of a reasonable number of macroparticles can be (at least initially) even quieter than the real beam distribution.

Several cases based on the Fermilab Main Injector (FMI) have been calculated to explore the effects of computational parameters; some results are given in Table 1. The maximum growth rate occurs near 60 GHz regardless of time/bandwidth scaling, much below the 120 GHz predicted by Hardt's model. The two orders of magnitude difference between the growth for 0.2 and 0.3 eVs is consistent with the threshold of just below 0.3 eVs calculated with Hardt's formulas. The amount of emittance growth is comparable but less with scaling factor $\lambda > 1$, a discrepancy not observed in previous uses of time scaling. Fig. 3 shows the spectrum of the beam current over the frequency range 21 to 74 GHz taken at intervals of 88 μs. The rapid shift of the peak to lower frequency may relate to the discrepancy between the analytical prediction and the lower peak frequency observed, *viz.*, the shift may progress substantially before the instability is visible on a linear amplitude scale. Computers are now fast enough and big enough that a calculation could be done for a physically interesting case with the actual number of particles in a bunch, thereby eliminating any question about scaling or numerical noise.

TABLE 1. Emittance growth $\Delta\varepsilon/\varepsilon$ and frequency of strongest Fourier component $\hat{f}$ [GHz] in NMI of $4.5 \cdot 10^{10}$ proton FMI bunches of full emittance ε [eVs]. Model parameters are number of macroparticles $n_p[10^6]$, number of charge bins n_b, and time scaling factor λ. See text for other parameters.

ε	n_p	n_b	λ	$\Delta\varepsilon/\varepsilon$	$\hat{f}$
0.2	128	4096	1	0.222	58.3
0.2	16	2048	2	0.112	58.3
0.3	128	4096	1	0.003	42.4
0.5	128	4096	1	0.000	none

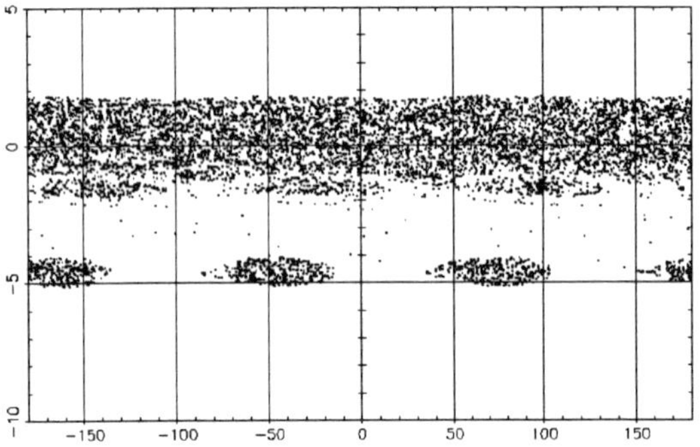

FIGURE 1. Coasting beam response with parameters like Los Alamos Proton Storage Ring with added passive cavity tuned 1% *above* the third harmonic of the circulation frequency; see text for parameter details .

PARTING THOUGHTS

Particle tracking of the evolution of a distribution is not in general superior or inferior to solving the Vlasov equation — simply different. Numerical modeling in dynamics is especially helpful to look for gross oversights, carry practical calculations beyond threshold, evaluate the result of the simultaneous action of several processes, and to get a quick check on expectations. Graphical output from macroparticle models can communicate results easily to non-experts. Also it's nice to get an easy reality check during a protracted calculation; reasonably well-tested codes are a useful source of comparisons or benchmarks.

REFERENCES

1. J. MacLachlan, "Longitudinal Phasespace Tracking with Spacecharge and Wall Coupling Impedance", Fermilab FN-446 (February 1987)
2. J. A. MacLachlan and Z. Nazario "Scaling for faster macroparticle simulation in longitudinal multiparticle dynamics", PRST-AB **3**, 114401 (November 2000)
3. J. Wei, "Longitudinal Dynamics of the Non-Adiabatic

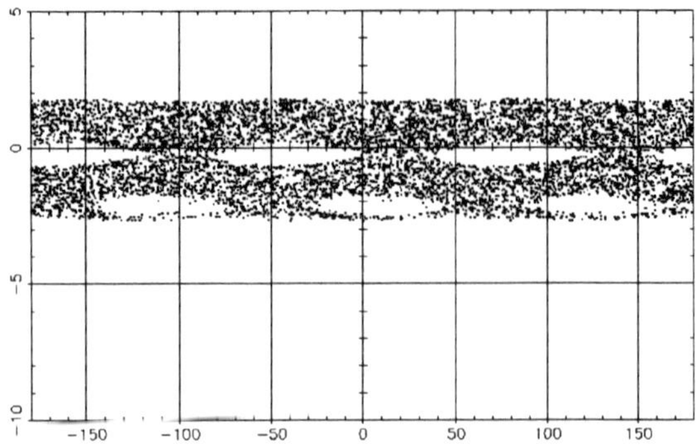

FIGURE 2. Coasting beam response to the same cavity used for the Fig. 2 result, except tuned 1% *below* the third harmonic of the circulation frequency.

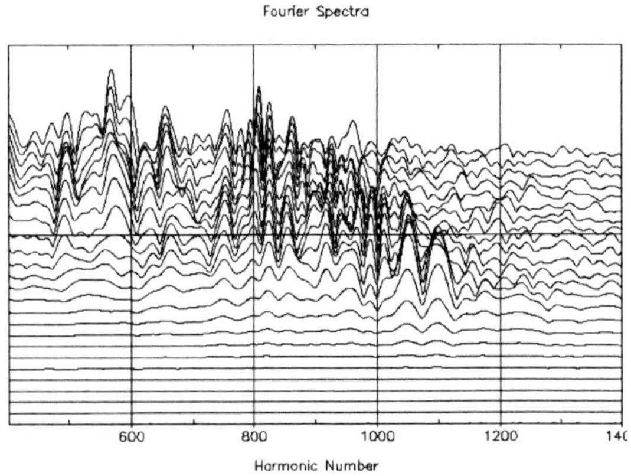

FIGURE 3. Fourier spectrum at times separated by 88 μs for 0.2 eVs FMI bunch of $4.5 \cdot 10^{10}$ protons just above transition ($\eta = 8.9 \cdot 10^{-5}$) The abscissa is labeled in harmonics of the 53 MHz rf; the frequency span shown is 21 – 74 GHz.

Regime on Alternating Gradient Synchrotrons", PhD dissertation for State University of New York at Stony Brook (1990)
4. J. A. MacLachlan, "Particle Tracking in E-φ Space for Synchrotron Design and Diagnosis", in Proc. 12th Int'l Conf. on the Application of Accelerators in Research and Industry, Denton TX, Jerome L.Duggan, ed., AIP Conference Proceedings (1992) also available as FERMILAB-Conf-92/333 (1992)
5. W. Hardt, "GAMMA-TRANSITION-JUMP SCHEME OF THE CPS", Proc. 9^{-th} Int'l Conf. on High Energy Acc., Stanford (1974)

Development and Applications of the UAL-based SNS Ring Simulation Environment

N.Malitsky, P.Cameron, A.V.Fedotov, J.Smith, J.Wei

BNL, Upton, NY 11973, USA

Abstract. The SNS Ring off-line parallel simulation environment based on the Unified Accelerator Libraries (UAL) has been implemented and used for extensive full-scale beam dynamics studies arising in high-intensity rings. The paper describes the structure of this environment and its application to various high-intensity topics and diagnostics modeling.

INTRODUCTION

Designs and parameters of modern high-intensity machines, such as the SNS accumulator ring, impose new expectations on the beam dynamics studies. One of the major scientific and technical challenges is the extremely strict requirement on uncontrolled beam loss at 10^{-4} level. In order to describe and analyze such low-level losses, one should closely reproduce all actual effects of a realistic machine. For addressing this task, we have employed the open simulation environment of the Unified Accelerator Libraries (UAL[1-2]). When we started this project, UAL had been already applied to several accelerators. However, unlike other UAL applications, the SNS Ring required a new combination of collective physical effects and dynamic processes. Then the new SNS-specific package has been implemented. Its development included the integration of the UAL infrastructure with three accelerator libraries (ORBIT, ACCSIM, and MB), deployment of this infrastructure on the MPI-based parallel facility, and the implementation of the injection framework. In the paper we describe the structure of the SNS simulation package and its application to complex beam dynamics studies and diagnostics modeling.

UNIFIED ACCELERATOR LIBRARIES

The Unified Accelerator Libraries (UAL) is an accelerator simulation environment aiming to develop the realistic beam dynamics models including an unlimited combination of physical effects and dynamic processes. The goal has been achieved by introducing the open infrastructure where diverse accelerator approaches and programs are implemented as collaborative C++ libraries connected together via Common Accelerator Objects (such as Element, Twiss, Particle, *etc.*). At this time, the UAL 1.x off-line simulation environment joins six object-oriented accelerator programs (see Fig.1). A universal homogeneous interface based on the Perl extension mechanism allows scientists to combine and manage all these C++ and Perl components in the single development and research environment. The high-level applications are built on the top of the project-specific Shell (or Façade) that consolidates and facilitates access to numerous low-level UAL interfaces and data containers.

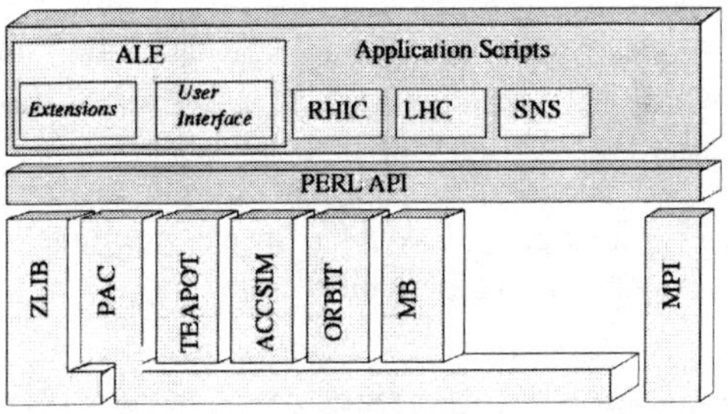

FIGURE 1. Unified Accelerator Libraries Environment.

The same UAL infrastructure fits to the MPI-based parallel environment and has been recently deployed on the parallel cluster for efficient high-intensity beam dynamics modeling. Parallel versions of the time consuming algorithms (such as collective space charge effects) are implemented as ordinary C++ shared libraries and can be mixed together with other UAL sequential and/or parallel modules.

SNS RING SIMULATION ENVIRONMENT

According to the UAL approach, the SNS simulation environment has been organized as an additional package integrating together SNS-specific modules and extensions. The overall architecture of the SNS environment is illustrated in Figure 2. A physicist interacts with the SNS::Shell interface that encompasses a set of high-level MAD-like or TEAPOT-like commands. Most of these commands are project-independent, and inherited from the common class ALE::Shell. The SNS Ring multi-turn injection scenario is considered as an

CP642, *High Intensity and High Brightness Hadron Beams: 20th ICFA Advanced Beam Dynamics Workshop on High Intensity and High Brightness Hadron Beams*, edited by W. Chou, Y. Mori, D. Neuffer, and J.-F. Ostiguy

extension of the present UAL environment and implemented directly in the SNS::Shell class.

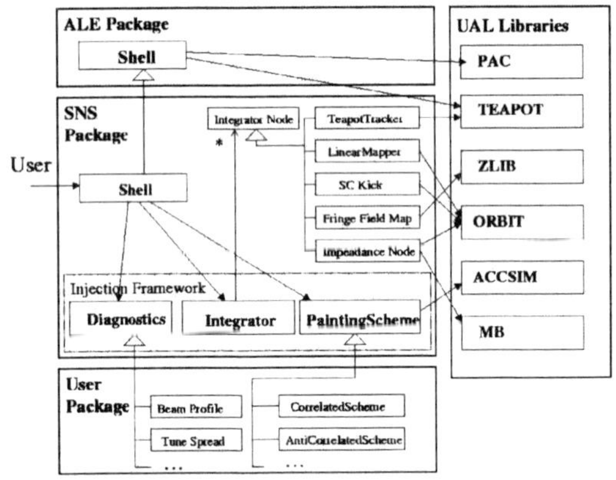

FIGURE 2. Structure of the SNS ring simulation package.

To facilitate the development and study of the SNS Ring applications, we have implemented the injection framework with three basis classes: SNS::Integrator, SNS::PaintingScheme, and SNS::Diagnostics. Each basis class originates an open collection of alternative approaches that can be selected in any combination with the instances of two other collections.

BEAM DYNAMICS APPLICATIONS

One of the major advantages of the SNS package is its capability to support comprehensive studies of various effects, as well as a complex combination of several effects in the single simulation environment. Below, we list some of the beam dynamics topics to which the SNS package was applied:

- Single-particle tasks, including such effects as kinematic non-linearity, non-linear tune-spread, dynamic aperture, and resonance driven diffusion maps [3]-[4].
- Effect of space charge during transverse painting [5].
- Optimization of painting bump functions [6].
- Combined tune spread due to the space charge, chromaticity and other nonlinearities [6]-[7].
- Imperfection resonance crossing in the presence of space charge with corresponding choice of working points and intensity limitation [7]-[8].
- Effect of ½ coherent resonance crossing in the presence of high-order resonances.
- Coherent resonance crossing of coupling resonances.
- Collective instability due to the transverse impedance [9].

An ability to study a complex combination of several effects enables us to have a realistic expectation for beam losses and intensity limitation. Here, we demonstrate the resonance driven beam loss for the two working points of the SNS: (6.23, 6.20) and (6.4, 6.3). The imperfection errors are excited at a level slightly higher than expected to get a conservative estimate. The full 1060-turn injection is then performed for each of beam intensities with beam losses at the end of accumulation recorded for a specific acceptance. Figures 3-4 and 5-6 show the tune spreads and corresponding resonance driven loss curves for both working points.

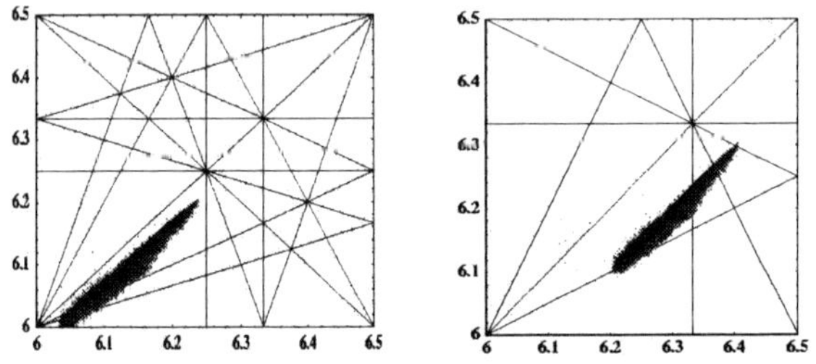

FIGURE 3-4. Tune spreads for working points (6.23, 6.20) and (6.4, 6.3), respectively.

The working point (6.23, 6.20) is essentially free from resonance losses apart from some low loss due to the resonances above the working point and chromatic tune spread. Here, the first-turn losses due to injection and H^- stripping are omitted from the loss curve. For high beam intensities the tune are effectively depressed by the space charge. The only resonances up to the 4th order, which beam crosses during accumulation, are the difference resonances. As a result of the space charge coupling, the beam with approximately similar transverse beam emittances is not susceptible to the difference resonance. The intensity limitation for this working point is associated with the coherent beam response near the tune of 6.0, which limits beam intensity to slightly above $N=2*10^{14}$ at the energy of 1 GeV [10]. However, this limitation is due to the structure resonances and thus is very strict.

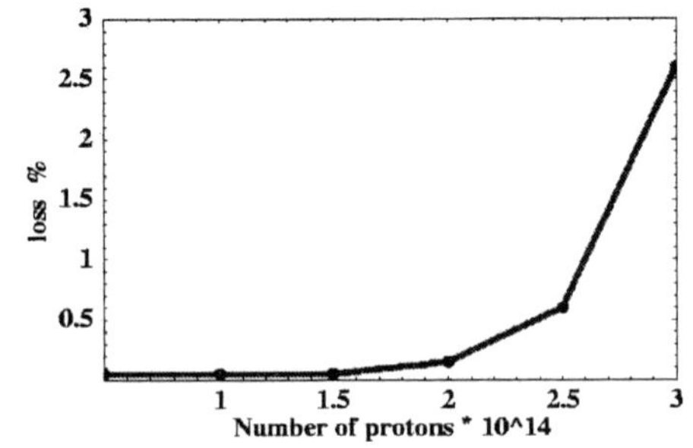

FIGURE 5. Loss curve for the working point (6.23, 6.20)

For the working point (6.4,6.3), the loss curve demonstrates impact of each individual resonance crossed during accumulation. The strong loss at low intensity is due to the sum sextupole resonance at which the working be slightly adjusted to avoid this strong resonance loss. Other loss peaks are due to the 3rd and 4th order resonances, which are crossed for higher beam intensity. The losses due to these resonance are very strong, however, they are all imperfection resonances. With appropriate correction schemes, which are available in the SNS, one can attempt to compensate such resonances. As a result, with successful compensation, the real loss for this working point happens only at a very high intensity due to the coherent beam response to the structure resonances, as shown in Fig. 6.

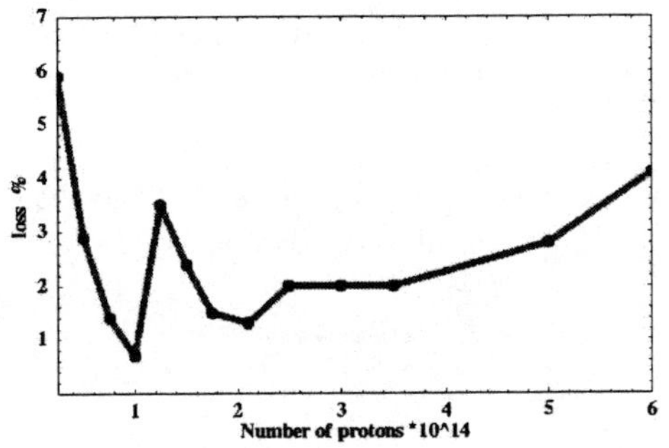

FIGURE 6. Loss curve for the working point (6.3, 6.4)

TUNE MEASUREMENT MODELING

Minimization of beam loss and activation in the Spallation Neutron Source Ring is highly dependent on proper control of the tune footprint. The major contributors to tune spread are space charge, chromaticity, and uncompensated fringe fields. In addition to the challenge of accurate measurement in the presence of large tune spread, large dynamic range will be required to permit measurement through the accumulation cycle. There are many possibilities for measuring coherent and incoherent tune and tune shift in the SNS Ring. These include [11]:

- Coherent (dipole) tune/tune shift from impulse excitation
- Incoherent tune from injection oscillations.
- Incoherent tune from Schottky.
- Incoherent tune from quadrupole mode oscillation.
- Incoherent tune from resonance crossing.
- Incoherent tune from Beam Transfer Function.

To facilitate the design of tune measurement systems, we have been employing the same simulation model used in the SNS Ring beam dynamics studies. The Beam Transfer Function (BTF) approach is its first application. Following the UAL generic scheme, the BTF measurement has been implemented as another tracking scenario with two new diagnostics elements: kicker and BPM. The kicker excites the propagated bunch during each turn of the injection painting and delegates it to the next tracking module. The BPM element calculates the center of bunch and writes the turn-by-turn data into the external file. This application allows one to measure the BPM response to the kicker deflection at a single predefined frequency. Then the tune footprint measurement is represented by a set of several simulation runs, each with a different predefined excitation frequency. Values of FFT signals determine the boundary of the tune footprint. Figure 7 shows preliminary FFT results for the tune spread due to chromaticity.

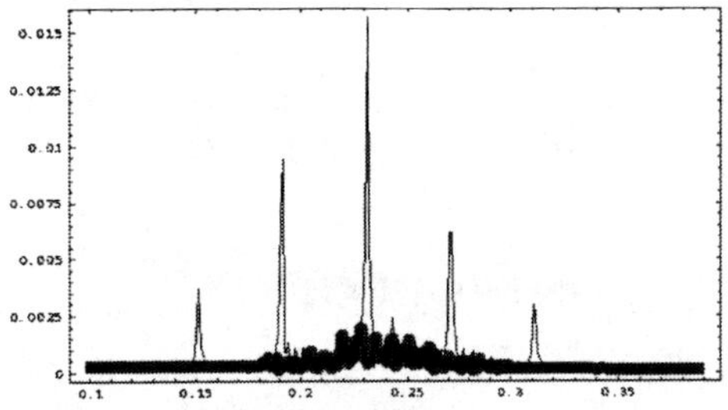

FIGURE 7. Model-based 'measurement' of the tune spread due to chromaticity.

These preliminary results confirm the measurement approach. The method is now being applied to measure full non-linear tune footprint with space charge. To facilitate future activities, the model will be enhanced with the new very efficient tracking engine and additional diagnostics extensions. These calculations were done at baseband with a non-resonant pickup, when in fact the kicker and pickup will operate at about 50MHz, and the pickup Q will approach 100. Kicks in the above modeling were about a factor of 100 stronger than what will be required with the resonant pickup. The observed BTF at 50MHz will show broadening of the revolution line due to a comparatively large relativistic slip factor η, as well as asymmetric chromatic broadening of the betatron sidebands.

REFERENCES

1. N.Malitsky and R.Talman, AIP 391 (1996).
2. N. Malitsky and R.Talman, ICAP 98 (1998).
3. A.V. Fedotov *et al.*, Proc. of EPAC'00, p.1492.
4. I. Papaphilippou, Proc. of PAC'01 (Chicago), p.462.
5. A.V. Fedotov *et al.*, Proc. of PAC'01, p. 2851.
6. A.V. Fedotov, SNS project ASAC reviews (2001-02).
7. A.V. Fedotov *et al.*, Proc. of PAC'01, p. 2848.
8. A.V. Fedotov, N. Malistky and J. Wei, BNL/SNS Tech Note 086 (2001).
9. A.V. Fedotov et al., "Exploring transverse beam stability in the SNS in the presence of space charge", EPAC'02 (to be published).
10. A.V. Fedotov and I.Hofmann, "Half Intger resonance crossing and space charge limit", EPAC'02 (to be published).
11. P.Cameron, A.Fedotov, and N.Malitsky, "Tune Measurement in the SNS Ring", EPAC'02 (to be published).

Longitudinal Dynamics and RF Hardware

Masahito Yoshii

High Energy Accelerator Research Organization (KEK)
Oho 1-1, Tsukuba 305-0801, JAPAN

Abstract. The longitudinal requirements to minimized beam losses are discussed from a longitudinal dynamics point of view. To achieve high field gradient RF system, the magnetic alloy loaded cavity system has been considered. In order to secure the RF system reliability/ reproducibility in operation, the original and novel designs arc proposed.

INTRODUCTION

The JKJ accelerator complex consists of the 400MeV linear accelerator, the 3GeV rapid cycle synchrotron (RCS) and the 50GeV main synchrotron (MR). The 1MW beam power in the RCS and a high intense proton beam of 15μA in the MR are designed as an intensity goal. Longitudinal beam manipulation is the key issue to minimize beam losses during acceleration. Employing the second harmonic RF system will control the bunching factor to mitigate beam loss due to the space charge tune shift. Stability under heavy beam loading is the most important issue on the RF system.

LONGITUDINAL DYNAMICS

In order to minimize the beam losses due to the space charge tune shift, the bunching factor must be controlled and be kept greater than 0.4 at the injection of the RCS, and 0.3 at the injection of the MR. Also, to avoid the single bunch instability, the longitudinal emittance should be enlarged to 10 eV·sec per bunch before the top energy in the MR. Those numbers are important designed values to achieve an acceleration of 6.7μC per bunch of proton beam. To control the longitudinal emittance is essential and required. The longitudinal particle simulations have been done carefully to optimize the parameters for (1) RCS injection, (2) RCS extraction for matching with the MR and (3) MR injection.

Longitudinal Issue in RCS

The linac beam pulse of 500μsec (nominal) with a momentum spread of ±0.1% is chopped by an RF clock signal of RCS injection and injected into the moving RF bucket. There is no flat bottom/top magnetic field for injection/extraction. This injection scheme is to avoid undesired capture losses during the adiabatic process. For optimizing the parameters to keep the required bunching factor, the pulse width, peak current, RF-chopping rate of the linac beam and injection timing and off-momentum injection are carefully searched. And also, the second harmonic RF system is going to be additionally employed.

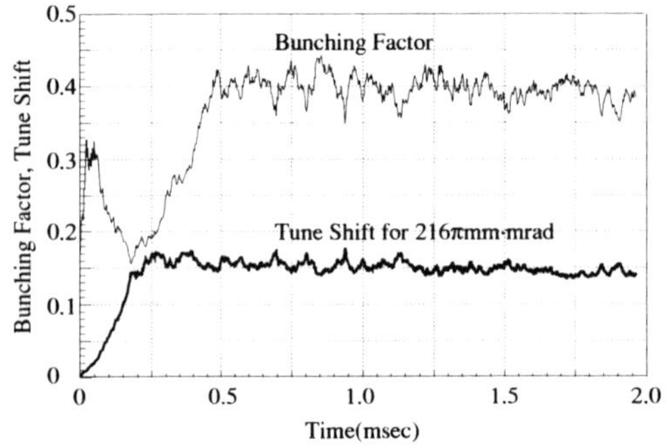

FIGURE 1. Time variations of the bunching factor and the calculated tune shift during the early time of the RCS injection.

The synchronization and the bucket matching are also the important issues. It should begin to manipulate right after the middle of the acceleration, because of low synchrotron tune. For the MR injection, the bunching factor is required to be more than 0.3. The extraction RF voltage must be reduced to 60kV for the longitudinal matching. The heavy beam loading becomes severe. The digital feedforward system is planned to compensate the beam loading effect over the whole cycle.

Longitudinal Issue in MR

Four transfers from the RCS fill the eight of nine RF buckets in the MR. One empty bucket is the space for the injection kicker pulse. The periodic transient effects must be suppressed during the injection and the acceleration. It is also necessary to keep the bunching factor more than 0.3 at the injection, which corresponds to the space charge tune shift of 0.17. The quadrupole oscillation due to the longitudinal mismatch should be avoided. And also, using both amplitude and phase modulation during the acceleration will allow the controlled emittance blow-up. The multi-particle tracking has been undergoing in

CP642, *High Intensity and High Brightness Hadron Beams: 20th ICFA Advanced Beam Dynamics Workshop on High Intensity and High Brightness Hadron Beams*, edited by W. Chou, Y. Mori, D. Neuffer, and J.-F. Ostiguy

order to optimize the parameters of RF manipulation in the MR.

BEAM LOADING

The stability of the RF system under the heavy beam loading is the most important issue. To keep the stability condition over the cycle, the relative loading factor is selected to be smaller than 1 as a design criteria and non-dynamic tuning system is considered [2].

The periodic transient effect is an important issue in the MR. In order to moderate the effects, it is preferable to select a low/ or high Q-value of the RF system. The most dangerous condition occurs when $Q \sim h/2$ [3]. During the MR injection, the effect becomes most severe when the beam fills in a half of the MR buckets. Also, even during the acceleration, the potential distortion caused by circulating sharp bunches must be taken into accounts.

The feedforward compensation is considered to suppress the heavy beam loading effects. In case of the RCS RF system, the bandwidth of the RF cavity covers the frequency of the second RF harmonic, so that the cavity can be driven by composite signals. The stability analysis of such a wide-band system is also the important issue [1]. The analytical approach might become essential to discuss the stability condition for the multi-harmonic RF system under the heavy beam loading.

TABLE 1. The RF Parameters

**imaginary*

	3GeV RCS	50GeV MR
Injection/Top energy (GeV)	0.4 ~ 3GeV	3 ~ 50 GeV
Cycle	25 Hz	3.64 sec
Intensity (ppp)	$8.3*10^{13}$	$3.3*10^{14}$
Accelerating voltage (kV)	450 kV	280 kV
Frequency range	1.23 ~ 1.67 MHz	1.67 ~ 1.72 MHz
Harmonic number	2	9
γ_t	9.14	-j31.6*
Number of RF systems	11	6 + 3
Acc. Voltage per system	41 kV	47 kV
Optimum Q-value	2 ~ 3	10 ~ 20
Ave. power dissi-pation per core	5 kW	9kW

HARDWARES

The high field gradient acceleration system is required in both the RCS and MR rings. The magnetic alloy is the only material [1], which can withstand to both high magnetic field and high power dissipation, in contrast to ferrite materials [Figure 2].

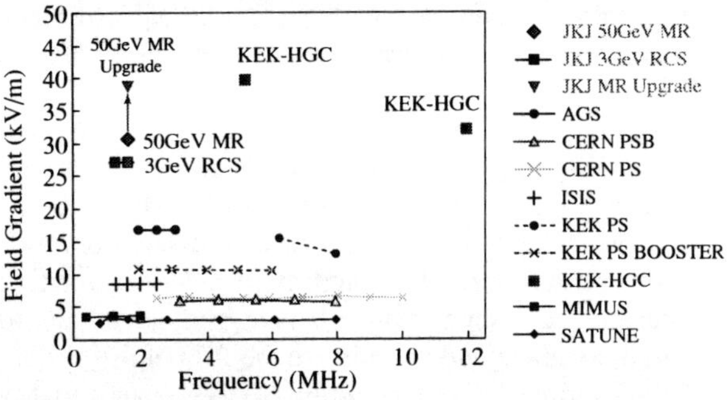

FIGURE 2. Field gradient of the RF cavities for various proton synchrotrons.

The RF power source must have the ability to supply the beam power and the RF generating power to obtain the accelerating voltages. The peak beam power amount is 2.8MW in the RCS and 1.5MW in the MR. The output stage of the power amplifier is a push-pull circuit with two tetrodes (TH558). Each three gaps of cavity is connected in parallel and directly driven by the amplifier. The plate voltage is decoupled from the cavity by using dc-blocking capacitors.

The anode dc-power supply for the amplifier is specified for (1) low ripple voltage, (2) low reactive power and (3) reliable operation. It is also necessary to minimize a drop in primary electric due to pulsed operation and harmonic voltage components. Moreover, the anode power supply should provide a fast cut-off circuit to protect tetrodes from accidental arcing inside.

An Integrated Gate Bipolar transistor, *IGBT*, inverter power supply is able to realize the above requirements. We have designed the 1.2MW anode dc-power supply for the MR RF system. The ripple voltage of ±0.1% was realized in a resistive load test.

RF Cavities

The RF systems in the RCS and the MR have basically the same design. The bandwidth of the cavity is identical in both. In case of the RCS, the cavity should cover the fundamental and second harmonic frequency. The Q-value of 2 ~ 3 is chosen. The combined RF signals can excite the same cavity. On the other hand, the Q-value of the MR cavity is selected 10~20 in order to minimize the transient effects. The cut-core configuration is applied to change the effective R/Q according to requirements [6].

The power dissipation in a core is 5kW and 9kW, respectively, in the RCS and the MR. Effective cavity cooling is the key to develop and to realize the high gradient cavity. Two different cooling methods have been employed. In case of "*direct cooling*" [6], each core is put into the water-sealed tank and *directly* cooled by water. This is the most effective way, however, the impedance reduction due to high dielectricity of surrounding water should be taken into account. Coating the surface of cores is also necessary to prevent corrosion. In case of "*in-direct cooling*", each core is *in-directly* cooled by means of contacting water-cooled Copper plates. It is a great advantage to be able to adjust and to optimize the R/Q of the system later. Thermal contact between Copper-plate and Ma core is the most important engineering issue. To provide thermal contact of two metallic materials while insulating them electrically, a thermal conductive material of a few mm is necessary to avoid impedance reduction and higher mode resonances

Digital RF Control

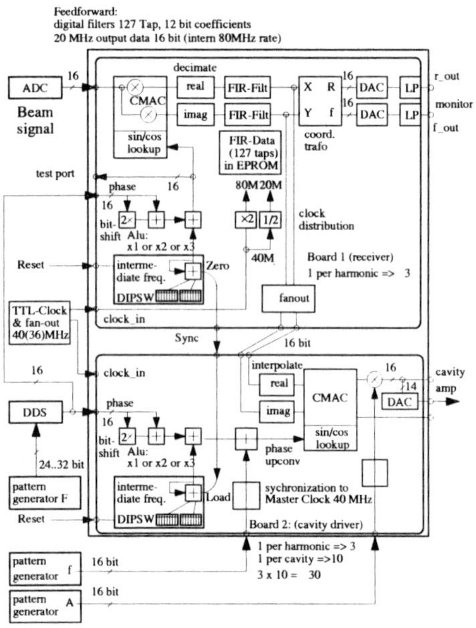

FIGURE 3. Block diagram of Feedforward digital circuit.

Since the MA cavity is a passive load and the system has no dynamic-tuning system, the digital low-level control is the best way to generate RF signal. It is possible to secure system *reliability*, *redundancy* and *reproducibility* in operation. The DDS based precision RF signal generator has been developed. Figure 3 shows the block diagram of digital feed forward circuit modules. This is also one of the key developments to compensate the heavy beam loading. In case of the RCS, The feedforward module must cover the harmonics of 2 ~ 4. A digital finite impulse filter is used as a low path filter. To achieve 400kHz narrow band, the input signal is re-sample with a 127-tap FIR filter running at a half of sampling frequency. The feedforward module has been designed to verify its specifications by testing at KEK-PS. The module consists of commercial filter (L3320 & HSP43320) in digital filtering section, and HSP45116 chips are used as complex multipliers.

SUMMARY

The longitudinal emittance control in both the RCS and MR is necessary to minimize beam losses due to space charge effects. Longitudinal full tracking in RCS shows that the *bunching factor* control in RCS seems to possible by adding 2nd harmonic system, adjusting *momentum*-offset at injection. And, the 2nd harmonic system in the MR is necessary to suppress the quadrupole oscillation at the injection. For the bucket matching to the MR, the extraction RF voltage must be reduced to 60kV in RCS. The feedforward system as well as feedback is indispensable to suppress the heavy beam loading. Cavity cooling is the most important issue to realize the high gradient RF system with magnetic alloy material. Both *direct* and *in-direct* cooling cavities are studied. The IGBT inverter power supply has been decided to use as an anode dc-power supply. Low ripple voltage of ±0.1% has been achieved, and the total appraisement is much better to compare to a conventional power supply with simple rectifiers and capacitor bank. The DDS based RF signal generator and the feedforward module have been designed and manufactured. The digital techniques in the feed-forward module can also be used to develop the digital phase detection.

REFERENCES

1. Mori, Y., et.al., *A new type of RF cavity for high intensity proton synchrotron using high permeability magnetic alloy*, Proceedings of EPAC 1998, , p.299-301.
2. Souda, M., et.al., *Stability analysis using EMTP for JHF systems*, Proceedings of APAC 1998, , p.405-407.
3. Yamamoto, M., et.al., *Beam loading effects on high gradient MA-loaded cavity*, Proceedings of the 1999 PAC, New York, p.860.
4. Tamura, F., et.al., *Digital RF Feedforward Systems for Beam Loading Compensation in JKJ Synchrotrons*, Proceedings of EPAC2002, Paris, France, p.2106-2108.
5. Uesugi, T., et.al., *Beam injection and longitudinal emittance control in the JKJ 50GeV synchrotron*, Proceedings of EPAC2002, Paris, France, p.1311-1313.
6. Yoshii, M., et.al., *Ma RF Cavity for the KEK 12 GeV PS*, Proceedings of EPAC2000, Vienna, Austria, p.984-986.

Diagnostic Investigation of Tune and Tune Shift in the IPNS RCS

J. C. Dooling, F. R. Brumwell, and G. E. McMichael

Argonne National Laboratory

Abstract. The Intense Pulse Neutron Source (IPNS) Rapid Cycling Synchrotron (RCS) accelerates 50 MeV protons to 450 MeV 30 times per second for spallation neutron production. Average current from the RCS has recently exceeded 16 μA with peak instantaneous current approaching 15 A. The RCS makes efficient use of 21 kV of RF accelerating voltage and uses phase-modulation between the two rf cavities to damp vertical instabilities. Split-ring electrodes in the ring suggest an anomalous tune shift that increases with time in the acceleration cycle. Based on a background gas pressure of 1 μTorr, the neutralization time for the beam is approximately 0.5 ms at injection suggesting the beam becomes fully neutralized relatively quickly in the cycle. Over-neutralization of the beam can lead to a positive tune shift that is presumably incoherent. Studies are underway to characterize the ionization within the RCS using the existing Profile and Position System (PAPS) and a newly installed Retarding Field Analyzer (RFA). Also a newly installed fast, deep-memory digitizing oscilloscope allows the entire history of a single acceleration cycle to be recorded from all four components of the split ring electrodes simultaneously at a rate of 250 MS/s.

INTRODUCTION

The Intense Pulsed Neutron Source (IPNS) Rapid Cycling Synchrotron (RCS) accelerates $3\text{x}10^{12}$ protons from 50 MeV to 450 MeV, 30 times per second. The bare x and y tunes are Q_x~2.2 and Q_y~2.35. Observations of spectra recorded from capacitive pick-up electrodes within the RCS suggest that for a portion of the beam, the tune may be increasing for part of the accelerator cycle. In 1999, looking at data from these "Pie" position electrodes[1] using a gated spectrum analyzer, a beam loss phenomena referred to as "pre-spill" was clearly seen to be the result of a resistive wall (vertical) instability as had been previously surmised[2]. The instability causes a 50 percent loss of beam at the end of the acceleration cycle if left unchecked. The resistive wall instability is controlled through phase modulation (PM) of the rf accelerating voltage. The phase modulator or "scrambler," varies the position of one of the two rf accelerating cavities by sinusoidal amplitude of 100 mrad (6°) near twice the synchrotron frequency. The scrambler circuit also introduces amplitude modulation (AM) at the same frequency. "Slow-Q" (captured charge) signals with and without the scrambler are presented in Fig. 1.

RESISTIVE WALL INSTABILITY

Since the mid-1980s, the scrambler has been used to stabilize the bunch during the last few milliseconds of the acceleration cycle. Figure 2 shows two spectra taken just prior to extraction, 14 ms after injection, from a vertical pie electrode using an HP4396B spectrum analyzer (SA). The spectrum presented in Figure 2a is that with the scrambler modulation on from 9.5 ms to 11.5 ms, and that in 2b is with the scrambler off. Significant side-bands are in evidence in the latter spectrum and are believed to reflect the vertical tune, Q_y. Amplitudes of the sidebands are a strong function of circulating charge.

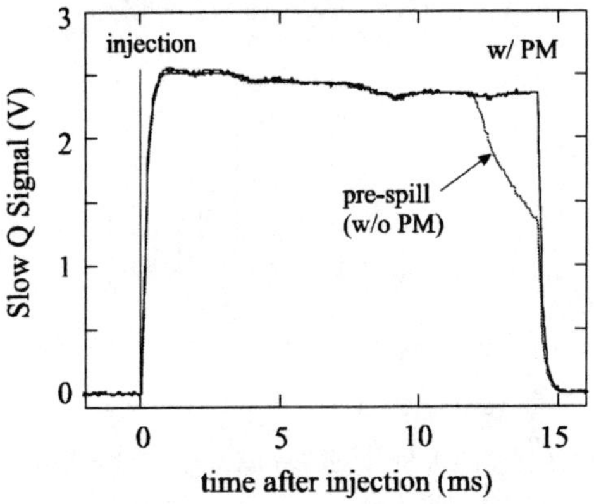

FIGURE 1. Captured RCS charge with and without PM.

The sidebands correspond to $(n\pm\nu_y)f_o$ for $n > Q_y$, in agreement with a resistive wall instability[3]. The scrambler damps the transverse instability presumably by increasing the energy spread of the bunch. One way this can occur is by establishing a parametric oscillation[4], where the bunch rotates in longitudinal phase space while maintaining a constant emittance.

FAST RING DATA

One concern regarding the SA is the degree to which it averages and possibly distorts the actual bunch signal during the cycle. To reduce noise, 10 sweeps are averaged to generate a single spectrum; each spectrum requires approximately 6 minutes to produce at a BW of 30 kHz. Thus at 30 Hz, we are sampling over 10,000 acceleration cycles; i.e., 10,000 bunches. The present generation of fast, deep-memory

CP642, *High Intensity and High Brightness Hadron Beams: 20th ICFA Advanced Beam Dynamics Workshop on High Intensity and High Brightness Hadron Beams,* edited by W. Chou, Y. Mori, D. Neuffer, and J.-F. Ostiguy

oscilloscopes makes it possible to produce fast spectra from a single 14 ms acceleration period. Using a Tektronix TDS7254 oscilloscope with 16 MB of fast memory, the entire RCS cycle can be recorded simultaneously on four channels at a rate of 250 MSamples per second. The penalty for speed is significantly higher noise with respect to the SA due to the greater oscilloscope bandwidth (2.5 GHz).

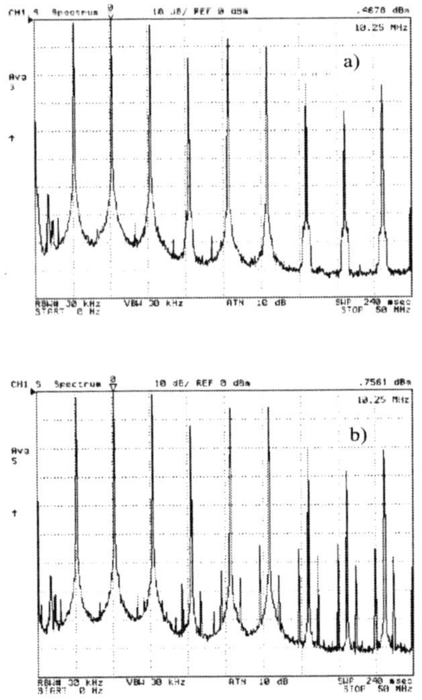

FIGURE 2. Bunch Spectra near extraction a) with scrambler and b) without scrambler.

To evaluate the above concern, a comparison has been made of spectra obtained from both the SA and oscilloscope with the data given in Figure 3. Approximately 0.3 ms after injection, the top pie electrode signal recorded with the SA exhibits strong sidebands as shown in Figure 3a. The sample period is 0.1 ms. Likewise, sideband structure can also be seen in the spectra generated from TDS7254 oscilloscope data and presented in Figure 3b. The time data was recorded at the same point in the cycle from an adjacent pie electrode during an 80 microsecond period (20 kS interpolated to 2^{15} values for the FFT). The individual bunch shape recorded with the oscilloscope differs slightly from that of SA but both represent an average. Note that the noise floor is several orders of magnitude higher for the oscilloscope generated spectrum. The noise level in the SA data is typically in the range of 10^{-12} W. Sidebands indicate a vertical fractional tune of 0.28; the horizontal tune is 0.20. Irrespective of the noise level, the two techniques do indicate similar spectral features.

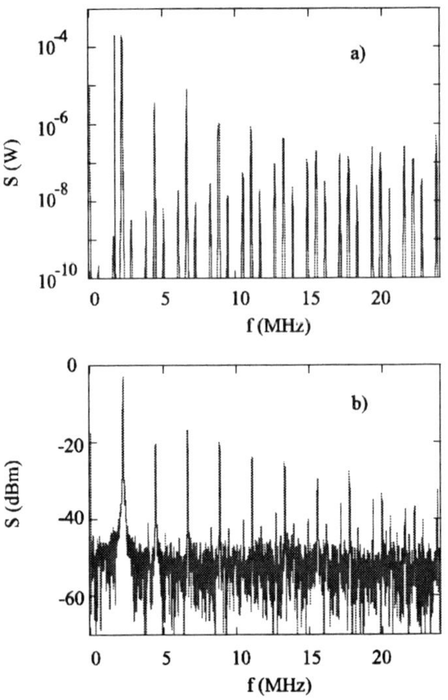

FIGURE 3. Comparison of top pie electrode data recorded with a) SA and b) oscilloscope 0.3 ms after injection.

SCRAMBLER STUDIES

The scrambler helps to damp oscillations that are indicative of destructive instabilities; these oscillations can appear as spectral features such as the sideband structure shown in Figure 1b. While the scambler removes one type of narrow band (coherent) structure from the spectra, it introduces other features which are broader in nature. Figure 4 presents pie spectra at two times in the RCS cycle, 10 ms and 11 ms after injection. The latter curve has been shift downward two orders of magnitude to make it more visible. The scrambler begins at 9.4 ms after injection; however, only the primary bunch harmonics are visible at 10 ms. One millisecond later, the spectrum has changed with a pair of broadband, transverse modulation features present on either side of the bunch harmonics. The modulation features are thought to be transverse because they do not appear on longitudinal diagnostics such as the beam current toroid (CT) or resistive wall monitor (RWM) in the RCS. Thus, coupling from longitudinal motion into the transverse plane may also help to damp the resistive wall instability. The broad sidebands are not present when the scrambler is off.

Knowing that the scrambler generates sidebands, an effort was made to see if the these spectral features could be observed elsewhere in the acceleration cycle. The scrambler electronics allows for two modulation periods at different times in the cycle. Keeping one at

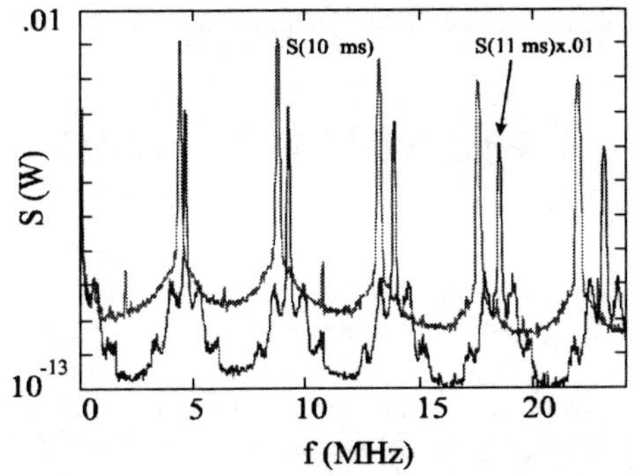

FIGURE 4. Bunch spectra from top pie electrode 10 ms and 11 ms (x0.01) after injection. Scrambler begins at 9.4 ms.

the normal time for stability while varying the other, it was possible to stimulate the bunch and observe sidebands. The scrambler study was first conducted in December 1999 and repeated again in October 2001. The second test generally repeated the results of the earlier study. Scrambler study measurements are given in Fig. 5. Background gas pressure in the RCS is on the order of 1 μTorr. Assuming a background of N_2; the neutralization time, τ_n = 0.5 ms at injection and 1.0 ms at extraction; therefore, the beam becomes fully neutralized. The tune shift introduced by partially compensated space-charge, may be expressed as[5],

$$\Delta\nu(\tau) = -\frac{I_p(\tau)R\left(1-\gamma^2(\tau)f_e(\tau)\right)}{I_0\varepsilon_n\beta^2(\tau)\gamma^2(\tau)} \qquad (1)$$

where f_e is the fractional neutralization. If over-neutralization is allowed and $f_e(t) = t/\tau_n$, where t is time measured from injection, then the tunes predicted by Eq. 1 are as presented in Fig.5. Some of the structure in the measured tunes may be the result of the pulsed quadrupole magnets in the RCS, the current waveforms of which are given at the bottom of Fig.5.

DISCUSSION AND FURTHER WORK

Neutralization through ionization means electrons and ions are generated in the background gas, perhaps leading to the formation of a plasma. A plasma sheath forming around the beam would both shield the space charge potential of the beam from the wall as well as carry image current. Plasma formation would be enhanced in the combined-function magnet sections of the RCS which cover just over half of the beam path. Charge confinement times in the magnets should be longer then in the field-free straight sections. Liberated charge is restricted to follow magnetic field lines which lead quickly above and below the beam to a stainless steel liner. Secondary emission from the liner produces more electrons which are again forced to follow field lines back to the beam. An electron

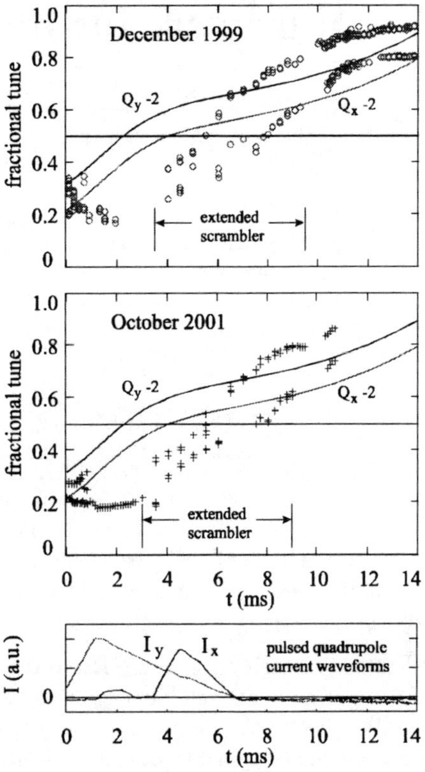

FIGURE 5. Extended scrambler tune measurements

collection grid at the bottom of one straight section, clearly shows the generation of electrons in the beam. We have recently installed a Retarding Field Analyzer (RFA) of the type used in the Proton Storage Ring at LANL and the Advanced Photon Source here at ANL[6]. With it we hope to learn about the electron density and energy spread near the beam. We look forward to having RFA data during our June run.

ACKNOWLEDGMENTS

Many thanks to the IPNS Accelerator Operations Group that makes this work possible.

REFERENCES

1. A. V. Rauchas, et al., IEEE Trans. Nuc. Sci., **28**(3), 2338 (1981).
2. C. Potts, et al., IEEE Trans. Nuc. Sci., **32**(5), 3107 (1985).
3. L. J. Laslett, V. K. Neil, and A. M. Sessler, Rev. Sci. Instrum., **36**(4), 1965(436).
4. M. Bai, et al., Phys. Rev. ST Accel. Beams **3**, 064001(2000).
5. M. Reiser, *Theory and Design of Charged Particle Beams*, Wiley, New York, 1994, p. 274.
6. R. A. Rosenberg, K. C. Harkay, NIM A, **453**, 507(2000)

Hardware for Beam Injection and Extraction

Y. Shirakabe, Y. Ishi, Y. Mori, I. Sakai, A. Takagi, and M. Tomizawa

KEK, Tsukuba, Japan

Abstract. The injection and extraction beam-lines of the 50 GeV synchrotron in the JHF project are described. The 50 GeV ring is equipped with one injection line from the preceding 3 GeV ring and three extraction lines for experimental facilities and for beam aborts. Some of the characteristic features on the injection/extraction hardware components, particularly on the kickers, are mentioned. These features are intended to fulfill two important demands on the kicker design, namely, a reduction of reflected waveforms and a bipolar fast extraction by the fast reverse of the kick direction.

INTRODUCTION

A newly designed synchrotron usually needs to fulfill novel demands that have not been achieved in older generation machines, and those demands may include, not only a higher energy or a higher intensity, but also new kinds of injection and extraction technologies.

In the 50 GeV synchrotron in the JHF project, there are a number of new technological tasks to be solved regarding the injection/extraction hardware. Since the ring is aimed to achieve an unprecedented beam power, the whole system must be designed to counter an unparalleled level of radiation damages in order to guarantee reliable operations of the machine. The injection and extraction components must similarly face with these radiation safety problems.

In the following sections of this paper, the injection and extraction systems of the 50 GeV ring is briefly outlined. Some of the characteristic features on the injection/extraction hardware components, particularly on the kickers, are then mentioned.

INJECTION AND EXTRACTION BEAM- LINES

The injection and extraction beam-lines of the 50 GeV synchrotron are illustrated in FIGURE 1. The ring employs one injection line and three extraction lines. In the fast half of the Insertion A, the injection line from the preceding 3 GeV synchrotron is situated. The second half of the insertion A provides the abort line at 3 GeV. This line is supposed mainly for the use in the commission stage in order to abort the beam before acceleration. These two lines are designed in a symmetric manner to each other with respect to the center of the Insertion A. In this way, the septum magnet designs can be made common between the insertion line and the 3 GeV abort line. As for the kicker magnets, however, the designs are not similar, since the rise-time of the abort kickers is much longer than that of the injection kickers. If kickers can be slower, the kicker volume can be made bigger, resulting in a smaller number of kicker units and costly power supplies.

The Insertion B is for the slow extraction line. This contains one electrostatic septum, 8 septum magnets, and 4 bump magnets. The second half of the Insertion C is for the fast extraction line for the neutrino oscillation experiment. A special demand to the fast extraction devices is to achieve a bipolar extraction by reversing the polarities of the extraction magnets. Its detail will be described later.

The injection and fast extraction scheme of the proton bunches in the 50 GeV ring is shown in FIGURE 2. The 3 GeV ring and the 50 GeV have the harmonic numbers of 2 and 9, respectively. The required time structures of the kickers --- rise-times,

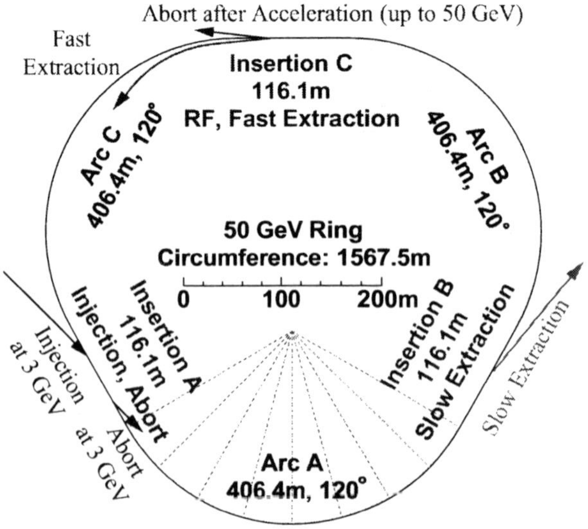

FIGURE 1. Injection and extraction beam-lines of the 50 GeV synchrotron of the JHF project. The ring has a three-fold symmetry, hence there are three straight insertion sections to which the injection and extraction functions are scattered, as are shown in the figure.

CP642, *High Intensity and High Brightness Hadron Beams: 20th ICFA Advanced Beam Dynamics Workshop on High Intensity and High Brightness Hadron Beams,* edited by W. Chou, Y. Mori, D. Neuffer, and J.-F. Ostiguy

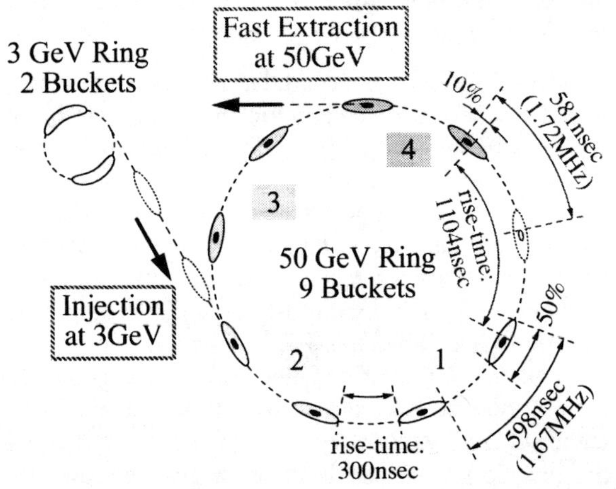

FIGURE 2. Injection and fast extraction scheme of the 50 GeV ring. Two successive bunches are injected from the 3 Gev ring to the 50 GeV ring at one injection cycle. After four injection cycles, numbered 1 to 4 in the figure, the 50 GeV ring is filled with eight bunches and starts acceleration. Since the harmonic number of the ring is nine, one bucket is left empty, which allows a longer rise-time for the fast extraction kicker.

flat-top lengths, and fall-times --- can be derived from the figure. Here, the bunch-to-bunch intervals are assumed to be 50% of one RF period at the injection energy, and 90% at the extraction energy. The rise-times for the injection kickers, the abort kickers, and for the fast extraction kickers are 300nsec, 900nsec, and 1100nsec, respectively.

REFLECTION TRAPPING METHOD FOR INJECTION KICKERS

In a kicker magnet, a magnetic field is generated with PFN (pulse forming network) pulses. The PFN system is connected to the kicker through a long transmission cable (FIGURE 3). The input pulses to the kicker will make reflection pulses repeatedly at the kicker. When the kicker is used for beam injection, these reflected pulses cause seriously damaging perturbation on the circulating beams. A reduction of reflected waveforms is an important design issue for the injection kickers.

In the 50 GeV synchrotron with a high beam power, however, conventional reduction techniques may not be satisfactory. It is because a slight mismatching of a PFN system is always unavoidable, and even a small fraction of reflected pulses caused by this mismatching may produce an prohibited level of the beam loss.

In order to keep the reflected waveforms within a harmless level, a new type of evading method is being designed for the 50 GeV ring injection kickers. Its principle is shown in FIGURE 4. Here, the pulse reflection time, t_{ref}, is adjusted to be 3 x t_{rf}, where t_{rf} is the RF period.

In the 50 GeV case, t_{rf} is 600nsec (FIGURE 2).

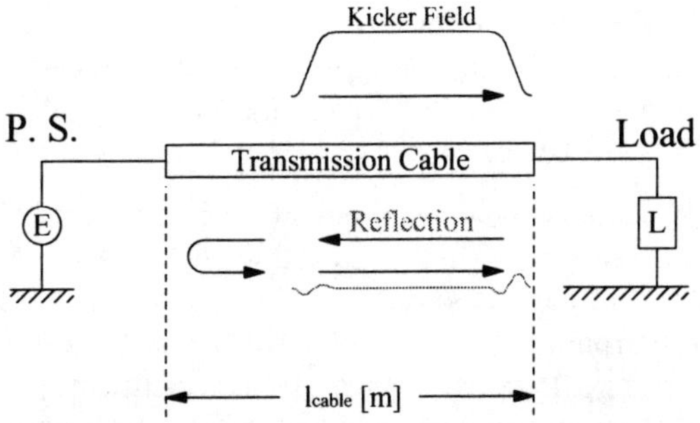

FIGURE 3. A simplified scheme of the kicker pulse propagation. A kicker pulse is generated at the power supply (left-end) and propagates through the transmission cable to the kicker load (right-end). The pulse is reflected at the kicker and returns to the power supply. Some part of the pulse energy is reflected again at the power supply and causes the reflected waveform at the kicker.

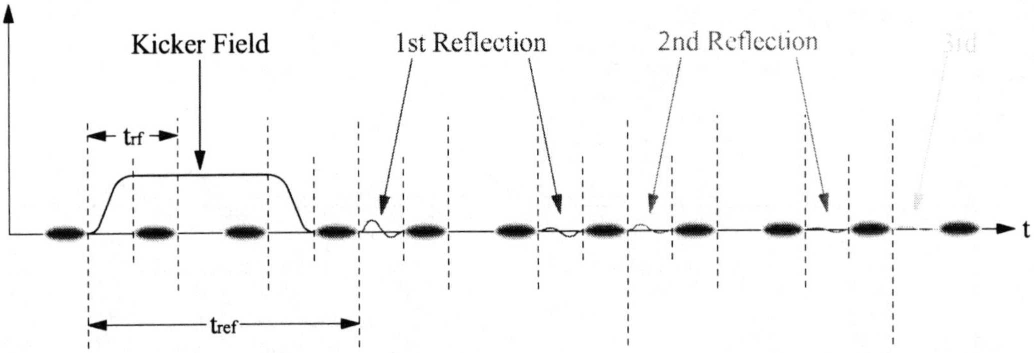

FIGURE 4. A principle of the reflection trapping method in order to make the reflected waveforms harmless. When observed at one kicker, the proton bunches (red ellipses) appear periodically with the RF period (t_{rf}). The reflected waveforms appear with a different period of t_{ref}, which is the period to propagate back and forth through the transmission line. If the ratio t_{ref} / t_{rf} is tuned to be an integer n (in the above case, n = 3), the reflected waveforms will always be trapped at the bunch intervals.

Then t_{ref} becomes 1800nsec. If the transmission cable length is l_{cable} [m], t_{ref} equals to 5 x 2 x l_{cable} [nsec]. Here, 5 means the pulse propagation time, 5[nsec/m], and 2 means that the pulse passes the length l_{cable} twice, i.e. backward and forward. From this relation, l_{cable} can be derived as t_{ref} / 10 = 180m. If the transmission cable length is selected to this length, all the reflection pulses will be trapped in the bunch intervals, therefore, they will become harmless to the circulating bunches.

The length, l_{cable} = 180m, is comparable to the distance between the injection kicker location in the 50 GeV tunnel and the power supply building. Eliminating the harmful effect by the reflection pulses will be realized with this reflection trapping method. More detailed design study is now underway.

BIPOLAR FAST EXTRACTION BY REVERSIBLE POLARITY KICKERS

The fast extraction line is situated at the downstream of the insertion C. The seven kicker units, each one having 1.5m in length and 100mm*100mm in aperture size, are designed to produce the strong enough kick with 0.1 T of the kicker field to extract the 50 GeV beam.

Recently a new requirement emerged to the fast extraction line, which can be called as the 'fast reverse' of the kick direction. This requirement comes mainly from the fact that the downstream beam line of the fast extraction, which leads to the neutrino oscillation experiment line, is designed to employ superconductive magnets. These magnets might get quenched at anytime, even just very short time --- milliseconds or shorter --- before the fast extraction of the 50 GeV beam. In order to avoid damaging the quenched magnets and the downstream ones, the fast extraction line should have the function of the fast reverse. Then the beam can be extracted in the reverse direction, where another beam line, an abort line at 50 GeV, is provided. The fast extraction kicker systems, therefore, are now requested to achieve the reverse of the kick direction within a possible shortest time.

One of the possible candidates of the PFN scheme to realize the fast reversible kicker is shown in Figure 5. This scheme is based on the Blumlein configuration, which is widely used, well proved scheme. In order to achieve the reversible polarity at the kicker, two switches are installed at the both ends of the Blumlein PFN lines. One switch corresponds to the normal direction kick, and another corresponds to the reverse direction kick. Since this PFN scheme is composed in a completely symmetric manner, a precisely equal kick with a reversed direction should be achieved. The choice of the switching between the normal switch and the reverse switch can be done at anytime, therefore the decision to abort the beam should be possible until even the very moment of the turn-on of the normal switch.

Another merit of employing the Blumlein system is that the PFN charging voltage can be reduced by a factor of 2. The fast extraction kicker requires 0.1 T of field strength at the gap height of 100mm, which amounts the necessary kicker current of 8 kAT. If the normal PFN impedance is 10 Ω, The PFN charging voltage becomes 80 kV, which is close to the possible maximum operating voltage. If the Blumlein system is chosen, the PFN impedance can be reduced to 5 Ω with the same time constant of the circuit system. The charging voltage then can be reduced to 40 kV. This value will allow a sufficient safety margin for the high voltage operation. A demerit of this system might be that the kicker magnet must be boosted to the high voltage while the PFLs are charged. High voltage troubles may occur more frequently with a longer high voltage operation time, but the reduction of the maximum voltage from 80 kV to 40 kV will farther relax the whole high voltage stress.

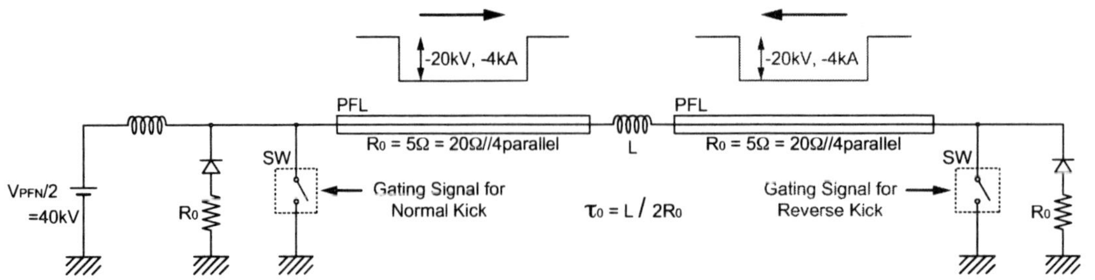

FIGURE 5. A symmetric Blumlein PFN configuration in order to achieve the fast reverse of the kicker field for the bipolar fast extraction kickers. At the both ends of the kicker (center, L), two PFLs are connected as in the same way as the conventional Blumlein PFN. The difference is that, on the farther sides of the PFLs from the kicker, two switching units are connected symmetrically.

Magnets for High Intensity Proton Synchrotrons

J.-F. Ostiguy, V. Kashikhin and A. Makarov

Fermi National Accelerator Laboratory, Batavia, IL, USA[1]

Abstract. Recently, there has been considerable interest at Fermilab for the Proton Driver, a future high intensity proton machine. Various scenarios are under consideration, including a superconducting linac. Each scenario present some special challenges. We describe here the magnets proposed in a recent study, the Proton Driver Study II, which assumes a conventional warm synchrotron, roughly of the size of the existing FNAL booster, but capable of delivering 380 kW at 8 GeV.

INTRODUCTION

One of the principal considerations in designing a high intensity proton synchrotron is to limit losses to prevent activation. Typically this translates into a requirement on the maximum fractional particle losses: on the order of 10^{-4} to as low as 10^{-6}. In order to provide sufficient dynamic stability, it is necessary to limit both space charge induced tune shift and tune spread. This is accomplished in two ways: by keeping the machine circumference small and by spreading out the proton distribution transversely. The former strategy implies rapid cycling, the latter implies large aperture; both have an impact on magnet design.

Aperture is a principal cost driver since magnet overall size, fabrication, power consumption and power supply hardware costs are directly proportional to stored magnetic energy. Magnet physical aperture, as opposed to available beam aperture, is determined not only by space charge considerations but also by the type of vacuum pipe employed. Because of the very substantial eddy current induced losses associated with rapid cycling, a conducting beam pipe generally cannot be used. One possible solution is a ceramic beam tube with thin conducting strips disposed on its inner surfaces to minimize beam impedance. This is costly in terms of physical aperture because for mechanical reasons, ceramic walls have to be considerably thicker than metallic walls. The Proton Driver magnets sidestep the difficulty by putting the entire magnet inside an evacuated enclosure an approach employed for the Fermilab Booster magnets. One drawback is that due a larger desorption area, achieving satisfactory vacuum in order to prevent gas scattering induced losses requires special care.

The presence of a large energy dependent space charge tune shift and tune spread dictates the need for tight tune control during the entire acceleration cycle. For this reason, dipole/quadrupole tracking errors are a special concern. A tracking error is equivalent to momentum offset error and results in a tune shift of magnitude

$$\Delta\nu = \xi_{\text{uncorrected}} \left[\frac{\Delta(G/B)}{(G/B)}\right] \qquad (1)$$

where $\frac{G}{B}$ is the ratio the gradient to main dipole field. Note that the tune variation is proportional to the *uncorrected* chromaticity because, in the context of a focusing error, there is no closed orbit error and the chromaticity correction sextupoles have no effect. The magnitude of the tolerable tune shift is debatable. At ISIS (RAL, U.K.), the ability to control the tune to within a part in 100 proved necessary, mostly to stay clear of specific resonances at extraction. While it is conceivable that this criterion can be relaxed, it provides a sense of what needs to be achieved.

Some rapid cyclic machines (e.g. the Fermilab Booster) use combined function magnets. This economizes space and naturally provides good tracking as long as the operating field is kept below ~ 1 T. The Fermilab Proton Driver study II assumes an aggressive 1.5T bending field and separate function magnets. The field strength is determined by two requirements. First, the circumference ratio between the Proton Driver and the Main Injector should be a simple rational fraction to allow synchronous beam transfers. Second, the total circumference should be as small as possible in order to minimize the space charge tune shift. Dipoles and quadrupoles are on a common bus; the residual tracking error (on the order of a percent) is handled by an independently powered active quadrupole correction system capable of compensating for the tune shift dependence on energy during acceleration.

DIPOLES

The Proton Driver dipole is a conventional H-magnet design. Field homogeneity is preserved at high excitation by profiled pole edges and by the presence of circular

[1] Work supported by the U. S. Department of Energy under contract number DE-AC02-76CH03000.

CP642, *High Intensity and High Brightness Hadron Beams: 20th ICFA Advanced Beam Dynamics Workshop on High Intensity and High Brightness Hadron Beams,* edited by W. Chou, Y. Mori, D. Neuffer, and J.-F. Ostiguy

TABLE 1. Proton Driver II Main Dipole Magnet Parameters

Peak Dipole Field	1.5	Tesla
Good Field Aperture	101.6×152.4	mm^2
Physical Aperture	101.6×273.1	mm^2
Field Homogeneity	± 0.0005	
Magnet Length	5.72	m
Cycle Frequency	15	Hz
Peak Current	5170	A
Conductor Dimensions	20.2×15	mm^2
Conductor cooling hole diameter	10	mm
No of turns/pole (3 conductors & top/bottom coils in parallel)	12	
Lamination Thickness	0.35	mm
Lamination Material	Si-Fe M17	
Inductance	18	mH
DC Resistance	4.7	mOhm
Stored Energy	0.063	MJ
Coil Losses	115	kW
Core Losses	16.3	kW
Core mass	37,000	kg
Peak Terminal Voltage	4.85	kV
Water Pressure Drop	10	bar
Water Flow	1.7	l/s
Water Temperature Rise	17	deg C

holes in the center of the poles. The magnet cross-section is shown in Figure 1; a list of relevant parameters is presented in Table 1. The dipoles are excited so as to produce a magnetic field strength of the form

$$B(t) = B_0 - B_1 \cos(\omega t + \phi) + 0.125 B_1 \sin(2\omega t + 2\phi) \quad (2)$$

where B_0 is the injection field, B_1 is the magnitude of the fundamental component, $\omega/2\pi = f = 15$ Hz and ϕ is a constant phase factor. The second harmonic component is introduced to reduce the maximum value of dB/dt, which determines the peak RF accelerating voltage and the overall RF system costs.

In an iron dominated dipole, good field homogeneity can be achieved over the entire extent of the physical vertical aperture g, that is, all the way to the pole surfaces. However, field homogeneity over the horizontal extent of the aperture requires a certain amount of pole overhang w. For a given field homogeneity, the required horizontal extent of the physical aperture is minimized by shimming the pole pieces edges. The optimality of a design can be accessed by comparing the achieved physical horizontal extent to a theoretical estimate developed by Klaus Halbach [2]. Figure 2 presents calculated field homogeneities achieved by the Proton Driver II dipole magnet.For this magnet, the ratio $g/w \simeq 0.6$ and we note that the achieved homogeneity is in good agreement with the prediction from Halbach's formula.

Eddy Currents

In a rapid cycling magnet, the presence of eddy currents is a source of technical difficulties. Eddy currents are induced both in the magnetic core and in the conductors. In the core, they are largely suppressed by a laminated core construction which impedes their flow in the longitudinal direction. As long as the lamination thickness is smaller than the skin depth, their is little impact on field quality and principal effect is to increased losses.

Eddy currents induced in conductors are potentially more problematic. If the current distribution is non-uniform, resistive losses can be substantial; furthermore, the field homogeneity can be affected. The latter problem is largely eliminated by avoiding to place conductors into or near to the magnet mid-plane.

For a rectangular conductor immersed in a uniform time-varying magnetic field $B_0 \sin(\omega t)$, it can be shown that the power losses are given by

$$P = \omega^2 B_0^2 A a^2 / 16\rho \quad (3)$$

where A is the conductor area, a is the conductor width and ρ is the resistivity. Clearly, eddy currents can be reduced either by reducing conductor cross-section or the magnetic field in which the coils are immersed. Reducing individual conductor area increases the number of turns N and therefore the magnet terminal voltage, since the inductance scales like N^2. To keep the voltage at a reasonable level, it becomes necessary to connect multiple turns in parallel. The Proton Driver II magnet assumes groups of three conductors connected in parallel in each coil and the top and bottom coils also connected in parallel. The electrical and mechanical connections in the end regions are shown in detail in Figure 1. Note that the conductors are not transposed. Although transposition would reduce losses slightly, it would also render connections in the end region very complex and cumbersome.

In the Proton Driver Study I, eddy current power loss problems were completely side-stepped by the use of a special water cooled stranded conductor. Aside from the technology required to produce reliable electrical and mechanical joints, the main drawback of the stranded conductor is its high cost (about an order of magnitude more expensive than conventional copper conductor). For study II, the possibility of using conventional solid water cooled conductors has been revisited. Figure 3 presents the result of an eddy current computation.

Eddy current losses reach a substantial level in the conductors closest to the edge of the pole, in the fringe field region. To take advantage of the fact that the magnetic field is predominantly vertical in that region, a rectangular (as opposed to square) conductor is employed. Relatively high losses affect approximately 20% of the total coil cross section; however, localized heating should be prevented by good thermal contact between conductors. Conventional water-cooled solid copper coil is a well-understood technology. Compared with stranded conductors, the trade-off is reduced operational efficiency vs reduced up-front fabrication costs.

Quadrupoles

The Proton Driver quadrupoles are four-fold symmetric magnets. Both horizontal and vertical focusing

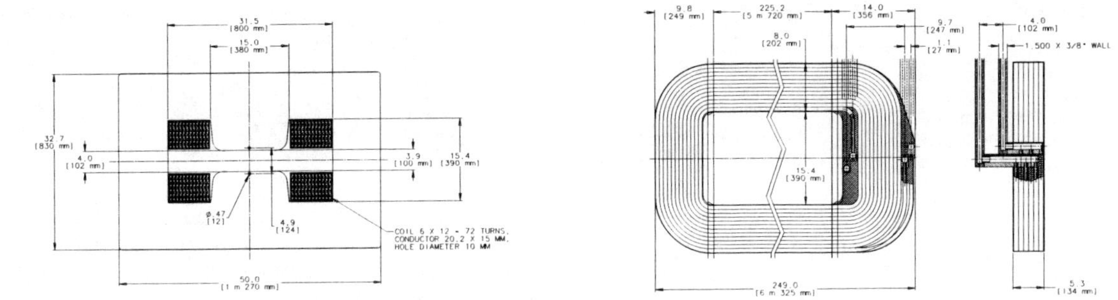

FIGURE 1. Proton Driver dipole cross-section and coil detail. Note the circular holes in the center of the poles. Note also the parallel connections at one of the coils extremities.

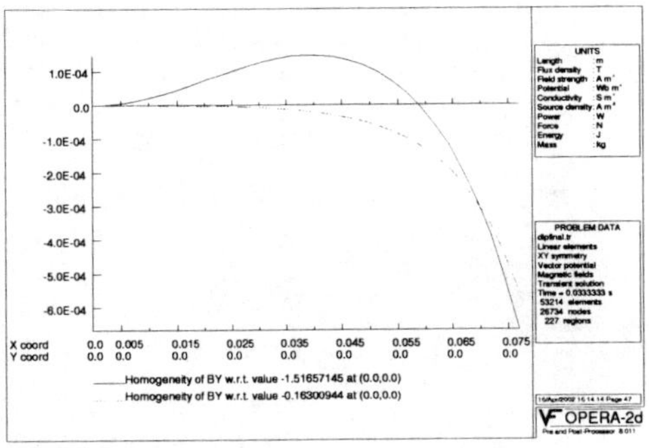

FIGURE 2. Proton Driver dipole fi eld homogeneity at minimum and maximum excitations.

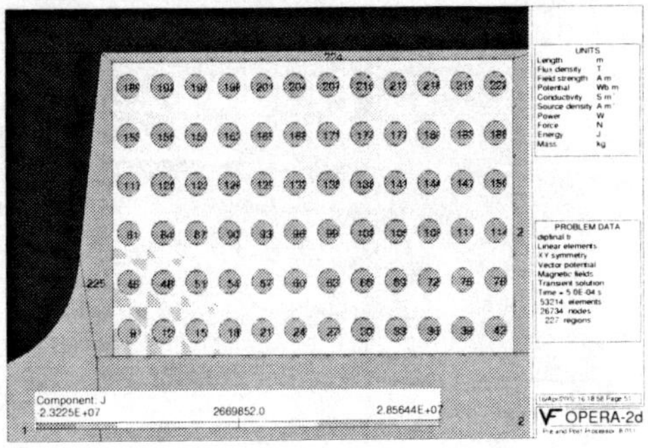

FIGURE 3. Eddy current distribution in the conductors. Note that the distribution is most uneven in conductors located near the pole edge.

quadrupoles are identical and the aperture radius is set to accommodate a rectangular good field region of the same size as the dipoles'. A common current bus provides dynamical tracking between the quadrupole gradient and dipole field. The number of turns and length of the quadrupole are selected so as to match optical and physical requirements at injection. At higher energies, the gradient/dipole strength ratio must not deviate by more than a percent or two; this effectively limits the maximum achievable gradient. For the required aperture, when the quadrupole pole tip field reaches 0.84 T, the field at the edges of the truncated hyperbolic profile reaches approximately 1.5 T, and saturation begins to affect the linearity of the relation between quadrupole strength and excitation. Note that saturation in backleg region also affect nonlinearity; however, this effect can be minimized by adjusting the backleg width. Note that for a four-fold symmetric quadrupole saturation does not adversely impact field quality since all harmonics except those of order 4n (8n-pole) are suppressed: the first allowed harmonic is the 12-pole. At 8 GeV, the deviation is on the order of 2.0%.

CONCLUSIONS

High voltage operation is a serious concern. The dipole magnets have a maximum terminal voltage on the order of 5 kV; in the proposed resonant configuration, the maximum voltage to ground reaches approximately 3.3 kV. Another source of concern in the fact that the entire magnet is be placed inside an evacuated enclosure. Special precautions will be needed during the assembly process to avoid excessive degassing.

An interesting avenue for future R&D would be an investigation of super-ferric magnet technology. In recent years, a new generation of Nb-Ti superconducting cable has been developed for applications in power generation and transmission. The cable has insulated filaments with a diameter as small as 0.1 μm embedded in a Cu-Ni alloy matrix. Superconducting coils would result in substantial savings in overall magnet size and power consumption; these savings may be substantial enough to offset the additional costs and complexity engendered by the cryogenic system.

REFERENCES

1. The Proton Driver Design Study, Fermilab TM-2136, December 2000.
2. A.W. Chao and M. Tigner Eds., *Handbook of Accelerator Physics and Engineering*, World Scientifi c, 1999.

WORKING GROUP II

LINEAR ACCELERATORS

Summary Report

J.M. Lagniel * and R. Garoby†

*CEA, Bruyères-le-Chatel, France
†CERN, Geneva, Switzerland

Linear hadron accelerators of high intensity/power/energy are at the core of many contemporary projects in the world. They cover a broad range of applications which reflects in the large span of their characteristics and their very different level of advancement, some of them being already approved and in construction, others being simply contemplated as long term possibilities. Based on an original table by T. Wangler (LANL), the compendium given in Table 1 was elaborated during the workshop with the help of the participants. Issues related to these machines were presented either in plenary talks or during the two sessions of working group 2. They all use superconducting RF structures.

The FNAL 8 GeV linac (W. Foster) stands out as the highest energy project, relying on TESLA technology to keep cost acceptable, using a high RF frequency (1.2 GHz) and a very high gradient (~ 23 MV/m). On the opposite , the 2.2 GeV CERN SPL proposal (R. Garoby) makes use of existing cavities recuperated from LEP, with a low RF frequency (352 MHz) and a modest gradient (~ 9 MV/m). The other projects assume the construction of accelerating structures optimised for their own use, with intermediate RF frequencies (600-800 MHz) and gradients (~ 16 MV/m). Considerations about RF adjustment and losses in the SNS linac were presented by S. Nath (LANL). M. Ikegami described the status of the linac accelerator for the JKJ project in Japan. The plans and studies for ESS and IFMIF (R. Ferdinand), CERN SPL and TRASCO (C. Pagani) were the subject of talks. A. Takagi showed the results obtained with a fast beam chopper modulating energy in front of an RFQ. These presentations and the active discussions they triggered clearly demonstrate that high performance hadron linacs is nowadays a very active field where significant progress can be expected in the coming years.

TABLE 1. High Power Linacs Survey

Name	Ion	Pulse Length (ms)	FRep (Hz)	Duty Factor (%)	IBunch (mA)	IAverage (mA)	K.E. (GeV)	PAverage (MW)	Start Date
LANSCE	H+/H-	0.625	100/20	6.2/1.2	16/9.1	1.0/0.1	0.8	0.8/0.08	On
SNS	H-	1.0	60	6.0	38	1.4	1.0	1.4	2006
CERN SPL	H-	2.8	50	14	22	1.8	2.2	4.0	?
ESS S Pulse	H-	1.2	50	6.0	114	3.75	1.33	5	2010
ESS L Pulse	H-or H+	2/2.5	16.67	4.2	114/90	+3.75	1.33	+ 5	2010
FNAL 8 GeV	H+/H-/e	1.0	10	1.0	25	0.25	8.0	2.0	?
JKJ 400 MeV	H-	0.5	50/25	2.5	50	0.7	0.4	0.28/0.14	2006
JKJ 600 MeV		25	1.25			0.35	0.6	0.21	?
TRASCO	H+	∞	0	100	30	30	≥ 1.0	≥ 30	?
IFMIF	D+	∞	0	100	2×125	2×125	0.04	10.0	2010

CP642, *High Intensity and High Brightness Hadron Beams: 20th ICFA Advanced Beam Dynamics Workshop on High Intensity and High Brightness Hadron Beams*, edited by W. Chou, Y. Mori, D. Neuffer, and J.-F. Ostiguy

High-Intensity Proton Linac for the KEK/JAERI Joint Project

Masanori Ikegami*, Takao Kato*, Kazuo Hasegawa† and Yoshishige Yamazaki†

*KEK, High Energy Accelerator Research Organization, 1-1, Oho, Tsukuba, Ibaraki, 305-0801 Japan
†JAERI, Japan Atomic Energy Research Institute, Tokai, Naka, Ibaraki, 319-1195 Japan

Abstract. The high-intensity proton accelerator facility in Japan, comprising a 600-MeV linac, a 3-GeV RCS (Rapid-Cycling Synchrotron), and a 50-GeV synchrotron, has been officially approved for construction of its Phase I. Phase I includes a 400-MeV normal-conducting linac as an injector, which consists of a 3-MeV RFQ, a 50-MeV DTL, a 190-MeV SDTL (Separate-function DTL), and a 400-MeV ACS (Annular Coupled Structure linac). The basic design of the injector linac has nearly been completed. In parallel with the design study, the construction of its low-energy part, which is up to 60 MeV, has been started at KEK to develop and establish the linac system before moving to the JAERI site. In this paper, we present the recent progress in the R&D and construction work together with the outline of the latest linac design.

INTRODUCTION

A high-intensity proton accelerator facility, comprising a 600-MeV linac, a 3-GeV RCS (Rapid-Cycling Synchrotron), and a 50-GeV synchrotron, has been proposed in Japan [1]. To reduce the initial construction cost, a phased approach has been adopted in this project, namely; we plan to build a 400-MeV linac, a 3-GeV RCS, a 50-GeV synchrotron, and a part of experimental facilities in Phase I, and we will upgrade the linac energy up to 600 MeV and reinforce the experimental facilities in Phase II. The output beam of the injector linac will be shared by the 3-GeV synchrotron and an experimental facility for ADS (Accelerator-Driven Systems) R&D.

Phase I of the project has been officially approved for construction. Phase I includes the construction of a 400-MeV linac which consists of a 3-MeV RFQ, a 50-MeV DTL, a 190-MeV SDTL (Separate-function DTL), and a 400-MeV ACS (Annular-Coupled Structure linac). In this paper, we present the main design features of this injector linac.

In parallel with the design study, the construction of the low-energy part of the injector linac has been started at KEK site, which includes a 3-MeV RFQ, a 50-MeV DTL, and the first two tanks of the SDTL. Although the whole facility will be constructed at JAERI Tokai campus, we perform the beam test of the low-energy part in KEK in order to develop and establish the linac system before moving to the JAERI site. The recent progresses of the beam test and construction work are also presented in this paper.

TABLE 1. Requirements for the injector linac

Particle	Negative hydrogen
Energy	400 MeV
Peak current	50 mA
Average current	1.25 mA (before chopping)
Pulse width	500 μsec
Repetition rate	50 Hz
Chopping ratio	56 %
Momentum spread*	$\pm$0.1% (100%)
Transverse emittance	3-5 πmm·mrad (99%)

*At injection point of RCS.

REQUIREMENTS

The requirements for the injector linac are summarized in Table 1. While the output beams of the linac are shared by a 3-GeV RCS and an ADS experimental facility, these requirements mainly come from the ring injection. It should be noted here that the requirement for the momentum spread is particularly severe, which necessitates precise rf control for linac tanks and an elaborate debuncher system in the beam transport line between the linac and the RCS. Furthermore, the macro-pulse is needed to be chopped to minimize the beam loss in the beam extraction process from the RCS, which requires a fast chopper system in the low-energy part of the linac. In addition, the linac tunnel is designed assuming uncontrolled beam loss of 0.1 W/m for most part of the linac, which imposes a strict condition for beam loss minimization.

FUNDAMENTAL DESIGN

To satisfy the above mentioned requirements, we have determined to adopt a combination of four types of linac structures, i.e., RFQ, DTL, SDTL, and ACS. The layout of the linac is schematically shown in Fig. 1. The SDTL

CP642, *High Intensity and High Brightness Hadron Beams: 20th ICFA Advanced Beam Dynamics Workshop on High Intensity and High Brightness Hadron Beams*, edited by W. Chou, Y. Mori, D. Neuffer, and J.-F. Ostiguy

TABLE 2. Main parameters of each part of the linac

	RFQ	DTL	SDTL	ACS	
Output energy	3	50	191	400	MeV
Frequency	324	324	324	972	MHz
Total length	3.1	27.1	91.2	108.3	m
Structure length	3.1	26.7	65.7	68.2	m
Num. of tanks	1	3	32	46	
Num. of cells		146	160	690	
Num. of klys.	1	3	16	23	
Accel. field E_0		2.5-2.9	2.5-3.7	4.3	MV/m
Vane voltage	82.9				kV
Synch. phase	-30	-30	-27	-30	deg
Wall loss	0.34	3.3	16.6	33.3	MW
Beam loading	0.15	2.4	7.0	10.5	MW
Total rf power	0.48	5.7	23.6	43.8	MW

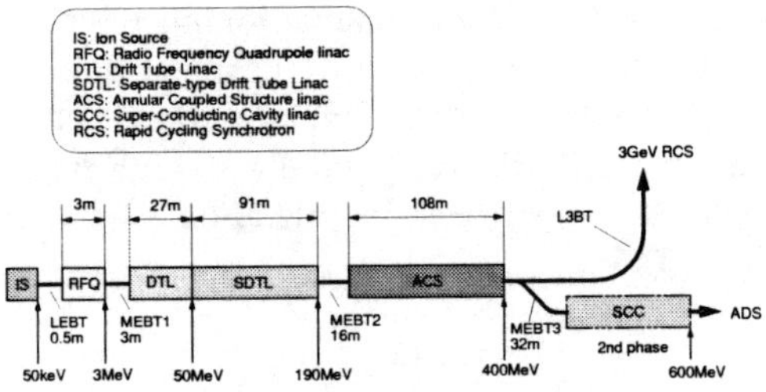

FIGURE 1. Layout of the injector linac

is basically the same with the DTL, except that the tank length is much shorter and the quadrupole magnets are placed between tanks instead of inside drift tubes [2]. Without quadrupoles inside drift tubes, we can optimize the drift-tube shape more freely to improve the effective shunt impedance. Furthermore, the shorter tank length reduces the effects of higher order modes, and, hence, facilitates field tuning greatly. The ACS is a variety of coupled-cavity linacs operated with the $\pi/2$-mode [3]. A distinctive feature of the ACS is its coupling cell geometry. In ACS, the coupling cell has a doughnut shape having the cylindrical symmetry with respect to the beam axis. The coupling slots between a coupling cell and an accelerating cell are placed with the four-fold symmetry in our design. Our primary goal in the linac design is to minimize the beam loss and the beam quality deterioration due to space-charge effects. To avoid excess beam loss, the stability of operation is indispensable. From this point of view, we choose stable structures for each part of the linac. Namely, we adopt PISL (π-mode stabilization loop) [4] for the RFQ. In the DTL, carefully designed post-couplers have been adopted together with the high-quality electroplating method. In the SDTL, its short tank length improves the rf characteristics, such as the magnitude of field tilts. The ACS has symmetrically placed coupling slots, which is expected to improve the symmetry of the accelerating field.

Because the required output beam quality for this linac is quite severe, precise beam tuning is strongly required. Having sufficient tuning knobs is essential to realize the precise beam tuning. From this point of view, we have adopted electromagnet quadrupoles for the DTL. Because there is technical difficulty in fabricating small electromagnetic quadrupoles with sufficient strength, it imposes the upper limit for the operation frequency of the DTL. On the other hand, higher frequency is preferable to improve shunt impedance and to suppress space-charge effects. Therefore, we determined to adopt the highest possible frequency for the DTL which enables the adoption of electromagnetic quadrupoles. Accordingly, the rf frequency of 324 MHz has been chosen for the low-energy part, which includes the RFQ, the DTL, and the SDTL. The choice of the rf frequency also enables us to utilize klystrons as rf power sources. The strength of the quadrupoles is chosen to enable us to satisfy the equipartition condition.

For high-energy part, we need to introduce three-fold frequency jump to realize the required output energy in the limited construction site at JAERI Tokai campus. To absorb the effects of the frequency jump and smoothly connect between the SDTL and the ACS, we have a 16 m long beam matching section after the SDTL. The smoothness of the beam envelope, and hence the external focusing force strength, is essential in suppressing space-charge-induced beam-quality deterioration.

Together with the smoothness, the strength of the focusing force is also important to ease space-charge effects. Because the space-charge effects are especially strong in low-energy part, we have adopted FODO lattice with period length of $2\beta\lambda$ for the DTL with β being the particle velocity scaled by the speed of light and λ the rf wave length. In the SDTL and the ACS, we have adopted doublet focusing lattices with relatively short period length. The main parameters of each part of the linac are summarized in Table 2.

R&D STATUS

As mentioned earlier, we plan to perform beam test up to the second SDTL tank at the KEK site before moving to the JAERI site. The RFQ and the MEBT1, which is the beam transport line between the RFQ and the DTL, have been installed and the beam test has been started. In the beam test, we are currently using a filament-discharge type ion source without cesium, and the peak current of 20 mA has been achieved at the exit of the RFQ. Figure 2 shows a typical beam waveform observed with current transformers installed in LEBT and MEBT1. In parallel with the beam test at KEK, a cesium-seeded ion source has been developed and tested at JAERI. Figure 3 shows the experimental setup for the ion source beam test. The beam current of 72 mA is achieved, and the

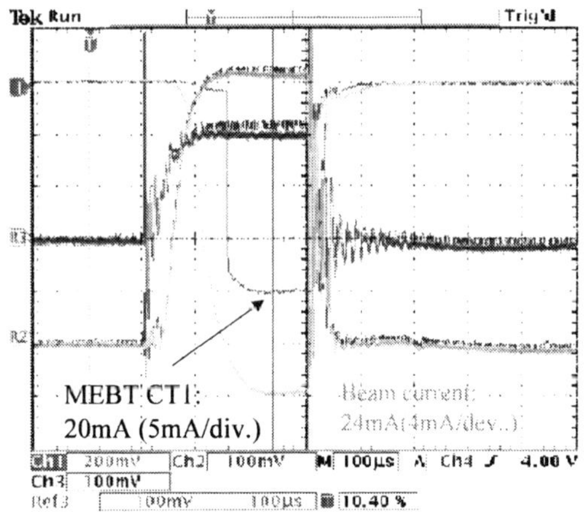

FIGURE 2. Waveform observed with current transformers in LEBT and MEBT1. Other signals, such as arc votage, are also shown.

FIGURE 3. Experimental setup of the cesium-seeded ion source at JAERI.

normalized rms emittance of 0.15 πmm·mrad has been obtained with 60 mA. The R&D study to increase the lifetime of the ion source is now in progress. We have not determined yet on the use of cesium-seed, because its long-term effect on the RFQ is still unclear. The type of ion source we adopt will be determined after reviewing the outcome of R&D work now in progress. The first DTL tank has been assembled in KEK as seen in Fig. 4, and the low-level rf measurement is currently performed. The first two tanks of the SDTL has been fabricated, and the high-power test has been completed. The rest of DTL and SDTL tanks have already been ordered. The ACS is still in the R&D stage. The cold model test with half-scale aluminum models and brazing tests with full-scale copper models have been performed.

TIMETABLE

The linac building will be completed in February, 2005. The low-energy part of the linac will be relocated from KEK to JAERI during 2005. After installation of the SDTL and the ACS, beam commissioning will be started in March, 2006. Then, the first beam is delivered to the RCS in September, 2006. Finally, whole accelerator complex, including the RCS and the 50-GeV main ring, is scheduled to start beam acceleration by the end of March, 2007.

FIGURE 4. The fi rst DTL tank assembled at KEK.

SUMMARY

The design of the injector linac for the KEK/JAERI joint project has nearly been completed. In the design, the emphasis has been put on the flexibility for beam tuning, stability of operation, and the minimization of beam-quality deterioration due to space-charge effects. In parallel with the design study, the beam test of the low-energy test has been started at KEK to establish the linac system. A 3-MeV RFQ and MEBT1, which is the beam transport line between the RFQ and the DTL, has been installed, and 20 mA beam has been successfully accelerated.

ACKNOWLEDGMENTS

The design and the R&D study of the linac have been performed by the linac group of the KEK/JAERI joint project.

REFERENCES

1. Mori, Y., "Accelerator Complex of Joint Project in Japan," in *these proceedings*, 2002.
2. Kato, T., Proposal of a Separated-type Proton Drift Tube Linac for a Medium-Energy Structure, Tech. Rep. KEK report 92-10, KEK (1992).
3. Kageyama, T., Morozumi, Y., Yoshino, K., and Yamazaki, Y., "Development of Annular Coupled Structure," in *Procs. of 1994 International Linac Conference*, 1994.
4. Ueno, A., and Yamazaki, Y., *Nucl. Inst. and Meth. A*, **300**, 15 (1991).

ESS SC Linac Design Overview

Robin Ferdinand

CEA-Saclay
DSM/DAPNIA/SACM bldg. 124, 91191 GIF-sur-Yvette cedex, FRANCE

Abstract. The ESS project (European Spallation Source) aims to produce high power beams for condensed matter studies. One option consists in having both H^+ and H^- beams. They have to be accelerated and guided to the different spallation targets. Beam power up to 5 MW at 50 Hz plus 5 MW at 16.2/3 Hz, will be delivered on respectively the Short Pulse Target Station and the Long Pulse Target Station. Two 50 mA H^- branches are funnelled with one 100 mA H^+ beam at around 20 MeV. A chopper line between two RFQs and DTLs constitutes the H^- front end. The H^+ front end is composed of one RFQ and one DTL. After the funnel, the two species are transported through the same linac up to 1.334 GeV. This common part is composed of a SDTL and a CCL from 20 to 185 MeV and followed by a SuperConducting Linac (SCL) to reach the final energy. Up to recently (ESS Volume III 1996), a NC version of the linac had been extensively studied and constituted the reference option. However, given the fantastic progresses made on SC technology and the many potential advantages, a SC alternative was developed. We concentrate on this option.

INTRODUCTION

The European Spallation Source (ESS) [1] Project represents the culmination of a decade of intense research, development and design by Europe's leading scientists. The single goal is to provide Europe with the world's most powerful neutron facility. Some 20 laboratories, universities and research organizations plus a large number of scientists from all over Europe, have joined together to develop the science case and the technical design, and to work on the planning and the realization of the ESS project.

The ESS will go one step further compared to the sources under construction in the US and Japan. Not only will its power be higher, which translates directly into more intensity, but the ESS Council has also opted for a source with two complementary targets, a 50 Hz Short Pulse (SP) and a 16 2/3 Hz Long Pulse (LP) target (both target stations being considered with equal priority).

Beam power	10 MW	
Linac beam energy	1.334 GeV	
	SP	**LP**
Linac average current	3.75 mA	3.75 mA
Linac peak current	114 mA	90 mA
Linac repetition rate	50 Hz	16 2/3 Hz
Linac beam pulse duration	2 × 0.48 ms (gap=0.1ms)	2.5 ms
Chopping efficiency	70%	no chopping
Linac beam duty cycle	4.8%	3.3%

With the latest development, the SC linac option shows the potential to provide a clearly superior solution to the old reference NC solution. A Technical Advisory Committee (TAC) comprising experts from a large number of major facilities all over the world has expressed its confidence in the proposed technical design and the ability of the ESS partners to build it. A dominant requirement is to minimise the uncontrolled beam losses to be below 1 Watt per meter and to localise the controlled beam losses in carefully designed serviceable collimators or beam dumps.

Early in the linac design we had also to consider the most efficient and reliable option compatible with the ambitious goals.

SC LINAC OVERVIEW

The 5MW Short Pulse (SP) is obtained by compressing two 480μs, 114mA-peak H^- pulses in two 600ns, 62.5A-peak proton pulses in two rings. H^- ions are used rather than protons for injection efficiency reasons. Since H^- sources are for the moment not able to produce 100 mA within all the other requirements (emittance, repetition rate, life-time…), two 50 mA sources are used and both beams are funnelled at upper energy (~20 MeV). The 2.5ms-100mA Long Pulse (LP) is directly sent to the target. Both H^- ions and protons can be used. The simpler solution consists in using H^- because the SP is already H^-. Nevertheless, the source capability of producing the LP is doubtful. A fallback solution consists in using protons. The main advantage is the already existing ECR proton sources (SILHI, LEDA) fulfilling all the requirements. The main difficulty is to demonstrate that it is possible to transport two high quality beams with different charges and current in the same linac.

This proton line is inserted at the funnel cavity input end. The first step of the beam dynamics studies consists in showing that the beam will fulfil the ESS requirements (losses <1W/m -short and long pulses-, $\varepsilon_{T,\,norm,\,rms}$<0.5 π.mm.mrad, $\varepsilon_{L,\,norm,\,rms}$<2.3 π.mm.mrad -SP only-). Linac sections have been matched and

CP642, *High Intensity and High Brightness Hadron Beams: 20th ICFA Advanced Beam Dynamics Workshop on High Intensity and High Brightness Hadron Beams*, edited by W. Chou, Y. Mori, D. Neuffer, and J.-F. Ostiguy

multi-particle simulations were done from the first RFQ input to the linac exit with the Saclay code package TOUTATIS-TRACEWIN-PARTRAN.

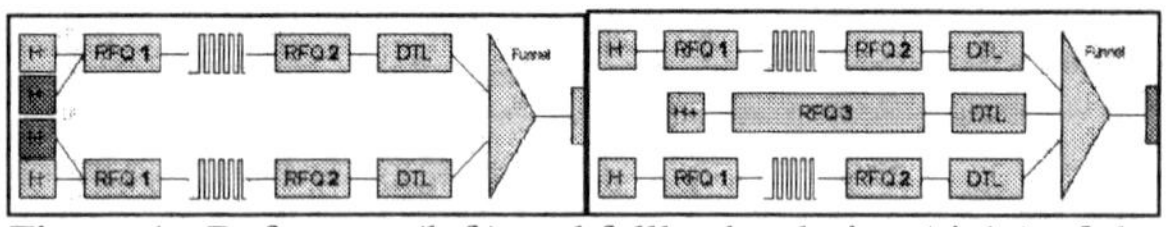
Figure 1 : Reference (left) and fallback solution (right) of the low energy Room Temperature Linac.

Front Ends

The front ends (i.e. at low energies below 20 MeV) of the accelerator require state-of-the-art performances from H^- ion sources, choppers and funnel branches. Between 20 MeV and ~200 MeV, normal conducting accelerating cavities are used.

Sources and RFQs

In order to meet the SP and LP ESS beam power specifications, much is required from the ion source(s). All the best performance levels achieved in the world are simultaneously required. This is considered to be challenging. Recently, development strategies have been drawn up on the basis of expert opinion. On the bases of present ion source operational experience and of development programmes already launched, it is likely that H^- ion sources will be produced to meet ESS short and long pulse specifications within 2–3 years (ISIS Penning source (RAL), volume ion source at IAP-Frankfurt, ECR H^- source at CEA-Saclay).

As the first accelerating components, the next RFQs play a significant role in determining the quality of the beam in the rest of the accelerator. A great care was taken in their design. RFQ1 accelerates the H^- from 60keV to 2MeV, allowing the insertion of the chopper line. 2MeV is chosen to avoid activation in RFQ1 and the chopper line. RFQ2 accelerates particles from 2MeV to 5MeV, lowest energy compatible with high duty cycle EM quadrupole in the DTLs. For the proton branch, RFQ3 accelerates protons from 95keV to 5MeV. RFQ1 minimizes the transverse emittance growth, with large acceptance (0.3 π.mm.mrad) and maximizes transmission (>99%). RFQ2 reshapes the beam, minimizes the emittance growth and gives the same zero current phase advance at exit as RFQ3, with 100% theoretical transmission. RFQ3 minimizes the losses above 2.16 MeV. The frequency choice of 352.2 MHz is consistent with present LEDA and IPHI experiences [2, 3]. At this frequency, four-vane geometries are used.

Chopper section

Time structure is required in the H^- beam to enable low loss multi-turn injection into a pair of accumulator rings and single turn fast extraction afterwards. A fast chopper is inserted after RFQ1 and must chop the beam cleanly between the RFQ bunches (2.84 ns) so as not to generate beam loss on the downstream linac structures. The emittance growth is kept below 3% with a great beam separation at the beam stop location.

DTLs

DTL tanks are used to increase the energy of the beams from 5 to 21.9 MeV where they can be funnelled much more easily. The ESS-DTL design is based on the IPHI R&D program. A power of 1 MW (1.3 MW klystron) per tank has been chosen. The use of conventional electromagnetic quadrupoles may allow restricting the halo development. They preserve the beam quality but at the expense of lower shunt impedance (bigger tubes). AFD focussing scheme is chosen. The phase advance per meter is conserved between RFQ output and DTL input. The H^- peak beam current is 57 mA while the H^+ peak current is 90 mA. It requires 2 tanks for the H^- lines and 3 tanks for the H^+ line.

Funnel

The funnel line is one of the most complicated sections of the linac. The energy at which the funnel line operates is a compromise between lower beam rigidity at low energy and lower space-charge effects along the beam transport at high energy. The compromise for ESS sets the funnel energy at 21.9MeV. The main difficulty is to transport the beam without emittance growth and to prevent halo development (obtained by a smooth change of the focusing force between structures). A big effort is made to insert the H^+ beam between the two H^- lines. This is made possible by the development of a new deflecting cavity allowing ± 5 degrees deflecting angle [4], ± 10.9° being obtained with quadrupoles.

SDTL to 90.5MeV, CCL to 185MeV

The frequency is now equal to 704 MHz. At 20 MeV, a SDTL was preferred for its doublet period, more convenient for the matching with CCL and SCL cavities. SDTLs require less matching than DTLs and are less complex than CCDTLs. They have higher shunt impedance. Above 100 MeV, the effective shunt impedance of the SDTL structure decreases rapidly. A switch is made to the Coupled Cavity Linac (CCL) structure, more efficient at high energy. A CCL is used from 90 MeV to 185 MeV.

Since the H^+ bunch repetition is 352.2 MHz and the H^- beam bunch repetition is 704.4 MHz, the proton beam charge per bunch is 1.58 bigger than that of the

H^- beam. The part of the linac common to both beams is matched for H^-. However, since the phase advance per meter and the focusing lattice are kept continuous along the linac, the proton beam is almost matched in the linac (current independent design).

Superconducting structure

Below ~200 MeV, the energy gain per metre of real estate of SC cavities is actually lower than 2MeV/m, a gradient which can be easily reached by warm NC structures. Furthermore, the design of SC cavities for beta<0.6 is complicated by stiffness and microphonics issues. The transition energy between NC and SC structures is then fixed at 185 MeV. The different cavity types and transition energies are obtained from length and cost optimisation. As a result, two 5-cell cavity families with geometric "beta" values of 0.66 and 0.85 are well suited for the acceleration between 185 MeV and 1330 MeV. The length of the SCL is 290m, with limited maximum peak field (Bp = 50 mT -about Ep = 27.5 MV/m), and reasonable power coupler capability (2 couplers per cavity each of 800kW max). From the beam dynamic point of view, the constrain of keeping the phase advance per cell below 90°, has been fulfilled in order to avoid structure instability.

Field control – energy stability

The one klystron per cavity scheme, each cavity with its own feedback/feed forward RF control system allows great flexibility and simplest operation procedure. Lorentz forces were carefully studied (welding rings between cells) and microphonics were simulated. The maximum deviations remain small (±0.2MeV and ±0.5°). Random errors of uniform distribution, simultaneous cavity field errors (1% amplitude, 1° phase) lead to a rms energy fluctuation of 0.8 MeV and phase fluctuation of 1.2° at linac exit (10 000 simulations).

END-TO-END SIMULATIONS

Linac sections were matched and multi-particle simulations were done from the first RFQ input to the linac exit. The input distribution is a 4D water-bag with rms transverse emittance of 0.3 π.mm.mrad to relax the H^- sources. Most of the emittance growth is visible in the funnel line. This is due to the non-perfectly smooth line and the deviation, which depend on the particle phase. Nevertheless, the final emittances are much smaller than the required ones.

In order to evaluate the beam losses arising from the beam dynamics in space charge in a real linac, typical static and dynamics errors on linac element have been defined and simulated. The errors are different from one section to the other, taking into account the element specification. More than 1000 linacs with 100 000 particles were simulated with no losses after the RFQs. The minimum ratio between the bore radius and the 10^{-5} level is about 1.15 in NC linac for H^+ and 2.55 in SC linac for H^-. The correction scheme allows the transport of both particles type with a rms residual orbit < 200µm (beam size < 15 mm).

ESS linac fulfils its requirements. Estimated beam losses are lower than 1 W/m, especially at high energy. Most of the losses are coming from the H^- stripping in the residual gas.

ACKNOWLEDGMENTS

The ESS project acknowledges all the European teams involved in the linac design. It also thanks the SAC and TAC committees, the very open and mutually beneficial collaboration with institutions and colleagues working on the US and Japanese spallation source projects, financial support from the European Commission through the neutron round-table.

REFERENCES

1. The ESS project, Vol III, Technical report, May 2002. http://www.ess-europe.de
2. H.V. Smith. and J. D. Schneider, "Status report on the low-energy demonstration accelerator (LEDA)", proceedings of Linac 2000, Monterey, USA, p581
3. P-Y. Beauvais et al, "Status report on the Saclay high-intensity proton injector project (IPHI)", Proceedings of EPAC 2000, Vienna, Austria, p283.
4. N. Pichoff et al., "The ESS-CONCERT Funnel Line", Proceedings of PAC 2001, Chicago, USA, p390

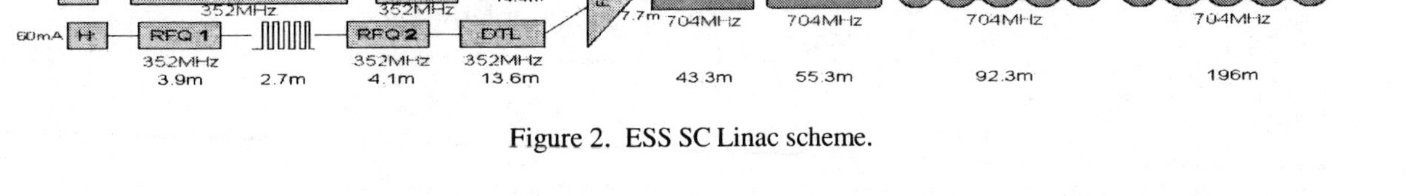

Figure 2. ESS SC Linac scheme.

Status and Plans of the SPL Study at CERN

R. Garoby for the SPL Study Team

CERN, Geneva, Switzerland

Abstract. The study of the SPL (Superconducting Proton Linac), a 4 MW / 2.2 GeV H^- linac, began at CERN in 1999. Based on the large inventory of RF equipment decommissioned from the LEP (Large Electron Positron collider), it was originally proposed as an upgrade to the injector complex for the high energy accelerators. Since that time, the proposal has attracted the interest of an increasing number of physics communities and the design has evolved in consequence. The present design of the SPL is presented in this paper, together with a proposal for a staged realization.

INTRODUCTION

The SPL [1,2] was first proposed as a way of profiting from the decommissioned LEP RF equipment, to upgrade the characteristics of the beam delivered by the CERN complex of proton accelerators at low cost. The SPL would then replace the 50 MeV proton linac (linac 2) and the 1.4 GeV PS Booster, and inject directly into the PS. Moreover, thanks to the high flux of protons potentially available from such a superconducting linac, other users could easily be accommodated. For a second generation ISOLDE-like facility [3], the SPL beam is directly usable. For neutrino physics (first a "Neutrino super-beam" facility [4] and ultimately a "Neutrino Factory" [5]), accumulator and bunch compression rings have to be added. The design of the accelerator set-up has consequently evolved.

SPL DESIGN

The proton driver of a neutrino factory is the most demanding application for the SPL [5]. For this, a 150 m radius storage ring is used to accumulate the linac beam. After accumulation, a 3 μs beam burst of 2.27×10^{14} protons is sent to a bunch compression ring before being delivered onto the pion production target. The resulting requirements for the SPL are shown in Table 1, and its overall layout is sketched in Figure 1.

TABLE 1. Main SPL characteristics.

Accelerated ions	H^-	
Kinetic energy	2.2	GeV
Mean beam power	4	MW
Repetition rate	50	Hz
Pulse duration	2.8	ms
Mean current during the pulse	13	mA
Number of H^- per pulse	2.27×10^{14}	
Bunch frequency	352.2	MHz
Chopping duty cycle	61.6	%
Bunch pattern (nb. of bunches / nb. of buckets)	5/8	
Norm. r.m.s. transverse emittances	0.4	π mm mrad
Longitudinal r.m.s. emittance	0.3	π deg MeV

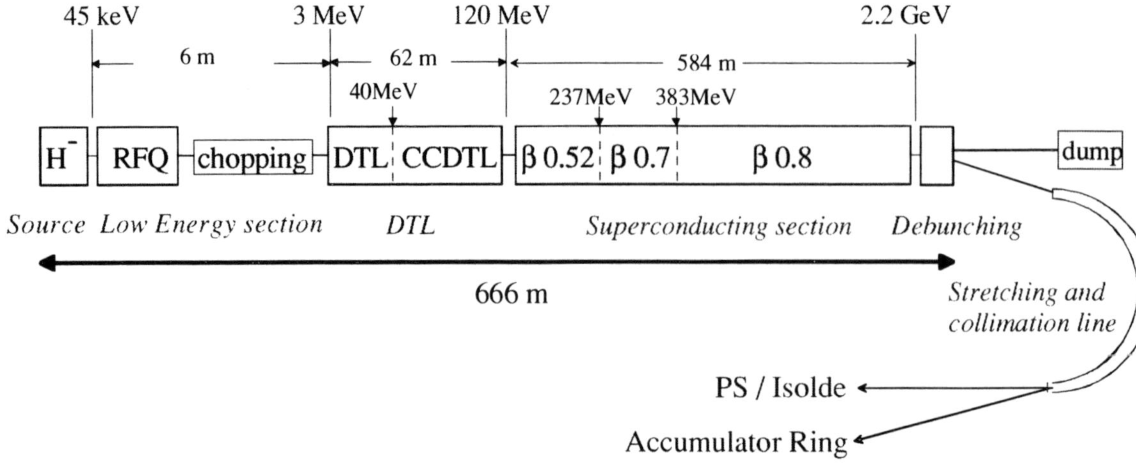

FIGURE 1. SPL synoptic

CP642, *High Intensity and High Brightness Hadron Beams: 20th ICFA Advanced Beam Dynamics Workshop on High Intensity and High Brightness Hadron Beams,* edited by W. Chou, Y. Mori, D. Neuffer, and J.-F. Ostiguy

Room temperature accelerating structures are used in the first 69 m of the linac (120 MeV kinetic energy). Above this energy and for most of the length of the accelerator, superconducting structures are employed. The parameters of the sections are detailed in Table 2.

TABLE 2. SPL sections parameters.

Section	Final Energy (MeV)	Nb. of cavities	Peak RF Power (MW)	Kly-strons	Te-trodes	Length (m)
Source	0.045					1
RFQ	3	1	0.5	1		2.4
Chopper	3	3	0.06		3	3.6
DTL	120	13	11.8	15		62
β=0.52	236	42	1.5		42	101
β=0.7	383	32	1.9		32	80
β=0.8	1111	52	9.5	13		166
β=0.8	2235	76	14.6	19		237
Debunch.	2235	4		1		13
Total		223	39.9	49	77	666

RFQ and Chopping line

The RFQ will be of the four vane type, and its precise design is now expected to benefit from the development of the IPHI device actually in construction for the CEA and IN2P3 [6].

The proposed SPL chopper structure consists of a pair of deflecting plates with a meander delay-line printed on alumina (Fig. 2), without separating ridges [7]. Attenuation and dispersion measured on prototypes are in good agreement with computation. The printed meander on alumina has very good vacuum properties, easy implementation and good radiation resistance and heat transfer (water cooled metal ground plane). The high permittivity (ε) of the ceramic permits a meander width below 25 mm for particle velocity of β=0.08, and gives the possibility to install the deflectors inside quadrupoles. The 100 Ω characteristic impedance helps reduce the driver power. A prototype 500 V chopper pulse amplifier has been realised, and tests are in progress [8].

The chopper line is composed of two 1 m long double FODO sections, for transverse matching between a fast phase advance in the accelerators and a slow one in the chopper, plus a 1.6 m long FODO period in between. The latter houses the chopper inside the quadrupoles and provides the 90° phase advance at the dump placed at its end, needed for the separation of chopped and unchopped beam.

A test of the chopping line using a 3 MeV segment of the 352 MHz IPHI RFQ, presently being assembled, is in preparation with CEA and CNRS-IN2P3 (France).

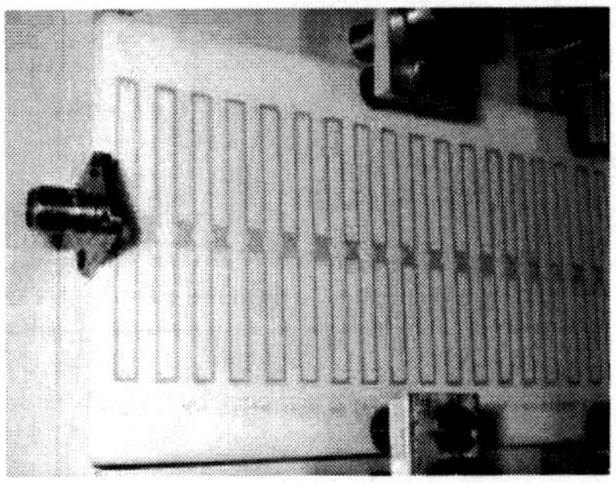

FIGURE 2. Chopper prototype: double 100 Ω meander-line on ceramic substrate.

By subsequently adding the second RFQ segment, acceleration will proceed up to 5 MeV. Beam tests up to 100 mA are foreseen, in pulsed mode with chopping and in CW without chopping. Beam measurements are planned to fully characterise chopping efficiency and halo development. The test stand will be located at CEA-Saclay.

Drift Tube Linac

The DTL section is based on a conventional Alvarez structure up to 40 MeV. From this energy, the focusing period is increased, keeping the longitudinal phase advance below 65° to avoid emittance exchange. A Cell-Coupled DTL design (CCDTL [9]) at 352 MHz is adopted, characterised by identical coupling cells. A 12-cell cold model (Fig. 3) has been realised and tested.

FIGURE 3. 3D drawing of a fraction of the CCDTL cold model.

The advantages of this structure are easy access and alignment for the quadrupoles, low construction cost, stable $\pi/2$ mode operation, continuous focusing lattice, and simple RF distribution with one klystron per tank. However, at this relatively low frequency, the real-estate shunt impedance of the CCDTL remains similar to that of a conventional DTL.

Superconducting Linac

The superconducting part of the SPL begins at 120 MeV kinetic energy. Up to 383 MeV, multi-cell cavities optimised for β=0.52, 0.7 are used. To ease stabilization of the field in the cavities and minimize the energy fluctuation of the beam, each cavity is driven by its own tetrode amplifier. Above 383 MeV, β=0.8 5-cell cavities are used. Unmodified LEP cavities are no longer employed, even at the highest energy. The additional cost is compensated by the higher accelerating gradient and transit time factor of the new cavities which allow to reduce the linac length. Housed in LEP cryostats, these 5-cell β=0.8 cavities are prepared at minimal cost. Four of them are driven by a single LEP klystron.

The beam dynamics design of the SC section is optimised for minimum emittance exchange, maximum stability against mismatch and simplified layout for minimum cost [9]. The length of the focusing periods, each containing a quadrupole doublet of 120 mm aperture diameter, increases along the linac to a maximum of eight cavities (two cryostats) per period above 1.1 GeV. This corresponds to 13 to 21 $\beta\lambda$ per period and keeps the maximum longitudinal phase advance below 65°. The relatively low longitudinal phase advance allows the full current tune ratio (σ_l/σ_t) to be kept below 0.8 thus avoiding emittance exchange between the longitudinal and the transverse planes [10]. At the same time, the maximum transverse phase advance can be held below 85° to avoid particle lattice instabilities. A smooth phase advance per metre in both planes ensures a minimum mismatch in the transition areas between sections, which are matched with existing beam line elements. Mismatch simulations with 50 M particles show only moderate emittance growth even for strong initial mismatch (30 % radial) [9].

A recent study having underlined the difficulty to properly control 4 cavities simultaneously [11], high power phase and amplitude modulators are now felt to be necessary to stabilize the field in each of the β=0.8 cavities individually (specifications in Table 3). The ferrite loaded waveguides and their external bias system are the only components that must be bought (Figure 4). Prototypes have been ordered and the first experimental results are expected by the end of 2002.

TABLE 3. Modulators specifications.

RF frequency	352.2 MHz
Peak power	350 kW
Phase modulation depth	± 25 degrees
Amplitude modulation depth	± 10 %
Rise-time (10 – 90 %)	1 ms

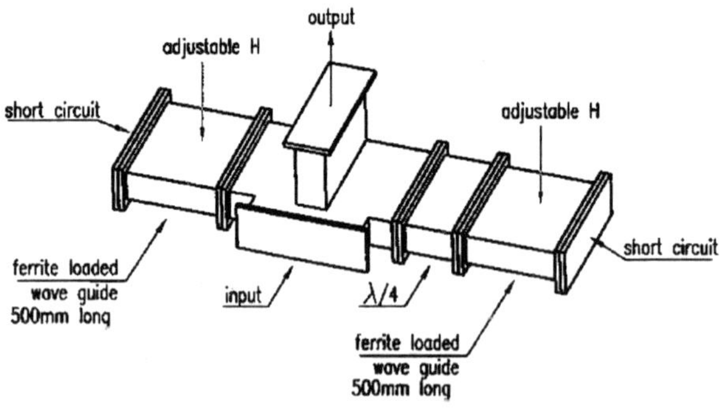

FIGURE 4. Phase/Amplitude modulator

STAGED REALISATION

A staged realisation is necessary to comply with the limited resources available at CERN during the construction of the LHC. Effort is concentrated on the low energy part of the machine (up to a few MeV) in close collaboration with the CEA and IN2P3 [6]. In a second stage, the realisation of the 120 MeV room temperature front-end is envisaged as an improved injector for the PSB.

REFERENCES

1. R. Garoby, M. Vretenar, "Proposal for a 2 GeV Linac Injector for the CERN PS", CERN PS/RF/Note 96-27.
2. M. Vretenar (editor), CERN 2000-012.
3. http://www.ganil.fr/eurisol/index.html
4. http://muonstoragerings.cern.ch/NuWorkshop02/
5. R. Garoby, CERN/PS 2001-055 (RF).
6. J.-M. Lagniel et al., "IPHI, the Saclay High-Intensity Proton Injector Project", PAC 97, Vancouver, pp. 1120-1122.
7. F. Caspers, A. Mostacci, S. Kurennoy, CERN/PS 2002-027 (RF).
8. M. Paoluzzi, CERN/PS 2002-026 (RF).
9. F. Gerigk, M. Vretenar, R.D. Ryne, "Design of the Superconducting Section of the SPL Linac at CERN", PAC 2001, Chicago, pp. 3909-3911.
10. F. Gerigk, I. Hofmann, "Beam Dynamics of Non-Equipartitioned Beams in the case of the SPL Project at CERN", PAC 2001, Chicago, pp. 2872-2874.
11. Tückmantel, "Mathematical analysis of spontaneous symmetry breaking in a multi cavity RF system with vector sum feedback and Lorentz detuning", CERN-SL-Note-2001-023.

The TRASCO Project

Carlo Pagani*

University of Milano and INFN Milano LASA
Via Fratelli Cervi, 2001, I-20090 Segrate (Milano), Italy

Abstract. TRASCO is a joint INFN/ENEA program, started in 1998, aiming at the design and the technological investigation of the main components of an accelerator driven system (ADS) for nuclear waste transmutation. The proposed 30 mA, 1 GeV, proton linac (TRASCO-AC) is briefly outlined, together with the ongoing R&D work on several key components.

INTRODUCTION

The disposal of radioactive waste represents a problem that is not yet fully solved, especially in terms of environmental and social acceptability. By using partitioning and transmutation technologies the most hazardous materials could be separated (partitioning) from the nuclear wastes and, then, converted to shorter-lived elements (transmutation). Transuranic elements would be transmuted by fission and Fission Products by neutron capture and beta decay.

Accelerator Driven Systems (ADS) are generally believed to be particularly suited for assessing the transmutation of nuclear waste and have become a major R&D topic in Europe [1].

Important ADS activities are also going on or are being planned outside Europe e.g., in Japan (the JAERI/KEK Joint Project), in USA (the AAA project) and in Korea (the KOMAC Project).

TRASCO PROGRAM OUTLINE

On the basis of a growing interest on the ADS concepts in Italy, the basic R&D program TRASCO has been funded in 1998 by the Ministry of the University, Scientific and Technological Research, and has been followed by an industrial program, in which major Italian research institutions and industries are involved [2]. TRASCO aims to study the physics and to develop the technologies needed to design an ADS for nuclear waste transmutation.

The TRASCO program consists of two main parts, regarding, respectively, the accelerator and the sub-critical system. The first part falls under the competencies of INFN and the second under those of ENEA, which are the two institutions jointly responsible of the TRASCO activities, which include also Italian industrial partners.

The program covers all the main subsystems of an ADS (accelerator, window/target, sub-critical reactor). However, due to the limited available financial resources, all the efforts are concentrated on starting a few significant and qualified experimental activities, in view of the goal of a future participation in an International project dedicated to the study and construction of an ADS prototype, like, for example, the PDS-XADS program of the 5th framework programme of the European Commission.

Reference Accelerator Parameters

The TRASCO accelerator program [2] is focussed on the conceptual design and the realization of prototypical components for a 30 MW (1 GeV, 30 mA) CW proton linear accelerator, which is the candidate for an industrial size transmuter with a thermal power in the range of 1500 MW.

The coupling of the accelerator with the subcritical system leads to rather tight requirements on the accelerator reliability and availability. This reliability issue therefore is at the base of the linac design, leading to conservative choices on all components with respect to the "state of the art" technologies, and to an appropriate amount of redundancy in the system, in order to increase the tolerance to component faults and to achieve beam delivery to the target in the presence of faulty components, in the time between regularly scheduled maintenance periods.

General layout

The linac consists in a microwave discharge proton source, operating at 80 kV, capable of providing 35 mA of CW proton beam, followed by a 352.2 MHz RFQ up to 5 MeV. A 352.2 MHz superconducting linac with independently phased resonators brings the energy to approximately 100 MeV, and the beam is finally brought to the nominal energy of 1 GeV by a

* On behalf of the TRASCO Linac Group

CP642, *High Intensity and High Brightness Hadron Beams: 20th ICFA Advanced Beam Dynamics Workshop on High Intensity and High Brightness Hadron Beams*, edited by W. Chou, Y. Mori, D. Neuffer, and J.-F. Ostiguy

three section superconducting linac that uses multicell elliptical cavities at 704.4 MHz.

The main objectives of the TRASCO program are:

- The conceptual design of the whole linac;
- The design and construction of the proton source and of the RFQ pre-accelerator;
- The design, construction and test of prototypes of the superconducting cavities, mainly for the high-energy linac.

A schematic layout of the TRASCO linac is presented in Figure 1.

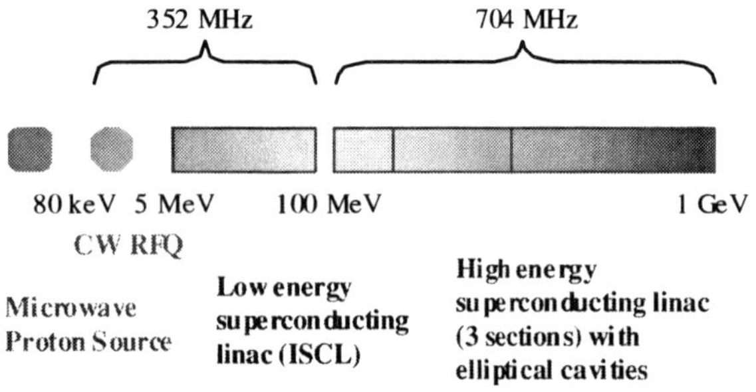

FIGURE 1. The schematic layout of the TRASCO linac.

In the following sections the status of the design of the components of the TRASCO linac will be outlined, together with the R&D ongoing activities.

TRIPS: TRASCO Intense Proton Source

High intensity proton sources of several tens of mA already exist but additional R&D efforts are still necessary for ensuring the availability and reliability required by the ADS system. TRIPS design is based on the principle of microwave discharge, off-resonance, and it is a modified version of the source SILHI operating at CEA/Saclay since several years.

With respect to that source some things have been changed and some new features have been added. Mainly the efforts have been concentrated on the design of a new extraction system with the aim to increase the source availability and reliability, in order to meet the requirement of a driver for an ADS system.

RFQ and low energy linac

The TRASCO RFQ has been designed with the constraints of low beam losses (less than 4%), the use of a single LEP type klystron (1.3 MW nominal power) and a Kilpatrick factor of 1.8 (corresponding to a 33 MV/m surface electric field). Table 1 summarizes the main parameters of the TRASCO RFQ.

The RF structure is split in three resonantly coupled segments (similar to the LEDA design). The vane tips have a fixed transverse radius of 2.93 mm, and the vane voltage is kept constant along the structure, to reduce power dissipation. A cold 3 m aluminium model has been built and has demonstrated the capability of field stabilization. Tests of the fabrication technology have been performed on a short model and the full RFQ structure is now under fabrication.

TABLE 1. TRASCO RFQ Parameters.

Beam Current	30	mA
Beam Emittance	0.2 π	mm mrad
Output Energy	5	MeV
Total Length	7.13	m
RF Input Power	850	kW
Peak Electric Field	33	MV/m

For the low energy linac several options have been investigated: a normal conducting DTL, and a few superconducting linac options based on different types of independently phased resonators (single gap reentrant cavities, half wave resonators and more recently spoke cavities, with one or two spokes).

The choice of a superconducting linac with independently phased cavities seems the most attractive both in terms of power consumption and in terms of reliability. Spoke cavities can allow a large beam aperture and can be divided in few families, as done for the elliptical cavities in the high energy part

The high energy linac

From energy of about 100 MeV the technology of elliptical superconducting cavities represents the most efficient and cost effective one to accelerate the beam to higher energies. The RF frequency has been set to 704.4 MHz (twice that of the RFQ and low energy linac) to use the bulk Nb technology for cavity fabrication, which can now guarantees outstanding performances, thank to the work done in the past years in the framework of the TESLA/TTF Collaboration. In order to cover the desired energy range (from ~100 MeV to 1 GeV), three different linac sections are required and sufficient, with the cavity betas set respectively to the values of: 0.47, 0.65 and 0.85. These beta choices, while efficient, leave a large margin for energy ugrade (ut to 2 GeV). Table 2 summarizes the main parameters of the three sections.

TABLE 2. TRASCO 704.4 MHz SC Linac Params.

Section #	1	2	3
Section Geometrical β_g	0.47	0.65	0.85
Input Enrgy [MeV]	90	190	480
Output Energy [MeV]	190	480	1000
Lattice Period [m]	4.2	5.8	8.5
# Lattice Periods	14	16	12
# Cells/Cavity	5	5	6
Cavity Eacc [MV/m]	8.5	10.3	12.3
# Cavities/Cryomodule	2	3	4

A set of multicell cavities for all three linac sections has been designed, taking into account both the optimization of electromagnetic performances (in terms of peak fields on the surface) and mechanical performances (stability under vacuum load, stiffness for microphonics and Lorentz forces excitation). Conservative values with respect to the state of the art technology of the TTF cavities have been used in terms of maximum peak design electrical and magnetic fields on the surface (set to < 30 MV/m and < 50 mT), in order to increase the reliability of the proposed technology.

Beam dynamics analysis and simulations

The whole linac has been designed using high current beam dynamics criteria aimed at minimizing the occurrence of beam emittance growth and particle losses. In particular structure and tune resonances, and strong tune depression, have been avoided in the design and a smooth variation of all linac parameters has been guaranteed. Proper adiabatic matching procedures, have been developed and employed at the section interfaces [4]. The validity of the design has been assessed by means of multiparticle simulations with standard beam dynamics codes.

R&D ON LINAC COMPONENTS

In addition to delivering a conceptual design report for an ADS driver, the TRASCO program foresees the realization of prototypical activities on critical components of each section of the proposed linac. The most representative examples of these activities are described in the following paragraphs. More information on the status of the different components can be found in [4] and in the references there quoted.

TRIPS operation

The TRIPS source has been assembled in LNS-Catania in May 2000, commissioned in January 2001 with a 20 mA beam at 60 kV and finally achieved 55 mA of beam current with 90% of proton fraction during beam tests in August 2001 [4].

Fabrication of the TRASCO RFQ

The RFQ design has been engineered, in collaboration with the industrial partners and the full structure is now in the fabrication stage [4]. Special attention during the design has been dedicated to the optimisation of the cooling channels.

SC cavities for the high energy linac

A single gap SC re-entrant cavity (at the 352.2 Mhz frequency) has been designed and manufactured. A report of the performances obtained is found in ref [4].

A total of four β=0.47 single cell cavities have been built with the TRASCO industrial partners, two of which using low grade niobium sheets (RRR>30) and two using high purity sheets (RRR>250). The vertical tests were very successful in all four cases. Both the cavities produced with low quality, and cheap, niobium exceeded the design parameters, showing a hard quench at an accelerating field value of 10-11 MV/m. The best of the other two cavities (Z104) reached the record accelerating field of 22 MV/m, the Q staying above 10^{10} up to > 20 MV/m. The other (Z103) was limited to 20 MV/m by field emission.

In Figire 2 the test results on the two best cavities are presented, showing also the corresponding values of the peak electric and magnetic fields. Treatments and tests were performed at TJNAF and CEA-Saclay.

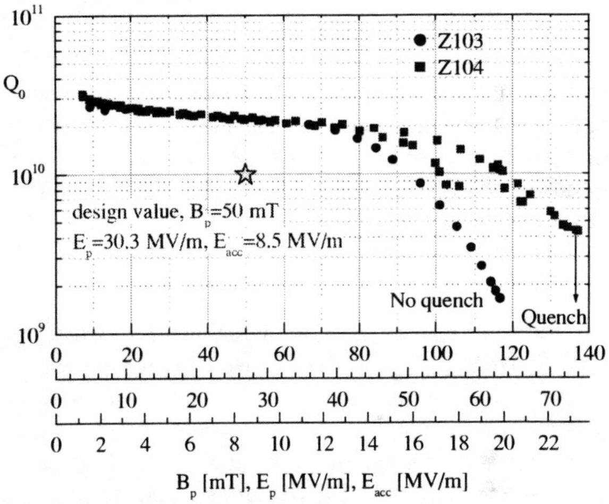

FIGURE 2: RF tests of the Z103 and Z104 cavities, made with high quality niobium (RRR > 250).

On the basis of this experience the construction of two complete five cell cavities has been launched. A more complete description is given in ref [4].

REFERENCES

[1] EU Tech. Working Group on ADS, "A European Roadmap for Developing Accelerator Driven Systems (ADS) for Nuclear Waste Incineration", April 2002, http://www.neutron.kth.se/TWG22/default/index.html.

[2] TRASCO Partners for the accelerator program are: INFN and the Italian companies HITEC, CINEL, Saes Getters, CESI and ZANON.

[3] The TRASCO_AC Group, "Status of the High Current Proton Accelerator for the TRASCO Program", INFN Report INFN/TC-00/23, December 2000, available from http://wwwsis.lnf.infn.it/pub/INFN-TC-00-23.pdf.

[4] Papers THBLA002, TUPLE121, WEPLE109, THPDO022, THPDO023, THPLE083 are being presented at EPAC2002 and will be in its Proceedings.

IFMIF - A Challenging High-Intensity Accelerator

Robin Ferdinand

CEA-Saclay
DSM/DAPNIA/SACM bldg. 124, 91191 Gif-sur-Yvette cedex, FRANCE

Abstract. The International Fusion Materials Irradiation Facility (IFMIF) employs an accelerator based D-Li intense neutron source as defined in the 1995-96 Conceptual Design Activity (CDA) study under the direction of the IEA's Executive Committee on Fusion Materials. Full performance operation (2 MW/m^2 @ 500 cm^3) allows to obtain engineering data for potential DEMO materials under irradiation up to 100-200 dpa and a systematic search for high performance materials. The linac design is reviewed and described.

INTRODUCTION

Materials required for the fusion reactor must be able to survive irradiation in a high intensity neutron field with an energy 14 MeV and annual damage doses of the order of 20 dpa (displacements per atom). Concepts for an irradiation test facility suitable for identifying and validating such materials have been explored through a number of studies over the period of the last several decades. An accelerator-based neutron source using the Deuteron-Lithium (D-Li) stripping reaction has been selected as the basis of the International Fusion Materials Irradiation Facility (IFMIF) studies [1]-[3].

The main specifications for the IFMIF facility are summarized in Table I. An intense flux of high energy neutrons will be produced within sufficient irradiation volume to enable realistic testing of candidate materials and components up to about a full lifetime of their anticipated use in DEMO and beyond.

Table 1: IFMIF top-level specifications

Neutron Flux	$\geq$ 2 MW/m^2 (@ 500 cm^3)
Operation Availability	70 %
D+ Beam Current	250 mA (CW, 2×125 mA)
D+ Energy	32, 40 MeV
D+ Beam Size	200 mm (width)×50 mm (height)
Li Jet Thickness	19, 25 mm (resp. for 32, 40 MeV D+)
Li Jet Width	260 mm
Li Jet Velocity	10-20 m/s

The technological approach adopted in the CDA design of the IFMIF requires a relatively modest extension of the current state of the art of the technology for the radio frequency (RF) linear accelerators and Li loops. In particular, the need to achieve the necessary neutron fluences within reasonable time and to maintain the required long term uniformity and stability of the neutron flux imposes high availability and reliability on the entire system (80.7% for the total facility, 88.0% for the accelerator facility). In order to assure the high availability and reliability required for IFMIF, its key technology elements like the 125 mA D$^+$ linac and a continuously operating liquid Li system require design and fabrication of suitable prototypes for performing the necessary endurance tests.

The basic approach is to provide two linacs modules, each delivering 125 mA to a common target. This paper describes the IFMIF linac modules.

COST AND STAGE APPROACH

The IFMIF CDA study was conducted during 1995-1996 [1]. In January 1999, IFMIF design focused on potential for cost reduction. In addition, staged deployment was to be examined as an option offering a potential reduction of the annual expenditures during construction. The total cost estimate was reduced to 61% of the CDA estimate, from 797.2 MICF to 487.8 MICF [4] (1 MICF ≈ \$1M US in January 1996).

The study of cost reduction and staged deployment considered the following major items:

1) The potential for a future upgrade to four accelerators with irradiation capability twice that of the current user requirements, has been eliminated.

2) The building volume was reduced in accelerator systems and lithium loop components.

In addition, the original CDA specifications were critically reviewed with the objective of improving component design and eliminating non-essential items. For example, the energy dispersion cavity, the buncher cavities in the High Energy Beam Transport (HEBT) lines and the associated RF power source were removed. The HEBT 90-deg beam turning lines were eliminated. The transition energy between the RFQ and the DTL was reduced from 8 MeV to 5 MeV.

The IFMIF deployment was assumed to proceed in three stages, each addressing a specific materials

CP642, *High Intensity and High Brightness Hadron Beams: 20th ICFA Advanced Beam Dynamics Workshop on High Intensity and High Brightness Hadron Beams*, edited by W. Chou, Y. Mori, D. Neuffer, and J.-F. Ostiguy

development issue as follows:

1st Stage: One accelerator with a maximum of 50 mA operation, to be used for material selection of the ITER breeding blanket test modules, fusion-fission data correlation and generic damage studies.

2nd Stage: One accelerator for 125 mA operation (i.e. 1 MW/m^2 @ 500 cm^3), to be used to demonstrate materials performance of a reference alloy for DEMO-relevant fluences.

3rd Stage: Two accelerators for 250 mA operation (i.e. 2 MW/m^2 @ 500 cm^3), used to obtain engineering data for potential DEMO materials under irradiation up to 100-200 dpa and a systematic search for high performance materials for fusion reactors.

In the 2nd stage, more RF stations are added to the accelerator to increase the current from 50 to 125 mA. In the 3rd stage, the second 125 mA accelerator is installed to bring the total beam on target to 250 mA.

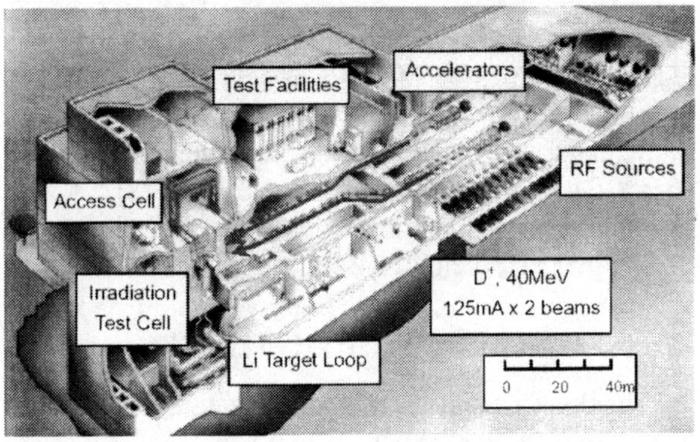

ACCELERATOR DESCRIPTION

The IFMIF facility requires generation, by a linear accelerator (LINAC), of 250 mA continuous current of deuterons at a nominal energy of 40 MeV, with provision for operation at ~32 MeV. The basic approach is to provide two linacs modules, each delivering 125 mA to a common target. This approach has availability and operational flexibility advantages. A quasi-continuous operation is mandatory. Annealing times of point defects would introduce unacceptable uncertainties. Consequently RAM is a major concern in the facility and accelerator design.

Acceleration of high current cw D^+ beams (125 mA) has not yet been demonstrated, although recent experiments with the 100 mA cw proton beams at the LEDA in Los Alamos [5] represent a significant step towards this goal.

A cw 155-mA deuteron beam is extracted from the ion source at 95 keV. A low energy beam transport (LEBT) guides the deuteron beam from the source to an RFQ. The RFQ bunches the beam and accelerates 125 mA to 5 MeV. The 5 MeV RFQ beam is injected directly into a Room Temperature (RT), ramped-gradient DTL of the conventional Alvarez type with post couplers, where it is accelerated to 32 or 40 MeV. A high-energy beam transport from the accelerator to the lithium target must perform a variety of functions, complicated by the presence of strong space-charge forces within the beam. Very low beam losses along the accelerator and transport lines is a prerequisite to perform maintenance without requiring remote manipulators (no halo development).

Sources

The IFMIF ion injector must deliver sufficient current to the RFQ to achieve a 125mA RFQ output current. This implies that the ion source will have to produce an estimated 155mA D^+ of which 140 mA will be transported through the LEBT. The injector has to provide excellent beam quality (transverse emittance). In addition to high performance, the ion injector will also have to provide high operational availability. There are two possibilities for the IFMIF deuterium ion source. The first one is the ECR source of the type operated successfully at Chalk River, Los Alamos, and at Saclay for the IFMIF program. The second one is the RF or filament driven volume ion source initiated at LBL and under development for IFMIF at the Institute for Applied Physics (IAP) of the University of Frankfurt and JAERI. Ion source availability may be one of the limiting operational considerations for the IFMIF accelerators. Therefore, the ion source design will emphasize preventive maintenance, rapid change out of failed parts and rapid restoration of service.

Recently the ion sources showed significant progress. The CEA-Saclay ECR source was able to demonstrate 95 kV, 114 mA with 99.8% availability over 160 hours (about 1 spark per day). The ECR source has no intrinsic lifetime limitations, and is power efficient (RF window is located behind a bend and was never changed since 1996). The source worked also in pulse deuteron mode (2ms/s) in order to minimize the neutron production (d,D reaction) in the LEBT. A coherent set of measurements gives a D^+ current of 130 mA (@100 kV) with a D^+ fraction over 96%, LEBT transparency of 75 % and rms beam noise of 1.2 % (mainly 19 kHz lines coming from the magnetron power supply). Up to 170 mA (267 mA/cm^2) were extracted.

IAP Frankfurt Source reached also IFMIF design current. The volume source produced 200 mA protons (corresponding to 140 mA D^+) in cw mode as well as in pulsed mode (1 msec pulse length, 50 Hz repetition rate) with excellent beam quality and low noise.

JAERI is testing in parallel three candidate ion sources on the same test stand (same beam extraction system and instrumentation).

RFQs

The RFQs accelerate the beams from 95 keV to 5 MeV. They are four vanes structures to ensure steady operations (based on LEDA [5] already achieved performances and IPHI developments) and minimum power consumption compared to coaxial RFQs. To achieve this in a stable, tuneable cavity, the RFQ will be composed in three segments resonantly coupled (damping of field errors [6]). The transmission is about 97.9%. IAP-Frankfurt evaluates the 4-rod type RFQ. In both cases, the RFQs lengths are about 12.5m long, requiring 3 RF systems each.

DTLs

The IFMIF DTL design is based upon conventional Alvarez technology with post couplers for field stabilization. The 5 MeV RFQ output energy allows the use of conventional electromagnetic quadrupoles which may help restricting the halo development. They preserve the beam quality but at the expense of lower shunt impedance (bigger tubes). A FD focussing scheme is chosen. The phase advance per meter is conserved between RFQ output and DTL input. The DTLs consist in 5 tanks for a total length of 28.9m. They require the use of 9 RF tubes. A 4-cell hot model prototype is being built to test the cw, high current DTLs technological feasibility of quadrupole magnet design and fabrication, vacuum problems, cooling, mechanical aspects, etc. Two drift tubes incorporating two conventional quadrupole magnets have been designed by AES [7] and CEA-Saclay and built. They are of two different types. Tests will be done soon at CERN. IAP is exploring the less common IH DTL type of structure.

Linac sections were matched and multi-particle simulations were done from the first RFQ input to the linac exit. The input distribution is a 4D water-bag. Without errors, there is no emittance growth and no losses in the DTL. In order to evaluate the beam losses arising from the beam dynamics in high space charge conditions in a real linac, typical errors on linac element have been defined and simulated (the machining errors are different between the RFQ and the DTL). As a result, the minimum ratio between the bore radius and the 10^{-6} level is about 1.25.

RF system

The RF power system for the IFMIF accelerator assumes the use of tetrode or diacrode technology with an output power level of 1.0 MW and a frequency of 175 MHz. The same relatively low frequency in both the RFQ and DTL is a conservative approach for delivering the high current deuteron beam with low beam loss in the accelerator. This will facilitate the achievement of hands-on maintainability without remote manipulators. It also provides operational simplification.

IFMIF needs 12×2 1 MW RF station. Since at the time of the design no such source was existing, development and testing of a RF system was identified as the highest impact development item. As a consequence, a follow up of the progress in the world was organised. THALES diacrode TH628 tube is a candidate tube for IFMIF. This diacrode has delivered 1 MW cw @ 200 MHz. Monitoring of a long test (1000h, 1MW cw) is on-going with success.

High power beam diagnostics

IFMIF beam power is about 15 kW after the source, 625 kW after the RFQ and 5 MW in the HEBT after the DTL. Any interceptive diagnostics would immediately melt. New types of diagnostics mostly based on light analyses are developed [8]. As an example, the profile and position measurement based on Doppler effect emitted light can be mentioned.

ACKNOWLEDGMENTS

Thanks to many participants in many countries: Bob Jameson from USA, Horst Klein and his IAP-Frankfurt team, Masayoshi Sugimoto and his JAERI Japan team, AES team in US and CEA-Saclay team.

REFERENCES

1. IFMIF CDA final report, ENEA Frascati Report, RT/ERG/FUS/96/11 (1996) edited by M. Martone.
2 FZK Report, FZKA 6199 (1999) edited by A. Möslang.
3 A. Möslang et. al., "Suitability and feasibility of the International Fusion Materials Irradiation Facility (IFMIF) for fusion materials studies", Fusion Energy 1998 (Proc. of 17th Int. Conf. Yokohama, 1998) p.1203, IAEA, Vienna (1999)
4 JAERI-Tech 2000-014 (2000) compiled by M. Ida
5 H.V. Smith. and J. D. Schneider, "Status report on the low-energy demonstration accelerator (LEDA)", proceedings of Linac 2000, Monterey, USA, p581
6 L. Young, "Segmented Resonant Coupled RFQ", proceeding PAC 1993, Washington, D.C..
7 Advanced Energy Systems, Inc., "Conceptual design development of the 5 MeV drift tube for the IPHI drift tube linac - Phase I final report", April 23, 1999.
8 R. Ferdinand, "High current ECR source for protons and deuterons at Saclay", this conference.

RFQ Designed to Accept Beam from a Weak Focusing LEBT*

L. M. Young

Los Alamos National Laboratory, Los Alamos, New Mexico 87545

Abstract.
The LEDA RFQ is a 350-MHz continuous-wave (CW) radio-frequency quadrupole linac. LEDA was designed as the full power front-end prototype for the accelerator production of tritium (APT) linac. This machine has accelerated a 100-mA CW proton beam from 75 keV to 6.7 MeV. The 8-m-long RFQ accepts a dc, 75–keV, ~110-mA H^+ beam from the LEDA injector, bunches the beam, and accelerates it to full energy with ~94% transmission. Output beam power is 670 kW. This RFQ consists of four 2-meter-long RFQs joined with resonant coupling to form an 8-meter-long RFQ.

INTRODUCTION

The RFQ [1-5] receives a continuous stream of 75-keV protons from the H^+ injector, [6,7] forms it into bunches with a high capture efficiency (~ 94%), and then accelerates these bunches to an energy of 6.7 MeV. Figure 1 shows the coupled RFQ structure mounted in the tuning laboratory. Figure 2 shows the RFQ structure configuration including tapered RF power feeds, vacuum-port placement, and section nomenclature. Figure 3 shows a photograph of the completed RFQ assembly in the LEDA tunnel with the injector pulled back. The array of vacuum manifolds, water-cooling manifolds, and RF waveguide almost completely hides the accelerating structure.

Design features

With output energy of 6.7 MeV the LEDA RFQ [1,8] is the highest energy and highest power RFQ in the world [3, 5, 9-11]. The beam power is 670 kW when operated with the design-value 100-mA CW proton beam, making it the second-most powerful linear accelerator (after the LANSCE 800-MeV linac). Some of its unique design features are as follows:

- Transverse focusing strength at the RFQ entrance is reduced for easier beam injection. This allows placement of the final focusing solenoid in the low energy beam transport (LEBT) at the optimum distance from the RFQ for input matching.
- It employs resonant coupling [12,13] between the four 2-m-long segments, providing high RF field stability throughout the entire structure length.
- It has a significantly larger aperture and gap voltage in the accelerating section than previously designed RFQs at this frequency.
- Transverse focusing at the RFQ exit is reduced to match the focusing strength in the next accelerating structure.
- RF power from three 1-MW klystrons is coupled to the RFQ through six waveguide irises. The structure itself combines the RF power.

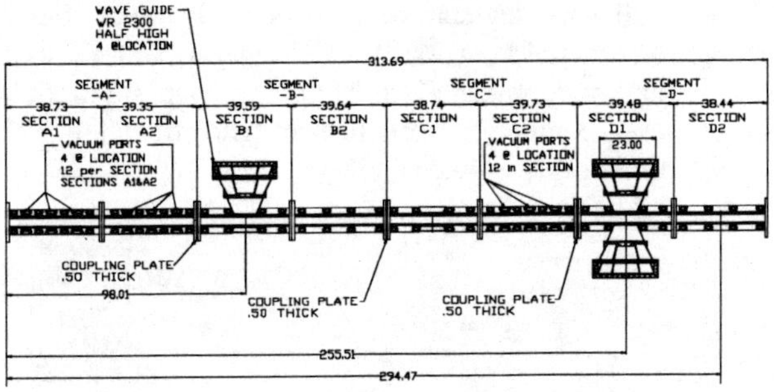

FIGURE 2: RFQ layout, showing RF feeds, vacuum ports, and segment nomenclature.

Resonant Coupling

In a typical RFQ that has constant focusing strength and constant gap voltage, as vane modulation increases to accelerate the beam, the aperture shrinks and beam can be lost on the vane tips. As the energy rises the cell length increases, and for a given modulation, the accelerating gradient decreases inversely with cell length. Since the maximum practical modulation is about 2, the RFQ would become very long if the gap voltage remained constant. To reduce beam loss and shorten the RFQ, we maintain a

Figure 1. Eight-meter-long RFQ in the tuning laboratory. The adjustable slug tuners can be seen in this picture.

* Work supported by the US DOE, Defense Programs.
† lyoung@lanl.gov

CP642, *High Intensity and High Brightness Hadron Beams: 20th ICFA Advanced Beam Dynamics Workshop on High Intensity and High Brightness Hadron Beams*, edited by W. Chou, Y. Mori, D. Neuffer, and J.-F. Ostiguy

large aperture, and increase the vane voltage to partially counter the decrease in transverse focusing as the vane modulation increases.

The increased gap voltage substantially increases the accelerating field, thus shortening the RFQ. However, even with this increased gap voltage, eight meters of length is required to accelerate the beam to 6.7 MeV. A conventional 8-m-long, 350-MHz RFQ would not be stable. Small perturbations would distort the field distribution intolerably [12,13]. Therefore, four 2-m-long RFQs (labeled A, B, C and D in Figure 2) are resonantly coupled to form the 8 m-long LEDA RFQ. This is implemented by separating the four 2-m RFQs by coupling plates. An axial hole in the coupling plates allows the vane tips to nearly touch. The capacitance between the vane tips of one RFQ and the next provides the RF coupling between the 2-m-long segments. The gap between the vane tips at the coupling joint is 0.32 cm. To minimize the effect of this gap on the beam, the gap position corresponds to a zero crossing of the RF electric field when the bunch passes the gap. The RF field is in phase in all four segments. The "coupling mode" has a strong electric field across the 0.32-cm gap and has one longitudinal node in each 2-m RFQ. The coupling mode's longitudinal component of electric field transmits RF power, providing the field stability.

FIGURE 3. A photograph of the entrance end of the RFQ mounted in the support/alignment frame. The accelerating structure is buried within the array of water-cooling manifolds, vacuum manifolds, and RF waveguide.

RFQ ELECTROMAGNETIC DESIGN

The RFQ was designed with the code PARMTEQM (Phase and Radial Motion in Transverse Electric Quadrupole; Multipoles) [14]. The code includes the effect of higher-order multipoles in the RFQ fields, which are important in accurately predicting beam loss. The earlier code version, PARMTEQ, used only the first two terms in the expansion of the RFQ fields. PARMTEQM uses the first eight terms. In addition, a realistic description of the input beam is required to accurately simulate beam losses in the RFQ. Simulations of the beam transport through the LEBT [15] with PARMELA [16,17] produce a more realistic distribution of particles for input into the RFQ codes than the ideal input distributions.

RFQ Entrance

To implement the reduced focusing strength at the entrance of the RFQ and have adequate focusing in the interior of the RFQ, the transverse focusing parameter increases smoothly from 3.088 to 6.981 over the first 32 cm of the RFQ. The focusing parameter is proportional to V/r_0^2 where V is the voltage between adjacent vane tips and r_0 is the average aperture. The voltage is held constant in this region and the aperture is reduced to increase the focusing parameter. On entry, the beam is not yet bunched, allowing the use of weak transverse focusing. By the time the beam starts to bunch, the focusing is strong enough to confine the bunched beam.

The low focusing strength at the RFQ entrance means that the matched beam size is relatively large, allowing some space, as shown in figure 4, between the second LEBT solenoid and the RFQ entrance. Without this feature, the solenoid would be right at the RFQ entrance.

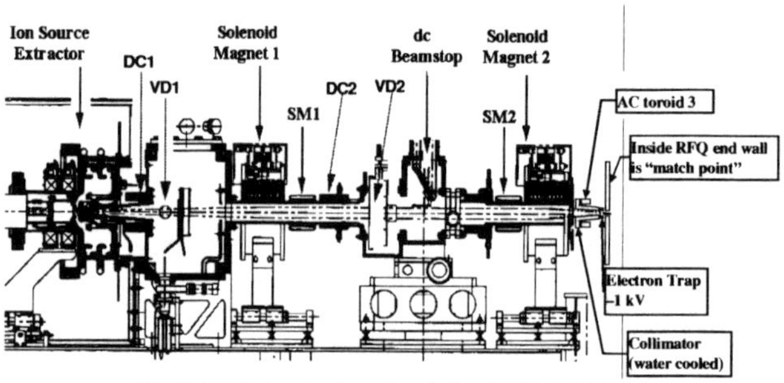

FIGURE 4. A sketch of the LEDA LEBT.

TRANSMISSION THROUGH THE RFQ

The code PARMELA simulates the LEBT beam with 95% space-charge neutralization [18]. These simulations showed that proper matching would be possible with an electron trap at the RFQ entrance, and solenoid-to-RFQ distance of 15 cm. The electron trap is a metal ring at the entrance of the RFQ. A ring voltage of –1 kV blocks low-energy plasma electrons, but does not affect the 75-keV protons. The electron trap performs two essential functions. One, it improves the space charge neutralization in the LEBT. Two, it prevents electrons from streaming into the RFQ through the torrid and corrupting the measurement of the beam current.

Using the simulated beam, two RFQ codes predict 93% transmission with the RFQ operating at design field levels. The codes are PARMTEQM and TOUTITIS [19] that use respectively 2D and 3D space charge effects. The measured transmission has been as high as 94% at 100 mA when the RFQ fields are 10% above the design field strength.

POSSIBLE ION TRAPPING IN RFQ

Figure 5 shows the time dependence of RFQ transmission in a 300-μs-long beam pulse with the RFQ fields at the nominal design value. At about 150 μs into the pulse, the transmission suddenly drops by about 10%. As the RFQ field is increased above the design value, transmission remains high for increasingly longer periods.

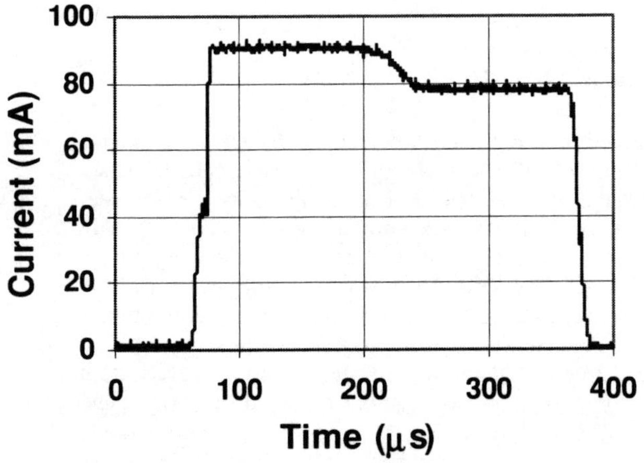

FIGURE 5. RFQ output beam current vs. time for a 300-ms-long pulse at ~97% of the design RF field level.

With fields ≥ 105% of design, the transmission drop is no longer observed, even for long pulses and CW operation. Along with the transmission drop, higher-than-expected activation is measured at the high-energy end of the RFQ, indicating significant beam loss at that location. Operating the RFQ with fields about 10% above the design value greatly reduces the magnitude of this beam loss. More work is needed to determine unambiguously the cause of the time-dependent transmission anomaly. At present, considerable evidence points to the possibility that it is caused by low-energy H^+ ions trapped near the axis by the RFQ fields [20]. The extra positive charge from the trapped ions causes the beam size to increase, reducing the RFQ transmission, and also increasing the beam size at the end of the HEBT. This hypothesis is consistent with the observation that the collimator ring in front of the beam stop glows, presumably from being struck by incident beam, when the RFQ pressure exceeds 1-2 x10^{-7} Torr. The low-energy H^+ ions can be produced by beam collisions with background gas (H_2) near the RFQ axis, or by beam collision with the vane tip surfaces. At fields ≤ the design value, the beam may be sufficiently large that its fringes strike the RFQ vane tips, creating H^+ ions that get trapped temporarily in the beam channel. As the trapped charge accumulates, the beam becomes larger still, until the transmission drops suddenly.

REFERENCES

1 D. Schrage et al., "CW RFQ Fabrication and Engineering," Proc. LINAC98 (Chicago, 24-28 August 1998) pp. 679-683.

2 L. M. Young et al., "High Power Operations of LEDA," Proc. LINAC2000 (Monterey, 21-25 August 2000) pp. 336-340.

3 L. M. Young et al., "Low-Energy Demonstration Accelerator (LEDA) Radio-Frequency Quadrupole (RFQ) Results," Proc. 2nd ICFA Advanced Accelerator Workshop on the Physics of High-Brightness Beams (Los Angeles, CA) pp. 488-503.

4 H. V. Smith, Jr. and J. D. Schneider, "Status Report on the Low-Energy Demonstration Accelerator (LEDA)," Proc. LINAC2000 (Monterey CA, 21-25 August 2000) pp. 581-3.

5 L. J. Rybarcyk et al., "LEDA Beam Operations Milestone and Observed Beam Transmission Characteristics," Proc. LINAC2000 (Monterey CA, 21-25 August 2000) pp. 584-6.

6 T. Zaugg, C. Rose, J. D. Schneider, J. Sherman, R. Stevens, Jr., "Operation of a Proton Microwave Source in Pulsed Mode," Proc. LINAC98 (Chicago, 24-28 August 1998) pp. 893-95.

7 J. D. Sherman et al., "Status Report on a dc 130-mA, 75-keV Proton Injector," Rev. Sci. Instrum. 69 (1998) 1003-8.

8 J. D. Schneider, "Operation of the Low-Energy Demonstration Accelerator: the Proton Injector for APT," Proc. PAC99 (New York, 29 March - 2 April 1999) pp. 503-507.

9 H. V. Smith, Jr. et al., "Update on the Commissioning of the Low–Energy Demonstration Accelerator (LEDA) Radio–Frequency Quadrupole (RFQ)," Proc. 2nd ICFA Advanced Accelerator Workshop on the Physics of High-Brightness Beams (Los Angeles, CA) pp. 466-479.

10 D. E. Rees et al., "Design, Operation, and Test Results of 350 MHz LEDA RF System," Proc. LINAC98 (Chicago, 24-28 August 1998) pp. 564- 566.

11 K. F. Johnson et al., "Commissioning of the Low- Energy Demonstration Accelerator (LEDA) Radio- Frequency Quadrupole (RFQ)," Proc. PAC99 (New York, 29 March - 2 April 1999) pp. 3528-3530.

12 M. J. Browman and L. M. Young, "Coupled Radio- Frequency Quadrupoles as Compensated Structures," Proc. of the 1990 Linear Accelerator Conference, (Albuquerque, 10-14 Sept. 1990) LA-12004-C, pp. 70-2.

13 L. M. Young, "An 8-meter-long Coupled Cavity RFQ Linac," Proc. 1994 Int. Linac Conf., (Tsukuba, 21-26 Aug. 1994) pp. 178-180.

14 K. R. Crandall et al., "RFQ Design Codes," Los Alamos National Laboratory report LA-UR-96-1836 (Revised August 21, 1998).

15 H. V. Smith, Jr. et al., "Comparison of Simulations with Measurements for the LEDA LEBT H+ Beam," Proc. PAC99 (New York, 29 March - 2 April 1999) pp. 1929-31.

16 L. M. Young, "PARMELA," Los Alamos National Laboratory report LA-UR-96-1835 (Revised February 27, 2001).

17 L. M. Young, "Simulations of the LEDA LEBT With H+, H2+, and e- Particles," Proc. of the 1997 Particle Accelerator Conference (Vancouver, 12-16 May 1997) pp. 2749-2751.

18 Robin Ferdinand Joseph Sherman, Ralph R. Stevens, Jr., Thomas Zaugg, "Space-Charge Neutralization Measurement of a 75-Kev, 130-mA Hydrogen-Ion Beam," Proc. PAC97 (Vancouver, B.C., Canada 12-16 May 1997) pp. 2723-2725.

19 R. Duperrier, R. Ferdinand, J-M, Lagniel, N Pichoff, "Toutatis, The CEA-SACLAY RFQ Code," Proc. LINAC2000 (Monterey, 21-25 August 2000).

20 M. S. deJong, "Background Ion Trapping in RFQs," Proc. 1984 Linac Conf. (Seeheim, Germany, 7-11 May 1984), pp. 88-90.

Wide Dynamic-Range Beam-Profile Instrumentation for a Beam-Halo Measurement: Description and Operation•

J. Douglas Gilpatrick*, D. Barr*, L. A. Day*, D. M. Kerstiens*, J. F. O'Hara*, M. Stettler*, R. Valdiviez*, M. Gruchalla#, J. H. Kamperschroer!

*Los Alamos National Laboratory, MS H808, LANL, Los Alamos, NM, 87545
Honeywell FM&T/NM, PO Box 5250, Albuquerque, NM, 87185
! General Atomics, Los Alamos, NM, 87544

Abstract. Within the halo experiment conducted at the Low Energy Demonstration Accelerator (LEDA) at Los Alamos National Laboratory, specific beam instruments that acquire horizontally and vertically projected particle-density distributions out to approximately 10^5:1 dynamic range are located throughout the 52-magnet halo lattice. We measure the core of the distributions using traditional wire scanners, and the tails of the distribution using water-cooled graphite scraping devices. The wire scanner and halo scrapers are mounted on the same moving frame whose location is controlled with stepper motors. A sequence within the Experimental Physics and Industrial Control System (EPICS) software communicates with a National Instruments LabVIEW virtual instrument to control the motion and location of the scanner/scraper assembly. Secondary electrons from the wire scanner 0.033-mm carbon wire and protons impinging on the scraper are both detected with a lossy-integrator electronic circuit. Algorithms implemented within EPICS and in Research System's Interactive Data Language subroutines analyze and plot the acquired distributions. This paper describes this beam profile instrument and describes our experience with its operation.

INTRODUCTION

At LEDA a 100-mA, 6.7-MeV beam is injected into a 52-quadrupole magnet lattice (see Fig. 1). Within this 11-m FODO lattice, there are nine wire scanner/halo scraper (WS/HS) stations, five pairs of steering magnets and beam position monitors, five loss monitors, and three pulsed-beam current monitors [1]. The WS/HS instrument's purpose is to measure the beam's transverse projected distribution. These measured distributions must have sufficient detail to understand beam halo resulting from upstream lattice mismatches [2,3]. The first WS/HS station, located after the fourth quadrupole magnet, verifies the beam's transverse characteristics after the RFQ exit. A cluster of four WS/HS located after magnets #20, #22, #24, and #26 provides phase space information after the beam has debunched. After magnets #45, #47, #49, and #51 reside the final four WS/HS stations. These four WS/HS acquire projected beam distributions under both matched and mismatched conditions. These conditions are generated by adjusting the first-four quadrupole magnetic fields so that the RFQ output beam is matched or mismatched in a known fashion to the rest of the lattice. Because the halo takes many lattice periods to fully develop, this final cluster of WS/HS are positioned to be most sensitive to halo generation.

FIGURE 1. The 11-m, 52-magnet FODO lattice includes nine WS/HS stations that measure the beam's transverse projected distributions.

As the RFQ output beam is mismatched to the lattice, the WS/HS actually observe a variety of distortions to a properly matched Gaussian-like distribution [2,3]. These distortions appear as distribution tails or backgrounds. It is the size, shape, and extent of these tails that predict specific types of halo. However, not every lattice WS/HS observes the halo generated in phase space because the resultant distribution tails may be hidden from the projection's view. Therefore, multiple WS/HS are used to observe the various distribution tails.

• Work supported by the US Department of Energy

CP642, *High Intensity and High Brightness Hadron Beams: 20th ICFA Advanced Beam Dynamics Workshop on High Intensity and High Brightness Hadron Beams*, edited by W. Chou, Y. Mori, D. Neuffer, and J.-F. Ostiguy

WS/HS DESCRIPTION

Each station consists of a horizontal and vertical actuator assembly (see Fig. 2) that can move a 33-μm-carbon monofilament and two graphite/copper scraper sub-assemblies. The carbon wire and scrapers are connected to the same movable frame. Attached to this movable frame is a linear encoder that provides the wire and scraper edges' relative position to within a typical rms error of 5 μm, and an additional linear potentiometer provides an absolute approximate position for LEDA's run permit systems. A stepper motor coupled to a ball lead screw is used to drive the moveable frame. A motor brake along with microswitches limit the frame's movement.

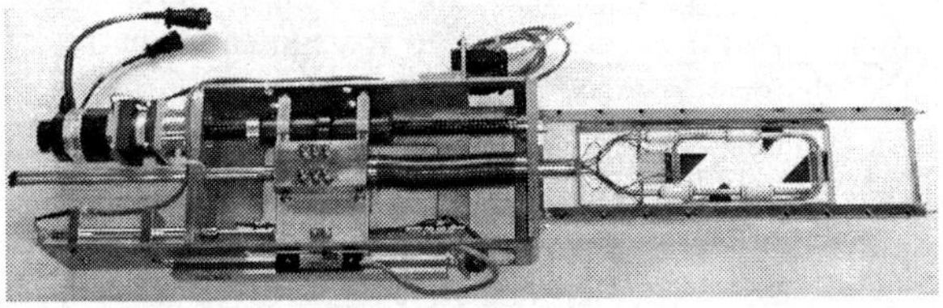

FIGURE 2. The WS/HS assembly contains a movable frame on which a 0.033-mm carbon wire resides between two water-cooled graphite scrapers.

The carbon wire, which senses the beam's core, is cooled by thermal radiation. If the beam macropulse is too long, the wire temperature continues above 1800 K resulting in the onset of thermionic emission. Thermionic emission causes an inaccurate appearance to the distribution by exaggerating the core's current density. To eliminate these effects for the halo experiment, the maximum pulse length and repetition rate is limited to approximately 30 μs and 1 Hz, respectively.

The halo scrapers are composed of a 1.5-mm thick graphite plate brazed to a water-cooled 1.5-mm thick copper plate. Since 6.7-MeV protons average range in carbon is approximate 0.3 mm, the beam is completely stopped within the graphite plate. Cooling via conduction lowers the average temperature of the scraper sub-assembly and allows the scraper to be cooled more rapidly than the wire. The lower average temperature and faster cooling allows the scraper to be driven in as far as 2 rms widths from the beam distribution peak without the peak temperature increasing above 1800 K.

The movement and positioning of each wire and scraper pair is controlled by a motion control system that contains a stepper motor, stepper motor controller, a linear encoder, and an electronic driver amplifier. The controller's digital PID loop controls the speed and accuracy at which the assembly is moved and placed.

The target position, as defined by the WS/HS operator, is relayed from the EPICS control screen via a database process variable to a National Instruments LabVIEW Virtual Instrument (VI). The VI also calibrates the relative position of the linear encoders based on the measured position of the limit switches, and provides some error feedback information. The total error between the target wire position and the actual wire position attained is within a total 4% range of a typical 1-mm rms-width beam.

As the wire is moved through the beam, it senses the projected beam core distribution. A small portion of the beam's energy is imparted to the wire causing secondary electron emission to occur. The secondary electrons leaving the wire are replaced by negative charge flowing from the electronics. This current flow for both axes is connected through a bias battery to an electronic lossy integrator circuit and followed by an amplification stage.

The integrator capacitance and amplifier gain are set to allow a very wide range of values of accumulated charge. Data are acquired by digitizing the accumulated charge through the lossy integrator at two different times within the beam pulse. This charge difference, acquired by subtracting the two values of charge, provides a low noise method of relative beam charge acquisition. The wire and scraper accumulated charge signals are digitized using 12- and 14-bit digitizers, respectively. The analog noise floor has been measured to be 0.03 pC, a noise level slightly lower than the scraper digital LSB noise level of 0.15 pC using the highest gain settings within the detection electronics.

The front-end electronic circuitry, mounted on a daughter printed circuit board, is connected to a motherboard that has all of the necessary interface electronics to communicate with EPICS via a controller module within the same electronics crate. A software state machine sequence was written within EPICS to control and operate WS/HS instrumentation. The state machine instructs the VI to move the wire and scraper to a specific location, acquire synchronous distribution data from either the wire or scraper, trigger the IDL routine to normalize the acquired charge with a nearby toroidal current measurement, graph the normalized data, and write the distribution to a file. The sequence also instructs IDL to calculate the first through fourth moments, fit a Gaussian distribution to the wire scanner data, and calculate the point at which the beam distribution disappears into the background noise.

To plot the complete beam distribution for each axis, the wire scanner and two scraper data sets must be joined. To accomplish this joining, several analysis tasks are performed on the wire and scraper data including,, scraper data are spatially differentiated and averaged, wire and scraper data are acquired with sufficient spatial overlap, and differentiated scraper data are normalized to the wire beam core data.

The scraper data need only be normalized in the relative charge axis since the distances between each wire and scraper edge are known to within 0.25-mm. In addition, the first four moments and the point at which the beam distribution disappears into the noise are also calculated for the combined distribution data.

ACQUIRED DISTRIBUTIONS

Fig. 3 shows data from WS/HS #26 with some slight mismatch generated by increasing the field above nominal by 5% in the first matching quadrupole magnet. These typical WS/HS profiles show distributions with a dynamic range of ~ 10^5:1 and provide distribution information to > 5X to 7X times typical rms widths of the beam. The calculated rms widths are 1.10 and 1.13 mm for the horizontal and vertical distributions, respectively.

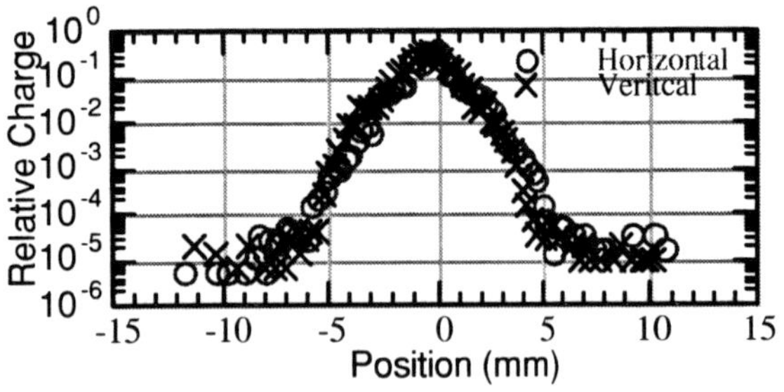

FIGURE 3. WS/HS distributions, such as #26 shown here, have a typical dynamic range of > 10^5:1.

WIRE AND SCRAPER PHYSICS

The WS wire is biased negative to optimize secondary emission (S.E.) yield, where the yield is defined as the ratio of the emitted secondary electron current and the proton beam current intercepted by the wire. All of the wires in the halo lattice WS are configured with a 33-μm, carbon monofilament. The HEBT WS is configured with a 100-μm SiC wire. The choice of bias potential was determined by measuring the wire and scraper currents as a function of bias potential. The resulting data showed that the wire is optimally biased at -6 to -12 V and the scraper is optimally biased at +20 to +30 V [4].

As the wire bias is positively increased from 0 V to > +100 V, the wire secondary electron emission is inhibited and the net wire current reduces to very near zero. As expected, a large positive bias reduces the wire detection signal. Furthermore, it appears that the wire collects positive ions with < -25 V bias potentials well after the beam pulse. This ion collection additionally limits the amount of negative bias that is applied to the wire for proper secondary emission operation.

The scraper detection goal is to inhibit secondary emission and detect only 6.7-MeV protons. With approximately +25 V bias applied to the scraper, the secondary emission is almost entirely inhibited and the net current reduces to the nominal proton current.

SUMMARY

A wire scanner and halo scraper have been integrated into a beam profile instrument capable of 10^5:1 dynamic range. This WS/HS combination was used extensively to acquire wide dynamic rage data in order to understand beam halo generation. The scanner and scraper V-I curves show that the wire and scraper are optimally biased at –12 V and +25 V, respectively.

REFERENCES

1. J. D. Gilpatrick, et al., "Experience with the Low Energy Demonstration Accelerator (LEDA) Halo Experiment Beam Instrumentation," *Proceedings of the 2001 Particle Accelerator Conference*, June 18-22, 2001, pp.2311-2313.
2. T. Wangler, et al., "Linac Based Proton Drivers," this workshop.
3. P. L. Colestock, et al., "Measurement of a Beam Halo Generation in an Intense Proton Beam," this workshop.
4. J. D. Gilpatrick, et al., "Biasing Wire Scanners and Halo Scrapers for Measuring 6.7-MeV Proton-Beam Halo," *Proceedings of the 2002 Beam Instrumentation Workshop* held at Brookhaven National Laboratory, on May 6-9, 2002.

Sources of Halo and Beam Losses in a High Intensity Hadron Linac

Nicolas Pichoff[1]

CEA-DSM/DAPNIA/SACM, 91191 Gif-sur-Yvette cedex

Abstract. Different possible sources of emittance growth, halo formation and eventually beam losses in a high intensity hadron linac are listed.

INTRODUCTION

Problem of beam losses are of main interest in high power protons or H^- linacs. They can activate the structure, complicating considerably the maintenance of the machine. This is a crucial point for many projects like SNS, ESS, JHF, SPL, APT ... running with protons and/or H^- ions. Last years, a big effort has been done in a large number of laboratories to investigate to possible causes of emittance growth and halo formation able to cause beam losses. The main sources of halo formation in a high beam-intensity linacs are listed in this paper. I apologize if I did not quote some of them; it is either because I consider that they are probably non-significant, or I simply forgot them. One can add that the only way to identify surely the main contributor of beam losses is the full scale experiment.

BEAM MISMATCH

The main cause of emittance growth is known as being the beam mismatch. A beam is matched when the beam phase-space distribution has the same period as the external focusing force. In a mismatch beam, the non-linear forces, especially the space-charge force, are responsible of this emittance growth through 2 phenomena: The beam filamentation and the resonant interaction between particle individual motion and beam core oscillation.

Beam Filamentation

In linear forces, particles are moving in the phase-space around ellipses, associated to the focusing channel, defined by the Courant-Snyder parameters. All the particles have the same phase advance per meter, i.e. they rotate around the concentric ellipses with the same angular speed. When the force is non-linear, these ellipses transform into close curves corresponding to curves on with the motion Hamiltonian is constant. Moreover, the particles do not rotate with the same speed: There is a spread in particles phase-advance. In the presence of space-charge, the force in the beam core is "compensated" by the space-charge force reducing the particle phase advances. Core particles are rotating more slowly than these of the halo. When the beam is matched, particles leaving one area of the phase-space are replaced by others with the same Hamiltonian: the emittance pattern stays the same, no rms emittance growth is observed. When the beam is mismatched, particles leaving a region are not replaced. The emittance pattern is changed and rms emittance growth is observed (see. FIGURE 1). Particle amplitude is not amplified by more than the beam mismatch factor.

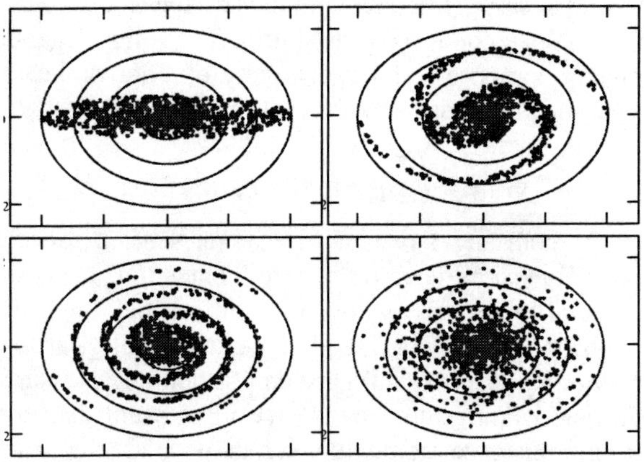

FIGURE 1. (x, x') phase-space plot of a mismatched beam in non-linear forces. External particles are rotating faster than core particles. Apparent emittance growth occurs.

Parametric resonance with core

The core oscillation of a matched beam is periodic with the same period as the external force. In these conditions, avoiding a parametric resonance consists mainly in keeping the zero current phase-advance lower than 90° (avoiding the strong fourth order resonance). Nevertheless, the core oscillation of a mismatched bunched beam is modulated by 3 other

[1] nicolas.pichoff@cea.fr

CP642, *High Intensity and High Brightness Hadron Beams: 20th ICFA Advanced Beam Dynamics Workshop on High Intensity and High Brightness Hadron Beams*, edited by W. Chou, Y. Mori, D. Neuffer, and J.-F. Ostiguy

lower frequency modes (mismatched modes [1]). Unfortunately and unavoidably, some beam particles have oscillation frequencies twice at least one mismatched mode. Second order parametric resonance with the core can give large amplitude to the concerned particles [2].

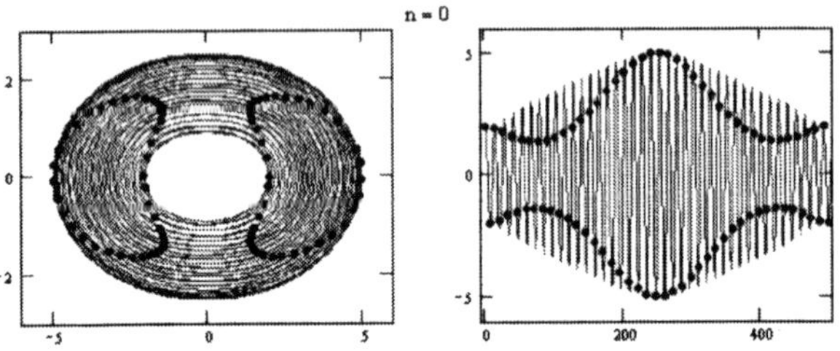

FIGURE 2. Trajectory (in red) of a particle in a second order parametric resonance with a mismatched mode. In blue, the particle position in a stroboscopic (Poincaré) plot with the mode periodicity. Left: (x, x') phase-space plot, Right: (z, x) plot.

INTERACTION WITH RESIDUAL GAS

In a hadron linac linac the residual gas pressure varies from 10^{-5} hPa (close to the source) to 10^{-7}-10^{-8} hPa in copper structures and can be lower than 10^{-11} hPa in superconducting structures. The residual gas is mainly composed of large variety of light or heavy gas. The partial pressure for each variety is important.

Scattering on Residual Gas

By colliding a residual gas atom, a beam particle can gain a significant transverse momentum. The gain probability is given by the differential nuclear-collision cross-section and depends on the particle energy and the residual gas composition and pressure. Particle gaining more transverse momentum than the beam transverse momentum spread populate the beam halo. If the momentum gain is sufficiently large, the particle can be directly (after a quarter of betatron period) lost. Beam tail at a given position of a linac can be easily estimated (FIGURE 3, from [3]). They are generally not dense but populate far halo.

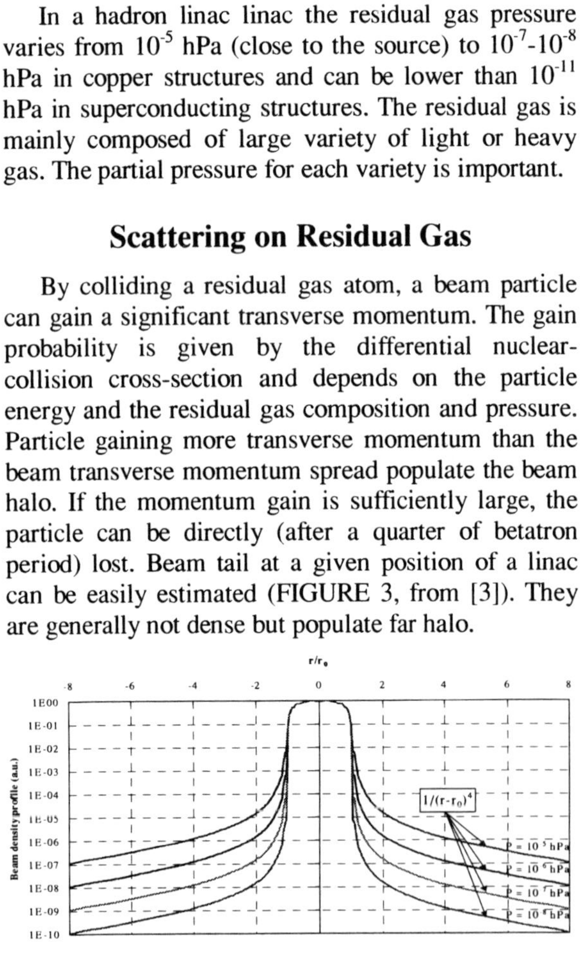

FIGURE 3. Beam profile at the end of a 20m channel for different Nitrogen gas pressure.

Charge Exchange with Residual Gas

Some particle (like H^-) can be stripped on residual gas. From the stripping cross section a rate of beam loss can be estimated depending on the gas composition, pressure and the beam energy (FIGURE 4, from [4]).

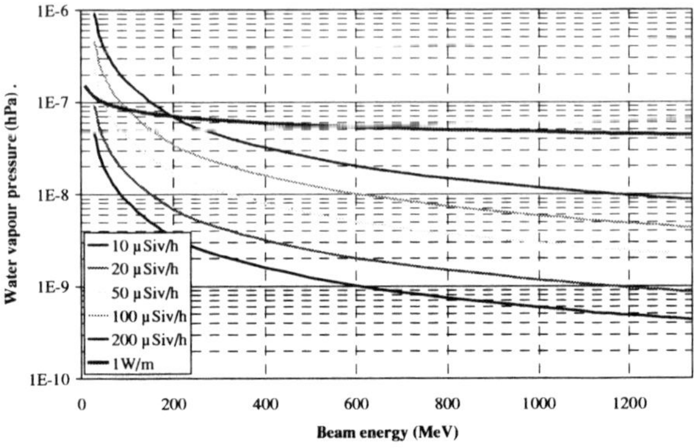

FIGURE 4. Pressure limit for a given structure activation (at 1 feet after 3 hours).

Transient Space-Charge Compensation

At low energy, because of the high ionization cross section combined with the relatively high gas pressure, the beam captures particles with opposite charge leading to a space-charge compensation regime. The effect of this compensation is benefic as, at stationary regime, it reduces the space-charge force non linearity and the induced emittance growth. Nevertheless, the main problem arises from the transient regime in which the space-charge force varies, inducing a transient mismatch of the beam in the channel. Order of magnitude of the transient time is about 10 μs, negligible in case of cw beam but concerning 1% of a 1ms pulse! This phenomena has to be study carefully to estimate the induced emittance growth and beam losses.

EMITTANCE EXCHANGE

The space-charge force induces a coupling between transverse and longitudinal directions (i.e. the longitudinal force depends on the transverse position, and the transverse force depends on the longitudinal position). This coupling induces an exchange between the particles transverse and longitudinal energies for equal phase-advances [5]. This could be dangerous as it changed the matching conditions (which depends on the beam emittance in space-charge driven beam) and it can transfer particles from the longitudinal halo to the transverse one.

MAGNETIC STRIPPING

An electron can be stripped from a H^- ion in a magnetic field through an electro-quantum effect. The probability per unit length to strip a H^- ion in a magnetic field B is:

$$P_{m^{-1}} = 9.53 \cdot 10^5 \cdot B_T \cdot \exp\left(-\frac{16.4}{\beta\gamma \cdot B_T}\right) \quad (1)$$

This effect is more important at high energy and depends exponentially on the magnetic field. As a consequence, this limits the maximum magnetic field and then quadrupole gradient at high energy.

INTRABEAM SCATTERING

The Coulomb collisions in beam induce energy transfer between particles and directions. The effect of this phenomenon has been estimated in high power protons linacs in [6] and [7]. Because of the long time-scale of this process compared to the beam life-time in linac, its effect is marginal (about 10^{-11} of a 100 mA beam in halo). Nevertheless, the effect varies with the square of the beam current and the induced halo extension can be large if the beam is far from energy equipartition.

TRANSPORT ERRORS

A real accelerator is made of components that can be badly aligned, powered, calculated or measured. Some of these errors can be corrected using an appropriate diagnostic-correction scheme, some cannot. Multipolar terms in quadrupoles or cavities contribute to the beam filamentation but generally at second order compared to space-charge force. Elements misalignment induces mainly a beam c.o.g. transverse displacement, increasing the probability to find a particle far from the beam axis. This error can be compensated with steerers minimizing the beam displacement in BPMs. Unfortunately fast errors like vibrations cannot be easily compensated. The errors on quadrupole gradients are even more difficult to compensate. They induce residual mismatch that could be estimated and corrected only from beam size measurements, not easy with such high power beam. The last important problem is the limited control of the RF phase and field in the cavity. It induces c.o.g. displacement in longitudinal phase-space, including longitudinal emittance growth and displacement of the beam final energy and phase.

An estimation of the effect of these errors has to be calculated in the linac design step. The efficiency of the correction scheme has to be evaluated (as example, for ESS [8]).

CONCLUSION

Many phenomena can be responsible of emittance growth, halo formation and beam losses in a linac. To minimize the risk, precautions have to be taken already from the design phase:

- keep both transverse and longitudinal phase advances below 90°, avoid the same transverse and longitudinal phase advances,
- minimize the number of transitions, make them smooth,
- take care to the residual gas,
- Estimate the safe element tolerances and associated correction scheme,
- Avoid reduction of linac acceptance where localized loses can take place,
- Estimate the mismatch of the pulse front-end and, if needed, chop it at low energy.

REFERENCES

1. Pichoff, N., "Envelope Modes of a Mismatched Bunched Beam" , Internal note, DAPNIA/SEA 98/44, 1998.
2. Lagniel, J.-M., *N.I.M.* **A346**, 46-53 (1994).
3. Pichoff, N. et al. *Particle Accelerators,Vol. 63, pp211-233.*
4. Pichoff, N. et al., "Beam losses in ESS from H^- stripping on residual gas", ESS Linac Technical Note ESSLIN-TN-0202-01, 2001.
5. Hoffman, I., "Emittance Coupling in high Intensity beam applied to the SNS linac", in Particle Accelerator Conference, edited by P. Lucas et S. Webber, Chicago, 2001, pp. 2902-2904.
6. Fedotov, A.V., Gluckstern, R.L., "Coulomb Scattering within a Spherical Beam Bunch in High Current Linear Accelerators", in Particle Accelerator Conference, edited by A. Luccio et W. McKay, New-York, 1999, pp. 1755-1757.
7. Pichoff, N., "Intrabeam scattering on Halo Formation", in Particle Accelerator Conference, edited by A. Luccio et W. McKay, New-York, 1999, pp. 3277-3279.
8. The ESS Project, Vol. III – Technical Report, Edited by F.H. Bohn et al., 2002. http://essnts.ess.kfa-juelich.de/

WORKING GROUP III

BEAMLINES AND TARGETS

Summary Report

Craig Moore and Alberto Marchionni

Fermilab, Batavia, IL, USA

INTRODUCTION

How to handle intense beams safely and how to target them safely were the main foci of this working group. The talks were evenly divided between the two subjects.

HIGH INTENSITY BEAMLINES

Beam transfer lines for the SNS
(D. Raparia, BNL)

The SNS talk was concerned with the physics and diagnostic capability of the beamlines into and out of the Main Ring. The injection line (HEBT) is more than a transfer line with very good diagnostic capability in both longitudinal and transverse dimensions. The line also has the capability of cleaning the beam in both longitudinal and transverse dimensions. The line to the target (RTBT) is noteworthy for its immunity to kicker failure; for one or even two failures the beam motion on target is minimal.

Sophisticated beam permit system
(R. Ducar, Fermilab)
Radiation protection utilizing electronic berms
(John Anderson, Fermilab)
Automatic beam line tuning (AUTOTUNE)
(T. Kobilarcik, Fermilab)

At Fermilab the upcoming neutrino experiments will deliver more intensity in a single year than was delivered in the seventeen years of running the Fixed Target Program. In MiniBooNE's case the beam will be almost continuous (18,000 pulses/hour). The intensity and nature of these beams have led to the development of control mechanisms for the care and handling of these high intensity beams. To begin with a "Sophisticated Beam Permit System" will check a large number of status bits and analogue information before permitting a pulse of beam. Then an automate intensity checking system will determine how much beam was lost from the beginning of the line to the end of the line (electronic berm system). Finally an automated tuning program (Autotune) will find and correct minor beam wandering before there are sufficient losses that the electronic berm system would stop operation.

Tails of beam distribution
(A. Marchionni, Fermilab)

The NuMI project is designed to handle an intensity of up to 4×10^{13} protons delivered in 8 (s pulses every 1.9 s, corresponding to a beam power of 0.4 MW.

For the NuMI beam line the normal concern about ground water activation is exacerbated due to the fact that the sloping beamline traverses the aquifer.Groundwater protection considerations severely constrain the allowable continuous proton losses. Fractional proton losses below 10^{-4} are required in the sloped transport region, with a tighter limit of about 10^{-6} in a limited 140 ft region at the interface between two geological formations. The primary proton beamline has been carefully designed to take into account these constraints, with input from recent measurements of transverse and longitudinal emittances of the Main Injector beam. A measurement program of beam tails is presently ongoing.

HIGH INTENSITY TARGETS

Mercury target
(K. McDonald, Princeton U.)

Neutrino factories demand targets capable of at least a few MW beam power, optimized for the production of soft pions (50 - 500 MeV/c). A free mercury jet target located inside a solenoidal magnetic field, with a field adiabatically falling off along the axis to provide focusing of the produced pions, is a promising design for 4 MW of beam power. The goal of BNL experiment E951 is to test key components of this design in realistic single pulse beam conditions. In particular the viability of a liquid metal jet target in intense, short proton pulses and in strong magnetic fields has to be investigated.

Simulations of jet distortions due to beam energy deposition and/or eddy currents have been performed. Experimental studies with 2×10^{12} protons on a Hg jet target of 1 cm diameter show no destructive effects. Com-

CP642, *High Intensity and High Brightness Hadron Beams: 20th ICFA Advanced Beam Dynamics Workshop on High Intensity and High Brightness Hadron Beams*, edited by W. Chou, Y. Mori, D. Neuffer, and J.-F. Ostiguy

bined tests of a Hg jet in a proton beam and in a 20 T pulsed magnet are being planned.

Target development for the SINQ high-power neutron spallation source
(W. Wagner, PSI)
Carbon and beryllium targets and beam windows at PSI
(Gerd Heidenreich, PSI)

PSI has an ongoing program of neutron spallation operation and a target development program for higher power. The SINQ facility has a layout for vertical targeting. This is used for the present Target Irradiation Program but in fact was set up for the proposed liquid metal spallation targets in order to utilize gravity for flow enhancement purposes. The MEGAPIE target is predicted to have a greatly enhanced neutron flux. The present solid targeting arrangement allows many different samples to be irradiated at the same time.

The beamline from Target-E to SINQ (or the dump) is very cleverly designed from a maintenance point of view. The vertical access and in particular the inflatable seal assembly are very important concepts. The rotating targets are well designed to handle the pattern of power deposition.

NuMI target
(James Hylen, Fermilab)

The NuMI target has been designed for reliable operation for at least 1 year ($\sim$ 107 beam pulses) and optimized for neutrino event rate. It has to withstand a 0.4 MW primary proton beam with a spot size of $\sigma_x\sigma_y = 0.7 \times 1.4\ \mathrm{mm}^2$ and must be capable of handling dynamic thermal stresses due to single-turn extraction. Graphite grade ZXF-5Q from Poco Graphite, Inc. is the chosen target material. A prototype target has been exposed to an accumulated dose of about 2×10^{18} protons/mm^2, corresponding to about 5in the NuMI operation have been induced in the prototype target by the use of a higher flux beam, with 10^{13} protons/pulse over a spot size of $\sigma_x\sigma_y = 0.2 \times 0.2\ \mathrm{mm}^2$. No visible damage to the target segments was detectable after this irradiation test.

Targetry issues for the Fermilab 2 MW Neutrino superbeam
(M. Kostin, Fermilab)

A study was presented for a solid target suitable for a neutrino "superbeam", with a primary proton beam power of 2 MW (5 times the NuMI intensity) and optimized for neutrino energies of about 3 GeV, that is pions around 7 GeV. . A beam spot size of (x (y = 3 Œ 3 mm2 on a graphite target, with a radius of about 2.5 (, would give an acceptable energy deposition of 300 J/g and still provide a good yield of $\sim$ 0.3 pions/proton.

Target Development for the SINQ High-Power Neutron Spallation Source

Werner Wagner

Paul Scherrer Institut, Spallation Neutron Source Division, 5232 Villigen PSI, Switzerland

Abstract. SINQ is a 1 MW class research spallation neutron source, driven by the PSI proton accelerator system. In terms of beam power, it is, by a large margin, the most powerful spallation neutron source currently in operation worldwide. As a consequence, target load levels prevail in SINQ which are beyond the realm of existing experience. Therefore, an extensive materials irradiation program (STIP) is currently underway which will help to select the proper structural material and make dependable life time estimates accounting for the real operating conditions that prevail in the facility. In parallel, both theoretical and experimental work is going on within the MEGAPIE (MEGAwatt Pilot Experiment) project, to develop a liquid lead-bismuth spallation target for a beam power level of 1MW.

INTRODUCTION

The PSI accelerator complex provides a proton beam of 590 MeV used for a wide range of applications. The upgrade program over the years took the proton current from its original design value of 100 μA to its present value of 1.8 mA, and further upgrading is in progress [1].

After penetrating two pion production targets, named M and E, about 1.25 mA (70%) of proton beam are transported to the spallation target of SINQ with a computed loss of as little as 10 ppm on the last 25 m of beam line. Injection into the SINQ-target is from underneath, keeping the whole circumference of the target shielding block free for neutron utilization [2].

SINQ is optimized for high time average neutron flux. Thermal neutron beams are extracted by beam tubes that end near the maximum of the thermal flux. A cold moderator system containing 20 l of liquid D_2 at 25 K [3] serves to slow neutrons down below thermal energies and to supply "cold" neutrons for seven supermirror coated neutron guides [4]. The primary mission of SINQ is to supply a suite of neutron scattering instruments serving a variety of research fields of current interest.

THE SINQ-TARGET DEVELOPMENT PROGRAM

In terms of beam power, SINQ is, by a large margin, the most powerful spallation neutron source currently in operation world wide. As a consequence, target load levels prevail in SINQ which are beyond the realm of existing experience, demanding a careful approach to the design and operation of a high power target. While the best neutronic performance of the source is expected for a liquid lead-bismuth eutectic target, no experience with such systems exists. For this reason a staged approach has been embarked upon, starting with a heavy water cooled rod target of Zircaloy-2 and proceeding via steel clad lead rods towards the final goal of a target optimized in both, neutronic performance and service life time.

In its current form the SINQ target is an array of D_2O-cooled lead rods in steel cladding. The rod array, shown in figure 1, is contained in a double walled Al-shell and suspended from a heavy shielding plug.

FIGURE 1. The rod array of the target Mark 3 used in SINQ.

Neutronic calculations for a very detailed model of the moderator tank and the target system were made using the LAHET-MCNP code system [5]. Comparing the results of the flux calculations for different target types envisaged (figure 2) one finds, as expected, that Zircaloy tube cladding gives the highest neutron flux for a solid lead target. For a liquid Pb-Bi-target, a further, quite significant flux increase is expected.

The SINQ Target Irradiation Program

Many of the questions relating to the life expectancy of different target concepts can only be answered by irradiation under realistic conditions. Therefore materials testing samples from different materials and of different shapes (tensile test, bending

CP642, *High Intensity and High Brightness Hadron Beams: 20th ICFA Advanced Beam Dynamics Workshop on High Intensity and High Brightness Hadron Beams*, edited by W. Chou, Y. Mori, D. Neuffer, and J.-F. Ostiguy

fatigue, TEM-disks, SANS samples, Charpy test etc.) were embedded in the SINQ targets (Mark 2 and 3) in the framework of an international collaboration (STIP) [6]. The aim of the program is to examine the effect of the realistic SINQ operating conditions on a variety of different candidate structural and solid target materials for SINQ and other future spallation targets.

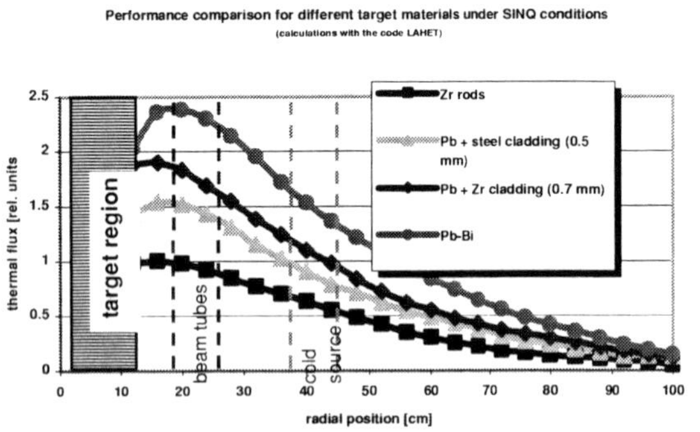

FIGURE 2. Calculated radial flux distribution in the SINQ moderator for different options of rod targets and a liquid PbBi target.

In target Mark 2 there were 10 rods holding test specimens (altogether about 1500), some containing thermocouples to monitor the irradiation temperature. Also dosimetry packages were placed with the test specimens. The temperatures of most of the test specimens were in the range of 250 to 400 °C. During its operation period (June 1998 – Dec. 1999), target Mark 2 has been charged with 6800 mAh of protons. The peak radiation damage and helium production were 12.5 dpa and 850 appm in steel, respectively.

The follow-up target (Mark 3 cf. figure 1), was equipped in a similar way, including more than 2000 specimens. This target received an accumulated proton dose of above 10 Ah, which gives a maximum of about 22 dpa in steel. The maximum irradiation temperature has risen up to 480°C. Among the samples are some irradiated in contact with stationary liquid metal (PbBi and mercury).

The analysis of the Al-Mg(3) safety-hull has been performed on several discs cut from the beam window (figure 3) [2]. The engineering strain-stress curves shown in figure 4 demonstrate: a) significant hardening has been introduced by the irradiation (as expected); and b) the material is still ductile even at the highest fluence of 3.1×10^{25} p/m^2. The latter is an important result for safety considerations in relation with the target enclosure.

There were several rods inspected by neutron radiography. One interesting result is shown in Figure 5.

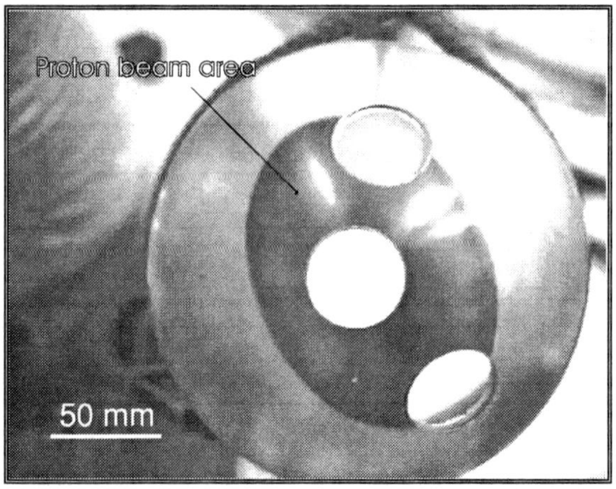

FIGURE 3. The window of the aluminium safety-hull of target Mark-2 after cutting several discs from it. Adapted from [2].

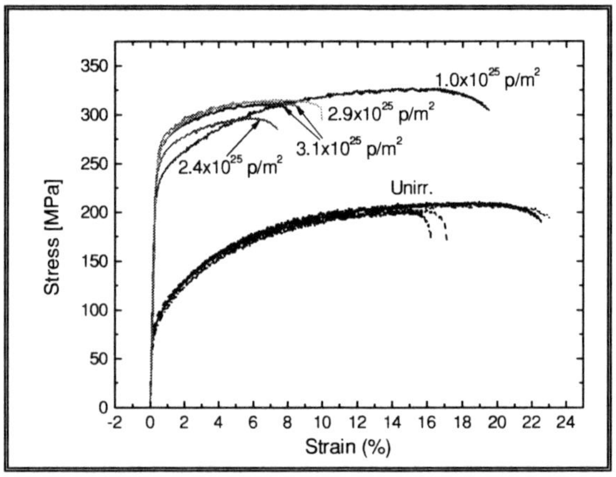

FIGURE 4. Tensile test results of samples cut from the centre and edge area of the proton beam and unirradiated material. Adapted from [2].

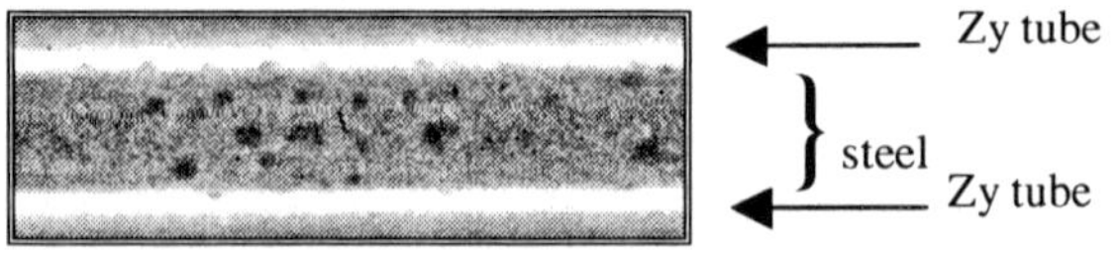

FIGURE 5. A picture of neutron radiography showing the middle part of a Zircaloy clad martensitic steel (F82H) sample. The black spots are believed to be hydrides formed in the Zircaloy cladding.

WORK TOWARDS A LIQUID METAL TARGET FOR SINQ

It is now generally acknowledged that, for high beam power and in particular for high beam power density as required for efficient neutron flux generation, liquid metal targets are the concept of choice. The sensitivity of SINQ to neutron absorption in or near the target essentially limits the selection of available target materials to lead and some of its low melting alloys, in particular lead-bismuth.

Corrosion and Liquid Metal Technology (The LiSoR experiment)

One of the major unknowns in liquid metal target development is related to the question, whether liquid metal-solid metal reactions (LiSoR) in the presence of (static or cyclic) stress are enhanced under irradiation. Since this is a problem that must be solved before a liquid metal target can be irradiated in a proton beam for an extended period of time, an experiment has been initiated to use PSI's 72 MeV Philips cyclotron (Inj. 1) to irradiate stressed steel specimens in contact with flowing liquid metal. Scoping calculations showed that, while much less radioactivity is produced, the damage levels and gas production in thin specimens by 72 MeV protons are, within reasonable limits, comparable to those on the inside of the proton beam window at 590 MeV. Also, the beam parameters can be adjusted in such a way that relevant heating rates at the solid-liquid interface are obtained. The irradiation program will start in June 2002. The initial program comprises four samples of a 9Cr-1Mo martensitic steel (T91), the most promising material to be used for the enclosure hull. The envisaged irradiation times are 10 and 20 days, respectively, at a temperature of 300 °C.

MEGAPIE

The **Mega**watt **Pi**lot Target **E**xperiment, MEGAPIE [7] is a joint initiative by six European research institutions (including PSI) and JAERI (Japan), DOE (USA) and KAERI (Korea). The aim of this initiative is to demonstrate the feasibility of a liquid lead-bismuth target for spallation facilities at a beam power level of 1 MW. The goal of this experiment is to explore the conditions under which such a target system can be licensed, to accrue relevant materials data for a design data base and to gain experience in operating such a system under the conditions of present day accelerator performance. Furthermore, design validation by extensive monitoring of its operational behaviour and post irradiation examination of its components are integral parts of the project. The activities described above (STIP and LiSoR) are part of an extensive materials R&D program which is carried out in order to maximize the safety of the target and to optimize its layout.

Currently (May 2002), the MEGAPIE project finished the engineering design phase, envisaging a target date of January 2005 to be installed at SINQ, followed by an approximately 9 months operational period in the year 2005.

CONCLUDING REMARKS

The world's most powerful spallation neutron source is currently operated at the Paul Scherrer Institut. While there is a potential for further increase of the neutron flux in SINQ, which will be exploited to improve the facility's competitiveness, running a Megawatt class spallation facility in itself is an important cornerstone for many development projects in the field. Examples are MEGAPIE and PSI's input to the target concepts of the next generation spallation neutron sources, like ESS, SNS and JNSC. The SINQ target development program will also benefit most other projects in the field of accelerator application in nuclear technology.

ACKNOWLEDGEMENTS

The participation and collaboration of my colleagues Y. Dai, F. Gröschel, H. Glasbrenner and E. Lehmann is greatly acknowledged. Special thanks go to Günter Bauer (now FZ Jülich) for his initiatives for the SINQ target development program.

REFERENCES

1 P. Schmelzbach "The PSI cyclotron and its extrapolation to a 10 MW driver", this volume (2002)

2 G.S. Bauer, Y. Dai, W. Wagner „SINQ Layout, Operation, Applications and R&D to High Power", Ann. Chim. Sci. Mat. 27 (2002)

3 H. Spitzer, G.S. Bauer, T. Hofmann "First Operation Experience with the Cryogenic Moderator at the SINQ Spallation Neutron Source" Proc. Int. Workshop on Cold Moderators for Pulsed Sources, Argonne, Ill. (1997)

4 W. Wagner, J. Duppich, Proc. ICANS XII, Rutherford Appleton Laboratory, Report No. 94-025 (1994) I-368-I-376

5 G.S. Bauer, A. Dementyev, E. Lehmann, "Target Options for SINQ - A Neutronic Assessment" in Proceedings ICANS XIV, ANL-98/33 (1998) 703-716

6 Y. Dai and G.S. Bauer, "Status of the first SINQ irradiation experiment, STIP-I", J. Nucl. Mater. 296 (2001) 43-53.

7 G.S. Bauer, M. Salvatores, G. Heusener, MEGAPIE, a 1 MW pilot experiment for a liquid metal spallation target J. Nucl. Mater. 296 (2001) 17-3

Carbon and Beryllium Targets at PSI

G. Heidenreich

Paul Scherrer Institut
Villigen PSI, Switzerland

Abstract. The Paul Scherrer Institut (PSI) operates a cyclotron facility for the production of a high intensity proton beam at an energy of 590 MeV. The majority of the beam is delivered to the two target stations M and E, mounted in series, to generate intense pion and muon beams for research in particle physics and muon-spin-resonance applications. Both targets consist of rotating wheels of polycrystalline graphite cooled by thermal radiation. About 40% of the proton beam is lost at the target-E station; the remainder of the beam is transported to the spallation neutron source SINQ or to a beam dump. Since the upgrade to the cyclotron, the target facilities have had to handle significantly higher beam current; both target stations were completely rebuilt (M in 1985 and E in 1990) to handle 2 mA. An overview of the meson production targets will be presented, concentrating on design and operational limits of the targets used at PSI.

INTRODUCTION

The 1 MW proton beam from the 590 MeV cyclotron is delivered to two meson production targets, "M" and "E", mounted in series, which generate intense pion and muon beams for research in particle physics and for muon spin-resonance applications (μSR). These targets consist of rotating wheels of polycrystalline graphite, cooled by thermal radiation. Some parameters for the targets are given in table 1. Target-M feeds two meson beam-lines, π-M1 a high resolution pion beam and π-M3, which is dedicated to μSR applications. The target-E complex provides five high intensity meson beam-lines; π-E1, π-E3 and π-E5 for pions; μ-E1 and μ-E4 for muons. After target E, the remaining proton beam (60% of typically 1.8 mA) is either delivered to the spallation neutron source SINQ or defocused and stopped in a high power beam dump. Before the meson production targets, there is an electrostatic splitter, which allows between 100 nA and 20 μA to be split from the main beam for use in the NA-Hall. This low intensity proton beam is used by PIF (proton irradiation facility) and Gantry (proton therapy facility). A degrader is available to reduce the energy of the proton beam to a few hundred MeV and an intensity of a few nA. Two new projects are planned in the NA - Hall:

(i) The project PROSCAN. This is a dedicated 250 MeV cyclotron for proton therapy, feeding two treatment rooms and one experimental area. (ii) The construction of an ultra cold neutron source (UCN), based on the spallation process. With the help of a fast kicker the whole beam (2 mA) will be directed onto the UCN spallation target for a few seconds every 10 minutes. This should produce a UCN density of 4000/cm^3 (orders of magnitude higher than any other UCN source in the world).

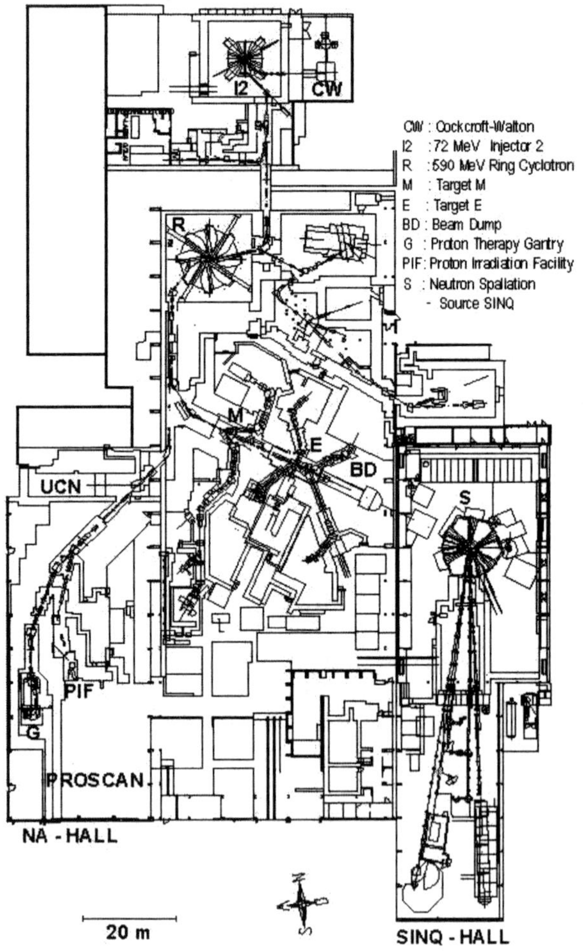

FIGURE 1. Layout of the PSI accelerator facility. Two new projects are planned in the NA-Hall: the project PROSCAN, a dedicated 250 MeV Cyclotron for cancer therapy and the Ultra Cold Neutron Source.

CP642, *High Intensity and High Brightness Hadron Beams: 20th ICFA Advanced Beam Dynamics Workshop on High Intensity and High Brightness Hadron Beams,* edited by W. Chou, Y. Mori, D. Neuffer, and J.-F. Ostiguy

TABLE 1. Some Parameters For The Targets

Meson Production Target	M	E
Mean Diameter (mm)	320	450
Target Length (mm)	5.2	60
Target Width (mm)	20	6
Graphite Density (g/cm^3)	1.8	1.8
Proton Beam Losses (%)	1.6	18
Power Deposition (KW/mA)	2.4	30
Irradiation Damage Rate (dpa/Ah)	0.11	0.1
Operating Temperature (K)	1100	1700
Rotational Speed (Turns/s)	1	1

TARGET STATION M

Target-M is designed as a horizontal insert. The rotation of the target is made using a long drive shaft equipped with a pair of ball bearings, the balls and rings are silver coated to achieve lubrication and to prevent adhesive wear in vacuum. The drive-motor is mounted outside the vacuum with the torque to the drive shaft transmitted by a permanent-magnet clutch. The present target-M unit has operated for more than 50'000 hours without failure since it was installed in 1991. The graphite of the target has been irradiated with total integrated beam current of 44 Ah, which corresponds to a radiation damage level of about 4 dpa.

FIGURE 2. Picture of the target-M unit. The arrow indicates the direction of the proton beam.

TARGET STATION E

In the year 1990 a new mechanical design of the target station E and of the beam dump was installed. The main features are:

i) All the beam line elements and their local shieldings are mounted on support stands, which are precisely positioned on ground plates, allowing a self-centering installation of the elements without any fastenings. All elements can thus be installed and removed exclusively in the vertical direction with the crane. There is no need for local mechanical work on the highly activated components. The rotating carbon target, beam monitors, beam collimators and the beam dump elements can be removed into a remotely controlled shielded flask and transported by a crane to the hot cell for maintenance.

ii) The connections between beam pipes and vacuum chambers are made by means of inflatable all-metal seals which do not require any clamping.

iii) All the power-, cooling- and signal connections are brought through the local shielding to a working platform 2.5 m above the proton beam axis, where the dose rate (with the beam off) is low enough to allow hands-on maintenance.

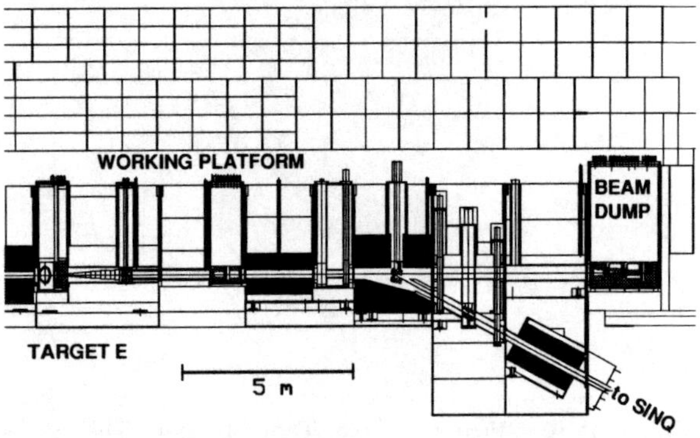

FIGURE 3. Design of the proton channel between Target E and the beam dump.

The target consists of a rotating cone of polycrystalline graphite, cooled by thermal radiation. The cone is attached to the wheel hub by six spokes. This design allows dimensional changes, such as thermal expansion during heating, to be taken up, but constrains the irradiation-induced anisotropic shrinkage of the polycrystalline graphite, which causes deformation of the shape and hence leads to a radial wobble. The radial displacement amplitude must be less then 2 mm during operation of the target. Figure 5 shows the measured radial displacement rate [mm/Ah] for the targets made from the graphite grades R6300P and R6400P [1]. Since 1997 the targets were made from R6400P, which is a more isotropic form of graphite. This has resulted in a significant improvement of the lifetime, which presently reaches 10 Ah (one operational year). The target is driven by a long vertically mounted drive shaft, so that the electric motor is mounted in a low radiation field region. The bearings contain steel balls; the rings and the balls are silver coated to act as lubricant. The lifetime of the bearings is about 3000 hours presently and causes up to three replacements of the target insert per operational year. A longer lifetime is expected by using commercially available bearings with silicon nitride balls [2]; these will be tried out this year.

FIGURE 4. Picture of the Target-E unit. The arrow indicates the direction of the proton beam.

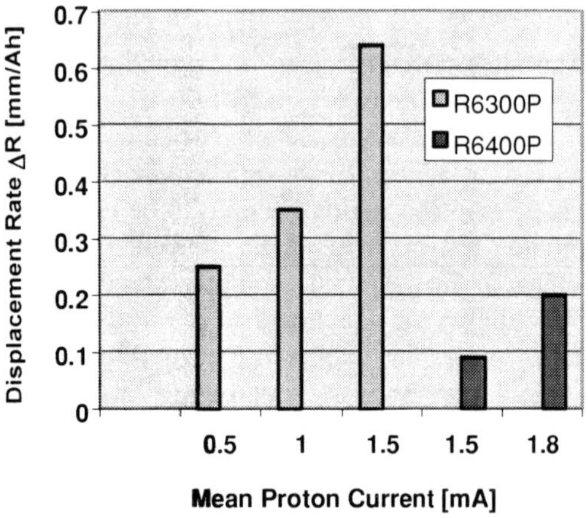

FIGURE 5. Measured irradiation induced radial displacement rate [mm/Ah] for the targets made from the graphite grades R6300P and R6400P [1].

The important operational concerns are mechanical reliability, mainly caused by thermal induced stresses, which can cause cracking of the target cone and spread of radioactivity from evaporation of the target material because of the high temperatures. Figure 6 shows the variation of the critical operational parameters (the temperature, the safety factor and the evaporation rate) for the beryllium and carbon targets as a function of $I/D\cdot\varepsilon^*$, which is the proton current I(mA), divided by the target diameter D(m) multiplied by the effective emissivity ε^* of the radiating surfaces. The current design for the graphite targets have a diameter of D=0.45 m and an effective emissivity of $\varepsilon^*=0.75$, which gives acceptable operational parameters for proton currents up to 3 mA ($I/D\cdot\varepsilon^*=8.9$). In the case of a beryllium target, a diameter ten times larger would be necessary. (e.g. for 3 mA 4.5 m diameter). In the early days of operation, beryllium targets were used with beam currents up to 150 μA. The targets with a diameter of 0.19 m and an effective emissivity of $\varepsilon^*=0.6$ failed at operating currents of 120 μA due to cracks of the target cone ($I/D\cdot\varepsilon^*=1.05$); with a diameter of 0.28 m they operated successfully at up to 150 μA ($I/D\cdot\varepsilon^*=0.89$).

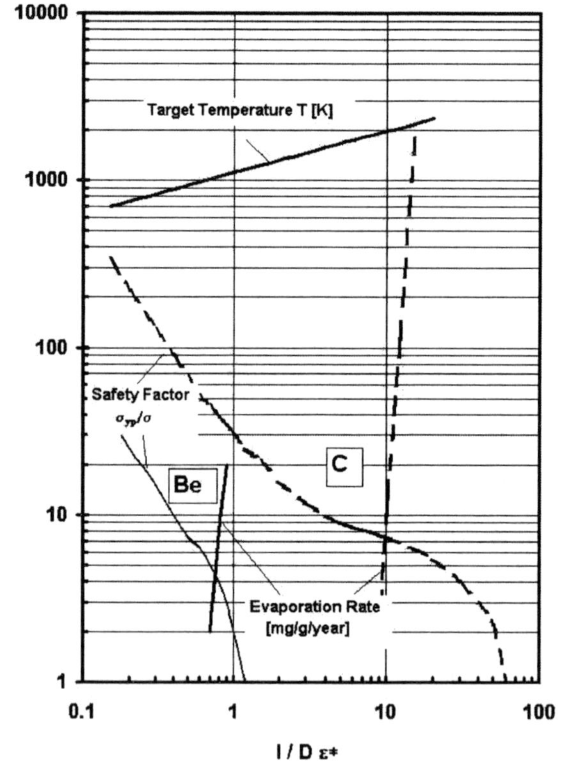

FIGURE 6. Variation of the critical operational parameters (the temperature, the safety factor and the evaporation rate) for the beryllium and carbon targets as a function of $I/D\cdot\varepsilon^*$, which is the proton current I(mA), divided by the target diameter D(m) multiplied by the effective emissivity ε^* of the radiating surfaces.

REFERENCES

1. SGL-CARBON, D-53170 Bonn, Germany
2. GMN, Paul Müller GmbH, D-90411 Nürnberg, Germany

Radiation Protection Utilizing Electronic Berms

John Anderson Jr. and Glenn Federwitz

Fermi National Accelerator Laboratory

Abstract. Excessive radiation losses from particle beams produce personnel, environmental, and component activation hazards. With the new high intensity fast repetition rate beam lines being constructed at Fermilab, small pulse to pulse losses are a significant concern. This paper presents the proposed design of a total loss monitoring system for the MiniBooNE Project.

INTRODUCTION

The 8 GeV Fixed Target Facility, being built at Fermilab, is used to transport beam from the Booster accelerator to a new area dedicated to external beam physics. MiniBooNE, the first experiment to use the new facility, has requested 5E20 protons per year. This equates to more beam per year than has been accelerated in the history of Fermilab. To minimize radioactivation of beamline components, groundwater, and areas outside the overburden; the project has specified a total beam loss monitoring system termed an Electronic Berm or E-Berm. This system is intended to monitor the amount of beam lost between two toroids and disable the beamline if losses exceed a predefined level.

The system described here will use two toroids and the associated integrating electronics. One toroid will be located at the upstream end of the 8 GeV beam line and the other will be located slightly in front of the target focusing horn. The comparator module will measure the absolute value of the difference of the two toroids on a pulse-to-pulse basis. Using the pulse-to-pulse difference along with summing the error of the last 10 pulses, both an instantaneous and integrated loss can be measured.

SYSTEM IMPLEMENTATION

Figure 1 shows a block diagram of the E-berm system. There are four major components to the system; the toroid and integrator electronics, the calibration module, the beam presence detector (resistive wall monitor), and the comparator module. The focus of this document is on the comparator module with a brief description of the other three components.

INPUT DEVICES

Toroid Integrator

The Toroid Integrator receives an event trigger to open a 10us integrate window. Upon completion of the integration window, the integrator sends the E-berm comparator module a 16 bit 2's compliment binary data word and a valid data pulse. There are 4 possible gain settings on the integrator electronics card; they are ±1.875E14, ±5.625E13, ±1.875E13, and ±5.625E12 protons at full scale. Using the lowest gain setting and taking into consideration that the signal is bipolar, the output resolution achieved is 1.717E8 protons per count.

The integrating electronics card buffers the output data word and the valid data pulse before being sent to the E-berm comparator module.

Beam Presence Detector

The beam presence detector utilizes a resistive wall monitor (beam sensing device) connected to a 53Mhz-tuned detector to provide a TTL level output pulse for each beam pulse. This signal is used as a coincidence check within the E-berm comparator module to verify the toroid integrator electronics is looking for beam at the correct time. The beam presence detector will detect beam signals of approximately 5E10 or higher.

Toroid Integrator Calibration Module

The Toroid Integrator Calibration Module is a part of the E-Berm total beam loss monitoring system. The module provides the ability to monitor the performance of the toroid and the integrator electronics, thus greatly improving the reliability of the beam to the experiment. The module is not part of the Fermilab Safety Systems, but rather a tool to improve the systematic performance of the E-Berm system.

CP642, *High Intensity and High Brightness Hadron Beams: 20th ICFA Advanced Beam Dynamics Workshop on High Intensity and High Brightness Hadron Beams,* edited by W. Chou, Y. Mori, D. Neuffer, and J.-F. Ostiguy

The intent of the calibration module is to perform a toroid calibration sequence in approximately 100ms. To do this, the module generates a family of calibration pulses sent through the toroid calibration winding, performs a least squares fit to the data measured by the toroid integrator module, and outputs the gain and offset values to the Fermilab control system in the form of five 8 bit digital data words.

In the normal operating mode, (not performing a calibration sequence), the module sits idle.

To initiate the calibrate mode, the module receives a 1 microsecond calibrate trigger from the Fermilab control system. A calibrate in progress bit is sent to the E-berm comparator module to trip the output permits to inhibit beam during a calibrate sequence.

COMPARATOR MODULE

The figure 2 shows the interconnecting logic functions performed by the comparator module. The logic functions are implemented in Altera™ large scale EPLD logic to eliminate the need for microprocessors. The diagram above shows both the "A" and "B" side comparison circuits. A description of each internal logic function is described below.

Intensity Compare

The 16-bit signal from the "A" integrator is compared to switch settings in the "A" side EPLD's of the module and if the signal from the integrator is higher, a trip occurs. The same function occurs for the "B" side with the "B" integrator. This allows setting of a maximum intensity limit for the beamline. This is not expected to be used in the MiniBooNE beamline but is included for future applications.

Difference Engine

This block takes the 16-bit 2's complement data values from Toroid Integrator "A" and Toroid Integrator "B", subtracts them and returns the absolute value of the result. The "A" side EPLD's perform an |"A-B"| function and the "B" side EPLD's perform an |"B-A"| function. The inversion of the two functions between the "A" and "B" sides is done to assist in preventing common mode errors.

Last 10 Sum

This block adds the current value of the difference engine to a running total and stores the value in a 10 deep by 16-bit shift register. The value that gets shifted out of the memory is then subtracted from the total. Thus generating a sum of the last 10 beam samples to be used for subsequent comparisons.

Pulse to Pulse Comparison

The absolute value of the difference between integrator "A" and integrator "B" is compared to a set of switches located on the comparator board. The value of the switches for MiniBooNE is set at approximately (~3E11). This equates to 6% of the total pulse-to-pulse beam of (~5E12) per pulse. If the difference value is higher than the switch setting a trip occurs.

Integrated Loss Comparison

The value of the sum of the last 10 beam pulse differences is compared to a set of switches again located on the comparator board. The value of the switches for MiniBooNE is set at approximately (~1E12). This equates to approximately a 20% loss of beam for the last 10 beam pulses. Looking at this from a pulse-to-pulse basis yields an average loss of less than 2%. Again, if the value is higher than the switch setting, a trip occurs.

Self Test Compare

Between beam pulses, the integrator is triggered to take a sample with no beam present. The absolute value is then compared to a small positive value to check if the integrator has drifted away from zero. If the value gets too high, the module is tripped. This value is defined inside the EPLD's. At present, the value is set to (2.7E9) and will be defined during the commissioning process after the noise floor for the toroid integrators has been established.

After completion of the zero test, a current pulse is sent through the calibrate winding of both toroids. This is to insure the integrity of all cables and the associated electronics. The absolute value is then compared against to a near full scale positive window. This checks for integrator gain drift and verifies cable integrity. If the value falls outside the window, the module is tripped. This value is defined inside the EPLD's

Beam Presence Detector Compare

The Beam Presence Detector compare logic is used as a beam coincidence check with the toroid integrator electronics. If either toroid integrator reads more than 1E11 beam, then a pulse indicating beam has passed is required from the beam presence detector. The absence of this pulse, when more than 1E11 beam is detected by either of the integrator electronics, will initiate a system trip.

Timing

The timing block takes the ACNET event strobe signals and passes the appropriate signal on to each integrator starting a beam event timing sequence. The event timing sequence requires that the beam pulse from the Beam Presence Detector occur between 2.5us and 8us after the event strobe occurs. Knowing that the Booster extraction pulse is 1.6us guarantees that both integrators were looking for the beam pulse at the correct time. This timing window is defined inside the EPLD's. Additionally a signal from the Beam Presence Detector at anytime outside the beam event will initiate a system trip.

Each Integrator Electronics module must provide a strobe signal indicating valid data is ready within 15us of an event strobe signal.

Event strobe signals may not occur at intervals of less than 40ms. This insures the previous event has been fully processes and the internal self test has completed before beginning the next event timing sequence.

Outputs

The five digital output words connected to an IRM consist of the buffered 16-bit 2's complement integrator values, the 16-bit absolute value of the difference between toroid A and B, the 16-bit sum of the last 10-beam pulse losses, the zero self test value, and the test pulse value.

Individual trip indicators are provided to the front panel for the self test zero and test pulse compare trips, intensity compare trip, single beam pulse loss trip, sum of last 10 beam pulse losses trip, strobe A or B errors, beam presence detector error, event timing errors, and clock errors.

Two 24vdc active high "A" and "B" output permits are provided for connection to the safety interlock system.

Two RS-485 connections for the Safety System Data Acquisition System are also provided.

Reset

There are three separate reset inputs for the comparator module. Each reset signal clears the data shift registers and resets the "A" & "B" output trip circuits. First, there is an internal power up reset for the board initialization. Second, there is a manual reset button located internal to the comparator module on the main board. Third, there is a reset in signal from the safety interlock system. To prevent undesired clearing of the shift registers; the reset-in signal from the safety interlock system is masked unless the system is in a tripped state.

Clock monitor

The comparator module uses two 10Mhz clocks for internal timing. There are two independent clock watchdog circuits. A failure of either clock will remove the output permit signals.

Proposed Use

The E-berm system is intended to be used in place of chipmunk radiation monitor arrays. The system is not intended to be used beyond allowing the same reduction in shielding as would be allowed by the use of chipmunks. The system does not and cannot prevent the initial beam loss. The major advantages of the E-berm over the standard chipmunk radiation monitors are that losses are detected from any direction. This will help to minimize ground water activation, component activation, and allow for a reduction in the required overburden.

Design Methodology

The design methodology used in the proposed design is consistent with the guidance provided in the Fermilab Radiological Control Manual (FRCM) Chapter 10. Two independent monitoring channels are provided inside the comparator module both capable of interrupting the output permit loops.

Failure Modes

During the design process, failure mode reduction principles utilizing redundant circuits were employed. Additionally, all signals to and from the comparator board are buffered. Monitoring two independent toroid integrators and comparing to a beam presence detector signal verifies event timing. The "A" side channel performs an Integrator |A-B| function while the "B" side channel performs an Integrator |B-A| function. The output permit signals are active high signals. Self test functions verify cable integrity. All data is echoed to the ACNET control system via an IRM for further analysis.

Testing

The calibration module built by Columbia University will be used to monitor the calibration drift of each toroid integrator between physical calibrations. A physical calibration of each toroid integrator will be performed at a minimum of every six months utilizing a pulse generator and oscilloscope. The gain and offset values for each integrator electronics module will be adjusted at that time to null out any errors since the last calibration.

The comparator module trip functions will be tested by inputting known data signals via an in-house built test unit. The test unit will verify the pulse-to-pulse difference trip settings, the sum of the last 10-event difference setting, the event timing functions, and the internal self-test functions.

Review Process

The review process established by the Beams Division Head is to produce a design with the joint inputs of the Fermilab ES&H Section, the RF&I Instrumentation Engineering Group, the MiniBooNE beamline Physicists, and the BD ES&H Interlock and Radiation Protection Groups. Upon completion of an approved design, the standard "Request for Interlock Modification" form, (Radiation Physics Form #42), will be submitted to the ES&H Section for a preliminary design approval. Final approval will be requested from the ES&H Section upon submission of completed test results.

Review Standards

The review process established for the proposed design is consistent with the Fermilab Work Smart Standards set and the FRCM.

ACKNOWLEDGMENTS

I would like to thank Glen Federwitz, Paul Czarapata, Craig Moore, Ralph Pasquinelli, James Crisp, James Zagel, Rick Ford, Peter Kasper, Kamran Vaziri, Michael Gerardi, and Daniel Johnson for serving as members of the project design team.

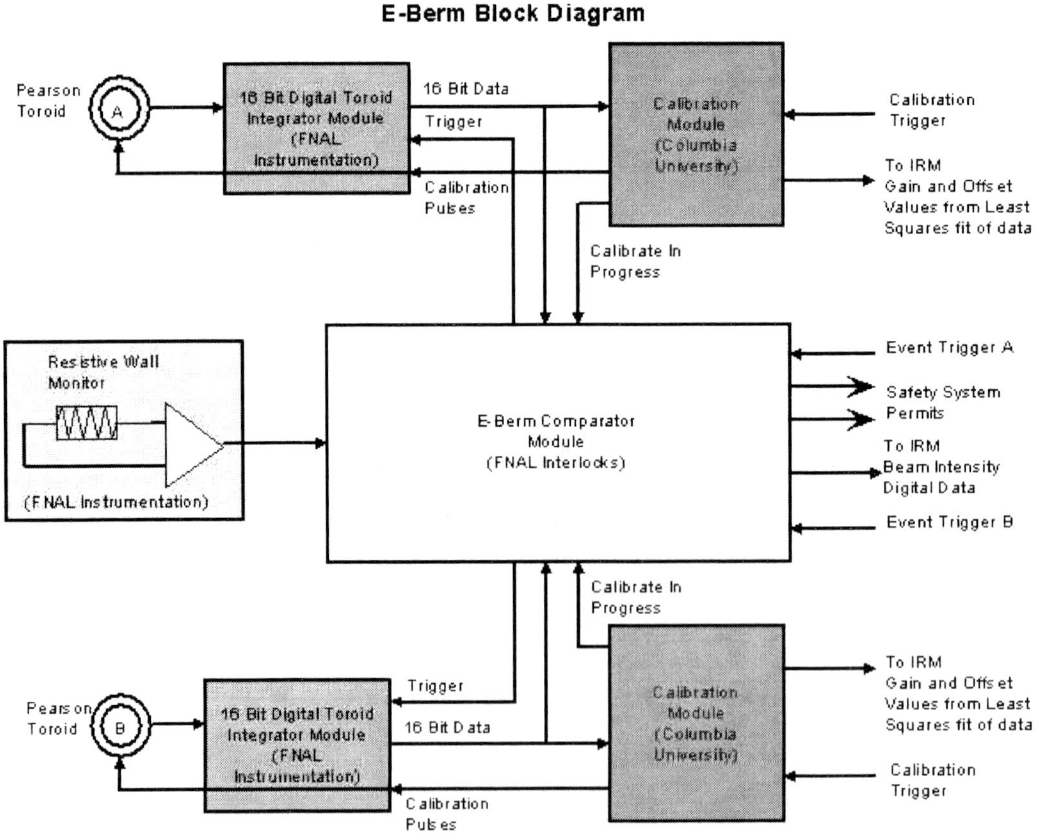

FIGURE 1. E-berm system block diagram.

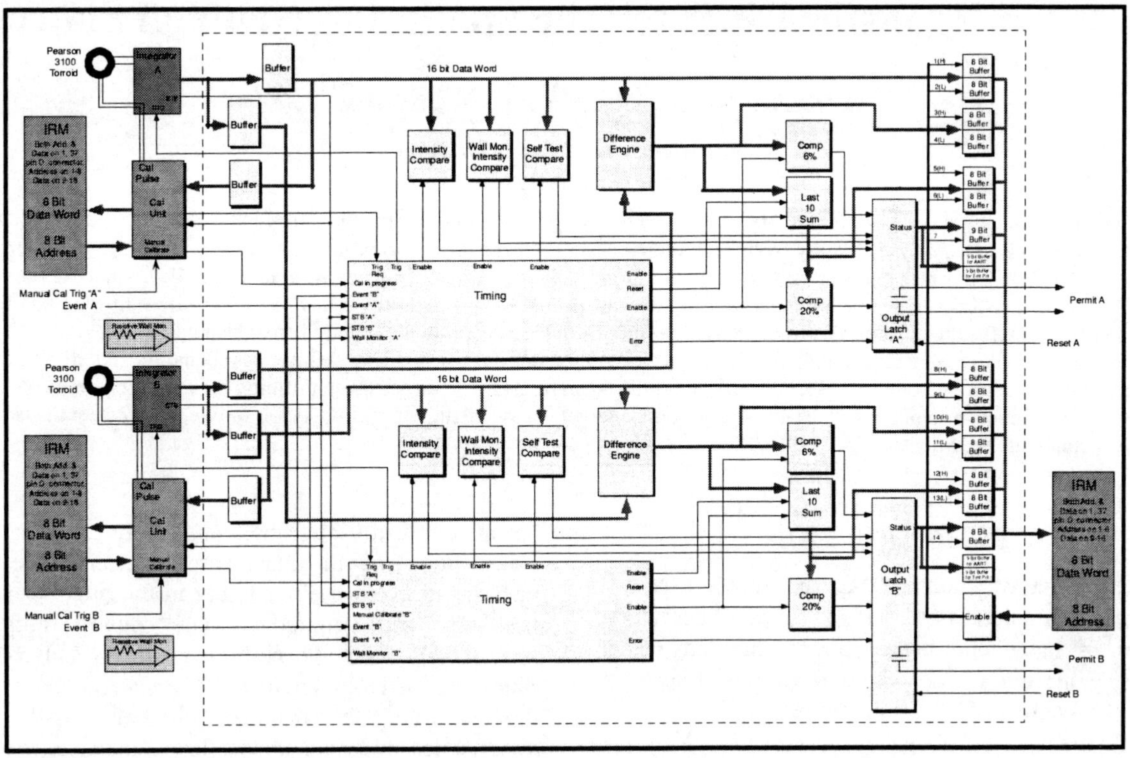

FIGURE 2. Interconnecting logic functions performed by the comparator module.

Beam Transfer Lines for the Spallation Neutron Source

D. Raparia, Y. Y. Lee, W. T. Weng and J. Wei

Brookhaven National Laboratory
P. O. Box 5000, Upton, NY 11793, Upton, USA

Abstract. Beam transfer lines for the Spallation Neutron Source (SNS) are designed to have low beam losses for hand on maintenance while satisfying the facility footprint requirements. There are two main beam transfer lines, High Energy Beam Transport (HEBT) line which connect super conducting linac to the accumulator ring and Ring to Target Beam transport (RTBT) which transfers beam from accumulator ring to the target. HEBT line not only transfer the beam from linac to ring but also prepare beam for ring injection, correct the energy jitter from the linac, provide required energy spread for the ring injection, clean the transverse and longitudinal halo particles from the beam, determine the linac beam quality, and provide the protection to the accumulator ring. RTBT line transport the beam from ring to target while fulfilling the target requirements of beam size, maximum current density, beam moment on the target in case of ring extraction kicker failure, and protect the target from the ring fault conditions.

INTRODUCTION

The 1.4 MW Spallation Neutron Source (SNS) is under construction in Oak Ridge National Laboratory, order of magnitude higher power than any other existing machines. The whole facility is design for low un-controlled beam losses for hand on maintenance. The SNS[1] consists 1.0GeV super conducting Linac[2] and an accumulator ring[3] and two main transfer lines. High Energy Beam Transfer (HEBT) line[4] connects super conducting Linac to the accumulator ring and Ring to Target Beam Transfer (RTBT) line[5] connects accumulator ting to the target. These lines are carefully designed to have low uncontrolled beam losses with several new features in these lines like energy corrector and spreader cavities, collimators etc.

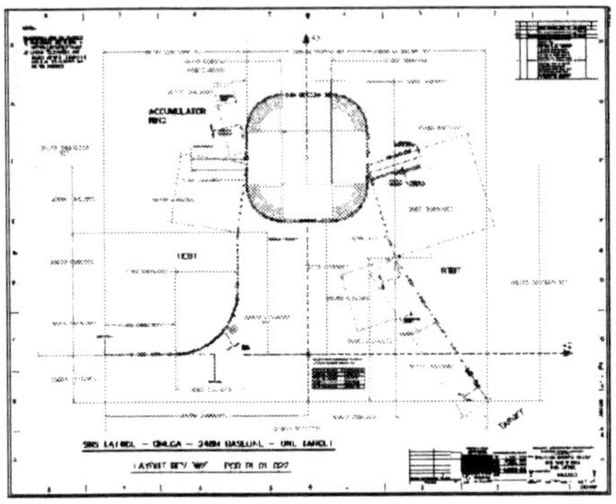

FIGURE 1. SNS transfer lines and ring layout.

HIGH ENERGY BEAM TRANSFER LINE

HEBT is about 180 meter long and carry H^- ion with peak average current of 38 mA in 1 ms long pulses at the rate of 60 Hz. The HEBT not only carry the H^- ion but also optically matches linac and ring, correct the energy jitter from the linac, increase the energy spread of beam to avoid beam stability in the ring, clean the transverse and longitudinal halo coming from the linac, characterize the beam from linac, and protect ring from the fault conditions. The general features of this line are, the magnetic filed in dipole and quadrupoles and vacuum are low enough to control Lorentz and gas stripping of H^- for energies 800 –1300 MeV[6]. The ratio of aperture to the rms beam size is kept greater than 10 through out the line except energy corrector , spreader and collimators. Figure 1 shows the layout of the HEBT, Ring and RTBT. For ease of operation and stability reasons the lattice of the is chosen is FODO which matches the linac and ring doublet-FODO lattices. Figure 2 shows the amplitude function (β_x ,β_y) and the dispersion function (η) along the HEBT. We can consider the HEBT as having three sections: Linac-Achromat Matching Section (LAMS), Achromat, and Achromat-Ring Matching Section (ARMS). In addition to the bend to the ring, there is a straight beam line used for linac beam characterization, as shown in Fig. 1. The first five cells (8 m/cell) after the linac (LAMS) are used to characterize the linac beam, match beam into the achromat, collimate beam halo. The energy corrector cavity is located in the last half cell of the LAMS. Following this, the four cell long achromat (14 m/cell) provides momentum selection by cleaning up the beam energy halo at the point of maximum dispersion (η_x=6.4m). The remaining six cells (8 m/cell) are used for matching the beam into the accumulator ring, diagnostics. The energy spreader cavity is located in the first cell following the achromat (in the ARMS), where the dispersion and its derivative are zero. There are eight horizontal and eight vertical dipole correctors placed in strategic positions of small apertures in the

CP642, *High Intensity and High Brightness Hadron Beams: 20th ICFA Advanced Beam Dynamics Workshop on High Intensity and High Brightness Hadron Beams*, edited by W. Chou, Y. Mori, D. Neuffer, and J.-F. Ostiguy

line. The line has following new features to control losses.

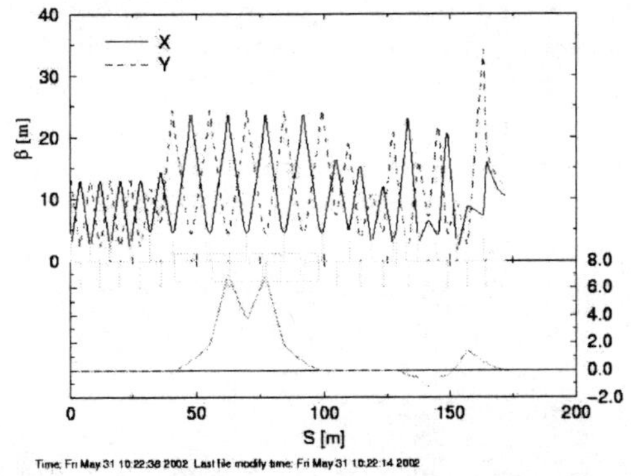

FIGURE 2. The β_x ,β_y and η_x along the HEBT.

Transverse and Longitudinal Collimators

To remove the transverse and longitudinal halo from the beam, we introduce adjustable carbon foils in the path of beam to intercept the tails. After the passage the H^- ion through the foil, the two electrons are removed and the H^- focusing lattice deviates the proton tails into large massive absorbers. There are a total of five set of such adjustable foil. Four transverse (2 each in x and y) are located just after linac and fifth one is located in the achromat at the high dispersion[7]

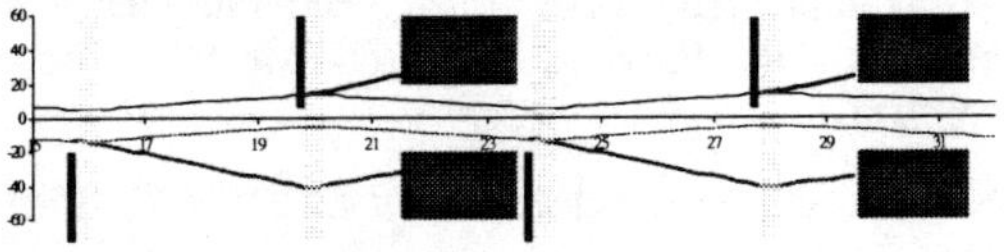

FIGURE 3. Carbon foils and absorber layout in HEBT.

Figure 3 shows the layout of the adjustable foils and absorbers. The phase advance between adjustable foils is 90° (in each plane) providing the most efficient cut of the phase space. For H^- collimation by charge exchange the interception of the secondary halo by the absorber is 100% efficient. Still, due to small impact parameters, the protons have a finite probability of escaping the absorber without being removed. We simulate the passage of protons through the material of the absorber with Monte Carlo code. The absorption efficiency for the secondary protons ranges 88 to 95 %.

For longitudinal halo collimation, a charge exchange foil is located in high dispersion region in the HEBT achromat to remove the longitudinal tails from the beam before entering the ring. After being stripped, the protons are deviated by the dipole filed into a large absorber outside the beam trajectory. The vacuum pipe geometry after foil has been adjusted to clear the dipole and drive the protons on the absorber. In this case, the absorption efficiency in the collimator is 100%, as protons are deviated from the main path and hit the absorber located outside the accelerator.

Energy Corrector and Spreader Cavity

Beam from the linac will have 2.0 MeV energy jitter and 3.5° for phase jitter due to 0.5° and 0.5 % amplitude control error in linac. To remove these phase and amplitude jitter an 805 MHz (linac frequency) RF cavity is place just before the achromat which is phase lock to the linac. The particle due to energy difference will suffer the phase slip with respect to design momentum given by

$$\Delta\phi_L \equiv \frac{\gamma}{\gamma(\gamma+1)}\frac{\Delta T}{T}\frac{L}{\beta c}2\pi f$$

where β,γ are the relativistic parameters, T is the design kinetic energy and f is the frequency and l is length for the phase slip. The required voltage is given by $V_0=\Delta E/\sin(\phi_{Slip})$. The energy jitter after the corrector cavity 0.2 MeV was achieved which is within the requirement of the ring.

SNS accumulator ring also need ± 4 MeV energy spread for stability reason. This is achieved by placing another RF cavity after the achromat which frequency is about 100 kHz different than the linac frequency. The linac bunches see different phase as they arrive at this cavity and gain/loss different energies depending on the cavity voltage, hence creating required energy spread without creating energy tails.

RING TO TARGET TRANSFER LINE

Ring to Target Beam Transfer (RTBT) line is about 150 meter long and carry the beam from the ring extraction region to the target and provide the desired footprint for the accelerator complex. The general features of this line are, the line is immune to one kicker failure and the ratio of acceptance to rms emittance is more than 20. Figure 4 shows the amplitude and the dispersion function along the RTBT.

The line has following function (a) extraction (b) beam dump (c) protect from device failure (d) beam spreader and (e) diagnostics. Following the extraction system beam can be dumped straight through a 16.8° dipole magnet. After the magnet two cell are used for the collimator system which protect line from the dipole failure. Following another six cells of transport, the last five quadrupoles in the line are used for final beam spreading to produced the beam sized required at the target.

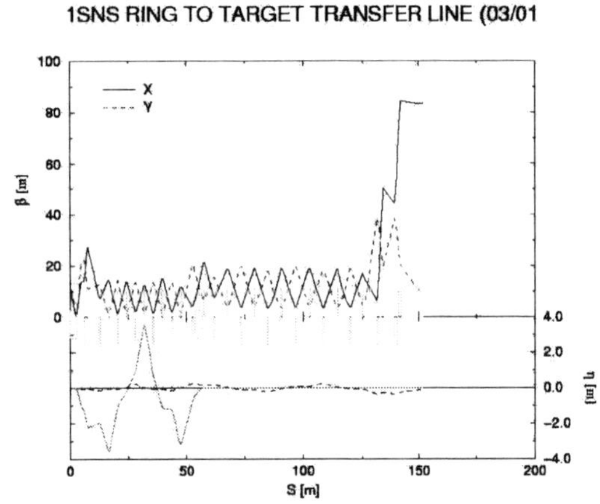

FIGURE 4. The β_x, β_y and η_x, η_y along the RTBT.

The phase advance between kickers and target is nπ, so that in event of a kicker failure beam will not move at the target. Figure 5 shows the closed orbit for different kickers failure.

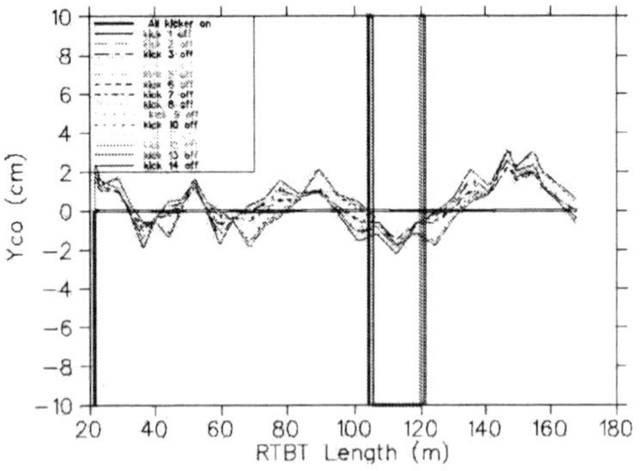

FIGURE 5. Close orbit due to kickers failure.

Extraction

The extraction of the beam is done in a single turn with full aperture at a pulse repetition frequency of 60 Hz. The extraction system consists of two sets (seven each) kicker magnets and a Lambertson magnet septum and a dipole magnet. The Lambertson septum magnet will receive the vertically kicked down beam and will provide large deflection angle (16.8°) to enable ejection horizontally from the accumulator ring. A dipole, which is 540° phase advance away from the Lamberton magnet, bends the beam horizontally in the same direction by 16.8°, making the extraction system achromatic. The Lambertson magnet is rotated 2.55° anti-clock wise to neutralize the vertical kick from the kickers and making the beam about 9 inches difference between the ring and RTBT beam height.

Beam Spreader

The beam spreader consists of five radiation hard quadrupoles near the end of the RTBT. These five 36 cm diameter aperture quadrupoles provide the desired beam size at the target. Due to thermal considerations of the target, the beam current density on the target must remain below 0.25 A/m^2 . This requirement results in a non-Gaussian beam distribution in space (with un-normalized rms emittance of 24 π mm mrad). The required current density distribution is achieved using painting scheme described elsewhere[8].

Beam collimation

The RTBT acceptance is 480 π mm mrad and the expected beam emittance (100%) from the ring is 240 π mm mrad. The line is design such a way that if one of fourteen kicker fail beam will pass through the line without scraping. To protect the line from device failures such as more than one kicker or dipole failure, there are two collimators 90° apart in phase advance are placed just after the 16.8° dipole. The acceptances of these collimators are 300 and 400 π mm mrad in horizontal and vertical plane respectively.

REFERENCES

1. N. Holkamp, "The SNS Linac and Storage Ring: Challenges and Progress Towards Meeting Them", EPAC 2002.
2. S. Nath, et al., "Selected Aspect of SNS Linac", EPAC 2002
3. J. Wei, et al., "Low-Loss Design for the High Intensity Accumulator Ring of the Spallation Neutron Source", PRST-AB, 3, 080101, pp. 1-19,2001
4. D. Raparia, et al., "The NSNS High Energy Beam Transport Line", Proc. 1997 Particle Accelerator Conference, p. 162.
5. D. Raparia, et al., "The SNS Ring to Target Beam Transport Line", Proc. 1999 Particle Accelerator Conference, p. 1297.
6. D. Raparia, et al., "Dependence of the SNS Transfer lines and Accumulator Ring on the Linac Energy", Proc. 2001 Particle Accelerator Conference, p. 3260.
7. N. Catalan-Lasheras, and D. Raparia, "The Collimation System of the SNS Transfer Lines, Proc. 2001 Particle Accelerator Conference, p. 3263.
8. J. Beebe-Wang, et al., "Transverse Phase Space Painting for SNS Accumulator Ring Injection", , Proc. 1999 Particle Accelerator Conference, p. 1743.

Targetry Issues for the Fermilab 2-MW Neutrino Superbeam

M.A. Kostin, V.I. Balbekov, N.V. Mokhov, Z. Tang

FNAL, Batavia,IL 60510, USA

Abstract. The possibility to use various target materials are studied for a 2 MW neutrino superbeam facility that can be built at Fermilab utilizing the Main Injector and a Proton Driver. A simple target solution found is a thick graphite rod taking a broad proton beam.

MOTIVATION

The neutrino oscillation is a hot topic nowadays. The oscillation is only possible when the neutrino is a massive particle. The Standard Model does not allow the neutrino to have mass, therefore observation of the neutrino oscillation opens window into the physics beyond the Standard Model. The resent results show that the neutrinos do oscillate. The SuperKamiokande experiment has provided a very strong evidence that the muon neutrino experience large flavor changing transition [1]. As shown ν_μ are transformed primarily to ν_τ. The SNO has reported that the solar electron neutrinos are changing their flavor to ν_μ and/or ν_τ [2] and the Large Mixing Angle solution is preferable.

Despite the fact that many of oscillation parameters will be determined more precisely by forthcoming experiments such as MINOS, MiniBooNE and others, some questions still will be out of reach of those. For example, the problems of neutrino mass hierarchy, CP violation in the leptonic sector and probably CPT violation can only be addressed by an experiment of the next generation, so-called neutrino superbeam facility. Such an experiment is expected to utilize a conventional neutrino beam of a very high intensity and a detector with fiducial mass of 20 kton and more. Whereas the detector and beamline issues are relatively well known, the target remains the most critical one. For instance, the FNAL Main Injector coupled with a Proton Driver would provide a proton beam of 2 MW on target. A target solution for such a facility is not obvious.

CRITERIA FOR A GOOD TARGET

A good target for a neutrino superbeam facility would satisfy the following conditions: it survives one spill; the steady state temperature is relatively low; the target life time is greater than 6 months; in the case of multiple choice for a target candidate, the optimal target is the one that provides the highest π^+ yield in the energy interval of interest. The first three clauses are quite obvious. The last one is driven by the physics case. The main goal of such a facility would be a measurement of the $\nu_\mu \to \nu_e$ oscillation parameters. For a ν_μ-beam coming from a NuMI-like facility, the ν_e appearance probability to leading order is given by

$$P_{\nu_\mu \to \nu_e} = sin^2(\theta_{23}) sin^2(2\theta_{13}) sin^2\left(\frac{1.27\Delta m^2_{32}(eV^2)L(km)}{E_\nu(GeV)}\right) \quad (1)$$

A detector must be placed at an appropriate distance from neutrino source to let the neutrino enough time to oscillate. For example, one of possible locations for the detector site is the Homestake mine at 1290 km from Fermilab.

The main source of ν_μ is the two body pion decay $\pi^+ \to \nu_\mu \mu^+$. The pions are produced in interactions of the initial proton beam with a target. Neutrino energy from the above decay is

$$E_\nu = \frac{0.43 E_\pi}{1+\gamma^2\theta^2}, \quad (2)$$

where γ is the pion relativistic boost and θ is the neutrino emission angle at decay. It is seen from (2) that in order to get a neutrino beam of the energy of 2.5 – 3.5 GeV in a detector placed on the facility axis, one would need to focus pions of 5.8 to 8.2 GeV energy. For an off-axis detector, the energy interval of pions should be extended up to 20 GeV [3].

MATERIAL CHOICE

What survives one spill ?

A number of commonly used materials was considered for a target. The target was simulated with the MARS14 code [4]. We assumed that a 120 GeV proton beam hits a rod target with a length of two interaction lengths. For the first trial, the beam was assumed to be a Gaussian with $\sigma_x = \sigma_y = 1$ mm (approximately NuMI parameters). The target transverse size was optimized for the maximal pion yield scanning the radius of target with a step of 0.5 mm starting from 2.5 mm (see Table 1). In order to achieve a power of 2 MW, the nominal NuMI

CP642, *High Intensity and High Brightness Hadron Beams: 20th ICFA Advanced Beam Dynamics Workshop on High Intensity and High Brightness Hadron Beams,* edited by W. Chou, Y. Mori, D. Neuffer, and J.-F. Ostiguy

TABLE 1. Density of peak energy deposition and pion yield in the 5.95 – 8.05 GeV interval for various targets. Also shown are the optimal target radii R_T for σ=1 mm.

Material	Peak ED density (J/g)	π^+yield (N/POT)	optimal R_T (mm)
Graphite	1581 ± 18	0.326 ± 0.005	5.0
Nickel	6520 ± 251	0.311 ± 0.004	2.5
Inconel	6011 ± 259	0.311 ± 0.004	3.0
Copper	6084 ± 216	0.310 ± 0.004	3.0
Indium	$5248 + 149$	0.336 ± 0.004	3.0
Mercury	10064 ± 293	0.324 ± 0.004	3.0

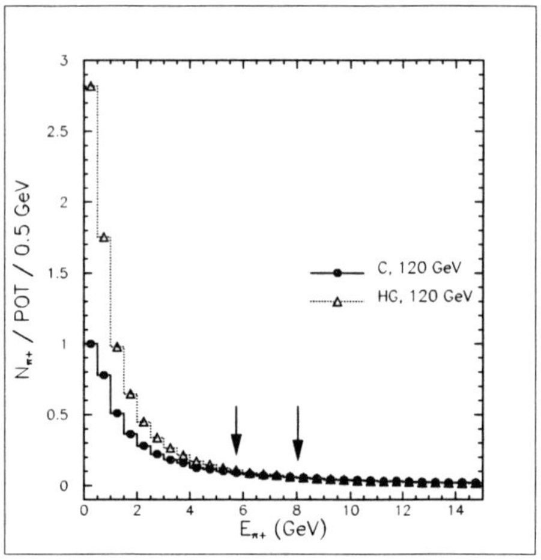

FIGURE 1. Pion yield versus pion energy. The arrows restrict the interval of our interest.

beam intensity is rescaled by a factor of 5 that corresponds to 2×10^{14} protons per spill. We also assumed that the Main Injector repetition period of 1.9 s does not change.

Calculated peak energy deposition densities on a beam axis are shown in Table 1 for C through Hg targets. For the beam conditions described above all the materials experience a significant thermal shock. The stress limits known for graphite, nickel and inconel are about 1000 J/g. The limit for copper is about 600 J/g. As one can see from the table, none of the solid materials can survive such conditions.

Despite the fact that the above conditions are not operable for the solid materials, it is interesting to notice that the pion yield in the defined energy interval does not vary significantly with the target material (Table 1). Fig. 1 shows the pion yield versus pion energy for graphite and mercury. Much more soft pions are produced in a mercury target. But what is important, the yield is about the same in the region of interest for on-axis beams, $5.95 \leq E_\pi \leq 8.05$ GeV. Pion spectra for the other materials studied behave similarly.

The most obvious idea of stress reduction is to increase

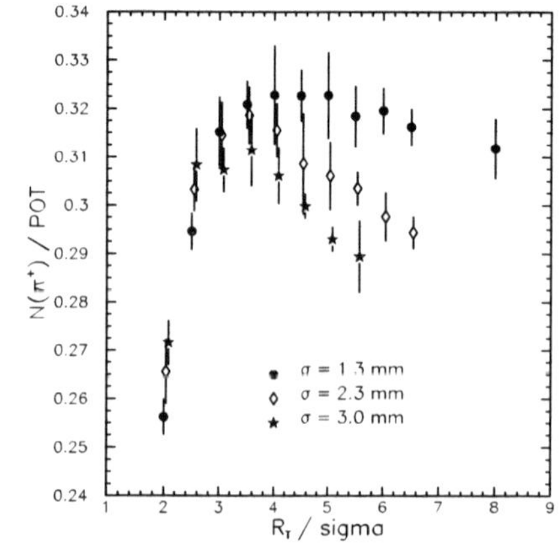

FIGURE 2. Number of π^+ coming from a graphite target in the interval 5.95 – 8.05 GeV against the value of target radius over beam sigma. Shown are dependencies for different beam RMSs. The distributions are normalized per the number of protons on target (POT).

the beam transverse spot size. In the study below, both the beam spot size $\sigma = \sigma_x = \sigma_y$ and target radius R_T are varied.

For a graphite target the pion yield as a function of R_T/σ is shown in Fig. 2 for different σ. As one can see the maximal yield does not change much with σ. This is due to the fact that pions leave the target at much larger angle than a specific angle at which the target collimates the pions, that is atan(R_T/λ_I) $\approx 1°$ (Fig. 3). Energy deposition in the hottest cell of target is acceptable at $\sigma > 1.5$ mm (Fig. 4). For example, for σ=3 mm and target radius R_T of 9 mm the peak energy deposition is 310 J/g that is below the limit. Thus the beam spot size and target radius for graphite can be substantially increased compared to the NuMI parameters without significant loss of the yield.

The same idea does not work well for solid dense materials. An attempt to bring the energy deposition safely below the limit by increasing σ and R_T results in a substantial reduction of the pion yield. For example, if one increases the beam σ up to 12.5 mm keeping the ratio to be optimal $R_T/\sigma = 2.5$, the pion yield reduces down to 0.22 pions per proton that is much smaller than the yield for an optimal graphite target of about 0.33. The density of peak energy deposition is 1250 J/g for inconel in this case that is still above the limit.

The studies performed show that graphite is the most convenient material for a FNAL neutrino superbeam facility. Indeed, the dense solid materials do not provide enough of pion flux at the acceptable beam and target radii. The use of light materials with large interaction lengths such as Li and Na will lead to a too long target, making focusing of secondaries problematic. A mercury jet target seems to be too complex, expensive and haz-

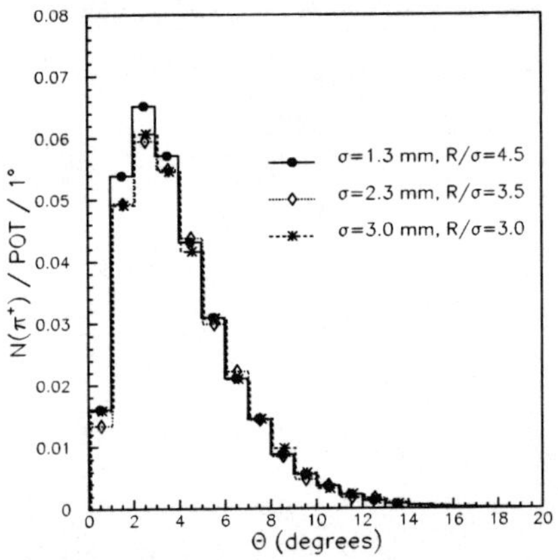

FIGURE 3. Angular distributions for pions coming off a graphite target. θ is an angle between the target axis and pion direction.

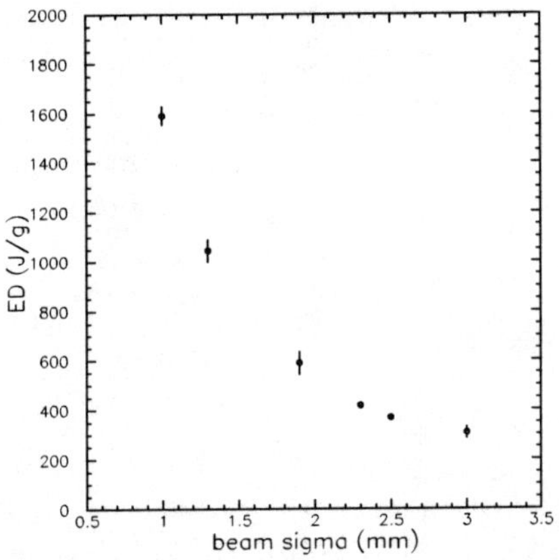

FIGURE 4. Energy deposition in the hottest cell of a graphite target versus beam spot size.

ardous device and at the same time it does not provide any advantages over a graphite target for the given experimental conditions.

Target life-time

One of the factors limiting the target life-time is radiation damage. The life-time determined here corresponds to the time when 5 dpa (displacements per atom) occur in the hottest cell of the target. The atoms are displaced due to interactions with hadrons with kinetic energy of >0.1 MeV. The 5 dpa limit corresponds to the integrated hadron flux of 5×10^{22} cm^{-2}. From the MARS simulations we have found that the limit of 5 dpa is reached within 5 years and 8 months assuming 10 months of operation per year at the full intensity.

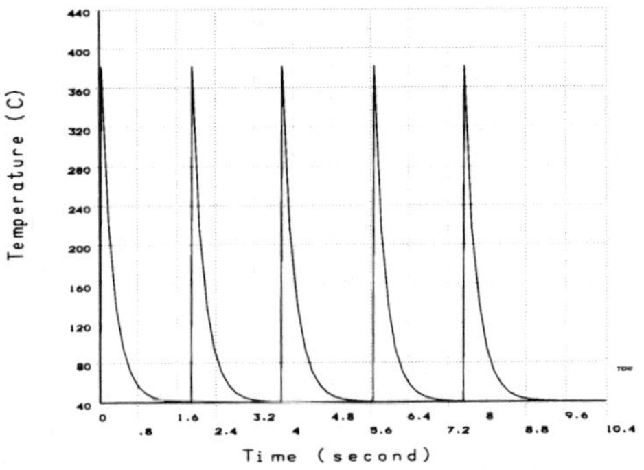

FIGURE 5. Temperature evolution in the hottest cell of a graphite target with 1.9 s repetition rate.

Temperature buildup

The temperature evolution has been investigated with the ANSYS code [5]. The temperature on the side target surface was fixed to 40° C. As Fig. 5 shows, the temperature in the hottest cell oscillates between 40° C and 380° C without any buildup. The tensile stress obtained from the simulation is about 20 MPa at the tensile strength for graphite of about 90 MPa.

CONCLUSIONS

A graphite target represents a simple and cheap solution for the FNAL neutrino superbeam facility. Such a target satisfies all the requirements applied: it survives one spill; radiation damage does not constrain the life-time significantly; there is no temperature buildup in target; pion yield is quite high.

REFERENCES

1. Y. Fukuda et al., Evidence for oscillation of atmospheric neutrinos. Phys. Rev. Lett. 81 (1998) 1562.
2. Q. R. Ahmad et al., Direct evidence for neutrino flavor transformation from neutral current interactions in the Sudbury Neutrino Observatory. e-Print Archive: nucl-ex/0204008.
3. A. Para and M. Szleper, Neutrino oscillations experiments using Off-axis NuMI beam. e-Print Archive: hep-ex/0110032.
4. N. V. Mokhov, "The MARS Monte Carlo", Fermilab FN-628 (1995); N. V. Mokhov and O. E. Krivosheev, "MARS Code Status", Fermilab-Conf-00/181 (2000); http://www-ap.fnal.gov/MARS/.
5. "ANSYS (rev.5.1)",Swanson Analysis System, Inc., SASI/DN-P511:51, Houston (1994).

SESSION A

LATTICES

Summary Report

Henry A. Thiessen

LA-UR-02-3315
Los Alamos National Laboratory
Los Alamos, New Mexico 87545 USA

Abstract. We discuss here the status of dynamic aperture calculations for the various machines discussed at this workshop. We recommend that these calculations be extended to include all magnet field errors and alignment errors as quickly as possible.

THE CONVENOR'S VIEW

At this meeting, we saw many new, innovative lattices proposed for machines either in design phase or in construction. These lattices have many essential features for operation at high current with low losses including explicit collimation schemes, avoiding transition, and intentional eta function manipulation to allow production of short pulses, etc. Most of these lattices have larger acceptance than their predecessors. Dynamic aperture is an important issue to be considered before committing to a final design of any of these machines.

We saw numerous studies of the dynamic aperture of the intrinsic lattices, *i.e.*, in the absence of magnet field and alignment errors. Tune issues, placement of chromaticity sextupoles, and nearby structural resonances all play important roles in limiting the dynamic aperture seen. Most of the dynamic aperture studies shown at this workshop did not include magnet errors, although all participants acknowledged that such calculations were planned for the future.

However, there was a general trend. The more magnet imperfections that were included, the smaller was the dynamic aperture. Indeed, the Fermilab Proton Driver talks showed that the differences in dynamic aperture among the various lattice designs tended to disappear when systematic errors of quadrupoles were included in the tracking.

The Los Alamos team found that the dynamic aperture of their lattices was marginal when all magnet errors - random, systematic, and alignment - were included in the tracking[1]. The solution needed at Los Alamos is to build better quadrupole magnets[2].

This reminds us of the early years of operation of the original Fermilab Main Ring. For this machine, the dynamic aperture was found to be very small at injection time, due to unexpected and unanalyzed remnant sextupole and decapole components in the dipole fields. The heroic efforts of many people and on the order of a year's time was needed to correct the dynamic aperture of the Main Ring. We do not want to repeat this exercise.

In contrast, the Fermilab Main Injector dynamic aperture was studied in advance of construction with all errors included. A decision was made to build better dipoles than in the Main Ring, and to operate these dipoles at a higher injection field. Many of the quadrupoles were reused. Tracking with worst-case errors showed that the dynamic aperture at injection time exceeded the physical aperture. Turnon of the Main Injector was rapid and uneventful, as we wish for future machines.

The message is clear – do a dynamic aperture study with all errors in advance of construction. Either provide sufficient quality magnets, or a suitably designed correction scheme such that the dynamic aperture can be made larger than the required aperture. In most cases, the critical time in the accelerator cycle is near injection time, when the required aperture is largest and magnet field errors may be largest.

REFERENCES

1. Filippo Neri, Martin Schulze, Dave Johnson, Peter Schwandt, "Transverse Tracking of the AHF Rings", in this workshop proceedings.
2. Martin Schulze, David E. Johnson and Ben Prichard, Filippo Neri and Arch Thiessen, "Magnets that Meet Tracking Requirements for AHF", in this workshop proceedings.
3. "The Fermilab Main Injector Technical Design Handbook" 1995, private communication.

CP642, *High Intensity and High Brightness Hadron Beams: 20th ICFA Advanced Beam Dynamics Workshop on High Intensity and High Brightness Hadron Beams*, edited by W. Chou, Y. Mori, D. Neuffer, and J.-F. Ostiguy

The JKJ Lattice

Kenta Shigaki*, Fumiaki Noda*, Kazami Yamamoto*, Shinji Machida#, Alexander Molodojentsev#, Yoshihiro Ishi@

*Japan Atomic Energy Research Institute, Tokai, Ibaraki 319-1195, Japan
#High Energy Accelerator Research Organization (KEK), Tsukuba, Ibaraki 305-0801, Japan
@Mitsubishi Electric Corporation, Kobe, Hyogo 652-8555, Japan

Abstract. The JKJ high-intensity proton accelerator facility consists of a 400-MeV linac, a 3 GeV 1-MW rapid-cycling synchrotron and a 50-GeV 0.75-MW synchrotron. The lattice and beam dynamics design of the two synchrotrons are reported.

INTRODUCTION

The JKJ is a joint project of Japan Atomic Energy Research Institute and KEK for a high-intensity proton accelerator facility [1-3]. Its accelerator complex consists of a 400-MeV[1] linac, a 3-GeV rapid-cycling synchrotron (RCS) and a 50-GeV synchrotron (MR). The RCS and MR are characterized with their design output power as large as 1 MW and 0.75 MW, respectively.

3-GEV RCS

The 3-GeV RCS receives a 400-MeV proton beam from the linac and accelerates it to 3 GeV with a repetition rate of 25 Hz. The design output beam power is 1 MW with 8.3×10^{13} ppp. Its output beam is time-shared between the pulsed spallation neutron and muon sources (95 %) and the MR (5 %).

Design Concepts

The design of the RCS primarily focuses on achieving the 1-MW output beam power. The ring acceptance is as large as 486 π mm mrad transversely and $\pm$ 1 % in momentum, to circulate the beam up to 324 π mm mrad and reduce the beam loss due to intra-beam scattering. The expected incoherent tune shift is -0.16 at the injection energy with the design bunching factor of 0.41. The loss is controlled and localized with transverse and longitudinal beam collimators in the ring. The uncontrolled beam loss is suppressed to a 1 W/m level except in the injection, collimation and extraction regions.

Ring Layout and Lattice Functions

The RCS has 27 FODO cells with three-fold and mirror symmetries, as shown in Figure 1, over its circumference of 348.333 m. The FODO structure eases requirements for the strength of the quadrupole magnets and also allows an effective arrangement of the secondary beam collimators (halo collectors) with the large and smooth phase advance.

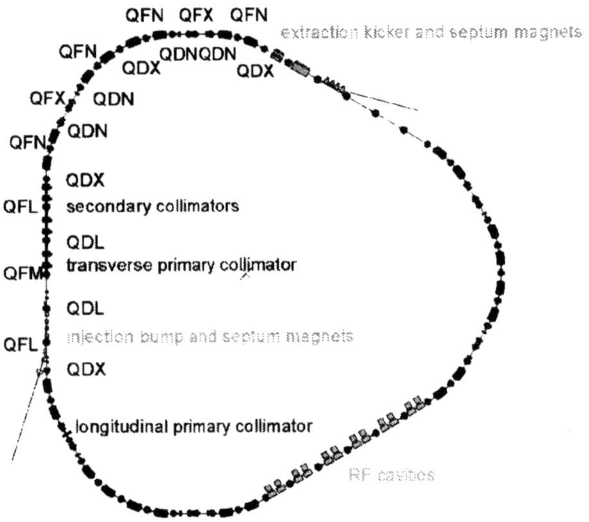

FIGURE 1. The ring layout of the 3-GeV RCS.

Figure 2 shows main lattice functions of a super-period of the RCS. The arc sections have so-called missing bends where the horizontal momentum dispersion function peaks. The raised transition γ at 9.14 makes the RF operation easier. The missing-bend sections are suitably used for longitudinal beam collimation and chromaticity correction. The three long dispersion-free insertions are utilized for injection and beam collimation, RF acceleration and fast extraction, respectively.

[1] A future extension to 600 MeV is planned for the linac.

CP642, *High Intensity and High Brightness Hadron Beams: 20th ICFA Advanced Beam Dynamics Workshop on High Intensity and High Brightness Hadron Beams*, edited by W. Chou, Y. Mori, D. Neuffer, and J.-F. Ostiguy

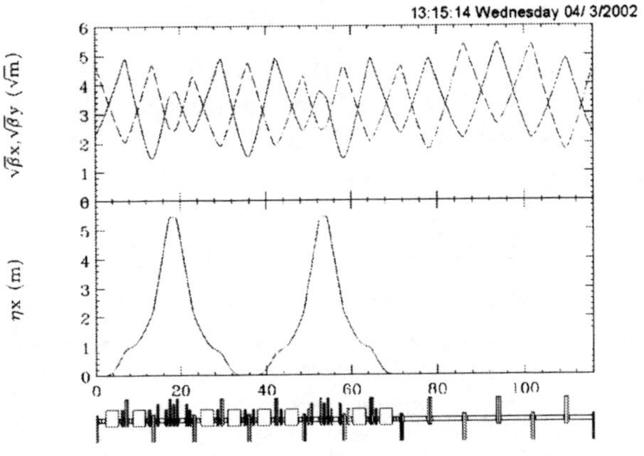

FIGURE 2. Lattice functions of a super-period of the 3-GeV RCS.

Working Point and Tunability

The nominal working point of the RCS is chosen at (ν_x, ν_y) = (6.72, 6.35) to avoid structure resonance and for small emittance blowup due to intra-beam scattering and β modulation with the injection bump orbit on. The lattice is flexible in the betatron tune space for alternate working points and stable enough to cover the expected incoherent tune shift and possible tune excursions during the acceleration.

Dynamic Aperture

The dynamic aperture of the RCS is designed to cover the physical aperture of 486 π mm mrad with $\pm$ 1 % momentum spread. Effects of field imperfection, *e.g.* higher order fringe field, on the dynamic aperture have been studied [4]. The dynamic aperture is comparable to the physical aperture even with the full chromaticity correction.

Closed-Orbit Distortion and Its Correction

The closed orbit will be distorted by up to 10 mm with expected magnetic and alignment errors without a correction. A beam position monitor and a correction dipole magnet are placed in each half-cell to correct the distortion. The expected residual closed-orbit distortion after correction is within $\pm$ 1 mm with the design resolution of the beam position monitor of 0.2 mm.

Injection

The RCS has a charge-exchange injection system with carbon stripping foils, located in the first half of the injection insertion. Approximately 300 pulses are injected from the linac in 500 μsec. The primary stripping foil is at the center of the system symmetrically sandwiched by four closed-orbit bump magnets. The system is designed to support multiple injection schemes, from center injection for commissioning to precise transverse painting injection for optimum operation. Four horizontal painting magnets in the ring and two vertical painting magnets on the injection line at a phase π upstream of the injection point are used for the two-dimensional painting. Specific schemes of transverse painting, *e.g.* correlated or anti-correlated in horizontal and vertical directions, are under study in terms of emittance deformation during the injection. The system is capable to adjust the painting area on the pulse-by-pulse basis.

Beam Collimation

Transverse and longitudinal beam collimators are in the RCS to control and localize the beam loss. The system consists of two primary collimators to scatter the halo particles and five secondary collimators to collect them. The secondary collimators are located in the second half of the injection insertion to collect halo particles directly after the injection without spreading them over the ring. The acceptance ratio between the collimators and the rest of the ring is adjustable with the movable collimator jaws and designed at 1.5 for the halo collection efficiency of 98 % in the collimator region.

Extraction

The RCS is equipped with a fast extraction system consisting of eight kicker and four septum magnets. The first half of the extraction insertion is allocated for the system. The system is designed to have the minimum acceptance of 324 π mm mrad, the collimator acceptance, for the one-pass extraction line and 1.5 times more for the circulating beam, at the nominal working point and for the momentum spread of $\pm$ 1 %. Note that the beam core at the extraction energy is 81 π mm mrad.

50-GEV MR

The 50-GeV MR accelerates protons from 3 GeV to 50 GeV. The design output beam power is 0.75 MW with 3.3×10^{14} ppp and a 0.3-Hz repetition rate. The MR is equipped with fast and slow extraction channels for particle and nuclear physics facilities [2].

Design Concepts

The design concept of the MR is basically on the same line as that of the RCS. The missing-bend technique further pushes the transition γ of the lattice to the imaginary region to avoid crossing it during the acceleration process. The achromatic arc sections are realized by setting the horizontal phase advance at $2n\pi$, where n is an integer (6).

Ring Layout and Lattice Functions

The MR has a circumference of 1567.5 m with a three-fold symmetry. Each super-period consists of an achromatic arc section (24 FODO cells), a dispersion-free insertion (3 FODO cells) and two matching sections (2 cells each). Main lattice functions of a super-period are shown in Figure 3. The missing-bend structure of the arc sections makes the momentum compaction factor flexible, with its nominal setting below zero leading to an imaginary transition γ. One of the three long dispersion-free insertions is dedicated for slow extraction. Another is for injection, beam abort at the injection energy and beam collimation, and the last for RF acceleration and fast extraction. A half-cell in the matching section has a relatively large β and a small α function for the slow extraction system.

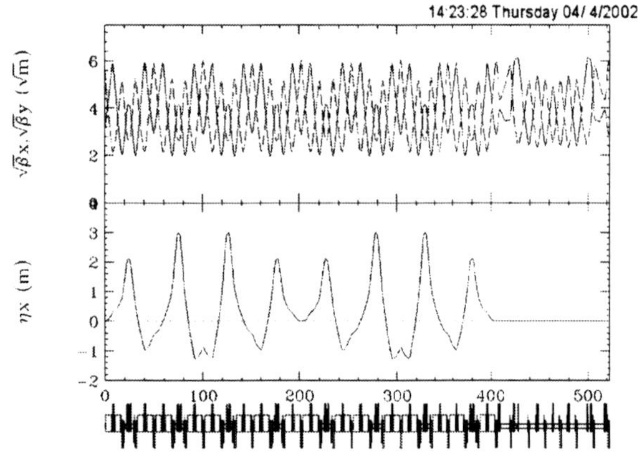

FIGURE 3. Lattice functions of a super-period of the 50-GeV MR.

Working Point and Tuning Knobs

The nominal working point of the MR is at (ν_x, ν_y) = (22.33, 22.28), with the horizontal tune constrained for the slow extraction. The integer part of the vertical tune is adjustable in the range from 17 to 22. The momentum compaction factor is another tuning knob of the lattice, adjustable from –0.0022 (transition γ at $21i$) to +0.0022 (transition γ at 21).

Slow Extraction

A major difference in the design of the MR from that of the RCS is the existence of the slow extraction channel. See Reference [5] for its design and discussions on the system.

Reference for Other Issues

See Reference [1] for other beam dynamics issues of the MR, such as the dynamic aperture, the closed-orbit distortion and its correction and the injection, beam collimation and fast extraction systems.

SUMMARY AND OUTLOOK

The lattice design of the 3-GeV RCS and 50-GeV MR of the JKJ high-intensity proton accelerator facility is basically completed, while it may still need minor final-stage tuning as the hardware R&D progresses. Most of the key issues and concerns with their beam dynamics design have been examined and cleared. See Reference [1] for further details. With the construction of the facility starting in 2002, the RCS expects its first beam in fall of 2006, and the MR in spring of 2007. Studies of their operation modes are under way on specific topics, *e.g.* the transverse injection painting of the RCS, halo formation and beam loss control and the commissioning scenarios.

ACKNOWLEDGMENTS

The project is supported by the Ministry of Education, Calture, Sport, Science and Technology of Japan.

REFERENCES

1. Accelerator Technical Design Report for High-Intensity Proton Accelerator Facility Project, edited by Y.Yamazaki *et al.*.
2. Imazato, J., "JHF Physics", in these proceedings.
3. Mori, Y., "Accelerator Complex of Joint Project in Japan", in these proceedings.
4. Molodojentsev, A., "Tracking Studies for the JKJ Lattice", in these proceedings.
5. Tomizawa, M., and Yokoi, T., "Radiation Handling in the Slow Extraction of the JHF 50 GeV Ring", in these proceedings.

Tracking Studies For JHF Lattice

Alexandre Molodojentsev, Shinji Machida

KEK, High Energy Accelerator Research Organization
1-1 Oho, Tsukuba, Japan 305-0801

Abstract. Dynamic aperture of the Rapid Cycling Synchrotron and the Main Ring of the JAERI-KEK project is simulated. The non-linear resonances, excited by the fringe fields of the magnets and the field nonlinearity of the sextupole magnets for the chromaticity correction, are studied. The operating point for both machines is optimized to provide maximum survived beam emittance. Finally, main misalignment errors, that introduce additional field non-linearities, are included in the simulations.

INTRODUCTION

Accurate tracking study for JHF [1] (JAERI-KEK Project) lattice should be performed to eliminate excitation of the resonances, which can limit the dynamic aperture. These resonances are determined by the intrinsic non-linearities of the magnetic field even without any magnetic field imperfections or alignment errors. In the case of RCS (Rapid Cycling Synchrotron) acceptance of the synchrotron and the aspect ratio of the magnets (inner diameter over magnet length) should be large to meet the multi-turn injection process. Then the fringe fields of the magnets can change the particle motion significantly. In the case of MR (Main Ring) another source of the magnetic field non-linearity should be taken into account, in particular, the field nonlinearity of the sextupole magnets used to correct the 'linear' chromaticity. Optimization of the working point position is performed to provide maximum available beam emittance survived in the collimator acceptance.

ANALYSIS OF LATTICES

The design working point on the betatron tune diagram for RCS is 6.72 and 6.35 in the horizontal and vertical planes, respectively. After the multi-turn injection the beam should have the full transverse emittances of 216 π·mm·mrad in the both phase planes with the momentum spread of $(\Delta p/p) = \pm 0.01$. The natural 'linear' chromaticities before the correction are −8.253 and −8.359 in the horizontal and vertical planes, respectively. To correct the momentum tune-spread two families of the sextupole are used. The collimator acceptance of RCS is 324 π·mm·mrad. In the area near the design working point the following structure resonance lines are located: $4Q_H=27$ and $2Q_V-Q_H=6$. Even without any magnetic field imperfections these resonances can be excited by the fringe fields of the quadrupole magnets ('octupole'-like effect) and the non-linear field of the sextupole magnets for the chromaticity correction in combination with the fringe field effect of the dipole magnets ('sextupole'-like effect), respectively.

The bending arc of MR consists of 8 identical modules with the betatron phase advance of 0.75 per module both in the horizontal and vertical phase planes. This phase advance in the horizontal phase plane is needed to provide the "zero-dispersion" condition outside of the arcs. The design working point of MR has the betatron tunes of 22.33 and 22.28 in the horizontal and vertical phase planes, respectively. The horizontal tune is determined by the slow extraction technique, based on the 3rd-order horizontal resonance. In the vicinity of the design working point the following structure resonance lines are located: $2Q_H-2Q_V=0$ and $2Q_H=45$. The natural linear chromaticity of MR is about -30 in the horizontal and vertical phase planes that should be corrected by the sextupole magnets placed in the arcs. For the design betatron phase advance per module of the arc all driving terms of the third-order resonances excited by the sextupole magnets and the average sextupole field components of the bending magnets are cancelled [2]. The beam emittance and the collimator acceptance are 54 and 81 π·mm·mrad, respectively.

Simulations of the dynamic aperture limitation are performed by COSY INFINITY [3]. This code is based on the Differential Algebraic and the symplectic integration. To assess the long-term stability of particles in a periodic accelerator structure in a reasonable amount of time, the code uses an approximation of the one turn map, and then track particles with the map. For the accurate presentation of the system, the maps have to be computed to a sufficiently high order.

Linear Detuning Effect

The fringe fields of the magnets induce a variety of amplitude/momentum dependent effects beginning from changing of the betatron tunes. The linear

CP642, *High Intensity and High Brightness Hadron Beams: 20th ICFA Advanced Beam Dynamics Workshop on High Intensity and High Brightness Hadron Beams*, edited by W. Chou, Y. Mori, D. Neuffer, and J.-F. Ostiguy

amplitude dependent detuning coefficients calculated for the design working points of RCS and MR are collected in Table 1.

TABLE 1. Linear amplitude detuning coefficients

	Fringe Field	Sextupole Chromaticity Correction	Combined: FF+SCC
		RCS	
a_{hh} [m^{-1}]	13.909	1.408	14.113
a_{vv} [m^{-1}]	12.824	25.625	33.981
a_{hv} [m^{-1}]	18.354	-22.997	-6.748
		MR	
a_{hh} [m^{-1}]	30.834	180.064	207.610
a_{vv} [m^{-1}]	35.813	135.128	161.758
a_{hv} [m^{-1}]	30.388	171.553	198.183

For RCS the detuning effect is determined by the combined effect of the non-linear fringe field of the ring magnets and the sextupole magnets for the chromaticity correction. In the case of MR mainly the sextupole magnets for the chromaticity correction determine this effect.

DYNAMIC APERTURE

The on- and off-momentum dynamic aperture for both machines is analyzed including the fringe field effects of the ring magnets and the sextupole magnets used for the 'linear' chromaticity correction. For these simulations the dynamic aperture (or the dynamic acceptance) of the machine is determined as the maximum initial beam emittance that remain circulating in the chamber, which has the horizontal and vertical sizes much bigger (10 times) than the physical aperture at the observation point. The physical limitation of the chamber size at the observation point is determined by the corresponding Twiss parameters and the collimator acceptance. Main goal of this study is optimization of the working point position to avoid, first of all, any limitations of the dynamic aperture caused by the nearest structure resonances. Moreover, the beam survival in the collimator acceptance for both machines will be discussed including main misalignment errors.

RCS Dynamic Aperture

For the simulation of the RCS dynamic aperture the 8th order Taylor map of the magnetic field has been used. In the case of RCS the nearest structure resonance excited by the sextupole field non-linearity of the sextupole magnets and the fringe field of the dipole magnets is the third-order difference resonance $2Q_V$-Q_H=6. Another limitation of the dynamic aperture of RCS is determined by the 4th order resonance $4Q_H$=27 excited by the fringe field of the quadrupole magnets.

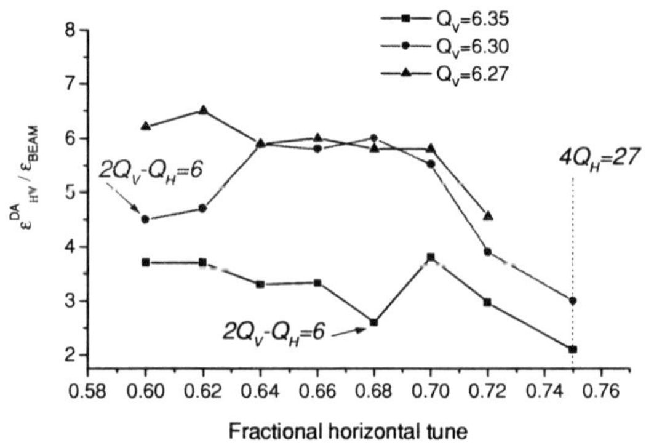

FIGURE 1. On-momentum dynamic aperture of RCS for different Q_H and Q_V around the design values.

In FIG.1 the initial beam emittances in the horizontal and vertical phase planes are equal to each other. The symplectic tracking was performed during 1000 turns. If the operation working point is Q_H=6.72, Q_V=6.35 the dynamic aperture is limited by $\varepsilon_{H,V}/\varepsilon_{BEAM}$~3. This value can be improved about 2 times if the operating working point is located in the area Q_H=6.64...6.70 and Q_V= 6.27...6.30.

From the practical point of view it is necessary to consider survival of the beam inside of the collimator aperture. Moreover, the misalignment errors of the ring magnets can lead to excitation of additional resonances, which are not structure ones. In the case of RCS we should study the influence of the coupling resonance Q_H+Q_V=13, excited by the skew quadrupoles, and the 3rd order resonances $3Q_V$=19 and $2Q_H$-Q_V=7, excited by the skew sextupoles. The axial shift of the sextupole magnets generates an additional source for the skew quadrupole field component.

The following misalignment errors are assumed for this simulation that correspond to the design requirements: σ_{TILT} = 0.5mrad and σ_{SHIFT}= 0.5mm. The Gaussian distribution of the misalignment errors has been generated with the maximum value 2.5σ. In the case of the shifted sextupole magnets the contribution to COD is negligible. In the case of the design Q_V value (Q_V=6.35) the survival of the beam is limited first of all by the structure resonance $2Q_V$-Q_H=6 and the 'linear' coupling resonance Q_H+Q_V=13. If the Q_V is changed (Q_V=6.27), the beam survival can be improved in the Q_H range of 6.64...6.70 (FIG.2).

Further study including the intrinsic magnet field nonlinearities (systematic and random) is needed to estimate the reduction of the dynamic aperture by the high-order resonances.

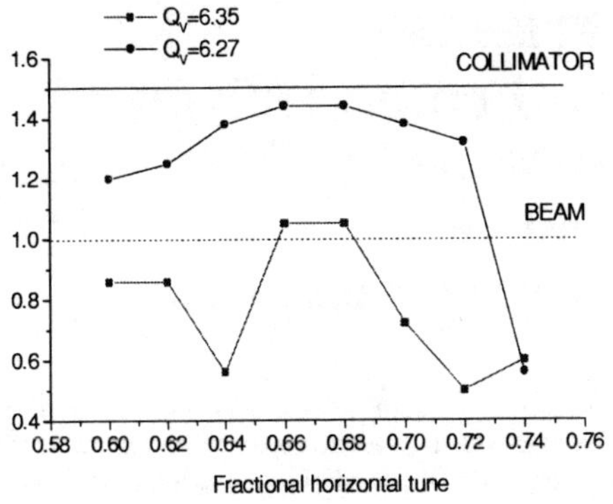

FIGURE 2. Survival of beam in the collimator of RCS for different operating working points including fringe field, sextupole field non-linearities and misalignment errors.

MR Dynamic Aperture

As was stressed above the MR lattice is designed to suppress all driving terms of the third-order resonances in the case of the design working point. From the other side, this working point is located near the DC 'linear' coupling resonance Q_H-Q_V=0. For comparison the MR dynamic aperture is calculated for the design and modified Q_V. The horizontal tune is fixed near the 3rd order horizontal resonance line. The fringe field of the ring magnets and the sextupole magnets for the 'linear' chromaticity correction are included in the tracking during 1000 turns by using the 9th order Taylor map.

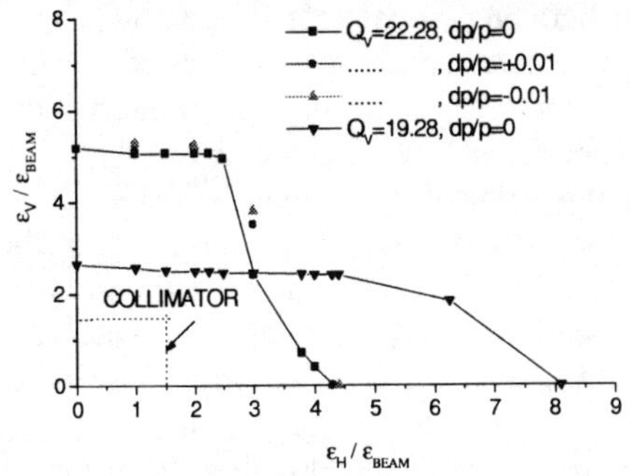

FIGURE 3. On- and off-momentum dynamic aperture in the units of the beam emittance for the design operating point and for the modified point.

The MR dynamic aperture (FIG.3) for the design working point (Q_V=22.28) for the on- and off-momentum particles is depicted. Additionally the on-momentum dynamic aperture is shown also for Q_V=19.28. For both cases the dynamic aperture without any errors is bigger than the physical acceptance of the collimator.

Survival of the beam in the collimator of MR depends on the betatron tune of the particles. In the case of the design and modified operation tunes the MR beam survival without any errors is determined by the coupling resonance $2Q_H$-$2Q_V$=nN_{super}, where N_{super} is number of the super-periodicity of MR (N_{super}=3). This resonance is excited by the 'octupole'-like fringe field effect of the quadrupole magnets.

If the misalignment errors are included in the consideration, the 'linear' coupling resonance is excited by the skew component of the skew quadrupole magnets and the shifted sextupole magnets.

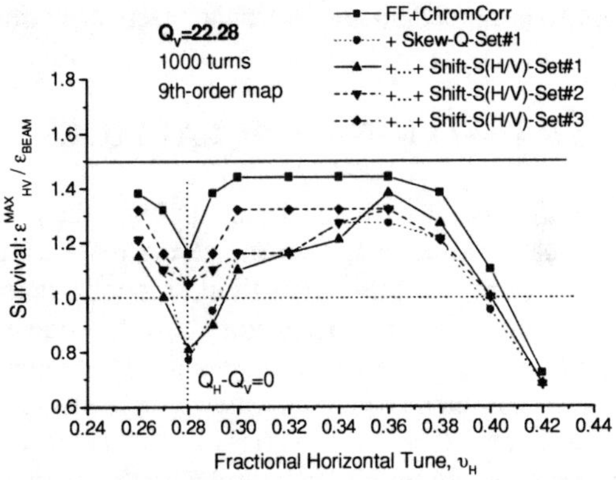

FIGURE 4. Survival of beam in the collimator of MR around the design working point including fringe field, sextupole field non-linearities and misalignment errors.

Survival of the beam in the collimator for Q_V=22.28 is shown in FIG.4 including the misalignment errors with the σ-value similar to the RCS design parameters. In the case of Q_V=19.28 the beam survival for Q_H=22.34...22.38 has the same limit ($\varepsilon^{MAX}_{HV}/\varepsilon_{BEAM}$ ~1.2). But in this case the width of the resonance $2Q_H$-$2Q_V$=6 becomes bigger than the width of the resonance Q_H-Q_V=0 in the case of Q_V=22.28.

ACKNOWLEDGMENTS

The authors would like to thank Y.Mori and S.Ohnuma for many useful discussions, and M.Berz, K.Makino and B.Erdelyi for their support of COSY simulations.

REFERENCES

1. K.Shigaki, F.Noda, K.Yamamoto, S.Machida, A.Molodojentsev, Y.Ishi, "The JHF Lattice", this proceedings.
2. "The proton driver design study", FERMILAB-TM-2136, December 2000.
3. K.Makino and M.Berz, COSY INFINITY version 8, Nuclear Instruments and Methods, A427:338-343, 1999.

Tracking of Three Variants of Transition-Free Lattices for a Proton Driver

Bela Erdelyi* and Carol Johnstone*

*Fermi National Accelerator Laboratory, P.O. Box 500, MS 220, Batavia, IL, 60510

Abstract. Transition-free lattices are favored as possible realization of proton drivers. Several variants have been proposed, some of which have considerably different behavior. One of the main quantities used to characterize this behavior is the short term dynamic aperture (DA). In this note we study three different variants of such lattices, and show that the differences in DA among the lattices essentially disappear as soon as magnet multipole errors are included in the simulation. The tracking results can be understood in terms of the normal form based amplitude dependent tune shift and resonance strength coefficients.

TRANSITION-FREE LATTICES

Transition is defined as the point during acceleration where there is no deviation in the revolution period as a function of momentum. Transitionless machines avoid transition either by injecting beam above the transition energy; that is, the pathlength of the higher momentum particles is sufficiently long that they always arrive later than the synchronous particle (relative to the rf waveform), or, more commonly, by shortening the pathlength of high-momentum particles via lattice design until a shorter revolution period is maintained throughout the acceleration cycle. In general, control over the transition energy is achieved by changing the dispersion function and the location of dipoles with respect to the dispersion function. This is easy to see, since the higher the dispersion, the larger the orbit is offset radially from the on-momentum orbit and, consequently, the longer the traversal time for higher momentum particles (conversely, it becomes shorter for low-momentum ones).

In general, transition is higher in energy the shorter the arc cell length when the phase advance per cell is kept constant. For the PD energies studied here, the cell length would require an inter-quadrupole spacing of only about 4 m which is clearly not an option after considering injection and extraction. The other approaches to controlling transition are "missing-dipole" and "strong focussing" techniques. The missing dipole is based on the standard FODO arc module, but dipoles are removed from high dispersion locations. The PD arc modules studied here are formed from a triplet of FODO cells with dipoles removed from the center FODO cell. The effect is to create a dispersion function across the three cell unit which, when integrated with $1/\rho$ across the length of the remaining dipoles, reduces the slip factor, η when compared with a normal three cell unit. If enough empty space is available, the dispersion function can be driven negative for some fraction of the module length. A second way to control the dispersion function is by increasing the quadrupole focussing strength. One example of increased quadupole strengths has already been mentioned–the shortened standard cell. A second way to increase focussing strength is to create a horizontal lower-beta region at the center of a focussing quadrupole by inserting a quadrupole tripolet or doublet and a short drift. Outside of the low-beta region, the beta functions are matched to normal FODO cell ones, making the insertion transparent in the linear optics (especially if the phase advance across the insertion is close to π). The peak beta functions are the same as the standard FODO cell, and the fractional cell phase advances can be made similar; hence the optical properties are also not strongly different. In this approach long regions of negative dispersion can be created, and, by propitiously locating dipoles, the transition can often be driven imaginary (transition does not occur at any energy indicating pathlength decreases with momentum).

The missing dipole is the simplest structure of the two, but generally has lower dipole packing so that the arcs must be longer, and realizes a more limited range in γ_t. Dispersion suppression for straights is standard or induced with a dispersion wave by instituting an integer phase advance across the arcs. The strong-focussing lattice has the more complicated low-beta insertion, but exhibits standard FODO cell dipole packing and arcs which are more compact. Also, dispersion suppression is usually more efficient in terms of length requirements. For similar peak beta functions, the strong-focussing lattices have higher transitions (unless significant bend is removed in the missing dipole case). PD lattices have been designed with both types of transition control, but the studies which follow concentrate on the "missing dipole" types which have so far been proposed for the new generation of PDs, mainly because of their close resemblence to conventional synchrotron lattices.

However, in the following, an abnormal sensitivity to

CP642, *High Intensity and High Brightness Hadron Beams: 20th ICFA Advanced Beam Dynamics Workshop on High Intensity and High Brightness Hadron Beams,* edited by W. Chou, Y. Mori, D. Neuffer, and J.-F. Ostiguy

the positions of the chromatic-correction sextupoles was discovered in the transition-free lattices proposed which resulted in a local nonlinear cancellation of geometric aberrations and seeming enhancement of dynamic performance. This artifact associated with the specific layout of the arc module was verified by constructing the same lattice using conventional FODO cells (which were not transition-free), and placing the chromatic-correction sextupoles as in the transition-free case. Hence the effect could be pinpointed to the particular placement of the chromatic-correction sextupoles. (For example, standard FODO cells with sextupole correctors every half cell do not show the same sensitivities and depowering of a sextupole family always results in an increase in dynamic aperture for an ideal, or error-free, lattice.) When standard field errors are introduced, the advertised properties (particularly the large dynamic apertures) of these sensitive lattices were found to deteriorate substantially, making the different versions almost identical in performance. The details leading to these conclusions are presented below.

RESULTS

As mentioned in the previous section, we tracked three versions of transition-free lattices for proton drivers. The lattices are termed hereafter: FODO, R. FODO, and PD which represent an arc module composed of 3 standard FODO cells, an identical three-cell FODO module, but with rearranged sextupoles to match the PD locations, and the PD arc module, which has missing dipoles, respectively. (The standard FODO module has chromatic correction sextupoles located near each quadrupole, while the rearranged FODO and PD modules have sextupoles in the center FODO cell only). The ideal lattices were tracked with the full complement of chromatic correction sextupoles, then with with horizontal sextupoles turned off, and, finally, with a realistic set of multipole errors.

The code COSY Infinity was used to track [1, 2, 3]. We note that COSY is a single particle map code, in which the correct ODEs are used (including the kinematic effects and thick elements) with a high precision integration algorithm. The fringe fields were turned off (sharp cutoff approximation). The tracking was performed with an order 18 one-turn transfer map, by first employing a quick search for the approximate DA using the truncated Taylor part, followed by fine tuning with the optimally symplectified EXPO map [4, 5]. It is worth mentioning that, while order 7 effects were still strong in the case of the ideal lattices, order 9 and higher effects were almost always negligible. With errors turned on, higher order effects became important.

The initial conditions for tracking were set up on a grid in polar coordinates in the $x-y$ plane with steps of 0.5 cm along seven directions with polar angles: $0°, 15°, 30°, 45°, 60°, 75°$, and $90°$. The momenta were set to zero and tracking preformed for 1000 turns. As DA along each angle the largest value of radial distance was taken, for which all particles with smaller radii survived.

DA of the ideal lattices

First, we tracked the ideal lattices. In the following, the results are presented along three angles only, because the DAs in other directions follow from an interpolation of the values shown in the tables below.

As shown in Table 1, the ideal lattices show widely different behavior: from the ridiculously large horizontal DA of the R. FODO to the small vertical DA of the FODO. One is tempted to conclude from this table that the R. FODO would be the best choice.

TABLE 1. The dynamic aperture of the three ideal lattice variants. The DA is shown in units of cm along the horizontal, diagonal, and vertical directions in the $x-y$ plane.

Lattice / Direction	0°	45°	90°
FODO	17.5	10.0	6.5
R. FODO	≈ 150	19.0	15.5
PD	30.5	13.5	15.0

However, we will see in the next section that this is misleading. The first clue that some "accidental cancellation" is causing this behavior is obtained by looking at the dynamic aperture of the ideal lattices with the horizontal sextupoles turned off. The results are shown in Table 2. The table is structured in the same way as Table 1. One expects that in the absence of the nonlinearities due to the sextupoles, the DA would increase or at least stay unchanged. This is indeed the case in the horizontal direction, but in the case of the vertical direction of the PD lattice the DA decreases by 33%, which implies cross-talking of the horizontal and vertical planes due to the horizontal sextupoles. Also, a small reduction in the diagonal DA of the R. FODO can be observed. This in turn implies that some accidental cancellations are going on in the ideal lattices, which might not be robust against systematic and/or random multipole errors. This is the case studied in the next section.

TABLE 2. The dynamic aperture of the three ideal lattice variants with horizontal sextupoles turned off. The DA is shown in units of cm along the horizontal, diagonal, and vertical directions in the $x-y$ plane. To be compared with Table 1.

Lattice / Direction	0°	45°	90°
FODO	≈ 110	10.5	8.0
R. FODO	≈ 182	18.5	15.5
PD	≈ 131	14.0	10.0

Lattices with Errors

A set of realistic multipole errors were included in the simulation of all three ideal lattices (with the horizontal sextupoles now turned back on), taken from the Fermilab's Main Injector Technical Design Handbook (see Table 3) [6]. Only normal systematic dipole and quadrupole body errors have been used, the dipole errors up to 14-poles, and quadrupole errors up to 20-poles. We note that the conclusions of these studies are not very sensitive to the magnitude of the errors, so one can think of the conclusions as being valid for any generic set of multipole magnet errors.

TABLE 3. Set of multipole magnet errors used in the tracking, taken from the Fermilab's Main Injector Technical Design Handbook.

Multipole order	Error $< b_n >$	
	Dipole	Quadrupole
4	0.06	-
6	-0.4	0.5
8	0.04	5.85
10	0.33	-0.1
12	-0.01	-1.82
14	-0.03	0.21
16		1.41
18		-0.03
20		-0.8

Under these conditions the DAs of all three lattices change dramatically compared to the ideal lattices. Now essentially all lattices have the same, reduced, DA, as presented in Table 4, with the notable difference of the horizontal DA of the PD, which might be significantly above the noise level. Clearly, the errors now dominate the DA, and any cancellation or point solution that was present in the ideal lattices vanished. Since errors in practice are unavoidable, it follows that realistically there is no clear best lattice in terms of dynamic aperture.

TABLE 4. The dynamic aperture of the three lattice variants with multipole errors turned on. The DA is shown in units of cm along the horizontal, diagonal, and vertical directions in the $x-y$ plane. To be compared with Table 1.

Lattice / Direction	0°	45°	90°
FODO	3.5	1.5	1.5
R. FODO	3.5	1.5	1.5
PD	6.0	2.0	1.5

ANALYSIS OF THE RESULTS

Since the tracking was performed with the one-turn map, the order-by-order amplitude dependent tune shifts and resonance strengths can be analytically extracted using Differential Algebraic normal form techniques [7]. Their relative magnitude sheds light on the reasons behind the widely varying dynamic apertures. We exemplify the process through a few examples.

For example, why is the horizontal DA of the ideal rearranged FODO so large compared to the other two? Looking at first and second order amplitude dependent tune shifts we obtained Table 5. The reason is clear now; the FODO has very large higher order amplitude dependent tune shifts.

TABLE 5. Comparison of first and second order amplitude dependent coefficients of two ideal lattices.

Tune shift coef.	R. FODO	FODO
First order	-0.2	-8
Second order	-2	11828

Or one might wonder why does the vertical DA stay the same for the rearranged FODO and is decreasing for PD, when the horizontal sextupoles are turned off? Again, a quick look at tune shifts solves the puzzle (see Table 6).

TABLE 6. Comparison of first and second order amplitude dependent coefficients of two ideal lattices with horizontal sextupoles turned on and off, respectively.

Tune shift c.	R. FODO		PD	
Horizontal	HS ON	HS OFF	HS ON	HS OFF
First order	-2	-2	-7	-1
Second order	6548	6385	-28516	-8144
Vertical				
First order	0.1	-0.1	2	-6
Second order	154	188	-19	654

Finally, the question arises why are the properties of the three lattices essentially the same when magnet errors are included? For example, the largest tune shift coefficients are roughly equal, as seen in Table 7. Likewise, the dominating resonance driving terms, included in Table 8 are also the same and of the same order of magnitude. However, notice that at third order the (1,0) resonance is not excited in the case of PD, which is the reason why PD has a larger horizontal dynamic aperture.

Finally, we checked also higher order tune shifts and resonance strengths, and the same qualitative aspects persist to very high orders.

TABLE 7. Comparison of the largest first and second order amplitude dependent coefficients of the three lattices with multipole errors on.

Tune shift coef.	FODO	R. FODO	PD
First order	144	94	150
Second order	4161514	3317155	2346097

TABLE 8. Comparison of the dominating resonance driving terms of the three lattices with multipole errors on.

D. t. c.	FODO	R. FODO	PD
3rd order	of (1,0):: 28	of (1,0):: 22	of (3,0):: 12
4th order	(2,-2):: 18217	14605	33198
5th order	(1,0):: 1160894	1174164	946206

CONCLUSIONS

Three variants of a Proton Driver lattice have been investigated for dynamic aperture with emphasis on robustness studies with respect to multipole errors. Ideal lattices showed a wide array of properties, mainly due to the positioning of the sextupoles. However, the "accidental" cancellations of the ideal lattices in general did not survive under the presence of generic magnet errors, and the lattices with errors showed essentially the same behavior (with PD performing a tad better for the set of errors employed in this study).

Although no systematic study has been done, it seems that the choice of the tune alters the results quantitatively, but qualitatively the same conclusions can be drawn.

REFERENCES

1. Makino, K., and Berz, M., *Nuclear Instruments and Methods*, **A427**, 338–343 (1999).
2. Berz, M., COSY INFINITY Version 8 reference manual, Tech. Rep. MSUCL-1088, National Superconducting Cyclotron Laboratory, Michigan State University, East Lansing, MI 48824 (1997), see also http://www.beamtheory.nscl.msu.edu/cosy.
3. M. Berz et al., The COSY INFINITY web page, http://www.beamtheory.nscl.msu.edu/cosy.
4. Erdelyi, B., and Berz, M., *Phys. Rev. Lett.*, **87**, 114302 (2001).
5. Erdelyi, B., Ph.D. thesis, Michigan State University (2001).
6. Fermilab Main Injector Department, The Fermilab Main Injector Technical Design Handbook, http://www-fmiinternal.fnal.gov/MI_Technical_Design/index.html (1994).
7. Berz, M., "Differential Algebraic Formulation of Normal Form Theory," in *M. Berz, S. Martin and K. Ziegler (Eds.), Proc. Nonlinear Effects in Accelerators*, IOP Publishing, London, 1992, p. 77.

AHF Synchrotron Lattices

P. Schwandt*, D.L. Friesel* and F. Neri#

*Indiana University Cyclotron Facility
#Los Alamos National Laboratory

Abstract. Lattice designs for a high-energy, moderate-intensity proton synchrotron complex under study for an Advanced Hydrotest Facility are presented.

INTRODUCTION

The conceptual design of an accelerator complex to provide 50 GeV proton beam pulses for an Advanced Hydrotest Facility (AHF) is underway at the Los Alamos National Laboratory [1]. This complex consists of a 4 GeV, 5 Hz Booster synchrotron, injected from a 157 MeV H^- linac, feeding a 50 GeV slow-cycling (3.3 sec acceleration ramp) Main ring synchrotron. This paper presents design criteria and general properties of the 50 GeV Main ring and 4 GeV Booster ring lattices. Two possible Main ring lattices presently being evaluated are described: a conventional, regular FODO lattice in which the beam crosses transition (real $\gamma_t < \gamma_{max}$) and a transitionless (imaginary-γ_t) lattice.

50 GEV MAIN RING LATTICES.

The intended application of this proton synchrotron in the AHF context demands robust, stable and highly reliable operation but also operational flexibility. Principal design criteria thus call for a simple lattice of conventional and conservative design. Operational simplicity mean few "knobs" and low installation cost

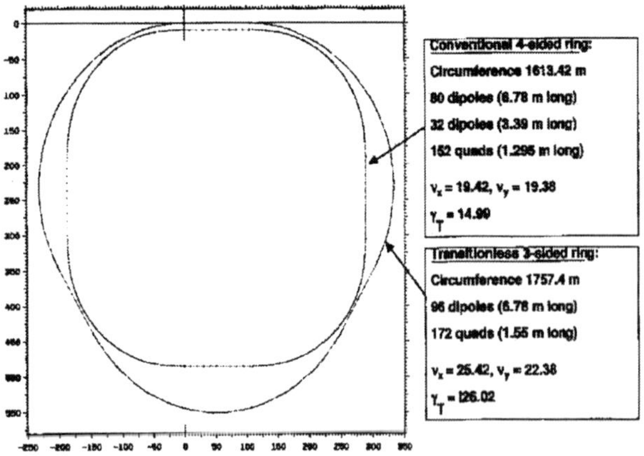

FIGURE 1. Layout of two 50 GeV main ring lattices (m)

requires repetitive cell structures involving few types of identical, room-temperature magnets. The lattice must have moderately large dynamic aperture (at least 60π) and momentum acceptance (at least ± 0.4%), and must accommodate at least 25 200-ns beam bunches of up to 2×10^{12} protons per bunch. The ring should incorporate at least 3 long (100m), dispersionless straight sections for fast kick-injection, 0.2 MV of RF and single-bunch fast extraction (with option for future slow-spill extraction).

Two lattice designs are presently under study. The principal feature separating these two lattices is the transition energy. In the conventional, regular FODO lattice option (A), the beam is expected to pass through transition (at about 13 GeV) with acceptably small beam loss and emittance growth using only the requisite fast RF phase jump, as in the Fermilab Main Injector (FMI). The acceleration rate $d\gamma/dt$ for the AHF ring is an order of magnitude lower than for the FMI and may set an intensity limit. The fall-back option for high-intensity operation is the addition of a bipolar γ_t-jump capability at transition using local dispersion inserts of fast, pulsed jump quad triplets [2].

The second lattice option (B) being explored is a transitionless lattice tuned to imaginary γ_t to avoid transition crossing altogether, as in the proposed 16 GeV Fermilab Proton Driver [3] or the 50 GeV Main ring of the Japan Hadron Facility [4]. Note, however, that in contrast to these machines the AHF ring is a very-slow-cycling machine for lower-intensity beams (by an order of magnitude) of smaller emittance and momentum spread.

The ring layout for these two 50 GeV AHF lattice options is presented in Fig. 1.

Option A: Conventional FODO Lattice

This 4-sided ring geometry employs a regular FODO cell structure repeated throughout the ring, in both arc and straight sections, using identical 92°/ 92° cells. This produces a very simple, straightforward lattice with perfectly smooth lattice functions around the ring (no beta-modulation) but places $\gamma_t = 15$ within the operating range. The design allows for the later addition of pulsed γ_t-jump quad insertions mentioned

CP642, *High Intensity and High Brightness Hadron Beams: 20th ICFA Advanced Beam Dynamics Workshop on High Intensity and High Brightness Hadron Beams*, edited by W. Chou, Y. Mori, D. Neuffer, and J.-F. Ostiguy

above if needed for improved transition crossing. Dispersion-suppressing sections at the ends of arcs require half-length dipoles in addition to full-length (6.8m, 1.64T) arc dipoles. A single family of identical quads (1.3m, 18.5T/m) is employed.

The present choice of operating point in betatron tune space is shown in Fig. 2. These tunes were chosen such that they are not close to integer multiples of the ring periodicity (4) and lie in a region free of systematic and structural sum resonances. Because of negative tune shifts from beam space charge, the working point is located below the octupole coupling resonance $2\nu_x - 2\nu_y = 0$, but since the horiz. & vert. tunes are essentially unsplit, some emittance growth in the presence of strong space-charge forces may occur.

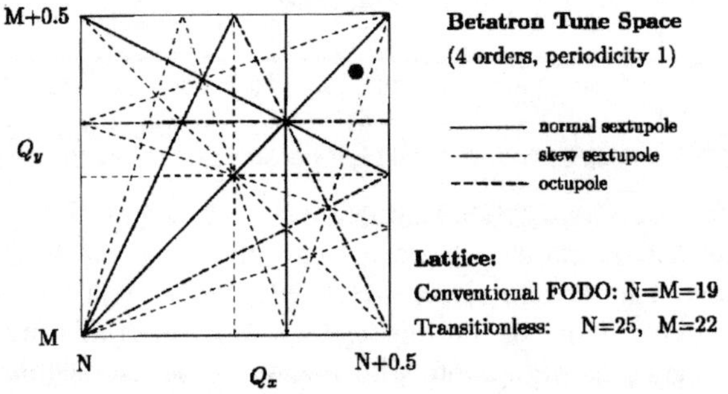

FIGURE 2. Nominal working point in tune space for main ring lattices and low-order resonances.

Option B: Transitionless Lattice.

This 3-sided ring geometry (cf. Fig. 1) uses an arc structure incorporating missing-dipole cells to reduce the value or change the sign of the momentum compaction factor $\alpha = 1/(\gamma_t)^2$, leading to large real or imaginary γ_t. Each arc module consists of 3 DOFO cells, with bend dipoles removed from the center cell. Such a module, as illustrated in Fig. 3, exhibits fairly smooth horizontal and vertical beta-function amplitudes but a bipolar dispersion function D which is either very small (on the average) or negative in the regions occupied by bend dipoles, hence resulting in either very small or negative α, i.e., imaginary γ_t in the latter case. Eight such modules per arc, each tuned to a horiz. phase advance of $7/8 \times 2\pi$, produce an achromatic arc (i.e., with integer phase advance, here $Q_x = 7$) which results in essentially zero horiz. dispersion in each of the 3 long straight sections. Fig. 4 shows 1/3 of a complete ring (straight + arc) with its bipolar dispersion wave through the arc which, for the particular ring betatron tunes of $Q_x = 25.42$, $Q_y = 22.38$, results in a transition $\gamma_t = i26$.

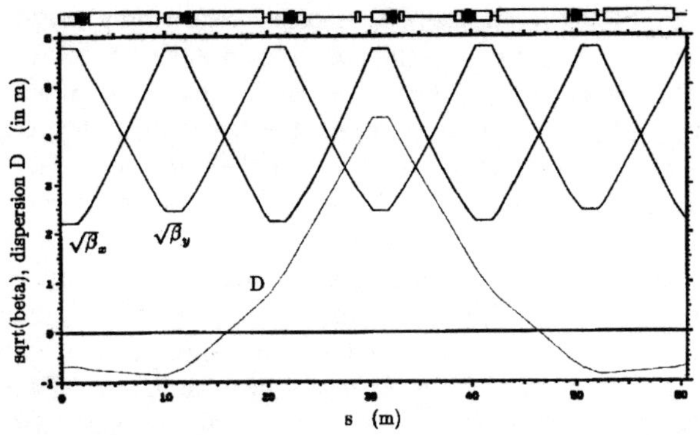

FIGURE 3. Transitionless lattice functions in one arc module.

The fractional tunes result in the same nominal working point location in betatron tune space (see Fig. 2) as for the conventional FODO lattice (A), but now the horiz. & vert. tunes per superperiod, $Q_{x,y}$ = 8.47, 7.46, are separated by one unit, hopefully rendering harmless any effects of the nearby octupole coupling resonance $2\nu_x - 2\nu_y = 6$.

Although proposed as a transitionless ring lattice for fixed operation at imaginary γ_t, this same lattice can also be retuned for operation at either (a) real $\gamma_t > \gamma_{max}$ (by simply switching quad polarities from DOFO to FODO, without any change in cell phase advances), or (b) real $\gamma_t < \gamma_{max}$ with horiz. cell phase advance adjusted to 90 deg which is optimum for addition of an effective γ_t - jump system (if needed) producing only minor, localized perturbation of lattice functions.

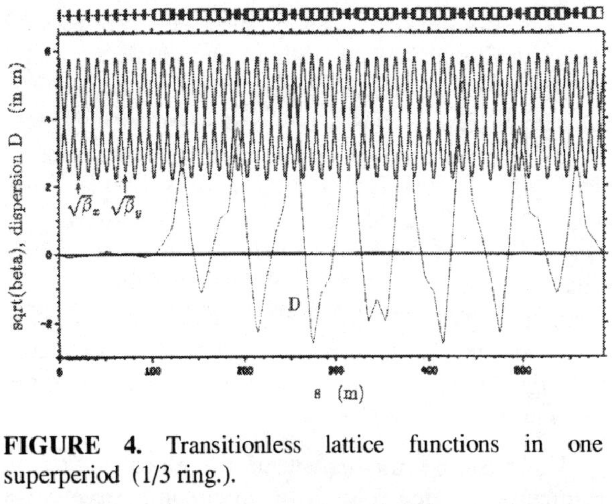

FIGURE 4. Transitionless lattice functions in one superperiod (1/3 ring.).

This single, versatile lattice design can thus provide a flexible choice for as yet unidentified future needs or as yet unforeseen problems or limitations with imaginary γ_t operation (while imag. γ_t has been adopted in several recent high-energy, high-intensity lattice designs, there is as yet no operating experience with such a lattice).

This lattice design (B) is based on a single length of each magnet type (dipole, quadrupole, sextupole). The resulting economy of scale (many identical magnets, few types) reduces the total cost of both magnets and associated power supplies. In comparison to the conventional FODO lattice A, however, lattice B has about 10% greater circumference and requires about 10% more quadrupoles (each 20% stronger) because of the missing-dipole arc cells. Lattice B also has 3 times larger maximum dispersion in the arcs and 40% larger natural chromaticity than lattice A. Two families of chromaticity-correcting sextupoles can reduce ring chromaticity to zero, but the ring will more likely be operated at a chromaticity of around -6; at this level the chromatic tune spread for a beam momentum spread of ±0.2% is only ±0.015.

Particle tracking calculations through both lattices A,B (described in detail in ref. 5) have shown both lattices capable of meeting the dynamic aperture and momentum acceptance criteria quoted earlier. Longitudinal dynamics calculations with space charge [6] have verified the desired evolution of beam bunch time structure during acceleration in both lattices and have shown longitudinal emittance growth and beam loss at transition crossing in lattice A to be manageable at half the design beam intensity in the ring (5×10^{13} protons). The final choice of Main ring lattice will be made after further comparative studies (of collective effects and coherent instabilities in particular).

4 GEV BOOSTER LATTICE

This Booster ring shares many of the general design criteria of the Main ring lattice: robust conventional and conservative design, operational simplicity (few "knobs"), low cost (compact ring, few magnet types) and avoidance of transition crossing. The latter is readily achieved in this small, "low-energy" ring by simply raising the lattice γ_t above γ_{max}. Other design features include H^- strip injection from a linac at either 157 or 800 MeV, and acceleration of a single bunch of 3-4 $\times10^{12}$ protons at a 5 Hz cycle rate, a rate low enough to moderate RF power requirements but high enough to fill the main ring with 25 bunches in a reasonable time (5 sec).

Fig. 5 shows the layout of the Booster ring: a symmetric 9-sided ring, 261m in circumference, with nine 8.4m long straight sections. Each sector or superperiod (1/9 ring) consists of a DOFO bend cell

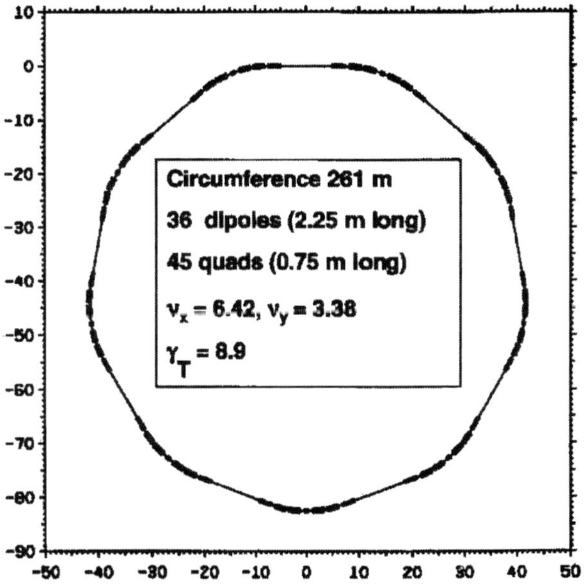

FIGURE 5. Layout of the 4 GeV booster ring lattice (m)

followed by a DOFO straight cell with central F-quad split and separated to create a straight of max. magnet-free length. This structure is equivalent to a 2-cell module with half the dipoles missing, i.e., a missing-magnet configuration which lowers the momentum compaction α and thus raises γ_t. Here, depending on the setting of the central F-quad in the bend cell, γ_t can be adjusted to any desired value above 9 (at fixed sector phase advances).

Figure 6 shows the lattice functions over one sector (center-to-center of adjacent straights) corresponding to γ_t = 9.4. Again, the horizontal dispersion D is a (positive) minimum in the region of the bend dipoles, resulting in a high real γ_t. Positive aspects of this lattice design are a sector horizontal phase advance of nearly 260° (odd multiple of nearly 90°) and a large

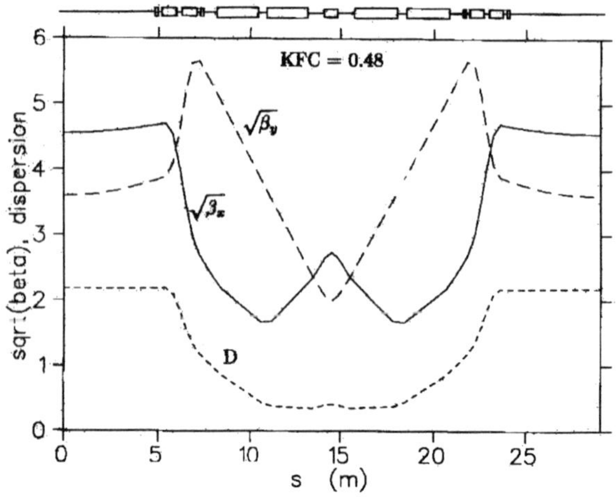

FIGURE 6. Booster lattice functions for one superperiod.

horizontal beta function in the straights. This combination allows (1) convenient placement of injection bumpers to locally a bump the circulating beam on/off the stripping foil, and (2) efficient kick-extraction of the beam where small-angle (2mrad) fast kicker in one straight produces a sizeable (4cm) beam displacement at a septum magnet in the next straight. Negative aspects of this lattice arrangement are (1) the dispersion is a maximum (about 2.2m) in the straights which contain all beam injection/extraction and beam manipulation insertions including RF cavities, and (2) the large vertical beta-function at the entrance of dipoles nearest each straight require a large dipole vertical gap (10cm). Quadrupole apertures are 15 cm diameter.

Again, the fractional ring tune chosen for the Booster lattice places the working point in the same region of betatron tune space, largely free of systematic and structural sum resonances, as for the 50 GeV Main ring lattices. The Booster tune diagram is shown in Fig. 7. Horizontal and vertical tune integers are widely split. Laslett coherent and incoherent space-charge tune shifts of order -0.15 at injection result in a bow-tie shaped tune spread region below this working point which encompasses several sextupole sum resonances, but preliminary beam simulations [7] using the space-charge code SIMPSONS show no dramatic ill-effects at full intensity with high-quality quadrupoles.

Two families of chromaticity-correcting short sextupoles (4 per sector) easily reduce the ring natural chromaticity from -10 to zero, if desired, with modest (0.1T) pole tip fields.

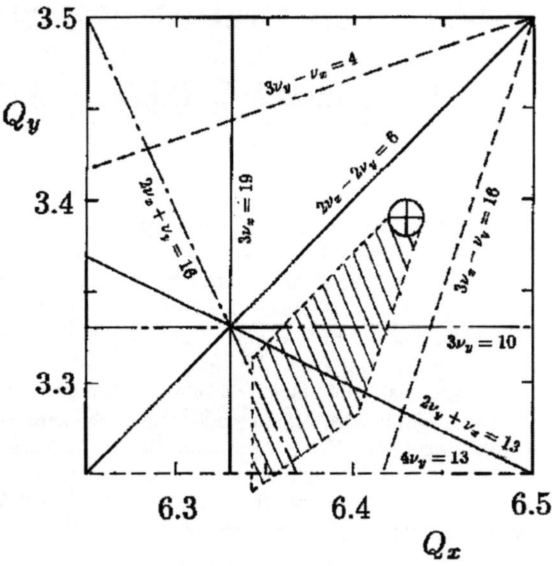

FIGURE 7. Booster tune diagram. Low-order resonances and nominal working point with space-charge tune shift tail at injection.

REFERENCES

1. A. Thiessen, "The AHF Project", these proceedings.
2. V. Visnjic, Phys. *Rev. Lett.* 73, 2860 (1994).
3. Proton Driver Design Study, *Fermilab Report* TM-2136, Dec. 2000.
4. K Shigaki, "JHF Lattice", these proceedings.
5. F. Neri, "Tracking AHF Lattices", these proceedings.
6. T.-S. Wang, LANL, private communication.
7. D.E. Johnson, Dallas, TX and LANL, private communication.

Magnets that Meet Tracking Requirements for AHF*

M.E. Schulze[1], B. Prichard[2], D.E. Johnson[2], F. Neri[3] and H.A. Thiessen[3]

[1]General Atomics, [2]SAIC, [3]LANL

Abstract. The magnet design and fabrication technology for the AHF 50 GeV synchrotron must be of sufficiently high quality to provide a dynamic aperture that does not result in unnecessary beam losses. Tracking studies indicate that the dynamic aperture is smaller than the physical aperture for some sets of systematic and random magnetic field errors in the quadrupoles. Magnet measurements of different quadrupole magnet designs have been analyzed to determine a representative set of systematic and random magnetic field errors that can be manufactured in industry and provide a dynamic aperture larger that the physical aperture. A 2D-error analysis was performed for different quadrupole magnet designs to ascertain the effect of magnet assembly tolerances and errors on the magnetic field harmonics. The analysis indicated that many systematic harmonics were correlated and that specific assembly tolerances and errors (coil placement tolerances and proximity to pole tip, half-core and quadrant assembly tolerances) result in significant random field errors for some magnet designs. Comparison of magnetic measurement data and the error analysis shows substantive agreement. These studies have allowed AHF to develop a set of quadrupole requirements that we feel can be constructed in industry and meet the beam dynamic requirements.

INTRODUCTION

The AHF synchrotron is designed to accelerate a proton beam from 4 GeV to 50 GeV. The dynamic aperture of the synchrotron depends greatly on the magnetic field quality at injection. Initial tracking studies were perform*ed using the systematic and random magnetic field harmonics presented in the FNAL MI Handbook. These studies indicated that the dynamic aperture of the AHF synchrotron is possibly smaller than the physical aperture [1]. The random multipole field errors in the FNAL MI quadrupole were found to be the cause of the small dynamic aperture.

Magnetic measurements of other quadrupole magnets were analyzed and compared with the systematic and random multipole errors presented in the FNAL MI Handbook. In addition, a systematic study on the FNAL MI quadrupole magnet was performed to understand the source and nature of the systematic and random multipole field errors. On the basis of these studies, a set of magnetic field errors was developed which meet the tracking requirements for the AHF synchrotron.

The magnet conventions used in this paper are that b_n and a_n are the normal and skew multipoles with n=1 for a dipole field and n=2 for a quadrupole field such that:

$$B_r(r,\varphi) = B_0 \sum_{n=1}^{\infty} \left(\frac{r}{r_0} \right)^{n-1} \left[b_n \cos(n\varphi) + a_n \sin(n\varphi) \right]$$

The harmonics will be expressed in units such that 1 unit is $10^4 b_n/B_0$ and all harmonics are quoted at a radius of one inch. For a group of magnets, the systematic multipoles are the average value of a given multipole and the random multipole is the standard deviation about the average.

* LA-UR:02-4027

MAGNET ANALYSIS

The FNAL MI quadrupole was studied using the magnet design code POISSON. The purpose of the study was to investigate the effect of mechanical assembly and fabrication tolerances on the harmonics. All POISSON calculations were performed in 2D only. The FNAL MI quadrupole consists of two laminated ½-cores assembled with four coils, each with four turns per pole. Figure 1 shows a 2D cross-sectional view of the FNAL MI quadrupole.

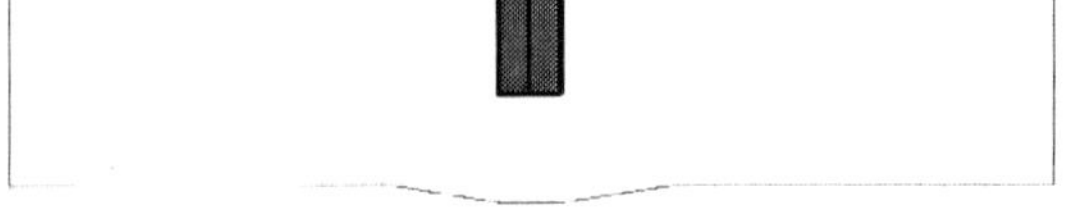

FIGURE 1. Cross Section of FNAL MI Quadrupole

The specific assembly tolerances considered includes gaps, tilts and shifts between the upper and lower half cores. These are the three degrees of

CP642, *High Intensity and High Brightness Hadron Beams: 20th ICFA Advanced Beam Dynamics Workshop on High Intensity and High Brightness Hadron Beams,* edited by W. Chou, Y. Mori, D. Neuffer, and J.-F. Ostiguy

freedom in the assembly of two half cores. In addition the effects of coil shifts, and tilts were considered. For each specific assembly tolerance a POISSON calculation was performed for a nominal offset. This difference between the predicted harmonics in the absence of any assembly tolerances was calculated.

In reality, the magnets will be constructed with a combination of all of these errors and the size of the errors will be different for all magnets. However, there will be distribution to these errors with an average value and a standard deviation. For example, the gap between the upper and lower half cores will be systematically positive with a distribution that can be described by a mean value and a standard deviation.

An analysis was performed in which to predict the measured random harmonic errors based on the above difference tables. The specific differences were added in quadrature with a multiplier indicating the fraction of the nominal error, F_n, (as described above). A second multiplier, S_n, was included to indicate the number of locations the error can occur ($S_n = 4$ for coil asymmetries and 1 for half core asymmetries). This total random error for a specific harmonic was compared to the measured random error. The sum of the squared errors (between measured and predicted random harmonic errors) was then minimized for the harmonics (sextupole through duodecapole only).

The quantities, σ_{An} and σ_{Bn}, representing the predicted random skew and normal multipole errors were calculated as below,

$$\sigma_{A_n,P} = \left(\sum_{i=1}^{n} (S_n F_n D_{An})^2 \right)^{1/2}$$

$$\sigma_{B_n,P} = \left(\sum_{i=1}^{n} (S_n F_n D_{Bn})^2 \right)^{1/2}$$

where,

$S_n = 1$ for steel errors and 4 for coil errors

F_n = Ratio of actual assembly error to calculated assembly error

D_{An} =Change in harmonic due to specific assembly error

The sum over n is the sum of the all assembly errors considered.

The sum of the squared errors, SSE, is then given by,

$$SSE = \sum_{A_3}^{A_6} (\sigma_{A_n,P}{}^2 - \sigma_{A_n,M}{}^2) + \sum_{B_3}^{B_6} (\sigma_{B_n,P}{}^2 - \sigma_{B_n,M}{}^2)$$

Here, $\sigma_{Bn,M}$ and $\sigma_{An,M}$ are the measured random errors.

The results are presented in Figure 2 below and show good agreement between the measured [2] and predicted random multipole errors. The analysis showed that the random normal octupole, B_4, results from a combination of the midplane gap and a tilt between the two half cores. The calculated standard deviation in the midplane gap and tilt was consistent about .001" for both of these errors. The rest of the random errors result from coil placement errors of about 0.020". All other assembly errors did not contribute significantly.

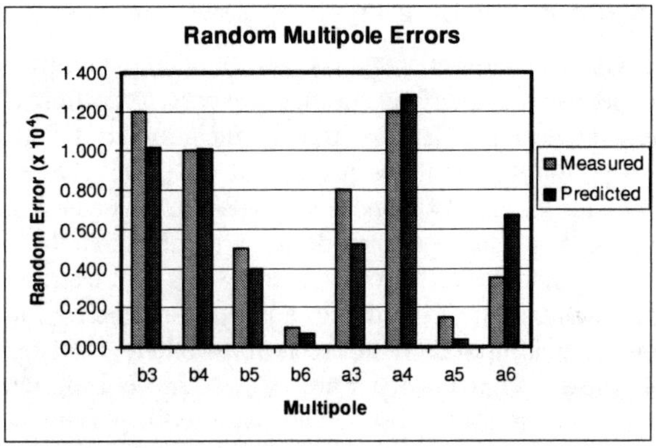

FIGURE 2. Comparison of Predicted and Measured Random Multipole Errors

Figure 3 shows that coil placement errors are the dominant source of random errors with the exception of the normal octupole as described above. This result implies that there will be a strong correlation between the measured harmonics for these multipoles. The strongest correlation is observed between the normal sextupole, b_3, and decapole, b_5, and between the skew octupole, a_4, and the duodecapole, a_6, as shown in Figure 4.

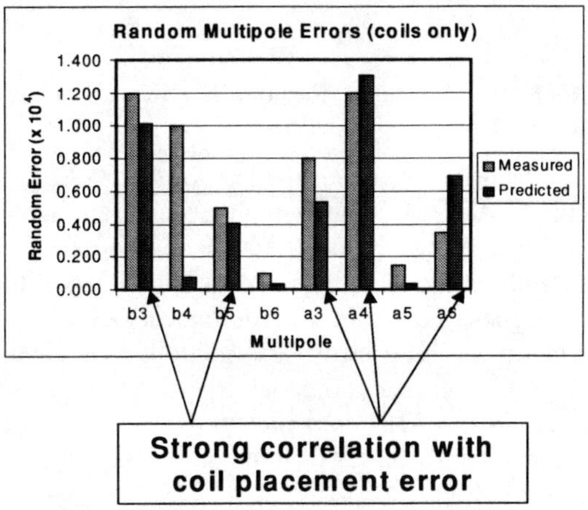

FIGURE 3. Contribution of Coil Placement Errors to Random Multipole Errors

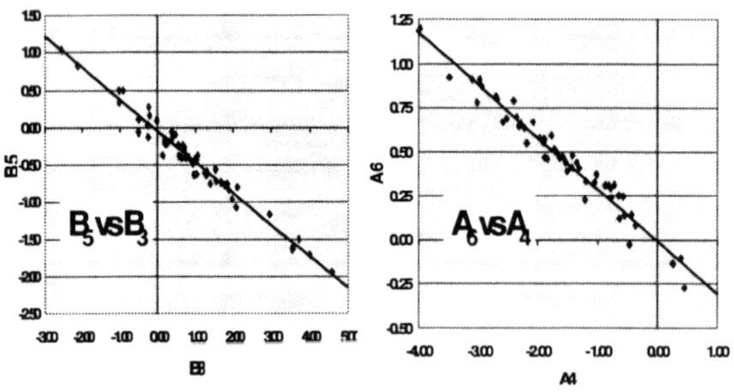

FIGURE 4. Correlation between Measured Harmonics

Other quadrupole magnets were studied to compare random and systematic multipole errors for different magnet designs. These magnets include the FNAL Antiproton Accumulator quads (SQC) [3], SLAC PEP-II quads (LER) [4], and the ANL ALS quads [5]. Figure 5 presents the random harmonics for these magnets and the AHF requirements. This figure shows that the random errors due to coil placement errors in the MI quadrupoles are not seen in the other. Analysis of these magnets does not show the correlation between harmonics associated with coil placement errors.

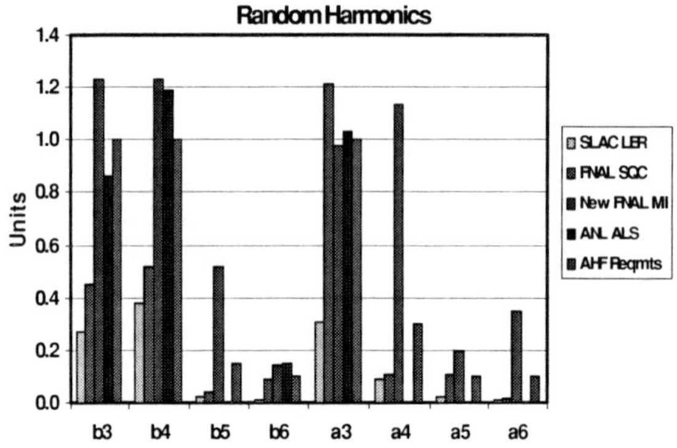

FIGURE 5. Measured Random Harmonics and AHF Requirements

The random errors are smallest for the FNAL SQC and the SLAP PEP-II quadrupoles, which are about a factor of three shorter than the AHF quadrupoles. The AHF random errors are thus expected to be larger than these magnets. To obtain the requirements for the AHF random errors these errors are multiplied by about a factor of two and an additional 30% is added. Using these errors the AHF dynamic apertures is calculated to be significantly larger than the physical aperture [1] as shown in Figure 6.

SUMMARY

We have determined values for the random errors for the AHF quadrupole magnets based on an analysis of the source of the random errors in the FNAL MI quadrupoles magnets. The analysis found that coil placement errors contributed significantly to the measured random errors in the FNAL MI quadrupole magnets. The AHF quadrupole magnets will be designed with the coils located away from the pole tip so that coil placement errors do not contribute to the measured magnet harmonics.

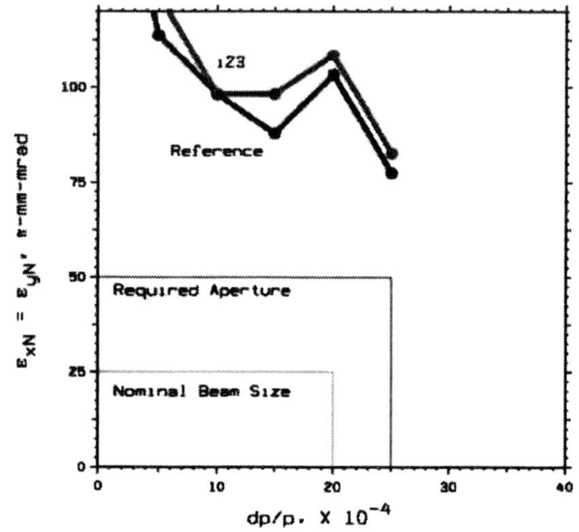

FIGURE 6. Calculated Dynamic Aperture for new AHF Requirements for Magnetic Field Errors

ACKNOWLEDGMENTS

The authors would like to thank D. Harding and B. Brown of FNAL for providing measurement data for the FNAL MI quadrupoles and N. Li, U. Wienands and J. Tanabe of SLAC for providing measurement data for the PEPII LER quadrupole magnets.

REFERENCES

1. Neri, F. et al., These proceedings.
2. David Harding, Private Communication.
3. D. Johnson, Private Communication.
4. Li, N et al., "Performance of the PEP-II Low-Energy Ring Quadrupoles", Proceedings 1997 Particle Accelerator Conference, Vancouver BC, 3318, (1998).
5. Kim, S.H. et al., "Statistical Analysis of the Magnet Data for the Advanced Photon Source Storage Ring Magnets", Proceedings 1995 Particle Accelerator Conference, Dallas TX, 1310, (1996).

Evolution of the Spallation Neutron Source Ring Lattice [1]

J. Wei*, N. Catalan-Lasheras*, A. Fedotov*, C.J. Gardner*, Y.Y. Lee*, Y. Papaphilippou*, D. Raparia*, N. Tsoupas* and J. Holmes†

*Brookhaven National Laboratory, Upton, NY 11973, USA
†Oak Ridge National Laboratory, Oak Ridge, TN 37831, USA

Abstract. Requirements of minimum beam loss for hand-on maintenance and flexibility for future operations are essential for the lattice design of the Spallation Neutron Source (SNS) accumulator ring. During the past seven years, the lattice has evolved from an all-FODO to a FODO/doublet hybrid, the circumference has been increased to accommodate for a higher energy foreseen with a super-conducting RF linac, and the layout has evolved from an α- to an Ω- geometry. Extensive studies are performed to determine working points that accommodate injection painting and minimize beam losses due to space charge and resonances. In this paper, we review the evolution of the SNS ring lattice and discuss the rationales.

TABLE 1. Major lattice parameters of the SNS ring.

Quantity	Value	Unit
Circumference	248.0	m
Kinetic energy	1	GeV
Repetition rate	60	Hz
Number of protons per pulse	1.6	10^{14}
Ring dipole field	0.792	T
Unnormalized full emittance, H+V	240	$\pi\mu$m
Betatron acceptance	480	$\pi\mu$m
Momentum acceptance (full beam)	$\pm$ 2	%
Number of super-periods	4	
Nominal tunes ν_x, ν_y	6.23, 6.20	
Transition energy, γ_T	5.3	
Maximum β_x, β_y	27.9, 15.7	m
Maximum β_x, β_y in arc	12.9, 13.3	m
Maximum ratio β_{max}/β_{min} (H, V)	11.6, 7.5	m
Maximum dispersion D_x	4.0	m
Natural chromaticity (ξ_x, ξ_y)	-7.9, -6.9	

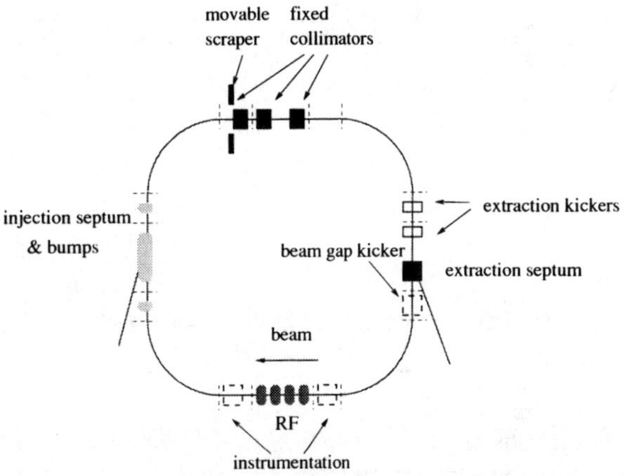

FIGURE 1. Functions of the SNS accumulator ring.

INTRODUCTION

The SNS project is designed to reach an average beam power above 1.4 MW for pulsed neutron production [1, 2]. The accumulator ring, operating at a fixed energy of 1 GeV, compresses at 60 Hz repetition rate 1 ms beam pulses containing 1.6×10^{14} protons to 650 ns bunches for delivery onto the target. Table 1 lists major lattice parameters of the ring.

DESIGN EVOLUTION

As shown in Fig. 1, the ring presently has a four-fold lattice symmetry containing four dispersion-free straight sections, each housing injection, collimation, radio-frequency (RF) system, and extraction [3].

Layout

At pre-construction stage, the lattice's four-fold symmetry [4] was chosen against a three-fold symmetry [5] for its separate, dedicated function of each straight section, reduced maximum dispersion, and fewer structure resonances. Since the start of construction in 1999, several iterations have been made to the general layout of the ring. The early α-geometry (Fig. 2) was later replaced by the Ω-geometry to avoid the crossing of injection and extraction lines for easy tunnel access and maintenance. The circumference of the ring was also increased from 221 m to 248 m to accommodate the potential increase of injection energy up to 1.3 GeV (longer injection-chicane magnets with lower field to avoid H^- stripping), which is made possible by the adoption of superconducting RF linac, and to reduce beam density and foil traversals.

[1] SNS is managed by UT-Battelle, LLC, under contract DE-AC05-00OR22725 for the U.S. Department of Energy. SNS is a partnership of six national laboratories: Argonne, Brookhaven, Jefferson, Lawrence Berkeley, Los Alamos, and Oak Ridge.

CP642, *High Intensity and High Brightness Hadron Beams: 20th ICFA Advanced Beam Dynamics Workshop on High Intensity and High Brightness Hadron Beams*, edited by W. Chou, Y. Mori, D. Neuffer, and J.-F. Ostiguy

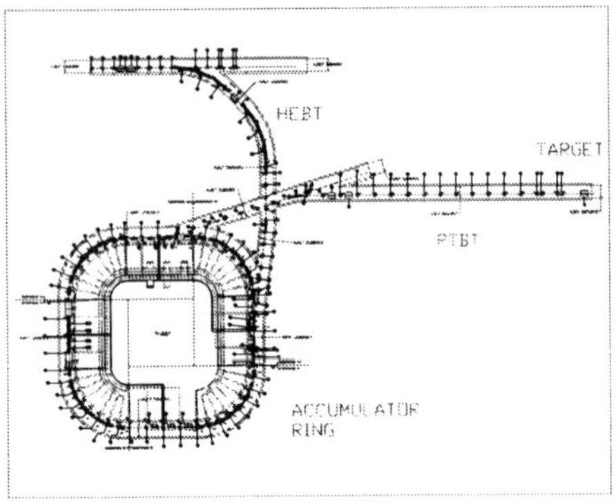

FIGURE 2. Original α-layout of the SNS ring.

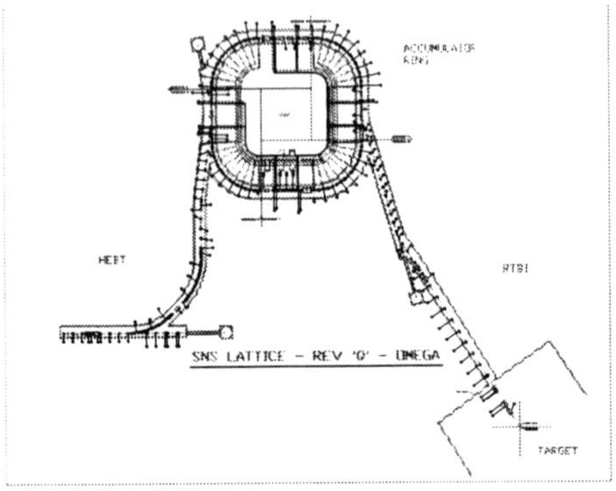

FIGURE 3. Present Ω-layout of the SNS ring.

Lattice

Each achromatic arc consists of 4 FODO cells with 90° horizontal phase advance [6]. In 1999, the original all-FODO structure (Fig. 4) was replaced a hybrid structure with matched FODO arcs and doublet straights (Fig. 5) [3]. The lattice combines the FODO's simplicity and ease of correction with the doublet's flexibility for injection and collimation. The 12.5 m-long uninterrupted straight section with a flexible phase advance further improves collimation efficiency, and eliminates magnets between secondary collimators [7]. Comparing with the original lattice, matching between the arcs and the straights increases the arc acceptance by 50% with the same magnet aperture. The dipoles were also centered in each half cell to maximize the acceptance.

Injection

Injection at a dispersion-free region allows independently adjustable painting in the transverse (with orbit bumps in the ring) and longitudinal (with an energy-spreading phase-modulated RF cavity in the HEBT) directions for a robust operation (Fig. 6) [8]. With the long straight section provided by the doublet lattice, the injection chicane is made independent from the lattice tuning

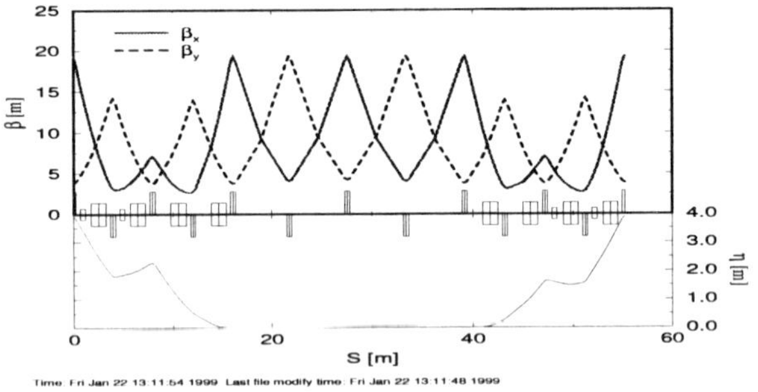

FIGURE 4. Original all-FODO lattice of the SNS ring.

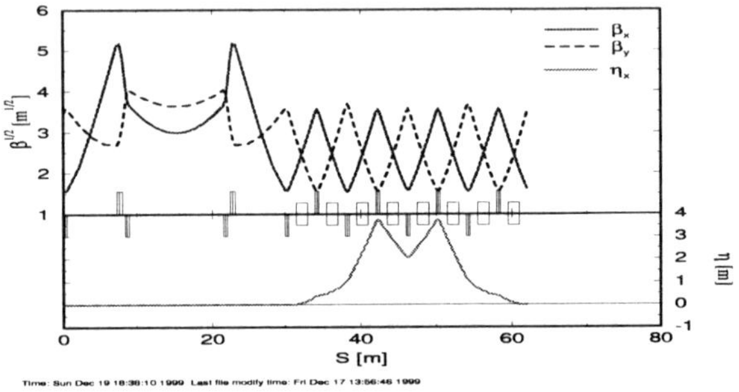

FIGURE 5. Present FODO-doublet lattice of the SNS ring.

(Fig. 7). The long straight section also accommodates possible future arrangement of laser-stripping injection [9, 10]. A symmetric layout of the dynamic bump further simplifies the set-up of transverse painting [11].

Tuning and Working Point

Tune adjustment is provided by five families of quadrupole power supply. Horizontally, the phase advance across the arc is fixed to satisfy achromatic condition. The tuning range is limited to about one unit (from 6 to 7) (Fig. 8). Vertically, the tuning range is about two units (from 5 to 7) without significant perturbation from the injection bump [12].

The nominal working point (ν_x, ν_y) is chosen to be (6.23, 6.2) near the coupling resonance $\nu_x = \nu_y$ (Fig. 9). With full coupling, a round beam is realized even with correlated painting [8]. The full tune spread is accommodated in the tune space without crossing lower-order structure resonances (Fig. 9). An alternative working (6.4, 6.3) is further away from the structure lines $2\nu_x = 12$ and $2\nu_y = 12$, although some third- and fourth-order

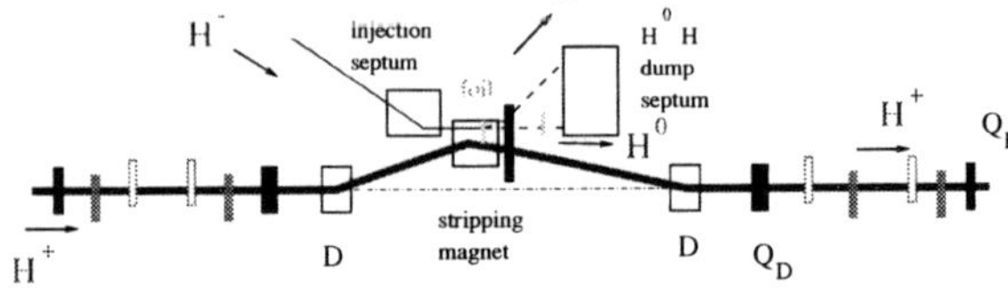

FIGURE 6. Original injection layout with a FODO straight.

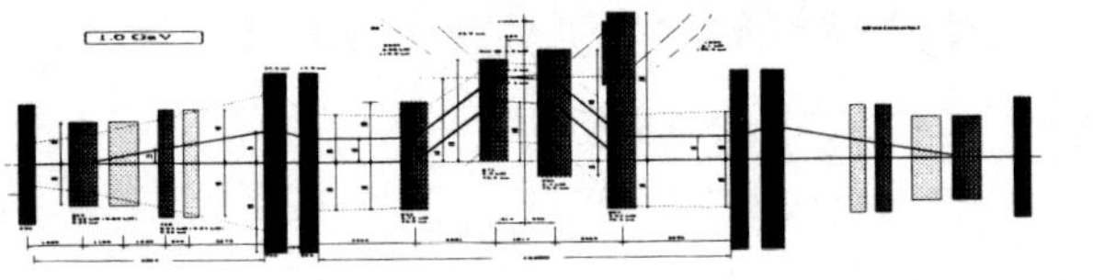

FIGURE 7. Present injection layout with a doublet straight.

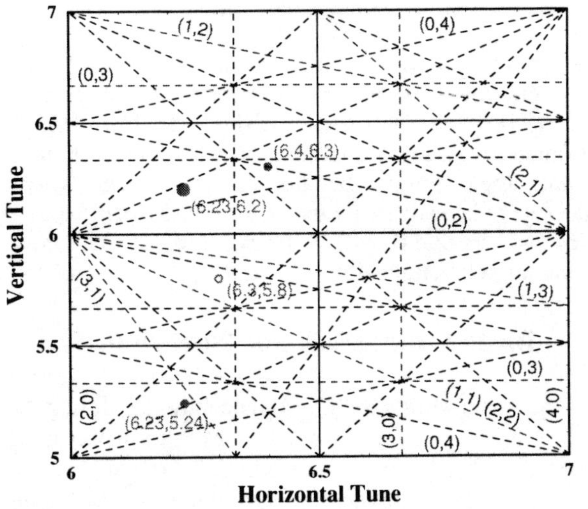

FIGURE 8. Nominal (6.23, 6.2) and alternative working points (6.4, 6.3), (6.23, 5.24) in the transverse tune space. Sum structure resonances up to fourth order are shown in red. Other resonances up to third order are shown in black.

non-structure resonances must be crossed. A split-tune working point (6.23, 5.24) can also be used for decoupled operation [14]. Old working points (5.82, 5.8) and (6.3, 5.8) were abandoned due to matching difficulty and proximity to structure resonance, respectively.

Perturbations and Correction

The injection chicane produces a perturbation of about 5% in $\Delta\beta_y/\beta_y$, and 0.1 m in D_x [12]. Together with the dynamic bump, the dispersions are perturbed by 0.25 m (H) and 0.15 m (V), respectively.

Correctors up to octupole order are used for orbit correction, decoupling, β-wave and vertical-dispersion correction, and resonance correction (Fig. 10) [16, 17].

SUMMARY

With four-fold symmetry, dispersion-free injection, collimation, RF, and extraction, and a matched FODO-arc/doublet-straight optics, the SNS ring lattice is optimized for maximum acceptance, maximized dispersion-free space, flexible injection, high-efficiency collimation, and easy corrections.

We thank G. Rees and the ASAC committee for advice, and Y. Cho, J. Galambos, N. Malitsky, S. Tepikian, D. Trbojevic, W. Wan, and W.T. Weng for discussions.

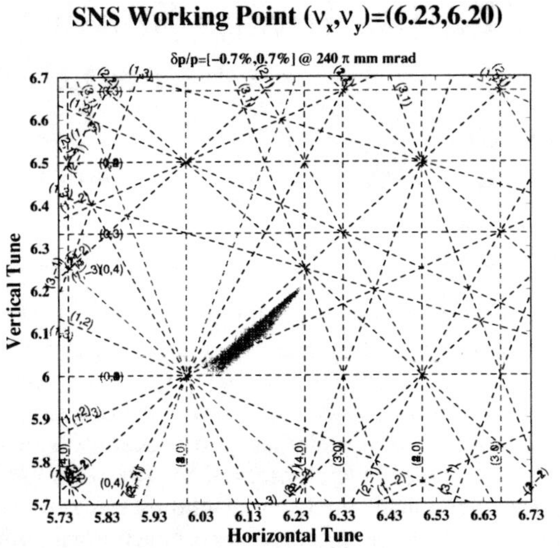

FIGURE 9. Tune spread of a beam containing 2×10^{14} protons in the SNS ring. The computer-simulation results are obtained with the Unified-Accelerator-Libraries (UAL) package [13]. Structure resonances are indicated in red.

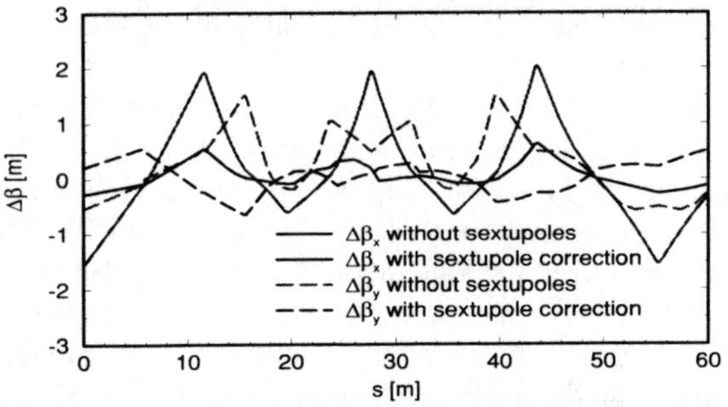

FIGURE 10. Correction of off-momentum ($\Delta p/p = -0.7\%$) optics using four-family sextupoles in a mismatched case. Data taken from [15].

REFERENCES

1. N. Holtkamp, EPAC (2002)
2. J. Wei, EPAC (2002)
3. J. Wei et al, Phys. Rev. ST-AB **3** (2000) 080101
4. Y.Y. Lee, BNL/NSNS/26 (1997)
5. A.G. Ruggiero, BNL/NSNS/1 (1997)
6. C.J. Gardner et al, PAC (1997) 962
7. N. Catalan-Lasheras et al, Phys. Rev. ST-AB **4** (2001) 010101; S. Cousineau et al, these proceedings
8. J. Wei et al, PAC (2001) 2560; EPAC (2000) 978
9. C.J. Gardner et al, PAC (1999) 3182
10. I. Yamane, Phys. Rev. ST-AB **1** (1998) 053501
11. J. Beebe-Wang et al, PAC (1999) 1743
12. J. Holmes et al, PAC (2001) 3618
13. N. Malitsky et al, PAC (1999) 2713
14. A.V. Fedotov et al., PAC (2001) 2848
15. N. Tsoupas et al, EPAC (2000) 1581
16. C.J. Gardner et al, BNL/SNS/31 (1997); BNL/SNS/37 (1997); BNL/SNS/40 (1997); BNL/SNS/82 (2000)
17. Y. Papaphilippou et al, PAC (2001) 1670

Lattice Design Studies for the CERN Proton Driver Accumulator and Compressor

B. Autin, C. Carli, M. Chanel, M. Giovannozzi, M. Martini, Ph. Royer

CERN PS Division, 1211 Geneva 23, Switzerland

Abstract. The CERN baseline scenario for a Neutrino Factory comprises accumulation and compression of the beam delivered by a 2.2 GeV kinetic energy H^- linac, to provide the time structure needed on the target. Originally, accumulation and bunch compression in one single ring was envisaged. To limit the RF voltage necessary for bunch rotation, quasi-isochronous lattices were studied. The low transition energy together with the long circumference, either leads to large dispersion or necessitates the addition of bending magnets with negative deflection angle. Several solutions for quasi-isochronous lattices were investigated. Finally, the quasi-isochronous lattices were abandoned, and a solution has been chosen with two separate rings, one for accumulation and the other for bunch-compression, and conventional lattices working far below transition.

INTRODUCTION

For a Neutrino Factory, a proton beam with sufficient intensity has to be delivered onto a target with a time structure appropriate for ionisation cooling of the secondary muon beam.

The CERN scenario [1] consists of acceleration in a Superconducting Proton Linac (SPL) [2], a 2.2 GeV kinetic energy H^- linac, followed by accumulation and bunch compression for a total beam power of 4 MW. The aim is to convert the long (about 2 ms) linac pulse into a sequence of very short bunches with a spacing appropriate for the pion/muon cooling channel, and a total length smaller than the muon decay ring.

QUASI-ISOCHRONOUS LATTICES

Originally, a solution consisting of one single ring for accumulation and bunch compression has been envisaged [3].

At that time it was felt advantageous to work with a very small momentum slip factor η, i.e. a machine close to transition. Then, neglecting direct space charge forces and potential deformations, modest RF voltages are sufficient to keep the beam bunched during the charge exchange injection and for the bunch rotation.

Furthermore, the circumference of the accumulator ring should be large, in order to limit the number of foil traversals during the charge exchange injection, but short enough that the secondary muon beam fits into the decay ring.

A ring of an appropriate circumference could be located in the existing Intersecting Storage Rings (ISR) tunnel. Thus, lattices should have a geometry which fits into this tunnel, i.e. with a mean radius of 150 m and a width of about 15 m.

Basic Limitations For Isochronicity

The main difficulty [3, 4] encountered during the search for quasi-isochronous lattices is to avoid an unacceptably large dispersion leading, in conjunction with the large relative momentum spread during bunch compression, to large horizontal beam sizes.

This can be analysed by considering the standard expression for the momentum compaction factor:

$$\alpha = \frac{1}{C}\int_0^C ds \frac{D(s)}{\rho(s)} = \frac{1}{\gamma_{tr}^2} \quad (1)$$

with C the machine circumference, $D(s)$ and $\rho(s)$ the dispersion and the curvature radius at the longitudinal location s respectively. In case of a quasi-isochronous lattice, γ_{tr} must be close to the relativistic γ of the beam, which is $\gamma = 3.45$ in our case. Equation (1) can be rewritten, assuming a constant radius of curvature inside one bending magnet (different magnets may have different curvature radii)

$$\alpha = \frac{1}{C}\sum \overline{D}_i \cdot \theta_i \quad (2)$$

with i an index running over all bending magnets, $\overline{D}_i$ and θ_i the average dispersion and the deflection angle of the i-th magnet respectively. Requiring isochronicity and assuming only positive deflections, one can deduce a lower bound for the maximal average dispersion $\max(\overline{D}_i) \geq C/(2\pi\gamma_{tr}^2)$, in the case under discussion $\max(\overline{D}_i) \geq 12.6$ m. The only way to achieve the necessary small γ_{tr}, with the given long circumference and reasonable values of dispersion, is to combine dipoles with both positive and negative deflection and, hopefully, positive and negative dispersion. These considerations are the motivation for the lattices presented in the following.

CP642, *High Intensity and High Brightness Hadron Beams: 20th ICFA Advanced Beam Dynamics Workshop on High Intensity and High Brightness Hadron Beams*, edited by W. Chou, Y. Mori, D. Neuffer, and J.-F. Ostiguy

LEAR-like Lattice

The lattice described in this section is similar to that of the old Low Energy Accumulator Ring (LEAR) [5], in the sense that focusing is mainly performed by doublets on either side of a bending section. However, one more horizontally defocusing quadrupole is added at the centre of the bending sections. The phase advance between centres of bending sections is somewhat larger than π as in LEAR. Whereas this led to a negative dispersion at the bendings (and thus a negative α) in LEAR, this leads to the appropriate behaviour of the dispersion for our application, since the bending sections have alternatively positive and negative deflection angle.

The geometry and the lattice functions along two periods are shown in Fig. 1. The total positive bending angle is 4π, while the total negative bending angle is 2π (to restore the necessary 2π overall deflection). The maximal dispersion is reduced to about 5 m. This type of lattice leads to relatively long straight sections without magnets, leaving space for equipment.

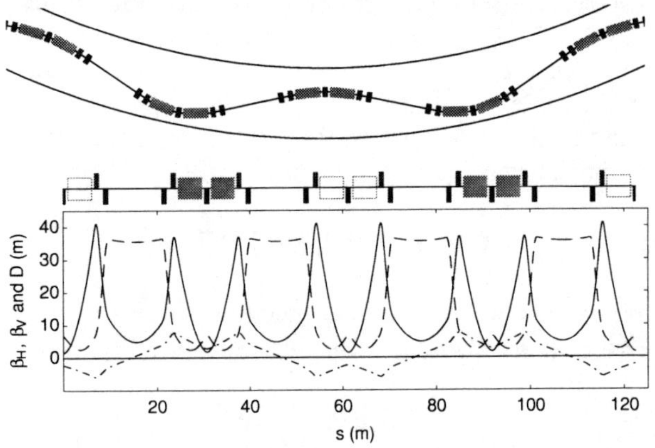

FIGURE 1. Geometry and Twiss functions for the LEAR-like isochronous lattice (β_H solid, β_V dashed, and D_H dot-dashed).

Modified FODO Lattice

The lattice presented in this section is similar to the one described above in the sense that the phase advance between sections with positive deflections and sections with negative deflections is somewhat larger than π. The focusing is done by a rather regular FODO channel with the bending magnets inserted at appropriate locations. Then, the gradients of the quadrupoles are only slightly readjusted to locate the transition exactly at $\gamma_{tr} = 3.54$.

The result is shown in Fig. 2. Again, the total positive and negative deflection is 4π and 2π respectively, and the maximum dispersion is about 5 m. The focusing is smooth and the betatron functions remain generally small.

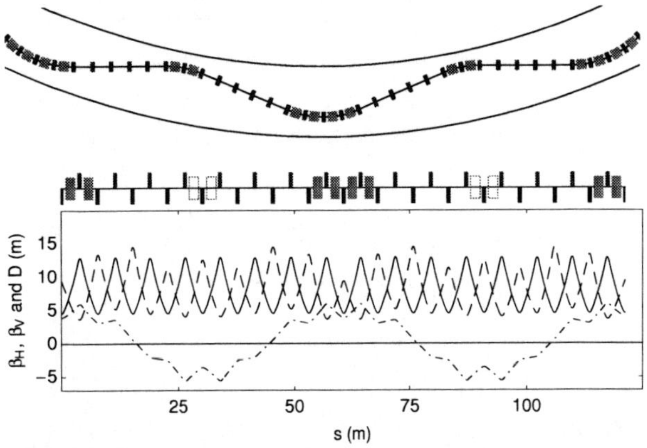

FIGURE 2. Geometry and Twiss functions for the modifi ed FODO lattice (β_H solid, β_V dashed, and D_H dot-dashed).

Lattice With Wigglers

The design philosophy of the lattice with *wigglers* differs from the lattices presented above. One starts from a piece of a regular FODO lattice (called arc in the following) and adds FODO cells with positive and negative bending angles, such that the optics in the arc is not affected by these wigglers [4]. Focusing is provided by a regular FODO lattice and, since this machine is modelled in thin element approximation, the dipoles do not affect the Twiss functions. The regular arc consists of one single FODO cell, extending from the centre of one defocusing quadrupole to the next (see Fig. 3) with two bendings in the centre between quadrupoles. One wiggler consists of three FODO periods with dipoles in between. The first attempt, placing the dipoles at the centres between quadrupoles, did not allow to satisfy all requirements. Thus additional bendings were placed at both ends of the wiggler structure, thus implying combined function elements, $3/2$ periods from the beginning and the end, and at the centre of the structure. Finally, four such modules were placed between regular arc cells. For a given focal length, the three bending angles are used as parameters to adjust the geometry (total deflection of wiggler vanishes), the dispersion in arcs (not affected by wigglers) and α. In thin lens approximation, this can be done analytically [4]. It turns out that the focal length of the lenses has to be rather short w.r.t. the distance between quadrupoles to find real solutions. This leads to large variations of the betatron functions and, thus, to rather large chromatic effects.

Lattice With Dispersion Suppressors And Low-β Insertions

Such a lattice is built by combining different modules in thin lens approximation. The optical properties are, as far as possible, expressed by analytic formulae. The

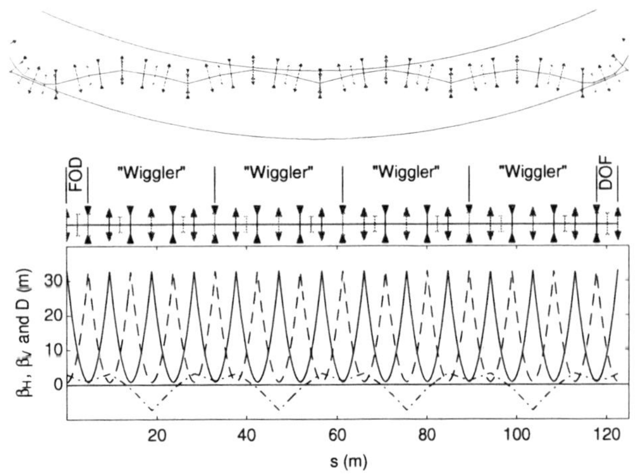

FIGURE 3. Geometry and Twiss functions for the lattice with wigglers (β_H solid, β_V dashed, and D_H dot-dashed).

modules (see also symbolic representation in Fig. 4) are:

- regular arc (made of 3 FODO cells),
- dispersion suppressors (made of 1 FODO on both sides of arc with 2 bendings between quadrupoles),
- matching section (doublet) and
- low-β insertion made of 3 FODO cells.

For a given focal length of the quadrupoles, three parameters (the deflection angles of both the arc dipoles and the two dipoles in the arc) allow to adjust the geometry, the dispersion (zero inside the insertion) and the momentum compaction α. Also in this case the focal length must be small to obtain real solutions for the deflection angles. Two solutions are found, one with a positive and one with a negative deflection angle in the arc. Only the latter is shown in Fig. 4, the first being similar apart from the geometry.

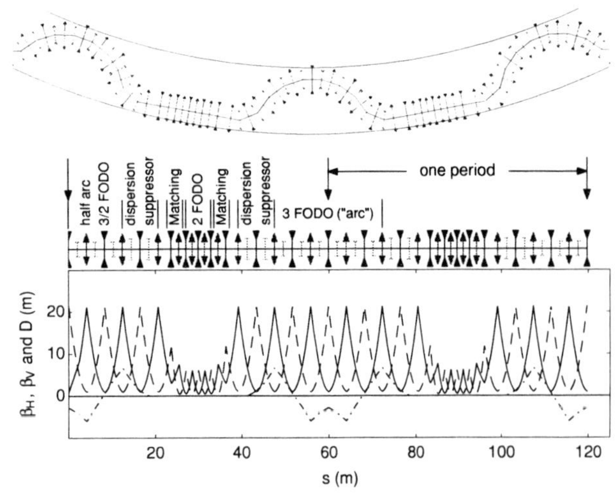

FIGURE 4. Geometry and Twiss function for the lattice with dispersion suppressors and low-β insertions (β_H solid, β_V dashed, and D_H dot-dashed).

CONCLUSIONS

In this paper, a number of ring lattices have been presented, having a large circumference combined with a low transition energy and rather small maximum value of the dispersion function. This has been achieved by using different schemes with various combinations of alternating sign dipoles. All these properties make the optical structures discussed in previous sections good candidates for a lattice of a single ring where both accumulation and bunch rotation take place.

However, plain optics considerations are not the only arguments to be considered in the choice of the final ring layout. In fact, the analysis performed in the framework of a Neutrino Factory study at CERN, showed that the very different physical conditions of the beam during accumulation and compression make it not feasible to perform both actions in the same ring.

The main problem is the beam loading in the high voltage RF system needed to perform the bunch compression. Therefore, the envisaged solution consists of two rings [6]: in the first one a low voltage RF system (about 0.5 MV needed to compensate the longitudinal space charge forces) runs during the 2 ms accumulation stage, while in the second ring the high voltage RF system (about 8 MV), performs the fast compression stage during a few machine revolutions.

ACKNOWLEDGMENTS

We would like to thank all members of the CERN Proton Driver Rings Working Group for stimulating discussions.

REFERENCES

1. Autin, B., et al., "Proton Drivers for Neutrino Factories: The CERN Approach" in *International Workshop on Muon Storage Ring for a Neutrino Factory*, edited by S. Chattopadhyay, Nucl. Instrum. Methods Phys. Res. **A 472**, 2001, pp. 504-10.
2. Autin, B., et al., "Conceptual design of the SPL, a high-power superconducting H^- linac at CERN", CERN-2000-012.
3. Autin, B., et al., "Analysis of Different Options For A High Intensity Proton Driver for Neutrino Factory", CERN PS/CA/Note 2000-001.
4. Autin, B., et al., "Application of Wigglers to Quasi-isochonous Transport Systems" in *7th EPAC Conference*, edited by J. Poole et al., Geneva: European Phys. Soc., 2000, pp. 1030-32.
5. Plass, G., et al., "Design Study of a Facility for Experiments with Low Energy Antiprotons (LEAR)", CERN/PS/DL 80-7.
6. Autin, B., et al., "Design of a 2.2 GeV Accumulator and Compressor for a Neutrino Factory" in *7th EPAC Conference*, edited by J. Poole et al., Geneva: European Phys. Soc., 2000, pp. 921-23.

SESSION B

BEAM LOSS, COLLIMATION, AND SHIELDING

Summary Report

N.V. Mokhov and R.C. Webber

Fermi National Accelerator Laboratory,
P.O. Box 500, Batavia, Illinois 60510, USA

PRESENTATION OVERVIEW

This group heard about beam loss, collimation, and shielding issues from a wide range of perspectives including design, simulation, and test and operational experience. The applications covered machines, systems, and processes as diverse as megawatt spallation sources, proton driver machines, high energy beam slow extraction, multi-megawatt spallation targets, colliding 340-MJ beams, and an operational rapid cycling synchrotron.

R. Webber (FNAL) provided a status report on Fermilab Booster operations. Difficulties and efforts to achieve Booster 8 GeV beam operation at 40 kW with acceptable losses were described. Measurement data on current residual radiation levels in the Booster was shown.

S. Cousineau (SNS) presented an efficient two-stage collimator design with performance simulation data for the SNS ring. A necessary design compromise was described in which a secondary collimator needed to be placed at a position of "less than ideal" phase advance in order to reduce the radiation load on a pair of quadrupole doublet magnets in a particularly sensitive location. A moderately complex scheme to clean beam out of the extraction kicker gap prior to extraction to reduce energy deposition on extraction components was described. The beam-cleaning scheme is complicated by the large tune spread in the intense SNS circulating beam. It was noted that a similar "beam in gap" problem exists in high energy collider rings where even a small amount of beam that has escaped RF bucket containment can be sufficient to quench superconducting magnets or damage sensitive particle detectors if partially kicker during the rise time of an abort or extraction kicker magnet. This has been a recent problem in the Tevatron.

C. Warsop (RAL) described collimator designs for the ESS ring. The importance of collimation in achieving acceptable performance at high beam powers was dramatically shown in a picture depicting 25% of the ESS accumulator ring circumference devoted to the collimation system. Another picture showed attention in early design stages paid to critical handling and maintenance issues including quick disconnect services and vacuum seals, modular component design, remote handling mechanisms, and importantly adequate working space in the vicinity of highly activated collimators.

M. Tomizawa (JHF) presented collimator design considerations for a slow extraction system to handle a 50-GeV 750-kW beam. Detailed modeling of collimator performance and interaction of beam with the electrostatic septum wires has been done with the MARS code.

A. Drozhdin (FNAL) informed the group of collimation system designs related to Fermilab Proton Driver proposal efforts. A comprehensive design philosophy for shielding and beam loss control was described. The impact of machine design parameters on collimation system design constraints and options was made apparent by comparison of the designs for an 8-GeV ring and for a 16-GeV ring. In one case, the collimation system could be placed almost completely within one straight section; in the other, it needed to be located in the arc of the ring. Sensitivities of collimation system performance to parameters like closed orbit deviation and betatron tune were shown.

J. Johnson (ORNL) described the impressive code library available at Oak Ridge for radiation shielding design and residual activity predictions. Examples of coupling between codes to take advantage of the best aspects of each while minimizing computing time were shown.

T. Gabriel (ORNL) impressed the group with some of the difficulties of developing and shielding a multi-megawatt neutron spallation target. Actual beam tests have been done at Los Alamos to understand target issues and tackle material deterioration problems caused by hydrodynamic cavitation of the liquid mercury target. The reality of the SNS project situation was "brought home" by pictures he showed of actual site construction. Any remaining SNS design issues must be resolved in a timely manner before they shift from being paper problems to being real problems. The second part of his talk was devoted to targetry and radiation issues of the muon collider and neutrino factory projects, studied in great detail with the MARS code at Fermilab.

J.B. Jeanneret (CERN) steered the session focus from high average power beam to extremely high peak power beams. He talked of the design issues associated with control of colliding 340-MJ beams in the LHC. The

CP642, *High Intensity and High Brightness Hadron Beams: 20th ICFA Advanced Beam Dynamics Workshop on High Intensity and High Brightness Hadron Beams,* edited by W. Chou, Y. Mori, D. Neuffer, and J.-F. Ostiguy

stored energy in each of the beams at 7 TeV is sufficient to melt 800 kg of copper in a single pulse. Yet this tremendous stored energy must be contained in a superconducting magnet ring. The LHC collimation system must protect against unavoidable particle losses that are 2 to 3 orders of magnitude above magnet quench limits. In addition, the LHC must survive a dump kicker failure that can spray 7-TeV particles in an uncontrolled manner - the process studied in detail at Fermilab with corresponding protective measures designed.

SUMMARY OBSERVATIONS AND IDEAS

There is much consensus and are many similarities in approaches to beam loss control, collimation, and radiation shielding on a worldwide basis despite the wide range of machines and initial perspectives:

- Two-stage collimation system designs are nearly universal. These typically utilize a thin primary collimator (scatterer) in each plane followed by one or more absorbers (collimators). The ESS design is an exception, proposing a two-stage system with a thick primary collimator.
- Uncontrolled losses limits at the level of 1 part in 10,000 of the total beam are typical objectives in new machine designs.
- Hands-on maintenance through the bulk of a machine is a goal in all cases with residual dose rates of 10 to 100 mrem/hr (at contact for 30 days irradiation and 1 day cool-down, or at 30 cm from the components for 100 days irradiation and 1 day cool-down) corresponding to about 1 Watt per meter uncontrolled beam loss.
- Some concerns were expressed about SNS and ESS modeling approximations that treat normal components in collimator regions as "black absorbers." Scattered particles, ignored in this approximation, may play a significant role in the beam loss distribution, energy deposition, and residual radiation in collimators and other components.
- Important code benchmarking work remains to be done for beam loss, collimation, and residual radiation prediction codes. Some open issues include energy dependence, beam tail modeling, and correlation between particle tracking codes and radiation codes. N. Mokhov suggests definition and development of a "toy model" accelerator, a generally accepted simple but realistic machine model, on which all codes can be run for benchmarking.
- R. Webber concluded the summary stating, "Personally, I am anxious for SNS startup as a true test of the validity and appropriateness of the predictions generated by the heroic efforts that have gone into machine performance and beam loss simulations and collimation system design on a worldwide basis".

Fermilab Booster Operational Status: Beam Loss and Collimation*

Robert C. Webber[†]

Fermi National Accelerator Laboratory, Batavia, IL 60510, USA

Abstract. Beam loss reduction and control challenges confronting the Fermilab Booster are presented in the context of the current operational status. In Summer 2002 the programmatic demand for 8 GeV protons will increase to 5E20/year. This is an order of magnitude above recent high rates and nearly as many protons as the machine has produced in its entire 30-year lifetime. Catastrophic radiation damage to accelerator components must be avoided, maintenance in an elevated residual radiation environment must be addressed, and operation within a tight safety envelope must be conducted to limit prompt radiation in the buildings and grounds around the Booster. Diagnostic and performance tracking improvements, enhanced orbit control, and a beam loss collimation/localization system are essential elements in the approach to achieving the expected level of performance and are described here.

INTRODUCTION

Efforts are underway to meet forthcoming demands on the Fermilab Booster. In addition to supplying 1E16 protons per hour (pph) for antiproton production for Collider Run II, Booster is requested to provide up to 9E16 pph for the MiniBooNE experiment [1][2]. The highest historical beam rate was about 2E16 pph, under much less restrictive radiation constraints than apply today. Booster currently operates near the 5E12 protons per pulse (ppp) expected by each user. The increased hourly beam throughput will be achieved by increasing average beam pulse rates to approximately 6 Hz. Recently Booster has provided beam at average rates less than 1 Hz. Reduction and control of beam loss and resulting radiation impacts is essential.

RECENT OPERATIONS

The Booster has operated for the past year in support of Tevatron Collider Run II operations and Recycler Ring commissioning. Figure 1 shows operation for the first 14 weeks of 2002 at typical peak pulse intensities of about 4.5E12 ppp and peak rates of 1.2E14 protons per minute (7.2E15 pph). These rates are typically sustained for 12-16 hours per day for antiproton production with much lighter demands the remainder of the time. During these 14 weeks, a total of 7.06E18 protons were accelerated corresponding to an overall time-averaged rate of 3E15 pph.

Several measures of beam loss performance have been recently established or improved to track the effectiveness of machine improvements. More than 50

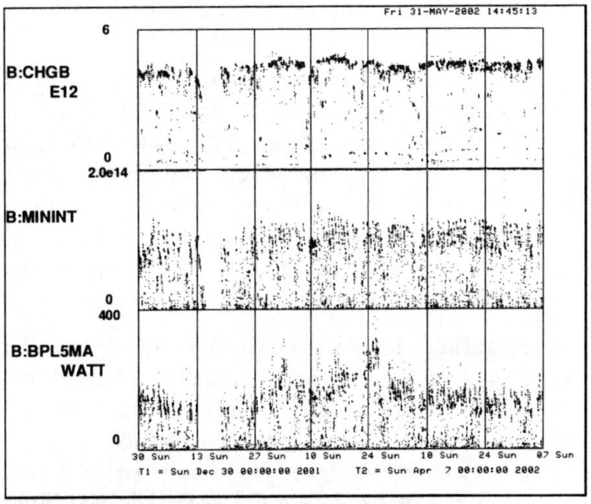

FIGURE 1. Recent 14 weeks of operation. Top: per pulse intensity. Middle: protons per minute accelerated. Bottom: average beam power lost.

radiation monitors (chipmunks), installed around Booster as radiation safety interlocks, provide useful average beam loss information that is data logged. Performance relative to interlock trip levels is tracked and used to project the potential trouble spots as beam rates are increased. This information is available at http://www-bd.fnal.gov/proton/booster/chipmunks/.

Improvements have been made in the data acquisition and data logging aspects of the Booster Beam Loss Monitor (BLM) system. The beam loss observed by each of >50 BLMs during each millisecond of each Booster cycle is measured and individually accumulated for each time interval. The accumulated values and the total accumulated loss are logged to provide long-term capability of tracking performance as a function of time in the cycle at each BLM location. Figure 2 is a typical BLM plot that can

Work supported by the U.S. Department of Energy under contract No. DE-AC02-76CH03000.

[†]webber@fnal.gov

CP642, *High Intensity and High Brightness Hadron Beams: 20th ICFA Advanced Beam Dynamics Workshop on High Intensity and High Brightness Hadron Beams*, edited by W. Chou, Y. Mori, D. Neuffer, and J.-F. Ostiguy

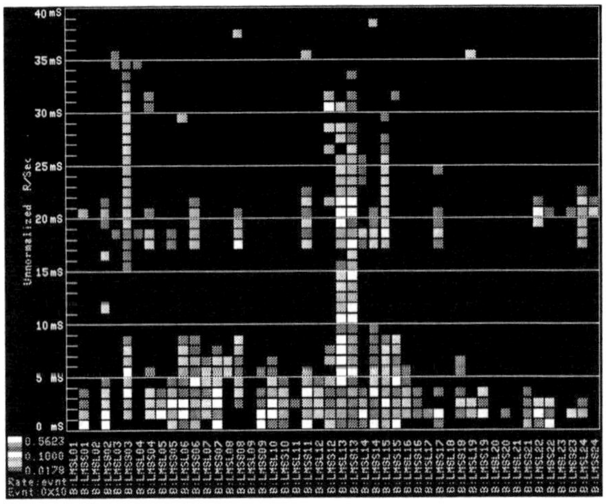

FIGURE 2. Color gradient BLM plot. Horizontal: BLM location. Vertical: time in cycle.

be generated from either current or logged BLM data showing color-coded loss levels around the ring vs. time in the acceleration cycle. The lower horizontal band in Figure 2 shows widely distributed losses during the first 5 msec of the cycle and the band mid-way up the plot shows losses at transition. The vertical bands near the left and center of the plot are losses due to apertures at the two Booster extraction regions.

A real-time 100-second running average of the total loss at each BLM location is also produced and data logged. These running averages are incorporated into the Booster Alarms and Limits system to alert machine operators and potentially inhibit beam should the loss at any location exceed a predetermined limit. This feature is expected to be the front-line defense in controlling component irradiation and residual radiation levels. Data on current machine performance relative to initial limit settings is currently at http://www-bd.fnal.gov/proton/booster/blms/.

Another real-time measurement of total beam energy lost is obtained from the Booster beam current monitor. That signal is effectively differentiated throughout each machine cycle and weighted by the energy at the time of the loss. The result is integrated throughout each cycle to yield a total beam energy lost value and then accumulated over all cycles to provide a running five-minute time-average of beam power lost. Typical levels of 150 to 250 watts during peak operations are shown in Figure 1. Given the 474-meter Booster circumference this represents an average power loss just less than 0.5 watts/meter.

Each shutdown, typically 2 to 8 hours after beam has been turned off, the radiation level at a fixed set of radiation survey points on or near the Booster beam line is measured and recorded to be compared and correlated with the real-time beam loss measures. Data has been accumulated since June 2001. Figure 3 shows the residual radiation levels for the last 9 months at "normal" long and short straight sections around Booster, i.e. excluding injection, extraction, and RF cavity locations. Figure 4 shows data from the same time period at the Booster RF cavity locations. The RF cavity data is particularly important because the power amplifiers, located directly on top of the cavities, are the highest maintenance items in the tunnel. Note that Figure 3 has a logarithmic vertical scale and that rates are "at contact" on or near the beam pipe, whereas Figure 4 has a linear scale and rates are at "one foot" from the cavities. This data corresponds to operation at a time-averaged beam rate of 3E15 pph.

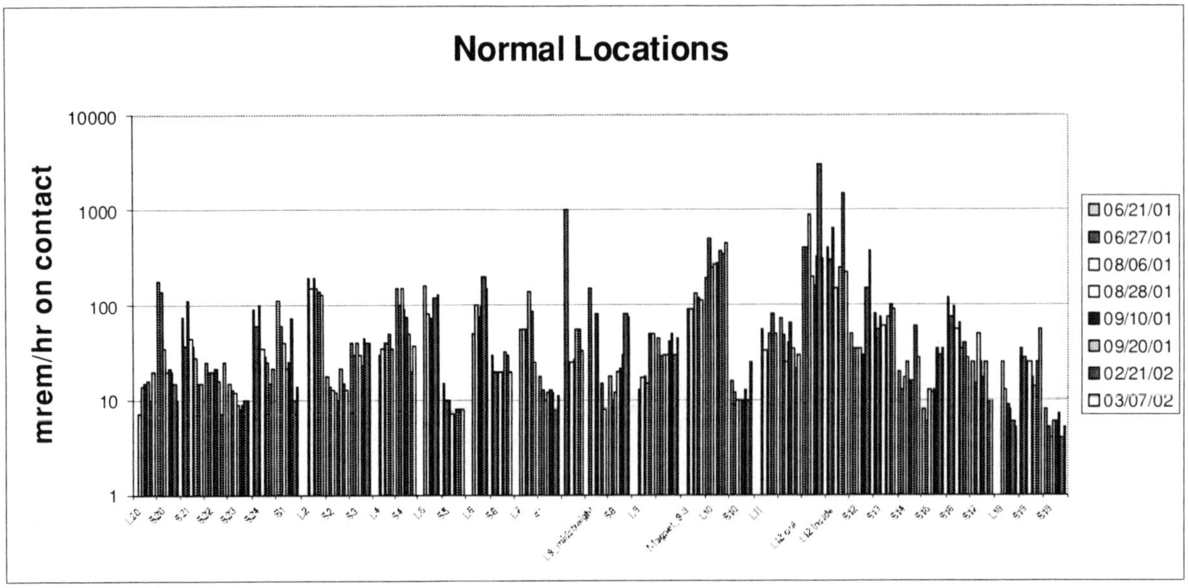

FIGURE 3. Residual radiation levels **on contact** at "normal" locations around Booster over recent 9 months.

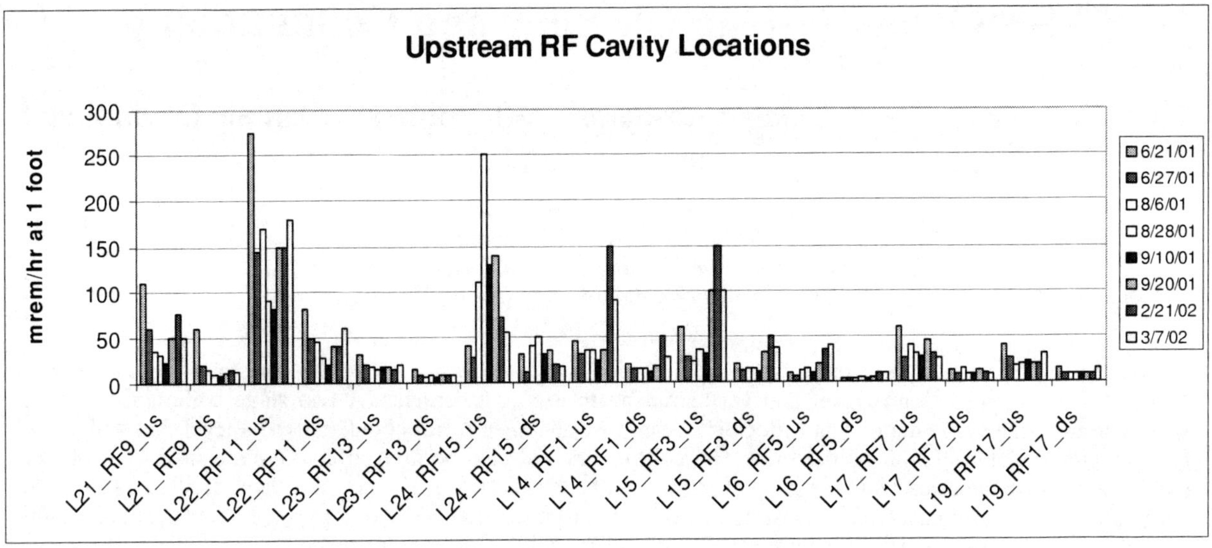

FIGURE 4. Residual radiation levels <u>**at one foot**</u> at RF cavity locations around Booster over recent 9 months.

IMPROVEMENT ACTIVITIES

One avenue to reduced beam loss is through improved orbit control with the addition of ramp capability to the Booster dipole corrector magnet system. Figure 5 shows first test results of ramped correctors attempting to hold position fixed throughout the cycle. Operational utility of the ramping capability will rely on soundly and robustly engineered software with a simple and friendly user interface to enforce and facilitate operational discipline.

A beam collimation system has been designed [3] and recently installed in Booster to localize unavoidable beam losses. The system spans two of Booster's 24 lattice periods and includes in each plane a thin carbon primary collimator followed by two 0.6 meter copper energy absorbing secondary collimators. See Figure 6. Initial tests of the collimation system are just beginning and an acceptable collimator shielding design is still to be completed.

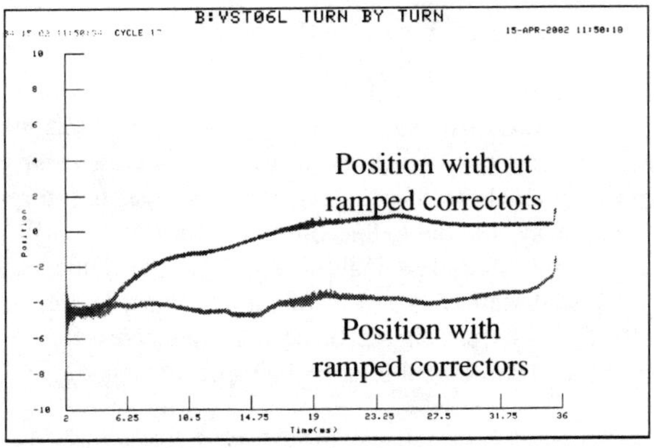

FIGURE 5. Ramped correctors in action.

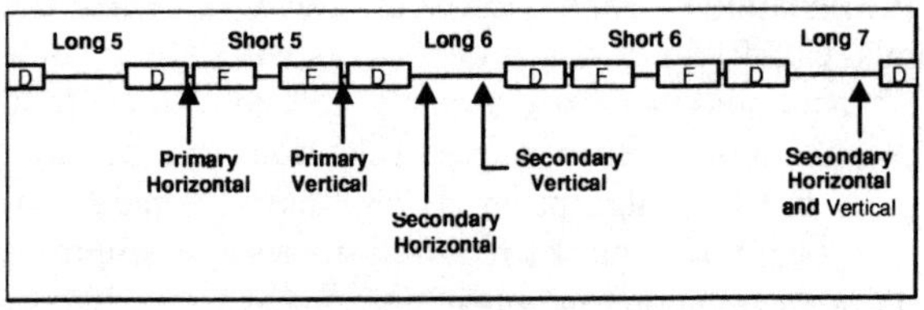

FIGURE 6. Collimator layout.

CONCLUSION

Booster is still far from achieving the requested 1E17 pph beam rate. Improved orbit control and clean collimation system operation will be key to improved performance. Operational loss limits must be established to avoid disastrous component radiation damage and unmanageable maintenance problems. Operational discipline with automated loss monitoring and limiting and data logging tools is important. In any case, Booster will face a new era in dealing with operations in a high radiation environment.

REFERENCES

1. MiniBooNE, http://www-boone.fnal.gov/.
2. R. Webber, "Challenges to the Fermilab Linac and Booster Accelerators", Proceedings of the 2001 Particle Accelerator Conference, IEEE Catalog # 01CH37268, pp. 2581-2583.
3. A. Drozhdin et al., "Beam Loss, Residual Radiation, and Collimation and Shielding in the Fermilab Booster", Proceedings of the 2001 Particle Accelerator Conference, IEEE Catalog # 01CH37268, pp. 2569-2571.

SNS Beam-in-Gap Cleaning and Collimation

Sarah Cousineau*, Nuria Catalan-Lasheras†, Jeff Holmes**, Daniele Davino‡ and Hans Ludewig‡

*Indiana University, Bloomington, Indiana
†CERN, Geneva, Switzerland
**Oak Ridge National Laboratory, Oak Ridge, Tennessee
‡Brookhaven National Laboratory, Upton, New York

Abstract. The Spallation Neutron Source (SNS) under construction at Oak Ridge National Laboratory will operate with very high beam power and very strict beam loss requirements. A two stage collimation system will be used in the accumulator ring to localize beam loss and keep residual radiation to a level that will allow for hands-on maintenance of the machine. The collimators will also be used in conjunction with a set of fast kickers to remove the beam from the gap before the rise of the extraction kickers. We have installed a full-scale collimation package into the ORBIT particle-in-cell tracking code, and have used it optimize both the collimation system and the beam-in-gap cleaning process. Results of the studies are presented here.

INTRODUCTION

The Spallation Neutron Source currently under construction at Oak Ridge National Laboratory will deliver 1.4×10^{14} to a liquid mercury target at a frequency of 60Hz to create the most intense source of neutrons in the world. Due to the very high beam power, unprecedented control of beam loss will be required in order to minimize radiation activation of the machine and allow for hands-on-maintenance in most areas. In the accumulator ring specifically, uncontrolled beam losses are not to exceed 1×10^{-4} of the total beam intensity, or 1 W/m around the ring. Loss levels of this order can not be achieved without the use of collimators that capture high emittance particles before they intercept the machine aperture. For this purpose, the SNS ring will employ a sophisticated two-stage collimation system composed of four adjustable primary scrapers and four fixed-aperture secondary absorbers. The system is designed to absorb either 1×10^{-3} of the beam intensity during every accumulation cycle, or two full beam pulses in the event of a failure scenario.

In addition to scraping the beam halo, the collimation system, together with a set of fast stripline kickers, will provide a means of cleaning the beam in the gap before the rise of the extraction kickers. The beam gap is expected to reach the 1×10^{-4} intensity level during accumulation due to effects such as nuclear scattering and energy loss at the foil, RF noise, and collimation inefficiency. Loss of these particles during the rise of the extraction kicker fields translates into over 100W of power deposition on the extraction region, far in excess of the 1W/m safety level. The proposed solution is to use the stripline kickers to excite the beam in the gap on the betatron resonance and deposit it into the collimation system before the rise of the kickers.

To optimize both the beam in gap (BIG) cleaning process and the collimation system in general, we have installed a complete collimation package into the particle-in-cell tracking code, ORBIT [2]. The structure of the collimation routine is modeled after a standalone collimation code K2 [3], with routines adapted specifically for the SNS energy range. Physical processes included are multiple coulomb scattering, ionization energy loss, nuclear inelastic absorption and nuclear elastic scattering, rutherford scattering, and a detailed treatment of the collimator edge. The ORBIT framework provides a realistic accelerator environment and a complete suite of diagnostics that, combined with the collimation package, allow for detailed studies of how the beam interacts with the collimation system. We have used this program to optimize several parameters of the collimation system, to determine uncontrolled loss distributions around the ring, and to test the full beam in gap cleaning process. We dedicate the first half of this paper to optimization work associated solely with the collimation system, and the balance to the beam in gap cleaning simulations.

COLLIMATION SYSTEM OPTIMIZATION

Much of the design work for the collimators was completed before installation of the ORBIT collimation package, and detailed descriptions of the system have been published [3]. The goal of our work was not to re-design the system but to fine-tune such parameters as primary scraper thickness and material, aperture shape, and collimator location. A few results are discussed here.

In a two-stage collimation system, particles first intercept the primary collimators (scrapers) where they are

CP642, *High Intensity and High Brightness Hadron Beams: 20th ICFA Advanced Beam Dynamics Workshop on High Intensity and High Brightness Hadron Beams*, edited by W. Chou, Y. Mori, D. Neuffer, and J.-F. Ostiguy

imparted a momentum kick. For betatron collimation, the momentum must be translated into displacement before absorption by the secondary collimators can occur. An analytic expression exists for the optimum phase advance between the primary collimator that induces the kick and the secondary collimator that performs the absorption:

$$cos(\Delta\mu) = \sqrt{(\varepsilon_1/\varepsilon_2)} \quad (1)$$

Where $\Delta\mu$ is the phase advance, and ε_1 and ε_2 are the emittance apertures of the primary and secondary collimators, respectively. This phase advance provides the cleanest absorption of the scraped beam by the secondary collimators. However, the equation neglects the existence of other limiting machine apertures. The multiple-coulomb-scattered beam is gaussian in nature and possesses a long tail that risks interaction with limiting ring apertures before optimum phase advance is reached. In the case of the phase-optimized SNS system, the scraped beam must negotiate two quadrupole doublets and over twelve meters of vacuum pipe in its passage from the primary scraper to the last of the secondary absorbers. Collimation simulations which include both beam pipe and quadrupole apertures show that about 7-8% of the scraped beam is lost in the vicinity of the second quadrupole doublet instead of the intended collimator. This motivates a change in location of the last collimator from after the quadrupole doublet to just before the doublet. The new arrangement represents a compromise between optimum phase advance and limiting geometric aperture. Simulations with the new layout result in almost no losses on the doublet, and an overall efficiency increase of about 7%. The separation of the last collimator from its optimum phase advance does allow a portion of the scraped beam to escape the collimation system, but it is below the limiting ring aperture therefore completes the turn without interaction and is collimated on a subsequent pass through the system.

The role of the primary collimators, or the "scrapers", is to kick the halo particles into large emittances without significantly changing the beam energy. The SNS primary collimator is comprised of four tantalum scrapers mounted on copper supports at angles of 0, 45, 90, and -45 degrees. Tantalum is chosen for its high melting point and good durability. The thickness of the scraper determines the impact parameters of particles on the secondaries, which in turn influences the absorption efficiency of these collimators. Figure 1 shows the dependence of the individual collimator efficiencies on the scraper thickness, where we invoke the definition $efficiency \equiv \frac{\#absorbed}{\#impacted}$, and normalize to the maximum efficiency (about 90%).

It is clear that the thicker scrapers provide better absorption efficiency. However, too thick of a scraper leads to scraped beam intercepting the quadrupole and beam

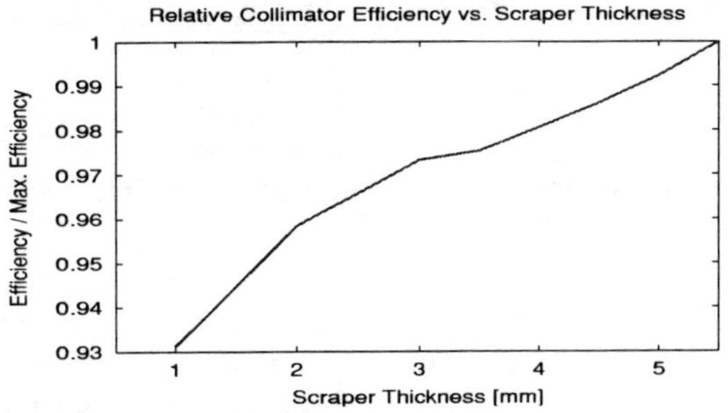

FIGURE 1. Individual collimator efficiency versus scraper thickness.

pipe apertures before reaching the collimators. Based on our studies, the optimum scraper thickness for the SNS system is ≈ 4.0mm.

Figure 2 demonstrates the effect of the fully optimized collimation system. The scrapers aperture was set to $200\pi mm \cdot mrad$, and about 15% of the initial gaussian beam distribution exceeded that value. The initial emittance profile of the beam is plotted along with emittance profiles after 10 turns and after 100 turns. After 10 turns around the machine, the beam halo is almost completely absorbed. Since betatron phasing ensures that at least few turns will be needed for all of the halo to intercept the scrapers, a 10 turn cleaning success implies almost single-turn collimation once particle intercepts the system. Typical loss distributions around the ring show 88-90% of particles being absorbed directly in the collimators, 3-4% impacting the radiation-hard quadrupoles, and about 7% impacting the vacuum pipe. Overall, 98% of all losses occur within the collimation straight section. In our simulations, both the beam pipe and the quadrupoles are treated as black absorbers, whereas in reality some fraction of particles landing there will scatter into downstream collimators. Also, it should be noted that the loss distribution depends heavily on the vacuum pipe radius which is still under design; a conservative estimate of 12.0cm radius was used for the simulation.

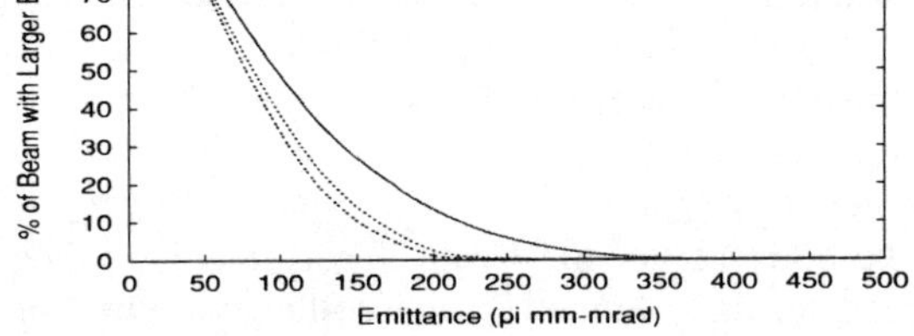

FIGURE 2. Beam emittance profiles before collimation (solid red line), after 10 turns (dashed blue line), and after 100 turns (dashed black line).

BEAM IN GAP CLEANING

Unlike the halo beam, gap particles do not necessarily have large transverse emittances and therefore will not reach the collimators without some type of intervention. Five vertical stripline kickers with a cumulative kicking strength of 1 mrad will be used to drive the gap beam into the collimation system. The kickers are capable of reversing polarity on a turn-by-turn basis according to a programmable polarity sequence, and therefore they can be made fire in resonance with the vertical betatron tune of the ring. To determine the exact sequence of kicks for any given tune, an initial kick of arbitrary polarity is imparted and the beam centroid is followed around the phase space for subsequent turns. On each turn kicks are assigned with the same polarity as the centroid momentum. Though the effect of a kick on each individual particle is sensitive to the particle's phase, using the centroid in this manner guarantees that the net effect is always to increase the magnitude of the momentum in the distribution.

For very small tune spreads, it is only necessary for a sequence of kicks to follow the main tune of the lattice. However, the SNS expects tune spreads on the order of $0.15-0.2$. Kicking on resonance with the bare tune alone will serve to excite only a small fraction of the beam in the gap. We account for the range of tunes by defining entire kicking sequences for several tunes within the spread and apply them sequentially to excite the entire beam.

The simulations included initial gaussian beam distributions of finite momentum spread. The five kickers were modeled with a single, 1 mrad kick, with kick polarity sequences defined by the aforementioned method. Second order transport matrices were used to induce chromatic tune spread in the beam. The full baseline collimation system was used, with the variable-aperture primary scrapers set to $200\pi mm \cdot mrad$. For SNS, the anticipated fractional momentum spread for the nominal (non-gap) beam is $\Delta p/p = 0.007$, the RF acceptance is $\Delta p/p = 0.010$, and the momentum aperture of the ring is $\Delta p/p = 0.020$. Results of simulations with the three different initial beam momentum spreads are shown in Figure 3, where the fraction of particles lost is plotted over several turns of the beam. For each of the three cases, the entire beam is absorbed in under sixty turns around the ring.

The efficiency of the beam in gap cleaning process is fundamentally limited by the collimation efficiency. Gap particles that arrive at the primary collimator before any other aperture will be collimated in the same manner as the non-gap beam. However, the drift speed of the beam in the gap is much greater than the anticipated non-gap halo drift, and therefore it is possible for particles of particular phases to bypass the primary collimators altogether and impact secondary collimators instead. In our simulations, about 75% of the gap particles initially impact the primary scraper, while the remaining 25% overshoot into the secondaries; no particles initially impacted any of the other limiting apertures. Particles that pass directly to the secondary collimators are less likely to be absorbed than those that pass through the primary collimator first. Figure 4 shows typical primary impact parameters on both the scrapers and on one of the secondaries. The primary impact parameters on the secondaries are large, but not as large as they would have been had the particles been scattered by the primary scrapers. This leads to an overall decrease collimation efficiency for beam in gap particles compared to non-gap particles.

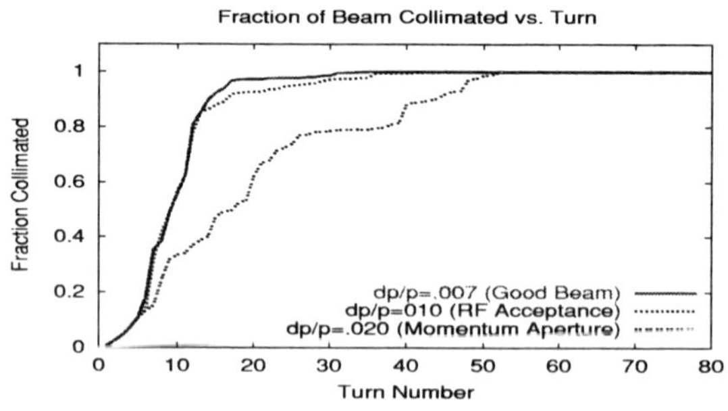

FIGURE 3. Fraction of particles lost versus turn for three different momentum spreads: the solid red line is for $\Delta p/p = 0.007$, the dashed blue line is for $\Delta p/p = 0.010$, and the dashed black line is for $\Delta p/p = 0.020$.

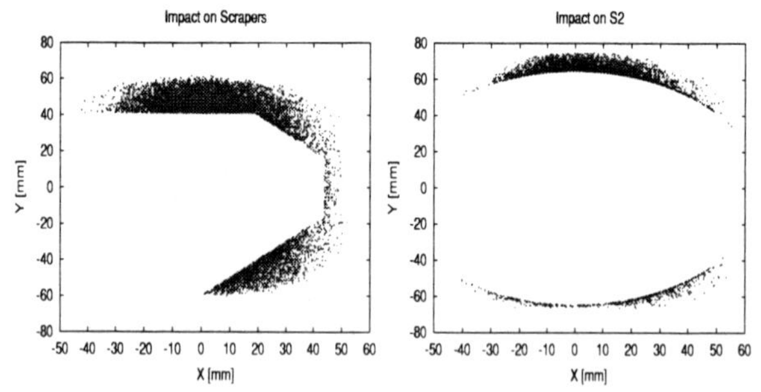

FIGURE 4. The figure on the left shows primary impact parameters of the beam in the gap on the four primary scrapers, and the figure on the right shows the primary impact parameters one of the secondary absorbers.

The two parameters that control the distribution of impacts among the collimators are the primary collimator aperture and the kicker strength. The primary aperture is fixed by the non-gap beam distribution and can not be varied to accomodate the beam in the gap. The other variable is kicker strength, which determines the drift velocity of the gap beam. A decrease in kicker strength would correspond to a slower, more controlled halo drift; however, this could also lead to considerably slower beam in gap cleaning.

CONCLUSIONS

A new collimation simulation package embedded in a particle-in-cell environment has allowed us to perform detailed studies of the SNS collimation system and beam-in-gap cleaning process. Results of the work have led to optimization of collimator location and scraper thickness. Additionally, we have shown that the collimation system can be used with a set of fast kickers to clean the beam in the gap before the rise of the extraction kickers.

This work is supported by SNS through UT-Battelle, LLC, under contract DE-AC05-00OR22725 for the U.S. Department of Energy. SNS is a partnership of six national laboratories: Argonne, Brookhaven, Jefferson, Lawrence Berkeley, Los Alamos, and Oak Ridge.

REFERENCES

1. See K2 code description in N. Catalan-Lasheras, "Transverse and Longitudinal Beam Colliamtion in a High-Energy Proton Collider" PhD Thesis, University of Zaragoza, Spain (1999).
2. J. Galambos et al., ORBIT User's Manual, SNS/ORNL/AP Technical Note 011, 1999.
3. N. Catalan-Lasheras et. al., "Optimization of the collimation system for the Spallation Neutron Source accumulator ring", Phys. Rev. ST Accel. Beams 4, 010101 (2001)).

Beam Loss Control on the ESS Accumulator Rings*

C M Warsop

Rutherford Appleton Laboratory,
Oxfordshire, UK

Abstract. The requirements for beam loss control on the 1.334 GeV accumulator rings of the European Spallation Source are summarised. The main features of the beam loss collector system design are described, along with the underlying aims. Use of a specially developed code to test the proposed system under most foreseeable loss conditions, with machine errors, is described. Predicted collection efficiencies, in terms of localisation and surviving halo, are given. Simulations indicate that the required uncontrolled loss levels of <1 W/m over most of the machine will be achievable.

INTRODUCTION

The ESS Accelerators [1] provide 5 MW of beam power for a short pulse length, 50 Hz spallation target. The ~1.0 ms pulse from the linac is compressed to ~1 μs by two accumulator rings, operating in parallel at 50 Hz. During the ~600 turn charge-exchange injection, 2.34×10^{14} protons are accumulated in each ring. The 1.334 GeV H^- linac provides a mean and peak current of 70 and 114 mA respectively, chopped at 70% duty factor, at the revolution frequency of the rings, 1.242 MHz. This allows lossless capture in the ring RF system, maintaining a gap for extraction. After fast extraction from each ring, beams are brought together in the target transport line.

The need for low and controlled loss has dominated the machine design. Key measures for loss control in the rings include 3D collimation in the injection line, highly optimised injection with 3D painting which minimises proton foil traversals. The concern here is control of lost protons, once they enter the acceptance of the ring [2]. To allow hands on maintenance, loss levels over most of the machine (uncontrolled loss) should not exceed 1 W/m levels. To achieve this, collimator systems localise most loss (controlled loss) in dedicated, well shielded regions of the machine.

Expected and Possible Losses

Regular losses, occurring operationally, 24 hours a day will dominate activation levels. There are unavoidable losses associated with foil interactions, expected at 0.01% levels, manifested as transverse emittance growth and energy loss. Unexpected effects, or non-optimal set up, may lead to some transverse and longitudinal loss. Important examples requiring precautions are possible space charge emittance growth and extraction loss. Also important are *fault losses*, where high loss levels trip off the beam. Up to a full beam pulse may require efficient removal, and fault scenarios may lead to loss in any plane. Highest priorities are for control of foil related losses, and transverse losses. For comprehensive protection, provision is also made for general longitudinal losses.

1. LOSS COLLECTION SYSTEMS

Betatron and Momentum Tail System

The main betatron collimation system is placed downstream from injection, in a dedicated, well shielded, dispersionless straight. For each transverse plane, the design is based on a standard two stage betatron collimation configuration [3], with long primary jaws followed by secondary jaws at relative betatron phases of 90° and 163°. The latter is given by $180-\mu$, with $\cos \mu = \sqrt{(\varepsilon_p/\varepsilon_s)}$. Painted emittance, primary (ε_p), secondary (ε_s) and aperture acceptances are 150, 260, 285 and 480 π mm mr, respectively. The secondary collimation limit is chosen on the basis of expected alignment, and also to protect the extraction system. Additional protective collimation is included at 20° and 32°. To enhance protection, jaws are double (both sides of the beam). This configuration is efficient over a wide range of loss modes. Placement of the primary jaw 180° from the injection point provides collimation of the low momentum tail generated at the foil.

General Momentum System

This consists of single stage collimation placed at the first dispersion peak after the betatron system. It is provided as a precaution against error conditions, and for removal of out-scatter from the betatron system.

Practical Features and Layout

To minimise the number of mechanical units, and enclose conveniently the most active surfaces,

* Work here based on parts of the author's Ph.D. thesis [2]

CP642, *High Intensity and High Brightness Hadron Beams: 20th ICFA Advanced Beam Dynamics Workshop on High Intensity and High Brightness Hadron Beams*, edited by W. Chou, Y. Mori, D. Neuffer, and J.-F. Ostiguy

horizontal and vertical jaws are combined into a single box construction. This tubular geometry is also beneficial for interception of secondary particles and protection of downstream components. Primary jaws are of copper and secondaries of graphite, lengths are ~0.5 and 1.0 m respectively (three interaction lengths). The straights are enclosed in concrete; active handling concepts are included for fast removal and installation.

Compromises are made in placement of betatron collimators for combination of horizontal and vertical jaws, and to protect quadrupoles. The final betatron system consists of combined horizontal and vertical jaws at 0°, 20°, 32° and 90°. The 163° horizontal and vertical collimators are split around the final quadrupole to achieve near optimal phases. Phases are within about 10° of ideal values. Simple collimator jaw designs are selected, flat in the longitudinal and transverse direction. The latter is strongly influenced by the rectangular machine apertures. A beam in gap kicker system could be added if required.

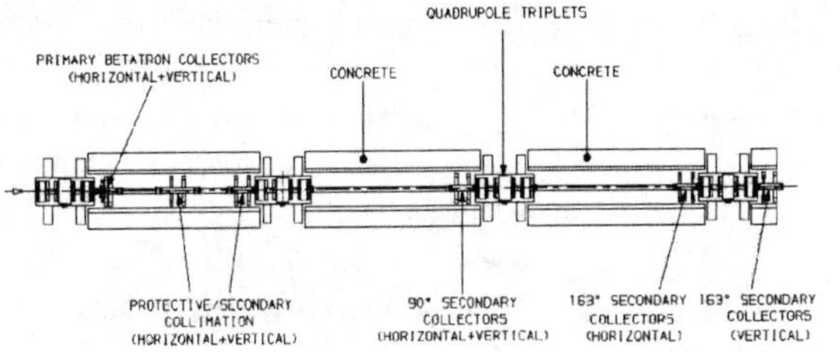

Figure 1.1 Main Betatron System Layout in Long Straight

2. MONTE CARLO CODE

Approach and Aims

Full understanding and control of activation on a high intensity proton machine would require complete models of (i) loss mechanisms (ii) proton loss control (iii) particle cascades and resulting distributions of unstable nuclei. The approach here is to concentrate on (ii) and model this realistically with some care. Loss mechanisms are not well known, therefore simple models of loss modes, for all expected planes and growth rates, are used. The aim here is to determine, and ensure control of, proton loss distributions: activation levels may be calculated from these.

Outline of Simulation

The code tracks protons around the machine, using a detailed lattice model with aperture geometries. Checks are made every half metre to see if a particle has been lost. Representative random magnet and alignment errors, and Q shifts are included.

Collimators are modeled as 3D objects, including interactions with inner faces, and appropriate treatment of out-scattering at boundaries. All important processes: ionisation energy loss, straggling, elastic and inelastic nuclear scattering, and multiple elastic scattering, are included. These have been checked against published data for the relevant regimes.

The aim is to ensure that the design works for most expected loss and machine error conditions. Therefore various loss conditions are defined, modelled, and loss distributions around the machine determined. The extent of surviving halos is also analysed. Basic behaviour is first assessed with no errors, and then observed over a number of runs with standard random error conditions.

3. SIMULATION RESULTS

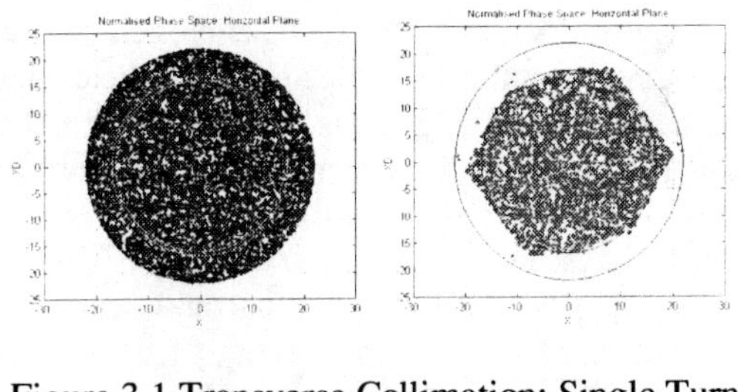

Figure 3.1 Transverse Collimation: Single Turn

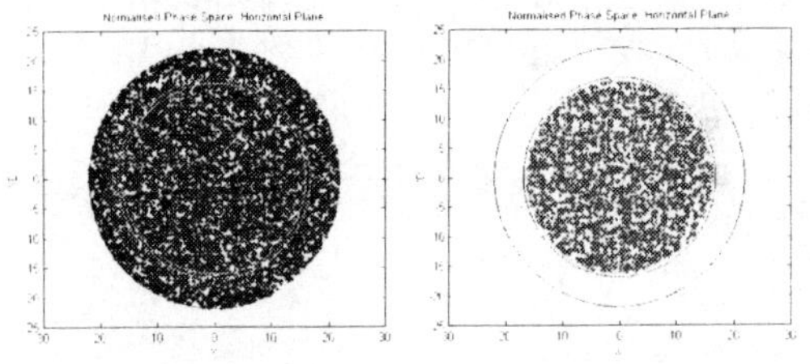

Figure 3.2 Transverse Collimation: Multiple Turn

Test 1: Basic Transverse Collimation

These tests checked the basic optical performance of the system, ensuring design compromises were reasonable. A matched beam distribution uniformly spanning the 480 π mm mr transverse horizontal and vertical acceptances was introduced at the primary jaw, and taken around the machine one turn. The momentum distribution was uniform over the nominal ±0.8% acceptance. The test was then repeated, but over multiple turns. Simulations were also repeated a number of times with randomly generated errors.

Simulation results for single and multiple turn collimation in the horizontal plane are shown in Figures 3.1 and 3.2. These show normalised transverse phase space (axes in units 10^{-3} $m^{1/2}$): input beam and collimated beam, with no errors. The circles indicate acceptances of 260, 285 and 480 π mm mr. The corresponding distribution of lost particles around the whole machine is shown in Figure 3.3, starting at the primary collimator. The lattice is shown schematically: darker elements are the main dipoles, indicating the main arcs of the three super period machine.

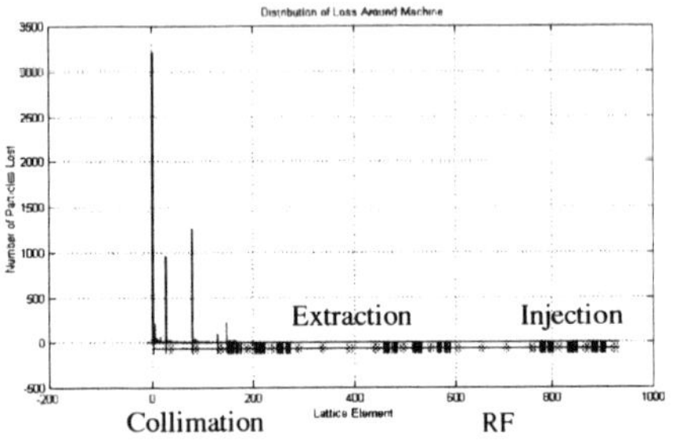

Figure 3.3 Loss Distribution for Test 1: Single Turn

These results are without machine errors: their inclusion had small effect, the most significant being an increase in surviving halo. The first figure shows the cuts made by the jaw system: in a single turn most beam is confined well within the machine acceptance, except that escaping due to out-scatter. The 99% surviving single-turn halos, without and with errors, were 344 and 374 ±3 π mm mr. Corresponding numbers for multi-turn collimation were 261 and 281 ±3 π mm mr. The single turn loss distribution is given in Table 1, case 1: in the worst case with errors, <1% escaped the collimator regions.

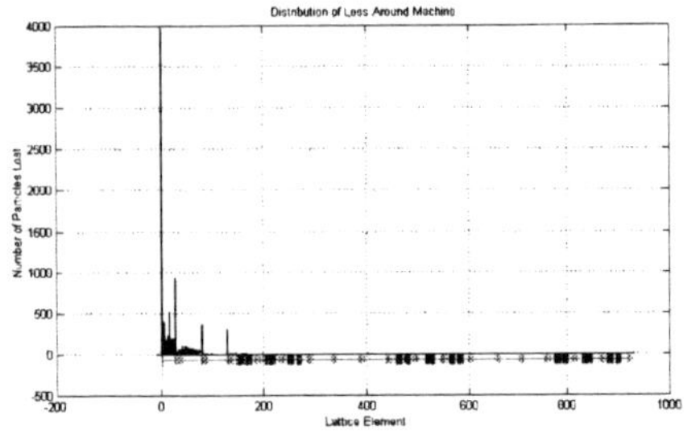

Figure 3.4 Loss Distribution for Test 2: Growth 10 μm/turn

Test 2: Transverse Growth Rate Tests

These tested the operation of the betatron system for a range of transverse beam growth rates, equivalent to 1-10^3 μm/turn at the primary jaws. A beam on a single matched emittance contour was introduced at the primary collimator, transported around the machine, and the emittance then incremented on each turn. The process continued over ~100's of turns until most beam was removed. Growth rates were tested in one plane at a time: in orthogonal planes beam uniformly populated the collimated acceptance. The resulting beam loss distributions for 10 μm/turn growth rates are shown in Figure 3.4, and for 10 and 1 μm/turn in cases 2 and 3 of Table 1. These are for the horizontal plane with no errors: vertical results were essentially the same. As would be expected, spread of loss down the collimation straight increased with slower growth rates. Simulations were repeated with errors. However, worst cases with errors still achieved overall control >95%.

TABLE 1. Percentage Loss in Sections of Machine (All results without errors, except* which are 'worst cases')

case	Betatron 1st Cell	Betatron 2nd & 3rd Cell	Momentum Collimator Region	Rest Of Machine
1	59.6±0.8	37.7±0.4	2.4±0.1	0.3 (0.4*) ±0.1
2	75.0±0.8	23.6±0.4	1.0±0.1	0.4 (1.4*) ±0.1
3	53.5±0.8	43.3±0.4	2.5±0.1	0.7 (4.6*) ±0.1

Test 3: Injection and Extraction Losses

Approximate simulations of the injection process confirmed foil associated losses at <0.01%, and indicated effective loss control. Similarly, efficiency of the general momentum system was satisfactory. Simulations also showed that if unexpected halo survives until extraction, use of correction elements should allow its controlled removal.

Test 4: Collector Options Studied

Simulations, including transverse and longitudinal angles on collimator jaws, showed significant enhancements in loss control. However, improvements did not justify the increased complexity. Similarly, simulations using higher atomic mass primaries also improved localisation of loss, but not to a level that justified the increased inherent radiation hazard.

4. CONCLUSIONS

Studies indicate that loss control to ≥95% levels should be possible. Larger halos observed when errors were included showed the importance of large machine acceptances. Simulated loss distributions suggested uncontrolled loss levels peaking at 0.1 W/m for a 1 kW total loss.

REFERENCES

1 The ESS Project, Volume 3, Technical Report, ISBN 3-89336-299-1, May 2002.

2 C M Warsop, Ph.D. Thesis, University of Sheffield, January 2002.

3 T Trenkler et al., "Principles of Two Stage Betatron and Momentum Collimation", CERN SL/95-03 (AP).

Radiation Handling in the Slow Extraction of the JHF 50 GeV Ring

Takeichiro Yokoi and Masahito Tomizawa

KEK Accelerator Laboratory

Abstract. In the slow extraction of the JHF 50GeV ring, about 1is expected to hit the ESS(Electro Static Septum). The beam loss in the scattering is about 1% corresponding to the beam power of 7.5kW. Considering the tolerable beam loss(0.5∼1W/m), careful study of the beam loss of the scattering beam is indispensable. To evaluate the radiation level around ESS, simulation study with MARS14(00), was carried out. In addition, the beam loss distribution in the ring was studied combining MARS14(00) and single particle tracking. The result shows that the radiation level around ESS is suppressed to the tolerable level. In addition, it was found that the actual beam loss is much smaller than the loss of 1% which hit the ESS wire .

INTRODUCTION

In the slow extraction of the JHF 50GeV synchrotron, whose beam power is 750kW(15μA, 50GeV), the beam hit rate on the ESS wire is expected to be suppressed down to about 1% [2]. Though the loss rate is suppressed so low, the absolute beam power is 7.5kW. Considering the tolerable beam loss from the point of view machine maintenance, ∼1W/m, the loss is quite large. Thus, a detailed study using computer simulation is indispensable. In following sections, the radiation level and loss rate in the ring are to be discussed.

SLOW EXTRACTION CHANNEL OF JHF MAIN RING

The instruments for the slow extraction system, such as ESS, septum magnets, are to be installed in a straight section of 116m long. The layout of the slow extraction line of the ESS channel and the beam envelope is shown in Fig. 1. In the slow extraction section, long straight sections are prepared for the beam collimation system of the slow extraction scattering.

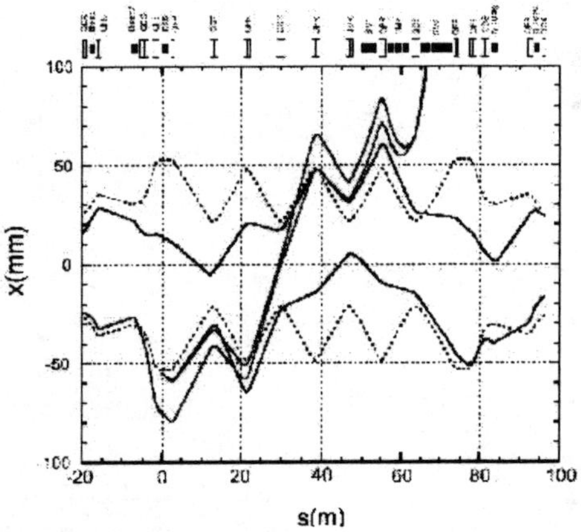

FIGURE 1. Layout of beam extraction devices and beam envelope

SCATTERING AT ESS

Beam loss at ESS is caused by direct hit of proton beam on the ESS wire. In such a case, the loss process and the amount are well defined and understood. In addition, the diameter of the ESS wire is 80μm. In such a thin target, most of the scattered particle go out with Moliere scattering.

In such a case, a detailed computer simulation of beam loss process is indispensable and possible with a good accuracy.

To study the beam loss at the ESS wire, a simulation study with MARS14(00)[1] was carried out. The setup of the simulation is shown in Fig. 2. In the simulation a uniform 50GeV proton beam with the size of 80μm(H) $\times$ 10mm(V) hits the ESS wire from the upstream.

Fig 3 shows the particle distribution at the exit of ESS. It indicates that most of the scattered proton go out with small energy losses, and that half of them keep circulating in the ring for turns. Then, what we have to evaluate are the radiation level around ESS due to the prompt scattering, and the loss map in the ring with multi-turn tracking.

RADIATION AROUND ESS

Using the setup shown in Fig. 2, radiation level around ESS was estimated using MARS. From the point of view of maintainability of the ring, life time of the elements and the residual dose are crucial problems. Among the element of the slow extraction channel, the quadrupole

CP642, *High Intensity and High Brightness Hadron Beams: 20th ICFA Advanced Beam Dynamics Workshop on High Intensity and High Brightness Hadron Beams,* edited by W. Chou, Y. Mori, D. Neuffer, and J.-F. Ostiguy

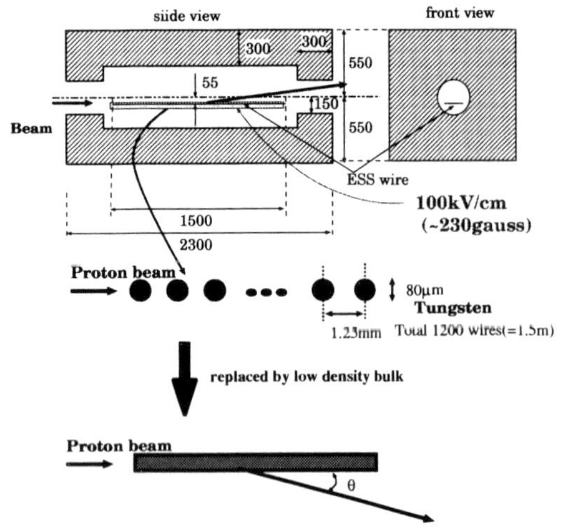

FIGURE 2. Simulation model of ESS

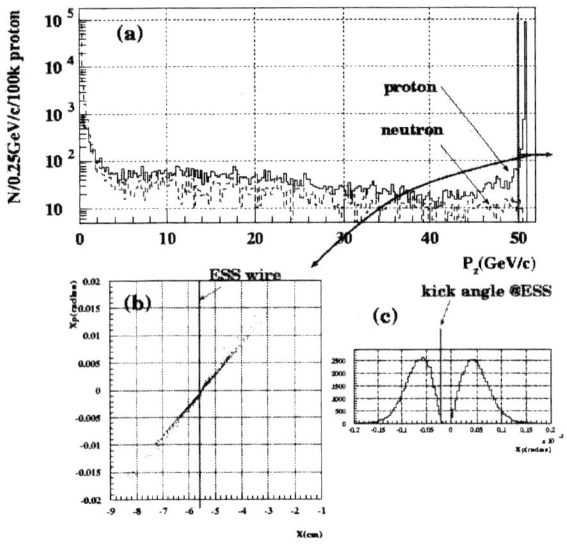

FIGURE 3. Distribution of Scattered particle at the exit of ESS (a)Momentum distribution,(b)phase space distribution of high energy region(>50GeV/c),(c)angle distribution

magnet installed at the downstream of ESS(QFP) is expected to suffer from the most serious radiation damage. Thus, by modeling the quadrupole magnet, radiation level there was estimated.

First, the absorption dose in the coil insulator was estimated. The result indicates that assuming 5000 hour operation per one year, absorption dose reaches $\sim$16MGy/y in maximum. Using polyimide film(Kapton) as the coil insulator, the life time is more than 25 year, which is sufficiently long.

Next, the residual dose rate was estimated. Assuming the operation with 30 day run and 1 day cooling, for the side of ESS and QFP, the residual dose is 15mSV/hour in maximum. However, for the duct between ESS and QFP,which is made of SUS, the residual dose reaches 150mSv/hour[1]!! Therefore, the maintenance of ESS would require special treatment.

However, in most of the ESS channel, the radiation level is tolerable.

BEAM LOSS ESTIMATION OF ESS SCATTERING

From the initial distribution of scattered particle(Fig. 3), half of the scattered particle go out in the first turn. Then, the remaining problem is the loss distribution in the ring for many turns. In order to study the issue, simulation study combining the result from MARS and single particle tracking(SAD) was carried out.

From ESS to the 1st septum magnet, all the scattered particles pass through the same field configuration. Thus, at first, the beam loss up to the 1st septum was examined. Fig. 4 shows the loss map in the straight section.

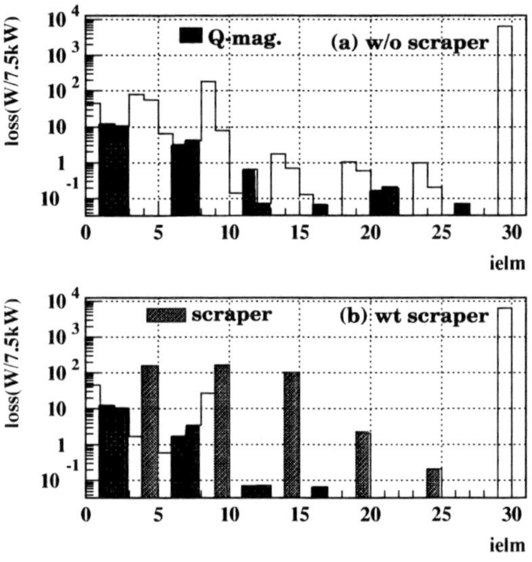

FIGURE 4. Beam loss distribution of ESS scattering event in the slow extraction line; (a)without scraper, (b) with scraper

In the study, the case with scraper was also examined. The scraper aperture was set to match the extraction orbit and injection orbit(acceptance:81πmm·mrad.). It indicates that without scrapers, the next quadrupole magnet of QFP suffers from severe beam loss of about 150W. However, once the scraper is installed, the beam losses of the magnets except QFP are reduced to less than 10W.

The next step is to carry out the multi-turn tracking and make the loss map in the whole ring. Fig.5 shows the beam loss map up to 12 turns. In the study, the injection

[1] If the duct material is replaced to Ti, the residual dose is reduced to 1/3

scraper is also installed. Without scrapers, beam power of 15W is expected to dissipated in a element in maximum. However, with scrapers, it drops to 2W. It is acceptance level considering the length of the elements(3~5m).

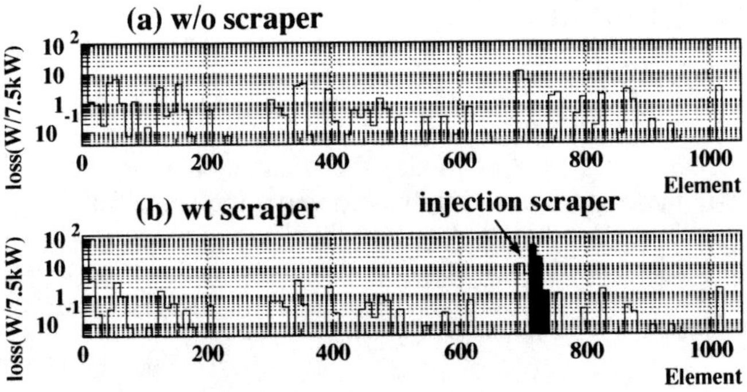

FIGURE 5. Beam loss distribution of ESS scattering event in the JHF 50GeV ring; (a)without scraper,(b) with scraper

Fig. 6 shows the horizontal phase space distribution at ESS for the first 3turns. It indicates that certain amount of scattered beam is extracted eventually; 0turn: ~ 50%,2turn:~ 2%,3turn:23%.

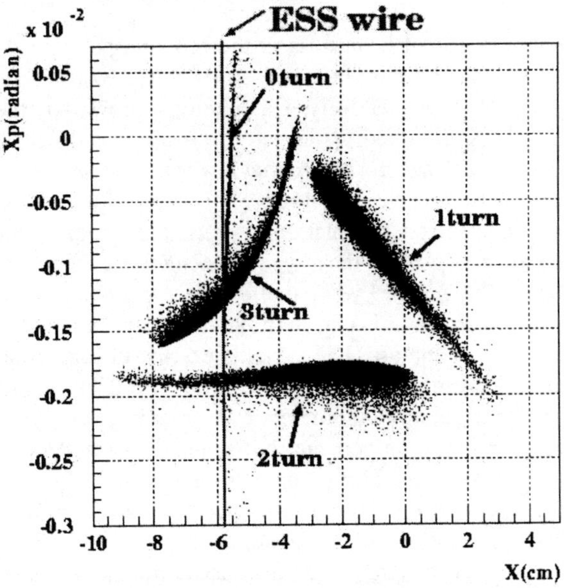

FIGURE 6. Horizontal phase space distribution of the scattered particle at ESS

Multi-turn tracking shows that such tendency lasts for turns. Fig. 7 summarizes the beam loss of the scattered particles. From the figure, the following features can be seen. (1) After 8turns, actually no beam loss occurs in the ring. (2) In every 3turns, 20~ 30% of circulating beam is in the side of extracted beam at the ESS wire. Thus, if the tendency lasts furthermore, after 60turns, the remaining scattered particle in the ring is expected to reduced to 1/1000. Thus more than 90% of scattered beam off the ESS wire is extracted eventually. It means the extraction efficiency of the slow extraction is quite high.

From these results, it can be said that as long as the scattering at the ESS wire is concerned, the loss in the ring can be controlled by employing the scrapers.

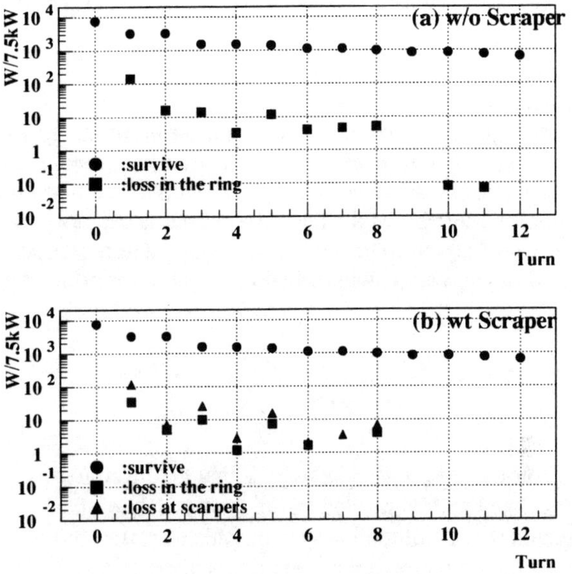

FIGURE 7. Beam loss summary of ESS scattering event(up to 12 turn) (a)without scraper, (b)with scraper

SUMMARY

In the slow extraction of JHF 50GeV main ring, about 1% of beam hit the ESS wire, which corresponds to the beam power of 7.5kW. For the beam loss, the simulation combining MARS14(00) and single particle tracking(SAD) was carried out. From the radiation level estimation, the radiation level around ESS is tolerable level. Multi-turn tracking of scattered particle off the ESS wire shows that the scraper is effective to reduce the uncontrolled beam loss in the ring, and that most of the scattered beam was extracted eventually without losing in the ring. Thus, as long as the scattering at the ESS wire is the main source of the primary beam loss in the slow extraction, the extraction efficiency is quite high.

REFERENCES

1. N.Mokhov, MARS13(98) code system User's guide
2. "Design of small-beam loss slow extraction in a high intensity 50-GeV proton synchrotron", Masahito Tomizawa et al. Proc. of EPAC'02, p2267~2269

Beam Collimation and Shielding in the Fermilab Proton Driver

A.I. Drozhdin, N.V. Mokhov

FNAL, Batavia,IL 60510, USA

Abstract.
A high beam power in the proposed Fermilab Proton Drivers – 1.2 MW in 16-GeV PD-I and 0.48 MW in 8-GeV PD-II – implies serious constraints on beam losses in these machines. Only with a very efficient beam collimation system can one reduce uncontrolled beam losses in the machine to an allowable level. The entire complex must be well shielded to allow acceptable hands-on maintenance conditions in the tunnel and a non-controlled access to the outside shielding at normal operation and accidental beam loss. Collimation and shielding performances are calculated and compared for both Proton Drivers.

BEAM LOSS AND SHIELDING STRATEGY

The Proton Driver design strategy is that the beam losses are localized and controlled as much as possible via the dedicated beam collimation system. This way, the source term for the radiation analysis is a derivative of the collimation system performance. A high loss rate is localized in the collimation section with components locally shielded to equalize prompt and residual radiation levels in the tunnel and drastically lower uncontrolled beam loss rates in the rest of the lattice [2, 3]. The radiation transport analysis is fundamentally important because of the impact on machine performance, conventional facility design, maintenance operations, and related costs. Results of this paper are based on detailed Monte Carlo simulations with the STRUCT [4] and MARS [5] codes.

The Fermilab regulatory requirements imply that: 1) prompt dose rate in non-controlled areas on accessible outside surfaces of the shield is $\leq$0.05 mrem/hr at normal operation and $\leq$1 mrem/hr for the worst case due to accidents; radionuclide concentration of 20 pCi/ml for ^{3}H and 0.4 pCi/ml for ^{22}Na in any nearby drinking water supplies are not exceeded; and anywhere in the machine, residual dose rate $P_\gamma \leq$100 mrem/hr=1 mSv/hr at 30 cm from the component surface, after 100 day irradiation at 4 hrs after shutdown (averaged over all the components, $P_\gamma \leq$10-20 mrem/hr=0.1-0.2 mSv/hr.

The radiation analysis for the arcs and long straight sections is performed both for normal operation and for accidental beam loss. The maximum shielding thickness from the both cases is put into the design as the tunnel shielding in that part of the machine. In normal operation, the source term is based on the beam loss distributions calculated with a beam collimation system described in the next section. For accidental beam loss, a *credible* accident is considered: a point-like loss of 0.1% of the full beam intensity during one hour. This is about 10^{15} protons. Once such an accident happens, the machine is shut down within 1 second to analyze the cause and undertake appropriate measures.

COLLIMATION SYSTEM

Assuming 1% of beam loss at the top energy, one gets total beam power of 11.5 kW and 4.8 kW lost in PD-I and PD-II, respectively. Calculations show that the peaks (at some quadrupoles) in the beam loss distribution can reach several kW/m that is a few thousand times higher than the tolerable levels [2, 3]. Therefore, multi-component beam collimation systems are designed for the projects. In PD-I, the system is located in a dedicated long straight section, while in the 8 GeV machine, due to space constrains, it is placed in the drifts available in the beginning of the arc. The systems consist of horizontal and vertical primary collimators, and several secondary and supplementary collimators as shown in Fig. 1. Secondary collimators generate out-scattered particles lost later in the lattice. This component is reduced with a *3-stage collimation system* positioning three (PD-I) or four (PD-II) secondary collimators close to the beam to deal with protons scattered in the primary collimator and five (PD-I) or two (PD-II) *supplementary* collimators farther from the beam to catch particles out-scattered from the main secondary collimators.

Secondary collimators need to be placed at phase advances which are optimal to intercept most of particles out-scattered from the primary collimators during the first turns after the halo interaction with the primary collimator. The optimal phase advances are around $k \cdot \pi \pm 30^o$. The horizontal and vertical primary collimators are placed at the edge of the beam after painting, with secondary collimators father from this position by an offset d. Beam loss distributions at injection and

CP642, *High Intensity and High Brightness Hadron Beams: 20th ICFA Advanced Beam Dynamics Workshop on High Intensity and High Brightness Hadron Beams*, edited by W. Chou, Y. Mori, D. Neuffer, and J.-F. Ostiguy

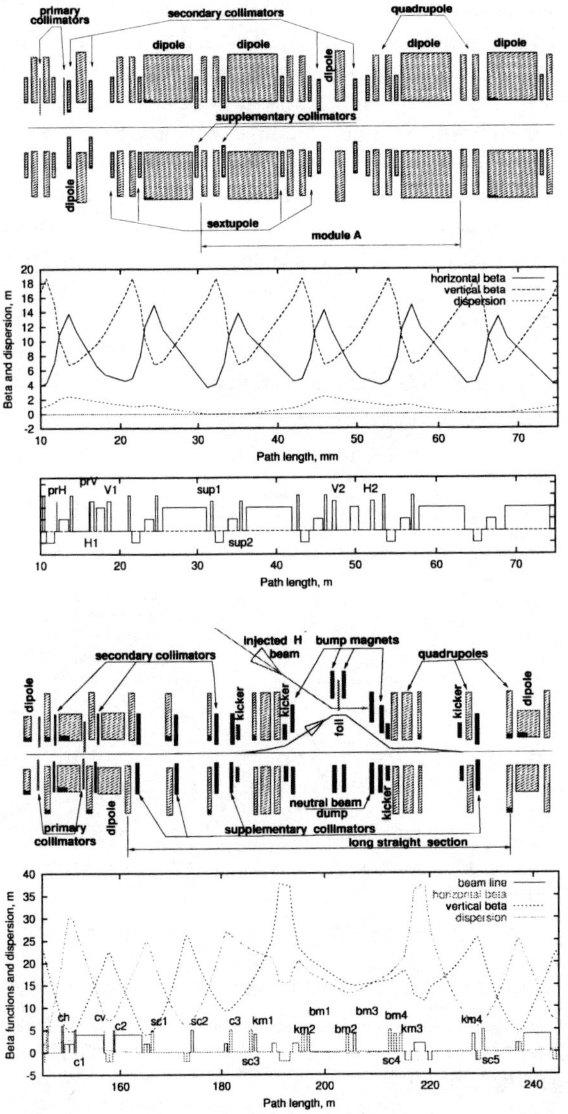

FIGURE 1. Beam collimation system layout, beta functions and dispersions in the 8 GeV (top) and 16 GeV (bottom) synchrotrons.

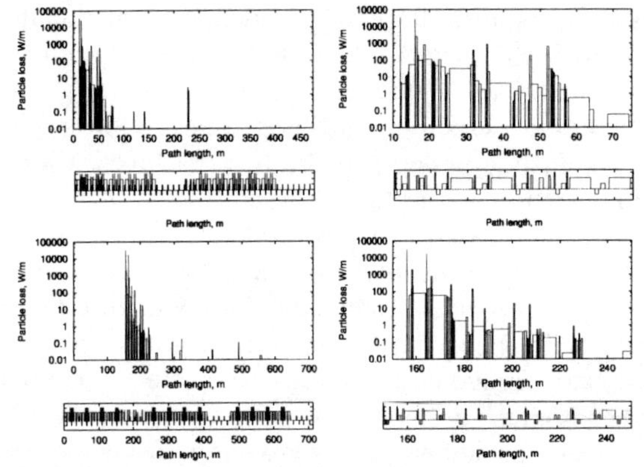

FIGURE 2. Beam loss at injection in PD-II at 0.6 GeV (top) and PD-I at 0.4 GeV (bottom).

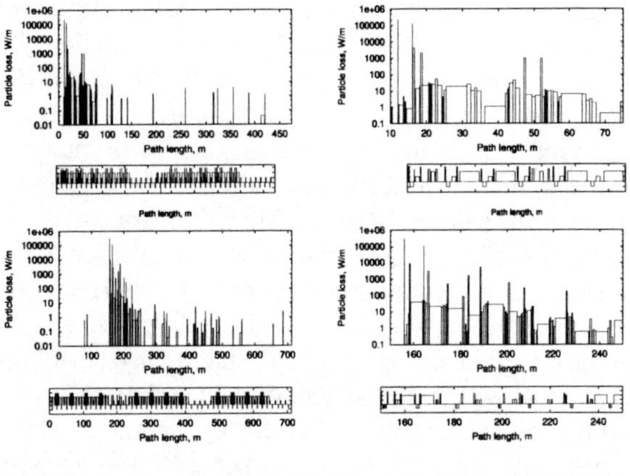

FIGURE 3. Beam loss in PD-II at 8 GeV (top) and PD-I at 16 GeV (bottom).

top energies are shown in Figs. 2 and 3 for the systems with 0.3-mm thick tungsten primary collimators, four secondary collimators (0.5-m long stainless steel or copper) positioned at d = 2 mm and two 0.3-m long supplementary collimators at d = 4 mm. The right sides of the Figures show details of beam loss in the collimation regions. It is assumed in calculations that 10% of the beam is lost at injection and 1% at the top energy, and 2/3 of these amounts interact the horizontal primary collimator (a half for off-momentum protons with $\Delta p/p = \pm 0.002$ and a half for on-momentum protons) and 1/3 the vertical primary collimator. The β-function varies along the secondary collimators, therefore the collimator apertures are tapered to follow the beam envelope after painting.

With the proposed system, ~99% of the beam halo is intercepted in the collimation section. About 1% is lost in the rest of the machine with mean rate of 0.2 and 0.12 W/m in PD-I and PD-II, respectively. At several locations the beam loss is noticeably higher (~2 W/m), exceeding the tolerable rates. Such "hot" locations need special care. Beam loss rates in the collimation system section itself are very high requiring a special shielding design.

A practicality of a rapid cycling proton synchrotron dictates a stationary collimator approach with collimator jaws in a fixed position with respect to the beam orbit during the entire cycle. In an ideal case, the circulating beam should be kept close to the collimators edge during the cycle. This requires rather complicated horizontal and vertical bumps, created by several fast magnets for each direction. To simplify the system, we propose to keep the beam at the edge of the primary collimators and close to the first secondary collimators using only three fast magnets for each direction. Most of the particles scattered out of the primary collimators are intercepted then by these secondary collimators, with other collimators intercepting the larger amplitude and off-momentum protons. Such a scheme allows to localize a majority of the beam loss in a short region.

A detailed sensitivity analyses were performed for the collimation systems in the two machines. Closed orbit deviations during the cycle and from cycle to cycle change a secondary collimator offset with respect to the primary one. In the worst case, the beam can hit initially a secondary collimator. This will result in a lower collimation efficiency and can cause damage of the collimator. Positioning secondary collimators 1 to 3 mm farther from the beam increases slightly beam loss rates in the ring, but allows larger closed orbit deviations (up to ± 3 mm) at these locations. The tune causes change of phase advances between the collimators and distance to the resonances. As betatron amplitudes of protons after interaction with primary collimators are large, the second factor can cause collimation efficiency degradation. We found that tune deviations affect mostly the beam loss peaks increasing them by up to a factor five.

The mechanical design of the secondary collimators is similar to that of those already built and installed in the Tevatron for Collider Run II. The collimator jaws consist of two pieces 30-40 mm wide welded together in an 130-mm "L" configuration. Primary collimators are made of tungsten 1 mm thick. Secondary and supplementary collimators are made of stainless steel or copper (choice will be the subject of further thermal analyses) 0.5 m (secondary) and 0.3 m (supplementary) long. These dimensions will accommodate the full beam size, after painting, as well as maximum impact parameters. Machining and assembly tolerances of 25 μm are easily met for the collimator jaws. All collimators will be in a fixed position during the machine cycle, but motion control is required in order to adjust collimators to their optimum position. The collimator assembly is welded inside a stainless steel box with bellows on each end.

RADIATION ANALYSIS

MARS calculations show that residual dose rates on the collimators and magnets of the collimation system significantly exceed the hands-on maintenance limits (Fig. 4). To reduce these levels and protect ground water outside the tunnel walls, the entire region needs to be shielded. The configuration found for PD-II, consists of steel shielding uniform in two sections of the 58-m region: first, 5-m long, starts 0.5 m upstream of the secondary collimator H1 and second is in the remaining downstream region. The first section is 1 m (vertically) and 1.3 m (horizontally) thick on each side of the secondary collimators and 0.6-m around magnets. The second section is 0.65-m (vertically) and 0.95-m (horizontally) thick on each side of the collimators, 0.25-m around dipoles, and 0.4 m (vertically) and 0.7 m (horizontally) around quadrupoles. This reduces residual dose rates below the limits, provides adequate protection of cables and other components in the tunnel and ground water around the tunnel, and equalizes (to some extent) the dirt shielding needed around the entire machine. Therefore, the same external shielding design in the arcs and straight sections is applied. Taking maximum of the normal operation and beam accident cases, the thickness of dirt shielding above the tunnel (with a safety factor of three) is 5.8 m or 19 feet. The maximum dose accumulated in the collimators and hottest spots of the magnet coils reaches 200 Mrad/yr. The maximum yearly dose at cable locations is about 150 krad per year.

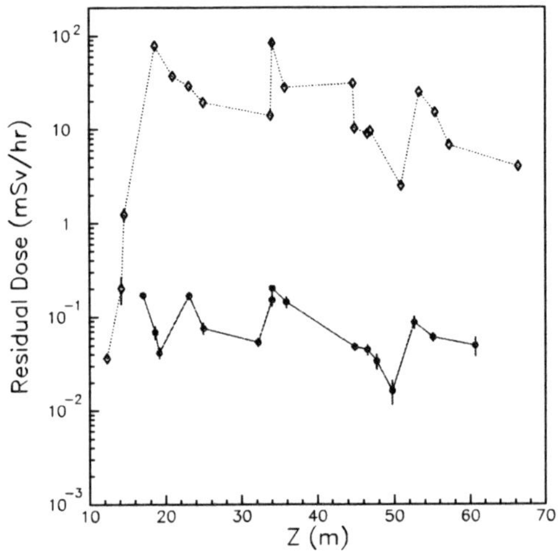

FIGURE 4. Maximum contact residual dose on site surfaces of the PD-II collimation section components (diamonds) and shielding (circles).

REFERENCES

1. "The Proton Driver Design Study" FERMILAB-TM-2136, December 2000.
2. A.I. Drozhdin, O.E. Krivosheev, N.V. Mokhov, "Beam Loss and Collimation in the Fermilab 16-GeV Proton Driver", Proc. 2001 Particle Accelerator Conference, Chicago, p. 2572 (2001); Fermilab-Conf-01/128 (2001).
3. N.V. Mokhov, A.I. Drozhdin, O.E. Krivosheev, "Radiation Shielding of the Fermilab Proton Driver", Proc. 2001 Particle Accelerator Conference, Chicago, p. 2578 (2001); Fermilab-Conf-01/132 (2001).
4. I.S. Baishev, A.I. Drozhdin, N.V. Mokhov, "STRUCT Program User's Reference Manual", SSCL–MAN–0034 (1994); http://www-ap.fnal.gov/~drozhdin/.
5. N. V. Mokhov, "The MARS Code System User's Guide", Fermilab-FN-628 (1995); N. V. Mokhov and O. E. Krivosheev, "MARS Code Status", Fermilab-Conf-00/181 (2000); http://www-ap.fnal.gov/MARS/.

Shielding Calculations in Support of the Spallation Neutron Source (SNS) Proton Beam Transport System

Jeffrey O. Johnson, Franz X. Gallmeier, and Irina Popova

Oak Ridge National Laboratory, P. O. Box 2008, MS 6363
Oak Ridge, Tennessee 37831-6363 U.S.A.

Abstract. Determining the bulk shielding requirements for accelerator environments is generally an easy task compared to analyzing the radiation transport through the complex shield configurations and penetrations typically associated with the detailed Title II design efforts of a facility. Shielding calculations for penetrations in the SNS accelerator environment are presented based on hybrid Monte Carlo and discrete ordinates particle transport methods. This methodology relies on coupling tools that map boundary surface leakage information from the Monte Carlo calculations to boundary sources for one-, two-, and three-dimensional discrete ordinates calculations. The paper will briefly introduce the coupling tools for coupling MCNPX to the one-, two-, and three-dimensional discrete ordinates codes in the DOORS code suite. The paper will briefly present typical applications of these tools in the design of complex shield configurations and penetrations in the SNS proton beam transport system.

INTRODUCTION

The Department of Energy is on schedule in the design and construction of the Spallation Neutron Source (SNS) facility at Oak Ridge National Laboratory with commissioning operations planned for FY06. The SNS initially will consist of an accelerator system capable of delivering an approximately 0.5 microsecond pulse of 1 GeV protons, at a 60 Hz frequency, with 1 MW of beam power, into a single target station. The SNS will eventually be upgraded to a 2 MW facility with two target stations (a 60 Hz station and a 10 Hz station). The proton beam is accelerated in a linac, consisting of a drift tube linac (DTL), a coupled cavity linac (CCL) and a super-conducting linac (SCL), bunched in an accumulator ring, and directed to a target station.

This paper briefly discusses some of the shielding calculations performed for the design of the SNS proton beam transport system based on a hybrid Monte Carlo and discrete ordinates particle transport method. This methodology relies on a newly developed high energy neutron cross-section library and new coupling tools that map boundary surface leakage information from the Monte Carlo calculations to boundary sources for one-, two-, and three-dimensional discrete ordinates calculations.

BACKGROUND

Determining the bulk shielding requirements for accelerator environments is generally an easy task compared to analyzing the radiation transport through the complex shield configurations and penetrations typically associated with the detailed Title II design efforts of a facility. Although there exist a couple of simple hand calculation formulas[1,2] developed over the years from the expertise gained at accelerator facilities, these are limited to simple duct arrangements through shields. Monte Carlo transport codes are able to model radiation fields in complex geometries. However, they consume a significant amount of resources, and may give solutions with poor statistics even when applying variance reduction methods.

In the design of the SNS proton beam transport system, hybrid Monte Carlo and discrete ordinates methods were applied to many of the complex shielding tasks. In almost all cases the front-end of the multi-step calculations was performed by MCNPX[3], modeling the proton beam interactions with the accelerator components, beam dumps, or building/tunnel structural elements. At some distance from the primary interaction area, the high-energy protons and secondary charged particles have interacted or ranged out, and neutrons and photons become the dominant radiation. At this distance neutron and photon boundary leakage terms are scored and passed as boundary sources to the discrete ordinates suite in the DOORS code system[4] to continue the radiation transport through the bulk shields and/or penetrations. If necessary, further coupling between successive discrete ordinates calculations allows the analyst to treat multiple legged duct arrangements and mazes.

The successful application of the discrete ordinates methods to the SNS accelerator shielding design was made possible by the development of two new tools. One is a series of coupling tools that serve as the interface between the different transport codes by generating boundary sources, and the second is the new 75-neutron-group/22-photon-group HILO2k

CP642, *High Intensity and High Brightness Hadron Beams: 20th ICFA Advanced Beam Dynamics Workshop on High Intensity and High Brightness Hadron Beams,* edited by W. Chou, Y. Mori, D. Neuffer, and J.-F. Ostiguy

cross-section library[5] with P_5 Legendre polynomial expansion of the scattering cross sections and an upper neutron energy of 2 GeV.

Coupling Tools

The coupling tools utilize the event-wise Monte Carlo boundary crossing information into energy bins, angular bins, and spatial bins (if applicable) of the corresponding discrete-ordinates mesh, angular quadrature, and multi-group energy structure[6]. This procedure maps the Monte Carlo surface crossing information into a boundary source definition to be used by the discrete ordinates codes.

The first coupling tool is Monte Carlo to ANISN (MTA), which prepares a boundary source for the one-dimensional discrete-ordinates transport code ANISN[4] from surface crossing information written primarily by MCNPX. MTA can prepare sources for planar, cylindrical and spherical problems. ANISN allows two types of sources, a distributed volume source that is restricted to isotropic angular distributions, and a shell source that is implemented as a directional flux step condition at a specified right mesh boundary. The second source type enables any angular distribution within the discrete angular quadrature set. Both source types are supported by MTA.

The second tool, MTD (Monte Carlo to DORT), prepares a boundary source for the two-dimensional discrete ordinates code DORT[4] from the surface crossing information written by the Monte Carlo code MCNPX. At the current time, MTD is limited to the preparation of surface sources for the (r,z)-cylindrical-geometry option meaning that sources are allowed on the top and bottom surfaces and on the radial surface at the outer boundary of the problem, but also at any internal right mesh boundary of constant-z planes or constant-r cylinder radii.

The third tool, MTT (Monte Carlo to TORT), which completes the suite of Monte Carlo to discrete-ordinates coupling tools, prepares a similar boundary source for the three-dimensional discrete ordinates code TORT[4]. MTT is similar to the MTD code except it has only been tested in Cartesian (x,y,z) coordinates. MTT prepares surface sources on all six sides at the user prescribed boundaries in the problem.

Similar coupling codes are available in the DOORS package to couple two-dimensional DORT to three-dimensional TORT problems, or two TORT problems. The coupling codes are TORSED[4] and TORSET[4], respectively. Of particular value for the analysis of the proton beam transport system penetrations has been the DTD (DORT to DORT) code[7] that couples two cylindrically symmetrical problems with arbitrarily aligned cylinder axes. The code performs a spatial and angular transformation to equate the angular fluxes at each radial, azimuthal, and axial boundary mesh point of the "perturbed problem to the angular flux at the nearest mesh point and angular direction of the base problem.

Geometry Models

Detailed geometry models for use in the MCNPX code were developed for the linac, high-energy beam transport (HEBT) line, the accumulator ring, and the ring to target building transfer (RTBT) line of the SNS proton beam transport system. The linac model integrated the details for the drift tube linac (DTL), the coupled cavity linac (CCL) and the super-conducting linac (SCL) into a single model. The MCNPX models were highly representative of the design and included the details of the proton beam transport line components (magnets, collimators, etc.), tunnels, buildings, and earth berms. Auxiliary models of the penetrations (oxygen deficiency/helium (ODH) vents, klystron wave-guides, etc.), personnel and truck egresses, beam dumps, and shield wall mazes were modeled in both MCNPX and DORT geometries for subsequent analyses using the coupling methodology described above. An example MCNPX model of the CCL is shown in Figure 1.

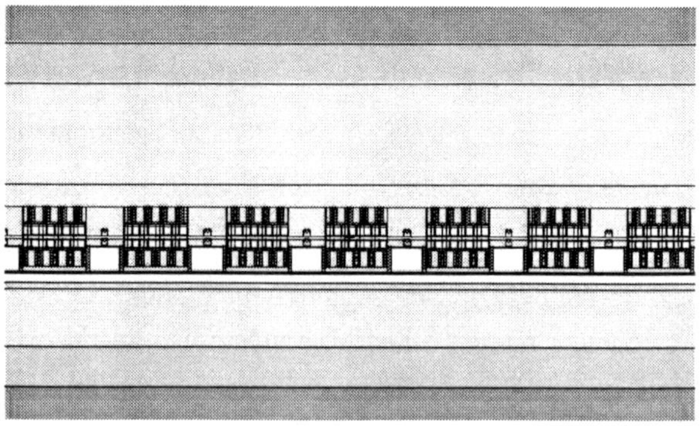

FIGURE 1. A section of the CCL model showing the complex cavity structure.

Representative Applications

The methodology and models described above were applied to a number of Title II shielding design problems associated with the SNS proton beam transport system. In addition to the standard bulk shielding design calculations, analyses were performed to design the various penetrations through the shielding including, ODH vents, HVAC ducts, personnel and truck egresses, and klystron wave-guides. The methodology was also applied to determine the shielding requirements for the DTL sections in the SNS Front End Building, the HEBT shield wall maze to allow work in the ring during linac tuning, the design of the linac tune, ring injection, and ring extraction beam dumps, and the required RTBT shielding for backstreaming from the mercury target. In conjunction with these analyses, the radiation environment during operation and residual activation after shutdown was determined for normal operational line losses as defined by the SNS accelerator physics group[8]. Furthermore, credible accident scenarios were examined to determine the adequacy of the shielding. A typical result is shown in Figure 2. Detailed descriptions of some of these calculations can be found in References 9 and 10.

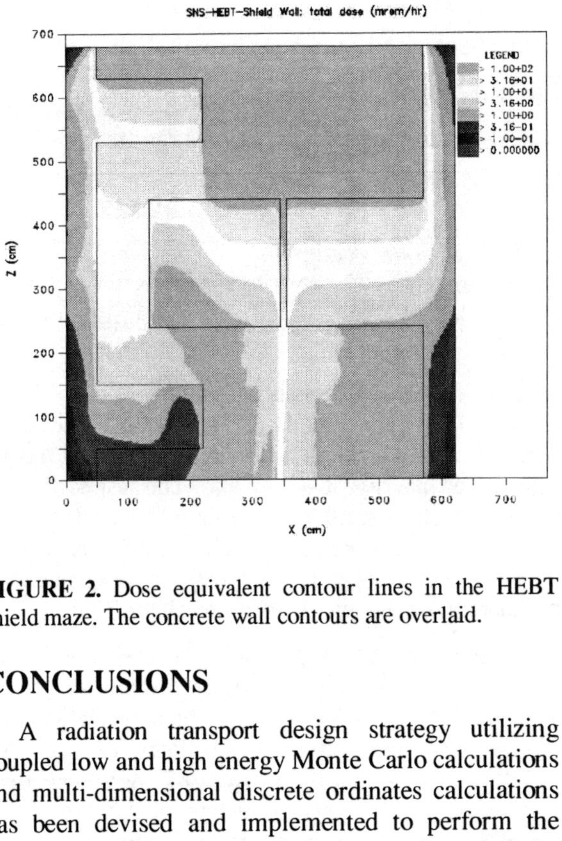

FIGURE 2. Dose equivalent contour lines in the HEBT shield maze. The concrete wall contours are overlaid.

CONCLUSIONS

A radiation transport design strategy utilizing coupled low and high energy Monte Carlo calculations and multi-dimensional discrete ordinates calculations has been devised and implemented to perform the requisite radiation transport design analyses of the SNS proton beam transport system and to characterize the neutronic performance of the SNS. The impact of this strategy has allowed the radiation transport design team to stay current with the design changes and provide timely feedback to the SNS design engineers. Design modifications are constantly being addressed to ascertain the impact the modifications have on the radiological safety of the SNS. Future analyses will continue to refine the models to account for the details of the advanced Title II design that affect the radiological safety or impact environmental concerns.

ACKNOWLEDGMENTS

This work was supported by the U.S. Department of Energy through the Spallation Neutron Source (SNS) Project. SNS is managed by UT-Battelle, LLC, under contract DE-AC05-00OR22725 for the U.S. Department of Energy.

REFERENCES

1. Sullivan, A.H., "A Guide to Radiation and Radioactivity Levels Near High Energy Particle Accelerators," Nuclear Technology Publishing, Ashford, England, (1992).

2. Dinter, H., Dworak, D., and Tesch, K., "Attenuation of the Neutron Dose Equivalent in Labyrinths through an Accelerator Shield," Nucl. Instrum. Meth. **A333**, 507-512, (1993).

3. Hughes, H. G. et. al., "MCNPX for Neutron-Proton Transport," International Conference on Mathematics & Computation, Reactor Physics & Environmental Analysis in Nuclear Applications, American Nuclear Society, Madrid, Spain, September 27-30, 1999.

4. DOORS3.2, One-, Two-, And Three-Dimensional Discrete Ordinates Neutron/Photon Transport Code System, RSIC Computer Code Collection, CCC-650, (1998).

5. Lillie, R. A., and Gallmeier, F.X., "HILO2k: A coupled Neutron-Photon Transport Cross-section Library for Neutron Energies up to 2000 MeV," Fourth International Topical Meeting on Nuclear Applications of Accelerator Technology, American Nuclear Society, Washington D.C., (2000).

6. Gallmeier, F. X. and Pevey, R. E., "Creation Of A Set Of Interface Utilities To Allow Coupled Monte Carlo/Discrete Ordinates Shielding Analysis," Third International Topical Meeting on Nuclear Applications of Accelerator Technology, American Nuclear Society, Long Beach California, (1999).

7. Lillie, R. A., "DTD: A Coupling Code for Two-Dimensional R-Z Cylindrical Geometries," Trans. Am. Nucl. Soc. **61**, 381, (1990).

8. Catlan-Lasheras, N. et. al., "Accelerator Physics Model Of Expected Beam Losses Along The SNS Accelerator Facility During Normal Operation," ANA/AP Technical Note 07, UT-Battelle, LLC, Oak Ridge (March, 2001).

9. Gallmeier, F. X., "Shielding The SNS Klystron Building From The Linac," *Fourth International Topical Meeting on Nuclear Applications of Accelerator Technology*, American Nuclear Society, Washington D.C., (2000).

10. Bucholz, J. A., Gallmeier, F. X., and Popova, I., "Egress Shielding Studies for the SNS Accelerator System," *Fifth International Topical Meeting on Nuclear Applications of Accelerator Technology*, American Nuclear Society, Reno, Nevada. (2001).

Development of High Powered Target Systems for the Spallation Neutron Source and the Muon Collider/Neutrino Factory

T. Gabriel*, J. Haines*, B. Riemer*
P. Spampinato*, T. McManamy*, N. Mokhov#

*Spallation Neutron Source, Oak Ridge, TN 37830
#Fermi National Accelerator Laboratory, Batavia, IL 60510

Abstract. The purpose and requirements of the Spallation Neutron Source (SNS) and the target area of the Muon Collider/Neutrino Factory are presented. Parts of the technologies that are being utilized to design these facilities are discussed. Emphasis is given to the technology issues that present the greatest scientific challenges.

SPALLATION NEUTRON SOURCE

In many areas of physics, chemistry, biology, materials, and nuclear engineering, it is extremely valuable to have a very intense source of neutrons so that the structure and functionality of materials can be studied. Discussions on all the Spallation Neutron Sources under consideration or development are given in Refs. 1–3. One facility under construction at ORNL for this purpose is the Spallation Neutron Source (SNS). This facility will consist basically of three parts: 1) a high-energy (~1 GeV) and high-powered ($\lesssim$ 2 MW) proton accelerator (60 Hertz, <1.0 μs/pulse, <33 kJ/pulse), 2) a target/moderator/reflector/shielding/shutter/utility [Target Systems (TS)] assembly, which converts part of the proton beam power to low-energy ($\lesssim$ 2eV) neutrons through spallation and delivers them to the third part, 3) the neutron scattering instruments. A picture showing the overall facility is given in Fig. 1. LBNL is responsible for the front end, which is currently being shipped to ORNL for installation; LANL/JLAB, the linac [parts of the Drift Tube Linac (DTL) and klystrons have been shipped to ORNL for installation]; BNL, the high-energy beam transport system and accumulator ring; ORNL, Target Systems and Conventional Facilities; and ANL, the neutron scattering instruments.

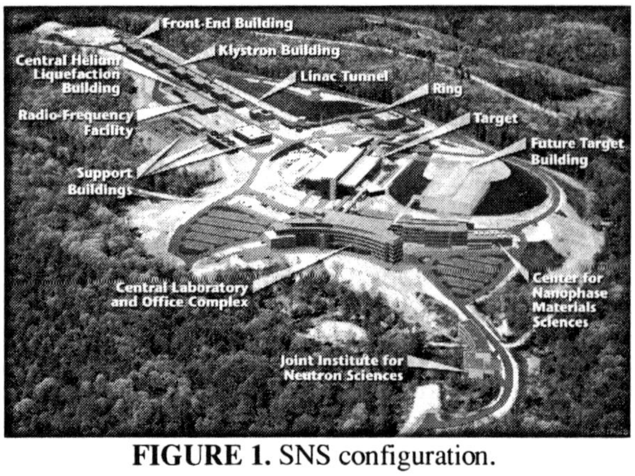

FIGURE 1. SNS configuration.

Accelerator Components

The SNS is composed of several sequential accelerator systems with a resultant proton energy of ~1.0 GeV delivered to the neutron production Hg target.

Beginning the process is a Cesium enhanced H- ion source, the Front End (FE) system. The next accelerator unit is the RFQ, which accelerates the H- beam to an energy of ~2.5 MeV. The last section of the FE system is the Medium Energy Beam Transport (MEBT). In the MEBT, the proper bunch structure of the beam is developed using buncher cavities to match the rotation frequency of the accumulator Ring.

After the front end, the H- beam enters the first of four distinct accelerating structures within the Linac. The first two parts are a DTL which accelerates the beam to ~87 MeV and a Coupled Cavity Linac (CCL) which accelerates the beam to ~187 MeV. The final two sections are Superconducting Radio Frequency (SRF) structures with a mechanical (relativistic) beta of 0.61 (~325 MeV) and 0.81 (~1.06 GeV).

At the injection region of the Ring, the H- beam traverses thin carbon foils that strip the two electrons from the H- beam and allow the resultant protons to be circulated and stored in the Ring. After ~1000 turns are accumulated in the Ring, a series of fast rise-time dipole elements (kickers) are triggered to extract the stored $\lesssim 2\times10^{14}$ protons circulating in the Ring for delivery to the Hg target.

CP642, *High Intensity and High Brightness Hadron Beams: 20th ICFA Advanced Beam Dynamics Workshop on High Intensity and High Brightness Hadron Beams,* edited by W. Chou, Y. Mori, D. Neuffer, and J.-F. Ostiguy
2002 American Institute of Physics 0-7354-0097-0

Scope and Requirements of Target Systems

The scope of TS is to provide low-energy neutrons from high-energy spallation reactions for short-pulsed neutron scattering instruments and to develop three-proton beam dumps, one for the linac and two for the storage ring (injection and extraction).

The first requirement for TS is to safely and reliably receive a proton beam in a flowing mercury target with the characteristics defined in the Introduction. As a second requirement, TS must be able to convert part of the proton beam power into short, high-intensity pulses of low-energy neutrons (both ambient and cold) which can be used by up to 24 neutron beam lines, and which meet the requirements of the neutron scattering instruments.

Building a TS for the SNS requires the development of a target, in this case a flowing-mercury target, which can give maximum neutron yield; a reflector/moderator assembly to trap, reflect, and thermalize (ambient and cryogenic temperatures) the neutrons; a vessel system to contain the reflector/ moderator assembly and provide support and alignment for the start of the neutron guides; bulk shielding and shutter assemblies to shield the personnel from the neutrons and to allow the closing of the low energy neutron pathways so samples can be replaced in the neutron scattering instruments; utilities (light and heavy water, vacuum and He) to help with the cooling and functions of the various systems; remote handling to accommodate the change out of the target, inner reflector assembly, neutron guides and shutters, proton beam window, etc.; and Instruments and Controls for the majority of the subsystems mentioned above. Currently, the detailed design of the TS is almost complete and many procurements have been placed.

Target Development R&D

Five areas have been defined as needing R&D to develop a successful Target Systems[4]. These are (1) steady-state power handling, (2) remote handling and operation, (3) radiation damage, (4) material compatibility, and (5) thermal shock. All of the issues have been addressed or solved except for the erosion through cavitation induced pitting brought on by the thermal shock of the beam. After testing at the WNR in 2001, pitting (both large and small) was detected. Additional tests are planned in 2002 to assess possible solutions based on geometry, mitigation methods, and materials.

MUON COLLIDER/NEUTRINO FACTORY: DEVELOPMENT OF THE TARGET AREA, REMOTE HANDLING, AND GRAPHITE TARGET FOR THE μ/ν

The development of a high-intensity source of muons that can be used for collider experiments or for the production of high-energy neutrinos opens the door for a broad range of physics experiments. A large effort is underway in this country to develop such a source. The concept is to use a high-intensity proton beam (~24 GeV, ~1MW, ~15 hz) incident on a Hg jet or graphite target to produce pions which decay to give the muons. These muons will be magnetically captured and then accelerated in a collider ring or in a "race track" system. The "race track" itself can be pointed at a detector located many hundred kilometers away. This part of the paper describes some of the target remote handling needs, which are necessary at such a facility, and also on some of the R&D that is underway on the graphite target.

Remote Handling Associated with the Target Area

Facility design concepts have been developed for graphite and mercury target systems based on the requirements to operate and maintain these different targets. A picture of a possible layout of the target

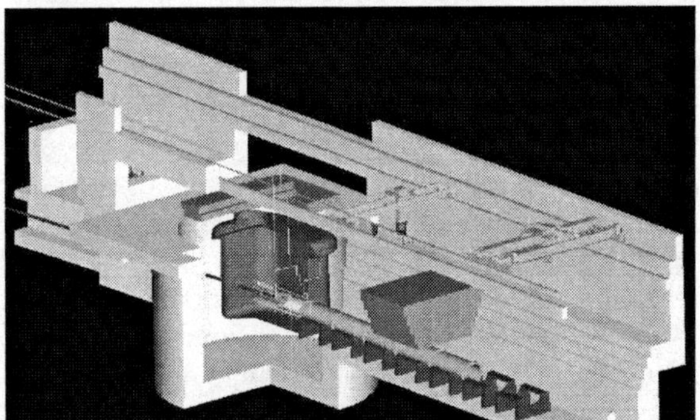

FIGURE. 2. Layout of the target area, decay channel, nuclear shielding, and remote handling for the muon/neutrino factory.

area, decay channel, nuclear shielding, and remote handling devices is given in Fig. 2. Both systems have highly activated components that must be replaced periodically using remote handling equipment and tools. For the case of a graphite target, overhead access was the preferred approach and the hot cell was located above the target area. For the mercury target, overhead access combined with below-grade access was necessary in order to maintain the mechanical components of the system, i.e., pumps, valves, and mercury storage. A more detailed writeup can be found in Refs. 5 and 6.

Graphite Rod Test at the Lansce-WNR and ORNL

Graphite rods were instrumented with fiber optic strain gages to measure the dynamic response from the LANSCE 800 MeV proton beam pulses. The energy deposition produced by the protons in the graphite is similar to that anticipated at a Mu/Nu facility. Structural analysis simulations of the response were performed with the finite element code, ABAQUS/ Explicit.[7] The predicted and measured strains at the mid length location compare favorably.

Sublimation of the radiatively cooled graphite target is a limiting feature of the target's lifetime. Tests have been carried out in a vacuum environment to determine the useful lifetime at about 2000ºC. The data, which is in reasonable agreement with theoretical data, indicates that the lifetime is short, but acceptable. This experiment is described in detail in Ref. 8. Tests are also underway to determine the lifetime in a 1 atm He environment, which should increase the lifetime.

Neutronics and Shielding

Substantial analysis based on the MARS code [9] simulations have been carried out [10] which includes pi/mu production and collection, energy deposition and residual dose levels, shielding and component lifetimes. The optimized system would provide 0.37 captured muons per proton at 36 meters from the tilted mercury jet impacted by 24-GeV protons. If a graphite target had been used, this number would be reduced by about a factor of two. Peak energy deposition and dynamic heat loads in the superconducting coils of a 20-T hybrid solenoid are kept below the tolerable limits via carefully designed shielding made of water-cooled tungsten-carbide balls. With the current setup, the anticipated lifetime of main components exceeds 15 years. The inner shielding is very radioactive, with residual dose rate up to 1 kSv/hr. This will require remote control and robotics for the inner parts of the system. This dose drops by two orders of magnitude after several weeks. The residual dose outside the cryostat is significantly lower, of the order of 100 mSv/hr.

REFERENCES

1. K. N. Clausen, "The European Spallation Source (ESS) Project," *ICANS-XV 15th Meeting of the International Collaboration on Advanced Neutron Sources*, Proceedings–Volume 1, November 6–9, 2000, Tsukuba, Japan.
2. Y. Oyama, S. Ikeda, and JAERI-KEK Joint Project Team, "Status of Spallation Neutron Source Program in High Intensity Proton Accelerator Project," *ICANS-XV 15th Meeting of the International Collaboration on Advanced Neutron Sources*, Proceedings–Volume 1, Nov. 6-9, 2000, Tsukuba, Japan.
3. T. E. Mason, T. A. Gabriel, R. K. Crawford, K. W. Herwig, F. Klose, and J. F. Ankner, "The Spallation Neutron Source: A powerful tool for materials research," 20th International Linac Conference, August 21-25, 2000, Monterey, California, published on the LANL e-print server.
4. T. A. Gabriel, "An Overview of the SNS," published in the proceedings of the "Golden Anniversary" of Nuclear Engineering at North Carolina State, Raleigh, NC, Dec. 2–4, 2000.
5. P. T. Spampinato, J. B. Chesser, D. L. Conner, T. A. Gabriel, F. X. Gallmeier, J. R. Haines, and T. J. McManamy, "Support Facility for a Graphite Target Neutrino Factory," ORNL/TM-2000/153 (August 2000).
6. P. T. Spampinato, J. B. Chesser, D. L. Conner, T. A. Gabriel, F. X. Gallmeier, J. R. Haines, and T. J. McManamy, "Support Facility for a Mercury-Jet Target Neutrino Factory," ORNL/TM -2001/124 (September 2001).
7. ABAQUS Finite Element Analysis Software, Hibbitt, Karlsson & Sorensen, Inc., Pawtucket, RI.
8. J. R. Haines and C. C. Tsai, "Graphite Sublimation Tests for the Muon Collider/Neutrino Factory Target Development Program," ORNL/TM-2002/27.
9. N. V. Mokhov, "The MARS Code System User's Guide," Fermilab-FN-628 (1995); N.V. Mokhov, O.E. Krivosheev, "MARS Code Status," Fermilab-Conf-00/181 (2000).
10. N. V. Mokhov, "Particle Production and Radiation Environment at a Neutrino Factory Target Station," Proc. 2001 Part. Accel. Conf., Chicago, p. 745 (2001); Fermilab-Conf-01/134 (2001).

Beam Loss and Collimation at LHC

J.B. Jeanneret*, I.L. Ajguirei†, R. Assmann*, I.S. Baichev†, D. Kaltchev**, T. Kurtyka* and A. Wroblewski‡

*CERN, Geneva, Switzerland
†IHEP, Protvino, Russia
**TRIUMF, Vancouver, Canada
‡Cracow University of Technology, Cracow, Poland

Abstract. Beam loss and collimation issues at LHC can only be briefly outlined in three pages, but some emphasis is given to the bibliographic material.

TABLE 1. Some LHC nominal parameters related to beam losses

Luminosity	10^{34}	$cm^{-2}s^{-1}$
Normalised emittence	3.75	μm
β^* at crossing	0.5	m
σ^* at crossing	16	μm
Bunch population	$1.05\ 10^{11}$	protons
Stored bunches	2800	
Injection energy	450	Gev
Collision energy	7000	Gev
Injected intensity	$12 \times 2.5\ 10^{13}$	protons
Stored intensity	$3\ 10^{14}$	protons
Stored energy	(per beam)	
Injection	12×2	MJ
	$\equiv 12 \times 4$	kg melted Cu
Collision	304	MJ
	$\equiv 800$	kg melted Cu

INTRODUCTION

In high energy proton colliders, the beam squeezing at the collision point is limited by the aperture and the strength of the final focus quadrupoles. High luminosity is therefore reached with high stored intensity. This is the case of LHC, see Table 1. The combination of high momentum and high intensity results in high injected and stored energies, while the main magnets are superconducting. Equipment integrity and quench prevention are therefore major issues. Several systems will be installed to protect the ring and ensure reliable operation. Many single pass low-Z absorbers will be installed (e.g. injection and dump insertions), while safe and strict injection procedures will be implemented [1, 2]. A multi-turn collimation system will capture multi-turn transient and steady losses. Finally, fast beam loss monitors installed both near the collimators and around the ring will trigger the machine protection system for beam dump action. We restrict the scope of this paper to the collimation system.

BEAM LOSS ISSUES

Quench limits

High energy protons impacting the vacuum chamber of a s.c. magnet develop a hadronic shower and deposit energy over an effective length of about one meter in particular in the coils of the s.c. magnets [3]. The maps of energy density obtained with simulation codes [6, 7, 8] are compared to both the heat reserve in the cables between the working and the critical temperature (transient losses) and to the heat flow allowed by the same temperature difference (steady losses). In the first case, the heat reserve is time dependent. In LHC magnets, if the energy deposition occurs in $\Delta T < 2.5$ ms, only the heat reserve of the metal can be used, because of a limitation of energy transfer at the interface metal/helium. Full use of the heat reserve of the helium occurs if $\Delta T \approx 50$ ms [3]. Quantitative values expressed in proton losses per meter of vacuum chamber are given in Table 2.

Regular losses

Regular losses can be split into three basic cases, quantified in Table 2.

Injection errors Even with collimation in the transfer lines and in the kicker sections, losses might occur during the first few turns, for example because of a failure of the damping of the injection oscillation. We consider losses up to 5% of an injected batch.

Ramping Because of phase errors, noise or uncontrolled instabilities during the long injection process, protons lying out of RF buckets will be lost at the beginning of the ramp. The flash of losses will last about one second [9]. We consider an upper limit of 10% of longitudinal losses.

Steady losses at collision A few hours will be spent between the dump of the stored beams and the restora-

CP642, *High Intensity and High Brightness Hadron Beams: 20th ICFA Advanced Beam Dynamics Workshop on High Intensity and High Brightness Hadron Beams*, edited by W. Chou, Y. Mori, D. Neuffer, and J.-F. Ostiguy

TABLE 2. Expected regular transient (top) and steady (bottom) losses compared to quench limits. Transient quench limits are time-dependent, see text.

Case	ΔN [p]	ΔN_q [p/m]	$\eta = \Delta N_q/\Delta N$ [1/m]
Injection	$1.25\ 10^{12}$	10^9	$8\ 10^{-4}$
Ramping	$3\ 10^{13}$	$2.5\ 10^{10}$	$8\ 10^{-4}$
	$\dot{N}$ [p/s]	$\dot{N}$ [p/m/s]	$\eta = \dot{N}_q/\dot{N}$
7 TeV	$8\ 10^{10}$	$8\ 10^6$	10^{-4}

tion of colliding conditions. Efficient production therefore requires large beam lifetimes, i.e. $\tau > 10$ hr, but poorer performance, e.g. $\tau \approx 1$ hr must be allowed during trial periods.

Above the loss rates quoted in Table 2, it may be necessary to dump the beams to avoid magnets quenches. More elaborated specifications can be found in [4, 5]. The ratio $\eta = \Delta N_q/\Delta N$ or $\eta = \dot{N}_q/\dot{N}$ (Table 2) clearly indicates that protons must be captured before reaching the vacuum chamber. This is the task devoted to the collimation system. Its efficiency must be larger than $1/\eta$. This is described in Sect. .

Accidental losses

While regular losses can induce magnet quenches, accidental losses can damage machine components in the absence of adequate protection schemes. Apart from the specific protections evoked in Sect. , the collimators which must be the aperture limit of the ring are inherently exposed to capture fast accidental losses and must therefore be adequately resistant to high energy deposition. The worst identified case is related to a particular dump failure, whenever one of the 15 kicker modules is self triggered. The other modules are triggered soon after ($\sim 1\mu$s), but the effective slower rise of the kick results in the impact of many bunches at the edge of at least one collimator. The density of protons may be as high as 6 nominal bunches over $200 \times 200\mu\text{m}^2$ [5]. Only low-Z materials can survive this [10]. We performed a preliminary stress analysis of a beryllium jaw exposed to the impact of 10 bunches at 7 TeV in the condition discussed in Sect. . The energy density map was obtained with the MARS code [11] and a time-dependent stress analysis was performed with the ANSYS FE code. The maximum dynamic stress calculated for this case within linear elastic analysis is quite local but quite high, $\sigma_{max} = 1.6$ GPa, and is well above the ultimate tensile strength of beryllium $\sigma_{uts} \simeq 1.1$ GPa. Further analysis is going on to establish the extent of plastic deformation or local rupture. We also consider improvements to the retriggering scheme of the kicker and the use of sophisticated low-Z materials (graphite family) for the few collimators which will be exposed to this kind of events.

TWO-STAGE COLLIMATION

In order to meet a collimation inefficiency $\eta < 10^{-4}$ m^{-1}, a two-stage collimation system is needed [13, 14]. The impact parameter at the edge of a primary collimator is small ($\sim \mu$m) with multi-turn losses, leaving the protons with a high probability to be scattered out of the jaw. Secondary collimators are therefore necessary to absorb this secondary halo. Provided adequate phase advances between collimators (Sect.), the size of the secondary halo $A_{sec-halo}$ can be kept smaller than the normalised geometrical aperture of the ring A_n. The inefficiency is then calculated with using the phase-space dilution of the tertiary halo (emitted by the secondary collimators) above $A_{sec-halo}$.

Optics

The optics of a two-stage two dimensional collimation system is designed by considering that protons are scattered out of a collimator in all directions. This 'stochastic coupling' imposes the use of several secondary jaws per primary collimator (in LHC four of them). Their location is optimum for well defined and correlated transverse betatron phase advances [15]. In the case of ramping, momentum collimation must be used. Conflicting optics requirements imply to use separate insertions for betatron and momentum collimation, even if the optics principles are similar in both cases [15]. In an insertion of finite length, the best correlation of the phase advances can only be approached. The location and the transverse tilt of the jaws are calculated numerically [16, 17]. The optics of the two insertions of LHC are discussed in [18].

Geometrical aperture

For cost optimisation, the aperture of the magnets is kept small and must be carefully optimised and checked. The normalised aperture must satisfy everywhere the condition

$$A_n(s) > A_{sec-halo}\ . \tag{1}$$

The normalised primary aperture n_1 delimited by the collimators was chosen to $n_1 = 7$. With the secondary jaws set at $n_2 = 8.2$, the size of the secondary halo is $A_{sec-halo} = 10$. A code coupled to MAD [19] takes into account mechanical, closed orbit and optics errors to match the optics with aperture constraints by respecting the condition 1, see Fig. 1.

Scattering in jaws and inefficiency

A Monte-Carlo simulation of nuclear and electromagnetic scattering near the edge of a jaw is coupled to

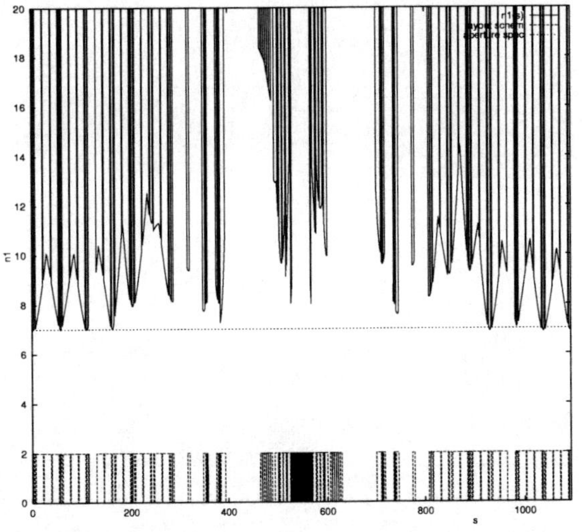

FIGURE 1. The normalised aperture of the insertion 1 of LHC after matching with aperture constraints, expressed in terms of equivalent primary aperture $n_1(s)$.

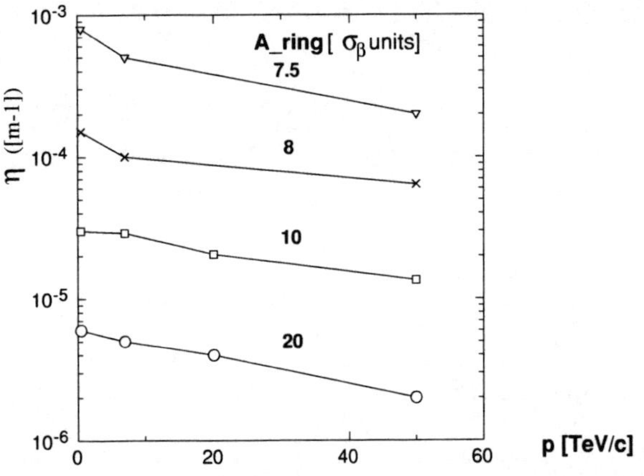

FIGURE 2. The inefficiency η of a two-stage betatron collimation system simulated with K2 for different beam momenta. Here three primary collimators are used (delimiting an octagonal normalised aperture, with primary and secondary retraction of the jaws $n_1 = 6$ and $n_2 = 7$ respectively.

multi-turn tracking around the ring to compute collimation efficiency. Up to recently two codes were used. The K2 code [20] allows high statistics but the dilution of the tertiary around the ring must be analysed separately. The STRUCT code [21] can produce a map of losses along the ring. Both codes have limited chromatic properties, making their use sometimes delicate. The K2 code was successfully checked in an experiment with coasting beams at 120 GeV in the SPS ring at CERN [22]. Used for LHC (and VLHC), K2 indicates inefficiencies somewhat better than the the specified $\eta = 10^{-4}$ m^{-1}, provided that the aperture is $A_n > 10$, see Fig. 2. Calculations with linear an non-linear optics errors and imperfect collimators are underway, by integrating the K2 and the STRUCT scattering module in MAD/SIXTRACK and DIMAD [23].

MORE ABOUT SIMULATION CODES

At this workshop, Weiren Chou advocated for more certification of the K2 and STRUCT codes. We can only agree with him, but it must be understood that a beam loss experiment in a ring is delicate and requires substantial machine time, while specific instrumentation is sometimes needed. On the other hand these codes need consolidation, see Sect. . This can only be envisaged with adequate resources. Beam loss and collimation are now essential systems in high intensity proton machines but this branch of accelerator science is nevertheless lacking resources and adequate recognition. Upgrades must thus be envisaged in parallel of further certification. A generic code combining low and high energy scattering and true 6D-tracking with linear and non-linear errors would be much useful. This might the subject of a session in a future workshop of this series.

REFERENCES

1. O. Bruening et al., CLPR 291 and PAC99,1999.[1].
2. R. Schmidt and J. Wenninger, CLPN 287,2002.
3. J.B. Jeanneret et al., CLPR 44, 1996.
4. R. Assmann et al., CLPN 277,2002.
5. R. Assmann et al., EPAC 2002,to be issued.
6. I. Ajguirei, I. Baichev and J.B. Jeanneret, CLPR 398 and EPAC2000,2000.
7. I. Baichev, CLPN 240,2000.
8. J.B. Strait and N.V. Mokhov, CLPR 43 and EPAC96,1996.
9. I. Baichev et al., CLPR 309,1999.
10. R. Assmann et al., CLPN 282,2002.
11. I. L. Azhgirey et al., in: *Materials of XV Work. on Charged Part. Acc.*, Vol.2,p.270,Protvino,1996.
12. Arauzo A. et al., CERN-SL-2001-027-BI.
13. Design study of the Large Hadron Collider, CERN/91-03,Chap.9,1991.
14. The Large Hadron Collider, Chap. 4, CERN/AC/95-05(LHC),Chap.4,1995.
15. J.B. Jeanneret, Phys. Rev. ST-AB,**1**,081001,1998.
16. D.I. Kaltchev et al., CLPR 134 and PAC1997,1997.
17. D.I. Kaltchev et al., CLPR 194 and EPAC98,1998.
18. D.I. Kaltchev et al., CLPR 305 and PAC1999,1999.
19. J.B. Jeanneret and R. Ostojic, CLPN 111,1997.
20. N. Catalan Lasheras et al., *Proc.Symp. 'Near Beam Physics',Fermilab,1997* and CLPR 156, 1998.
21. I. Baichev, et al., SSCL-MAN-0034, Dallas, 1994.
22. N. Catalan-Lasheras, CERN-THESIS-2000-019,2000.
23. R. Assmann, D.I. Kaltchev and J.B. Jeanneret,EPAC 2002,to be issued.

[1] CLPR or CLPN stand for CERN-LHC-Project Report or Note

JOINT SESSION C AND G

FFAGs AND CYCLOTRONS

Summary Report

P. Schmelzbach*, Y. Mori† and G. Dutto, R. Baartman**

*Paul Scherrer Institut, Switzerland
†KEK, Tsukuba, Japan
**TRIUMF, Vancouver, Canada

There were about 25 attendees in this session. Papers presented were as follows:

- PSI Cyclotron
  (P. Schmelzbach)
- High Intensity Cyclotrons at IBA
  (Y. Jongen)
- Isochronous and Scaling FFAGs
  (R. Baartman)
- FFAG Development at KEK
  (Y. Mori)
- Beam Dynamics in the FFAG
  (M. Aiba)
- Non-Scaling FFAGs for Muon Acceleration
  (C. Johnstone and S. Koscielniak)
- Longitudinal Motion in Nearly-Isochronous FFAGs
  (J.S. Berg)
- Stochastic Cooling in FFAGs
  (M. Wakasugi)

The first 2 speakers emphasized that cyclotrons are indeed high intensity accelerators. Many of the "Cyclone" series of cyclotrons built and commercially available from IBA routinely deliver 10mA of protons for isotope production. The PSI cyclotron (Villigen, Switzerland) routinely delivers >1MW of beam power (2mA at 590MeV). The TRIUMF cyclotron (Vancouver, Canada) routinely delivers 0.2MW beam power. New developments include "self-extraction" and use of molecular hydrogen. A non-intuitive aspect for synchrotron physicists is that at high space charge fields, such as occur in the PSI injector cyclotron, the space charge force in an isochronous machine "curls" up the bunches, reducing their length to equal their width. With sufficient rf voltage, extraction from machines such as those at PSI can be very clean, with loss on the order of 1:104. Machines of the PSI type have been studied to reach yet higher power and energy. An extrapolation to 1 GeV and 10 mA (10MW) is considered feasible. In fact, it is believed that up to this amount of beam power, cyclotrons are a more economic choice than linacs. Energies higher than 1GeV become expensive because isochronism demands that the radial tune be equal to γ (Baartman). This means that for large energy, many low-order resonances are crossed. To avoid low-order intrinsic resonances, requires the cyclotron be composed of many sectors. Intensity larger than 10mA becomes expensive because space charge fattens the turns, making clean extraction only possible for very large energy gain per turn. At some point the rf becomes sufficiently expensive that one might as well save the cost of the magnet and go with a linac. Between the regimes of the cyclotron and the sychrotron, there is the synchrocyclotron. With the addition of strong focusing, this has come to be know as the FFAG. Strictly speaking however, cyclotrons with sector focusing (this includes practically all modern cyclotrons) are FFAGs, since they use strong focusing and are "fixed field". So there are two types: isochronous FFAGs on the one hand, and pulsed beam FFAGs on the other. If we release the isochronism requirement, the optics can scale from inner radii to outer and the radial tune can be made constant. This allows for extremely large acceptance (Aiba). On the other hand, if the machine is isochronous, beam intensity is large because it is cw, but acceptance is small because of the resonances crossed, as noted above. As Mori pointed out, pulsed FFAGs have an advantage over synchrotrons besides their large acceptance: Since the magnetic field is fixed, the ramp rate and therefore intensity is limited only by the ability of the rf to ramp quickly. A major development in this regard is the magnetic alloy (MA) rf cavity. The MA of choice (FINEMET) has good μQf up to very high rf excitation, and has a high Curie temperature. Both these features make possible much higher ramp rates and rf voltage as compared with conventional ferrite core cavities. The Mori group has built and commissioned a 0.5MeV "proof of principle" proton FFAG. This machine has demonstrated both high ramping rates and large acceptance. Both of these features make such machines ideal for accelerating secondary particles such as muons. For the particular application of the (-factory, a goal is to reach about 20GeV in a short time of about 1ms. FFAGs seem a more natural choice than linacs (recirculating or otherwise) because they have such large acceptance. This has the potential of reducing the very difficult beam

CP642, *High Intensity and High Brightness Hadron Beams: 20th ICFA Advanced Beam Dynamics Workshop on High Intensity and High Brightness Hadron Beams,* edited by W. Chou, Y. Mori, D. Neuffer, and J.-F. Ostiguy

cooling requirements. Ideas were presented by Koscielniak, Johnstone and Berg on how to achieve the acceleration in the required time. The extremely short acceleration time implies need for very large rf voltage, which may only be possible with high Q cavities. Since this makes ramping difficult, an idea is to make the machine close enough to isochronous that phase slip over a few turns can be tolerated. The 20GeV FFAG would have a circumference of 2km and to avoid excessive phase slip, acceleration would need to be complete in the order of 10 turns. Wakasugi presented RIKEN's ideas on using an FFAG and in particular, stochastic cooling inside the FFAG. The application here is to radioactive isotope beams (RIBs). A novel feature of the design is the use of the ASTOR scheme to convert the cw RIB to a pulsed beam before transfer to the FFAG.

The PSI High Intensity Cyclotron and its Extrapolation to a 10 MW Driver

P.A. Schmelzbach, Th. Stammbach, S. Adam, A. Mezger, and P. Sigg

Paul Scherrer Institute, CH-5232 Villigen PSI, Switzerland

Abstract. The acceleration of a 2 mA, 590 MeV proton beam has been demonstrated at the PSI High Intensity Cyclotron. The peculiarities of the machine, the operation mode for high currents, and the ongoing upgrade program are discussed. The conceptual design of a 10 MW driver based on the extrapolation of the present performances is presented.

INTRODUCTION

The PSI accelerator facility was built in 1974 with the goal of reaching a beam intensity of 100 μA at an energy of 590 MeV. The main stage of the accelerator chain, the PSI Ringcyclotron is a separated sector cyclotron, specially designed for high beam intensities [1]. The concept proved to be successful and an upgrade was undertaken to achieve a beam current of 1.5 mA. In 1985 a new injector cyclotron was commissioned [2] and in 1991-95 the RF systems of the Ringcyclotron were rebuilt in order to provide the necessary RF power. In the course of this upgrade the peak voltage in the four accelerating cavities was raised from 450 kV to 730 kV. According to a law described by W. Joho in reference [3], the beam current limit set by longitudinal space charge forces in the Ringcyclotron is proportional to the third power of the acceleration voltage. The predicted beam intensity was reached in 1995, thus validating this law. With stronger bunching, better beam matching and other minor improvements the beam loss was further reduced and the beam current for routine operation could be raised to 1.8mA. In a short beam test in July 2000 a maximum beam current of 2 mA was accelerated and extracted from the Ringcyclotron. The layout of the accelerator complex is shown in figure 1.

To achieve 1 MW of beam power on the PSI spallation source SINQ, 2.7 mA have to be extracted from the Ringcyclotron and transported to the last Pion production target, after which about 60-70% of the scattered beam is accepted for further transportation. An upgrade of the accelerators to a current of 3 mA is under way. It bases on the experience made with the special operation mode of the Injector 2 and on the facts established during the continuous development of the Ringcyclotron RF system. The demand for stronger neutron sources and applications in transmutation technologies call for ever more powerful drivers. The feasibility of cyclotrons to generate a beam power in the order of 10 MW has therefore to be discussed, also. A tentative layout of a cyclotron facility with a beam current of 10 mA at 1 GeV has been presented [4]. The extrapolation of the beam performance from 2 mA up to 10 mA is a realistic issue and will be also discussed in the present report.

The understanding of the limitations due to space charge effects is the key to the development of high intensity cyclotrons. Some aspects will be summarized here, a more detailed discussion is presented in [5].

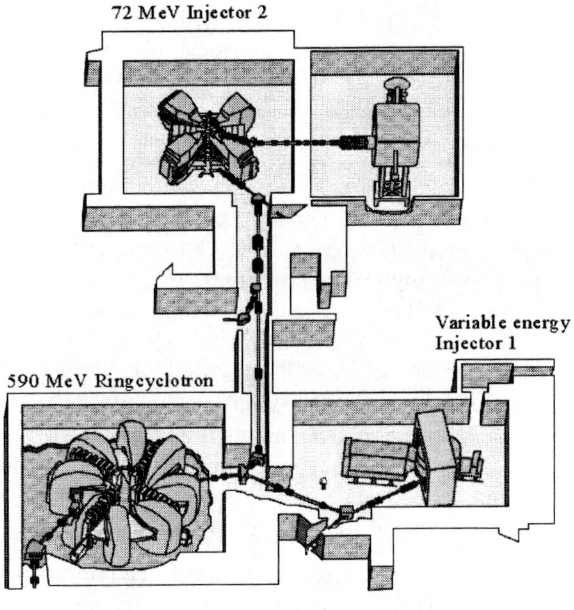

FIGURE 1. Layout of the PSI accelerators. The Injector 1 is presently only used for low energy applications.

THE INJECTOR CYCLOTRON

For the high intensity beam production the so-called Injector 2 is used. It can presently deliver a beam of up to 2.2 mA.. The preinjector is an 870 keV Cockcroft-Walton generator equipped with a cusp ion

CP642, *High Intensity and High Brightness Hadron Beams: 20th ICFA Advanced Beam Dynamics Workshop on High Intensity and High Brightness Hadron Beams*, edited by W. Chou, Y. Mori, D. Neuffer, and J.-F. Ostiguy

source operated at DC currents of 8 to 12 mA. For a 10 MW facility 50 to 60 mA would be necessary. Sources delivering such currents are already available. The corresponding beam power and the importance of a high beam stability is an important issue for the choice of a future preinjector. The possible use of a RFQ as preaccelerator has to be considered, also.

A major aspect in the 6-dimensional matching of the phase space of the beam to the injector cyclotron is a strong bunching. A double gap buncher is installed at a distance of 4.9 m from the injection point on the first turn. It is operated at a peak voltage of 7.2 kV. At first glance one would expect that this high buncher voltage introduces an undesirable energy spread at the time focus at the injection point. However, with a proper combination of the buncher voltage and the DC beam current, the space charge forces can be employed to reduce this energy spread [6].

The special features of a beam bunch of equal dimension in radial and longitudinal directions have already been noted by Chabert [7] in 1981. The PSI Injector 2 is, however, the first machine taking advantage of the properties of such a beam. In order to operate a cyclotron in this regime the beam has to be longitudinally matched. The ideal phase width at injection is about 20º in our case. The centre region of the Injector 2 with phase-defining and cleaning collimators is descibed in ref. [8].

Under space charge dominated conditions in a cyclotron, such a circular bunch has a stable configuration [7, 9, 6 and references therein], especially at high beam intensities. The bunch remains circular and of equal size during acceleration, and has therefore an extremely narrow phase width at extraction. In the case of Injector 2 it is about 2º. Consequently, a flattop system is not needed. The originally installed two flattopping cavities are now used for additional acceleration. In the upgrade programme to reach 3 mA it is planned to replace at least one of them by a true accelerating system [10].

The Injector 2 cyclotron has well separated turns up to the extraction radius, where the separation amounts to 23 mm. This corresponds to about 7σ of the beam width at 1.9 mA. The beam loss at extraction is correspondingly low and the extraction rate is about 99.98%. The beam width depends on the beam intensity as shown in figure 2.

The extrapolation to the case of a 10 MW facility is straightforward. With an extraction energy of 120 MeV the injector cyclotron would be larger, but it could be operated in the same regime. The beam size is expected to increase with beam current, but using four instead of two accelerating cavities enough energy gain per turn can be provided in order to assure separated orbits. A code to perform full six dimensional simulations is being developed [9, 11] in order to optimise the injection process and to further improve the capture rate.

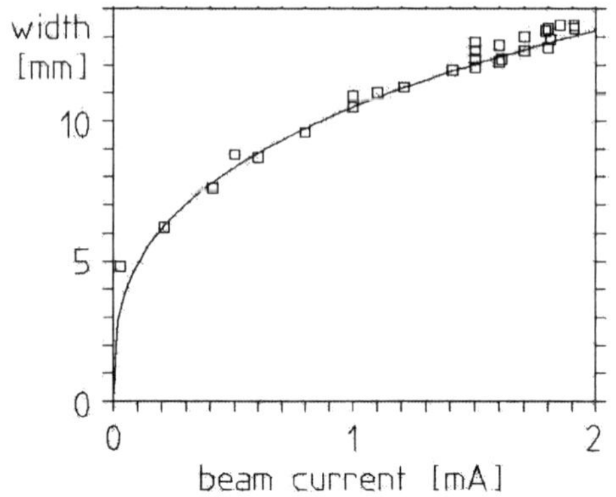

FIGURE 2. Beam size at the extraction from Injector 2. The width of the beam profile (4σ) in function of the extracted beam current averaged over the last seven turns is shown. The width is proportional to the cubic root of the intensity

THE RINGCYCLOTRON

The beam bunches in the Ringcyclotron are not circular as in the case of the Injector 2, but very elongated. Hence, space charge forces and especially longitudinal effects [3, 9, 6 and references therein] have to be accounted for. Due to the strong coupling between longitudinal and radial motion the longitudinal components generate a tilt of the elongated bunch in the radial direction. This tilt can be compensated by using a flattop system and properly adjusting its phase against the accelerating cavities.

Based on simple models W. Joho [3] predicted that the maximum beam current should be proportional to the third power of the average energy gain per turn. The limit reached when each year one cavity was upgraded from 450kV to 730kV followed exactly the prediction as shown in figure 3. With the installation of new, 1 MV cavities now in development [10] the space charge limit of the Ringcyclotron will be increased to 4 mA.

The Ringcyclotron also has well separated turns at extraction. The separation due to the acceleration only amounts to 6 mm. It is doubled by a coherent betatron oscillation induced by off-centre injection on the first orbit in the cyclotron. The resulting separation of 12 mm corresponds to about 7σ of the observed beam profile width at 1.8 mA. As for the Injector 2 the beam loss at extraction is low and the extraction rate is 99.98%. The predicted performance of the 1 GeV

Cyclotron beam bases on the experience from the upgrade of the PSI Ringcyclotron as shown in figure 3. It is assumed that longitudinal space charge effects set

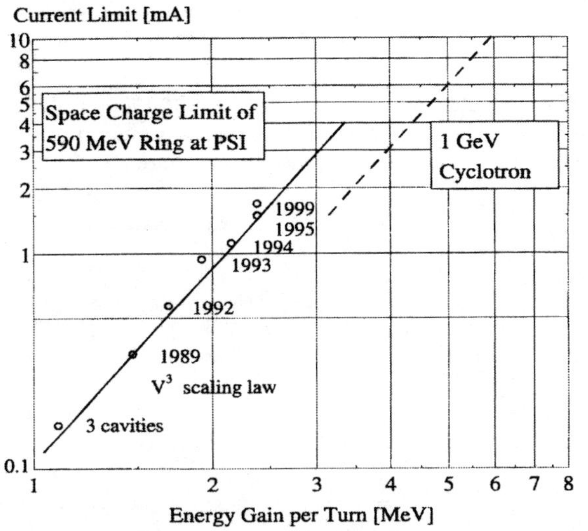

FIGURE 3. Maximum beam current extracted from the Ringcyclotron as a function of the average energy gain per turn as established during the upgrade of the cavity voltage. The solid line is the dependence due to the longitudinal space charge. Also shown is the extrapolation to a tentative facility with a 1 GeV, 10 mA cyclotron (dashed line).

the intensity limit. With eight accelerating cavities operated at a peak RF voltage of 1MV [10], the proposed 1GeV cyclotron has an averaged energy gain of 6.3 MeV/turn, compared to 2.4 MeV/turn in the PSI Ring with four accelerating cavities at 730 kV. Taking into account the different size and the different final energy a maximum beam current of 10 mA is possible, as shown by the dashed line in figure 3. A turn separation of 7σ similar to the situation today in the Ringcyclotron is expected under the following assumptions: acceleration of the beam into the fringing field where ν_r drops to 1.5, increased beam emittance from the injector cyclotron as extrapolated from figure 2 and no limitation from other sources. A rebuncher in the injection transport line might be necessary if the energy spread of the beam from the injector cyclotron becomes too large. A choice of parameters of the PSI and of the 10 MW facility are summarised in table 1.

CONCLUSION

Available technologies and experience at PSI are a realistic basis for the design of a cyclotron as a 10 MW proton driver. While scaling laws are seen to apply in a wide range, further progress in the simulation of space charge dominated beams is needed to estimate the ultimate performances of this type of accelerators.

TABLE 1. Comparizon of some parameters of the PSI cyclotrons with design values of a 10 MW driver.

	PSI Injector 2	**120 MeV**
Energy	72 MeV	120 MeV
Beam Intensity	2.2 mA	10 mA
Cavities	2 + 2 FT(see text)	4 (=2-2.5 MeV/t)
dR/dn	23 mm	28-35 mm
ΔR(4σ)	13 mm at 1.9 mA	23 mm
	PSI Ring	**1 GeV**
Magnete	8	12
Cavities	4 (730 kV)	8 (1000 kV)
Flat top	1 (460 kV)	2 (650 kV)
R_{av}	4462 mm	5677 mm
Number of turns	215	140
E_{gain} at extr.	2.44 MeV/turn	6.3 MeV/turn
Number of turns	215	140
$dR/dn = (R_{av}.\gamma/(\gamma+1)\nu_r^2)(E_{gain}/E)$	6 mm	11 mm
x 2 with betatron oscillation	12 mm	
ΔR(4σ)	7 mm at 1.8 mA	
Turn separation	7 σ	7 σ
Space charge lim	2 mA	10 mA
Beam power	1.18 MW	10 MW

REFERENCES

1 H.A.Willax, "Status Report on SIN", Proc. *7th Int Conf. on Cycl. and their Appl.*, Zürich 1975, p.33.

2 U.Schryber et al., "Status Report on the New Injector at SIN", *Proc. 9th Int Conf. on Cycl. and their Appl.*, Caen 1981, p. 43.

3 W.Joho, "High Intensity Problems in Cyclotrons", *Proc. 9th Int. Conf. on Cycl. and their Appl.*, Caen 1981, p.337.

4 Th.Stammbach et al., "The 0.9 MW Proton Beam at PSI and Studies on a 10 MW Cyclotron", *Proc. 2nd Int. Conf. on Accel.-Driven Transm. Tech.*, Kalmar 1996, p. 1013.

5 Th.Stammbach et al., "The PSI 2 mA Beam and Future Applications", *Proc. 16th Int. Conf. on Cycl. and their Appl.*, East Lansing 2001, p. 423.

6 J.Stetson et al., "The commissioning of PSI Inj.2 for high intensity, high quality beams", *Proc. 13th Int. Conf. on Cycl. and their Appl.*, Vancouver 1992, p. 36.

7 A.Chabert et al., *Proc. 7th Int. Conf. on Cycl. and their Appl.*, Zürich, 1975, p245 and *IEEE Trans. NS* 22-3, 1975, p. 1930.

8 U.Schryber et al., "High Power Operation of the PSI Accelerators", *Proc. 14th Int Conf. on Cycl. and their Appl.*, Cape Town 1995, p. 32.

9 S.Adam, "Step to enhance the knowledge on space charge effects", *16th Int. Conf. on Cycl. and their Appl.*, East Lansing 2001, p. 428.

10 P.K.Sigg et al., "Upgrade Concepts of the PSI Accelerator Systems for a Projected 3 mA Operation", *Proc. 16th Int. Conf. on Cycl. and their Appl.*, East Lansing 2001, p. 300.

11 A.Adelmann et al., Contribution to this workshop.

Isochronous and Scaling FFAGs

R. Baartman

TRIUMF 4004 Wesbrook Mall, Vancouver, BC, Canada V6T2A3

Abstract. We explore the differences between isochronous FFAGs (i.e. cyclotrons) and non-isochronous FFAGs (i.e. where the beam is necessarily pulsed). The general formula for tunes is derived. This is used to understand the acceptances and also the suitability of such machines as high energy and/or high power accelerators.

INTRODUCTION

If an FFAG can be made isochronous, acceleration can occur without ramping the rf frequency. But this is of course simply a cyclotron, of which many examples exist worldwide: 2 of the largest are PSI (580 MeV p, up to 2 mA beam current[1]), TRIUMF (500 MeV H^-, up to 0.2 mA beam current). We would like to know how the beam dynamics change if an FFAG is isochronous. In particular, can we still achieve large acceptance?

ISOCHRONISM

Let us define θ to be the angle of the reference particle momentum with respect to the lab frame. Orbit length L is given by speed and orbit period T:

$$L = \oint ds = \oint \rho d\theta = \beta c T.$$

The local curvature $\rho = \rho(s)$ can vary and for reversed-field bends even changes sign. (Along an orbit, $ds = \rho d\theta > 0$ so $d\theta$ is also negative in reversed-field bends.) Of course on one orbit, we always have

$$\oint d\theta = 2\pi.$$

What is the magnetic field averaged over the orbit?

$$\overline{B} = \frac{\oint B ds}{\oint ds} = \frac{\oint B\rho d\theta}{\beta c T}.$$

But $B\rho$ is constant and in fact is $\beta\gamma m_0 c/q$. Therefore

$$\overline{B} = \frac{2\pi}{T}\frac{m_0}{q}\gamma \equiv B_c\gamma = \frac{B_c}{\sqrt{1-\beta^2}}.$$

Remember, β is related to the orbit length: $\beta = L/(cT) = 2\pi R/(cT) \equiv R/R_\infty$. So

$$\overline{B} = \frac{B_c}{\sqrt{1-(R/R_\infty)^2}}.$$

Of course, this means the field index is $k = \frac{R}{B}\frac{dB}{dR} = \frac{\beta}{\gamma}\frac{d\gamma}{d\beta} = \beta^2\gamma^2 \neq$ constant.

So isochronism implies non-scaling.

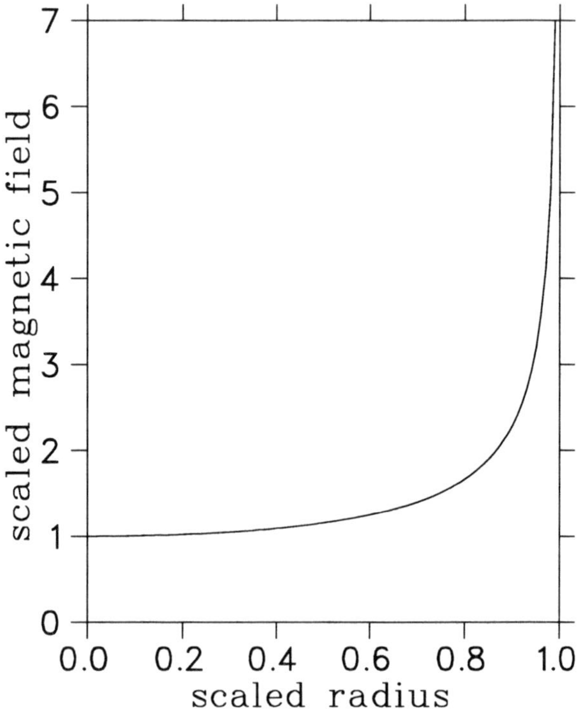

FIGURE 1. Magnetic fi eld law for an isochronous macine.

FOCUSING

Flat Field

For the moment, let us imagine that there are no sectors, no azimuthal field variation, just radial variation with field index $k = \beta^2\gamma^2$.

We know the transfer matrix in such a dipole, and can write directly:

$$\begin{aligned} \nu_r^2 &= 1+k & \nu_z^2 &= -k \\ &= 1+\beta^2\gamma^2 & &= -\beta^2\gamma^2 \\ &= \gamma^2 & &< 0 \\ \nu_r &= \gamma & \nu_z &= \text{imaginary} \end{aligned}$$

So why do such cyclotrons work at all? Lawrence's first cyclotrons, and those built before the early 50s, had no sector focusing. To obtain vertical focusing, k was made slightly negative. This resulted in phase slippage,

CP642, *High Intensity and High Brightness Hadron Beams: 20th ICFA Advanced Beam Dynamics Workshop on High Intensity and High Brightness Hadron Beams,* edited by W. Chou, Y. Mori, D. Neuffer, and J.-F. Ostiguy

so the rf voltage was made as large as possible to achieve the final energy before an accumulated phase slip of $\pi/2$. In addition, some vertical focusing was achieved by accelerating on the falling side of the rf voltage. In this way, such cyclotrons achieved maximum energies of 20 to 30 MeV (protons).

To reach higher energy with simple magnet poles, it was necessary to release isochronism, ramp the rf frequency, and pulse the machine. Unfortunately, this meant that the maximum intensity was 2 or 3 orders of magnitude lower than for the (cw) cyclotron.

Azimuthally-Varying Field

Strong focusing was invented/developed in the 50s. This had implications for

- **cyclotrons**: they could now be isochronous AND vertically focusing
- **synchro-cyclotrons**: tunes need be no longer near 1, so beams were smaller, space charge limits higher
- and **synchrotrons**: same advantages as for synchro-cyclotrons

The application was by far simplest for synchrotrons. For FF machines, the extended nature of the field was a calculational headache. That is why synchrotrons developed rapidly in the 50s (when large computing power was unavailable), while synchro-cyclotrons of the FFAG type did not. Nowadays, large computing power is freely available.

For Azimuthally-Varying Field (AVF) or the special case of Alternating Gradient (FFAG), let us use **Mathematica** to calculate the tunes. To make it transparent, let us consider all identical dipoles and drifts; no reverse bends. We have drifts d, dipoles with index k, radius ρ, bend angle ϕ, and edge angles $\phi-\theta$:

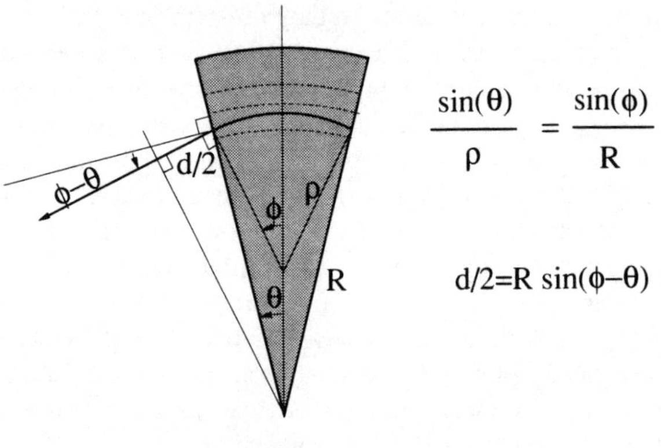

FIGURE 2. FFAG sector magnet with no spiral. Notice that the (blue) orbit does not follow a line of constant field.

In addition, imagine that the edges are inclined by an extra angle ξ, not shown in the figure. This is called the "spiral angle".

In this hard-edged case, the "flutter" $F^2 \equiv \langle (B-\overline{B})^2\rangle/\overline{B}^2 = R/\rho - 1$. Aside: Notice that the particle trajectory (blue curve) does not coincide with a contour of constant B (dashed curves). This has large implications for using existing transport codes to describe FFAGs. The

```
s = ρ * ϕ;
θ = ArcSin[(ρ / R) Sin[ϕ]];
f1 = ρ / Tan[ϕ - θ + ξ];
f2 = ρ / Tan[ϕ - θ - ξ];
d = 2 R * Sin[ϕ - θ];
R = ρ (1 + F^2);
k = κ * ρ / R;
kx = Sqrt[1 + k] / ρ;
ky = Sqrt[k] / ρ;
Mz := (( Cosh[ky s]        Sinh[ky s] / ky )
       ( ky Sinh[ky s]     Cosh[ky s]      )) .
      (( 1      0 )    (( 1  d )    (( 1      0 )
       ( -1/f2  1 )) .  ( 0  1 )) .  ( -1/f1  1 )) .
      (( Cosh[ky s]        Sinh[ky s] / ky )
       ( ky Sinh[ky s]     Cosh[ky s]      ))
νz^2 = FullSimplify[
   (ArcCos[Series[
       (Mz[[1, 1]] + Mz[[2, 2]]) / 2,
       {ϕ, 0, 3}]]) / (2 ϕ)]^2
(-κ + F^2 (1 + 2 Tan[ξ]^2)) + O[ϕ]^2
```

```
Mr := (( Cos[kx s]         Sin[kx s] / kx )
       ( -kx Sin[kx s]     Cos[kx s]      )) .
      (( 1     0 )    (( 1  d )    (( 1     0 )
       ( 1/f2  1 )) .  ( 0  1 )) .  ( 1/f1  1 )) .
      (( Cos[kx s]         Sin[kx s] / kx )
       ( -kx Sin[kx s]     Cos[kx s]      ))
νr^2 = FullSimplify[
   ArcCos[Series[
       (Mr[[1, 1]] + Mr[[2, 2]]) / 2,
       {ϕ, 0, 3}]] / (2 ϕ)]^2
(1 + κ) + O[ϕ]^2
```

FIGURE 3. `Mathematica` code to generate FFAG tunes.

resulting expressions (see Fig. 3) are identical to those originally derived by Symon et al. in the original 1956 Phys. Rev. paper about FFAGs[2].

But beware! Since there is now a distinction between local curvature (ρ) and global (R), the definition of field index is ambiguous. The local index, used in the dipole transfer matrix, is

$$k = \frac{\rho}{B}\frac{dB}{d\rho},$$

while the Symon formula uses

$$\kappa = \frac{R}{B}\frac{dB}{dR} \approx k\frac{R}{\rho}$$

As we proved, it is in fact this latter quantity which must be equal to $\beta^2\gamma^2$ for isochronism. We therefore still have

$$\nu_r = \gamma \qquad \text{(isochronous)}$$

But

$$\nu_z^2 = -\beta^2\gamma^2 + F^2\left(1+2\tan^2\xi\right) \qquad \text{(isochronous)}$$

Aside: We could have **Mathematica** print out the next higher order, but it would be in error because the trajectories are not circular arcs.

Example: TRIUMF cyclotron

See for example [3] for some details of the magnetic field configuration.

TABLE 1. Vertical tune at a few radii in the TRIUMF cyclotron.

Energy	R	ξ	$1+2\tan^2\xi$	F^2	ν_z
100 MeV	175 in.	0°	0.0	0.30	0.28
250 MeV	251 in.	47°	3.3	0.20	0.24
505 MeV	311 in.	72°	20.0	0.07	0.24

A peculiar feature of the TRIUMF cyclotron is that the peak field is only 0.5 T. The reason for this is that the accelerated particle is H^-, and higher fields at the 500 MeV energy needed for efficient meson production would strip the loosely bound extra electron. This makes the flutter at large radii very small and forces reliance on spiral angle instead. The extreme spiral angle used was considered quite daring at the time the design was conceived.

Irregular FFAG cyclotrons

By adjusting the flutter F and spiral angle ξ as functions of R, we can arrange to make ν_z constant.

But what about ν_r? Can we change it by departing from the regular N-cell lattice? Perhaps, but no one has ever tried it.

In synchrotron language, isochronism means $\gamma = \gamma_t$. But we know that for a "regular" lattice, $\gamma_t = \nu_r$. Hence, $\nu_r = \gamma$. For any lattice[4],

$$\frac{1}{\gamma_t^2} = \frac{\nu_r^3}{R}\sum_n \frac{|a_n|^2}{\nu_r^2 - n^2}$$

where a_n is the Fourier transform of $\beta_x^{3/2}/\rho$. For N identical cells where $N \gg \nu_r$, this gives $\gamma_t = \nu_r$ because the $n = 0$ term gives by far the largest contribution to the sum.

FIGURE 4. TRIUMF cyclotron. This is the lower pole, shown during construction (1972).

If the lattice is arranged to have a superperiodicity $n = \pm S$ where $S \sim \nu_r$, γ_t can be moved away from ν_r:

$$\frac{1}{\gamma_t^2} = \frac{1}{\nu_r^2} + \frac{2|a_S|^2}{R}\frac{\nu_r^3}{\nu_r^2 - S^2}$$

This is the approach used to make γ_t imaginary in e.g. the JHF main ring[5]. Whether this can be used to fix ν_r in a cyclotron where necessarily $\gamma_t = \gamma$, is not clear, since it prescribes a peculiar variation of a_S with R.

ACCEPTANCE

The chief virtue that FFAG scaling machines have over non-scaling isochronous machines is the constancy of the tunes. This results in huge acceptance ($> 1000\pi$ mm-mrad[6]), especially in the bending plane. But because they cannot be isochronous, they must be pulsed.

The chief virtue of isochronous FFAGs is that they can run cw and reach very high intensity. However, because $\nu_r = \gamma$, many resonances must be traversed to reach high energy. This reduces acceptance to zero: all beam would be lost if energy gain per turn were as low as it is in non-isochronous pulsed FFAGs. However, since the rf frequency is fixed, one can use high Q cavities and very high energy gain per turn. For example, a 9 GeV proton cyclotron with 30 sectors was studied[7] and found to have an acceptance of less than 1 πmm-mrad normalized even with an energy gain of 25 MeV per turn. This was due to the intrinsic 30/3 resonance crossed at $\gamma = 10$.

CONCLUSIONS

FFAG synchrocyclotrons can be designed with very large acceptance and repetition rate and so can be used for accelerating very large time-averaged current or very large phase space populations as is the case for example with uncooled secondary beams. If an FFAG is isochronous, then the rf can have constant frequency (and the beam can be cw), but the acceptance advantage over synchrotrons is lost because isochronism implies that the horizontal tune cannot be constant (is actually equal to γ). Isochronous FFAGs already exist, accelerating protons up to $\gamma = 1.6$. It seems possible to extend such designs to energy of a few GeV.

ACKNOWLEDGMENTS

TRIUMF receives federal funding via a contribution-agreement through the National Research Council of Canada.

REFERENCES

1. Schmelzbach, P., Stammbach, T., Adam, S., Mezger, A., , and Sigg, P., *these proceedings (FFAG)* (2002).
2. Symon, K., Kerst, D., Jones, L., Laslett, L., and Terwilliger, K., *Phys. Rev.*, **103**, 1837–1859 (1956).
3. Craddock, M., and Richardson, J., *(PAC 1969) IEEE Trans. Nucl. Sc.*, **16**, 415–420 (1969).
4. Courant, E., and Snyder, H., *Annals of Physics*, **3**, 1–48 (1958).
5. Mori, Y., *these proceedings (JHF)* (2002).
6. Mori, Y., *these proceedings (FFAG)* (2002).
7. Baartman, R., Mackenzie, G., and Lee, R., *(PAC 1983) IEEE Trans. Nucl. Sc.*, **30**, 2010–2012 (1983).

Beam Dynamics of FFAG Accelerator

Masamitsu Aiba

Department of Physics, University of Tokyo

Abstract. Beam dynamics of FFAG Accelerator is described, focusing on the non-linearity. The tracking simulations were performed, in order to understand the non-linear motion in the guiding field of FFAG accelerator.

INTRODUCTION

A large acceptance of FFAG is ideal as a phase rotator and a secondary particle accelerator, not only as a high current proton accelerator. On the other hand, lattice magnets of FFAG are full of non-linearity. It is not obvious why FFAG has large transverse acceptance and what limits the acceptance. We started systematic study to answer these questions, both analytically and by tracking.

NON-LINEAR MOTION IN FFAG

In FFAG accelerators, for the sake of achieving the zero the chromaticity condition, the higher order components are introduced into the guiding field. Then, the betatron tune has no dependence upon momentum but has a dependence upon its oscillation amplitude. Then the bare tune (the tune of vanishingly small amplitude) should be selected, taking into account the non-linear motion at a large amplitude.

HORIZONTAL MOTION

Relation between the Horizontal Acceptance and the Phase Advance

The tracking simulations (Runge-Kutta integration) using two-dimensional field of "hard edge model" [2] were performed to calculate the horizontal acceptance. Such a simulation is one of the effective methods to analyze the non-linear motion. Figure 1 shows the simulation results for the cases of the periodicity of 16 and 32 with triplet (DFD) lattice. In Fig.1, the horizontal axis corresponds to the phase advance per cell. Comparing the rings having the different periodicity, it is general and convenient to use the phase advance per cell. Two common tendencies in two rings can be seen. One is that the acceptance becomes smaller as increasing the phase advance. The other is that the acceptance becomes small rapidly around the structure resonance corresponding to the phase advance of 2π/3, 2π/4, 2π/5.

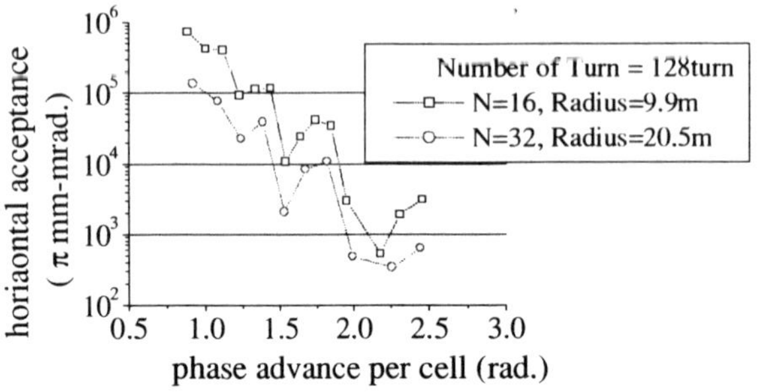

FIGURE 1. Horizontal Acceptance vs. Phase Advance

First tendency is explained by the large k value. The k value is main knob of the horizontal phase advance; the phase advance becomes larger as increasing the k value. Equation 1 shows the radius dependence of FFAG guiding field and its expansion. Here, B_0 is the field strength at the radius r_0.

$$B_z = B_0\left(\frac{r}{r_0}\right)^k = B_0 + B_0\frac{k}{r_0}x + B_0\frac{k(k-1)}{2!\,r_0^{\,2}}x^2 + \cdots$$

$$(Taylor\ \ Expansion\ \ around\ \ r_0, r = r_0 + x). \qquad (1)$$

As shown in Eq.1, a large k value enhances the non-linearity. It seems a good solution to take a small phase advance so that the acceptance becomes large; however, the phase advance, in other word the k value, should be kept certain large in order to keep the orbit excursion in a feasible value.

Second tendency is explained by the separation between the bare tune and the structure resonance line. Figure 2 shows the tune shifts for the cases that the phase advance is around 2π/3.

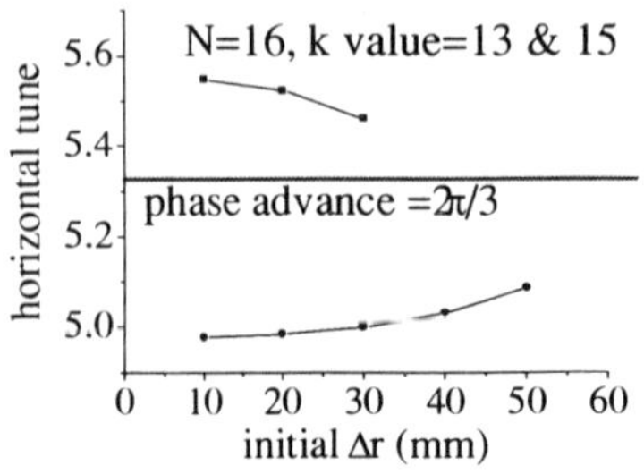

FIGURE 2. Horizontal Tune Shift around Phase Advance of 2π/3

CP642, *High Intensity and High Brightness Hadron Beams: 20th ICFA Advanced Beam Dynamics Workshop on High Intensity and High Brightness Hadron Beams,* edited by W. Chou, Y. Mori, D. Neuffer, and J.-F. Ostiguy

The direction of tune shift faces to the structure resonance line. Therefore, "the valley of acceptance" appears and naturally the acceptance becomes zero on the structure resonance.

Normalization of Horizontal Acceptance

In order to obtain more general result on the relation, a normalization of horizontal acceptance is introduced as follows.

With an approximation of the large k value, the low order terms of Eq.1 can be transformed into Eq.2.

$$B_z \cong B_0\left[1+\left(\frac{k}{r_0}x\right)+\frac{1}{2!}\left(\frac{k}{r_0}x\right)^2+\frac{1}{3!}\left(\frac{k}{r_0}x\right)^3+\cdots\right]. \quad (2)$$

In Eq.2, the quantity kx/r_0 can be treated as normalized coordinate of the distance. On the other hand, the emittance is expressed in Eq.3,

$$W = x^2/\beta, \quad (3)$$

where β is the beta-function. Using approximations of the beta-function and the betatron tune,

$$\because \beta \cong r_0/\nu_h,\ \nu_h \propto \frac{k}{N^2}\times N = \frac{k}{N}, \quad (4)$$

where ν_h is the horizontal tune and N is the periodicity, Eq.3 can be transformed into Eq.5, which includes normalized coordinate.

$$W \propto x^2\left(\frac{k}{r_0 N}\right) = \frac{r_0}{kN}\left(\frac{k}{r_0}x\right)^2. \quad (5)$$

Then, the remaining factor is only kx/r_0 and that is the normalization factor of the acceptance. As the result of the normalization, Fig.3 is obtained.

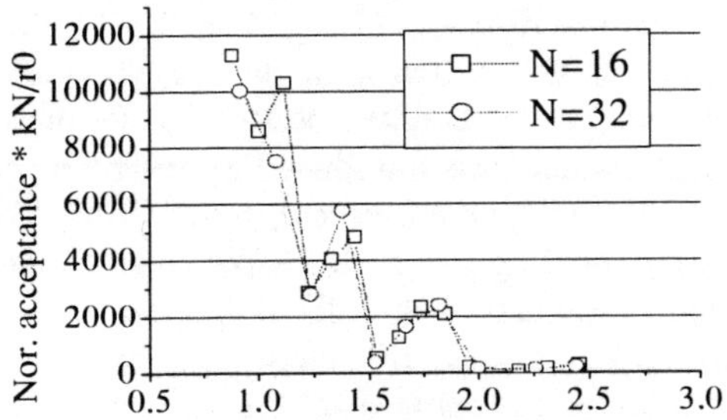

FIGURE 3. Normalized Horizontal Acceptance vs. Phase Advance

Three parameters in the normalization factor are the main parameters of the ring. And the horizontal phase advance is almost determined only by the k value and the periodicity. Therefore, using Fig.3, it is possible to estimate the horizontal acceptance of a ring with any parameters.

VERTICAL MOTION

Soft Edge Model of FFAG Focusing Field

In order to study the vertical motion including its non-linear term, "soft edge model" of three-dimensional field is introduced in the followings.

First, three components of the cylindrical coordinate are expressed by the polynomial of the vertical coordinate as shown in Eq.6.

$$B_i = B_{i0}(r,\theta) + B_{i1}(r,\theta)z + B_{i2}(r,\theta)z^2 + \cdots \quad i = r,\theta,z \quad (6)$$

Inserting these equations into Maxwell's Equation, the coefficients of the polynomial can be solved, using the boundary condition on the median plane shown by Eq.7 and Eq.8,

$$B_{r0}(r,\theta) = B_{\theta 0}(r,\theta) = 0, \quad (7)$$

$$B_{z0}(r,\theta) = B_0\left(\frac{r}{r_0}\right)^k E(\theta), \quad (8)$$

where B_0 is the field strength at the radius r and $E(\theta)$, which changes from 1 to 0, represents the distribution of the fringing field.

Finally, equations from 9 to 11 are obtained as "soft edge model",

$$B_r(r,\theta,z) = E\frac{kB_0}{r_0}\left(\frac{r}{r_0}\right)^{k-1} z - \frac{1}{3!}(Ek^2+E^{(2)})\frac{(k-2)B_0}{r_0^3}\left(\frac{r}{r_0}\right)^{k-3} z^3 + \cdots, \quad (9)$$

$$B_\theta(r,\theta,z) = E^{(1)}\frac{B_0}{r_0}\left(\frac{r}{r_0}\right)^{k-1} z - \frac{1}{3!}(E^{(1)}k^2+E^{(3)})\frac{B_0}{r_0^3}\left(\frac{r}{r_0}\right)^{k-3} z^3 + \cdots, \quad (10)$$

$$B_z(r,\theta,z) = EB_0\left(\frac{r}{r_0}\right)^k - \frac{1}{2!}(Ek^2+E^{(2)})\frac{B_0}{r_0^2}\left(\frac{r}{r_0}\right)^{k-2} z^2 \cdots. \quad (11)$$

Furthermore, first term of the radial and vertical field can be expanded as shown in Eq.12 and Eq.13 respectively.

$$E\frac{kB_0}{r_0}\left(\frac{r}{r_0}\right)^{k-1} z = E\left(\begin{array}{l} B_0\frac{k}{r_0}y + B_0\frac{k(k-1)}{r_0}xy \\ + B_0\frac{k(k-1)(k-2)}{2!r_0^{2}}x^2 y\cdots \end{array}\right), \quad (12)$$

$$EB_0\left(\frac{r}{r_0}\right)^{k} = E\left(B_0 + B_0\frac{k}{r_0}x + B_0\frac{k(k-1)}{2!r_0^{?}}x^2 + \cdots\right), \quad (13)$$

where the notation is changed; the coordinate y represents the coordinate z. Then the natural result appears; the guiding field of FFAG accelerator can be expressed by the summation of multipole components.

Vertical Tune Shift

The vertical tune shift is mainly occurred by the component of normal octupole. The effect of higher order components more than octupole is weak, unless the vertical tune is very close to the value affected by these components. Then, the direction of tune shift is one-side. The tracking simulation with "soft edge model" was carried out for various bare tunes. The results shown in Fig.4 indicate that the direction of the vertical tune shift is upward.

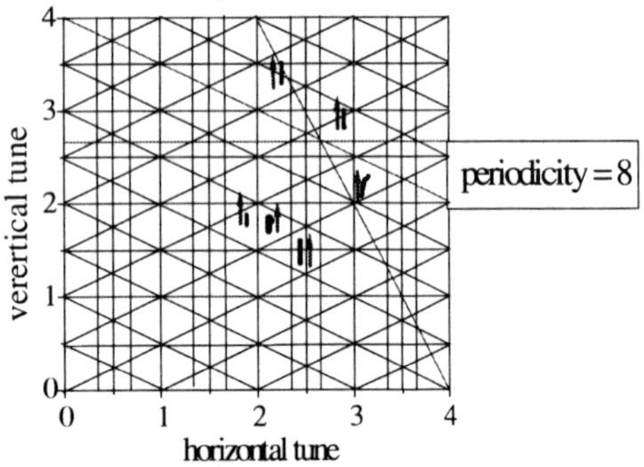

FIGURE 4. Vertical Tune Shift with "soft edge model"

Then, the criterion of the tune selection is obtained; the vertical tune should not be set just bellow the strong resonance line.

Non-linear Coupling

Non-linear coupling induced by normal sextupole is observed with the simulation as shown in Fig.5. When the pair of tunes is very close to the non-linear resonance line ($\nu_x+2\nu_y=q$), the motions in each phase space are not bounded. As shown in Eq.2, the guiding field has normal sextupole, intrinsically. Especially, the non-linear structure resonance line should be avoided.

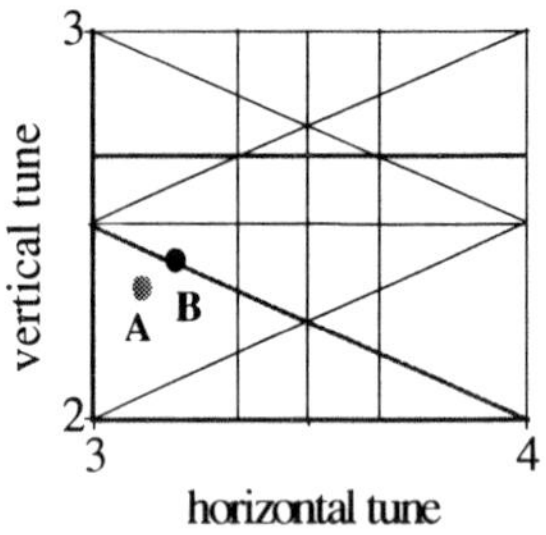

(a) Two Simulation Conditions in the Tune Diagram

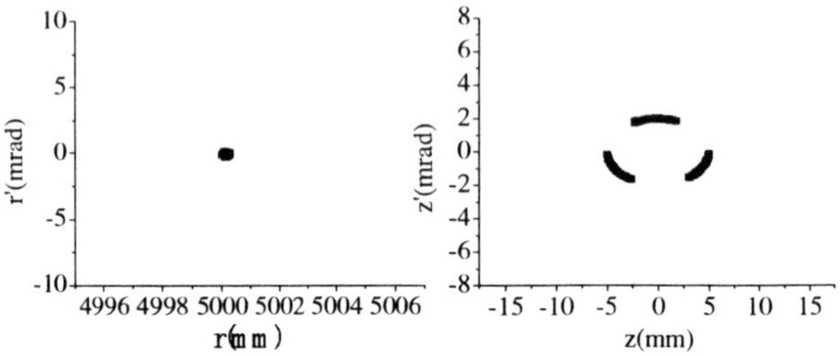

(b) Bounded motions with avoiding the non-linear coupling (A)

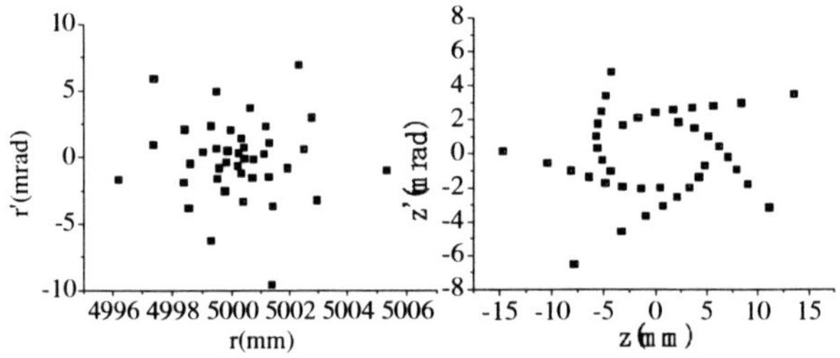

(c) Unbounded motions due to the non-linear coupling (B)

FIGURE 5. Non-linear Coupling with "soft edge model"

CONCLUSION

The beam dynamics of FFAG accelerator was studied to obtain a large acceptance. We defined normalized coodinates. Using that the horizontal acceptance is described as a function of phase advance per cell, independently of the periodicity. Furthermore, "soft edge model" was introduced to study the vertical motion. It shows the vertical tune shift and the nonlinear coupling in FFAG. We found that the direction of the tune shift is fixed and has to be taken into account to choose bare tune.

REFERENCES

1. Y. Mori, 'Neutrino Factory in Japan: Based on FFAG Accelerator', Proc. of EPAC02, p278-p280.
2. M. Aiba et al., 'Study of Acceptance of FFAG Accelerator', Proc. of EPAC02, p1226-p1228.

Rapid Acceleration in an FFAG using High-Frequency RF

C. Johnstone * and S. Koscielniak †

*Fermi National Accelerator Laboratory, P.O. Box 500, MS 220, Batavia, IL, 60510
†TRIUMF, Vancouver, B.C., V6T 2A3 Canada

Abstract. When large transverse and longitudinal emittances are to be transported through a circular machine, extremely rapid acceleration holds the advantage that the beam becomes immune to nonlinear resonances because there is insufficient time for amplitudes to build up. Uncooled muon beams exhibit large emittances and require fast acceleration to avoid decay losses and would benefit from this style of acceleration. The approach here employs a fixed-field alternating gradient or FFAG magnet structure and a fixed frequency acceleration system. Acceptance is enhanced by the use only of linear lattice elements, and fixed-frequency rf enables the use of cavities with large shunt resistance and quality factor.

INTRODUCTION

Acceleration of large emittance beams, particularly those with simultaneously large transverse and longitudinal emittances, present a challenging new direction in accelerator design. Conventional accelerators such as synchrotrons or linacs cannot support acceleration of ultra-large emittances for either case. *Scaling* FFAG accelerators, such as the radial or spiral sector[1], display an almost unlimited momentum acceptance, but transverse acceptance is sensitive to design and is generally restricted. The approach described here is development of a *nonscaling* FFAG[2] wherein the ideal optics demonstrate strong linearity; that is, the acceptance is not sensitive to energy or amplitude. The discerning feature of the nonscaling version, in keeping with the condition of linearity, is that the optics are not held constant, as in the scaling machine, but change slowly with momentum.

However, a signature of fixed field acceleration is that orbit length unavoidably changes with energy; it can be substantial and can result in a significant phase-slip relative to the accelerating waveform[2][3]. For relativistic beams and rapid acceleration, this poses a non-standard problem which must be addressed by the rf system. Recent workshops have focussed on the phasing problems of FFAGs and a number of solutions are being advanced[4]. These include: (i) momentum-dependent chicanes[5] to correct pathlength differences, (ii) broad-band RF that can be phased quickly, or (iii) a frequency low enough (25 MHz, for example), to make the phase errors ineffectual. Outside of the obvious solutions of broadband rf, or very low frequency rf, this paper outlines alternative approaches using high-Q, high-frequency rf systems. This paper reports significant progress on both a lattice and rf acceleration system for a high-energy FFAG in the context of rapid acceleration.

FFAG LATTICES

A circular accelerator system can be designed with magnetic fields that remain constant during acceleration by adopting an alternating gradient focussing lattice. The arcs of such machines, composed of large aperture magnets, can be designed to accommodate the large energy range in acceleration. The beam centroid orbit is not fixed as in a ramped machine, but rather moves across the magnet aperture during acceleration. Lattices have been developed which can contain up to an energy change of a factor of four[6][7][8].

In the nonscaling FFAG, not only do the central orbits move across the aperture, but also the optics functions vary with the central momentum. When acceleration occurs so rapidly that the beam experiences only a few turns in the machine one does not have to avoid resonances or control lattice parameters as a function of momentum. Instead, one has the freedom to choose parameters optimal for the application such as minimizing circumference and supporting a large transverse dynamic aperture[2].

The example, which provides the focus of this paper, is a 6-20 GeV non-scaling FFAG cell optimized for ultra-rapid, stable acceleration, as is required for intense muon sources. The entire lattice is comprised solely from 314 simple arc FODO cells. (The periodicity is, therefore, 314). The rf system is assumed distributed over most of the ring with 3 cavities filling the 3 m straight in each half cell. (To keep field gradients to conventional values, $\approx 3MV/m$, the rf needs to occupy about 300 of the cells, given a few number of acceleration turns, although much higher-gradient cavities are currently under construction and testing.) The remaining empty cells are filled with the kickers necessary for injection and extraction, processes which require several cells to complete. The simplicity of lattice design; i.e. its unbroken periodic structure, is responsible for its stability over a large range

CP642, *High Intensity and High Brightness Hadron Beams: 20th ICFA Advanced Beam Dynamics Workshop on High Intensity and High Brightness Hadron Beams,* edited by W. Chou, Y. Mori, D. Neuffer, and J.-F. Ostiguy

in momentum despite changing optical conditions. The lattice components, parameters, betatron functions and peak orbit excursions are given in Table 1 and plotted in Figures 1 through 4. The central energy has been arbitrarily defined as the energy at which the beam centroid is centered in the focussing quadrupole (where B=0). The plots of peak orbit excursions show the radial difference in the beam centroid positions at injection and extraction relative to the orbit described by the central energy.

TABLE 1. Parameters of a 6-20 GeV *Nonscaling* FFAG. Where two sets of values are given, they represent superconducting/normal conducting magnets, respectively.

General (SC/NC)	
Central Energy	16.5 GeV
Circumference	2041/2355 km
Magnetic Rigidity	55 T-m
Poletip Field	6T/2T
Basic Cell	
Number	314
Length	6.5/7.5 m
Bend/cell	0.02 rad
Quad Gradient	75.9/25.3 T/m
"F" length	0.15/0.45 m
"F" strength	1.38/0.46 m^{-2}
"D" length	0.35/1.05 m
"D" strength	0.59/0.20^{-2}
Cell Tunes	
6 GeV	0.45 (162 deg)
20 GeV	0.08 (29 deg)
Maximum Displacements	
6 GeV	-7.5 cm
20 GeV	7.1 cm

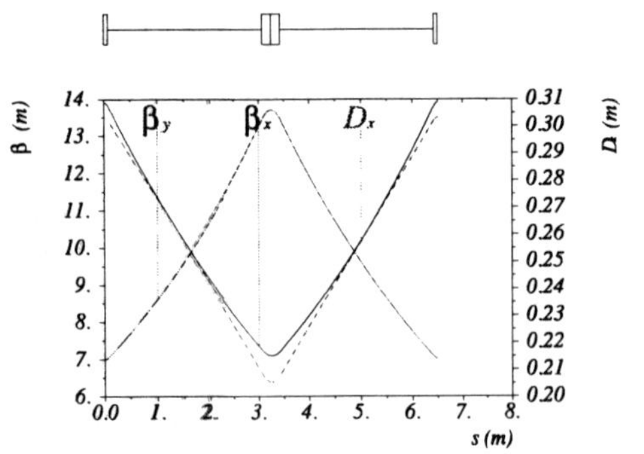

FIGURE 1. Lattice functions at 16.5 GeV for a 6-20 GeV non-scaling FFAG.

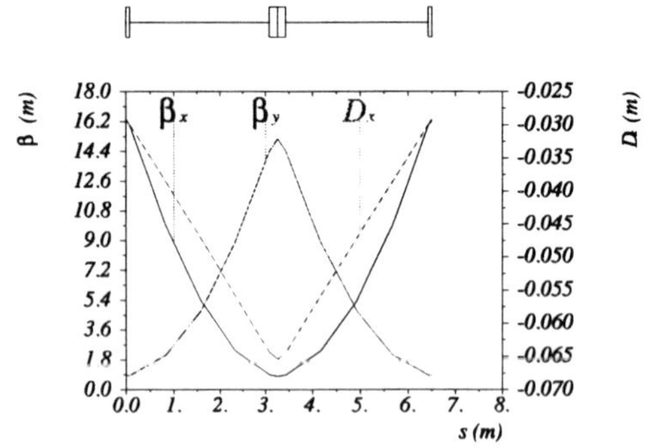

FIGURE 2. Lattice functions at 6 GeV for a 6-20 GeV non-scaling FFAG.

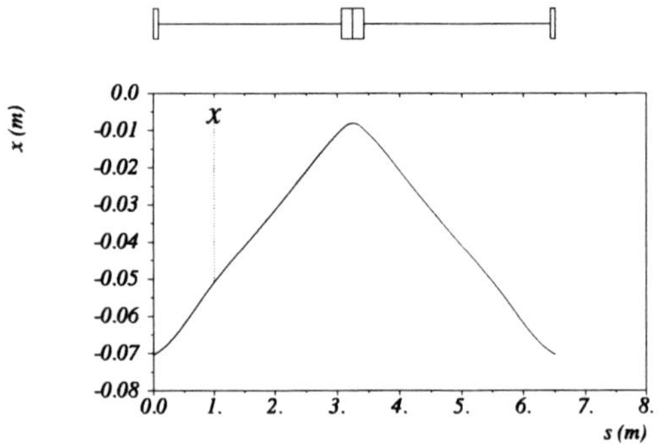

FIGURE 3. 6-GeV orbit excursion for a 6-20 GeV non-scaling FFAG.

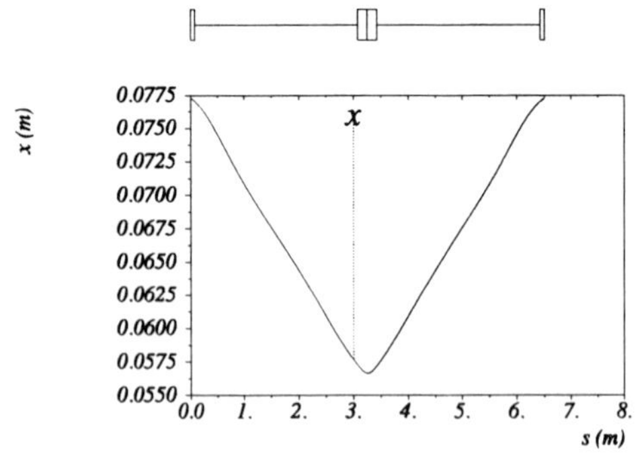

FIGURE 4. 20-GeV orbit excursion for a 6-20 GeV non-scaling FFAG.

Pathlength Dependencies in FFAGs

A main drawback to FFAGs in both the scaling and the non-scaling versions is the large changes in path-

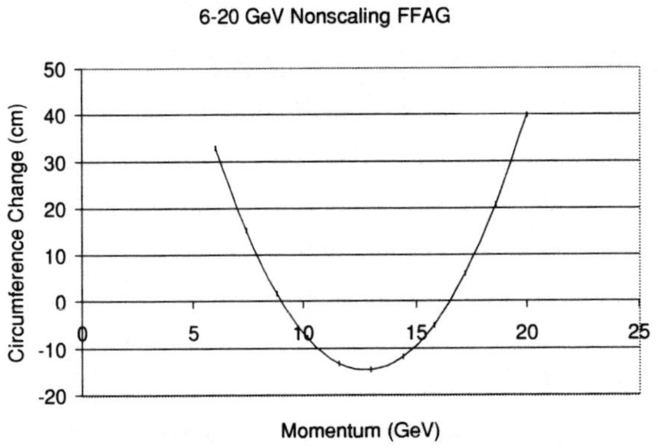

FIGURE 5. Circumference change as a function of momentum

length as a function of energy. The pathlength dependence is clearly linear with momentum for radially-staggered, parallel orbits as in the radial-sector FFAG, but it is parabolic in *nonscaling* FFAGs. (The following discussion of *scaling* FFAGs refers to the radial-sector FFAG explicitly.)

Pathlength and therefore traversal time change with the reference energy. In a regime of slow acceleration this variation is distributed over many turns, but for rapid acceleration non-negligible changes occur from cell to cell. Of course, the cell traversal times must be synchronized with the waveforms in the RF cavities responsible for acceleration. This consideration leads to the following discussion of acceleration modes and the rf systems that are compatible with them.

FFAG ACCELERATION

In a circular machine, particles make repeated passages through the same cavities; and so on every revolution of the machine the frequency and phasing of each cavity must be readjusted. To make this possible, the demanded phase change per cavity filling time should be small; and this leads to the condition $2\pi(\Delta C/C) < 1/Q$ where Q is the quality factor and $\Delta C/C$ is the relative fractional change in circumference per turn. If no attempt is made to adjust the cavity frequency or phase, then errors accumulate linearly with time and inversely as harmonic number. A fixed-field machine can be operated in several modes, classifiable in terms of the timescale for acceleration; for example, conventional and rapid, which range over tens of milliseconds and tens of microseconds, respectively. Between these extremes, ies moderate acceleration: theoretically, a few hundred acceleration turns (hundreds of microseconds) can be achieved in either a *nonscaling* or *scaling* FFAG.

Rapid Acceleration

When acceleration is to be completed in a few turns, the large energy gain per turn forces one to consider on-crest acceleration as in a cyclotron. In this case, the rf is used almost entirely to provide acceleration. One may envisage use of low-Q ($\leq 10^3$) cavities and *on-crest* operation with phase shifting of accelerating stations to make up for non-isochronous behaviour of orbits in the FFAG. Even when the phases of the cavities are readjusted, this regime can be sustained only for a few tens of turns, due to the near absence of phase focusing.

A more affordable approach for rapid acceleration employs high-Q ($\geq 10^6$) and high frequency ($\geq$ 100 MHz). Because high-Q cavities cannot be rephased on a rapid acceleration timescale (microsecond), the phase relation of the beam to the acceleration waveform is completely determined by the changes in the orbital pathlength during acceleration, and these changes are substantial at high cavity frequency. (For the example in this paper, the total pathlength changes by $\approx$ 50 cm which is 1/3 the wavelength at 200 MHz). Thus acceleration does not remain fixed at the crest of the rf waveform, but rather crosses over it one or more times. The *nonscaling* FFAG demonstrates at least two operational modes for rapid acceleration, termed cross-crest and near-crest acceleration, with the difference being a common initial phase *versus* independently-optimized starting phases for individual cavities.

FFAG RF system

The present work investigates the simplest approach: the application and optimization of a single high-frequency, high-Q rf system to operate in the *near-crest regime*; since this allows lower voltage and may furnish a less distorted phase space than *cross-crest regime* when an equal number of turns are employed. Modest elaborations, such as second harmonic or a mixing of several fundamental frequencies to produce a waveform that better matches the beam traversal times, are also considered.

There are CERN[9, 10] designs available for 200 MHz normal T conducting (NC) and super conducting (SC) cavities, and for 400 MHz SC cavities that provide a starting point for extrapolation. A design with $R =$ 14 Mohm, $Q = 7 \times 10^4$ and 2 MV gap voltage is within reach of present NC technology and the peak rf power is some 250 MW distributed between 1800 cavities. However, the filling time of 350 μs is too long to allow rephasing. The modulation could be achieved by vector addition of a constant in-phase term and a varying quadrature term. Unfortunately, to sweep the phase correctly requires a quadrature modulation term comparable with the in-phase carrier. Thus the demands of the "quadrature-subsystem" are no different from the in-phase RF system. Given the large power requirement of NC cavities, a FFAG for rapid acceleration would clearly benefit from the adoption of superconducting cavities (SC) with quality factors ranging from 10^7 to 10^9. In either case, NC or SC, pure sinusoid operation is the only mode possible.

There exist many possible optimization strategies. We report only that one which aims to give the reference bunch the maximum acceleration on each turn. The optimization is based on reference particles and, strictly speaking, it cannot be used directly to maximize input acceptance or minimize output emittance. However, extensive studies demonstrated that the acceptance and performance of the FFAG depend critically on combinations of initial cavity phasing, rf overvoltage, and the time scale set for acceleration.

200 MHz acceleration

If one uses a single optimal frequency, and fixed, but different, initial cavity phases, then at least 5-turns of acceleration can be sustained with a modest over-voltage. (The amount of over-voltage represents the increase in rf voltage required relative to pure on-crest acceleration.) A 25% over-voltage yields a 1.5 eV.s admittance (Fig. 7). If one adds 2nd harmonic (Fig. 8), then one needs $(4/3) \times 1.25$ of the nominal voltage, and the admittance rises to 2.28 eV.s. The input admittances are similar in shape to that in figure 6 except that their time-width is halved. Again, when considering the entire region, the transport is nonlinear, but the emittance which corresponds to the central region is reasonably well conserved.

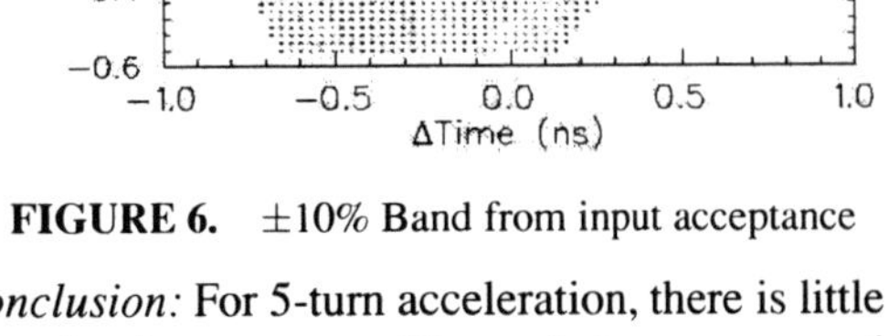

FIGURE 6. ±10% Band from input acceptance

Conclusion: For 5-turn acceleration, there is little difference in the output emittance between conventional acceleration where cavity phases are adjusted, *versus* using a combination of "best initial phases"which are optimized on an individual cavity basis and an over-voltage. When second harmonic is employed, the useful acceptance is greater when the "best phases" scenario is adopted, and is typically doubled. In all cases, the overall transport is non-linear with the useful, or preserved, phase space area comprising roughly one half of the full admittance. However, since most beams do not approach such large longitudinal emittances, nonlinear transport issues do not pose serious concerns with the operation of these machines in a rapid acceleration mode.

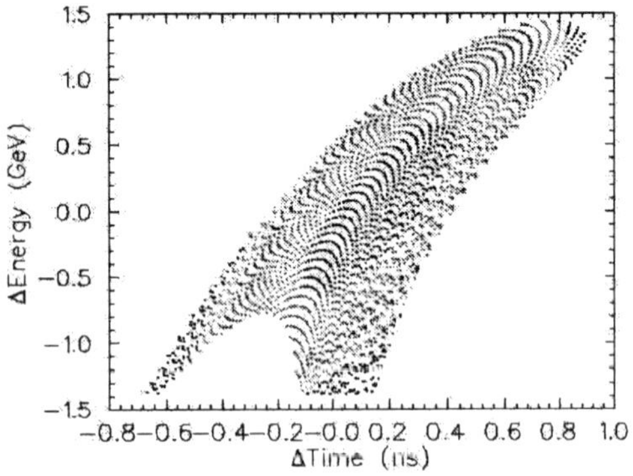

FIGURE 7. Output emittance, single harmonic.

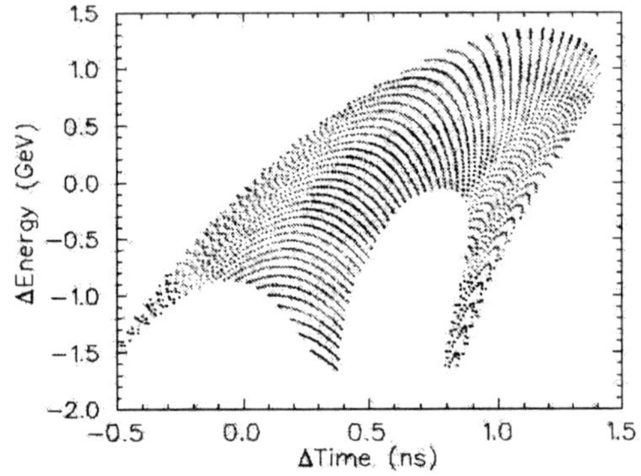

FIGURE 8. Output emittance, dual harmonic.

Acceleration with 100 MHz RF

Since the phase slips accumulate half as quickly when the RF is halved, one anticipates that 100 MHz acceleration will be less compromised by using a larger number of turns. Based on the results of the 200 MHz case studies, we have considered only the use of a "best frequency and phases" for 100 MHz rf. Certain general trends can be noted. The input admittance rises almost linearly from 0.2 to 4.1 eV.s as the number of turns is reduced from fourteen to six and the voltage-per-turn is raised from 1.25 to 2.92 GV. For example, 10-turn acceleration with a modest 25% over-voltage super-posed on the nominal voltage of 1.4 GV/turn, yields a 2.3 eV.s acceptance; see figures 9 and 10. If one adds second harmonic, the admittance rises to 4.2 eV.s. Varying the number of rf stations from 100 to 600 again has little influence; admittances vary from 2.22 to 2.33 eV.s.

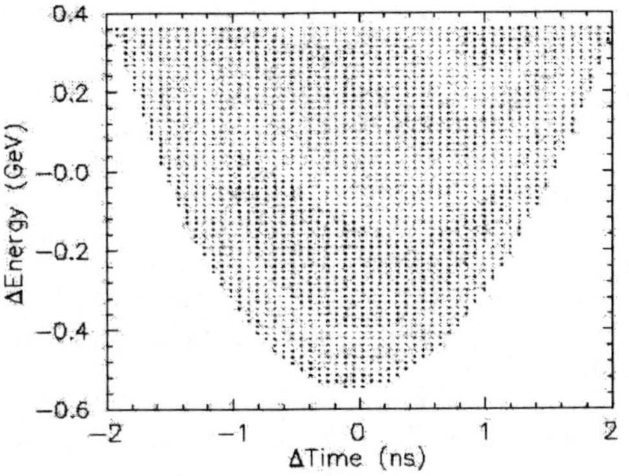

FIGURE 9. ±10% Band from input acceptance.

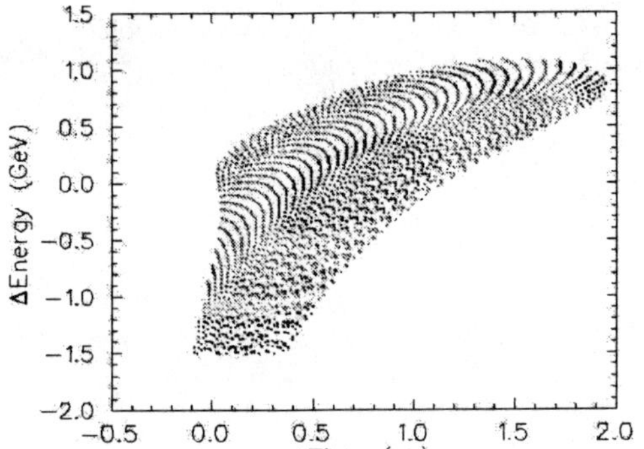

FIGURE 10. Maps to the output emittance

Conclusion: For acceleration with 100 MHz RF using a single frequency and fixed "best phases", the optimum admittance of the machine is achieved when the number of turns appears to be eight or nine combined with modest overvoltages. However, admittances comparable to the 200 MHz case can be achieved in 10,11 or even 12 turns with a corresponding and advantageous decrease in rf gradient. For 100 MHz rf, use of the second harmonic in combination with an overvoltage and best phasing produces a tremendous longitudinal acceptance in a FFAG for a rapid-acceleration application.

SUMMARY

Nonscaling FFAGs have a strong advantage in applications requiring rapid acceleration of large-emittance beams by providing a transverse admittance beyond conventional *scaling* FFAGs. The magnet layout (horizontally-focussing quadrupole and horizontally-defocussing combined-function magnet) with appropriate optimization represents an innovation in FFAG lattice design, and has been described in detail in earlier publications. This design approach produces the most efficient and compact linear machine design in terms of circumference and magnet aperture. (Circumference reduction approaches a factor of two from *scaling* FFAG designs developed for the same application.) The slowly-changing optics which characterize this *nonscaling* FFAG design have an additional advantage in the suppression of nonlinear resonances for rapid, or moderately-rapid acceleration applications. At least initially, it was felt that the circumference change, or phase-slip, posed a serious problem, but since then numerous solutions have been proposed. However, most of the previous solutions had the disadvantage of requiring impractical rf power by applying very low-frequency rf, or broadband rf to accomodate the phase-slip. This work represents the first successful study of the application of reasonably high-frequency and high-Q cavities to rapid acceleration in a fixed-field accelerator, thereby dramatically reducing the rf power required in previous solutions (that is, when a bunch train is being accelerated).

Rapid acceleration with no adjustments possible in the cavity phasing, however, show strict operational constraints in order to meet stablility requirements in the output phase space and energy distribution. For *near-crest* operation, which displays stable performance across a large longitudinal machine acceptance three factors were found to be critical and interrelated: optimal cavity frequency and starting phases, overvoltages, and the timescale for acceleration (the total number of turns in the machine). Only a few turns are supportable in the context of a large and practically stable machine admittance and, even then, it is only achieved by applying different initial phases on rf stations and allowing modest overvoltages (typically 20-40%). Even though little phase-space increase is evident in the final emittance, the transport, which tends to increase the momentum spread and decrease the bunch length, is still nonconventional,

Specifically, in the application of 200 and 100 MHz rf, the number of turns increases from 5 to 10, respectively, indicating the expected general dependence on the inverse of the frequency. In this approach, a hard limit occurs when the phase-slip exceeds 90° from crest at which point a sharp falloff of stable and useful transmission occurs. (One can force an increase in the number of turns, but at great expense in the overvoltage and at the cost of even greater convolution of the phase space.) Applying a second harmonic has a dramatic effect on the transmitted phase space area, but had only a minor impact on the number of turns.

In practical terms, when 200 MHz RF is utilized, useful admittances ($\geq$ 1 eV.s) can be achieved with acceleration in 6 or less turns using either either a phase-agile or a high-Q fixed-frequency RF system and a modest over-voltage in a FFAG. Surprisingly, there appears to be little advantage in using a phase-agile or low-Q cavity to the overall performance when operating in the on- or near-crest regime. For the case of 100 MHz RF, study of a fixed frequency system, shows that a useful output emittance of (≥ 2 eV.s) is realized for acceleration in 10 or less turns. In all cases, addition of second harmonic roughly doubles the phase space conserved within the defined cuts.

In conclusion, the *nonscaling* FFAG coupled with the rf approach developed here presents not only a solution to rapid acceleration, but also a new acceleration technique. Ultra-large emittances are successfully transported in a conventional machine with minimal complexity in components (when compared with the more design-intense magnets required for a *scaling* FFAG). The use of only linear fields form the basis for the tremendous transverse and longitudinal acceptance, one that is large in comparison with conventional acceleration, including traditional FFAG machines. With appropriate rf technique and design, this large acceptance was found to be preserved despite the problems introduced by phase-slippage. Specifically, this approach provides the necessary transverse and longitudinal acceptance match to high-energy muon beams with little or minimal cooling in place. Based on this work, it looks promising to build a chain of muon accelerators from FFAGs and replace the costly and somewhat restrictive RLAs, which so far have been the baseline accelerator for the feasibility studies of a Neutrino Factory in the U.S [11, 12].

REFERENCES

1. D. Kerst, K. Symon, L. Laslett, L. Jones, K. Terwilliger, "Fixed Field Alternating Gradient Particle Accelerators", pgs. 32-35, Proc. of CERN Symposium on High Energy Accelerators, Geneva, June 11-23, 1956.
2. C. Johnstone and S. Koscielniak, "Recent Progress on FFAGs for Rapid Acceleration," Proc. of the APS Snowmass 2001, The Future of Particle Physics, Snowmass, CO, June 28-July 21, 2001. C. Johnstone, "Recent Studies of FFAGs in the USA", Proc. of the 16th International Conference on Cyclotrons and their Applications, May, 2001, available at http://meetings.nscl.msu.edu//cyclotrons2001 C. Johnstone, W. Wan, and A. Garren, "Fixed Field Circular Acceleration Design", Proceedings of the 1999 Particle Accelerator Conference, New York, NY, Mar. 29-Apr. 2, 1999, pp. 3068.
3. F. Mills, "Linear Orbit Recirculators", pgs. 693-696, Proc. of the 4th International Conference on the Physics Potential and Development of the $\mu+\mu-$ Colliders", San Francisco, CA, Dec. 10-12, 1997. C. Johnstone,"FFAG design", talk, pgs. 696-698, Proc. of the 4th International Conference on the Physics Potential and Development of the $\mu+\mu-$ Colliders", San Francisco, CA, Dec. 10-12, 1997.
4. FFAG02, Tsukuba, Japan, Feb., 2002, http://hadron.kek.jp/FFAG/ffag02 FFAG Workshop, CERN, Geneva, Switzerland, July 3-4, 2000. FFAG00, KEK, Tsukuba, Japan, Oct. 11-13, 2000.
5. A. Garren, presented at the CERN FFAG Workshop, Geneva, Switzerland, July3-4, 2000.
6. C. Johnstone, W. Wan, and A. Garren, "Fixed Field Circular Acceleration Design", Proceedings of the 1999 Particle Accelerator Conference, New York, NY, Mar. 29-Apr. 2, 1999, pp. 3068.
7. S. Machida, et al, "Beam Optics Design of an FFAG Synchrotron", (MOP1B20), submitted to EPAC 2000, and Y. Sato, et al, "Development of a FFAG Proton Synchrotron", (MOP1B21), submitted to EPAC 2000.
8. K. Symon, MURA-KRS-6, (MURA-43) (1954); K. Symon, et al, Phys. ReV. 103 p. 1837, (1956)..
9. D. Boussard et al: "Design Considerations for the LHC 200 MHz RF System", LHC Project Report 368, January 2000.
10. E. Chiaveri & R. Losito: private communication.
11. N. Holtkamp and D. Finley, eds., A Feasibility Study of a Neutrino Source Based on a Muon Storage Ring, Fermilab-Pub-00/108-E (2000).
12. S. Ozaki, R. Palmer, M. Zisman, and J. Gallardo, eds., "Feasibility Study-II of a Muon-based Neutrino Source", BNL-52623, June, 2001, available at http://www.cap.bnl.gov/mumu/studii/FS2-report.html.

Longitudinal Dynamics in High-Frequency FFAG Recirculating Accelerators

J. Scott Berg

Brookhaven National Laboratory; Building 901A; PO Box 5000; Upton, NY 11961-5000

Abstract. A recirculating accelerator accelerates the beam by passing through accelerating cavities multiple times. An FFAG recirculating accelerator uses a single arc to connect the linacs together, as opposed to multiple arcs for the different energies. For most scenarios using high-frequency RF, it is impractical to change the phase of the RF on each pass, at least for lower energy accelerators. Ideally, therefore, the FFAG arc will be isochronous, so that the particles come back to the same phase (on-crest) on each linac pass. However, it is not possible to make the FFAG arcs isochronous (compared to the RF period) over a large energy range. This paper demonstrates that one can nonetheless make an FFAG recirculating accelerator work. Given the arc's path length as a function of energy and the number of turns to accelerate for, one can find the minimum voltage (and corresponding initial conditions) required to accelerate a reference particle to the desired energy. I also briefly examine how the longitudinal acceptance varies with the number of turns that one accelerates.

LATTICE DESCRIPTION

For the purposes of this paper, a recirculating accelerator consists of an alternating sequence of identical linacs and arcs. The arcs are identical in the sense that each has the same path length as a function of energy. The linacs are identical in the sense that they all have the same voltage and the same phase. By "the same phase," I first mean that the phase of the RF does not change from one turn to the next. Second, if there are M linacs in the recirculating accelerator, the phase of one linac differs from that of the previous linac by $2\pi k/M$ for some fixed integer k. If one has an FFAG arc with a particular characteristic variation of path length with energy, one can cause the path length at one particular energy to give the same phase in the linac following the arc as in the linac preceding the arc by a combination of adding an additional length to the arc and varying the aforementioned k. Changing k allows one to keep the required changes to the dynamics of the arc lattice small.

There are two extremes in this design: one is the racetrack design, where there are two long parallel linacs connected by arcs; the opposite extreme is a distributed RF system, where one has a sequence of short arcs with a single RF cavity between them. The racetrack design allows one to attempt to suppress dispersion in the linacs, eliminating longitudinal-transverse coupling. It is generally difficult to suppress dispersion over a large energy range, and in addition, the dispersion suppression may reduce the dynamic aperture of the system. The distributed RF system allows longitudinal-transverse coupling, but maintains a high degree of symmetry in the system, in principle giving a good dynamic aperture. The longitudinal-transverse coupling may not be so important, however, since we are on-crest, and the energy gain does not vary so strongly with time-of-flight (that variation is what causes the longitudinal-transverse coupling).

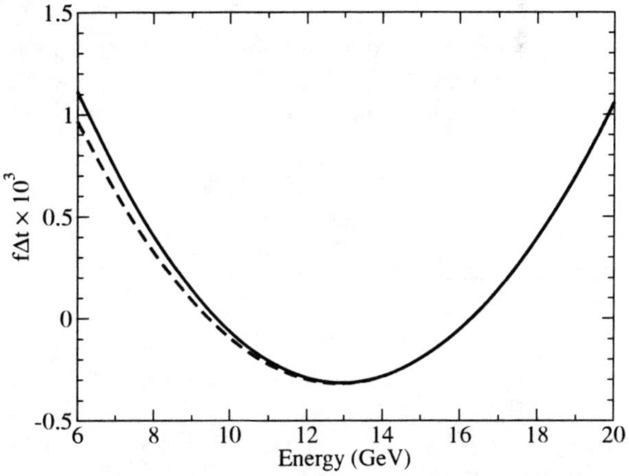

FIGURE 1. Path length as a function of energy for a single 6.5 m FFAG cell [1], calculated using COSY INFINITY [2]. Path length is given as a fraction of a 200 MHz RF period. The solid line is the exact path length variation, the dashed line is a quadratic approximation.

It turns out that the path length in an FFAG arc is generally well approximated as a quadratic function of energy (see Fig. 1 for an example). One would expect that it is desirable to minimize the total variation in the path length over the desired energy range, and so one generally adjusts the lattice design to place the minimum of the parabola in the center of the energy range of the arc. Thus, the path length as a function of energy takes the form

$$\Delta T\left(\frac{2E-E_{\max}-E_{\min}}{E_{\max}-E_{\min}}\right)^2-T_0. \qquad (1)$$

CP642, *High Intensity and High Brightness Hadron Beams: 20th ICFA Advanced Beam Dynamics Workshop on High Intensity and High Brightness Hadron Beams*, edited by W. Chou, Y. Mori, D. Neuffer, and J.-F. Ostiguy

We expect ΔT to vary very little as T_0 is adjusted over a small range. Adjusting T_0 changes the energies for which the particle will see the same phase betweeen subsequent linacs.

Time-of-flight variation with energy is ignored in the linacs. For the purposes of this study, and considering the relatively large energies that these recirculating accelerators are designed for, it is a very good approximation to distribute any path length variation with energy in the linacs into the adjacent arcs.

EQUATIONS OF MOTION

The equations giving the energy and time-of-flight at the entrance and exit of the linacs are

$$E_{n+1} = E_n + V\cos(\omega\tau_n) \tag{2}$$

$$\tau_{n+1} = \tau_n + \Delta T\left(\frac{2E_{n+1} - E_{\max} - E_{\min}}{E_{\max} - E_{\min}}\right)^2 - T_0. \tag{3}$$

Here E_n is the energy after the nth pass through a linac and τ_n is the time-of-flight relative to the crest in the nth linac. It is useful to change coordinates to $x_n = \omega\tau_n$ and

$$p_n = \frac{2E - E_{\max} - E_{\min}}{E_{\max} - E_{\min}}. \tag{4}$$

Then the equations become

$$p_{n+1} = p_n + v\cos x_n \tag{5}$$

$$x_{n+1} = x_n + \Delta\phi p_{n+1}^2 - \phi_0 \tag{6}$$

where $v = 2V/(E_{\max} - E_{\min})$, $\Delta\phi = \omega\Delta T$, and $\phi_0 = \omega T_0$.

Say we want to accelerate from $E_{\max}$ to $E_{\min}$ in N turns. Then $p_0 = -1$ and $p_N = 1$. The problem that we wish to solve is given these endpoint conditions, minimize v by varying x_0, the phase at which you enter the first linac, and ϕ_0. This solution will depend only on $\Delta\phi$ and N.

Note that this problem can be formulated for any voltage profile (e.g., something different from $\cos\phi$) as well as a different relationship of time-of-flight to energy. Changing the voltage profile will only change the results quantitatively, not qualitatively [3]. Changing the relationship of time-of-flight to energy will be discussed later.

Continuous Approximation

It is useful to make a continuous approximation to Eqs. (5-6) [3]. The continuous approximation indicates that for $N\Delta\phi$ large, $v \to \Delta\phi/12$, $\phi_0 \to \Delta\phi/4$, and $x_0 \to -\pi/2$. The reason that these results are interesting is that they give a good approximation to what will occur in the discrete case. In particular, since the continuous approximation has a solution for all N, we might expect that the discrete system does as well. In the case of the distributed RF system, $\Delta\phi$ is small and N is very large, and the discrete equations are in fact a very good approximation to the continuous ones. For a racetrack system, however, it is far from clear that the approximation is good. Since $\Delta\phi$ is relatively large, the change of x in one step can be large, making it questionable whether the contuous approximation is really very good. However, we will subsequently see that for a large number of turns, a large fraction of the steps occur at points where the change in x is small, and x_n is large and therefore the change in p is also small. The continuous approximation thus turns out to give the correct results for large numbers of turns both qualitatively and nearly quantitatively as well.

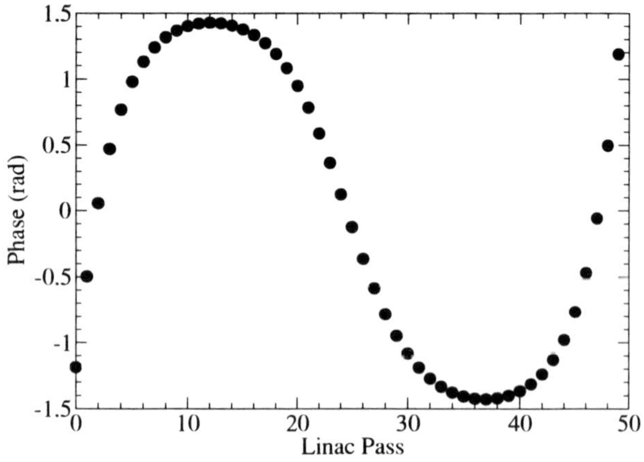

FIGURE 2. Phase as a function of linac pass when $\Delta\phi = 1$ and $N = 50$.

EXAMPLE

We now find the minimum-v solution of Eqs. (5-6) for $\Delta\phi = 1$. This is a relatively large phase swing: remember that the phase errors accumulate, and so after only 4 steps with this phase error, one would certainly be decelerating. This example is more appropriate for a racetrack configuration: the phase swing per arc would be orders of magnitude smaller for a distributed arc system (but there would be correspondingly more linac passes required).

If one makes 50 linac passes for this system and performs the aforementioned optimization, the phase as a function of turn number is shown in Fig. 2. Note that the reference particle crosses the crest three times; this is related to the parabolic shape of the path length. The particle spends most of its time at the two turning points in phase (turns 8-18 and 32-42); these are the points where the path length error is near zero. Due to the large phase at this point, the particle is not gaining very much energy, and so remains at the point where the path length is near

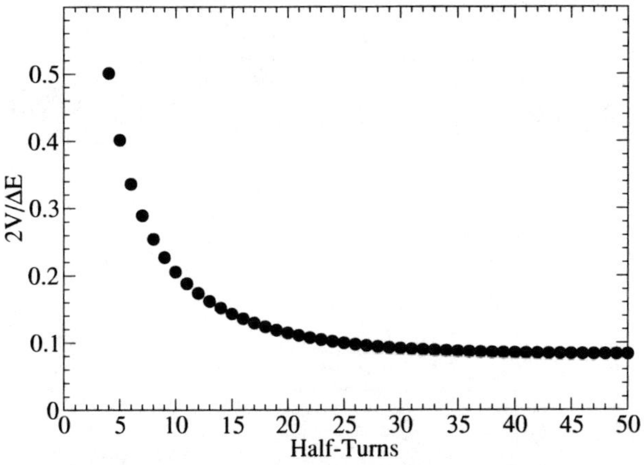

FIGURE 3. Linac voltage as a function of the number of linac passes.

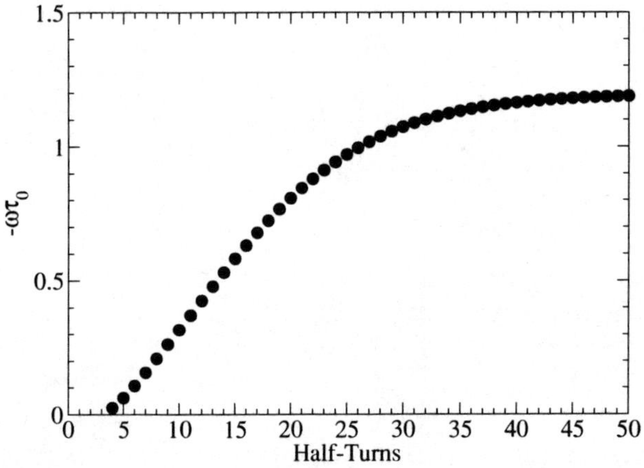

FIGURE 4. Initial phase as a function of the number of linac passes.

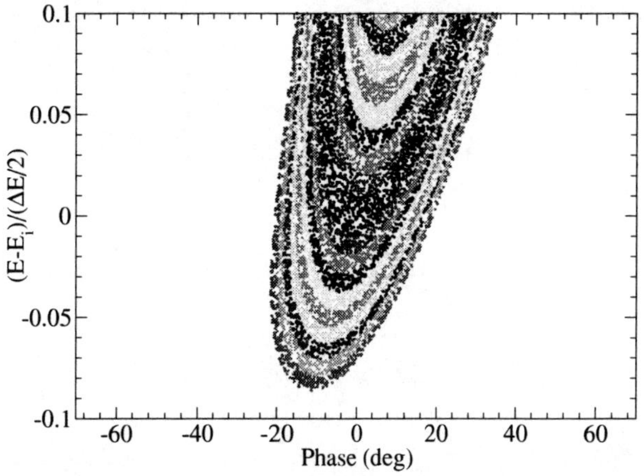

FIGURE 5. Initial phase space that is accepted in a 8 linac pass system. Each color represents a band of ± 0.01 in p_N.

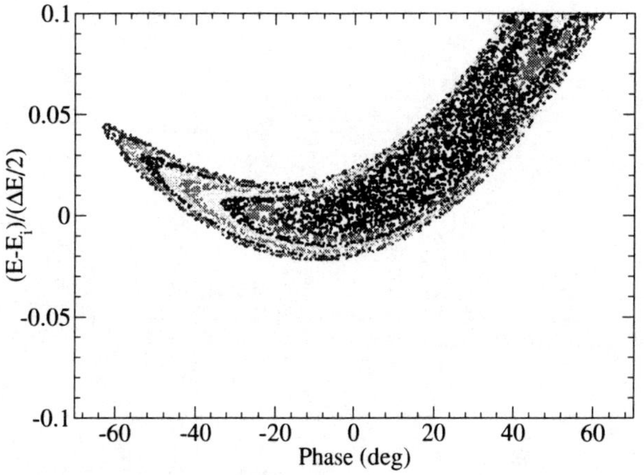

FIGURE 6. Initial phase space that is accepted in a 24 linac pass system.

zero for a long time. This is the mechanism by which the particle can spend an arbitrary number of turns in this system.

Figure 3 shows the voltage as a function of the number of linac passes, and Fig. 4 shows x_0. For large numbers of turns, these (including ϕ_0, not shown) do appear to be approching the large N limits given above. Note that while it is not clear that $x_0 \to -\pi/2$ in Fig. 4, from Fig. 2, one can see that the maximum phase swing is slightly larger than $-x_0$. The fact that $\Delta\phi$ is large causes this difference from the continuous approximation. It is really the regime near the turning points in phase that approaches the continuous approximation. It turns out that the voltage limit as $N \to \infty$ is very slightly less than what is found in the continuous approximation, and the difference is also due to the the finite $\Delta\phi$.

Finally, Figs. 5 and 6 demonstrate how the acceptance varies with the number of turns. A smaller number of turns seems to caus a large region of phase space to be accepted; what one sees at 8 linac passes is very close to what one expects from on-crest acceleration. For a large number of linac passes (24 here), one notes that most everything that is accepted ends up within a small band of the reference momentum at the final energy. In fact, while it appears that there is a smaller total acceptance for more turns, the phase space area ending up within a small energy band at the end is mmuch larger for more turns. The analysis of acceptance is still very preliminary, and must be studied more thoroughly.

REFERENCES

1. Carol Johnstone. Private communication.
2. K. Makino and M. Berz. COSY INFINITY version 8. *Nuclear Instruments and Methods*, **A427**, pp. 338–343 (1999).
3. J. Scott Berg. Proceedings of Snowmass 2001.

SESSION D

NEW IDEAS

Summary Report

Ken Takayama* and Charles Ankenbrandt†

*KEK, Tsukuba, Japan
†Fermilab, Batavia, USA

In Session D, 9 papers were presented. Their subjects are in a wide range of accelerator physics and technology. Their outlines are briefly summarized here. These papers may be classified under 5 topics.

Griffin has reviewed a historic "New Idea" on passive compensation of longitudinal space charge effects by insertion of inductors initiated by Sessler and Vaccaro and related works. In addition, he described milestone experiments that have been performed in the KEK 12GeV PS and LANL PSR with great success. He addressed the merits and demerits of this technique such as the inherent resonance leading to self-bunching instability at high intensity.

Ongoing plans for RF gymnastics in the Fermilab accelerator complex and a new idea, demonstration of which is expected in near future, have been covered by three papers. Bhat presented experimental results of beam handing with barrier buckets and pbar/proton beams in the Fermilab Recycler Ring. Most important features of longitudinal beam handling such as capture, stacking, and debunching were introduced. These experimentally obtained results will provide guidance for various succeeding applications. Ng presented results of simulations of a method for doubling the beam intensity in the Main Injector with the aid of barrier buckets. Koba discussed slip stacking that had been proposed in order to increase the primary proton-beam intensity for pbar production. A size of sufficient frequency separation to avoid undesired interference between two adjacent booster batches has been experimentally confirmed. Stacking with low intensity beams has been successfully demonstrated. She addressed the importance of beam loading compensation in a practical case at high intensity.

Use of beam echo as a means of beam diagnostics was discussed from both theoretical and experimental perspectives. Stupakov, one of the pioneers developing beam echo since early 90's, presented an instructive view of the theory. Colestock reviewed typical results experimentally obtained in the Fermilab Antiproton Accumulator Ring, CERN SPS, Fermilab Tevatron, and DESY-HERA. Both presentations have emphasized the importance of its application to measurement of small diffusion rates.

A novel concept of super-bunch acceleration, which is expected to be realized in Induction Synchrotrons, was explained by Kishiro, and the R&D status at KEK for required induction accelerating devices was introduced. He asserts that a crucial point of super-bunch acceleration is acceleration by long step-voltages induced in the induction gaps. A proof-of-principle experiment in the KEK 12GeV PS is anticipated to justify reliance on this type of device, because it will replace a conventional RF device.

An ambitious plan, where a post-linac accelerating section called "Linac Afterburner" is installed in the Fermilab 8GeV Booster tunnel, is presented by Popovic. He claims that increasing the injection energy by 200 MeV by means of additional 805 MHz side-coupled cavities could be helpful to mitigate transverse space-charge effects. The plan in near future seems to be quite practical with existing infrastructures and R&D work as well as a reasonable cost estimate.

CP642, *High Intensity and High Brightness Hadron Beams: 20th ICFA Advanced Beam Dynamics Workshop on High Intensity and High Brightness Hadron Beams,* edited by W. Chou, Y. Mori, D. Neuffer, and J.-F. Ostiguy

Using Beam Echo to Recover Transverse Emittance

G. V. Stupakov

Stanford Linear Accelerator Center, Stanford University, Stanford, CA 94309

Abstract. If a beam is injected into the ring with an offset a, it undergoes betatron oscillation. After the oscillation decoheres, the beam transverse emittance increases by $\Delta\varepsilon$. To avoid this emittance increase one typically uses a feedback (damper) that takes out the oscillation before it damps down. We show that using echo one can recover a fraction of $\Delta\varepsilon$ long after the beam oscillation decoheres.

INTRODUCTION

Beam echo has been introduced in accelerator physics in Refs. [1, 2]. The echo effect in a circular accelerator could be observed in a situation where the beam in the ring is deflected (or injected) off-orbit at time $t = 0$, causing its centroid to undergo betatron oscillation. After this oscillation has completely damped out due to a betatron frequency spread in the beam, the beam is excited again by a quadrupole kick at $t = \tau$. This kick does not produce any visible displacement of the beam at that time, but it turns out that close to time $t = \tau_{\text{echo}} = 2\tau$ the beam centroid undergoes transient betatron oscillations with an amplitude that is a fraction of the initial beam offset.

Originally the echo was predicted and calculated for the transverse betatron oscillation of the beam. It was pointed out later that the echo can also be observed in the longitudinal direction [4].

Experimentally, for the first time, the echo was measured in Fermilab on Antiproton Accumulator for a coasting beam [5]. Detailed studies of the longitudinal echo were later performed at CERN [6]. Theory of the longitudinal echo was further advanced in Refs. [6, 7].

The echo effect is based on the fact that the decoherence is not a true damping — it can be somewhat reversed and the initial perturbation of the beam can be partially restored. At the same time, it turns out that the echo is very sensitive to the diffusion in the phase space. In CERN experiments the diffusion coefficient of order of 10^{-13} s^{-1} was measured corresponding to the longitudinal emittance doubling time ~ 100 days!

In this paper we explore another aspect of the time-reversing property of the echo. If the beam is injected with an offset (or the offset arises due to a deflection of the beam), the beam emittance increases after the oscillation decoheres. This increase, however, is not fully irreversible, and we will show that part of it can be recovered. Although the fraction of the recovered emittance is relatively small, it is remarkable, that the recovering can be carried out long after the oscillation decoheres (assuming that there are no diffusion effects on the time interval of interest).

ECHO THEORY

Let us assume, for concreteness, that the beam offset a is due to beam deflection that happens at time $t = 0$. The offset is much smaller than the transverse beam size, $a \ll \sigma_0$. For description of particle's motion in the ring, we will use dimensionless variables x and p related to the transverse deviation from the equilibrium orbit X by the following relations:

$$x(s) = \frac{X(s)}{\sqrt{\varepsilon_0 \beta(s)}}, \qquad p(s) = \beta(s)\frac{dx(s)}{ds}, \tag{1}$$

where ε_0 is the beam emittance. The variables x and p are the conjugate variables of Hamiltonian motion; we will also use the action-angle variables J and ϕ: $J = (x^2+p^2)/2$, $\phi = -\text{ArcTan}(p/x)$.

We will assume that, due to nonlinearity of the lattice, the tune is a function of the amplitude of the betatron oscillations, or action, J:

$$\nu = \nu_0 + \Delta\nu J, \tag{2}$$

where ν_0 is the linear tune, and $\Delta\nu$ represents the tune spread in the beam.

The beam evolution is described in terms of the distribution function $\psi(x,p,t)$, or equivalently, in angle-phase representation, $\psi(J,\phi,t)$. The dimensionless beam emittance, measured in the units of the equilibrium emittance ε_0, is defined as

$$\varepsilon = \int_0^\infty dJ \int_0^{2\pi} J\psi(J,\phi,t)d\phi\,. \tag{3}$$

The equilibrium distribution function ψ_0 of the beam depends only on the action, and for a Gaussian distribution, $\psi_0(J) = (2\pi)^{-1}\exp(-J)$. The equilibrium beam emittance calculated for this distribution function with the help of Eq. (3) is equal to unity. To simplify the notation we will measure time in units of inverse betatron frequency ω_β^{-1}.

CP642, *High Intensity and High Brightness Hadron Beams: 20th ICFA Advanced Beam Dynamics Workshop on High Intensity and High Brightness Hadron Beams*, edited by W. Chou, Y. Mori, D. Neuffer, and J.-F. Ostiguy

The echo calculation is carried out in four steps.

1. Beam deflection. It is performed as a dipole kick that translates the beam distribution function along the p axes, $p \to p+\alpha$, where α is the dimensionless offset, $\alpha = a/\sqrt{\varepsilon_0 \beta}$. The kick changes the distribution function from ψ_0 to ψ_1,

$$\psi_1(p,x) = \psi_0(p-\alpha,x). \tag{4}$$

In terms of the action-angle variables, the distribution function after the kick, which we denote ψ_2, is given by the following relation

$$\psi_2(J,\phi) = \psi_1(p(J,\phi),x(J,\phi)). \tag{5}$$

Using Eq. (3) it is easy to calculate that for small α the emittance increase of the beam after the dipole kick is

$$\Delta\varepsilon = \frac{\alpha^2}{2}. \tag{6}$$

2. Free betatron oscillation during time τ. The oscillation conserves the action and advances the angle, $\phi \to \phi + \tau(\nu_0 + \Delta\nu J)$, and results in the new distribution function ψ_3 at time τ:

$$\psi_3(J,\phi,\tau) = \psi_2(J,\phi - \tau(\nu_0 + \Delta\nu J)). \tag{7}$$

Calculation of the average offset of the beam with this distribution function reveals decoherence of the betatron oscillation on the time scale equal to the inverse betatron frequency spread in the beam, $\Delta\nu^{-1}$.

3. Quadrupole kick at time τ. We assume that $\tau\Delta\nu \gg 1$, which means that the quadrupole kick is applied after the initial betatron oscillation decoheres. The kick generates a distribution function ψ_5 which we calculate in two steps: first transforming from $J-\phi$ to $x-p$ variables,

$$\psi_4(x,p,\tau) = \psi_3(J(p,x),\phi(p,x),\tau), \tag{8}$$

and then applying the transformation $p \to p + qx$:

$$\psi_5(p,x,\tau) = \psi_4(p-qx,x,\tau), \tag{9}$$

where q is the dimensionless strength of the kick equal to the ratio of the beta-function at the location of the quad to the focal length of the quadrupole, $q = \beta/F$.

4. The final step is to allow a free betatron oscillation for the time interval $\tau + \Delta t$. It gives the distribution function ψ_{echo} which is calculated via an intermediate function ψ_6

$$\psi_6(J,\phi,\tau) = \psi_5(p(J,\phi),x(J,\phi),\tau), \tag{10}$$

$$\psi_{\text{echo}}(J,\phi,\tau,\Delta t) = \psi_6(J,\phi - (\tau+\Delta t)(\nu_0 + \Delta\nu J),\tau). \tag{11}$$

In the analytical approach we used the following approximations that greatly simplify the analysis:

- We assumed that $\alpha \ll p,x$ and expanded ψ in Taylor series keeping terms of order of α and α^2 only. This assumption means that the initial beam offset is much smaller than the transverse beam size.
- We assumed that the quadrupole kick is weak, $q \ll 1$, and expanded ψ keeping only linear terms in q. Moreover, it turns out that the echo signal is proportional to the product $q\tau$, so we kept only those terms in which q is multiplied by τ.
- We used Fourier decomposition in angle ϕ and kept only terms $\propto e^{in\phi}$ with $n = 0, \pm 1$, because only those terms contribute to the beam offset ($n = \pm 1$) and the emittance ($n = 0$).

With these simplifying assumptions, calculations can be performed with the help of symbolic capabilities of Mathematica [10]; the notebook with the code can be obtained from the author's web site [11]. As a result, the part of the distribution function $\Delta\psi_{\text{echo}}$ that is responsible for the echo signal takes a very simple form:

$$\Delta\psi_{\text{echo}}(J,\phi,\tau,\Delta t) = \tag{12}$$
$$-\frac{a\sqrt{J}}{\sqrt{2\pi}} e^{-J} \sin(\nu_0\Delta t + \Delta\nu_0 J\Delta t - \phi) \mathrm{J}_1(Jq\tau\Delta\nu),$$

where J_1 is the Bessel function.

RECOVERING EMITTANCE

The idea of recovering beam emittance is based on the application, at some time Δt (measured from the echo time $\tau_{\text{echo}} = 2\tau$) another dipole kick of amplitude ξ and phase ϕ, so that $p \to p + \xi\sin\phi$ and $x \to x + \xi\cos\phi$. The parameters of the kick ξ and ϕ as well as the application time Δt should be chosen so as to minimize the beam emittance after the kick.

The contribution to the beam emittance due to this last kick comes from two sources. First, perturbation of the equilibrium part of the distribution function gives the emittance *increase* $\xi^2/2$ analogously to the original term Eq. (6). However, this increase is offset by the perturbation of the echo term Eq. (12), which, for a small kick can be calculated as

$$\delta\psi_{\text{echo}} = \xi\cos\phi \frac{\partial\Delta\psi_{\text{echo}}}{\partial x} + \xi\sin\phi \frac{\partial\Delta\psi_{\text{echo}}}{\partial p}. \tag{13}$$

As a result, the change of the beam emittance $\Delta\varepsilon_1$ is

$$\Delta\varepsilon_1 = \frac{\xi^2}{2} + \Delta\varepsilon_{\text{echo}}, \tag{14}$$

where $\Delta\varepsilon_{\text{echo}}$ is the change of the emittance generated by the distribution function Eq. (13),

$$\Delta\varepsilon_{\text{echo}} = \int_0^\infty dJ \int_0^{2\pi} J\delta\psi_{\text{echo}} d\phi\,. \tag{15}$$

Calculations show that the minimum value of $\Delta\varepsilon_1$ is negative, which means that emittance decreases as a result of the kick. It is achieved for $\Delta t = 0$, $\phi = \pi/2$, when

$$\xi = 0.51\alpha, \qquad q\tau = \frac{0.43}{\Delta\nu}, \tag{16}$$

and is equal to

$$\Delta\varepsilon_{1\,\text{min}} = -0.26\frac{\alpha^2}{2}. \tag{17}$$

The conditions $\Delta t = 0$, $\phi = \pi/2$ mean that the emittance restoring beam offset should be applied at the moment of the maximum echo signal τ_{echo}. Of course, one can achieve the same result by applying a dipole kick of proper magnitude a quarter of the betatron period earlier. The strength of the kick, as Eq. (16) shows, should be such that it would produce approximately 50% of the initial beam offset. In addition, the optimal value of the quadrupole strength given by Eq. (16) is related to the tune spread in the beam and the delay τ between the first (dipole) kick and the second (quadrupole) one.

The emittance decrease Eq. (17) constitutes approximately a quarter of the initial increase Eq. (6) due to original beam offset. Hence the beam is still left with three quarters of the initial $\Delta\varepsilon$.

CONCLUSION

In this paper we demonstrated another aspect of the time-reversing with the beam echo. We showed that using the echo effect which regenerates the beam dipole signal long after the original oscillation decoheres and applying a proper dipole kick at the time of the echo one can partially compensate for the initial beam emittance increase. Unfortunately, the compensation is apparently too small to be of practical interest. We hope, however, that this demonstration reveals new properties of the echo phenomenon which can find applications in the future.

ACKNOWLEDGEMENTS

The author is thankful to A. Chao for critical remarks.

This work was supported by Department of Energy contract DE–AC03–76SF00515.

REFERENCES

1. G. Stupakov, SSCL Report 579 (1992).
2. G. Stupakov, S. Kauffmann, SSCL Report 587 (1992).
3. G. Stupakov, A.W. Chao. PAC 97.
4. N. Mahale et al., SSCL-N-817, 1993.
5. L.K. Spenzouris, J.-F. Ostigy, P.L. Colestock, PRL 76, 620 (1996).
6. O. Bruning, CERN SL/95-83 (AP), 1995.
7. E. Shaposhnikova, CERN SL/Note 95-125 (RF), 1995.
8. O. Bruning et al, CERN SL-MD Note 217, 1996.
9. O. Bruning et al, EPAC 96, 1996.
10. S. Wolfram, *The Mathematica Book* (Wolfram Media/Cambridge University Press, 1999), 4th ed.
11. http://www.slac.stanford.edu/~stupakov/codes.shtml

Slip Stacking

Kiyomi Koba and Jim Steimel

Fermi National Accelerator Laboratory, Batavia, IL 60510, USA

Abstract. We have started beam studies for 'slip stacking'[1] in the Main Injector in order to increase proton intensity on a target for anti-proton production. It has been verified that the system for slip stacking is working with low intensity beam. For a high intensity operation, we are developing a feedback[2][3] and feedforward system.

INTRODUCTION

Fermilab Main Injector (MI) accelerates protons and extracts them to a target to produce anti-protons. In the operation cycle for the anti-proton production, 84 bunches are injected from Booster to MI. Bunches are accelerated from 8GeV to 120GeV and extracted to hit the production target. Figure 1 shows a typical operation cycle. The total beam intensity is $4.5*10^{12}$ particles per pulse (ppp) with a cycle of 1.5 sec. Main rf parameters are listed in Table 1.

Higher proton intensities are needed to get more anti-protons and higher stacking rates in the accumulator. In order to achieve this, we are planning to use a scheme called "slip stacking". With slip stacking, the intensity of the bunches can be doubled by injecting one bunch train at lightly lower energy, another train at lightly higher energy and bringing them together.

Since two bunch trains have different energies, MI must have an enough momentum aperture to accept both. The momentum aperture of MI is +/-0.7% at injection, that is, the rf frequency for each bunch train can be shifted by +/-3000Hz from the original value.

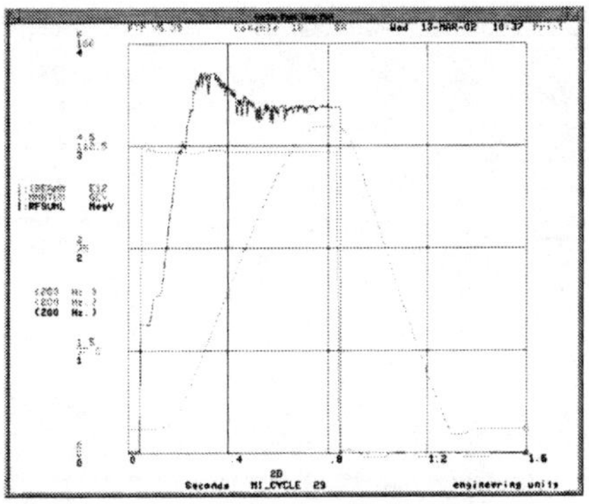

FIGURE 1. Typical operation cycle for anti-proton production. The green trace and red trace show a total intensity and momentum respectively.

TABLE 1. MI RF Parameters.

Rf frequency@njection	52.811400MHz
Number of rf cavities	18
Maximum total rf voltage @ injection	1.6MV
Proton energy @injection	8GeV
Harmonic number	588
Transition gamma	21.6
Mean radius	528.3m

STACKING PROCEDURES AND MEASUREMENT RESULTS

For slip stacking, we use three different rf systems and follow four steps. Step 1: the first bunch train is injected from Booster on the central orbit and captured by the first rf system. To make a room for the second bunch train, the first bunch train is then decelerated until it circulates inside of the central orbit. Step 2: the second bunch train is injected on the central orbit and captured by the second rf system. Step3: As bunch trains have two slightly different energies, they can move relative to each other as they accelerated. Step 4: when two bunch trains coincide at the same longitudinal location, they are captured by the third rf system.

Optimum For Frequency Separation

Since there are two rf voltages during stacking, they act on both bunch trains. The bunch shape has been measured to demonstrate that the frequency separation between the first and the second rf systems is adequate.

In this measurement, one bunch train was injected, then two rf voltages were raised and the frequency separation was increased from 400 to 1200Hz. Figure 2 shows the bunch shape at injection and at 150msec after injection. The signals, plotted here with 5nsec/div, were obtained with a wall current monitor. It is clear that the frequency separation of 1200Hz is enough to keep the bunch shape unchanged.

CP642, *High Intensity and High Brightness Hadron Beams: 20th ICFA Advanced Beam Dynamics Workshop on High Intensity and High Brightness Hadron Beams,* edited by W. Chou, Y. Mori, D. Neuffer, and J.-F. Ostiguy

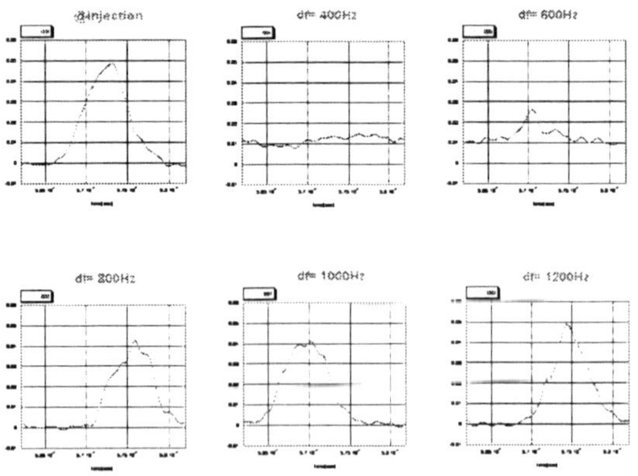

FIGURE 2. Comparison of the bunch shape, which is from a wall current monitor, at injection and after 150msec. The bunch on upper left was at injection. Other bunches were at after 2msec. The frequency separation was 400Hz(upper middle), 600Hz(upper right), 800Hz(Lower left), 1000Hz(lower middle), 1200Hz(lower right).

Simulation

A simulation study of the slip stacking has been curried out with a simulation code, ESME[4]. Two bunches each with emittance of 0.1 eV-sec were injected. An area corresponding to one bucket area of 0.1eV-sec was kept between lower and higher energy bunches. They then followed the four stacking steps and were captured by the third rf system which provided a bucket area of 0.3 eV-sec. At the end of these processes, we found a single bunch of emittance 0.3 eV-sec without beam loss, indicating an emittance growth of 50 %.

Measurement

The frequency and voltage as a function of time are shown in Figure 3 for the first and the second rf systems. Of 18 cavities at 53MHz, one was used for the first system, another for the second system. The frequency separation was kept at 1200Hz. The first bunch train was injected on the central orbit with nominal frequency at 0.13msec and captured by the first rf system of 62kV. The intensity was low, $0.4*10^{12}$ppp, in order to reduce the beam loading effects. At this time, the frequency of the second rf system was 1200Hz higher than the first rf system. The first bunch was then decelerated to the frequency 1200Hz lower than the original value. After one Booster cycle of 66.7msec, the second train was injected on the central orbit and captured by the second rf system. After slipping, both bunch trains were captured by the third system, all 18 cavities. Figure 4, a mountain range picture of the signals from the wall current monitor, reveals the progress of slip stacking from the beginning to the end.

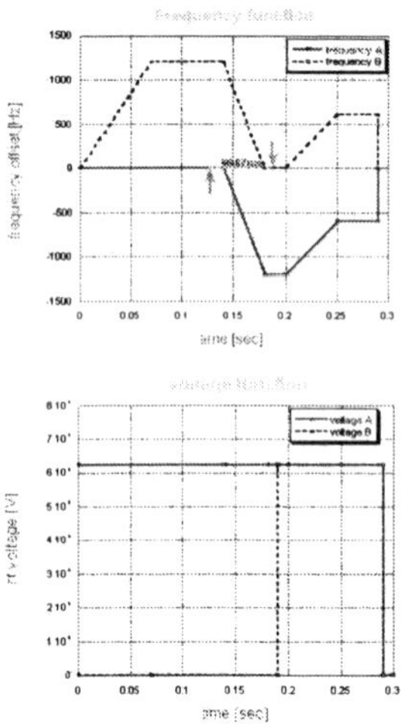

FIGURE 3. Frequency (upper) and voltage (lower) as a function of time. Red lines are for the first rf and blue line are for the second rf.

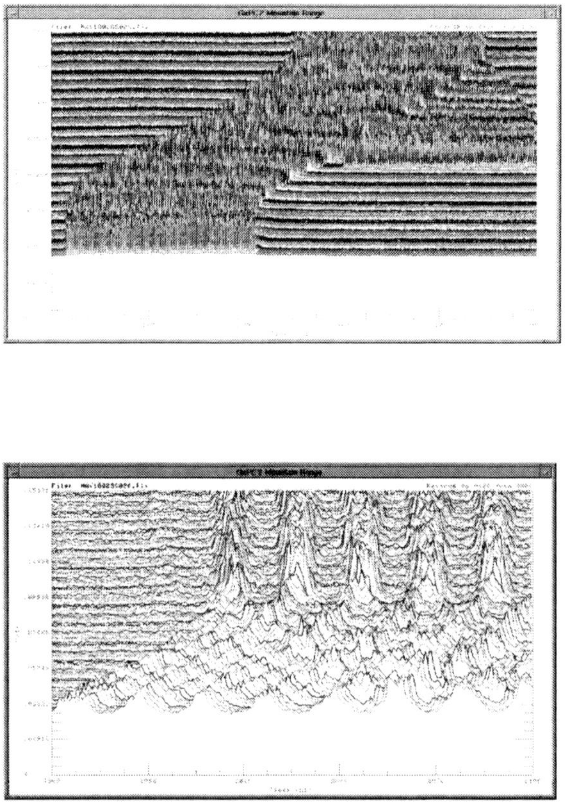

FIGURE 4. Measurement results of the whole slip stacking cycle. The lower figure shows the expanded signal at just after recapture by the third rf.

BEAM LOADING COMPENSATION

For high intensity operation, beam loading compensation is the key issue. A simulation was carried out to compare the beam with and without compensation of beam loading effects. Two trains of 84 bunches with the total intensity of $5*10^{12}$ went through the slip stacking process. At the end of stacking, with no compensation, the bunch shape became distorted and many particles were left outside of the bucket. When there was compensation of 40 dB of loop gain, there were no particle losses.

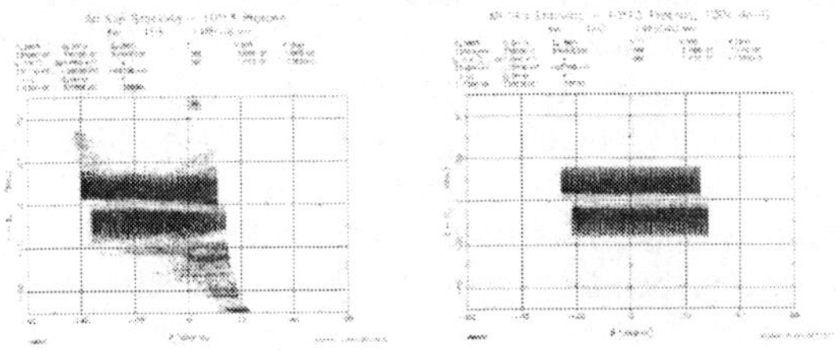

FIGURE 5. Simulation results comparing the beam after the slip stacking with (right) and without (left) a compensation of beam loading effects.

Development Of Beam Loading Compensation

We are developing feedback and feedforward systems for beam loading compensation. Figure 6, in which the center frequency is the fundamental, shows frequency spectra from a gap voltage monitor. The gain for feedback system was –20dB. One can see from this that the feedforward system reduces the revolution sidebands by 20dB.

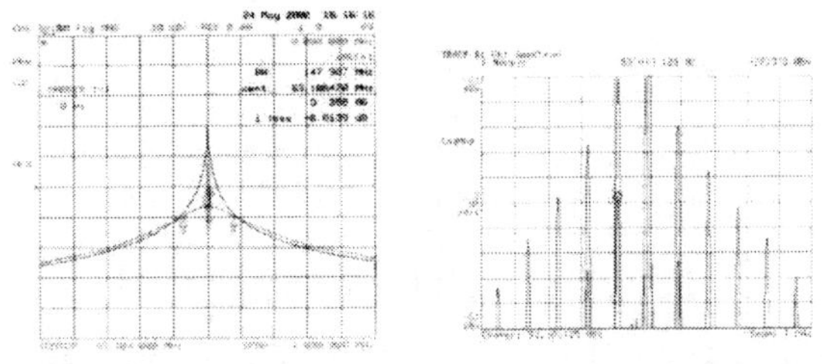

FIGURE 6. Measurement of the beam spectra from a gap voltage monitor. On the left graph, the green and red traces show without and with feedback system, respectively. On the right graph, the blue and green traces show without and with feedforward system, respectively.

SUMMARY

In order to achieve an increase in proton intensity, Fermilab Main Injector will use a stacking process called "slip stacking". The intensity will be doubled by injecting one train of bunches at a slightly lower energy, another at a slightly higher energy and then brining them together for the final capture. Beam studies have stared for this process and we have already verified that, at least for low beam intensity, the stacking procedure works as expected. For higher intensity operation, development work of the feedback and feedforward system is under way.

REFERENCES

1. S. Shukla et al, "Slip Stacking in the Fermilab Main Injector.", Snowmass'96, June 1996.
2. J.Dey and J.Steimel, "Improving the linearity of ferrite loaded cavities using feedback", PAC2001
3. J.Dey, J. Steimel and J. Reid, "Narrowband Beam Loading Compensation in the Fermilab Main Injector Accelerating Cavities", PAC2001
4. J. MacLachlan, "Users Guide to ESME", 2000.

Doubling Main Injector Beam Intensity using RF Barriers

K.Y. Ng

Fermi National Accelerator Laboratory, P.O. Box 500, Batavia, IL 60540

Abstract. Using rf barriers, 12 booster batches can be injected into the Fermilab Main Injector continuously, thus doubling the usual beam intensity. After that, adiabatic capture of the beam into 53 MHz buckets can be accomplished in about 10 ms. The beam loading voltages in the rf cavities are small and they can be eliminated by a combination of counterphasing and mechanical shorts.

THE INJECTION METHOD

Instead of slip-staking [1], a method to double beam intensity in the Main Injector by rf barriers without adiabatic bunch compression was proposed by Griffin [2]. A booster batch of protons of length T_b and full fractional momentum spread Δ, denoted by 1 in Fig. 1(a) is injected into the Main Injector at a *negative* momentum offset, so that the highest fractional momentum offset is δ_{i1} and the lowest fractional momentum offset is $\delta_{i2} = \delta_{i1} - \Delta$. The longitudinal position of the injection is so chosen that the right side of the batch just touches the left side of a square barrier B of width T_1 and magnitude V, which is moving to the left at the speed of $\dot{T}_2 = T_b/(2T_c)$ where T_c is the booster-cycle time. The momentum offset of the batch injection is so chosen that protons of highest energy on the left side of the batch will drift to the right at the speed of $\frac{1}{2}T_b$ per booster cycle or $-\dot{T}_2$. Thus, after one booster cycle, another batch marked 2 in Fig. 1(b) can be injected again with its right edge touching the left side of the barrier. In other words, at every booster cycle, a new booster batch can be injected with the injection point moved half a booster-batch length to the right. The linear beam density can therefore be doubled.

The only parameter here is the strength of the barrier,

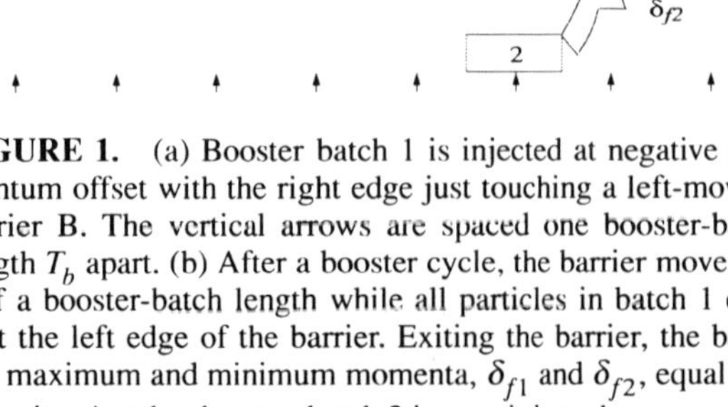

FIGURE 1. (a) Booster batch 1 is injected at negative momentum offset with the right edge just touching a left-moving barrier B. The vertical arrows are spaced one booster-batch length T_b apart. (b) After a booster cycle, the barrier moves by half a booster-batch length while all particles in batch 1 drift past the left edge of the barrier. Exiting the barrier, the batch has maximum and minimum momenta, δ_{f1} and δ_{f2}, equal and opposite. Another booster batch 2 is now injected.

which is so chosen that after the booster batch emerges from the barrier, the highest and lowest energies of the batch become symmetric about the nominal energy of the accelerator ring. This is necessary because we need to place another stationary barrier on the right side of the moving barrier in order to limit the longitudinal motion of the protons after passing through the moving barrier so as to guarantee empty spaces along the Main Injector for future batch transfer from the Booster. We use a rectangular barrier here just for simplicity, although the results depend only on the integrated strength of the barrier.

To derive the strength of the barrier, let us go to a more general scenario of having the barrier moving to the left at the rate of $\dot{T}_2 = xT_b/T_c$ ($x = \frac{1}{2}$ in above). Suppose the first proton with momentum spread δ_{i1} exits the barrier in the N_1 turns. Its phase drift in time towards the right is

$$\int_0^{N_1} \eta T_0(\delta_{i1} + n\Delta\delta)dn\,, \tag{1}$$

where η is the slip factor, T_0 is the revolution period, $\Delta\delta = eV/(\beta^2 E)$, e is the proton charge, E is the on-momentum proton energy with βc its longitudinal velocity and c the velocity of light. At this moment, the barrier has moved towards the left by the phase $N_1\dot{T}_2T_0$. Because the width of the barrier is T_1, we must have

$$T_1 - N_1\dot{T}_2T_0 = \int_0^{N_1} \eta T_0(\delta_{i1} + n\Delta\delta)dn\,, \tag{2}$$

from which N_1 can be solved. The exit momentum spread of this particle is therefore

$$\delta_{f1} = \frac{\dot{T}_2}{|\eta|} - \sqrt{\left(\delta_{i1} - \frac{\dot{T}_2}{|\eta|}\right)^2 - \frac{2T_1\Delta\delta}{|\eta|T_0}}\,. \tag{3}$$

The final fractional momentum offset δ_{f2} of the protons with the lowest initial momentum spread δ_{i2} can be obtained similarly. Equating the two, we arrive at the strength of the barrier

$$VT_1 - \frac{T_0T_b^2\beta^2E}{2|\eta|T_c^2}\left[1-\left(\frac{\Delta|\eta|}{2\dot{T}_2}\right)^2\right]\left[\left(1+\frac{x\Delta|\eta|}{2\dot{T}_2}\right)^2-x^2\right], \tag{4}$$

where use has been made of the fact that the protons with initial momentum spread δ_{i1} must drift to the right at a rate of $(1-x)T_b/T_c$ so that a next batch can be injected or $\delta_{i1} = -(1-x)\dot{T}_2/(x|\eta|)$.

CP642, *High Intensity and High Brightness Hadron Beams: 20th ICFA Advanced Beam Dynamics Workshop on High Intensity and High Brightness Hadron Beams*, edited by W. Chou, Y. Mori, D. Neuffer, and J.-F. Ostiguy

Obviously, a larger initial fractional momentum spread Δ of the beam will lead to a larger final fractional momentum spread. However, when the initial fractional momentum spread is large enough, protons with the highest momentum will acquire so much energy from the moving barrier that their drifting speeds to the left exceed the speed of the moving barrier. These particles will not be able to emerge from the moving barrier and this injection method fails. This happens when $\dot{T}_2 < |\eta|\delta_{f1}$, resulting in the critical strength of the moving barrier

$$(VT_1)_c = \frac{T_0 T_b^2 \beta^2 E}{2e|\eta|T_c^2}\ , \qquad (5)$$

which is independent of x, and the maximum allowable initial full fractional momentum spread of the beam is

$$\Delta_c = \left(\sqrt{x^2+\frac{1}{4}}-\frac{1}{2}\right)\frac{2T_b}{|\eta|T_c}\ . \qquad (6)$$

For a given full fractional momentum spread Δ of the beam, the speed of the moving barrier must be faster than xT_b/T_c where

$$x = \sqrt{\frac{|\eta|T_c\Delta}{2T_b}\left(\frac{|\eta|T_c\Delta}{2T_b}+1\right)}\ . \qquad (7)$$

EMITTANCE INCREASE

The position of injection along the accelerator ring slips to the left by xT_b for each injection or for each booster cycle. implying that a whole batch of length T_b of beam can be injected in the length xT_b. Therefore the ratio of final to initial emittances is

$$r_e = \frac{\delta_{f1}\times 1}{\frac{\Delta}{2}\times\frac{1}{x}} = 1+\frac{|\eta|T_c\Delta}{2T_b}\ , \qquad (8)$$

which is independent of x. Here, we assume circumference of the accelerator ring is infinitely long so that injection can be continued forever. On the other hand, the final linear density does increase by the factor $r_d = 1/x$. For this reason, we should choose the lowest barrier moving speed allowable by the initial momentum spread of the batch. In other words, given Δ, we should stick to the critical x given by Eq. (7) and the critical size $(VT_1)_c$ for the barrier given by Eq. (5).

APPLICATION TO MAIN INJECTOR

Simulations were performed for doubling the Main Injector intensity at the injection energy of $E = 8.938$ GeV at the critical condition for barrier movement of $\frac{1}{2}T_b$ per booster cycle ($x = \frac{1}{2}$) with the initial half critical half energy spread of $\Delta E_{\frac{1}{2}i} = 4.900$ MeV. Here, $T_b = 1.59$ μs, $T_0 = 7T_b$, $T_c = 0.667$ ms, $\eta = -0.00892$ and the critical barrier strength is $(VT_1)_c = 3.1421$ kV-μs. Particles with the highest energy are injected with the offset of $\Delta E_{i1} = -\dot{T}_2\beta^2 E/|\eta| = -11.8285$ MeV while particles with the lowest energy are injected with offset $\Delta E_{i2} = \Delta E_{i1} - 2\Delta E_{\frac{1}{2}i} = -21.6276$ MeV. After exiting the moving barrier, their final energy offsets are $\Delta E_{f1} = -\Delta E_{i1} = 11.8285$ MeV and $\Delta E_{f2} = -\Delta E_{f1} = -11.8285$ MeV. Thus, the energy spread has been increased by the factor $\Delta E_{f1}/\Delta E_{\frac{1}{2}i} = 11.8285/4.8995 = 2.419$ and this is also the ratio of the final longitudinal emittance to the initial longitudinal emittance. Figure 2 shows the situation one booster cycle after the injection of the 12th batch. The consecutive bunches are represented by different colors for easy identification. The vertical dashed lines separate the ring into 7 booster lengths and the first batch was injected at the location between 5.5 and $6.5T_b$. The reflecting barrier is represented by the solid square at location $7T_b$, while the injection barrier is represented by the hollow square with the arrow pointing to its direction of motion. Theoretical reasoning and simulations show that about 10 ms will be required to capture the beam into the 53-MHz buckets.

Barriers are usually used to confine a beam within a designated region along the accelerator ring. The integrated sizes of the barriers need not be accurate as long as they are large enough to confine the particles of the highest and lowest energies. However, this is not true here. Although the injection process does not depend on the shape of the moving barrier, its integrated size has to be accurate in order to have the final positive and negative energy spreads be equal and opposite. Otherwise, after reflection by the stationary barrier, the positive energy spread will be modified, resulting possibly in an increase in the longitudinal emittance. Here, we would like to study the effect of such an error. The left plot of Fig. 3 shows the situation when VT_1 is 10% larger than its critical value, 2 booster cycle after the injection of the 12th bunch. Actually the final momentum spread becomes smaller. However, many particles cannot emerge from the strong barrier resulting in a loss of 10.6%. The situation of a reduction of VT_1 by 10% from its critical value is shown in the right plot. Now all particles will emerge from the moving barrier earlier and their energies

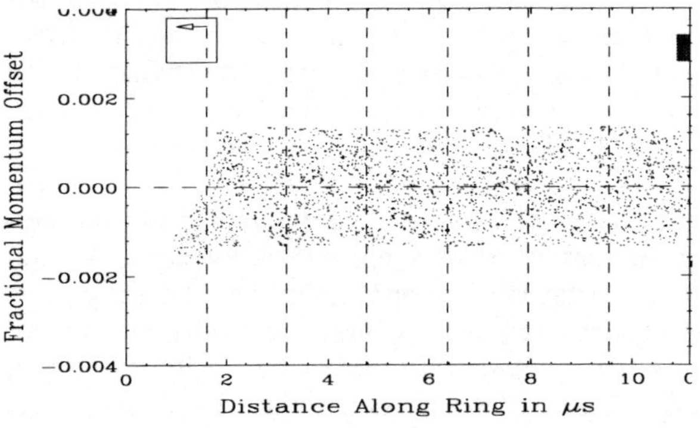

FIGURE 2. (color) Longitudinal phase space one booster cycle after the injection of the 12th batch between

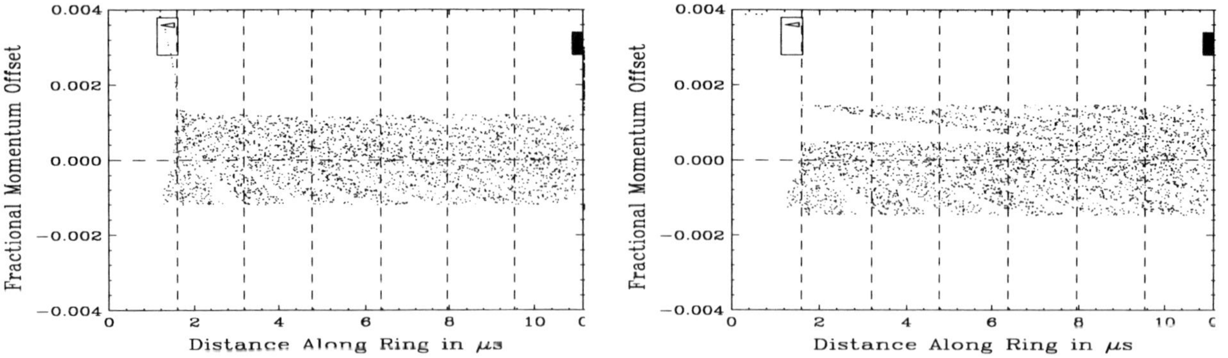

FIGURE 3. (color) Plots showing 2 booster cycle after injection of 12th bunch, with the strength of the moving barrier (left) increased by 10% and (right) decreased by 10%.

are therefore reduced, which results in an increase in momentum spread by 9.8% after reflection by the stationary barrier. A small amount particles have energies high enough that their left-drift velocities after reflection are larger than that of the moving barrier. Eventually, they may catch up with the moving barrier and get lost.

BEAM LOADING

Because of the high beam current, the problem of beam loading must be examined. At the passage across the $N_c = 18$ rf cavities, a very short bunch containing 6.0×10^{10} protons will induce a total instantaneous beam loading voltage $V_{b0} = qN_c\omega_r R_L/Q_L = 5.5$ kV, where $R_L = 500\,\text{k}\Omega$, $Q_L = 5000$, and $\omega_r/(2\pi) = 52.8$ MHz are, respectively, the loaded shunt impedance, loaded quality factor, and resonant frequency of each cavity. For a batch of 84 such bunches injected into the Main Injector, the beam loading voltage after the passage of the last bunch increases to 444 kV. Thus the difference in beam loading voltages experienced by the last and first bunches is 439 kV. Taking into account of the finite lengths of the bunches, this difference becomes 388 kV when steady state is reached. During normal operation, the total rf voltage at injection is 1.2 MV. If the designed synchronous phase $\phi_s = 0$ is synchronized to the middle bunch of the batch, the rf phase errors introduced become $\Delta\phi_s = \pm 9.18^\circ$ for the first and last bunches. Eventually, the longitudinal emittances of the bunches will be increased up to 18%, which is considered still tolerable.

During slip-stacking, although the bunch density is doubled, the beam loading voltage may not increase because the bunch spread out to almost fill the buckets at the very low rf voltage of ~ 64 kV, almost 7 times less than the difference in beam loading voltages seen by the first and last bunches. Such low rf voltage is required in order to allow two series of buckets to stay within the momentum aperture of the ring. The rf phase errors now become so large that the most beam particles will be driven out of the buckets. Feedforward and feedback have been planned to compensate for the beam loading.

In the present continuous multiple injection scheme using barriers, the booster bunches before injection will be made very long in the Booster so that the momentum spread can be reduced. These bunches start to debunch immediately after injected into the Main Injector. For a completely uniformly distributed beam, there is no 53 MHz rf component and therefore no beam loading at all. Here, although the distribution is not uniform (see, for example, Fig. 2), the 53-MHz rf component of the beam current at any moment of the injection is found to be less than 0.25% of twice the dc beam current, implying that the beam loading voltage will be reduced at least 400 times.

Actually no rf voltage is required in this injection scheme. The Main Injector is presently equipped with fast mechanical shorts, which can be inserted in 100 ms and removed in 50 ms [4]. To cope with beam loading and arrive at zero rf voltage, first, the rf drive of 16 cavities is turned off and shorts are inserted. Second, counterphasing is used for the two remaining cavities to arrive at zero rf voltage with the tiny beam loading voltage taken into consideration. Further fine adjustment can be implemented using a fast low-level feedback. Notice that counterphasing is not possible for slip-stacking because the two rf controls have been employed already to produce the two series of buckets at slightly different frequencies.

REFERENCES

1. C. Ankenbrandt, *"SLIP-STACKING": A New Method of Momentum-Stacking*, Fermilab Report FN-352, 1981; F.E. Mills, *Stability of Phase Oscillations under two Applied Frequencies*, BNL Report AADD 176, 1971.
2. J. Griffi n, *Momentum Stacking in the Main Injector using Longitudinal Barriers*, Fermilab Report, 1996.
3. K.Y. Ng, Continuous Multiple Injections at the Main Injector, Fermilab Report FN-715, 2002.
4. D. Wildman, *Transient Beam Loading Reduction during Multi-Batch Coalescing in the Fermilab Main Ring*, Proceedings of the 1991 IEEE Particle Accelerator Conference, San Francisco, May 6-9, 1991, p. 410.

Barrier Buckets in the Fermilab Recycler Ring

C. M. Bhat

MI/RR Department, Beams Division,
Fermilab, P.O. Box 500, Batavia, IL,60510, USA,*

Abstract. The Fermilab Recycler Ring is an 8GeV permanent magnet antiproton storage ring. It is equipped with a wide-band rf system to produce barrier buckets of any shape. We report here on some beam dynamics simulation studies and beam experiments carried out using the barrier rf system in the Recycler Ring.

INTRODUCTION

The Fermilab Recycler Ring (RR) [1] has been built with the goal of increasing ppbar collider luminosity for Run-IIb in the Tevatron from 5E31cm^{-2}sec^{-1} to 2E32cm^{-2}sec^{-1}. The RR will be the main pbar storage and cooling ring in the Fermilab accelerator complex. Hence, RR plays a vital role in future Tevatron collider operations.

Presently, we are commissioning the RR using protons from the Fermilab Booster or pbars from the Accumulator Ring (AR). The beam transfer in both cases has to be done by separating previously transferred beam from the new transfer. In future, unused pbars at the end of each collider run will be decelerated first in the Tevatron from 1 TeV to 150 GeV and then in the MI from 150 GeV to 8 GeV and, finally, transferred to RR. The emittance of the pbars from the Tevatron at the time of injection to RR is expected to be <16 eV-s in longitudinal plane and <20π-mm-mr in transverse planes. These pbars will have considerably larger emittance as compared to the pbars from AR. In RR, we have to separate the cold pbars and the hot pbars in azimuth. Thus, at any given time it is required that a maximum of three zones in RR azimuthally separated, *viz.,* transfer, stacked (cooled) and hot beam regions as indicated in Figure 1. All these specific needs can be met only by having barrier buckets[2] in the RR and be able to dynamically change their locations in the RR azimuthally.

The barrier bucket rf technology was invented at Fermilab[2] in 1983 and used successfully in the pbar Debuncher Ring. The shape of barrier pulses used in the Debuncher Ring was sinusoidal. In the RR we adopted bipolar square barrier pulses. Recently, a multi-particle beam dynamics 2-D simulation code[3] has been developed to study beam dynamics issues in RR and a comprehensive Hamiltonian formalism of particle motion in barrier rf buckets is presented in reference 4.

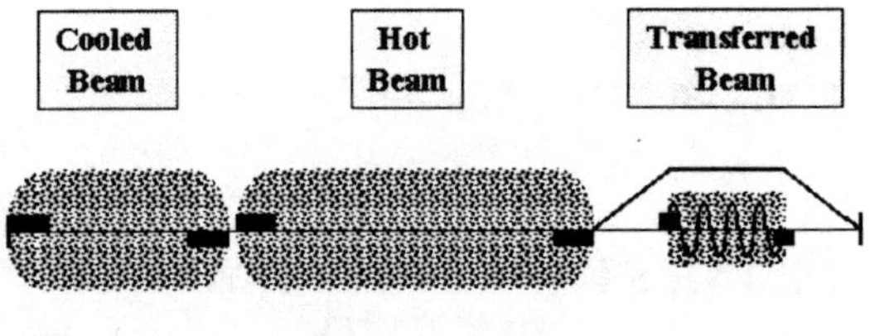

FIGURE 1. A schematic view of cooled, hot and transfer beam regions in RR separated using barrier buckets. The transfer region has 2.5MHz rf wave with rf bucket matched to the beam coming from MI.

BARRIER BUCKETS IN RR

A wide-band RF system capable of generating many barrier buckets of any shape has been built[5] and installed in RR. The maximum height of a barrier pulse is set to about +/-2kV each. The width of a pulse can be varied. The pbar beam from AR will be having 2.5MHz rf structure and that from RR to MI can have 2.5MHz or 7.5MHz rf structure. The rf system can produce 2.5MHz/7.5 MHz rf buckets in between barrier buckets. Figure 2 shows a schematic view of a barrier bucket in RR generated using square pulses.

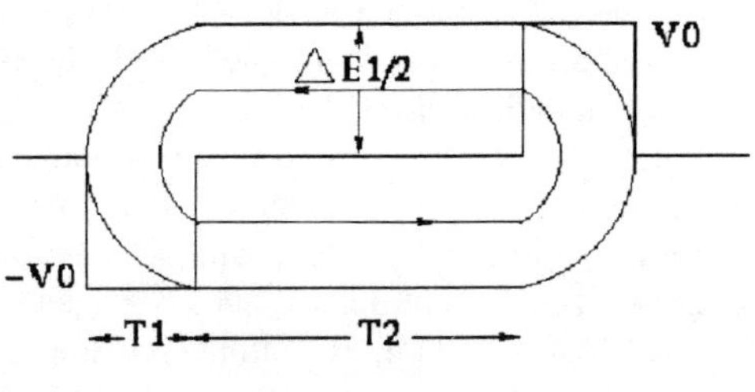

FIGURE 2. A schematic view of a barrier bucket in RR. Equal Hamiltonian contours are also shown.

*Work supported by the U.S. Department if Energy under contract No. DE-AC02-76CH03000.

CP642, *High Intensity and High Brightness Hadron Beams: 20th ICFA Advanced Beam Dynamics Workshop on High Intensity and High Brightness Hadron Beams,* edited by W. Chou, Y. Mori, D. Neuffer, and J.-F. Ostiguy

Half-height of a barrier bucket shown in Fig.2 is given by [1,4],

$$\Delta E_b = \left(\frac{eV_0 T_1}{T_0} \frac{2\beta^2 E_0}{|\eta|} \right)^{1/2}$$

and, bunch area is given by,

$$\mathcal{A} = 2T_2 \Delta\hat{E} + \frac{8\pi|\eta|}{3\omega_0 \beta^2 E_0 e V_0} (\Delta\hat{E})^3.$$

E_o is synchronous energy of the beam and η is slip factor (=-0.0087 for RR). $\omega_o = 2\pi f_o = 2\pi/T_o$. Where beam revolution period in RR $T_o \sim 11.12$ μsec. $\beta = v/c$. The quantity $\Delta\hat{E}$ is energy spread of the beam. The remaining parameters are indicated in Fig.2.

PBAR STACKING IN BARRIER BUCKETS

RF manipulations involved in beam stacking in RR have been simulated using the program ESME[3]. These simulations do not take into account any rf noise or other instrumental effects which may give rise to additional longitudinal emittance growths. Therefore we demand all rf operations carried out in simulation to be fully adiabatic *i.e.*, with no emittance dilution. However, from practical point of view we allow up to about 5% longitudinal emittance dilution.

Figure 3 shows phase-space distributions of particles in (ΔE,Δϕ)-space for four stages of beam stacking. The upper left figure shows beam in 2.5MHz rf bucket in RR. Next, the beam is de-bunched iso-adiabatically in about a second (see Fig.3 top right) in a barrier bucket with pulse width set to 491 nsec. This width was not fully optimized during the simulation. During the next two seconds the de-bunched beam particles are squeezed to match momentum spread of the stacked beam (notice that the shown here is a first transfer). There is a potential for emittance growth during squeezing and transferring the beam to stacked region. This emittance growth has to be minimized by optimizing speeds for squeezing and transferring the beam. The simulation showed that if beam squeezing is carried out faster than about 0.22 radian/sec in azimuthal plane we see emittance growth more than 5%. Some level of emittance growth may be acceptable for hot beam, but, not for cooled beam from AR. Fully squeezed beam is shown in bottom left side of Fig. 3. Finally, the new transfer along with the previously transferred beam is shown in bottom right of Fig. 3.

We have conducted series of experiments with protons and pbars in the RR barrier buckets to understand beam dynamics. We have used pbars from AR and protons from Booster.

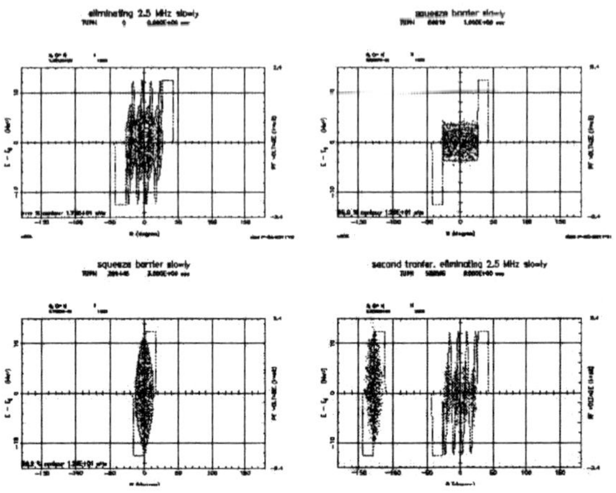

FIGURE 3. Longitudinal beam dynamics simulation results for beam stacking. Shown are distribution of beam particles in (ΔE,Δϕ)-space. The rf wave forms and buckets are also shown. For details see the text.

Experimental results for pbar beam stacking in RR are shown in Fig.4. The lower traces in each of the figures 4(a)-4(e) show RF cavity gap monitor voltage outputs. The upper traces are wall current monitor signals fed to a digital oscilloscope. The later results are similar to the projection of the phase space distributions shown in Fig.3 on ϕ axis (i.e, on horizontal axis). Descriptions of figures are as follows: a) four bunches on the left are newly transferred pbar beam and the long bunch on the right hand side is the stacked beam, b) beam immediately after debunching in barrier bucket. In this case, notice that the beam distribution is slanted indicating asymmetry in barrier pulse area. c) squeezed beam in barrier bucket, d) after transferring the newly stacked beam to originally stacked beam. e) rf waveform just before next beam transfer; this time notice the stacked bucket is slightly wider than in 4(a) to 4(d). So far, the maximum pbar stacked in RR was about 30E9 and that of proton beam was about 750E9 with a beam life time of a few hours.

We have carried out some preliminary experiments to un-stack the beam from a stacked beam and tried to transfer the beam to the MI. The sequences of this operation is essentially reverse of stacking.

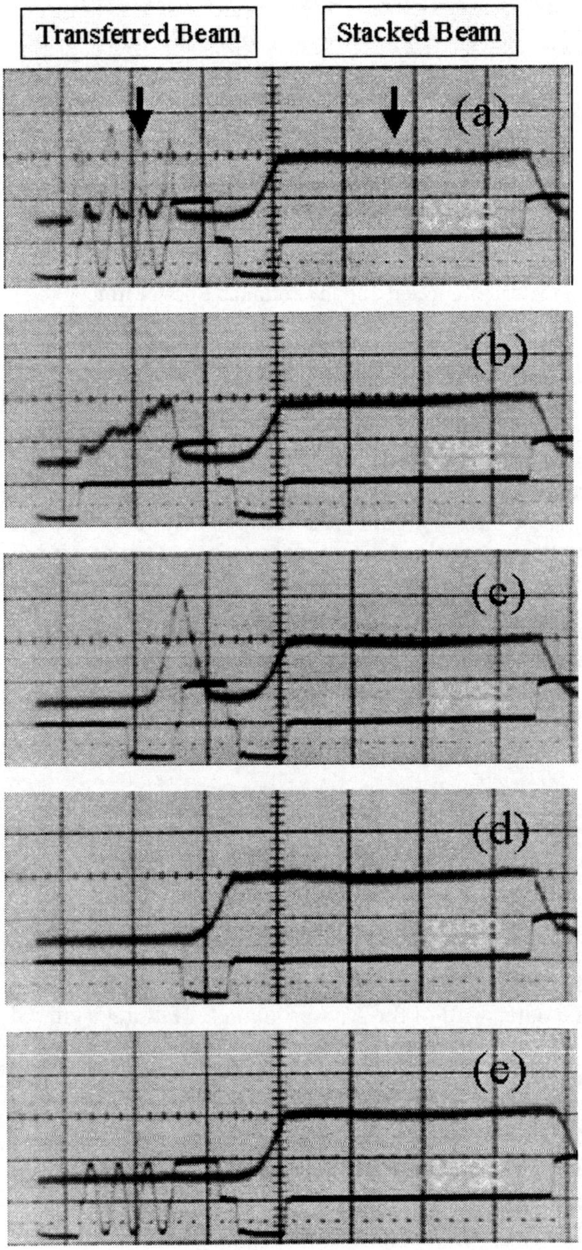

FIGURE 4. Barrier buckets in RR (lower traces). Transferred and stacked pbars in RR (upper traces). (a) and (e) also show 2.5MHz rf wave inside transfer bucket.

Measurement of beam intensity and the longitudinal emittance of the beam in the barrier bucket is vital in understanding beam life time and emittance growth in RR. Presently, we are developing methods to measure beam intensity by using a technique very similar to fast bunch integrator[6]. Longitudinal emittance measurements in barrier buckets is not trivial. We plan to measure it by use of schottky signals of the beam in the barrier buckets.

The author would like to thank J. MacLachlan during the beam dynamics simulations. Special thanks are due to Brian Chase, Joe Dey, Consolato Gattuso, Keith Meishner and John Reid for their help at various stages of this work

REFERENCES

1. G. Jackson, "The Fermilab Recycler Ring Technical Design Report" , FERMILAB-TM-1991 (1996).
2. G.E. Griffin et. al, IEEE, Trans. Nucl. Sci., NS-30, 3502 (1983).
3. J. MacLachlan, ESME modified to handle beam dynamics in barrier buckets (1996), (private communications 2001.
4. S.Y. Lee and K. Y. Ng, Phys. Rev. E, Vol. 55, 5992 (1997).
5. J.E. Dey and D.W. Wildman, IEEE, Proc. the 1999 Part. Accel. Conf. New York, NY, 869, 1999).
6. B. Fellenz and J. Crisp, a set of two timer gates are being developed for each of the barrier system. The system will integrate total area under wall current monitor signal to determine beam intensity (private communication 2002).

A Linac Afterburner to Supercharge the Fermilab Booster

C. Ankenbrandt, J. MacLachlan, M. Popovic, R. P. Johnson

Fermi National Accelerator Laboratory

Abstract. A Linac Afterburner is proposed to raise the energy of the beam injected into the Fermilab Booster from 400 MeV to about 600 MeV, thereby alleviating the longitudinal and transverse space-charge effects at low energy that currently limit its performance. The primary motivation is to increase the integrated luminosity of the Tevatron Collider in Run II, but other future programs would also reap substantial benefits. The estimated cost is $23M.

INTRODUCTION

The Fermilab Linac Upgrade Project, completed in 1993 at a total cost of $31M, increased the linac energy from 200 to 400 MeV. The Afterburner proposal emulates that successful project in several respects. The same kind of warm copper 805-MHz side-coupled cavities are used to accelerate the beam. Once again, existing enclosures and buildings are used to house the accelerator components and their supporting equipment. The installation and commissioning are planned to minimize their interference with Run II. Emulating the previous successful project allows a quick start and a rapid completion of the Afterburner Project. The luminosity of the Tevatron Collider increased by a factor of about 2.5 after the Linac Upgrade Project; the Afterburner is expected to provide a corresponding luminosity enhancement in Run II.

THE BOOSTER AND THE AFTERBURNER

When the Booster runs at $5 \cdot 10^{12}$ protons per cycle, as called for in the Main Injector era, several mechanisms of emittance growth and beam loss are evident. In the first few milliseconds after injection, transverse space-charge effects cause rapid emittance growth accompanied by substantial (~30%) beam losses. Meanwhile, longitudinal space-charge forces distort the potential well created by the rf focusing force, causing longitudinal emittance growth and failure to capture part (~10%) of the beam. Of course that uncaptured beam is swept radially inward out of the aperture as soon as the magnetic field ramps up significantly. These two low-energy space-charge effects are the dominant beam-loss mechanisms in normal operation, but they are not the whole story.

Although aggressive pursuit of measures to control higher-energy losses is certainly necessary to the success of the near-future physics program, only modest intensity gains are likely to result, because the higher-energy losses amount to less than ten percent of the beam.

The Booster is currently the bottleneck preventing higher-intensity operation of the whole facility, because both the upstream and downstream machines, the Linac [1] and the Main Injector/Tevatron complex, can handle higher intensities than the Booster.

AFTERBURNER TECHNICAL DESIGN CONCEPTS

To accelerate the beam between the Linac and the Booster, accelerating structures like those in the downstream end of the Linac will be installed along the outer wall of the Booster tunnel. (Putting a curved linac in the Booster tunnel was previously suggested in another context.[2]) The Afterburner is preceded by a longitudinal matching section and followed by a debunching section. The existing 400-MeV linac and beam transfer line to the Booster are left essentially untouched up to MH2, a horizontal bending magnet near the outer tunnel wall that currently bends the beam from the linac across the aisle toward the Booster injection system.

A. The Afterburner

The Afterburner is based on the 805 MHz warm copper side-coupled cavities installed at the downstream end of the existing linac during the Linac Upgrade Project [3]. There are several advantages to that choice. The technology is well understood and the costs well known, so no R&D is necessary prior to writing a definitive proposal. The transverse size of the side-coupled cavities is small enough to fit comfortably in the tunnel. Spare parts, including

CP642, *High Intensity and High Brightness Hadron Beams: 20th ICFA Advanced Beam Dynamics Workshop on High Intensity and High Brightness Hadron Beams*, edited by W. Chou, Y. Mori, D. Neuffer, and J.-F. Ostiguy

several 12 MW klystrons, exist and can be shared with the present system. Most importantly, the present system works very well, reliably producing high-quality beam with low beam losses. The decision to employ that technology allows the quick start and rapid completion that are essential in order to benefit Run II.

The Booster itself consists of 24 periods, with each period containing a long straight section, two combined-function magnets, a short straight section, and two more combined-function magnets. Figure 1 shows the layout of the Booster in its tunnel on the south side, where the Afterburner is to be installed. The length of the outer tunnel wall is 20.5 meters per period. The shape of the Afterburner and its transfer lines is regular and symmetrical, like the Booster and most of its tunnel. There are 7.5 degree bend centers opposite the middle of the short and long straight sections, separated by 10.25 meters. The 10.25 meter distance between bend centers is sufficient to accommodate one accelerating module along with the required bending and focusing magnets. In the present design there are actually five small magnets at the bend centers: a quadrupole triplet with two 3.75 degree dipoles interspersed. The dipoles are 0.5 meters long to reduce field stripping to negligible proportions.

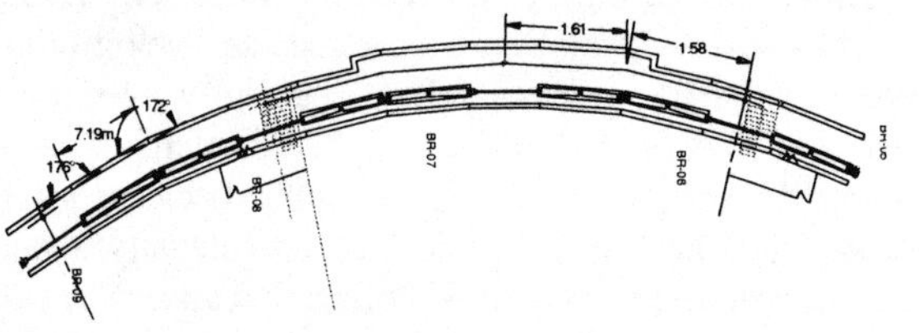

FIGURE 1.

The accelerating structure has five modules. Each module consists of four sections joined by bridge couplers to accommodate quadrupole focusing magnets. Each section contains twelve accelerating cavities. The choice of twelve cavities per section, rather than sixteen as in the existing linac, makes the modules short enough so that the Afterburner can match the shape of the outer tunnel wall. The rest of the parameters are the same as in the existing linac.

The maximal accelerating field is limited by the requirement that the electric field is less than 1.4 times the Kilpatrick value. The shunt impedance of the structure is taken to be 80% of the value calculated by Superfish; the same criterion was used in the design of the 400MeV energy upgrade. The transit-time factor, maximal field, and shunt impedance are extrapolated from the Superfish calculation for the Linac Upgrade Project for beta in the region from $0.57 < \beta < 0.71$.

Figure 2, taken from Trace3D output, shows the phase-space ellipses in the transverse and longitudinal dimensions. In the upper left corner are input beam ellipses with alphas and betas for three planes. The upper right corner shows the same beam parameters at the end of the Afterburner. The lower portion of the graph shows beam envelopes and accelerator elements. The displayed envelopes are the square root of five times their respective rms values. The two upper traces are the horizontal envelope and bunch length and the lower trace is the vertical envelope. The quad settings are adjusted to keep the beam envelope small along the linac. The full scale of 15 mm corresponds to the radius of the physical aperture.

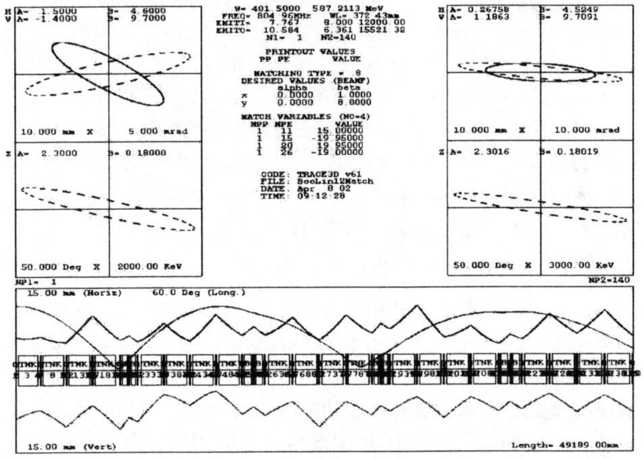

FIGURE 2.

It is interesting to note the similarity of Figure 2 with the corresponding Figure 15 of the Fermilab Linac Upgrade Conceptual Design Report [Ref]. In particular the beam envelopes in both cases are contained within about half of the physical aperture. That shows that the dispersion associated with the required bends is not a problem. The momentum spread contributes only a few mm of beam size in quadrature with the transverse beam sizes. The transverse sizes shrink adiabatically from their values at 116 MeV while the physical apertures remain the same size.

B. Longitudinal Matching

The Afterburner requires a buncher to match into its side-coupled structure (SCS) at the upstream end and a debuncher at the downstream end to reduce the momentum spread for injection into the Booster. (Transverse matching is of course also required, but it is not discussed here because it is routine.) The bunching system makes use of the existing 400 MeV debuncher but requires an additional bunching tank closer to the Afterburner.

The existing 805 MHz debuncher is located about 30 m downstream of the end of the 400 MeV SCS linac where the beam bunches have sheared to span about 80 degrees of rf phase. This is a good spread for rebunching with good linearity using a modest rf system. However, it is much too far upstream to permit the sharp focus (small longitudinal beta function) required to match to the Afterburner. This design includes a new buncher structure located about 25 m upstream of the Afterburner. To reduce the power required for the buncher, a twelve-cell SCS is used. It requires about 480 kW of rf power, which is not low enough to use the same kind of small TV klystron that powers the first buncher, so a shorter structure powered by one of the 12 MW klystrons might be preferred. The existing four-cell debuncher is excited to 2 MV/m to keep the bunch width below one hundred degrees. The new buncher is excited to an effective gradient of 1.7 MV/m so that the bunches entering the Afterburner have the same distribution in longitudinal phase space as the ones exiting the 400 MeV linac: 12 degrees full width and 2.6 MeV full energy spread.

The 600 MeV debuncher serves exactly the same purpose as the existing debuncher in the present transfer line between the 400 MeV linac and the Booster, but substantially more power is needed. It can be located 40 m downstream of the Afterburner and run at exactly the same gradient as the buncher. The energy spread of the 600 MeV beam is reduced from 2.8 MeV full width to 0.54 MeV at the Booster, corresponding to a momentum spread of 0.03%.

C Injection into the Booster

A new injection system to be located at Long Straight Section 11 is a part of this proposal. The existing injection system will be left where it is at Long Straight Section 1 so that it can continue to be used for injection of 400-MeV beam until the commissioning of the Afterburner is complete. Temporarily having dual injection systems will reduce the interruption to Run II, and a new injection system can incorporate the latest design ideas. Besides, the existing injection system probably would not work at 600 MeV anyway.

Figure 3 shows the layout of the new H^- injection system. The design borrows ideas from that of the injection system for the Proton Driver[4], but unlike that system it is local, that is, confined to a single straight section with no lattice elements interspersed among injection components.

Injecting a high-energy beam in a limited space is a design challenge because the field strength B of the magnets that transport the incoming ion beam must be limited; they must satisfy the constraint B<0.7 Tesla/p (GeV/c) in order to reduce the probability of field-stripping to about 10^{-3}. The parameter choices used in Figure FFF-INJ satisfy that constraint and comfortably accommodate 600 MeV beam in the available space. If necessary considerably higher incoming kinetic energies could be injected using this kind of design.

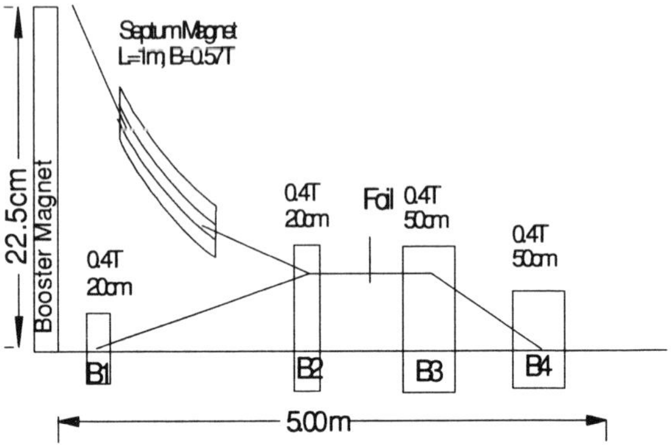

FIGURE 3.

SUMMARY AND CONCLUSIONS

A Linac Afterburner is proposed to supercharge the Fermilab Booster by raising its injection energy, thereby alleviating the transverse and longitudinal space charge effects that currently limit its performance. The Booster is the bottleneck in the whole Fermilab accelerator complex because all the other machines can handle higher intensity beams. Considerable leverage is involved because a modest improvement upstream of the Booster would benefit all components of the Fermilab experimental high energy physics program. Increasing the brightness and intensity of the Booster and Main Injector beams would be quite valuable to the US HEP program. However, the primary motivation for this proposal is to increase the integrated luminosity for Run II of the Tevatron Collider. CERN's remarkable effort to discover the Higgs in the last LEP run was impressive; now Fermilab must pick up the torch. A Linac Afterburner is the logical Fermilab response to the mandate to maximize the discovery potential of Run II. This document has demonstrated the feasibility of the Afterburner Project. The nature and scope of the project would be much like that of the highly successful Linac Upgrade Project. A quick start is possible because no R&D is necessary; we already know everything we need to know to write a credible proposal to the Department of Energy. A rapid completion is possible because the cost and scope are modest and because existing structures can be used to house the accelerating components and supporting equipment. Can Fermilab afford not to pursue these opportunities vigorously?

REFERENCES

1. LINAC Staff (M. Popovic for the collaboration), PRESENT, NEAR FUTURE AND FUTURE PERFORMANCE OF THE FERMILAB LINAC, FERMILAB-PUB-96-046, Mar 1996
2. M. Popovic, A. Moretti, R. Noble, and C.W. Schmidt, 1-GeV Linac Upgrade Study at Fermilab, 1998 Linac Conf
3. http://wwwlib.fnal.gov/archive/linac/FERMILAB-LU-ConceptualDesign.pdf
4. W. Chou, C. Ankenbrandt, and E. Malamud, Eds., The Proton Driver Design Study, FERMILAB-TM-2136, December 2000.

SESSION E

SPACE CHARGE SIMULATIONS

Summary Report

I. Hofmann* and R. Ryne†

*GSI, Darmstadt, Germany
†LBNL, Berkeley, USA

This session included the following 7 presentations:

- *ORBIT: A Code for Collective Beam Dynamics in High Intensity Rings,*
  (S. Cousineau and J. Holmes, ORNL)
- *Behavior of Intense Beams with Simpsons,*
  (S. Machida, KEK)
- *Coherent Transverse Resonance Effects,*
  (I. Hofmann, GSI)
- *Space-Charge Calculation with the code ORBIT,*
  (A. Luccio, BNL)
- *Simulation Studies with PATRASH,*
  (Y. Shimosaki, KEK)
- *Space-Charge Effects on Beam Loss due to Resonance Overlap,*
  (G. Francetti, GSI)
- *Status of the IMPACT Parallel Beam Dynamics Code,*
  (J. Qiang and R. Ryne, LBNL)

The main topic was space-charge simulation in rings. There was a general consensus that linac modeling was well in hand, if compared with ring simulation challenges. The presentations and discussions focused on simulation tools, new phenomena and issues of concern, and future tasks and challenges in space-charge simulations.

As is evident from the list of talks, several simulation tools were presented (ORBIT, Simpsons, PATRASH, IMPACT, Micromap, and others.) Some talks presented comparisons of the various codes, e.g. a comparison of simulation results for ORBIT, Simpsons, and PATRASH. Most of the ring modeling results that were presented used 2D or 2.5D space-charge solvers, in simulations of up to 1000 turns. There were also some reports of 3D simulations (e.g. using IMPACT).

In the area of new phenomena, an important issue was that of the coherent versus incoherent resonance condition. The main point is that the single particle (incoherent) resonance condition needs to be modified if coherent motion occurs. The shift of the resonance provides what I. Hofmann has called the "coherent advantage," meaning that it appears to be possible to propagate beams with higher intensity than would be predicted on the basis of the incoherent condition simply adjusted for space-charge tune spread. This leads to a 30% advantage in the PSR and SNS rings. It was pointed out that this is not a new idea (it was originally due to Sacherer), but it has now been confirmed by simulation in realistic lattices. A related issue that was discussed was the question of whether or not the coherent advantage still applies if resonance crossing occurs over many synchrotron oscillations? It is possible that the synchrotron oscillation might "wash out" the coherent response. This is a question that will be explored in future simulations. Another phenomena discussed were the possibility that space charge might have a beneficial effect on the beam dynamics. This is because space charge detuning in nonlinear resonances (e.g. 3rd and 4th order) leads to self-limitation, making the resonances less serious. In fact, it was mentioned during the discussions that an ideal barrier bucket might be an attractive option from a resonance point of view.

In addition to self-consistent multi-particle simulations, there were also presentations involving analytical and semi-analytical work. It was shown that, in very long ($\sim$1 million turn) simulations using an analytical space-charge model, repeated sweeping over resonances could lead to beam blow-up and particle loss. Given the numerical inaccuracies in self-consistent solvers, the question arose as to whether analytical models might provide a higher degree of confidence than self-consistent models over time spans involving hundreds of thousands or millions of turns. One advantage of the analytical models is that it is possible to gain insight from them due to the ability to make "clean" Poincare plots.

During the discussion period following the presentations there was much discussion devoted to the accuracy of Particle-In-Cell (PIC) codes and what they could be reliably predicted by PIC codes. Clearly one needs to pay attention to convergence of the results with respect to grid size, number of simulation particles, step size, number of space-charge kicks, etc. Furthermore, when testing such codes, one test is not enough to cover all problems and range of phenomena (i.e. all resonances, all intensity regimes, etc.). Numerical diffusion is an issue in long-

CP642, *High Intensity and High Brightness Hadron Beams: 20th ICFA Advanced Beam Dynamics Workshop on High Intensity and High Brightness Hadron Beams,* edited by W. Chou, Y. Mori, D. Neuffer, and J.-F. Ostiguy

time simulations, particularly given that it may obscure physical diffusion associated with the true dynamics. Numerical diffusion occurs at some level in all simulations, and has been shown to lead to a linearly growing emittance. This is a major issue for long-term simulations, since, e.g., a 1% emittance growth in a 1000-turn simulation would not be acceptable if one was interested in performing longer simulations of 100,000 or more turns. A related issue involves the smoothness of the forces in PIC simulations. If one is very interested in long-time simulations, it may be necessary to use models in which high derivatives of the self-force are continuous across grid boundaries.

In regard to the future, it was felt that the biggest challenges lie in the real of long-term ring simulations. We have reached the point where the current capability of 2D or 2.5D models over ~1000 turns need to be extended to 3D models for 0.1 to 1 million turns. This will require the use of massively parallel computers, and such work is already underway – Examples include the DOE/HENP SciDAC accelerator-modeling project, the parallelization of ORBIT, and the WARP suite of codes used in the heavy ion fusion community. Finally, the validation of such computer simulations will require sufficient diagnostics in dedicated experiments and sufficient characterization of the machine nonlinearities so that modelers and experiments have an adequate description to quantitatively compare simulations and experiments.

ORBIT: A Code for Collective Beam Dynamics in High-Intensity Rings

J.A. Holmes*, V. Danilov*, J. Galambos*, A. Shishlo*, S. Cousineau@, W. Chou#, L. Michelotti#, J.-F. Ostiguy#, J. Wei&

**Oak Ridge National Laboratory, Oak Ridge, TN 37831, USA*
@Indiana University, Bloomington, IN 47405, USA
#Fermi National Accelerator Laboratory, Batavia, IL 60510, USA
&Brookhaven National Laboratory, Upton, NY, 11973, USA

Abstract. We are developing a computer code, ORBIT, specifically for beam dynamics calculations in high-intensity rings. Our approach allows detailed simulation of realistic accelerator problems. ORBIT is a particle-in-cell tracking code that transports bunches of interacting particles through a series of nodes representing elements, effects, or diagnostics that occur in the accelerator lattice. At present, ORBIT contains detailed models for strip-foil injection, including painting and foil scattering; rf focusing and acceleration; transport through various magnetic elements; longitudinal and transverse impedances; longitudinal, transverse, and three-dimensional space charge forces; collimation and limiting apertures; and the calculation of many useful diagnostic quantities. ORBIT is an object-oriented code, written in C++ and utilizing a scripting interface for the convenience of the user. Ongoing improvements include the addition of a library of accelerator maps, BEAMLINE/MXYZPTLK; the introduction of a treatment of magnet errors and fringe fields; the conversion of the scripting interface to the standard scripting language, Python; and the parallelization of the computations using MPI. The ORBIT code is an open source, powerful, and convenient tool for studying beam dynamics in high-intensity rings.

INTRODUCTION

High-intensity proton ring beams are characterized by low energy, high intensity, and low loss requirements. Satisfying the beam-loss requirements necessitates a detailed understanding of beam dynamics in this regime. At high intensity, collective effects due to space charge and wakefields strongly affect the beam behavior, and single particle models do not suffice. Also, because of the complexity of collective phenomena for bunched beams in high-intensity rings, a computational approach is productive for theoretical studies and indispensable in solving detailed design and engineering problems.

Recognizing this, the SNS Accelerator Physics Group at ORNL, with help from colleagues at BNL, undertook the development of an object-oriented general-purpose code, ORBIT [1,2]. ORBIT began as a C++ rewrite of ACCSIM [3], developed under the SuperCode driver shell [4], but has since undergone extensive independent development, which will be described here.

ORBIT is a particle tracking code in 6D phase space. Its basic classes are herds, which are groups of particles, and nodes, which operate on the herds. ORBIT has been designed to simulate real machines: it has detailed models for transport through various lattice elements; injection foil and painting; rf and acceleration; 2.5D and 3D space charge; longitudinal impedance and space charge; transverse impedance; apertures: collimation; and beam diagnostics.

Present work on ORBIT focuses on the exploitation of some of the more recent and computationally demanding physics modules and on the enhancement of overall capabilities. With the completion of the transverse impedance, 3D space charge, and collimation routines, the physics for new classes of problems is available. However, meaningful calculations will require high performance computing techniques, such as the implementation of parallel computing. Also, because ORBIT has developed with an emphasis on collective effects, single particle transport models are still rudimentary. In order to assess losses to the accuracy demanded in recent and future high intensity rings, high order maps and a comprehensive treatment of magnet errors are necessary. Finally, although SuperCode is an innovative tool, it is neither standard nor supported. Comparable tools, based on the Python programming language, now are standard and present an attractive option.

Consequently, we are now developing parallel algorithms for collective beam dynamics, including higher order single particle maps, creating magnet error routines, and planning to implement a new driver shell based on Python. We now describe the models in ORBIT.

COMPLETED ORBIT MODELS

The coordinate representation in ORBIT utilizes the usual accelerator expansion. The horizontal direction is

CP642, *High Intensity and High Brightness Hadron Beams: 20th ICFA Advanced Beam Dynamics Workshop on High Intensity and High Brightness Hadron Beams,* edited by W. Chou, Y. Mori, D. Neuffer, and J.-F. Ostiguy
2002 American Institute of Physics 0-7354-0097-0

represented by x, x', the vertical direction by y, y', and longitudinal phase space by the phase angle ϕ, and energy deviation, dE, from a specified closed orbit reference particle. The independent variable is the machine location s.

ORBIT can construct lattices by reading output files from MAD [5] or DIMAD [6], by direct specification, or by specifying a uniform focusing channel. Linear transport is carried out through symplectic matrix multiplication. Nonlinear elements can be evaluated in the thin lens approximation, and higher order single particle transport terms, such as chromaticity, can be evaluated using second order transport matrices. There is no specific facility for the treatment of errors or fringe fields. Field errors in straight lattice elements can be calculated in MAD and included in the MAD output.

ORBIT can inject particles turn-by-turn or utilize a complete distribution specified at the start. A variety of distributions can be generated internally. Any externally generated distribution can be imported. Injection painting is treated with user-defined time-dependent closed orbit bumps. ORBIT contains an injection foil model that keeps track of foil hits and applies transverse kicks based on multiple Coulomb scattering. Particles that miss the foil at injection are removed from the beam.

The RF cavity model provides longitudinal kicks based on a specified time-dependent waveform with multiple harmonics. For nonaccelerating cases, the synchronous phase is assumed to be zero, and only the harmonics and time-dependent voltages need to be specified. For accelerating cases, the harmonics, voltages, and time-dependent dipole fields must be specified. The model uses this information to calculate the synchronous phase and the resulting kicks. The transverse phase space is also adjusted to conserve normalized emittance.

The 2.5D space charge model is implemented as a series of kicks separated by other transport operations. Particles are binned in a 2D rectangular grid using a second order distribution scheme. The potential for the distributed charges is then solved on the transverse grid using a fast FFT solver. Conducting wall (circular, elliptical, or rectangular beam pipe) boundary conditions are then imposed using a method described in Ref. [7]. Particle kicks are obtained from second order interpolation of the potential, completing what might be called a "quasi-symplectic" evaluation. Finally, the kicks are weighted by the local longitudinal density to account for bunch factor effects. This is the reason we call the model 2.5D. There is also an alternative direct force (momentum-conserving) solver without beam pipe that uses a method described in Ref. [8].

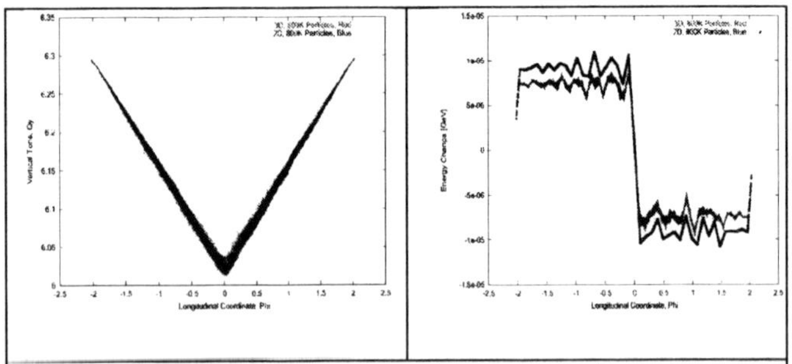

Figure 1. 3D space charge benchmark for triangular charge distribution: a) tune shifts vs. position, b) longitudinal force vs. position. 2.5D and 3D results agree with each other and with analytic calculations.

The 3D spacecharge model is a simple generalization of the 2.5D routine. Particles are distributed to a 3D rectangular grid using a second order scheme. Typically, for rings, the longitudinal spacing greatly exceeds the transverse spacing. The potential is solved as a 2D problem using the distributed charges and fast Fourier transforms on the transverse grid for each longitudinal slice. Conducting wall boundary conditions (circular, elliptical, or rectangular beam pipe) are used to "tie together" the transverse solutions into a 3D potential. Particle kicks are obtained by interpolating the potentials in 3D using a second order "quasi-symplectic" interpolation scheme.

ORBIT treats longitudinal impedances and/or space charge in a fashion similar to ESME [9]. The longitudinal impedance is represented as harmonics of the fundamental ring frequency. Particles are binned longitudinally and the binned distribution is Fourier transformed. The space charge contribution to the impedance is combined with the external impedance. The Fourier transformed distribution is multiplied by the impedance to give the longitudinal kicks to the particles. For many rings, the synchrotron period is much longer than a turn, and it is sufficient to evaluate the longitudinal impedance and space charge kicks once each turn. More frequent evaluations may be required for rings having higher synchrotron frequencies.

Transverse impedances are also treated as localized nodes in ORBIT. The validity of this approach requires the element length to be short compared to the betatron oscillation wavelength. Otherwise, multiple impedance nodes are required. As with the longitudinal impedance, the transverse impedance is represented by its Fourier components, but now the appropriate frequencies are the betatron sidebands of the ring frequency harmonics. For both impedance models, the formulation incorporates velocities below the speed of light. Particle kicks are obtained by convolution of the beam current dipole moment with the impedance. In evaluating the beam current it is assumed that the dipole moment evolves from the previous turn through a simple betatron oscillation.

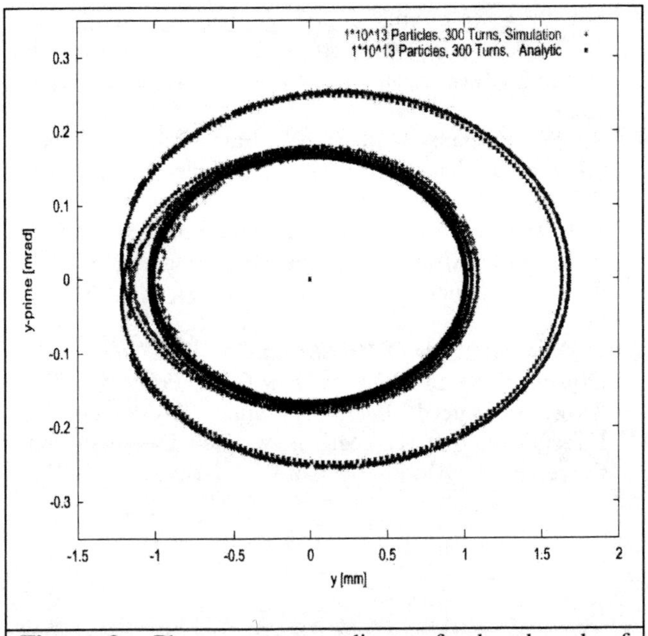

Figure 2. Phase space coordinates for benchmark of transverse impedance model with analytic calculation for pencil beam.

Circular, elliptical, rectangular, or racetrack apertures are defined in ORBIT. The apertures can be set either to allow particles to pass through and simply tabulate the hits, or to remove the particles from the beam and tabulate the loss locations.

ORBIT contains a detailed collimation model. Collimator shapes include the aperture shapes, single or multiple edges at arbitrary angles, and also a rectangular plate collimator that can be used for beam windows or foils. Physics includes multiple Coulomb scattering, ionization energy loss, nuclear elastic and inelastic scattering, and Rutherford scattering for a number of materials. Monte Carlo algorithms are used for particle transport inside the collimator, and step sizes are carefully adjusted near collimator boundaries.

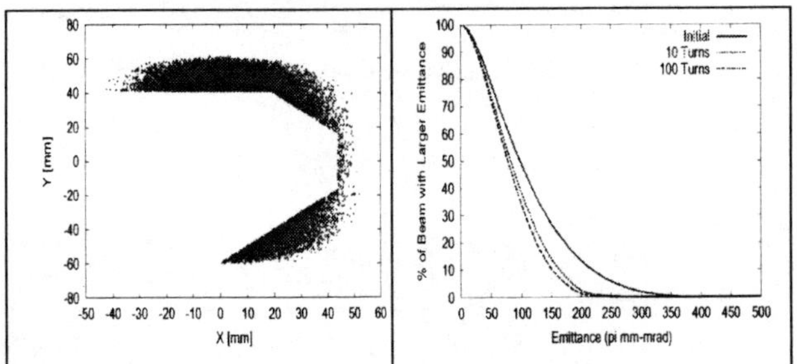

Figure 3. Results from SNS collimation study: a) particle impacts on beam scrapers, b) effects of collimation on emittance distribution.

A list of useful diagnostics in ORBIT includes the following: dumps of particle coordinates, tunes, or emittances at any point/node in the ring; histograms of particle distributions in x, y, phi, and emittance; rms emittances or beam moments versus turn or versus position; statistical calculation of beta functions; and longitudinal harmonics of the beam centroid. Because of the recent development of 3D models in ORBIT, we are extending our diagnostics to include more functions of longitudinal position.

ONGOING DEVELOPMENTS

The present emphasis in ORBIT development is on improvement of both physics and computational capabilities. In the physics category, we are significantly extending the single particle transport model by interfacing ORBIT and the BEAMLINE/MXYZPTLK library for maps and tracking [10]. In addition, we are including a comprehensive set of magnet error routines and a hard edge fringe field model. At this time, the ORBIT - BEAMLINE/MXYZPTLK interface has been partly completed with some options working, and the errors will be included this summer.

The most urgent computational improvement to ORBIT is the introduction of parallel processing algorithms. We are doing this using the MPI libraries, and at present the parallelization of the 2.5D space charge routines has been carried out and parallel versions of the 3D space charge routines are being developed. We anticipate near-term completion of this work, to be followed by creation of parallel versions of the diagnostic routines.

In the slightly longer term, we plan to convert the ORBIT user interface and driver shell from SuperCode to Python. The feasibility of and methods for doing this have been examined, but the work remains to be carried out.

OPEN CODE

Finally, we emphasize that ORBIT is an open code. The source code can be obtained by downloading from http://www.sns.gov//APGroup/Codes/Codes.htm and there are also a user manual, instructions for building ORBIT, and examples at that web site.

ACKNOWLEDGMENTS

We wish to thank our colleagues J. Beebe-Wang, M. Blaskiewicz, F. Jones, A. Luccio, N. Malitsky, and F. MacLachlan for their advice and collaboration during the development of ORBIT.

SNS is managed by UT-Battelle, LLC, under contract DE-AC05-00OR22725 for the U.S. Department of Energy. SNS is a partnership of six national laboratories: Argonne, Brookhaven, Jefferson, Lawrence Berkeley, Los Alamos, and Oak Ridge.

REFERENCES

1. J. Galambos, J. Holmes, D. Olsen, A. Luccio, and J. Beebe-Wang, *ORBIT Users Manual*, http://www.sns.gov//APGroup/Codes/Codes.htm
2. J. Galambos, S. Danilov, D. Jeon, J. Holmes, D. Olsen, J. Beebe-Wang, and A. Luccio, in *Proceedings of the 1999 Particle Accelerator Conference*, (New York, 1999) 3143.
3. F. Jones, *Users' Guide to ACCSIM*, TRIUMF Design Note, TRI-DN-90-17 (1990), http://www.triumf.ca/compserv/accsim.html
4. S. W. Haney, *Using and Programming the SuperCode*, http://www.sns.gov/APGroup/Codes/Codes.htm
5. H. Grote and F. Christoph Iselin, *The Mad Program, Version 8.19, User's Reference Manual*, CERN/SL/90-13, (Geneva, 1996).
6. See ftp://csftp.triumf.ca/pub/CompServ/dimad/
7. F. W. Jones, in *Proceedings of the 2000 European Particle Accelerator Conference*, (Vienna, 2000) 1381.
8. R. W. Hockney and J. W. Eastwood, *Computer Simulation Using Particles*, Institute of Physics Publishing (Bristol: 1988).
9. J. A. MacLachlan, Longitudinal Phase Space Tracking with Space Charge and Wall Coupling Impedance, Fermi National Accelerator Laboratory, FN-446, (1987).
10. L. Michclotti, *MXYZPTLK and BEAMLINE: C++ Objects for Beam Physics*, AIP Conf. Proc. No. 255 (Proc. Advanced Beam Dynamics Workshop on Effects of Errors in Accelerators, Their Diagnosis and Correction, Corpus Christi, Texas, 1992).

Behavior of Intense Beams Simulated with Simpsons

Shinji Machida

KEK, Accelerator Research Laboratory
1-1 Oho, Tsukuba-shi, Ibaraki-ken, 305-0801, JAPAN. (Shinji.Machida@kek.jp)

Abstract. We will show some examples of space charge simulation results using the code Simpsons [1]. We also discuss single and multi particle behavior as a function of simulation parameters.

INTRODUCTION

Space charge effects in a synchrotron are characterized with space charge detuning and resonance. In fact, space charge effects in a rapid cycling synchrotron (RCS) such as 3 GeV RCS in JKJ accelerator complex and 50 GeV MR show quite different behaviors. In a RCS, everything is transient: bunch shape as well as transverse emittance is being created during so-called painting process at injection. In addition, quick acceleration begins before the particle distribution is settled down. Here, the stability of the beam core becomes a primary issue. In MR, on the other hand, a well-shaped bunch is injected and any coherent oscillations of core are minimized, at least, a the beginning. The main concern of that type of machine is slow and continuous halo formation and resulting beam loss. In this paper, first we will show an example of space charge effects in a RCS. After raising a question of single and multi particle behavior in simulation, we will discuss beam behavior in MR.

ANTI-CORRELATED PAINTING

One typical example of space charge effects in a RCS emerges during phase space painting. In order to make K-V like particle distribution, anti-correlated painting was proposed. A beam is injected from the center in one transverse plane and from the maximum amplitude in the other plane. It works fine as expected when the space charge is not included (Fig. 1).

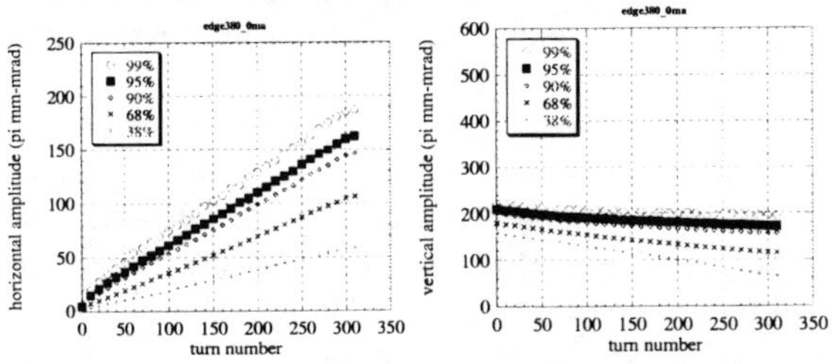

FIGURE 1. Emittance evolution during anti-correlated painting. Left is horizontal and right is vertical. Space charge is not included. There are five different emittance, namely 38%, 68%, 90%, 95%, and 99%. Those are defined as the amplitude within which the corresponding fractions of the particles are enclosed.

The space charge force during the injection, however, changes the resulting distribution. The asymmetric emittance at the beginning induces the coupling through space charge effects. Figure 2 shows the horizontal and vertical emittance as a function of turn number during anti-correlated painting with space charge effects. We observed five different emittance, namely 38%, 68%, 90%, 95%, and 99%. Those are defined as the amplitude in which the corresponding fractions of the particles are enclosed.

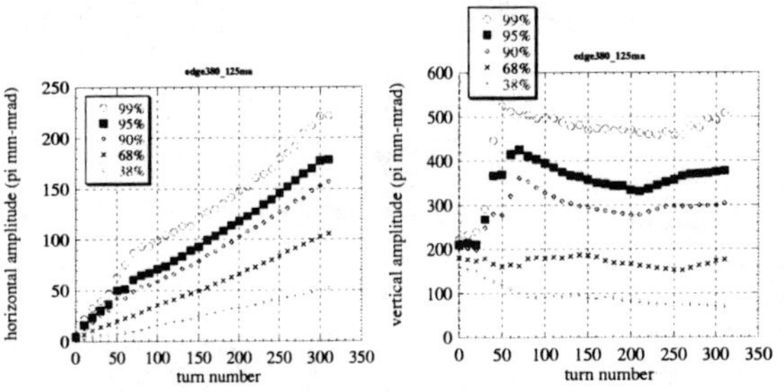

FIGURE 2. Emittance evolution during anti-correlated painting when the beam current is 50 mA.

SINGLE AND MULTI PARTICLE BEHAVIOR

In order to make sense of multi particle simulations, saturations of parameters have to be checked. In space charge calculations using a Particle in Cell (PIC) method, parameters to be optimized are number of macro particles, grid size, integration time step, and order of mode expansion if there is a smoothing process like Simpsons in frequency space. Although we may expect that smaller grid size, shorter time step, and higher order modes always make a calculation more precise, memory size and CPU time force compromised solutions.

Figure 3 shows a parameter dependence of emittance vs. the number of radial grids. Saturation is obtained when the number of radial grids is more than 50. Similarly, the macro particles should be more than

CP642, *High Intensity and High Brightness Hadron Beams: 20th ICFA Advanced Beam Dynamics Workshop on High Intensity and High Brightness Hadron Beams,* edited by W. Chou, Y. Mori, D. Neuffer, and J.-F. Ostiguy

5,000, the azimuth mode number is more than 16, and integration step is less than 2 m (In fact, Simpsons uses time as the independent variable, so that it is more appropriate to say that integration step time is less than 9ns.)

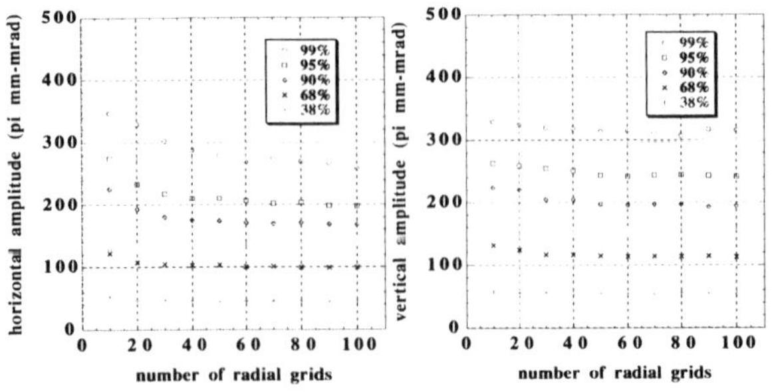

FIGURE 3. Emittance vs. the number of radial grids in horizontal (left) and vertical (right).

It is intriguing that single particle trajectory is not necessarily independent of the parameters even though the emittance defined by multi particles is saturated. An example is shown in Fig. 4 where the number of radial grids is changed from 60 to 80. According to the multi particle optimization, emittance does not change much when it is more than 50.

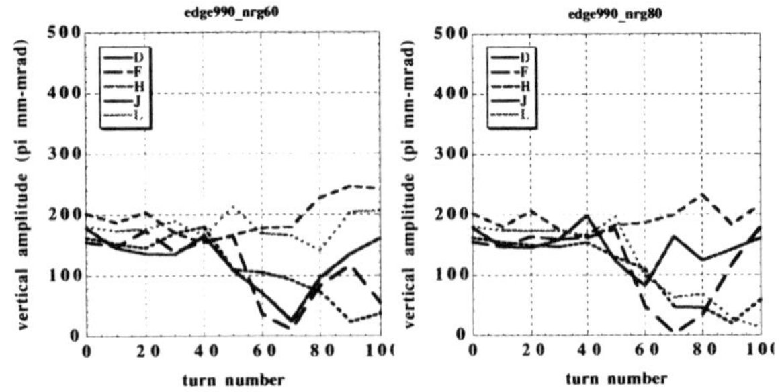

FIGURE 4. Single particle trajectory when the number of radial grids is 60 (left) and 80(right). There are five test particles with the name of D,F,H,J,L. They start from the same initial coordinates in both radial grids.

EMITTANCE EVOLUTION NEAR A RESONANCE

We have already known that, 1) Coherent tune shift is a relevant measure, not incoherent. 2) Ratio of the coherent and incoherent tune is 5/8 in a quadrupole mode. The higher the coherent order, the smaller the difference. Although those results are manifest, there are still some questions. First, those results are valid in a coasting beam, but not obvious in a bunched beam where incoherent tune is a function of longitudinal position as well as transverse amplitude. The synchrotron oscillations are also involved. Secondly, the main concern of a high intensity synchrotron is a small beam loss such as a few percents or even less. A question is then whether the coherent model of space charge effects helps understand those small beam losses.

In order to study beam behavior near a resonance, two bare tunes are picked up. The first tune is (6.566, 6.200) that is near a structure resonance line at ν_y=6.0. Since the model lattice we took here is 3 GeV RCS of JKJ with three fold symmetry, all the resonances located on ν_y=6.0 are structure. The second one is (6.850, 6.700) that is located just above ν_y=6.5. The resonance is non-structure and not excited until quadrupole imperfections are introduced.

As for the beam intensity, we tested three different levels. The lowest level intensity is low enough so that neither incoherent or quadrupole coherent tune shift does not reach the resonance. At the second level, the incoherent tune shift is below the resonance, but the quadrupole coherent is still above. At the third level, the intensity is high enough such that both incoherent and coherent ones are below the resonance.

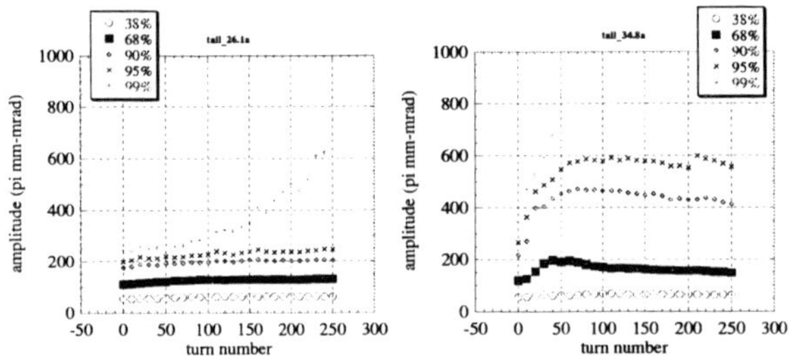

FIGURE 5. Emittance evolution when the tune is above non-structure resonance. Only incoherent tune is below the resonance (left) or both coherent and incoherent tune is below the resonance (right).

Behavior near Non-Structure Resonance

Figure 5 shows the results when the bare tune is chosen just above the non-structure resonance. Without any error fields, nothing happens for any intensity except a few growth of 99% emittance (not shown here). With random quadrupole errors of the width of 0.02, a significant growth of the beam core as well as the tail is observed when the both tune are below the resonance (Fig. 5, right). Only the growth of 99% emittance is observed when only the incoherent tune is below the resonance (Fig. 5, left). In other words, core growth is observed only when the quadrupole coherent shift is large enough and it satisfies the resonance, whereas the 99% tail particles start growing even below that intensity.

Behavior near Structure Resonance

Figure 6 shows the results when the bare tune is chosen just above the structure resonance. Now, the core growth is observed when only the incoherent tune is below the resonance without random errors. In fact, ν_y=6.0 is structure of all orders so that it is reasonable to conclude that the growth is not due to the incoherent tune, but by higher order coherent motions. The higher the order of coherent motion becomes, the smaller difference between the incoherent tune shift and the coherent tune shift. Practically speaking, the incoherent tune shift is a good measure for the structure resonance although the incoherent tune shift is not the true source of the growth.

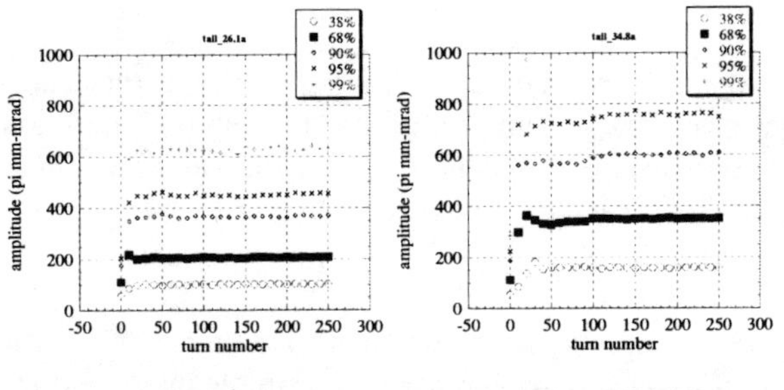

FIGURE 6. Emittance evolution when the tune is above structure resonance. Only incoherent tune is below the resonance (left) or both coherent and incoherent tune is below the resonance (right).

COHERENT OSCILLATIONS OF A BUNCHED BEAM

So far, the simulation results are all for coasting beams. It is interesting to see if there are coherent motions in a bunched beam. As a first step, instead of defining coherent motions as a whole beam, we observed the quadrupole motions of each longitudinally sliced beam. Figure 7 shows the tune spectra of each slice. As expected, a slice of the bunch center has the maximum tune and the tune shift becomes smaller near the bunch tail. It is also confirmed that the shift of the coherent quadrupole tune is the same as the one of a coasting beam with the same line density.

Figure 8 shows the emittance evolution when the bare tune is chosen near the non-structure resonance. At the all three intensity levels defined above, emittance growth is observed and it becomes larger for the tail. The clear threshold intensity disappears that was observed in the coasting beam simulation. Qualitatively, there is no difference whether the coherent quadruople tune is below and above the resonance.

Near the structure resonance, the behavior is similar to the coasting beam. Namely, the beam core starts growing when the incoherent tune is below the resonance. It is also reasonable to say that the growth is actually caused by higher order coherent modes instead of the incoherent one.

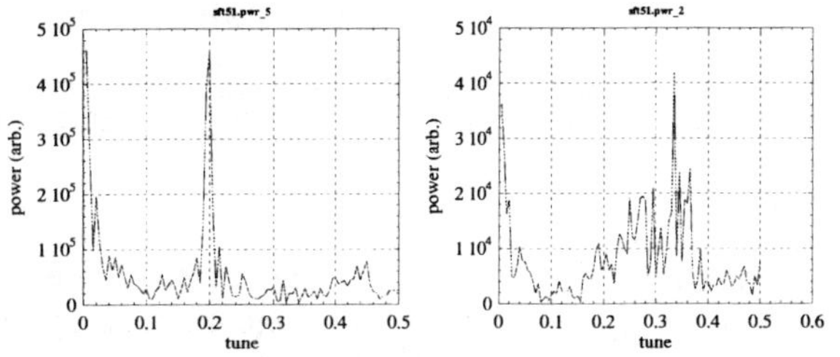

FIGURE 7. Spectra of coherent quadrupole tune at the center slice (left) and the tail one (right). Without space charge, the peak should be located at 0.44.

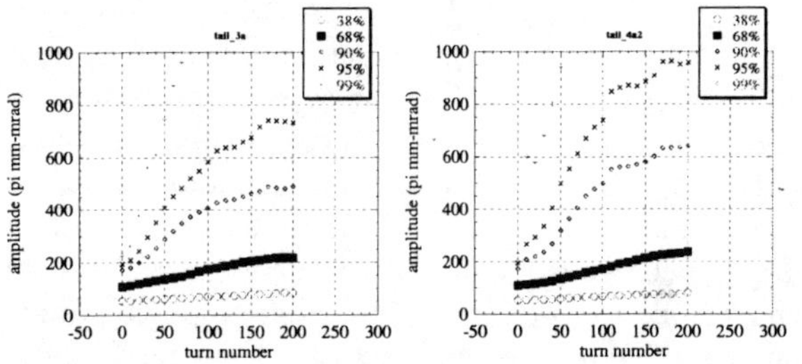

FIGURE 8. Emittance evolution of a bunched beam when the tune is above non-structure resonance. Only incoherent tune is below the resonance (left) or both coherent and incoherent tune is below the resonance (right).

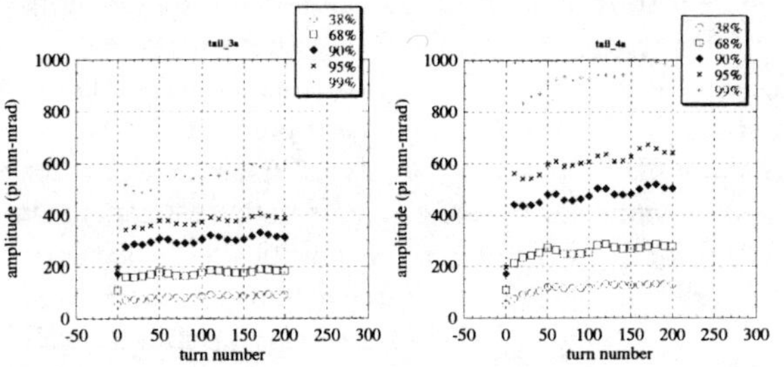

FIGURE 9. Emittance evolution of a bunched beam when the tune is above structure resonance. Only incoherent tune is below the resonance (left) or both coherent and incoherent tune is below the resonance (right).

ACKNOWLEDGMENTS

We would like to thank Prof. Y. Mori for his encouragement and continous support of the work.

REFERENCES

1. Machida, S., and Ikegami, M., "Simulation of Space Charge Effects in a Synchrotron" *in Workshop on Space Charge Physics in High Intensity Hadron Rings*, edited by A. U. Luccio and W. T. Weng, AIP Conference Proceedings 448, New York: American Institute of Physics,1998,pp.73-84.

Space Charge Resonances and Instabilities in Rings

I. Hofmann, G. Franchetti* and A. Fedotov†

*GSI, 64291 Darmstadt, Germany
†BNL, Upton NY 11973, USA

Abstract. We show that the influence of space charge on resonances of coasting beams in circular machines can be discussed in terms of a coherent frequency shift leading to a "coherent advantage", and of space charge induced resonant instabilities absent in the zero space charge limit. These mechanisms are exemplified for the SNS ring as well as the SIS18 of GSI.

INTRODUCTION

While resonances in circular machines driven by systematic or imperfection lattice terms have been widely explored since the early times of accelerators, the role of space charge has found comparable attention only in recent years [1, 2, 3, 4, 5, 6, 7]. The most obvious effect of space charge is the incoherent tune shift and spread, which leads to an extended foot print of single particle tunes in the tune diagram. Less obvious is the observation that the response on a resonance may be modified by the coherent motion of all or a large fraction of particles. This leads to an additional oscillating force, which must be added to the external forces and may cause a coherent shift of the resonance condition; furthermore a different response is obtained, depending on whether the resonance is crossed from above or below, which was first studied in Ref. [1]. The main point of this paper is to show that a systematic study of space charge effects needs to consider the existence of both, resonances *and* instabilities. Such structure instabilities have been studied earlier in the context of linear transport systems [8], but these findings can be applied also to circular accelerators. Our model is that of a coasting beam or, equivalently, a constant line current beam in an ideal barrier bucket. In usual bunched beams in an rf bucket the combined effect of synchrotron motion and line current variation leads to additional effects due to resonance crossing discussed elsewhere in this conference [9].

COHERENT SHIFT OF RESONANCE

The analytical basis for calculating coherent frequency shifts is the calculation of eigenfrequencies for charge density oscillations with arbitrary tune and emittance ratios, hence fully anisotropic beams, by using the dispersion relations derived in Ref. [3] for KV beams and in smooth approximation. These eigenmodes can be excited resonantly in a circular accelerator by appropriate external field perturbations (for round isotropic beams see also Ref. [5]); furthermore, they can be excited by space charge itself, or a combination of both. Hence, the coherent space charge contribution leads to a shifted condition of resonance of the *coherent* frequency ω with a field perturbation at harmonic n according to

$$\omega = m\nu_x + l\nu_y + \Delta\omega = n. \qquad (1)$$

Here we express the "mode tune" ω in units of the revolution frequency similar to single particle tunes, and introduce $\Delta\omega$ as coherent shift due to the resonant density oscillations. The importance of such a shift was first pointed out by Smith [10] for second order, and extended to any order by Sacherer [11]. Numerically, it was demonstrated in Ref. [1] – using particle-in-cell simulation – for an example of fourth order resonance crossing.

In order to more easily quantify the effect of the coherent shift, a coherent mode coefficient C_{mk} defined by the equation $n = m(\nu_0 - C_{mk}\Delta\nu)$, thus including the incoherent *and* coherent space charge effect, was proposed in Ref. [4] for isotropic beams and uncoupled modes, based on the dispersion equations of Ref. [3]. Here m is denoting an azimuthal, and k a radial mode number, while $\Delta\nu = \nu_0 - \nu$. The radial mode number reflects the order of the perturbed space charge potential; the azimuthal one gives the multiple of ν_0 for vanishing space charge. Note that for non-KV beams ν and $\Delta\nu$ must be understood as rms values. C_{mk} can thus be interpreted as coefficient of attenuation of the single particle space charge shift $\Delta\nu$ by the coherent motion, provided that $C < 1$. The intensity can thus be increased by the factor $1/C$ compared with the single particle based resonance condition, which expresses the "coherent advantage". For the symmetric breathing mode the result is $C = 1/2$, which reflects the strong density perturbation by the cross sectional breathing. This allows increasing $\Delta\nu$ by a factor 2 compared with the corresponding single particle resonance condition $n = 2\nu$; for the quadrupolar mode the coefficient is only 3/4, hence an allowed intensity increase by the factor 4/3. The general case for split horizontal and vertical tunes (also unequal emittances) leads to a coefficient C close to the arithmetic mean, e.g. 5/8, for both modes, hence a "coherent advantage" by the fac-

CP642, *High Intensity and High Brightness Hadron Beams: 20th ICFA Advanced Beam Dynamics Workshop on High Intensity and High Brightness Hadron Beams,* edited by W. Chou, Y. Mori, D. Neuffer, and J.-F. Ostiguy

tor 1.6. In fourth order we find that $C \approx 0.87$ is a good approximation for a broad variation of tune splitting and emittance ratios. This implies a weakening of the "coherent advantage" if fourth order resonances are expected to play a role. We note here that modes with $C > 1$ may appear, in principle, as solutions of the analytical KV-based dispersion relations. Following the discussion in Ref. [12] they should, however, be discarded as artifact of KV-distributions associated with "negative energy".

SECOND ORDER RESONANCES APPLIED TO SNS

We first discuss the appearance of an imperfection driven second order resonance for the fictitious un-split working point $(\nu_{0x}, \nu_{0y}) = (4.6, 4.6)$ driven by a harmonic $n = 9$ gradient error. The effect of the error resonance depends primarily on the tune and not the details of the focusing lattice, hence we can assume a constant focusing for simplicity. The expected resonance condition is

$$2\nu_x + \Delta\omega = 9. \quad (2)$$

An anti-symmetric error in x and y (as would be produced by a single quadrupole) then drives a quadrupolar mode resonance, whereas a symmetric error is needed to drive the breathing mode resonance. Note that for sufficiently split tunes a single quadrupole error will drive both modes. We assume equal emittances in x and y, and solve the envelope equations with a symmetric relative gradient error Fourier harmonic of 0.001 at $n = 9$ as well as a small initial envelope mismatch of 2%. If the single particle tune is exactly on resonance there is only a small beating of the envelope due to proximity of the coherent resonance (Fig. 1 top). The resonance of the in-phase mode should be expected at $\nu_{x,y} = 4.4$ based on the calculated value for C. At this value, in fact, the envelope response is significant though not yet maximum (Fig. 1 bottom). The complete picture is shown in Fig. 2 for both, symmetric and anti-symmetric gradient errors. Due to the nonlinear nature of the envelope response the maximum is reached for even slightly stronger tune depression, with a sharply dropping response beyond. As expected, the zero-current case has infinite response at $\nu_{x,y} = 4.5$. For the actual simulation the particle-in-cell ORBIT code was applied to the SNS lattice with a working point $(\nu_{0x}, \nu_{0y}) = (6.45, 4.6)$. The response to an error in a single quadrupole leads to a similar behavior of a shifted curve and steeply dropping maximum envelope slightly beyond the envelope resonance condition. Results for the out-of-phase mode excited by crossing the resonance condition in the y-plane are shown in Fig. 3, where intensity is plotted in terms of a normalized tune shift defined as actual space charge tune shift relative to the distance of the bare tune from the half-integer, i.e. $\Delta\nu_{sc}/\Delta\nu_{inc}$. The solid vertical line indicates the coherent resonance condition, where the normalized tune shift equals $1/C$, the long-dashed the r.m.s. incoherent resonance condition, and the short-dashed line the incoherent limit due to the small-amplitude particles in a waterbag distribution. As with the simple envelope model above, we find again that the resonance happens at significantly

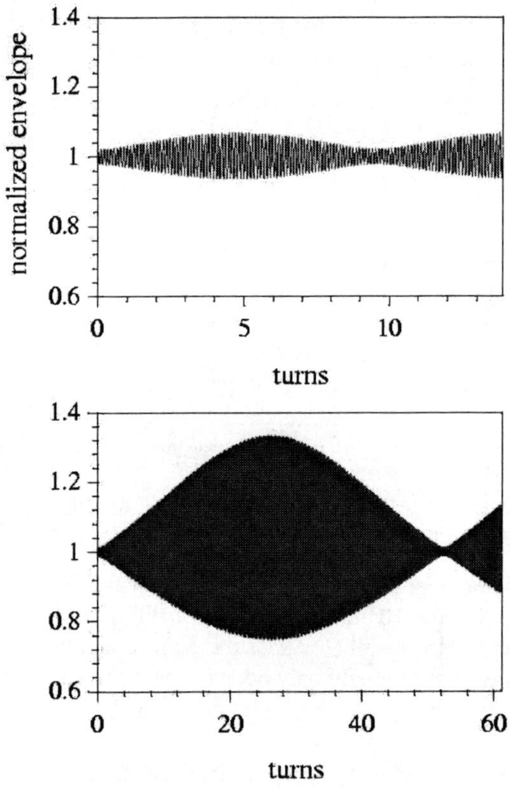

FIGURE 1. In-phase mode envelope evolution at the single particle resonance condition $\nu_{x,y}$=4.5 (top), and at the point of coherent resonance, $\nu_{x,y}$=4.4 (bottom) for $\nu_{0,x,y}$=4.6.

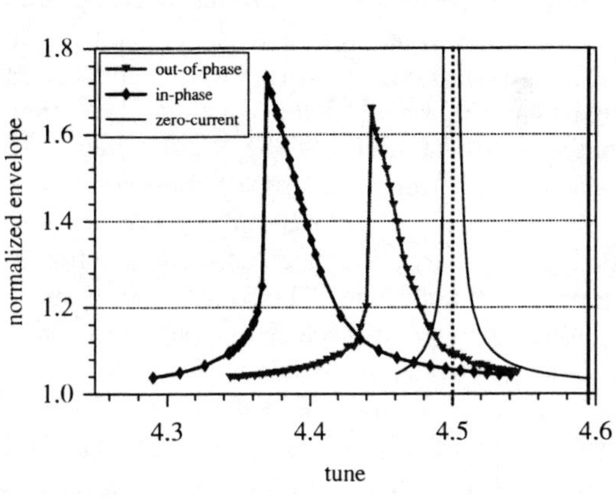

FIGURE 2. Maximum envelope response (normalized to initial envelopes) for both eigenmodes and bare tune $\nu_{0,x,y}$=4.6 as function of depressed incoherent tune.

higher intensity than for single particles in good agreement with the theoretical expectations. In this context the

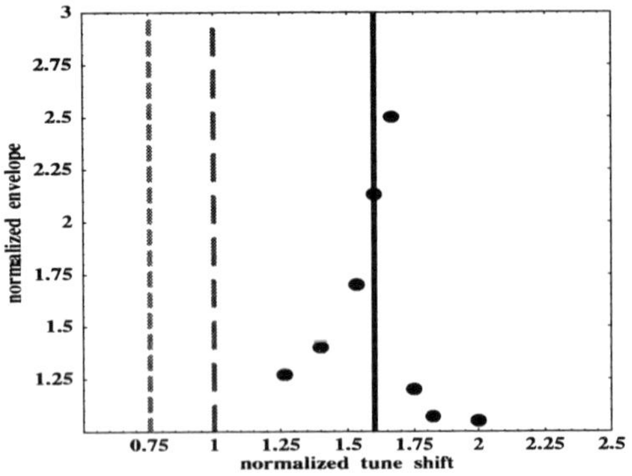

FIGURE 3. Maximum envelope response for SNS lattice as function of normalized tune shift and for (ν_{0x},ν_{0y})=(6.45,4.6).

question arises, if fourth order resonances might play a role over the 10^3 turns required for filling the ring. For the typical value of $C = 0.87$ the corresponding normalized tune shift would only be 1.15, hence nearly canceling the second order "coherent advantage".

RESONANCE AND INSTABILITY FOR THE SIS18

We next consider the working diagram of the SIS18 lattice with 12 super periods and triplet focusing. In Fig. 4 we indicate the usual fourth order systematic single-particle resonance line $4\nu_y = 12$, its coherent counterpart given by $4\nu_y + \Delta\omega = 12$ (dotted), which nearly coincides with the lower harmonic of the fourth order coherent resonance $2\nu_y + \Delta\omega = 12/2$; furthermore the coherent second order resonance condition $2\nu_y + \Delta\omega = 6$, which could be excited by a 6-th harmonic imperfection gradient error (not studied here). The coherent shifts are calculated for an incoherent space charge tune shift $\Delta\nu_y = 0.4$ and found practically independent of ν_{0x}.

In Table 1 we summarize analytically calculated coherent mode coefficients relevant for the resonances in Fig. 4. They are practically independent of both tunes provided that the incoherent shift is not comparable with or larger than the tune splitting between x and y, in which case corrections are needed.

TABLE 1. Some coherent mode coefficients C for different emittance ratios.

$\varepsilon_x/\varepsilon_y$	2nd	4th / resonance	4th / instability
4	0.66	0.88	0.87
1	0.61	0.85	0.89

For the simulation study we have employed the MICROMAP 2D particle-in-cell code employing $5x10^4$ simulation particles on a 128x128 rectangular grid with

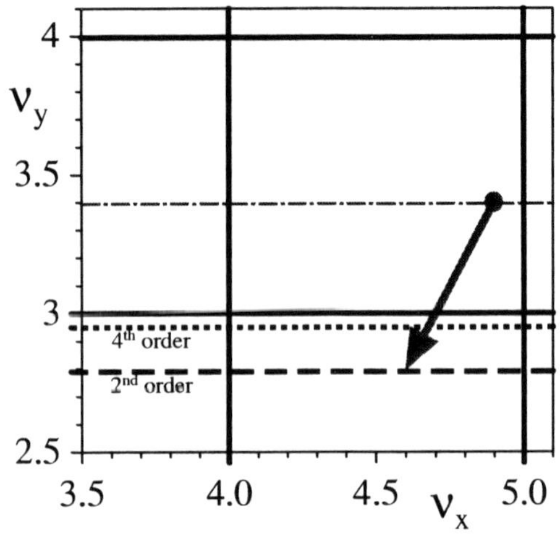

FIGURE 4. Incoherent and coherent shifted resonance lines of SIS18 for fixed ν_{0y}=3.4 and variable ν_{0x}.

conducting boundary conditions [13]. We have used rms matched waterbag distributions generated by filling a 4D hyper-ellipsoid uniformly and deforming it according to the rms matching. Carrying out the self-consistent simulation with $\nu_{0x} = 4.9$ (which is relatively arbitrary in this context) and $\nu_{0y} = 3.4$ we find in Fig. 5 that there is a significant emittance growth in y in the region $2.7 < \nu_y < 2.95$. In fact, we expect two distinct mechanisms to play a role and cause this broad a stopband. Firstly, the structure envelope instability described by the equation $2\nu_y + \Delta\omega = 12/2$ is expected. It is characterized by a phase advance of 180^0 of the underlying envelope mode per focusing period, hence a 1:2 (half-integer) relationship between mode and lattice periodicity; secondly we expect a structural fourth order resonance or instability. Such structure instabilities for sec-

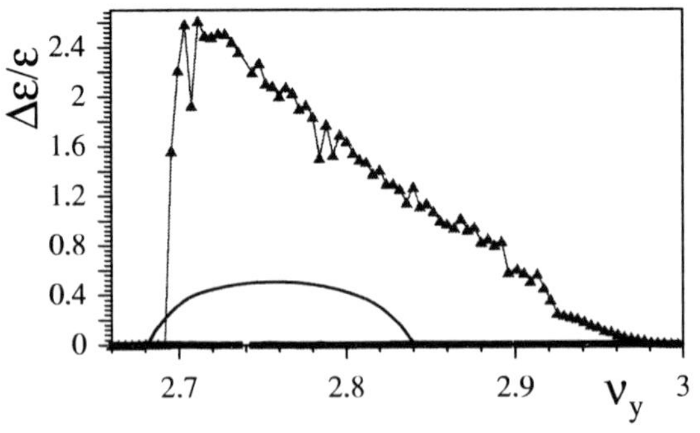

FIGURE 5. Saturated r.m.s. emittance effect after 200 turns in an ideal linear lattice with (ν_{0x},ν_{0y})=(4.9,3.4) as function of ν_y ($\varepsilon_x/\varepsilon_y$=4). Also shown is 10x($\lambda$-1) of the theoretical envelope instability growth factor λ.

ond and higher order have been systematically explored for periodic axi-symmetric transport systems in Ref. [8]. These findings apply equally to rings with different focusing in x and y with appropriately modified resonance conditions. Thus, it is sufficient for an instability of the envelopes to have the zero current phase advance per period sufficiently close to, but above 90^0 in at least one of the directions. In order to confirm the appearance of an envelope instability we have used the envelope eigenvalue solver KVXYG [14] and found that the present linear lattice with a zero-current phase advance per cell $\sigma_{0y} = 110^0$ is subject to envelope instability in the region $2.68 < \nu_y < 2.84$, with an associated envelope growth per cell by a factor λ . This permits us to associate the first part of the simulation stop-band with this mode, while the rest of the simulation stop-band can be only due to the structural fourth order resonance or instability as predicted analytically in Fig. 4 and Table 1. To verify the fourth order nature in this part of the stop-band we show in Fig. 6 the associated structure in phase space for $\nu_y = 2.86$ at an early and a more advanced stage. Note

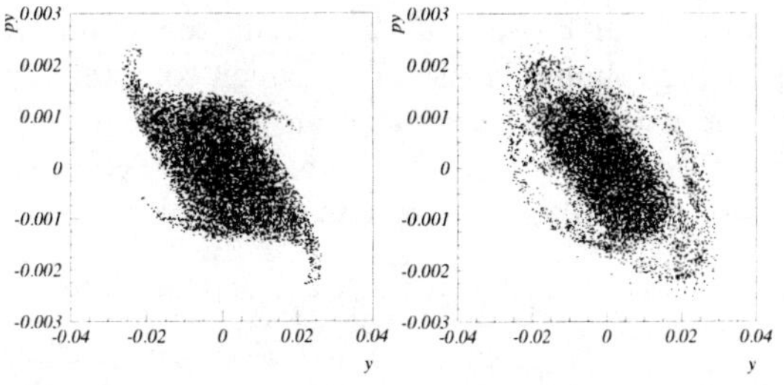

FIGURE 6. Phase space projections in y for fourth order structural response and ν_y=2.86 after turn 2 and 30.

that the location of the coherent fourth order resonance was calculated as $\nu_y \approx 2.95$ in Fig. 4, hence close to the right edge of the simulation stop-band . The asymmetry of the latter is a general nonlinear feature of resonances with significant space charge already identified in the nonlinear envelope simulation of Fig. 1. It is a combined effect of the tune spread existent in a waterbag simulation as well as the space charge detuning effect associated with emittance growth of the fraction of particles on resonance. There is still an effective "coherent advantage" of about 50% compared with the single-particle resonance picture, where one expects the small amplitude particles of a waterbag to be on resonance for $\nu_y < 3.1$.

In our case of strongly split tunes we find that the instability has almost no effect in the x-direction. In order to distinguish between resonances and instabilities we point out that in case of a resonance the driving term is assumed to be given by the lattice or by the space charge of the matched beam; for instabilities the driving term initially exists only on the noise level, but rises – under resonant conditions – exponentially in time. A characteristic feature of such instabilities is their complete absence in the zero-current case. In contrast, the second order resonance $2\nu_y + \Delta\omega = n$, which could be driven by a $n = 6$ imperfection gradient term would not be distinguishable from the envelope instability in our example, but it would be observable in the zero current limit, where the envelope instability vanishes. A more complete study of second order resonances and the envelope instability in rings (including a weak imperfection driven envelope instability) is found in Ref. [7]. For the fourth order mode above we postulate the simultaneous presence of resonance driven by the structural fourth order term in the waterbag space charge potential, and of structural instability – which may exist near $\sigma_{0y} = 90^0$ following Ref. [8] – similar to the self-consistent "Montague resonance" of the next section.

THE "MONTAGUE RESONANT INSTABILITY"

An important case of coupling resonance driven by space charge only is the well-known "Montague resonance", a fourth order difference resonance first studied in Ref. [15], It can be seen in a simplified way as a resonance driven by the zero-th harmonic of a quartic term in the space charge potential of a non-uniform beam. In machines with noticeably larger horizontal than vertical emittance it may lead to an exchange of emittances and possibly loss due to vertical acceptance limitation. It has been found in a number of synchrotrons and usually avoided by choosing a large enough tune split. For a recent detailed experimental study see Ref. [16].

The original analysis by Montague was based on a single particle approach using a "frozen-in" space charge potential defined by the initial Gaussian distribution, which neglects the effect of the induced time-varying collective space charge force. It is, however, adequately described only by a coherent resonance condition of the type $2\nu_x - 2\nu_y + \Delta\omega = 0$. The self-consistent simulation explored in detail in Ref. [17] shows, however, that this case is in reality a combination of instability and resonance: it develops as a pure instability for a KV distribution, which has no initial driving nonlinearity as is shown in Fig. 7 (top) for $\nu_y/\nu_{0y} = 0.8$ and $\varepsilon_x/\varepsilon_y = 2$. In this simulation the focusing ratio was re-defined for each data point such as to obtain a scan over a range of tune ratios. Note that for equal emittances the KV distribution leads to a symmetric loop around $\nu_x/\nu_y = 1$ as was recently shown in a study related to the SNS [18]. The initial growth for the KV-case is found to be exponential, starting from noise level (see Ref. [17]).

An initial waterbag distribution, instead, has all features of a space charge driven resonance due to the presence of a finite nonlinearity from the beginning (bottom of Fig. 7). The instability is, however, still superimposed on the resonance phenomenon. We conclude this by comparing it with a case, where the initial waterbag space charge distribution is "frozen-in": the maximum emittance exchange reaches only half the amount shown in Fig. 7 due to absence of a coherent response. Effective exchange occurs for specific tune ratios shifted

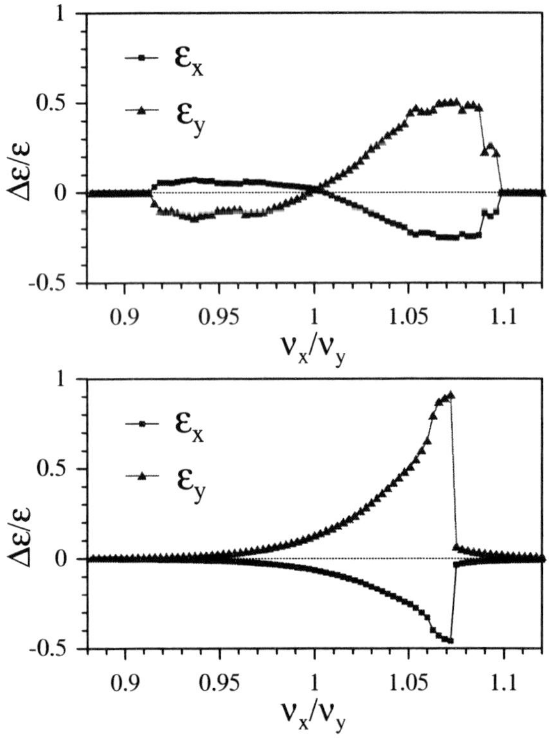

FIGURE 7. Saturated emittance exchange for KV (top) and waterbag (bottom) initial distributions, with ν_y/ν_{0y})=0.8, $\varepsilon_x/\varepsilon_y$=2, as function of ν_x/ν_y.

from the single particle condition $2\nu_x - 2\nu_y = 0$ due to space charge. It can be shown [17] that maximum growth rates as well as the width of the stop-band are roughly proportional to $\Delta\nu$, while the detailed shape of exchange profiles and maximum emittance exchange are nearly independent of $\Delta\nu$. This enables us to construct a stop-band within the usual tune diagram of a circular machine, which is shown in Fig. 8 for weakly split working points (arbitrary integer parts), using the data of the waterbag simulation in Fig. 7. The asymmetry of the stop-band with respect to the single particle condition $2\nu_x - 2\nu_y = 0$ (dotted line in Fig. 8) is due to the coherent space charge effect and the fact that $\varepsilon_x > \varepsilon_y$. Note that the width of the stop-band for the waterbag case shrinks to zero for equal emittances; in this limit there is no energy to exchange. The strongest growth of ε_y is actually closer to $2\nu_{0x} - 2\nu_{0y} = 0$ than to $2\nu_x - 2\nu_y = 0$.

CONCLUSION

We have found good agreement of space charge induced shifts of resonances between self-consistent simulation

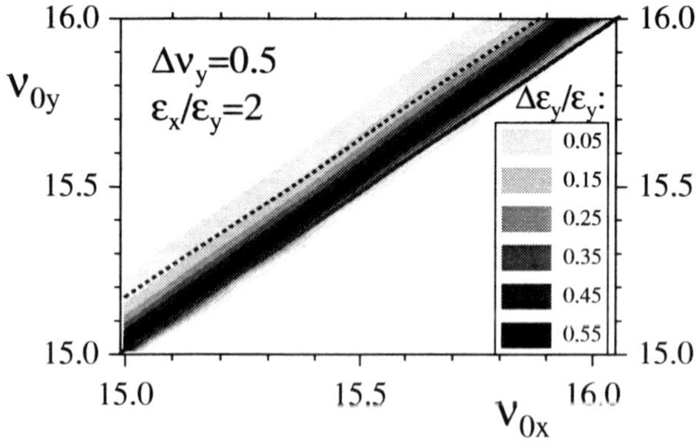

FIGURE 8. Stop-band of self-consistent Montague resonance: color scales indicate saturated growth of the originally smaller emittance in units of $\Delta\varepsilon_y/\varepsilon_y$.

and analytical theory results and confirmed the existence of resonant instabilities in second and fourth order. Both are shown to give rise to a significant "coherent advantage", which should be useful for bunch compression in SIS18 near $\nu_y = 3$. These studies also increase our confidence in the simulation codes and should be considered as a basis for further more systematic studies.

REFERENCES

1. I. Hofmann and K. Beckert, IEEE Trans Nucl. Sci. **32**, 2264 (1985).
2. S. Machida and M. Ikegami, AIP Conf. Proc. 448 (New York, 1998), p. 73
3. I. Hofmann, *Phys. Rev.* **E 57**, 4713 (1998).
4. R. Baartman, AIP Conf. Proc. 448 (New York, 1998), p. 56
5. M. Venturini and R.L. Gluckstern, *Phys. Rev. ST Accel. Beams* **3**, 034203 (2000).
6. S. Strasburg and R.C. Davidson, *Phys. Rev.* **E 61**, 5753 (2000).
7. A.V. Fedotov and I. Hofmann, *Phys. Rev. ST Accel. Beams* **5**, 024202-1 (2002).
8. I. Hofmann, L.J. Laslett, L. Smith and I. Haber, *Part. Accel.* **13**, 145 (1983).
9. G. Franchetti and I. Hofmann, these Proceedings
10. L. Smith , Proc. of the Conference of High Energy Accelerators, Dubna, 1963 (AEC, Oak Ridge, TN, 1965), p. 1235.
11. F. Sacherer, Ph.D. thesis, University of California (Lawrence Radiation Laboratory Report UCRL-18454, 1968, unpublished)
12. I. Hofmann, *Part. Acc.* **10**, 253 (1980)
13. G. Franchetti, I. Hofmann, and G. Turchetti *AIP Conf. Proc.* **448**, 233 (1998); ed. A.U.Luccio and W.T. Weng.
14. J. Struckmeier, private communication
15. B.W. Montague, CERN-Report No. 68-38, CERN, 1968.
16. I. Sakai et al., these Proceedings
17. I. Hofmann and O. Boine-Frankenheim, *Phys. Rev. Lett.* **87**, 034802 (2001).
18. A.V. Fedotov, J.A. Holmes and R.L. Gluckstern, *Phys. Rev. ST Accel. Beams* **4**, 084202 (2001)

Progress of Space Charge Calculation in the Code ORBIT [1]

A. U. Luccio and N. L. D'Imperio

Brookhaven National Laboratory, Upton,NY

Abstract. The code *ORBIT* [1] has been designed for PIC tracking of beam in a high intensity hadron synchrotron. In the code, space charge forces are continuously calculated and applied to the individual macro particles as momentum and energy kicks. Space charge solvers in the code, as developed at Brookhaven are described.

PIC TRACKING WITH SPACE CHARGE

A PIC [2] herd of macro particles is pushed through a lattice represented by maps [3] (external fields). At 'space charge nodes' in the lattice (internal) forces are calculated and applied to the macros as kicks. We use a Split Operator technique with independent treatment of maps and kicks. For space charge kicks the herd is binned on a grid according to $(x, y, c\Delta t)$ to find the charge density ρ, and according to $(p_x, p_y, \Delta p/p)$, to find the current density $\vec{j}$. Then, we solve the partial elliptic differential equations for Q(source point) $\rightarrow P$(field point)

$$\nabla^2\Phi(P) = -\frac{\rho(Q)}{\varepsilon_0} \quad ; \quad \nabla^2\vec{A}(P) = -\frac{\vec{j}(Q)}{\mu_0} \tag{1}$$

to find the scalar Φ and the vector potential $\vec{A}$.

For long bunches, the beam current is parallel to walls, electric repulsion and magnetic attraction partially compensate and we may approx. only solve for Φ (multiplied by a factor $1/\gamma^2$). In the following we will refer to this approximation.

Independent variable can be time t or the longitudinal coordinate s. Time is attractive because Eqs.(1) are solved with all macros at the same t. Space is convenient in periodic accelerators as particles cyclically pass through the same positions and maps are referred to s. Either way, we must apply relativistic transformations between space and time. *ORBIT* uses space as the independent variable.

3D TREATMENT OF SPACE CHARGE

Use a transverse grid terminated at wall boundary, and a longitudinal grid over the length of the beam bunch. For long bunches we can comfortably make s grid steps larger than transverse, since (i) the space charge distribution varies smoothly along the beam, and (ii) the motion along the beam is much slower than in the transverse.

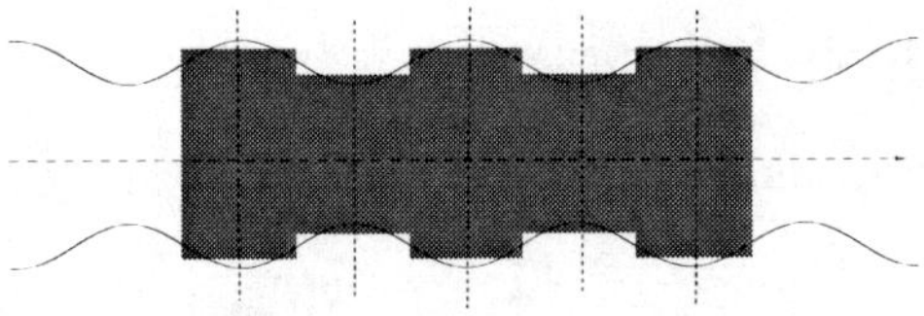

FIGURE 1. Slicing a beam. Wavy lines: envelope of the beam (β-wave). Dashed vertical lines: planes where to solve Poisson

For the longitudinal grid we cut the beam in slices, long enough that in each the average density, the transverse aspect ratio, and the surrounding wall configuration can be considered constant (Fig. 1).

$$\rho(x,y,z) = \rho_{\perp}(x.y)\,\rho_{\parallel}(z), \tag{2}$$

and solve the 2D transverse problem simultaneously in each slice, by parallel computing . As a slice length we use a fraction of the beam envelope wavelength. A slice is associated with the local wall configuration -A similar approach has been used by L.G.Vorobiev *et al* [4],[5]-

Each macro particle reach the same node at different time. To reconstruct the beam at a fixed time we project macros forward and backward using the maps for the bare lattice -more accurately, we should use the self-consistent lattice including extra focusing due to SC forces. It is an approximation, however consistent with the split operator technique-

POISSON SOLVERS

In its integral form, Poisson Eq.(1a) is

$$\phi(P) = \frac{1}{4\pi\varepsilon_0\gamma^2}\int\frac{\rho(Q)}{r}dQ, \tag{3}$$

with $r = |P-Q| + \varepsilon^2$, and ε a smoothing parameter to avoid poles. The equation can be solved by direct integration (BF), a slow but very transparent method good as a check, or by FFT, with the integral reduced to a convolution

While in the integral form the image on walls is part of the input, in differential Poisson solvers with boundary

[1] Work performed under the auspices of the U.S.Department of Energy

CP642, *High Intensity and High Brightness Hadron Beams: 20th ICFA Advanced Beam Dynamics Workshop on High Intensity and High Brightness Hadron Beams,* edited by W. Chou, Y. Mori, D. Neuffer, and J.-F. Ostiguy

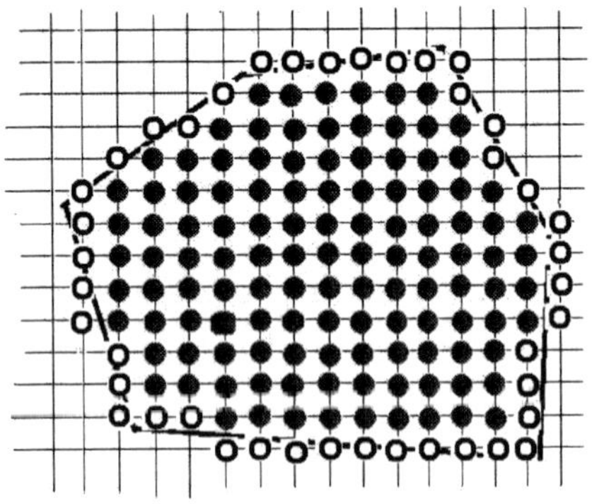

FIGURE 2. Solving with perfectly conducting walls

conditions the image is part of the answer

$$\nabla^2\Phi(P) = -\frac{\rho(P)}{\varepsilon_0}. \quad ; \quad \Phi(P_{wall}) = 0 \tag{4}$$

Two differential 2D solvers were implemented in *ORBIT*: (i) LU Decomposition plus matrix multiplication, and (ii) Successive Over Relaxation (SOR).

For (i), express the Laplacian operator ∇^2 in discrete form on a $M \times N$ grid that extends to wall

$$-4\pi\rho_{ij} = \mathscr{L}_{ij}^{kl}\Phi_{kl} \quad ; \quad \Phi(P) = -\frac{1}{4\pi}\mathscr{L}^{-1}\rho(Q) \tag{5}$$

$\mathscr{L}$ is an $M^2 \times N^2$ band-sparse matrix (δ is Kronecker's),whose inverse is unfortunately not sparse

$$\mathscr{L}_{ij}^{kl} = -4\delta_i^k\delta_j^l + \delta_{i+1}^k\delta_j^l + \delta_{i-1}^k\delta_j^l + \delta_i^k\delta_{j+1}^l + \delta_i^k\delta_{j-1}^l$$

and multiply the inverse by the $M \times N\ \vec{\rho}$.

For (ii), solve by iteration, starting with a guess. At step $k+1$ the discretized Poisson's is

$$\Phi_{i,j}^{k+1} = \frac{1}{4}\left(\Phi_{i-1,j}^k + \Phi_{i,j+1}^k + \Phi_{i+1,j}^k + \Phi_{i,j-1}^k - \rho_{i,j}\right).$$

Since the beam density evolves slowly from one space charge node to the next, iterative techniques show rapid convergence. Iterative procedures used were: Basic SOR (most efficient for small grids, $M,N < 128$), SOR with Chebychev acceleration (large grid), and Conjugate Gradient, that showed the most rapid convergence.

Walls are naturally introduced in differential P solvers. For a perfect conductive walls $\sum\rho_{beam} = \sum\rho_{image}$. As a schematic example, examine the explicit 2D calculation on the geometry of Fig.2. Walls are represented by n ∘, the interior by m •. The system of equations is exactly determined: $n+m$ known quantities, i.e. $\Phi = 0$ at the n ∘ and ρ at the m •; $m+n$ unknowns, i.e. m values of Φ to be calculated at the •'s and n values of ρ_{image} at the ∘'s.

Once we have Φ, the space charge force and the momentum kick on each macro are

$$\vec{F}(P) = \frac{e}{\gamma^2}\vec{\nabla}\phi \quad ; \quad \frac{\Delta\vec{p}}{p} = \frac{1}{p}\int\vec{F}\,dt \tag{6}$$

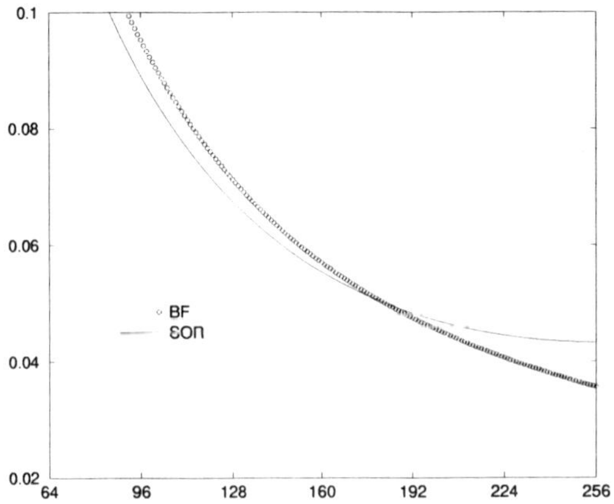

FIGURE 3. Comparison of field between 2-D BF and SOR,, with walls at bin 256

With $dt = \Delta t = L/\beta c$, the transverse momentum kick and the longitudinal energy kick are

$$\frac{\delta p_\perp}{p} = \wp\frac{\partial\phi}{\partial r}L_\perp \quad ; \quad \frac{\delta\Delta E}{E} = \beta^2\wp\frac{\partial\phi}{\partial z}L_\parallel \tag{7}$$

with the separation between successive transverse SC kicks $L_\perp$, the separation between longitudinal kicks $L_\parallel$, the perveance $\wp = \frac{4\pi\lambda q h r_0}{\Delta x\beta^2\gamma^3 m_0}$, the charge per unit length λ and the size of a square grid cell Δx.

COMPARISON OF 2D SOLVERS

We compared the SC field calculated with a BF integral and a SOR differential solver on a 256×256 grid, conductive walls, and a Gaussian random beam in the chamber center (Fig.3). The BF field goes to zero at large distance, while SOR ends at the walls with a finite value, where the image charge density is equal to the field and the field lines are perpendicular. The sum of the image charges equals the total of the real charges. In this case the images are uniformly distributed.

Fig.4 shows the image charge on the walls for an offset beam.

3D FORCES IN A LONG BEAM

Transverse kicks depend on the transverse aspect ratio of a slice. For the same $\wp$, the force is on the average larger where the value of the Twiss β function is smaller, and vice-versa. This is shown in Fig.5 and Fig.6 from a SOR simulation of a FODO lattice.

Fig. 7 shows the longitudinal kick calculated from the difference of potential between analogous (x,y) points in the median transverse plane of successive slices is compared with the traditional formula for a beam of radius a in a round pipe of radius b [6]. Z_0 is the impedance of

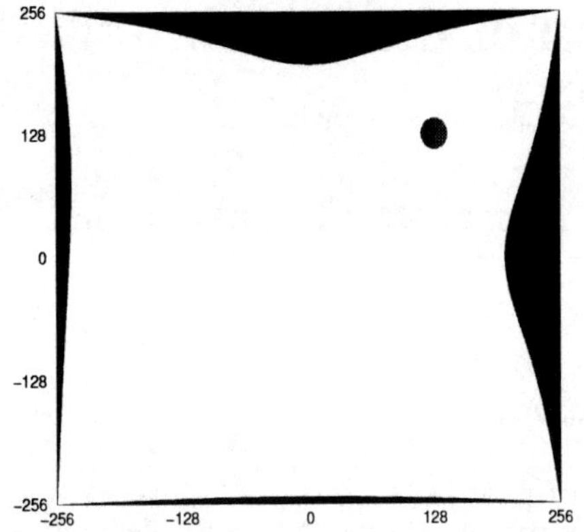

FIGURE 4. Beam offset in x and y in a square chamber. ρ and σ_{image}. Calculated by differential SOR

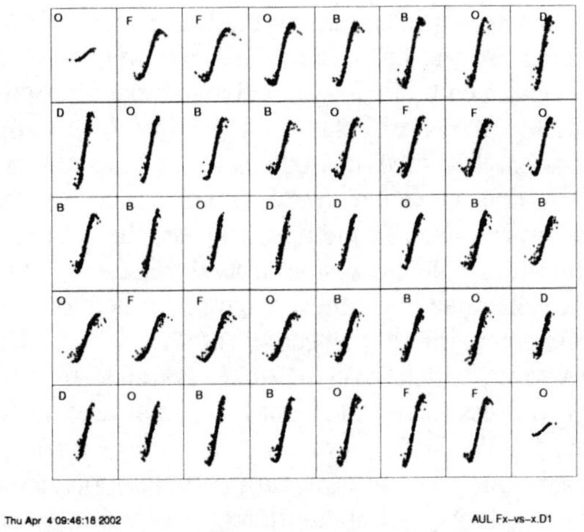

FIGURE 5. SC force F_x vs. x in each of a 40-slice beam whose central slice is in a defocusing lattice location

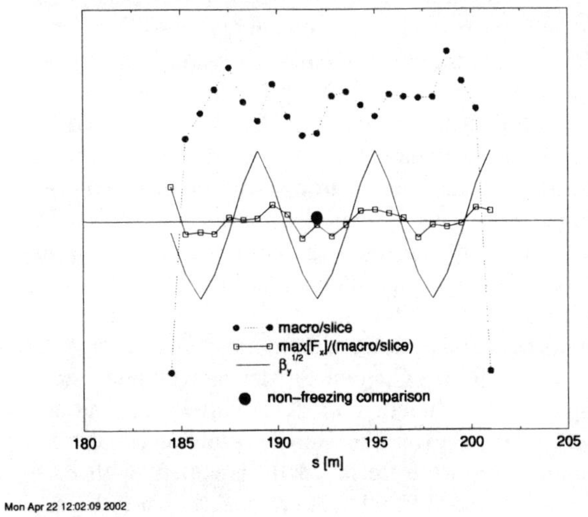

FIGURE 6. Maximum transverse kick in a 9-slice beam (AGS) compared with 2D calculation

free space and λ' the charge gradient along the beam.

$$(\Delta E)_{SC} \propto Z_0 \frac{\lambda'}{2\gamma^2}\left[1+2\ln\frac{b}{a}+f(r)\right]. \qquad (8)$$

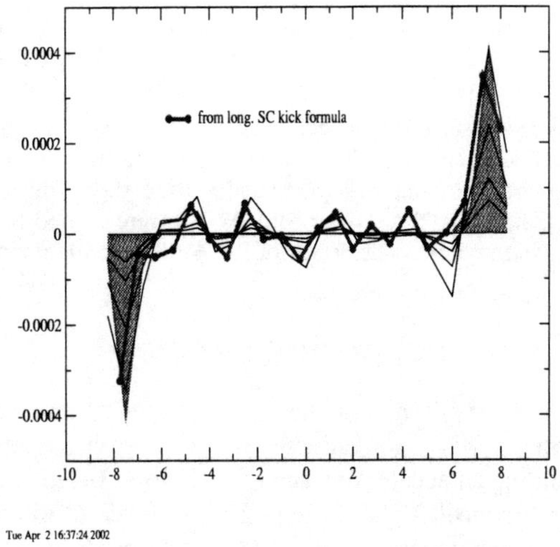

FIGURE 7. Longitudinal SC energy kick in a 9-slice beam (AGS). Each line: distribution of kick for various x,y. Thick line: simulation that uses Eq.(8)

REFERENCES

1. J.D.Galambos, J.A.Holmes, D.K.Olsen A.Luccio and J.Beebe-Wang: *Orbit User's Manual Vers. 1.10.* Technical Report SNS/ORNL/AP 011, Rev.1, 1999.
2. R.W.Hockney and J.W.Eastwood: *Computer Simulation Using Particles.* Adam Hilger, IOP Publishing, New York, 1988.
3. H.Grote and F.Ch.Iselin: *The MAD program, Vers.8.19.* Technical Report CERN/SL/90-13, Geneva, Switzerland, 1996.
4. L.G.Vorobiev and R.C.York: *Beam Dynamics Modeling By Sub-Three-Dimensional Particle-In-Cell Code.* In: *Proc. Particle Accelerator Conference 2001*, 2001. Paper RPAAH097.
5. L.G.Vorobiev and R.C.York: *Method Of Template Potentials To Find Space Charge Forces For High-Current Beam Dynamics Simulation.* In: *Proc. Particle Accelerator Conference 2001*, 2001. Paper RPAAH098.
6. A.W.Chao: *Physics of Collective Beam Instabilities in High Energy Accelerators.* Wiley, New York, 1993.

Simulation Studies for Inspection of the Benchmark Test with PATRASH

Y.Shimosaki*, S.Igarashi*, S.Machida*, M.Shirakata*, K.Takayama*, F.Noda#, K.Shigaki#

*KEK, Japan
#JAERI, Tokai, Ibaraki, Japan

Abstract. In order to delineate the halo-formation mechanisms in a typical FODO lattice, a 2-D simulation code PATRASH (PArticle TRAcking in a Synchrotron for Halo analysis) has been developed. The electric field originating from the space charge is calculated by the Hybrid Tree code method. Benchmark tests utilizing three simulation codes of ACCSIM, PATRASH and SIMPSONS were carried out. These results have been confirmed to be fairly in agreement with each other. The details of PATRASH simulation are discussed with some examples.

INTRODUCTION

One of the major issues in high-power hadron accelerators is activation of the environment surrounding an accelerator due to beam loss. Beam loss must be reduced to a sufficiently low level to allow hands-on-maintenance. In order to produce an acceptable design, a reliable simulation code is essential for estimating beam loss caused by space charge effects, nonlinear components of magnetic fields and machine imperfection. For the purpose to justify the simulation codes in our hands now, benchmark tests utilizing three simulation codes of ACCSIM, PATRASH and SIMPSONS were carried out by the 3GeV ring simulation group of the Joint Project. The machine parameters are based on the early version of the 3GeV Ring lattice for the Joint Project. The lattice is characterized by (1) three-fold symmetry, (2) high γ_t, (3) dispersion free in the straight sections, (4) beta function with low symmetry and (5) half integer tune split. The operating tune has been chosen in the region far from the major structure resonances, where the lattice functions are stable under perturbations and give a relatively small beam size. In this paper, the detail of PATRASH is described in the following subsections. The characteristics of each code are summarized in [1]. In the next section, the preliminary results for the inspection of the benchmark tests by using PATRASH are presented.

The electric field originating from the space charge is calculated by the Hybrid Tree method briefly explained below. Charge of the macro-particles, which are tracked in simulations, located in the beam core region is assigned as PIC-style charge, where the beam core is defined as a square where the length of one side is twice the rms beam size. Then, the transverse area including the above PIC-style particles on the grids and other macro particles outside the beam core is divided into small square-cells. In the Tree method, the electric field generated by PIC-style and macro particles placed in each cell is calculated in a way of multipole expansion. The Tree method consists of two processes, one of that is called "upward pass" and the other "downward pass". In the upward pass, the coefficients of the multi-pole expansion around the center of each cell are obtained; then, those are transformed to the new coefficients around the center of the square-cell of upper level, which involves the lower class four cells. Same process is repeated until the total area is filled with 24 cells. In the downward pass, the electric filed exerted on a particular tracking particle is obtained by using the above plural coefficients according to some rule.

BENCHMARK TEST OF CODES

As benchmark tests, 2D calculations were performed over 1000 turns at the injection energy of 400MeV and $\Delta P/P = 0$. Saturation of simulation results against the simulation parameters such as longitudinal step-size and number of macro particles had been checked by varying those parameters in the individual codes. Sextupole fields for chromaticity correction are included here. An initial distribution was prepared, which is generated by the anti-correlated painting method without space-charge forces and other non-linear magnetic fields.

In the benchmark test, the temporal evolution of 90%, 95% and 99% Courant-Snyder invariants were compared as a function of peak current I_{peak} in the bunched beam. Over the entire region of revolution, simulation results were in good agreement with each other within the 10% relative error as shown in Figs.1. We believe that ACCSIM, PATARSH and SIMPSONS

CP642, *High Intensity and High Brightness Hadron Beams: 20th ICFA Advanced Beam Dynamics Workshop on High Intensity and High Brightness Hadron Beams*, edited by W. Chou, Y. Mori, D. Neuffer, and J.-F. Ostiguy

are helpful tools to estimate emittance growth, halo formation and beam loss.

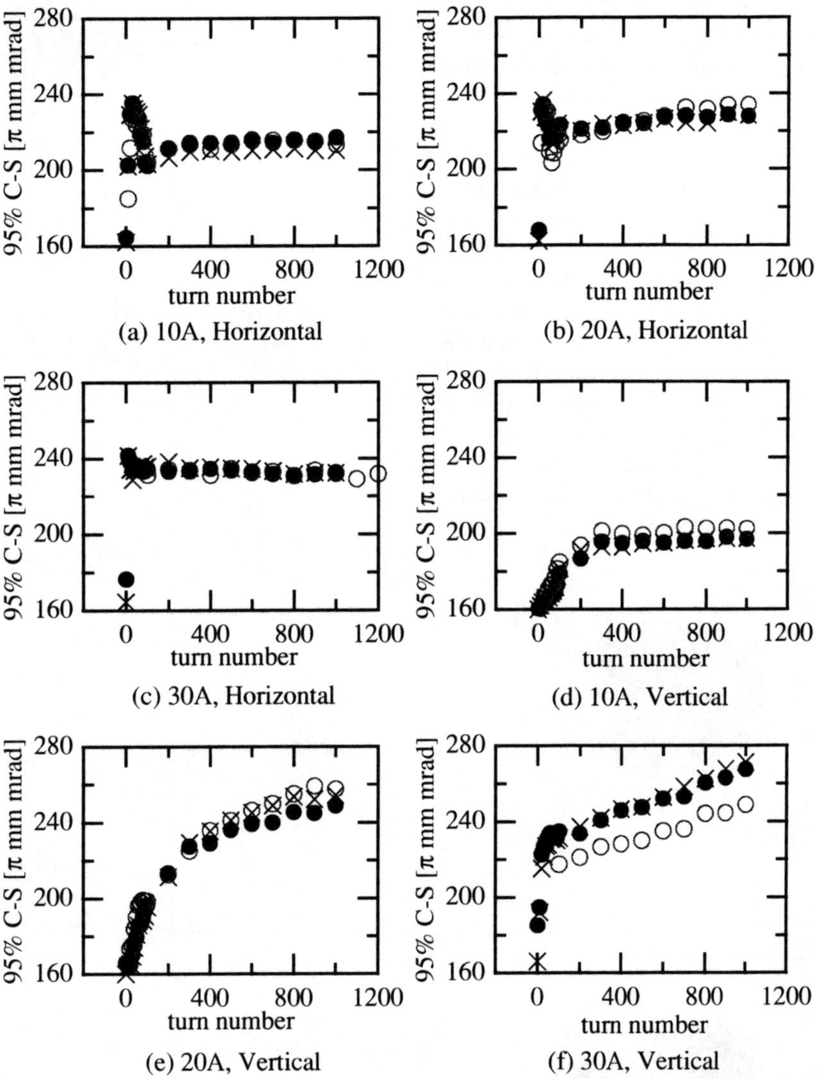

FIGURE 1. Time evolution of the (a)-(c) horizontal and (d)-(f) vertical Courant-Snyder invariant(C-S) as a function of I_{peak}. Black, blue and red represent ACCSIM, PATRASH and SIMPSONS, respectively.

EMITTANCE GROWTH AT EARLY STAGE JUST AFTER DEPOSIT OF THE INITIAL BEAM

As shown in Figs.1, the horizontal emittance rapidly blows up to 240π μmrad at the early stage (1-20turn) without the dependence on the peak current in contrast to the vertical emittance. Because this rapid emittance blow up independent of the beam intensity indicates that there is the possibility of the rapid halo generation independent of the beam intensity at the early stage, the inspection of the benchmark tests becomes important for making it clear that this phenomenon comes from the physics such as the resonance or the artificial error in these simulations. Therefore, the behavior of the particles at early stage has been carefully examined by PATRASH from a beam–dynamics point of view as a preliminary study for the benchmark test.

The beam distributions at the 4th turn are shown in Figs.2. It is clearly shown that the inner particles at the initial turn move to the outside in the vertical phase space. In the horizontal phase space, the structure can be also seen around the outer edge. These structures in the phase space were not seen at the initial turn. On the other hands, the distribution in the real space maintained the asymmetry in the vertical direction from initial turn.

For understanding what happened at the early stage, The Poincaré plot analysis was carried out. Its results are shown in Figs.3, where the positions of the test-particles were plotted turn by turn at the foil position until 20turn. The plots of the red and green particles in the horizontal direction clearly indicate the 2 islands. The black moves from the outside to the inside in horizontal and from the inside to the outside in vertical. The blue is in reverse of the black. The behaviors of the black and blue particles seem to be induced by some coupling resonances. For finding a trace of the coupling resonance, the histories of the each Courant-Snyder invariant are plotted as shown in Fig.4. The Courant-Snyder invariants of the black and blue particles clearly move along some straight lines. The canonical perturbation theory tells us that a combination of the horizontal and the vertical action variables become a constant if a coupling resonance is induced. We suspects that the emittance blow up at the early stage is induced by some coupling resonances. The specifications of the coupling resonances and the source of the resonances are now under analyzing.

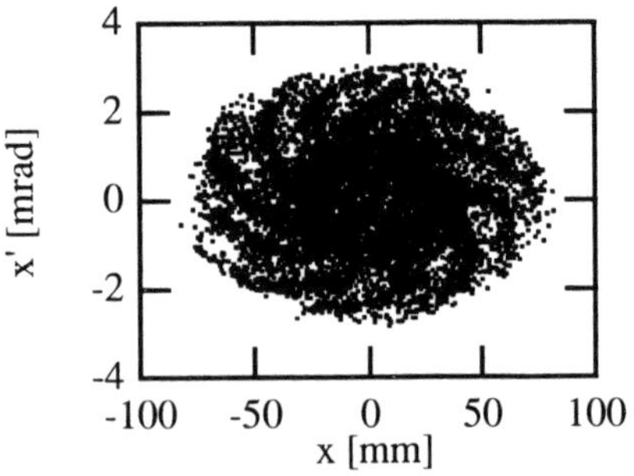

(a) horizontal

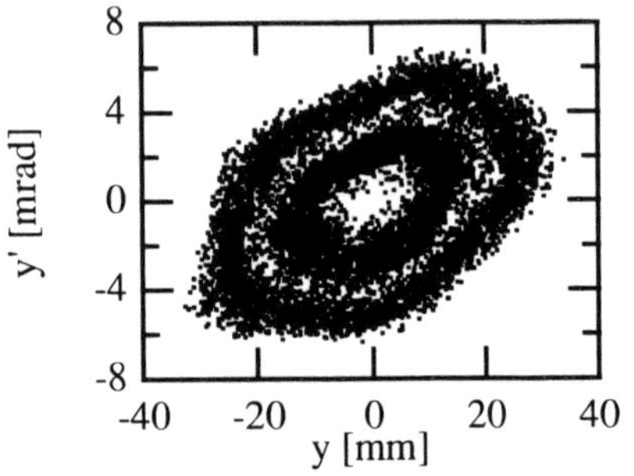

(b) vertical

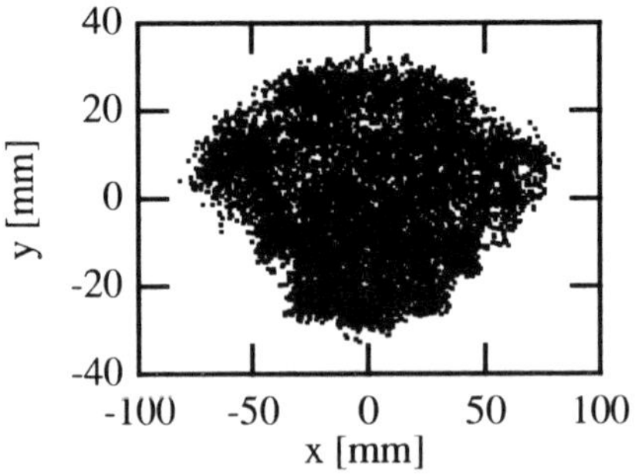

(c) y vs. x

FIGURE 2. Beam distributions (a), (b) in the phase and (c) in the real space.

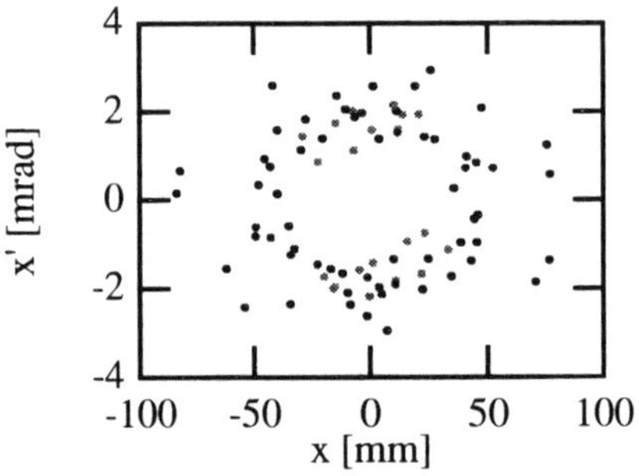

(a) Horizontal

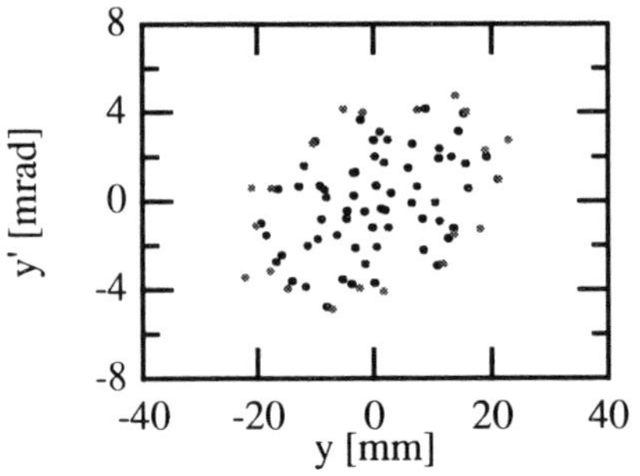

(b) Vertical

FIGURE 3. Poincaré plot of the characteristic 4 particles of the 18 test partices.

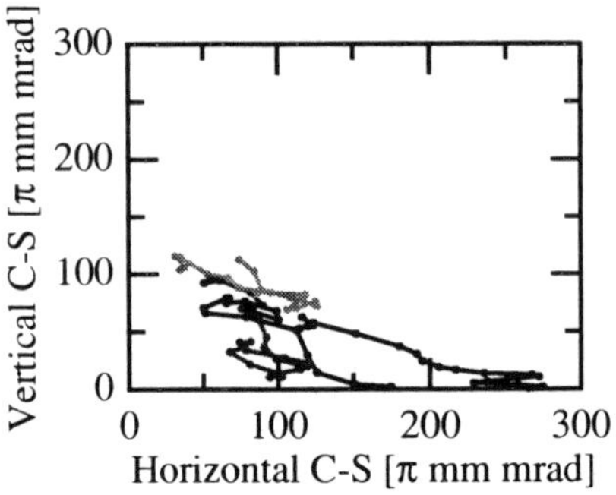

FIGURE 4. Temporal evolution of the Courant-Snyder invariant. The color corresponds to Figs.3.

SUMMARY

The results of PATRASH were fairly in agreement with ACCSIM and SIMPSONS. PATRASH can be said as one of the reliable codes in our hands.

The rapid emittance blow up independent of the beam intensity in the horizontal direction may be induced by the coupling resonances. The preliminary study is presented in this paper and a whole analysis will be given in the coming paper.

REFERENCES

1. http://hadron.kek.jp/member/onishi/tdr/index.html

Modeling of Space Charge Induced Resonance Loss

G. Franchetti and I. Hofmann

GSI, 64291 Darmstadt, Germany

Abstract. For high-current circular machines like the SIS100, which has been proposed as driver for a future radioactive beam facility at GSI, an exploratory study of space charge effects on resonances during long-term beam storage is indispensable. We discuss the effect of space charge on sextupole error resonances by a 4D model using an analytical expression for a stationary space charge. The effect of space charge induced resonance crossing in a bunch beam is approached by a simplified model.

INTRODUCTION

Beam loss control is an important issue for high-intensity heavy ion rings [1]. Long term bunched beam storage, typically 10^6 turns, and resolution of beam loss under the percent level are very demanding for PIC simulation. As pointed out in [2] numerical noise generated in PIC solvers is responsible for an artificial emittance growth. This effect may get controlled by increasing the number of macroparticles used in the simulations at expenses of larger CPU time. A study of this effect in 3D simulation was presented in [3]. Since small beam loss is of concern, a particle-core model provides a way to avoid numerical noise artifacts while retrieving important aspects of the dynamics during long term beam storage. If loss are found to be small and collective effects may be neglected a frozen core model may give reasonable predictions.

2D MODEL

The model is based on a coasting beam in a constant focusing lattice. We consider a 2D axisymmetric beam with frozen transverse Gaussian particle distribution of rms radius σ. The analytic space charge electric field [4] used in the particle equation of motion provides a particle-core model. In order to simplify the discussion we consider initially the particle dynamics only in the x-plane setting for any test particle $y = y' = 0$. The equation of motion becomes

$$x'' + \left(\frac{\nu_0}{R}\right)^2 x = K\frac{x}{r^2}(1 - e^{-\frac{r^2}{2\sigma^2}}) \quad (1)$$

where $()' = d/ds$, ν_0 is the ring bare tune, R is the ring radius, K is the perveance, $r = \sqrt{x^2+y^2}$. In this case $r = x$. Calling $C = 2\pi R$, and by using the coordinate transformation $x = \tilde{x}\sigma$, $s = \tilde{s}C$ we rescale Eq. 1 to the dimensionless equation

$$\ddot{\tilde{x}} + (2\pi\nu_0)^2\tilde{x} = 2(2\pi)^2(2\nu_0\Delta\nu - \Delta\nu^2)\frac{1}{\tilde{x}}(1 - e^{-\frac{\tilde{x}^2}{2}}) \quad (2)$$

where $\dot{()} = d/d\tilde{s}$. In Eq. 2 we use the tuneshift as a measure of the beam intensity, while the perveance is given by $K = 2(\sigma/C)^2(2\pi)^2(2\nu_0\Delta\nu - \Delta\nu^2)$. Equation 2 can be integrated through a micromap + nonlinear kick symplectic method. Dividing the ring in N integration steps the nonlinear transport map becomes

$$\begin{pmatrix} \tilde{x} \\ \dot{\tilde{x}} \end{pmatrix} = \begin{pmatrix} \mathscr{C} & \frac{\mathscr{S}}{2\pi\nu_0} \\ -2\pi\nu_0\mathscr{S} & \mathscr{C} \end{pmatrix} \begin{pmatrix} \tilde{x} \\ \dot{\tilde{x}} + f_n(\tilde{x}) \end{pmatrix} \quad (3)$$

with $\mathscr{S} = \sin(2\pi\nu_0/N)$, and $\mathscr{C} = \cos(2\pi\nu_0/N)$. The nonlinear kick f_n is $f_n(\tilde{x}) = \frac{2}{N}(2\pi)^2(2\nu_0\Delta\nu - \Delta\nu^2)(1 - e^{-\tilde{x}^2/2})/\tilde{x}$. The lattice nonlinearities can be added to $f_n(\tilde{x})$ at any integration step through the kick $A_m\tilde{x}^m$. Here A_m is the integrated strength of a m-th order nonlinear thin element in the frame $\{\tilde{s}, \tilde{x}\}$. The relation with its integrated strength K_m in the laboratory frame is $A_m = C\sigma^{m-1}K_m$.

RESONANCES AND SPACE CHARGE

With this model we study the effect of space charge on lattice nonlinearity induced resonances. Here the space charge acts as a perturbation leaving unchanged the main mechanisms of single particle resonances. We consider a ring with $\nu_0 = 13.36$, a beam with maximum tuneshift $\Delta\nu = 0.2$, and use 1060 integration steps to resolve the dynamics with space charge. With this choice the resonances $3\nu_0 = 40$, and $4\nu_0 = 53$ can be met if the depressed tune scans over $\nu = 13.16, ..., 13.36$. We show in Fig. 1a the nonlinear tune ν for particles with initial coordinates $\dot{\tilde{x}} = 0$. The calculation of ν is obtained with a FFT interpolated method (see in [5]) by using the particle coordinates in a fixed ring section over 2048 turns. As expected, at the center of the coasting beam $\nu = 13.16$, which moves asymptotically towards $\nu_0 = 13.36$ if the particle is taken very far from the beam center. Note that third and fourth order resonances are crossed at $x = 5.5\sigma$, and $x = 2\sigma$ respectively. We excite these two resonances by using 20 equal sextupolar kicks with $A_2 = 0.01$, and

CP642, *High Intensity and High Brightness Hadron Beams: 20th ICFA Advanced Beam Dynamics Workshop on High Intensity and High Brightness Hadron Beams,* edited by W. Chou, Y. Mori, D. Neuffer, and J.-F. Ostiguy

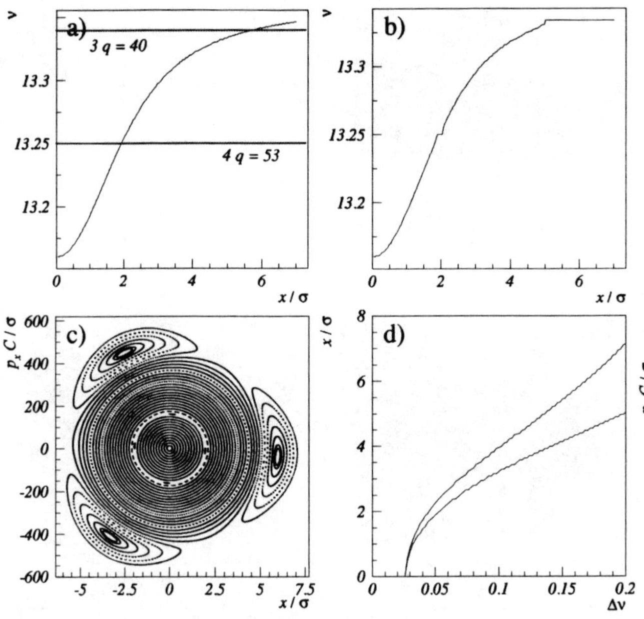

FIGURE 1. a) space charge induced nonlinear tunes, b) effect of resonances on nonlinear tunes, c) Poincare' section for $\Delta\nu = 0.2$, d) amplitude of 3rd order island as function of $\Delta\nu$.

53 equal octupolar kicks with $A_3 = 0.005$. In Figs. 1b,c we show the nonlinear tune and Poincaré section. The amplitude of the island may be measured by the size of the flat region in the nonlinear tune which crosses the resonance. The position of the fixed point is where $\nu_0 - \Delta\nu$ equals the harmonic of the nonlinearity driving term; the area of the islands seems to depend on the gradient of the nonlinear tune: the weaker the nonlinear tune gradient the bigger is the island. From the example in Fig. 1b it follows that large islands are obtained at 6σ. Setting properly the bare tune ν_0 and varying $\Delta\nu$ it is possible to create islands at any position. In Fig. 1d we plot the outer and inner separatrix of the third order resonance (at $\dot{\tilde{x}} = 0$) as function of the space charge intensity expressed in terms of $\Delta\nu$. The bare tune is still $\nu_0 = 13.36$. This picture shows consistently with the interpretation of Figs. 1a and b) that the third order islands move outward and increase their areas as the tuneshift increases. From this study we find that typically we may create big islands far beyond 3σ, whereas the islands within 3σ are much smaller.

Dynamics in a Bunch

The situation may be different for the single particle dynamics in a bunch. In fact synchrotron oscillations induce oscillations of the instantaneous current: when the particle is in the center of the bunch the local transverse current is maximum whereas on the bunch head or tail the current is minimum. The global effect is that a single particle experiences an oscillating instantaneous beam current with frequency much slower than the betatron oscillations. We consider first a linear current ramp which takes the tuneshift to $\Delta\nu = 0.2$. In first order the current ramp causes a migration of islands consistent with Fig. 1d. If the test particle has initial coordinates $x' = 0, x < 7.5\sigma$ the third order island eventually will cross the particle phase space orbit. The *speed* of the island at the crossing is crucial for the effect on a particle. In Fig 2 we explore this effect. In the left column we

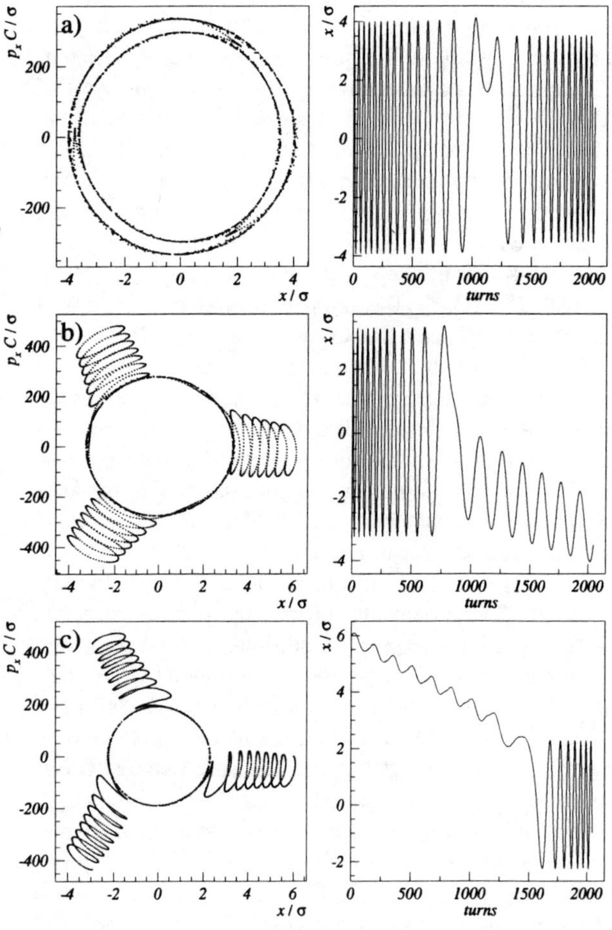

FIGURE 2. Phase space orbit (left) and time evolution of the particle amplitude plotted every third turn (right).

plot pictures of the single particle orbits. In the right column the time evolution of the $\tilde{x}$ coordinate. In the case a) (first two pictures at the top) a positive linear ramp brings the tuneshift to $\Delta\nu = 0.2$ in 2048 turns. The particle initial coordinates are $\tilde{x} = 4, \dot{\tilde{x}} = 0$. At the beginning the particle lies on an elliptic orbit, the frequency of the oscillations (picture on the right) decreases due to the nonlinear tune change for effect of the space charge ramp. The increasing space charge moves the third order islands outward. When the islands reach the particle orbit the third order resonance tries to capture the particle. This happen for half of an island revolution. Afterwards, the particle is closer the inner separatrix, and due to the outward motion of the island it crosses the separatrix again and gets unlocked from the resonance. The total effect is that the particle orbit gets kicked inward by an amplitude comparable with the island width, then its orbit amplitude remains unaffected. The orbit jump is visible after 1000 turns in the left picture. Simulations showed also the inverse effect for a decreasing current: in this case the orbit jump is outward. We conclude that the resonance crossing causes an orbit jump which has a direction opposite to the motion of the islands. The partial capture process depends strongly on the particle and island position in the phase space and on the speed of the island when approaching the particle. If a particle inside an island during half a revolution around the fixed point moves less than the island size, it may get trapped and follow the fixed point. In Figs. 2b,c we show two examples of particles trapped by the resonance. In b) the particle with $\tilde{x} = 3.3, \dot{\tilde{x}} = 0$ is trapped and moves outwards as the current increases linearly to reach $\Delta\nu = 0.2$. In c) a particle with $\tilde{x} = 6, \dot{\tilde{x}} = 0$ gets trapped and moves inward following the island. When the trapping condition gets broken, the orbit gets unlocked and follows the usual ellipse. From these simulations we conclude that a particle in a bunch may be subject to periodic resonance crossing: the islands oscillate back and forth periodically, kicking the particle orbit in and out. The single particle dynamics has then a nonlinear resonance induced stochastic regime. Over many synchrotron oscillations the particle may reach the trapping condition and get transported to the farest transverse position determined by the position of the island at the maximum intensity. Fig 3a shows the evolution of the $\tilde{x}$ coordinate of one particle with $\tilde{x} = 1.4, \dot{\tilde{x}} = 0$ over 10^6 turns. The space charge oscillations are simulated with $\Delta\nu = 0.2[1+\cos(2\pi\nu_s n)]/2$, where synchrotron tune is $\nu_s = 10^{-3}$. Note the diffusive-like process in the first $2\cdot 10^5$ turns while above it spikes bring the particle to $\sim 7\sigma$. Due to the reversible nature of the resonance trapping process the particle does not remain long at the maximum distance but falls within 3σ and later it gets trapped again and is brought to $6-7\sigma$. Note that $\sim 6\sigma$ is the position of the fixed point for $\Delta\nu = 0.2$ as shown in Fig 1d.

4D Tracking and Outlook

We repeat the same simulation for the 4D system (the condition $\tilde{y} = \dot{\tilde{y}} = 0$ is removed). The bare tunes are $\nu_{x0} = 13.91, \nu_{y0} = 13.58$, and the maximum tuneshift is $\Delta\nu_x = \Delta\nu_y = 0.3$. We excite now the harmonics 41 by using equal sextupolar errors with $A_2 = 0.01$. In Fig. 3b

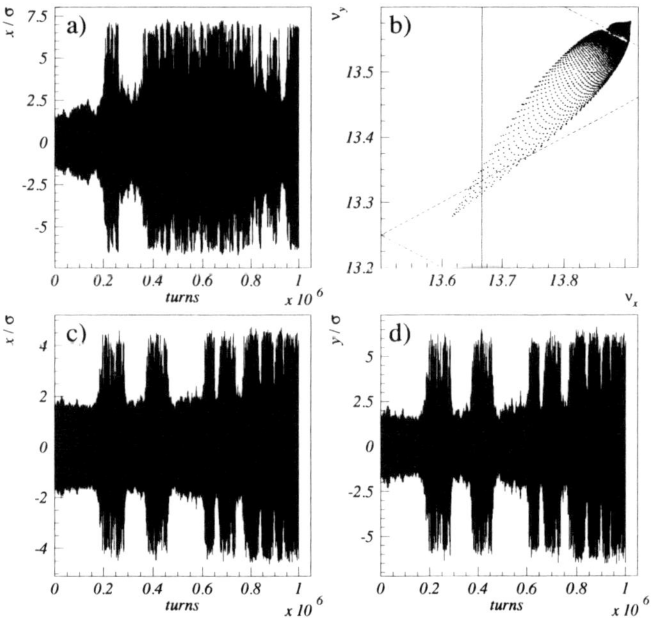

FIGURE 3. a) Particle evolution in a bunch (2D case), b) resonance lines and tune footprint for 4D case, c),d) particle evolution in the 4D case.

we plot the third order resonance lines and the tune footprint for test particles distributed over a larger area compared with the beam size. Note that the vicinity of the line $2\nu_y - \nu_x = 41$ is evacuated, which shows the existence of a stop band. The position of this resonance is chosen to be near the bare tunes. According to the 2D interpretation we expect again that the periodic resonance crossing will create a diffusion, and the resonance trapping would eventually bring the particle at the farest position. These effects are found in Figs. 3c,d for a particle with $\tilde{y} = \tilde{x} = 1.5, \dot{\tilde{x}} = \dot{\tilde{y}} = 0$. This study shows that in a bunched beam the most dangerous resonances appear to lie near the bare tune and not in the most densely populated part of the footprint corresponding to the bunch case. Consequently the single particle dynamics in a bunch is then very sensitive with respect to the particle synchrotron amplitude as it effects the tune oscillations. In order to quantify losses induced by the periodic resonance crossing a 3D self consistent space charge calculation along with a consistent modeling of synchrotron oscillation is needed.

REFERENCES

1. C. D. R. http://www.gsi.de/GSI Future/cdr/ .
2. J. Struckmeier, Phys. Rev. **E54**, 830 (1996).
3. J. Qiang and R. Ryne, these Proceedings.
4. M. B. Ottinger, T. Tajima, and K. Hiramoto, 1653 , 6th European Particle Accelerator Conference, Stockholm, Sweden, 22-26 Jun 1998.
5. R. Bartolini *et al.*, Part. Accel. **52**, 147 (1996).

SESSION F

REMOTE HANDLING

Remote Handling and Shielding at PSI

Eyke Wagner

Paul Scherrer Institut - PSI
CH-5232 Villigen / Switzerland

Abstract. The proton accelerator at PSI was originally designed to feed two target stations (with thin and thick targets respectively) with a 100 μA beam of 590 MeV protons for meson production. This report presents an overview of the evolution of remote handling and shielding concepts arising from the upgrade of the beam intensity (> 1 MW) and proposes guidelines concerning radioactive waste.

INTRODUCTION

The design intensity of 100 μA proton beam at the accelerator facility at PSI was achieved in 1980. After several years of stable operation, an upgrade program to reach 2 mA was started. An essential part of this work was the development of a new mechanical handling concept because of the increased activation of components (radiation hard components, remote handling design, local shielding, etc.); upgrading of the biological shielding was also needed. The actual job of replacing the proton channel beam-line gave practical experience in remote handling and radioactive waste conditioning.

DISMANTLING OF TARGET-E AND BEAMDUMP AREA

Before the new proton channel could be installed dismantling of the original (activated) system was necessary. This task was done in 1990 and can be summarised by the following statistics:

- Total mass of activated material: ≈ 500 tons
- Total amount of activity: ≈ 1.5 TBq
- Activation dose rates: < 4 Sv/h @ 1m
- Duration of dismantling: 6 months
- Collective dose (≈ 80 persons): ≈ 300 mSv

The most significant problem was contamination. This was caused mainly by water leaks leading to radiation-enhanced corrosion (the protective coating of component surfaces used was not effective).

To stop the spread of contamination a temporary bunker with airflow control and a remotely controlled power-manipulator unit was installed over the area. After any necessary de-contamination, the components were lifted out through an opening in the roof using the main crane of the experimental hall.

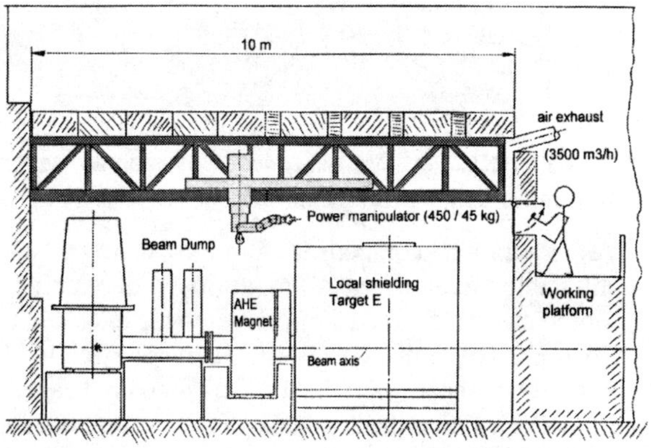

FIGURE 1: The improvised bunker with power manipulator, ventilation system and working platform.

Improvement Of The Target/Beam-stop Area For Higher Beam Intensity

The new system has been built using the following guidelines:

- High density (local) shielding around the beam line components,
- Radiation hard design of beam line components,
- Improved design of electrical, cooling-water and diagnostic connections for hands on operation at reasonable radiation levels,
- Vertical installation and removal for all highly activated units inside the proton-channel,
- Remotely controlled crane attachment with hook rotation,
- Self-alignment system for all the puzzle's pieces,
- Protection of mild steel surfaces against corrosion; galvanic coatings or 2-component painting for lower irradiation level
- Design of the components adapted to the packaging concept for radioactive waste
- Careful and complete documentation including material specifications for future declaration of radioactive waste

CP642, *High Intensity and High Brightness Hadron Beams: 20th ICFA Advanced Beam Dynamics Workshop on High Intensity and High Brightness Hadron Beams,* edited by W. Chou, Y. Mori, D. Neuffer, and J.-F. Ostiguy

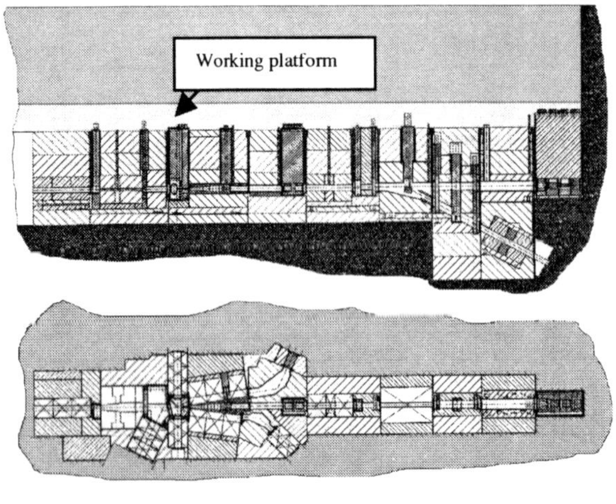

FIGURE 2: Vertical and horizontal sections through the new proton channel

An important component for this concept is the inflatable vacuum seal – an in-house development - which can be installed vertically into a gap of less than 30 mm. The unit is made of stainless steel to give radiation resistance. Tightening is based on differential pumping with controlled leakage at the two concentric inflated cushions on both sides of the seal.

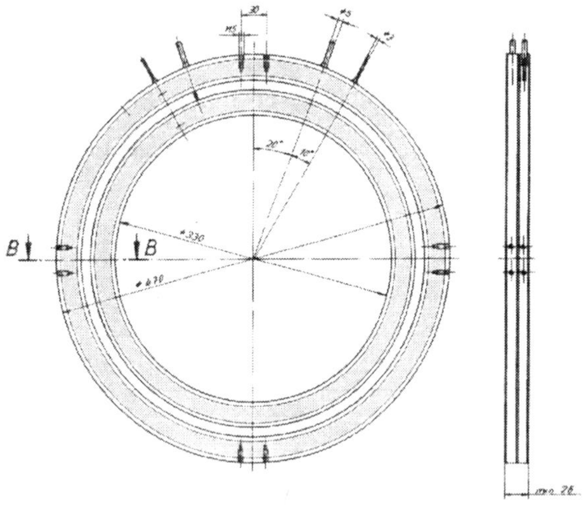

Detail for section B-B

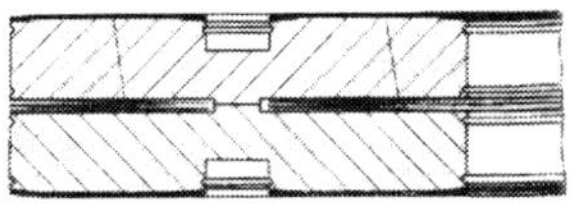

FIGURE 3: The PSI inflatable vacuum seal

Repairs Of Components Of The Beam Line:

After removal of the roof shielding of the proton-channel (4 to 4.5 m concrete beams) personnel access to a "Working platform" is allowed (average dose rate ≤ 1 mSv/h) and enables the disconnection of cooling lines, electrical feeding, etc by hand.

Highly activated components like targets, collimators or beam-stopper are transferred directly into a shielded transporting flask using contamination jackets with controlled airflow; all the other components can be transferred directly to the hot-cell with the crane.

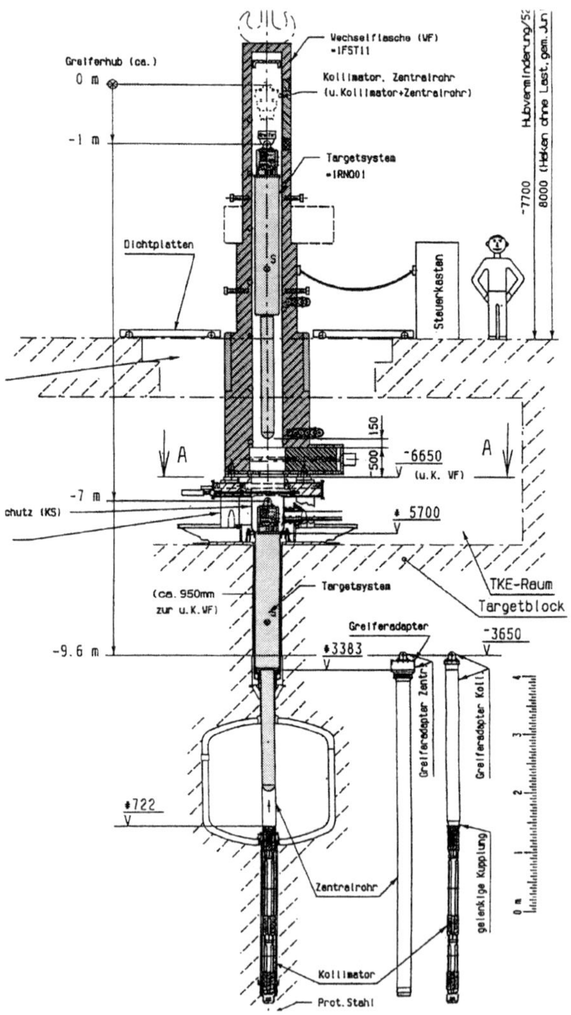

FIGURE 4: The handling-concept for SINQ targets. These are transferred into a shielded transporting flask containing a built-in crane.

For maintaining the highly activated components an adequate hot-cell under the same crane is provided. Corresponding to the size of the components the inside dimensions are 6 x 9 m and 6.5 m high. Beside the usual equipment of a hot-cell, a special working table is provided. This unit (2.6 x 1.6 x 0.5 m) is remote controlled, battery powered and can move loads up to 60 tons in all directions with high precision (it does not use wheels!). It can also be rotated in either direction to put components into an optimal working and viewing position in front of the lead-glass window.

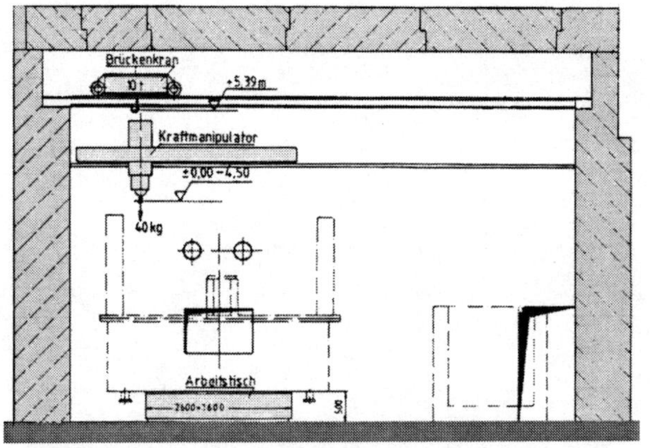

FIGURE 5: A vertical section through the hot-cell (ATEC) showing the power-manipulator, crane and remotely controlled working table for loads of up to 60 tons.

HANDLING OF RADIOACTIVE WASTE

In an ideal situation, a concept for the disposal of radioactive waste should be available before starting the design of an accelerator facility. When planning the upgrade of the target and beam-stop areas, no suitable packages for waste existed, so before starting we had to develop a conditioning concept for large components. This was done in the early 80-ies including obtaining a license for final disposal.

Corresponding to the size and the activity of the waste the following reinforced concrete containers were developed:

Type	Outer dimensions [m]	Wall-thickness [cm]	Maximum filled weight [kg]
KC-T12	1.5/1.5/2.0	12	20'000
KC-T30	1.5/1.5/2.0	30	20'000
GC-T24	4.0/2.0/2.0	24	60'000

The dimensions of the type KC are precise (± 1 mm) so that they can be used, for example, as shielding blocks – a sort of recycling so long as the facility is expanding.

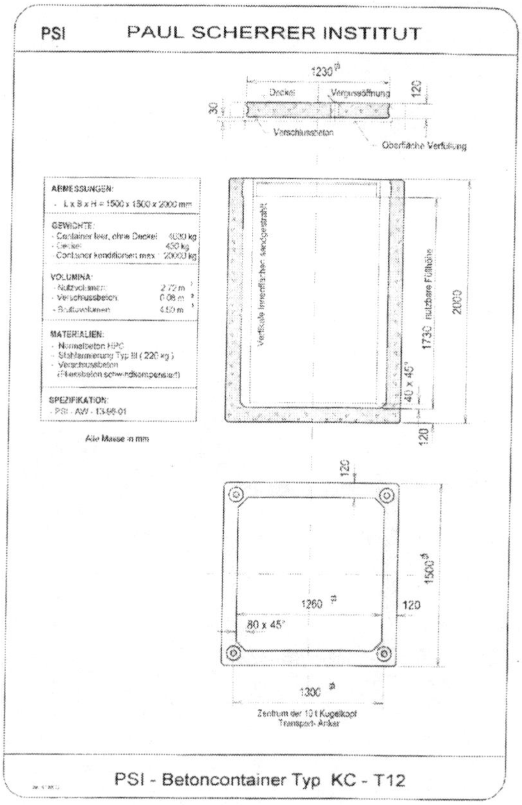

FIGURE 6: The most used concrete container for PSI radioactive waste. The components are potted into concrete with all voids filled. This makes them suitable for use as shielding blocks.

FIGURE 7: A 25 t cast-iron shielding block in process of being conditioned into a GC-T24 concrete container.

SESSION H

ION SOURCES

Summary Report

Horst Klein

Institut für angewandte Physik (IAP)
University of Frankfurt, Germany

In the ion source session, a total of 9 talks had been given, all of them were very interesting, presenting new results and generated a lot of lively discussions. It can be considered as a very successful session in this Workshop, putting together the scientists of the leading labs in ion source physics and technology all over the world.

It is remarkable, that most of the talks (7 out of 9) were devoted to the production of H^- ion beams, thus reflecting the increasing importance of such high current sources for many ongoing projects and proposed facilites like spallation neutron sources, accelerator for neutrino production, for high energy physics etc. All these accelerator-based facilities need intense H^- beams from the linacs for the non-Liouvillian injection by charge exchange into synchrotrons or accumulator rings. The requirements for the ion source become more and more stringent: beam currents up to 100 mA, high duty cycles up to 10 % or even 100 %, long pulse lengths (several msec), low noise and beam current fluctuations (1%), good beam quality, low transverse and longitudinal emittance. In addition: The lifetime, reliability, and the availability of the ion source must be very high. It is clear, that in most cases the use of cesium is inevitable. And all these call for more and intensify research work, since so far no H^- ion source worldwide can satisfy all the requirements.

The production of positive ions is much simpler than the prediction of negative ions, but even for them the requirements for many applications (e.g. transmutation, IFMIF, RIA, energy amplification) become difficult. So Richard Pardo, ANL showed that the ion source - it is the ECRIS of course, which is the proper and best choice - had no difficulty to produce the required beam current for the lighter ions (e.g. 600 pμA for mass 1, 29 pμA for mass 40). But for heavier ions, the ECR has to be improved (for example using higher frequencies), since for mass 209 only 5 pμA instead of 11 pμA have been achieved, and for mass 238 only 2 instead of 8 pμA have been obtained.

The ECR is unbeatable for heavy ions, but it can produce also protons in the range of 100 mA continuously, with higher reliability and long life time. This has been demonstrated by Robin Ferdinand, CEA Saclay: In long operations, the availability was measured up to 98 %! and currents up to 175 mA in pulsed mode have been achieved for deuterons (necessary for IFMIF). In addition to these remarkable experimental results, the Saclay group also develops nondestructive diagnostics, which is essential for the high intensity beams. Especially interesting is the use of the Doppler shift to get the beam profiles from the light produced by the interaction with the residual gas.

Now lets go to H^- sources. Alain Girard, CEA Grenoble, discussed the possibility to use the reliable ECR source for H^- -production also. In first tests in Saclay low current H^- beams could be extracted from a modified ECR, but the work has just started, and more optimization, especially in the magnetic field configurations have to be done.

Working horses for pulsed H^- beam production at low duty cycle are the magnetron sources. V. Dudnikov, Fermilab, gave a nice overview on the history and the experiences at Fermilab, and J. Alessi, BNL, presented the results achieved in Brookhaven. In both cases the magnetron sources are working very reliably, but the extrapolation to higher current and high duty cycle seems to be somewhat problematic; V. Dudnikov is more optimistic, recommending the Semiplanotron, whereas J. Alessi sees some difficulties regarding noise and stability.

B. Ellingboe, Dublin City University, comes from the plasma physics side. He showed among others an experiment where the influence of added noble gas on the H^- production is measured. Some surprising results need to be explained, and he pointed out, that the theory of the H^- sources in general has to be improved including the role of the cesium.

J. D. Sherman, LANL, gave an review of the Penning-H^- source. This type of source is very successfully working at ISIS, RAL, for many years. It is based on a Penning source operated in Los Alamos. A scaled Penning source produced later on in 1984 a 250 mA H^- beam, but at low duty cylce. J. Sherman is convinced, that using the applied scaling rule further the Penning source can achieve all the properties required by the new accelerator facilities including the high duty cycle.

Another powerful type of ion source is being used for

CP642, *High Intensity and High Brightness Hadron Beams: 20th ICFA Advanced Beam Dynamics Workshop on High Intensity and High Brightness Hadron Beams,* edited by W. Chou, Y. Mori, D. Neuffer, and J.-F. Ostiguy

the SNS, the volume source. Roderich Keller, LBNL, reported on the newest achievements. This volume source uses the rf (2 MHZ) for the plasma generation and is based on the developments of K.-N. Leung. The source now produces 50 mA with a duty factor of 6 % and normalized emittance between 0.2 and 0.3 πmm mrad rms. The reliability has substantially improved, especially by a new coating of the antenna. Moreover, the electrostatic LEBT system worked as expected, and the operation with the full RFQ could be done, resulting in very good transmission of more than 90 %. So the crew of R. Keller did an excellent job, achieving all the SNS requirements -congratulations.

J. W. Thomason, RAL, had a difficult task. He had to report on three different H^- sources at DESY, IAP Frankfurt and at Rutherford Laboratory and he did it in a fair and convincing way. In the volume source at DESY, the plasma is produced by rf also, but - in contrast to the Berkeley source - the antenna is now outside of the plasma volume, so the plasma can not damage the coating of the antenna. The source is operating at HERA, therefore the duty cycle is low (0.15 %, τ= 0.15 msec), but it produces 40 mA in pulse mode without cesium! The Frankfurt volume source is working with filaments. It fullfills nearly all the requirement of ESS: 120 mA at 6 % duty cycle, 50 Hz rep. rate, (= 1.2 msec. The beam quality is excellent. But it has not been operated with an accelerator, only at a test stand so far, and the question of realiability and life time have not been addressed yet. The ISIS source is a Penning source, which delivers the beam routinely to the synchrotron. The beam current is 35 mA with the nomalized emittance of 2 (mm mrad, pulse length 0.25 msec, rep. rate 50 Hz, and duty factor $\sim$ 1 %. The source is very reliable and can operate about 1 month between services.

The ion sources, and especially the H^- sources, are still somewhat a black magic. Therefore intense theoretical and experimental work has to be performed in different labs to achieve the new requirements. In Europe the NIS-network, supported by the EU, with its 8 partners will help to reach the goal. But also such a meeting as we have had in Fermilab is very helpful and intensifies the worldwide collaboration. Concerning the different types of ion sources, I think the most promising candidates for H(are the Penning ion source and the volume source.The ECR source may be a hope for the future.

High Brightness Negative Ion Sources with High Emission Current Density

Vadim Dudnikov

Fermi National Accelerator Laboratory & Brookhaven Technology Group, Inc.

Abstract Through the development of Charge Exchange Injection [1] and Surface Plasma Sources (SPS) with Cesium Catalysis [2,3,4,5] the possibility for the accumulation of a high brightness proton beam in circular accelerators was increased greatly, and now it is more than sufficient for all real applications. The combination of the SPS with charge-exchange injection improved large accelerators operation and has permitted beam accumulation up to space-charge limit and **overcome** this limit several times [6]. In this note an attention is concentrated on the seldom-discussed distinctive features of high brightness beam formation in noiseless regimes of negative ion source operation. Beam quality enhancement up to the level j/T>1 A/cm^{2}eV is possible by optimization of negative ion generation, extraction, and transportation in SPS with cesium catalysis. Advanced version of the SPS for accelerators will be described. Features of negative ion beam formation, transportation, space-charge neutralization- overneutralization, and instability damping will be considered.

INTRODUCTION

One practical result of development of high brightness negative ion source is accepting of the charge-exchange injection in circular accelerators for a routine operation. Now negative ion sources are "Sources of life" for gigantic accelerators complexes as FNAL, BNL, KEK, DESY,.. and an efficiency and reliability of these sources operation are determined a productivity of these big collaborations. Many results of the high energy physics were discovered with using of negative ion sources. Development of high brightness H^- sources was stimulated by first success of high current proton beam accumulation with using a charge-exchange injection [1]. Until 1971 a main attention was concentrated on the charge-exchange ion sources, because was no hope to extract from the plasma directly more than 5 mA of H^-. After first observation of a strong increasing of negative ion production following a small admixture of cesium into the gas discharges [2-4], and development of a first SPS for accelerators [5] has been start very fast development and adaptation of SPS in many USA laboratories, in Europe and in Japan, and International Symposiums for Production and Neutralization of Negative Ions and Beams has been established [7]. Now the Surface Plasma Method of negative ion production and SPS considered in many books (recently in book [8]). Good review of SPS for accelerators presented in reports of J. Peters [9-11]. Production of polarized negative ions by charge-exchange with a slow negative ion in SPS has been proposed and has been realized with a good success [12]. This development has permeate to use a charge exchange injection for accumulation of high intense beams of polarized ions in circular accelerators [13].

FEATURES OF SPS

Many versions of the SPS have been developed and optimized for different applications. Cesium admixture enhances negative ion formation in all types of discharges, but the most efficient negative ion production and highest beam quality is attained using an SPS that is optimized for each application. Some basic discharge configurations of SPS are presented in Fig.1. Compact SPS (CSPS) such as magnetrons (planotrons) are seen in (*a*) and (*b*). A Penning discharge CSPS is shown in (*c*). A semiplanotron is shown in (*d*), and (e) is a hollow cathode SPS using a cold cathode glow discharge in a crossed E x B field. These CSPS have high plasma density, high emission current density of negative ions (up to 8 A/cm^2), small cathode – emitter gap (1-5 mm) (2) and small extraction aperture in the anode (1). They are very simple and effective, have a high brightness and high pulsed gas efficiency. CSPS's are very good for pulsed operation but electrode power density is often too high for DC operation. The opposite situation exists in the Large Volume SPS (LVSPS), presented in Fig.1, *f*, *g*, *h*, and first developed in Lawrence Berkeley Laboratory (LBL) [14]. The gap between emitter, 5, and extractor aperture is very large (8-12 cm) and the plasma and gas density must be kept low to prevent negative ion destruction. LVSPSs use a hot filament, 7, an RF coil, 9, or microwave discharge and multicusp magnets, 8, for plasma confinement. LVSPSs have a low power density and can be used for DC operation. Emission current density is only about 10 mA/ cm^2 and the brightness is not so high. Some versions of LVSPS with emitter (5) were adapted for heavy negative ion production [15]. LVSPS with production of negative ions on the plasma grid surface (anode production on Fig.1*g*) were adopted for high

CP642, *High Intensity and High Brightness Hadron Beams: 20th ICFA Advanced Beam Dynamics Workshop on High Intensity and High Brightness Hadron Beams,* edited by W. Chou, Y. Mori, D. Neuffer, and J.-F. Ostiguy

current (up to 40 A) negative ion beam production for plasma heating [16].

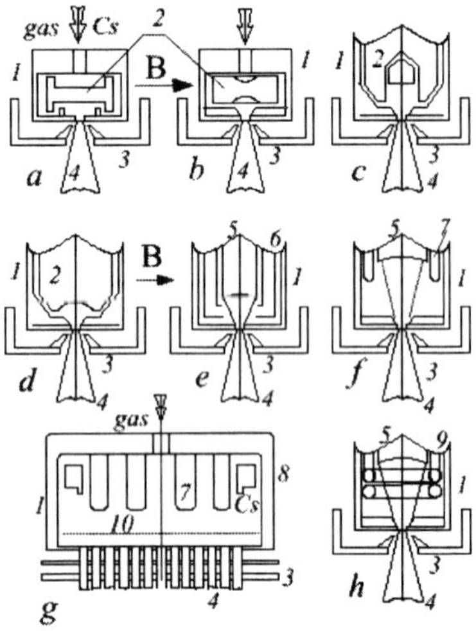

FIGURE 1. Schematic diagram of the basic versions of SPS: (a) planotron (magnetron) flat cathode; (b) planotron (magnetron) geometrical focusing (cylindrical and spherical); (c) Penning discharge SPS (Dudnikov type SPS); (d) semiplanotron; (e) hollow cathode discharge SPS with independent emitter; (f) large volume SPS with filament discharge and biased emitter; (g) large volume SPS with anode negative ion production; (h) large volume SPS with RF plasma production and emitter. 1- anode (gas discharge chamber); 2- cold cathode-emitter; 3- extractor with magnetic system; 4- ion beam; 5- biased emitter; 6- hollow cathode; 7-filaments; 8- multicusp magnetic wall; 9- RF coil; 10-magnetic filter.

The efficiency of negative ion formation depends very much on the catalytic property of the surface, mainly the work-function. For enhanced negative ion formation in the SPS a mixture of substances with a low ionization energy, such as alkaline or alkaline earth elements or compounds, are used. Most efficient is the addition of cesium. Still the surface work-function and catalytic properties of the surface for negative ion formation depends very much on many parameters such as surface-cesium concentration, admixtures of other compounds, such as oxides, halides, nitrides, and surface temperature. Small changes in the surface condition dramatically change the efficiency of negative ion formation. It is a fine art and some magic to optimize the surface and plasma condition for high efficiency of negative ion formation. Small changes in the surface condition can increase or decrease the intensity of a negative ion beam by large factors. The intensity of H^- beams can has a variation up to 10 times for the same discharge current. A stronger variation can has a beam brightness. An efficient ion temperature can has a variations from a part of eV to some keV.

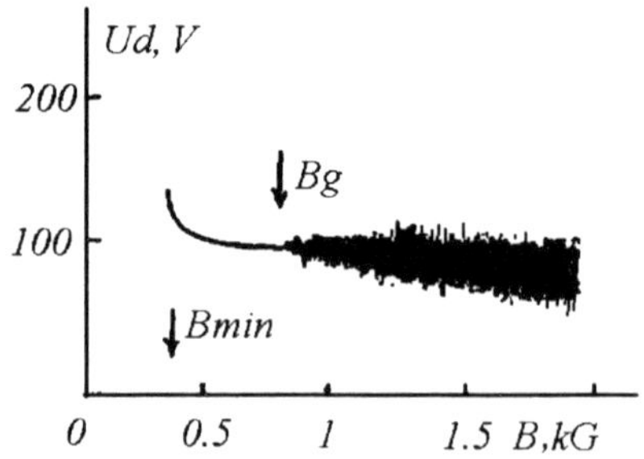

FIGURE 2. The discharge voltage and level of noise vs. magnetic field in SPS with Penning geometry.

It is easier to have stable operation with relatively low beam parameters such as intensity I ~30-50 mA, emission current density J ~0.5-1 A/cm^2, transverse ion temperature Ti ~5-10 eV. Present experience permits better optimization for longtime stable production of high-brightness high-intensity beams of negative ions (I ~0.1-0.15 A, B ~J/Ti > 1 A/cm^2 eV, lifetime N>10^8-10^9 pulses). Highest brightness could be realised only with noiseless operation. The level of discharge noise (hash) depends on many parameters. For stable discharge a surface properties should be in the stable conditions and frequency of electron scattering by plasma particles should be higher than Larmor frequency. A discharge noise could be suppressed by decrease of magnetic field as shown in Fig. 2 and by increase a gas or Cs density. Admixture of heavy gas could be useful for noise suppression, but it increase a sputtering. Examples of discharge voltages with a different level of noise are shown in Fig. 3. A transition from the noisy discharge (*a*) to noiseless one (*c*) increases a beam brightness at order of magnitude.

NEGATIVE ION SOURCES FOR ACCELERATORS

The first versions of the Surface-Plasma Sources (SPS) developed for charge-exchange injection of protons has an operating intensity I ~50 mA with pulse lengths of 0.05-1 msec, noisy discharges and a repetition rate up to 50 Hz [6-11]. H^- beam parameters of these SPS was sufficient for normal operation of large proton accelerator complexes during the past 25 years without significant modernization of ion sources. Now, new accelerator projects need an increase of the ion beam intensity and brightness. Some upgrading of existing SPS could achieve the necessary increase of intensity, duty factor and beam quality without degradation of reliability and availability of the achieved satisfaction level.

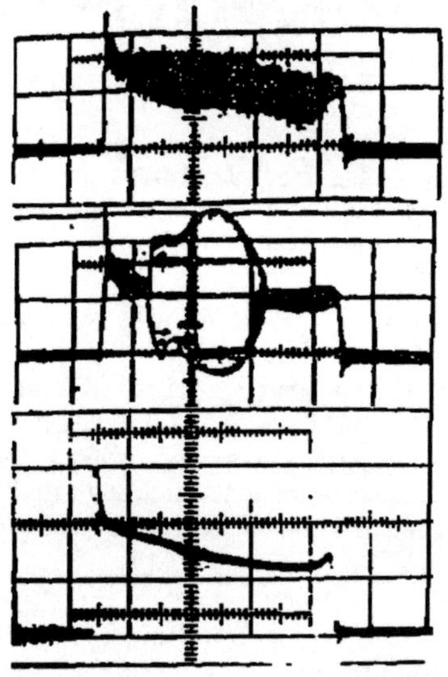

FIGURE 3. The examples of discharge voltages for different conditions in SPS. (*a*) a discharge with noise; (*b*) a discharge with RF generation: (*c*) noiseless discharge. Vertical scale is 100 V/div; Horizontal scale is 0.2 ms/div.

The Fermilab Magnetron SPS has been operational since 1978 [17]. The peak current of the H^- ion beam at the exit of the 750 keV accelerator column is I_b=65 mA with an extraction voltage Uex= 20 kV, and I_b ~70 mA with Uex = 25 kV with a beam pulse length T= 0.075 msec at 15 Hz. The pulse length could be increased with a new arc discharge pulser and adjusted parameters. It is useful for stable operation to have a discharge power supply as a current source with a high impedance (Z= 5-10 Ohm, now Z=1 Ohm) and corresponding higher voltage. After optimization of the discharge electrode configuration the intensity was increased above 0.1A without increasing the discharge power. A suppression of a back accelerating of the positive ion should suppress cathode and anode sputtering by accelerated positive ions - a main reason for the shorting of ion source lifetime. Improved cathode and anode cooling is necessary for increased discharge pulse length and intensity. The Semiplanatron version of the SPS is the best one for operation at higher duty factor. From previous experience it is possible to have reliable operation of a SPS with parameters: peak current after bending magnet I_b ~0.12-0.15 A with pulse duration of T ~0.25 msec, repetition rate F=100 Hz normalized emittance $\varepsilon(90\%) = 1\pi$ mm mrad. SPS with these parameters was tested with a relatively long run as 300hs [18].
Optimized cesium film recycling (deposition-desorbtion) could be used for shielding of electrodes from the sputtering and can reduce the sputtering to a very low level. Cesium in the SPS acts as an oil in an engine, increasing the operational lifetime. "Cold Start" of a discharge without cesium for a few minutes could be more destructive than many hours of low voltage operation. Electron emission from discharges without cesium is very high. Recently, new version of RF SPS has been developed in DESY, SPS with pulsed arc discharge in Frankfurt University and in KEK for high intense Proton Drivers[9-11].

LOW ENERGY BEAM TRANSPORT

The ion beam from a compact SPS has a very high current density (j~1-3 A/cm^2) and perveance. For transport of these beams it is necessary to use a deep space-charge neutralization (compensation) or very strong continuous focusing by electrostatic forces as in the RFQ. Partial compensation of space charge with magnetic focusing and nosy operation will create a strong variation of focusing and lead to an increase of emittance by ellipse rotation. Still, this mode of transport is used in almost all injectors, and until recently it was acceptable. A good solution could be a short LEBT with a fast beam over-neutralization by streams of noiseless plasma from a separate plasma source.

REFERENCES

1. Budker G., Dimov G, Dudnikov V., in Proc. Internat. Symposium on Electron and Positron Storage Ring, France, Sakley,1966, rep. VIII, 6.1 (1966).
2. Dudnikov V, The Method for Negative Ion Production, SU patent, C1.H013/04, No 411542, Appl. 3/3/72.
3. Dudnikov V., Rev. Sci. Instrum., 63(4),2660 (1992)..
4. Dudnikov V., Rev. Sci. Instrum., 73(2),992 (2002)
5. Dudnikov V., Proc. 4th All-Union Conf. On Charged Part. Accel.,Moscow,1974' V.1, p.323.
6. Dudnikov V., PAC01, tppho94, 2001. This Proc.
7. Schmidt C. W., Prod. Neutral. of Negative Ions and Beams, 8th Internat. Symp. AIP 1-56396-773-5. 1998.
8. Zhang H., Ion Sources, Springer,1999
9. Peters J., LINAC' 98, Chicago, 1998.
10. Peters J., Rev. Sci. Instrum., 71(2),1069 (2000).
11. Peters J., EPAC'2000.
12. Belov A., Dudnikov V., et al. Nucl. Inst. Meth. A333, 256-259 (1993).
13. Belov A., Derinchuk V., PAC01, 2001.
14. Leung K.N., and Ehlers K., R.S.I., 53, 803 (1982).
15. Ishikawa J., R.S.I., 67(3) 1410 (1996).
16. Okumura Y., et al., R.S.I.,71(2),1219 (2000).
17. Schmidt C.W., Curtis C., IEEE Trans. Nucl. Sci. NS-26,4120(1979).
18. Dimov G., Dudnikov V., Derevyankin G., IEEE Trans. Nucl. Sci. NS-24,1545 (1977).

Status of the SNS H⁻ Ion Source and Low-Energy Beam Transport System*

R. Keller,[@] R. W. Thomae,[@] M. P. Stockli,[#] and R. F. Welton[#]

[@] *E. O. Lawrence Berkeley National Laboratory,* [#] *Oak Ridge National Laboratory*

Abstract. The ion source and Low-Energy Transport (LEBT) system that will provide H^- ion beams to the Spallation Neutron Source (SNS)** Front End and the accelerator chain have been developed into a mature unit that will satisfy the operational needs through the commissioning and early operating phases of SNS. The ion source was derived from the SSC ion source, and many of its original features have been improved to achieve reliable operation at 6% duty factor, producing beam currents up to the 50-mA range. The LEBT utilizes purely electrostatic focusing and includes static beam-steering elements and a pre-chopper. This paper discusses the latest design features of the ion source and LEBT as well as some future improvements, gives performance data for the integrated system, and reports on commissioning results obtained with the SNS RFQ accelerator.

INTRODUCTION

Berkeley Lab has just completed building the linac injector (Front End, FE) for the Spallation Neutron Source project (SNS) and is commissioning the entire system. The main subsystems are the H^- ion-source, the low-energy beam-transport system (LEBT), the 2.5-MeV radio-frequency quadrupole (RFQ) accelerator, and the medium-energy beam-transport system (MEBT). Ion source and LEBT are the subject of this paper; their task is to create a 65-keV, 38-mA ion beam, to match and steer it into the RFQ, and to pre-chop it into mini-pulses of about 600 ns duration. The nominal duty factor is 6%, with 1-ms macro-pulse length and 60-Hz repetition rate.

Based upon the main design features of the SSC ion source [1], an R&D version of the SNS ion source was built first to demonstrate the viability of the chosen approach, utilizing an rf driven discharge inside a multicusp plasma generator with magnetic filter, cesium enhancement, and electron suppression at low energy [2].

For the LEBT, a purely electrostatic focusing system [3] was chosen, thereby avoiding time-dependent space charge compensation usually encountered with magnetic LEBTs. The production version of the ion source and LEBT [4] aims at generating and transporting a beam with 50-mA current, thought to be sufficient to satisfy the latest SNS design goal of 38 mA to be injected into the Linac while assuming a 20% beam loss in the RFQ. Actual RFQ transmission results [5], quoted below, indicate that a LEBT output current of about 44 mA should satisfy the SNS current goal.

A schematic view of ion source and LEBT is shown in Figure 1, below. By now, four plasma generators (including one "startup source") and one LEBT have been built and tested, and the specific design features and performance data of this production system are discussed in the following sections.

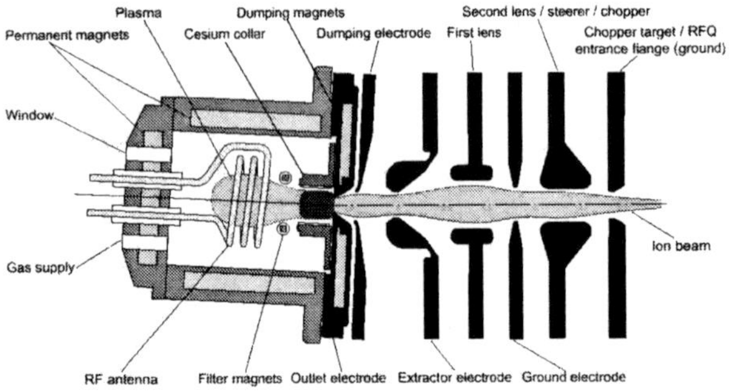

FIGURE 1. SNS Ion Source and LEBT schematic.

ION SOURCE

The production-version ion sources aim at generating H^- beams of up to 50-mA current. The goal for the normalized, transverse rms emittance is determined

---

* This work is supported by the Director, Office of Science, Office of Basic Energy Sciences, of the U.S. Department of Energy under Contract No. DE-AC03-76SF-00098.

** The SNS project is being carried out as a collaboration of six US Laboratories: Argonne National Laboratory (ANL), Brookhaven National Laboratory (BNL), Thomas Jefferson National Accelerator Facility (TJNAF), Los Alamos National Laboratory (LANL), E. O. Lawrence Berkeley National Laboratory (LBNL), and Oak Ridge National Laboratory (ORNL). SNS is managed by UT-Battelle, LLC, under contract DE-AC05-00OR22725 for the U.S. Department of Energy.

CP642, *High Intensity and High Brightness Hadron Beams: 20th ICFA Advanced Beam Dynamics Workshop on High Intensity and High Brightness Hadron Beams,* edited by W. Chou, Y. Mori, D. Neuffer, and J.-F. Ostiguy
2002 American Institute of Physics 0-7354-0097-0

by the requirement of 0.2 π mm-mrad for the beam exiting the LEBT.

Plasma Generation

The plasma is produced by a hydrogen discharge inside the multi-cusp vessel, sustained by tens of kW of 2-MHz rf power. The power is transmitted through an impedance-matching network to an antenna that consists of a 2-1/2 winding copper coil covered by a multi-layer porcelain coating. The efficiency of beam generation is about 1.0 mA per kW of rf power.

An antenna with 0.25-mm coating [6] underwent an endurance test at full duty factor, and the test was intentionally stopped after 107 hours while the ion source delivered 20 mA of beam current. An upgraded antenna with 0.8-mm thick 10-layer coating produces the same plasma density for a given rf power level as the thin version and should last significantly longer.

To reliably ignite the plasma at the beginning of every 1-ms pulse, a continuous, low-power discharge is sustained by an additional 13.56-MHz rf system. Not having to provide conditions suitable for ignition allows tuning the main discharge parameters towards optimum production of H^- ions.

H^- Creation

Negative hydrogen ions are preferentially created in the space confined by the cesium collar, the magnetic filter field, and the outlet electrode, see Fig. 1. The filter field keeps energetic electrons that would destroy the H^- ions away from the collar region. Volume production alone is sufficient only to generate about 15 mA of beam current; cesium enhancement is needed to reach the 50-mA level. For that purpose, the collar is fitted with eight cesium-chromate containers and is thermally isolated from the source body. The presence of a minute amount of cesium on the inner collar surface not only multiplies the abundance of negative ions in the discharge plasma by about a factor of three, but it also reduces the abundance of electrons in the extracted beam by one order of magnitude.

To best utilize the cesium, a freshly cleaned plasma generator is operated at full duty factor for about 15 min., heating the cesium collar to more than 500°C by forcing hot air through the collar wall and utilizing the rf power as a source of heat. After this initial conditioning, the collar is cooled down by room-temperature air and kept at about 280°C for optimal beam production. The cesium layer can then last for several days. Additional cesium reconditioning can be performed in-situ as needed.

Recently a new collar has been developed that seamlessly merges with the outlet aperture [7]. This design brings three major advantages: 1), it provides for keeping the surfaces around the outlet aperture at the same temperature as the collar; 2), with the help of an isolating centering ring, it allows precise aligning of the collar to the axis of the outlet aperture and at the same time biasing the entire unit to an optimal potential with respect to the source body [8]; and 3), it allows modifying the contour of the outlet aperture as discussed below, without having to build another main outlet flange. This integrated collar/outlet aperture has been fabricated but so far not yet been tested in the SNS ion source. Its design is shown in Figure 2.

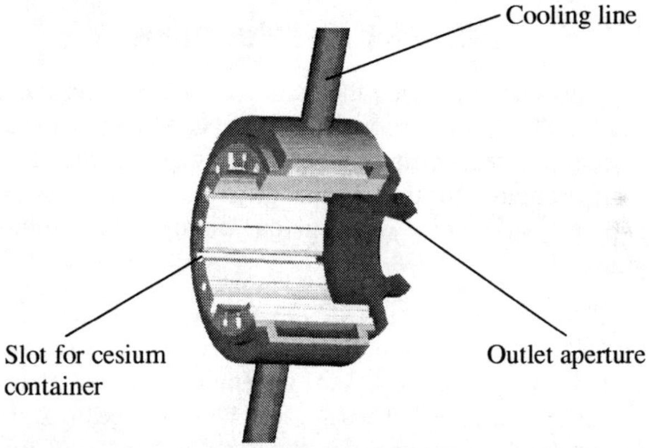

FIGURE 2. SNS Ion Source and LEBT schematic. Integrated cesium collar with slots for cesium containers (left) and outlet aperture (right). Heated or room-temperature air is conducted around the collar to stabilize the temperature.

LOW-ENERGY BEAM TRANSPORT

Electron Dumping

When a negative-ion beam is being extracted from a plasma, a substantial amount of electrons is extracted as well. This leads to the problems of increasing the space-charge density, especially near the outlet aperture, and of the power load to the structure where the electron beam is deposited. With the SNS ion source, the electrons are deposited on a dedicated 'dumping electrode' at moderate energies of about 5 keV, aided by a set of permanent magnets inside the outlet electrode. Usually some fraction of the electrons misses the dumping electrode entirely, and these electrons are accelerated to the full beam energy of 65 keV and hit either the extractor electrode or its support structure. A water-cooled shield has now been installed to absorb the associated heat load. It is expected that the new, integrated collar/outlet electrode described above will provide a more satisfactory solution to this problem.

Because of the steering action of the dumping magnetic field, not only on the electrons, but also on the ions, the entire ion source is tilted by an adjustable angle of ~3° with respect to the LEBT axis.

Beam Simulations

For the earlier design work, the positive-ion code IGUN [9] had been used, approximating the electron space charge by increasing the assumed ion current in the volume near the meniscus. The negative-ion version of the code PBGUNS [10] appears to better simulate the beam formation process [11] and was used to design the new outlet aperture contour.

Principal LEBT Functions

Apart from forming the beam, the main purpose of the LEBT is to transport it to the RFQ and match the injection requirements. To efficiently pump the gas load produced by the plasma generator the electrode support structures were given highly transparent shapes.

The focusing action of the two-lens electrostatic system, captured in a tuning matrix, works as predicted by simulations, but generation of less than nominal beam current results in a narrower beam size inside the first lens and effectively reduces its focusing power. To widen beams of less than 35 mA during RFQ commissioning, the extraction gap was increased by 4 mm as compared to the nominal size of 20 mm.

For pre-chopping and static steering, pulsed voltage signals of ±2.5 kV and, independently, dc potentials are applied to the four quadrants of the center electrode of the second lens. Not all of the chopped beam is intercepted by the ring target on the LEBT-exit/RFQ-entrance electrode; the remaining particles are deposited inside the RFQ cavities

BEAM RESULTS

The nominal beam-current goal of 50 mA pulse average measured at full 6% duty factor downstream of the LEBT was reached about a year ago, still with the nominal gap width installed. The peak current at the beginning of every pulse even reached 68 mA. With the extraction gap increased by 4 mm for RFQ commissioning, up to 36 mA were measured, and the pulse shape was much more uniform.

The emittances show pronounced distortions at the 10% intensity level, but after subtracting back-ground signals from the raw data, the normalized rms sizes are very close to or even better than the nominal values of 0.2 π mm mrad.

A round-the-clock endurance test of the ion source and LEBT was conducted over more than a week, continuously producing beam of about 25 mA current at 3% duty factor, with few interruptions. It proved that the beam-generating system is ready to support commissioning of the SNS accelerators and even SNS operations for the first few years.

In fact, the commissioning of the SNS RFQ at low duty factor [5] was aided by very stable performance of Ion Source and LEBT, and up to 33 mA were transmitted, out of about 36 mA that had been injected.

The functionality of the LEBT pre-chopper system has been tested as well, and rise and fall times of 25 ns were measured, better than the nominal requirement by a factor of 2. The beam signals were not clear enough to allow a precise determination of the pre-chopper attenuation factor, but a value around 1×10^{-3} appears quite plausible from extrapolations of results obtained at less than nominal chopping voltages.

ACKNOWLEDGMENTS

The authors would like to acknowledge the support by a large number of SNS staff at LBNL and ORNL who supported this work. Thanks are due in particular to R. Gough, R. Yourd, R. DiGennaro, A. Ratti, S. Lewis, T. Schenkel, D. Cheng, K. N. Leung, J. Greer, J. W. Staples, D. Syversrud, W. Abraham, T. Kuneli, N. Ybarrolaza, C. Lionberger, P. Cull, M. Hoff, J. Pruyn, R. MacGill, M. Monroy, M. Regis, J. Dougherty, D. Garfield, and K. Barat.

REFERENCES

1. Saadatmand, K., Arbique, G., Hebert, J., Valicenti, R., and Leung, K. N., *Rev. Sci. Instr.* *67* (3), p. 1318 (1996).
2. Leitner, M. A. et al., *Proc. PAC '99*, Paper WEA 13, New York (1999).
3. Staples, J. W., Hoff, M. D., and Chan, C. F., *Proc. Linac '96*, Geneva (1996).
4. Reijonen, R. J., Thomae, R., and Keller, R., *Proc. Linac 2000*, Paper MOD19, Monterey (2000).
5. Keller, R. et al., *Proc. EPAC '02*, Paper THPLE012, Paris (2002).
6. Welton, R. F., Stockli, M. P., Kang, Y., Janney, M., Keller, R., Schenkel, T., Thomae, R., and Shukla, S., *Rev. Sci. Instrum.* **73** (2), p. 1008 - 1012 (2002).
7. Welton, R., Stockli, M., Keller, R., and Thomae, R., *Proc. EPAC '02*, Paper THPLE019, Paris (2002).
8. Peters, J., *Rev. Sci. Instrum.* **73** (2), p. 900 - 902 (2002).
9. Becker, R., *EPAC'98*, Paper THP44G, Stockholm(1998).
10. Boers, J. E., PBGUNS Manual, available through Thunderbird Simulations, Garland, TX, 75042.
11. Welton, R. F., Stockli, M. P., Boers, J. E., Rauniyar, R., Keller, R., Staples, J. W., and Thomae, R. W., *Rev. Sci. Instrum.* **73** (2), p. 1013 - 1016 (2002).

Performance of the Magnetron H⁻ Source on the BNL 200 MeV Linac *

James G. Alessi

Brookhaven National Laboratory
Upton, NY, 11973 USA

Abstract. A magnetron surface plasma H^- ion source has been used at Brookhaven since 1982 for injection into the 200 MeV linac. Since 1989, this source has operated with a circular aperture, injecting into an RFQ. The source typically produces 90-100 mA of H^- at 35 keV, in 700 μs pulses at up to 10 Hz. The arc parameters are ~15 A, 150 V, and the extracted e/H- ratio is 1/2. Ions are extracted from a 2.8 mm diameter aperture (J=1.6 A/cm^2). The emittance is approximately $\varepsilon_{n,\,rms}$ = 0.4 π mm mrad. The performance is very reliable, operating continuously for ~6 months, with essentially no parameter adjustments required once the source is stabilized.

INTRODUCTION

Since 1982, a magnetron surface plasma H^- source has been used on the Brookhaven 200 MeV Linac. The linac typically runs at a repetition rate of 6.6-7.5 Hz, and a pulse width of ~ 500 μA, and the current at the exit of the linac is typically 35-40 mA. This beam is injected into the Booster synchrotron and then AGS, for high energy physics. Since the Booster typically takes only 4-6 pulses every 2.5-3.5 seconds, the rest of the pulses are switched to a second beamline, where this ~ 70-100 μA average current is used for isotope production.

The magnetron source was first developed in Novosibirsk (planotron) [1], but the basis of the version used at BNL is the very clever design developed by C. Schmidt at FNAL [2]. While changes have been made over the years to incorporate cathode focusing, circular aperture, extractor geometry changes, etc., there have been remarkably few changes over the years to his basic mechanical design, which is simple and reliable. From 1982–1989 the source operated in the dome of a 750 keV Cockcroft-Walton preinjector [3]. A slit extraction geometry was used at 20 keV, followed by a small 90 degree double-focusing dipole before the beam entered the high gradient accelerating column. In 1989, the Cockcroft-Walton was replaced by an RFQ preinjector. At this time, the magnetron was converted to circular aperture extraction at 35 keV, followed by a 2-solenoid matching into the RFQ [4].

MAGNETRON GEOMETRY, LAYOUT

Figure 1 shows the basic geometry of the magnetron. A transverse magnetic field of ~900 G produces a cyclic motion of electrons in the ~ 3 mm gap between the Mo cathode and anode, producing an intense hydrogen plasma in this region. Cesium vapor is fed into the source from an external oven containing metallic Cs, through a transfer tube kept at ~ 300C. An all metal valve between the source and the Cs oven allows the source to be let up to air without having to reload Cs. Hydrogen is fed into the source through a pulsed gas valve. The source has a very small discharge volume (~1 cm^3), so good gas pulsing is effective. The valve is placed close to the source to reduce the conductance and keep as sharp a gas pulse as possible.

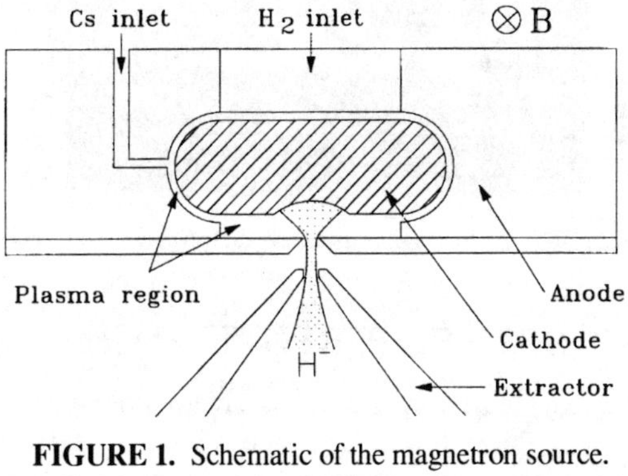

FIGURE 1. Schematic of the magnetron source.

Ions are extracted through 2.8 mm diameter apertures in both the anode and extractor, across a 4 – 4.5 mm extractor gap. The extractor tip is an insert made of tungsten, to reduce erosion coming from impingement by extracted electrons. We operate with pulsed extraction voltage. Figure 2 shows the source without the anode cover on, and Fig. 3 shows the source with extractor in place.

The source sits reentrant in the vacuum box. The only opening between the source vacuum and the

* Work supported under the auspices of the U.S. Department of Energy.

CP642, *High Intensity and High Brightness Hadron Beams: 20th ICFA Advanced Beam Dynamics Workshop on High Intensity and High Brightness Hadron Beams,* edited by W. Chou, Y. Mori, D. Neuffer, and J.-F. Ostiguy

LEBT line is the 2.8 mm diameter extraction aperture. This results in a rapid reduction in pressure along the beamline, reducing H^- stripping losses. The source chamber is pumped by a 2200 l/s turbopump, while only a 170 l/s turbopump is needed on the LEBT line.

FIGURE 2. Source with the anode cover removed. Permanent magnets are visible on the sides of the source.

FIGURE 3. Source with anode cover and extractor.

SOURCE PERFORMANCE

Typical source operating parameters are given in Table 1. One should note the very low extracted electron current. We often operate the source at higher arc current than is required to achieve the desired beam current. This is partly just to keep the source temperature high enough to allow good Cs distribution in the source, but also by operating the source in the space charge limited mode, the beam current pulse becomes somewhat flatter.

The good power efficiency of the magnetron has helped reliability. The power efficiency improved dramatically as we went from the conventional flat cathode to geometrical focusing of ions from a grooved cathode into the slit extractor, and subsequently spherical focusing from the cathode into a circular aperture. Table 2 shows how the power efficiency has now improved by a factor of 30 from what it was initially. This improvement leads us to believe that the source could easily operate at higher duty factors than those now used.

TABLE 1. Typical running parameters

H- current	90-100 mA
J (H-)	1.5 A/cm^2
Extraction voltage	35 kV
Electron/H-	0.5- 1.0
Arc voltage	140 – 160 V
Arc current	8 – 18 A (see note)
Rep rate	7.5 Hz
Pulse width	700 μs
Duty factor	0.5 %
rms emittance	~ 0.4 π mm mrad
Cs consumption	< 0.5 mg / hr
Gas flow	~ 2 sccm

(Note: often runs at 15-18 A to keep source temperature high)

TABLE 2. History of improvements in BNL magnetron power efficiency

Aperture/ cathode focusing	H- (mA)	Arc I (A)	Arc V (V)	Pwr effic (mA/kW)
Slit/flat	50	150	150	2.2
Slit/grooved	50	50	150	6.7
Circular/dimpled	100	10	150	67

OPERATING EXPERIENCE

With continuous 24-hour/day operation, times between source maintenance are ~ 6 months. (It can vary from 3-9 months; almost always shutting down when the program ends, rather than due to failure). The source can be shut down, a minor repair made or the spare swapped in, restarted, and be running well again in 8 hours. Once the source is started, we try to keep it running. That is, even for a 1-2 week linac shutdown, we would typically choose to leave the source pulsing, with extractor on. This is mostly for convenience, since the source almost always returns to normal operation following a shutdown or power outage. (On the test bench, a source may be turned on and off daily without difficulty). The attempt is made to keep the extraction voltage on whenever the source is running. Beam heating of the extractor tip tends to keep it clean, and if the extractor is turned off for a couple of hours, it may have to be reconditioned to get back to 35 kV operation.

We typically operate the source slightly gas- and Cs-starved, i.e. increasing either one would lower the arc impedance, but it has been found to be more reliable to keep these parameters reduced. The disadvantage is that it makes the discharge (and beam current) "noisy". Once stabilized after turnon, it's not unusual for the source to run for a month at constant current without a single adjustment. It is fairly common that a slow buildup of CsOH will gradually begin to block the hydrogen inlet, which then requires one to increase the gas to the source (voltage to the pulsed gas valve) every 1-2 weeks. This may start after a few months of operation.

When opening after a long run, the source is generally very clean inside (polished). Cs is not seen downstream of the extractor electrode. There is erosion of the extraction tip (from ions, electrons), anode aperture (electrons), cathode dimple (backstreaming ions), and cathode opposite the Cs feed (discharge). In spite of these very significant changes in source dimensions for the extraction geometry and cathode focusing, the performance (and output from linac) remains very constant. We often clean the source parts and reinstall them, even with the heavy erosion.

MATCHING TO THE RFQ

The 35 keV beam is matched from the source to the RFQ with 2 solenoids, over a distance of ~1.3 m. It is important for the beam to be space charge neutralized in this LEBT, and this neutralization typically occurs within 50 μs at a pressure in the line $\leq 10^{-5}$ Torr. The layout of the line is shown in Fig. 4. The pulsed dipole allows switching between polarized and unpolarized beam on a pulse-to-pulse basis.

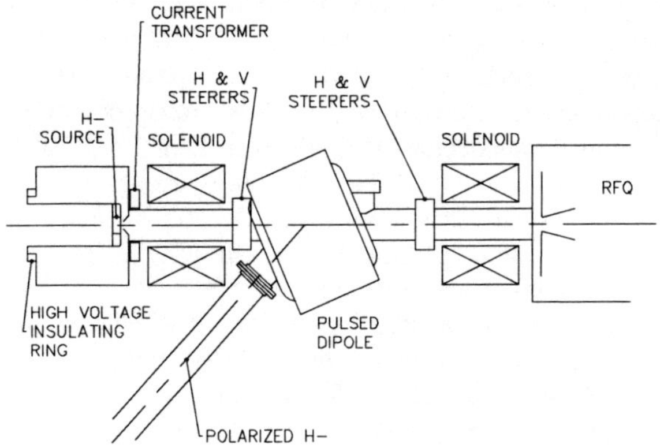

FIGURE 4. Layout of the matching line between source and RFQ. The distance from the source to RFQ is ~ 1.3 m.

The 200 MHz RFQ, built by LBL [5], accelerates the 35 keV input beam to 750 keV. Operation of the RFQ has been extremely reliable. By operating with space-charge limited extraction, and "flooding" the transport line, the transmission numbers suffer, but the current out of linac is high and stable. RFQ transmission is typically 80-90% (see Fig. 5), but at lower currents (40 mA) we have achieved essentially 100% transmission. Under normal conditions, the emittance out of the RFQ is ε(n, rms)~0.4 π mm mrad.

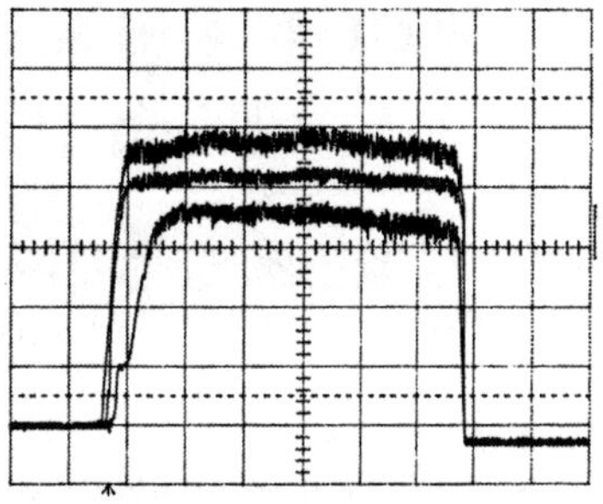

FIGURE 5. Source output, RFQ input, and RFQ output; 20 mA/div, 100 μs/div.

SCALING TO HIGHER DUTY FACTOR

In 1983, the magnetron source was operating at 150 V x 150 A x 5 Hz x 600 μs = 68 W average power. With the present power efficiency, simple scaling would imply that the source could now produce 100 mA at 4.5% duty factor. The source is presently uncooled, so one could gain some margin by adding even simple cooling. At reduced beam currents, one could either reduce the arc power, allowing even higher duty factors, or reduce the extraction aperture, resulting in a smaller emittance and reduced gas flow. In reality, the situation will be more complicated. Issues to be explored for higher duty factor would have to include the effects of the resultant higher gas flow out of the source, the increased erosion of source parts, and the increased heating of the extractor electrode.

ACKNOWLEDGMENTS

The very good performance of the source over the years is in large part due to the many excellent engineers and technicians in the BNL linac group who have operated and made improvements to this source.

REFERENCES

1. Belchenko, Yu.I., Dimov, G.I., and Dudnikov, V.G., Proc. 2nd Symp. On Ion Sources and Formation of Ion Beams, Berkeley, CA, VIII-1, LBL-3399 (1974).
2. Schmidt, C.W. and Curtis, C.D., Proc. Symp. Prod. and Neut. of Negative Ions and Beams, Brookhaven, BNL-50727 (1979) 123.
3. Barton, D.S. and Witkover, R.L., IEEE Trans. Nucl. Sci. NS-28 (1981) 2681.
4. J.G. Alessi et al., Proc. 1989 IEEE Part. Accel. Conf., Chicago, IEEE-89CH2669-0 (1989) 999.
5. R.A. Gough et al., Proc. 1986 Linear Accel. Conf., SLAC, SLAC Report No. 303 (1986) 260.

ECR Ion Sources for H^- Ion Production: First Results and Prospects*[*]

A. Girard*, D. Hitz*, G. Melin*, R. Gobin°, R. Ferdinand°, K. Benmeziane°, O. Delferrière°, J. Sherman**

*Département de Recherche Fondamentale sur la Matière Condensée, Service des Basses Températures, CEA Grenoble 17 rue des Martyrs 38054 Grenoble Cedex 9 FRANCE

°Département d'Astrophysique, de Physique desParticules, de Physique Nucléaire et de l'Instrumentation associée, Service des Accélérateurs, de Cryogénie et de Magnétisme, CEA Saclay 91191 Gif sur Yvette Cedex FRANCE

** Los Alamos National Laboratory, Los Alamos, N.M. 87 545 USA

Abstract. ECR Ion Sources are well known for their efficient production of Highly Charged Ions [1], and also for the production of intense proton beams [2]. Recently European laboratories have decided to join their efforts to develop and improve various plasma techniques for the production of intense beams of H^-, for a future application to the European Spallation Source, and possibly for other high power accelerators. Because of its great experience and skill in ECR plasmas and ion sources, CEA has decided to develop ECR ion sources for negative ion production. At first glance, this seems to be a real challenge, as negative ion production requires a very low electron temperature, incompatible with ECR heating. We will show that this contradiction can be solved. In this article we briefly summarize the present status of Negative Ion Sources (NIS). Then we describe ECR Ion Sources and how they can be of great interest for H^- production. Although this domain is still rather unexplored, some work has already been performed in the field of negative ion production with ECR plasmas. That work will be shortly summarized. Eventually promising preliminary results, obtained at CEA Saclay at 2.45 GHz, will be shown.

INTRODUCTION

Electron Cyclotron Resonance (ECR) plasmas and ECR Ion Sources (ECRIS) are now widely used in accelerators: heavy ions are usually produced with ECRIS delivering multiply charged ions [1]. Moreover high currents of protons can also be produced in ECR ion sources [2], although these sources are very different from the previous type, from the point of view of the magnetic configuration: ECRIS for highly charged ions are basically mirror confined plasmas, while high current proton ECRIS are moderate confinement plasmas (only an axial magnetic field is present). All these sources are able to produce both cw and pulsed beams of high reliability with emittances meeting the requirements of the accelerators. However the production of negative ions with ECR plasmas is a new domain: only few experiments have been performed and no ECRIS is still used to inject negative ions into an accelerator. Today there is a great interest for high current accelerators; in particular the production of high flux neutron beams for spallation reactions (ESS), neutrino and muon production for high-energy particle physics, are challenging new projects in Europe and in the world. These projects require intense beams of negative ions (in the range of 60 mA), with a long pulse duration (from 1 to 2 ms); moreover the pulses should be noiseless and highly reproducible, and the (normalized rms) emittance of the source should be less than 0.3 pi mm mrad. Today no existing source is able to fulfill all the requirements for ESS. Therefore nine European laboratories have recently joined their efforts to develop various techniques for an improved negative ion production. At CEA it was decided to develop an ECR Ion Source for Negative Ions. In section 1 the various existing types of NIS are shortly reviewed. Section 2 deals with the present status of ECRIS, and the main features of the plasmas of these sources; we also discuss the possibility of H^- production by an ECR plasma. In section 3 the first tests on the ECR NIS developed at CEA Saclay are described, and some prospects for new developments are given.

PRESENT STATUS OF NIS

NIS are usually divided in two main types: surface sources, and volume sources.

Surface Sources

In surface sources [3] the negative ion production occurs on electrodes in contact with the discharge plasma. This plasma is of very small size, and a very

* Work supported by European Commission under contract HPRI-CT-2001-50011

CP642, *High Intensity and High Brightness Hadron Beams: 20th ICFA Advanced Beam Dynamics Workshop on High Intensity and High Brightness Hadron Beams*, edited by W. Chou, Y. Mori, D. Neuffer, and J.-F. Ostiguy

large power (typically tens of kilowatts) is injected to sustain it. Hence the power density in the discharge reaches a very high value, and the resulting currents of Negative Ions are also very high. Penning SPS are widely used for accelerators, and further developments continue in that domain [4] to meet the requirements for ESS.

Volume Sources

In volume sources, the negative ions are produced directly in the plasma; however small amounts of cesium added in the discharge lead to a very significant improvement of the performances for NI production, which suggests that surface processes are also of importance in "volume" ion sources. For accelerator purposes the volume sources are usually of moderate size (typically 1 liter for the plasma region). Different techniques are used for the production and confinement of the plasma in volume sources: multicusps [5] are widely used for the confinement of the plasma, but an axial magnetic field may also be used; the plasma is excited by either filaments [6], or RF excitation [7,8]. Usually more than ten kilowatts are injected into such sources, which leads to a lower power density than in SPS.

PRESENT STATUS OF ECRIS

ECRIS For Multiply Charged Ions

ECRIS can produce beams of multiply charged heavy ions [1]. For that purpose a complex magnetic configuration is achieved, which confines the plasma so that atoms can be step-by-step ionized up to high charges. The plasma is produced by the resonant interaction of the electrons with the High Frequency (HF) wave at the electron cyclotron frequency $\omega = eB/m$; this interaction generates hot electrons (mean energy up to tens of keV), which are then able to ionize the atoms to high charges. It was found that the performances of these sources are related to the frequency of the wave: the density of the plasma produced is nearly proportional to the square of the HF frequency, which leads designers to use higher and higher frequencies for improved performances. 28 GHz operation has been recently achieved. Because of the production of the plasma via the HF wave, no cathode is necessary, which enhances the reliability of the source. This type of source is now the most commonly used for heavy ion production in accelerators.

ECRIS For Intense Proton Beams

After a pioneer work at Chalk River and Los Alamos, ECRIS are now widely used for intense proton beams [2]. As compared to the previous type of ECR sources, the magnetic configuration is much simpler: only an axial magnetic field is necessary; the frequency used is 2.45 GHz, with a resonant magnetic field of 875 G. The density reached is slightly below 10^{12} cm^{-3}, with an electron temperature of the order of 10 to 20 eV (much lower than in confined ECRIS). The SILHI source at Saclay has now proved to be efficient and reliable for proton production, and it will be used for the next generation of high power proton accelerators.

Because of these successes, it was decided to continue to develop ECRIS for negative ions.

ECRIS FOR NEGATIVE IONS

Main Characteristics, Previous Work

In ECRIS the dimension of the plasma is of the same order of magnitude as the wavelength of the HF wave. Therefore 2.45 GHz operation requires typically 1 liter of plasma volume. This shows that ECRIS for Negative Ions are necessary of the "volume" type. Although only few experiments have been performed so far, quite encouraging results have been obtained previously: at Argonne [9], a few mA of H^- were extracted from an ECR with an axial magnetic confinement. Similarly H^- had been extracted from multicusp ECR ion sources [10,11]. These results show that it is possible to eliminate the HF-generated hot electrons, which could destroy the negative ions. A special magnetic configuration is however necessary.

An ECRIS For Negative Ions at CEA Saclay

An ECRIS for negative ions has been designed recently [12]. The starting point was the SILHI magnetic field configuration. In order to prevent hot electrons to move to the extraction zone, a magnetic filter has been designed. A schematic drawing of the source is shown in figure 1.

As compared to the above reference, the following changes have been performed: the plasma chamber source is now biased at HV, while diagnostics are grounded.

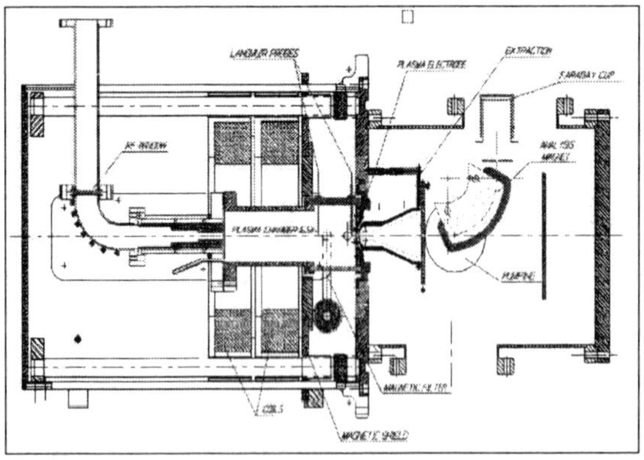

FIGURE 1. The CEA Saclay negative ion source.

A pulsed 90° dipole analysing magnet has been installed behind the 6 mm diameter hole collector, in the diagnostic box. The 10 mm central aperture dipole is made with non-zero entrance and exit to allow vertical focusing. Good transmission through the dipole magnet is predicted for extreme particle trajectories (at the limits of geometrical acceptance). As this magnet is not cooled, it operates in pulsed mode, typically at a repetition rate of 1 Hz. The timing of the source is tuned to extract a single pulse during the flat-top portion of the magnet current pulse. A shielded Faraday cup is located at the dipole exit to eliminate plasma effects and secondary electrons. An ACCT allows to measure the total extracted beam. Beam currents could be also measured on extractor and collectors. The Faraday cup measurements are done with an amplifier and its screen is biased by a few hundred volt tuneable power supply.

The dipole was calibrated with positive extracted charges (H^+, H_2^+, H_3^+). Then the source was negatively biased at –6.4 kV and fed by hydrogen and helium gas. 40 mA total current is easily extracted, and a peak characteristic of H^- ions can be clearly identified. This peak does not appear in helium plasmas. The presence of H^- ions has therefore been proved. However more precise measurements have still to be performed, to give a precise value of the H^- current extracted from the source. Nevertheless this result is very encouraging and demonstrates (if necessary) that ECR plasmas can produce H- ions.

Future Plans

This ECR-based NIS needs still many improvements, in particular for the diagnostics of the negative ion beam. Moreover the influence of the magnetic filter has to be optimized.

On the other hand, another ECR-based prototype will be developed at CEA Grenoble: a 10 GHz source will be developed, which is likely to bring some improvements: (i) working at a higher frequency may lead to a higher density, if the frequency scaling observed in ECRIS for heavy ions still holds in other types of magnetic configurations; (ii) the size of the plasma will be reduced, as the wavelength at 10 GHz is smaller than at 2.45 GHz. From this reduction of the plasma size we hope to be able to produce the same H^- current with a lower HF power.

CONCLUSION

We have proved that ECR ion sources can produce negative ions; however many developments are necessary to reach an optimized magnetic configuration and an optimum H^- production. We believe that ECR-based NIS will prove as efficient and reliable as ECRIS for heavy ions and protons. We think that these sources can solve all the source-related reliability problems for future high power accelerators.

REFERENCES

1. A. Girard and G. Melin, *Nucl. Instrum. Methods* **A 382**, 252-266 (1996).
2. R. Gobin et al, *Rev. Sci. Instrum.* **69**, 1009, (1998).
3. Yu. Belchenko, *Rev. Sci. Instrum.* **64**, 1385-1393 (1993).
4. J.W.G. Thomason, *Rev. Sci. Instrum.* **73**, 896-898 (2002).
5. K.N. Leung, *Rev. Sci. Instrum.* **65**, 1165-1169 (1994).
6. K. Volk et al , *Rev. Sci. Instrum.* **67** (1995).
7. J. Peters , *Rev. Sci. Instrum.* **69**, 992-994 (1998).
8. R. Keller, *in EPAC 2000,* Proceedings 1657-1659 (2000).
9. D. Spence, K.R. Lykke and G.E. McMichael, "Plasma Modified Production of High-current, high purity cw H+, D+ and H- beams from Microwave-driven Sources" in *Linac96*, 1996.
10. K. Jayamanna et al, *Rev. Sci. Instrum.* **67**, 1061- 1063 (1996).
11. O. Fukusama and M. Matsumori, *Rev. Sci. Instrum.* **71**, 935-938 (2000).
12. R. Gobin et al, *Rev. Sci. Instrum.* **73**, 983-985 (2002).

High Current ECR Source for Protons and Deuterons at Saclay

Robin Ferdinand

CEA-Saclay
DSM/DAPNIA/SACM bldg. 124, 91191 Gif-sur-Yvette cedex, FRANCE

Abstract. The SILHI (Source of Light Ions with High Intensities) source has been producing proton beams since 1996. The target objective is to produce up to 100 mA cw proton beams at 95 keV for IPHI (Injector of Protons for High Intensity) demonstrator or 155 mA cw deuteron beam for IFMIF. This prototype is developed by a CEA/DSM – CNRS/IN2P3 collaboration for applications such as Accelerator Driven Systems for nuclear waste transmutation, production of radioactive ion beams or secondary particles. To measure reliability of the installation, continuous five-day long to 30 day long runs have been performed. A peak of 157 mA cw beam has been extracted. In the framework of the IFMIF project, up to 170 mA-100 kV deuteron pulsed beams were produced. Recent SILHI results are also presented.

INTRODUCTION

For several years, in France, CEA and CNRS have undertaken an important R&D program on very high beam power accelerators. CEA is also implied in projects such as ESS (European Spallation Source) and IFMIF (International Fusion Material Irradiation Facility).

SILHI is developed to be the source of the IPHI prototype [1]. The SILHI main objective is to produce 100 mA proton or 140 mA deuteron cw beam currents at 95 keV with rms normalized emittances lower than 0.2 π.mm.mrad. An Electron Cyclotron Resonance (ECR) source has been chosen to reach these performances with a high reliability/availability. Experiments with SILHI are also devoted to the production of deuterons for IFMIF and a new test bunch is under study for H^- ion production.

Since 1996, SILHI has been regularly producing proton beams, in cw or pulsed mode, with performance close to the set objective [2]. A new extraction system has been designed to minimize beam losses on the electrodes by reducing the initial divergence. Beams of more than 130 mA total current are now currently extracted. New reliability tests were performed to analyze EMI-hardened device improvements as well as automatic procedures. The first deuteron pulsed beam measurements are briefly reported. The source is presently disassembled to be re-installed at its final location in the IPHI building.

PROTON SOURCE PERFORMANCE

General Layout

The ECR source operates at 2.45 GHz and the 875 Gauss ECR axial magnetic field is provided by 2 coils. The RF power is fed to the source through rectangular waveguides and a specific ridged transition. The source and its ancillaries are located on a 100 kV platform behind a protective cage. The source is running at 95 kV. A 5-electrode extraction system allows easy meniscus tuning to minimise beam losses and backstreamed electron limitation.

Different classical diagnostics (Faraday cup, insulated screens, cameras, current transformers, emittance measurement unit) allow beam characterisation in the 2-solenoid Low Energy Beam Transport (LEBT) line. As high beam density precludes from using interceptive diagnostics, specific optical diagnostic developments are in progress to analyse the beam at the RFQ matching point [3].

2 turbomolecular pumps (1000 l/s each) are used to take care of source and LEBT vacuum. The working pressure (with a 5 sccm hydrogen flow) turns out to be higher than 0.1 Pa in the plasma chamber and varies from 2 to 1 mPa in the LEBT.

Beam intensity and noise

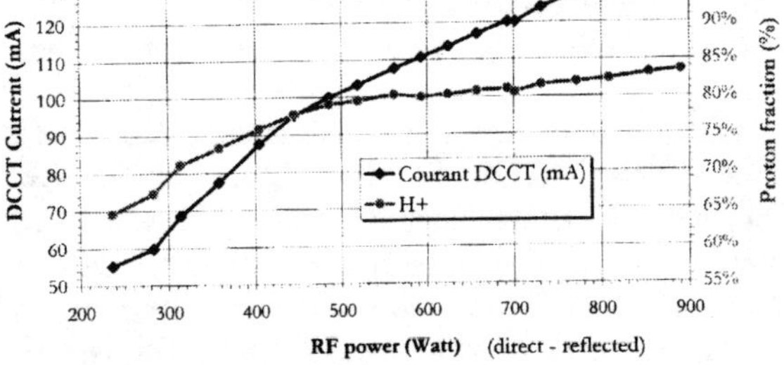

FIGURE 1: Extracted beam and proton fraction vs rf power

CP642, *High Intensity and High Brightness Hadron Beams: 20th ICFA Advanced Beam Dynamics Workshop on High Intensity and High Brightness Hadron Beams,* edited by W. Chou, Y. Mori, D. Neuffer, and J.-F. Ostiguy

The extracted beam intensity is checked by using DC or AC toroid as well as Faraday cup. The DCCT is located at the exit of the source and plots the total extracted current. A maximum of 157 mA has been observed and current as high as 130 mA are routinely extracted. **FIGURE 1** shows the total extracted beam and the proton fraction plotted versus the effective RF power transferred to the plasma.

The 6 MHz bandwidth ACCT installed between the 2 solenoids allows beam noise analysis and/or pulsed beam monitoring. 19 kHz oscillations are transferred to the plasma from the magnetron RF switched power supply (see the recorded spectrum obtained with a 120 mA total beam **FIGURE 2**). A 50 Hz ripple due to alternative magnetron filament heating is also observed. Nevertheless, rms beam noise lower than 1 % is currently achieved and is compatible with the RFQ requirements.

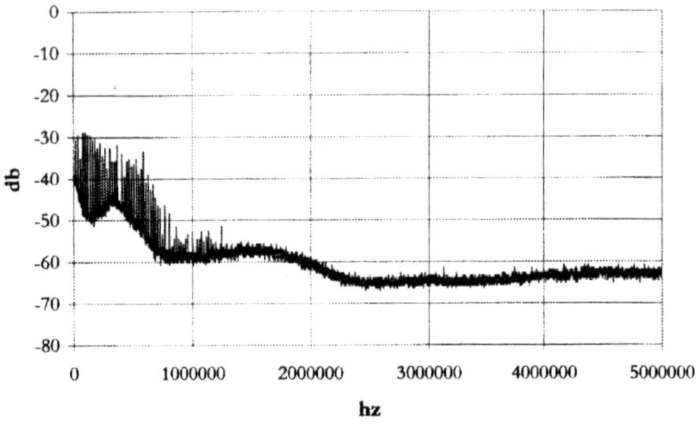

FIGURE 2: Beam noise spectrum for 120 mA extracted beam

Classical CCD cameras are used for beam position monitoring. Beam interacts with the residual gas in the LEBT and produces excited and ionised gas atoms. In the visible region, emitted light (mainly the Balmer lines of the atomic hydrogen spectrum) gives information on the beam position and current [4].

Emittance measurements

The emittance measurement unit could be placed in two different places either at the exit of the source (0.53 m from extraction) or 3.9 m further behind the 2 LEBT solenoids. First, the emittance has been analysed as a function of the intermediate electrode voltage which modifies the extraction electric field configuration (**FIGURE 3**). This experiment has been done with a 97 mA proton beam (120 mA total). Extraction simulations indicated a minimum emittance value for a 40 kV first gap extraction voltage which has been confirmed by the measurement. The r-r' rms normalized emittance value varies from 0.175 to 0.15 π.mm.mrad while the intermediate electrode voltage is going from 27 to 49 kV. The minimum value has been obtained for 43 kV.

Then the EMU was moved to the second location. Emittance measurements were performed by tuning the intensity, the LEBT magnetic configuration and the intermediate electrode voltage. The beam emittance at the end of the LEBT depends dramatically on the LEBT solenoids values and looks quite stable by varying the extraction electric field configuration. A minimum value of 0.23 π.mm.mrad has been observed downstream a cross over located at the future RFQ entrance for 70 mA transported proton beam.

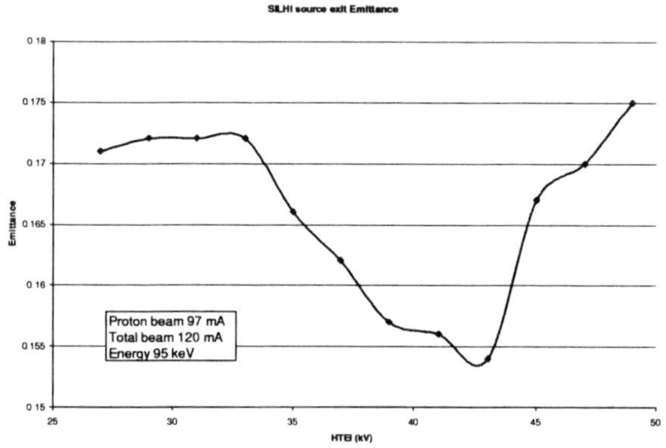

FIGURE 3 : Source emittance vs intermediate electrode voltage

Previous work [5] also indicated that the emittance is strongly dependent on the space charge compensation in the beam line. Heavy gas injection (Ar or Kr) in the LEBT creates a large amount of free electrons which allows space charge compensation and emittance improvements.

Reliability tests

Since SILHI delivered its first beams in 1996, several specific tests have been performed to analyse the source reliability-availability. To minimize possible breakdown and to optimise the reliability, different developments and technical choices were progressively adopted.

The following list presents several items developed in this framework : Quartz window protected behind a water cooled bend, electrode shape optimisation to minimize the electric field and the spark rate, large safety margins on all Power Supplies (HV and others), optimisation of Power Supplies air or water cooling, separate cable path and shielding for signals and power, galvanic insulation of analog and digital signals, use of EMI hardened devices especially for all sensitive electronics and PLC, development of beam current feedback, development of EPICS automatic start/restart procedures.

Table 1 summarises the 5 reliability runs. This

table shows that the reliability-availability can reach higher than 99.5 % with a very low number of beam off within a whole week. Since the source remote control is completely updated with the EPICS system, automatic procedures and home internet network connections allow us to leave the source working without any operator locally. In March 2001, an oil LEBT contamination led to a very high spark rate. To minimize new contamination risk, dry pump will be nearly installed.

Parameters	Déc. 97	Mai 99	Oct. 99	March 01	June 01
Energy (keV)	80	95	95	95	95
Intensity (mA)	100	75	75	118	114
Duration (h)	103	106	104	336	162
Beam off number	**53**	**24**	**1**	**53**	7
MTBF (h)	**1.75**	**4**	**n. appl.**	**≈ 6**	**23.1**
MTTR (mn)	**6**	5.3	2.5	**≈ 18**	2.5
Uninterrupted beam (h)	**17**	**27.5**	**103**	**25**	**36**
Availability (%)	**94.5**	**97.9**	**99.96**	**95.2**	**99.8**

TABLE 1. Reliability tests

Up to now, more than 800 hours continuous operation gave lots of information to optimise the source behaviour. Several weak points have been solved. No spark occurs without beam after specific accelerator column conditioning. The use of EMI hardened devices enhanced dramatically the source performance. Sparks now do not lead to power supplies failures or PLC reboots. Moreover, as the source is also dedicated to other experiments like diagnostic developments or EPICS control improvement, short experiments indicated a lower spark rate with a lower beam intensity as well as with pulsed beam. These results will have to be confirmed by specific runs.

DEUTERON SOURCE PRODUCTION

For the IFMIF project, CEA analysed the characteristics of the deuteron beam produced by such an ECR source (with the 120 mA proton extraction system). To minimise structure activation, the deuteron beam has been produced in pulsed mode (2ms/s) by modulating the 2.45 GHz magnetron power. A 135 mA-100 kV beam has been easily produced and guided through the beam line with a 75 % transparency (**Figure 4**). The deuteron fraction reached more than 96 % and D_2^+ was lower than 4 %. No D_3^+ or heavy masses were observed.

Coherent measurements (intensity, beam noise, species fraction) were checked for a 135 mA total extracted beam. As a result, the SILHI source looks as well adapted for deuteron production as for protons. As the (d,D) reaction at the surface of the copper target produces 2.45 MeV neutrons, leading to a progressive activation level increase, the experiment has been done within only 2 days. Appropriate shielding will be necessary for further experiments.

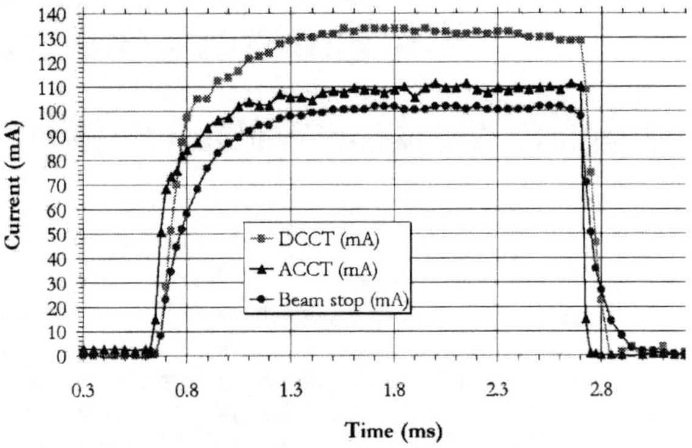

Figure 4. LEBT transparency for 135 mA deuteron beam.

CONCLUSION

The SILHI source, based on ECR plasma generation was built in 1996 and is, since then, regularly producing high intensity light ion beams. Table 2 summarises the beam characteristics either for Protons or Deuterons.

Particles	PROTON		DEUTERON	
Parameters	Requests	Status	Request	Status
Energy [keV]	95	95	95	100
Intermediate Electrode [kV]	55	56	?	50
Proton , Deuteron Current [mA]	100	108	140	129
Total Current [mA] **(I max)**	110	130 **(157)**	155	135 **(166)**
Proton, Deuteron Fraction [%]	> 90	83	> 90	96
Plasma electrode diameter [mm]	-	9	-	9
Current Density [mA/cm²]	140	204	243	212
Availability [%]	AHAP	> 99	AHAP	-
RF Forward Power [W]	< 1200	850	< 1200	900
Duty Factor [%]	100	100	100	0.2 *
H_2 , D_2 Gas Flow [sccm]	< 10	5	< 10	1
Beam Noise rms [%]	2	1.2	2	1.2
Source exit rms normalized emittance [π.mm.mrad]	0.2	0.15 @120 mA	0.2	-

* **Limited by neutron production**

TABLE 2. SILHI beam characteristics

SILHI has been recently stopped to be moved in the future IPHI building. A previous move showed that a minimum of 6 months will be necessary before to restart the source and recover the best results.

REFERENCES

1. P-Y. Beauvais: "Status Report on the construction of the French High Intensity Proton Injector (IPHI)", proceeding of EPAC2002, Paris, FRANCE
2. R. Gobin et al.: RSI, Vol.73, n°2, February 2002 (922)
3. P. Ausset et al.: Optical Transverse Beam Profile Measurements for High Intensity Proton Beam, proceedings of EPAC2002, Paris, FRANCE
4. B. Pottin et al.: "Optical beam profiler for high current beams", proceedings of EPAC 2000, Vienna, Austria
5. P-Y. Beauvais et al.: RSI, Vol.71, n°3, June 2000, (1413)
6. R. Ferdinand et al.: "Deuteron Beam Tests for IFMIF", proceedings of EPAC2002, FRANCE.

Review of Scaled Penning H⁻ Surface Plasma Source with Slit Emitters for High Duty Factor Linacs

Joseph D. Sherman, William B. Ingalls, Gary Rouleau, and H. Vernon Smith, Jr.

Los Alamos National Laboratory, Los Alamos, NM87545

Abstract. The Penning H^- surface plasma source (SPS) has been used at Rutherford Appleton Laboratory (RAL) for 20 years to provide the required H^- beams for charge-exchange injection into the 800-MeV proton synchrotron on the ISIS spallation neutron source. The RAL source is based on the first H^- Penning SPS operated at Los Alamos. Since that original technology exchange, Los Alamos has developed scaled-up versions of the Penning H^- SPS with the goal of extending the H^- beam duty factor (df) while maintaining high beam brightness. A 250-mA H^- beam with rms normalized emittance of <0.3 (πmm-mrad) in both transverse planes has been extracted from a 4X scaled Penning source at a discharge df of 0.5%. Using discharge scaling laws and the 250 mA H^- current results, it is predicted that a 4X Penning H^- SPS with a slit emitter would be capable of producing >100-mA, low emittance H^- beams in the 5% df range. A source with these parameters would be suitable for the European Spallation Source (ESS) and other high-power proton driver projects.

INTRODUCTION

Proposed accelerator drivers for short and long pulse spallation targets for neutron beam production place demanding requirements on the injector section[1]. The European Spallation Source (ESS) proposal, for instance, asks in the short-pulsed mode for 100-mA H^- beams in a linac operating at 5% duty factor (50 Hz, 1.0 ms)[2]. The purpose of this paper is to review the dimensionally and discharge-power scaled H^- Penning surface-plasma sources (SPS) using slit emitters in terms of high H^- current and comparatively long duty factor (df) requirements (5%). Previously published work on the 4X scaled Penning source at Los Alamos showed that 250-mA H^- current could be extracted at 29 keV from a slit emitter[3] at 0.5% df (5 Hz, 1 ms). This attractive beam current was somewhat overshadowed by asymmetric emittances of 0.15 X 0.29 (πmm-mrad) rms, normalized, which was thought to be difficult to match to a radio-frequency quadrupole (RFQ). However, over the past ten years RFQ design activities have shown that RFQs with a beam acceptance of 0.3 (πmm-mrad) are possible while still maintaining a high-quality linac design[2]. Further, design studies of matching beams with asymmetric emittances in the transverse planes to an RFQ using solenoid-lens[4] and quadrupole magnet focusing[5] low-energy beam transport (LEBT) systems indicate successful matches are possible. Thus there is a renewed interest and acceptance of the use of slit beams as injectors for RFQs.

The following section summarizes the Penning H^- discharge scaling laws as developed at Los Alamos National Lab(LANL)[3]. Historical comparison of the LANL Penning and the Rutherford-Appleton Labs (RAL) Penning SPS H^- source performances are made. The final section summarizes conclusions concerning the H^- Penning SPS scaling laws, and the possibility of attaining a slit-based Penning H^- SPS with beam characteristics of >100 mA, transverse plane emittance < 0.3 (πmm-mrad), and 5% df. For background literature for H^- production using cesiated cathode surfaces see[6], and other publications contained in this series.

PENNING H⁻ SPS SCALING LAWS AND HISTORICAL PERSPECTIVE

Application of discharge scaling laws to the SPS H^- Penning discharge for scaling source dimensional parameters up by the factor 4 in two of the three spatial dimensions has led to relationships among the discharge parameters summarized in the equations of Table 1. Here the subscript 4X refers to the Penning SPS H^- source with a cathode area four times the 1X source. These equations are based on the general theory of surface H^- production[6] and discharge scaling laws contained in standard texts[7] on discharge physics. The prospect for increased 4X source df (df_{4X}) is related to the decreased cathode power load, F_{4X}. A fundamental limit to the stability of the Penning discharge and hence stable H^- beam formation is the cesium equilibrium concentration in the discharge volume and cathode surfaces[8]. The pulsed temperature rise of a surface[9] is given by $\Delta T_{nX} = 2F_{nX}\mathrm{sqrt}(\Delta t/(\pi K\rho C))$ where ΔT_{nX}=pulsed temperature rise, F_{nX}=cathode power flux (W/cm^2)=$V_{nX}I_{nX}/(3CA_{nX})$, Δt=pulse length(s),

CP642, *High Intensity and High Brightness Hadron Beams: 20th ICFA Advanced Beam Dynamics Workshop on High Intensity and High Brightness Hadron Beams,* edited by W. Chou, Y. Mori, D. Neuffer, and J.-F. Ostiguy
2002 American Institute of Physics 0-7354-0097-0

K=thermal conductivity constant(W/cm°C), ρ=density(gm/cm^3), and C=specific heat (J/gm°C).

TABLE 1. Predicted relationships of the 4X and 1X ion source discharge parameters. The lower case parameters i and j refer to the extracted H^- beam. Upper case characters refer to the ion source discharge parameters. Typical units are indicated.

	DISCHARGE PARAMETER	4X – 1X RELATIONSHIP
1	Discharge pressure (Torr)	$P_{4X} = P_{1X}/4$
2	H_2 gas mass flow (sccm)	$Q_{4X} = Q_{1X}$
3	Discharge magnetic field (G)	$B_{4X} = B_{1X}/4$
4	Discharge Voltage (V)	$V_{4X} = V_{1X}$
5	Discharge Current (A)	$I_{4X} = I_{1X}$
6	Cathode area (cm^2)	$CA_{4X} = 4CA_{1X}$
7	Cathode power load (kW/cm^2)	$F_{4X} = F_{1X}/4$
8	Discharge current densities (A/cm^2)	$J_{4X} = J_{1X}/4$
9	Emission area (cm^2)	$EA_{4X} = 4EA_{1X}$
10	Extracted H^- beam current (mA)	$i_{4X} = i_{1X}$
11	Extracted H^- beam current density (mA/cm^2)	$j_{4X} = j_{1X}/4$
12	H^- beam production efficiency (mA/kW)	$\xi_{4X} = \xi_{1X}$

The index n equals one or four for the discussion in this paper. The factor (1/3) in the surface temperature increase equation comes from equipartitioning the total discharge power to each of the two cathodes and to the anode. The fundamental hypothesis of the 4X technology is that by decreasing the pulsed surface temperature rise the proper cesium balance can be maintained at longer duty factors. The beam current density defined as entry 11 in Table 1 is defined as $j_{nX} = i_{nX}/EA_{nX}$. The emission area EA_{nX} is equal to the emission slit length (y_{nX}) multiplied by its width (x_{nX}) – please see Table 2. The x_{4X} = 2.8mm is somewhat larger than the value of 2.0mm desired from the factor 4X scaling, and was the result of a fabrication error.

Table 2 shows dimensional comparisons of the LANL 1X source, the RAL source, and the 4X source as developed at LANL. It is noted the LANL 1X and RAL Penning sources have nearly identical dimensions, reflecting the earlier technology exchange between the two labs[10]. Thus, both the LANL 1X and RAL sources may be referred to as 1X sources. Further definition of the ion source and emission slit components are described in Fig. 2 of ref.[3].

TABLE 2. Comparison of the LANL and RAL 1X SPS dimensions with the 4X source.

COMPONENT	RAL 1X	LANL 1X	LANL 4X
Ion Source			
L (mm) Cath.-cath. gap	5	4.3	17
W (mm) Discharge depth	2	3	17
T (mm) Discharge length	10	12	16
Emission slit			
y (mm) Perp. to B field	10	10	11.4
x (mm) Parall. to B field	0.6	0.5	2.8

Figure 1 shows the measured cathode power load F_{nX} for the 1X and 4X sources plotted versus performance publication date. The RAL 1X source data are from ref.[11,12], and the LANL 1X source data are from ref.[13,14]. The i_{nX} and df for the various sources are shown in Fig. 1. An observation for the 1X sources is that H^- current is inversely rated to the duty factor with the product of $(i_{1X})(df_{1X})$ = 80 (mA%) being nearly constant. The exploratory experiment completed for the 4X slit emitter which produced 250 mA H^- did not probe the df_{4X} limit, although the 4X source has been operated at 2.3% df_{4X} with circular emission apertures with the instantaneous cathode power load of 2.5kW/cm^2[3], slightly more than the 2.2kW/cm^2 for the slit beam result shown in Fig. 1. It is readily apparent from previous 4X circular aperture measurements and the data contained in Fig. 1 that the 4X slit source df_{4X} could be increased.

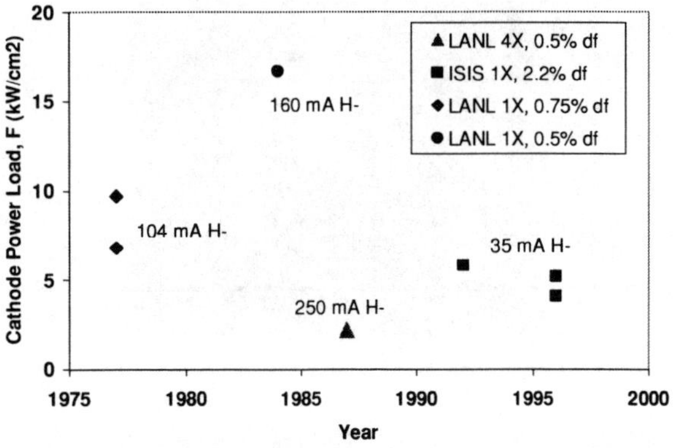

FIGURE 1. Measured cathode power loads for the 1X and 4X scaled H^- Penning SPS.

Another useful factor to consider in scaled H^- Penning SPS technology is the H^- beam production efficiency ξ_{nX} shown as entry 12 in Table 1[3]. The definition of $\xi_{nX} = j_{nX}/F_{nX}$, thus $\xi_{4X} = \xi_{1X}$ is expected from the scaling laws. The publication data used in

Fig. 1 is used to construct Fig. 2, where ξ_{nX} is plotted vs year for the 1X and 4X sources. ξ_{nX} can be thought of as the global efficiency of moving H^- ions from their cathode birthplace to the emission slit aperture by the resonant charge-exchange process on slow atomic hydrogen ions, and the subsequent H^- attenuation as they move to the emission slit[6,15]. The LANL 1X source ξ_{1X} is somewhat greater than the RAL 1X. This difference may be due to slightly different anode geometries, where i_{1X} are exponentially dependent on anode length in the direction of the emission slit[15]. The outstanding feature observed in Fig. 2, however, is the factor two increased cathode power efficiency of the 4X source compared to the 1X sources.

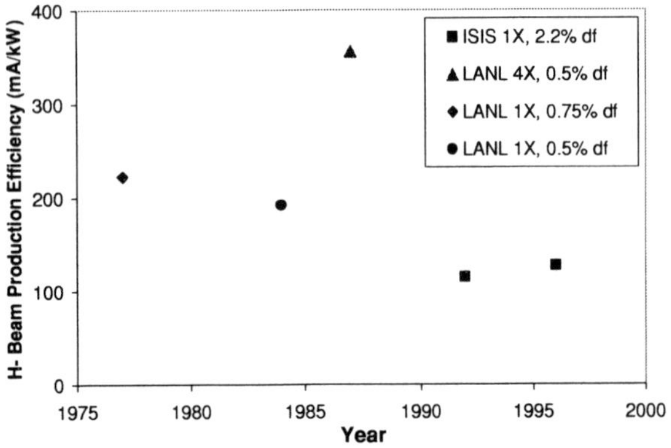

FIGURE 2. H^- beam production efficiency for the 1X and 4X sources plotted vs data publication year.

4X SCALING LAW RESULTS

The 250-mA H^- beam pulse at 29 keV extraction energy is shown in Fig. 3. The beam noise of +/- 1%

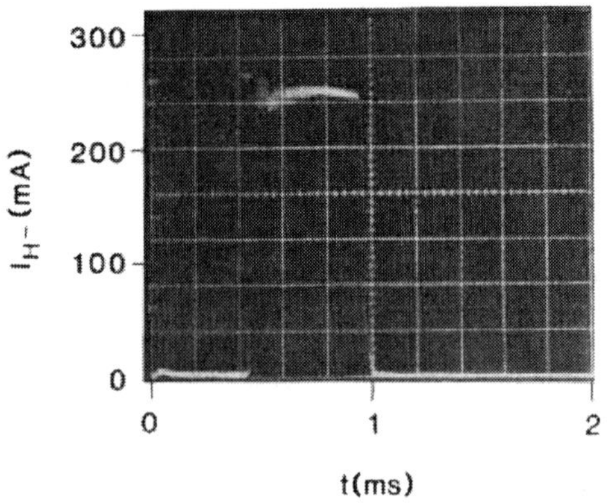

FIGURE. 3. 250-mA H^- beam pulse obtained from the 4X slit source.

is compatible with linear accelerator applications. Figure 4(A) shows the measured x-plane (narrow slit dimension) phase space and Fig. 4(B) shows the measured y-plane (long slit dimension) phase space measurement corresponding to the 250-mA H^- current. These phase-space measurements were made approximately 10 cm after the extraction electrodes with high resolution electric sweep phase-space scanners[16]. Corresponding x-y phase-space emittances at the 250-mA H^- current are ε_x X ε_y = 0.15 X0.29 (πmm-mrad) rms, normalized. The emittances are derived by extrapolating the measured rms emittances at beam fraction F (ε_{rms}(F)) to 100% beam fraction assuming a Gaussian beam model[17]. For the case of the x-plane phase-space distribution, the beam appears to be Gaussian to F = 0.95. For the y-plane distribution, deviations from the Gaussian model occur at F = 0.63. The y-plane non-Gaussian behavior at higher beam fractions is attributed to the slit-end aberrations seen in Fig. 4(B). These aberrations could

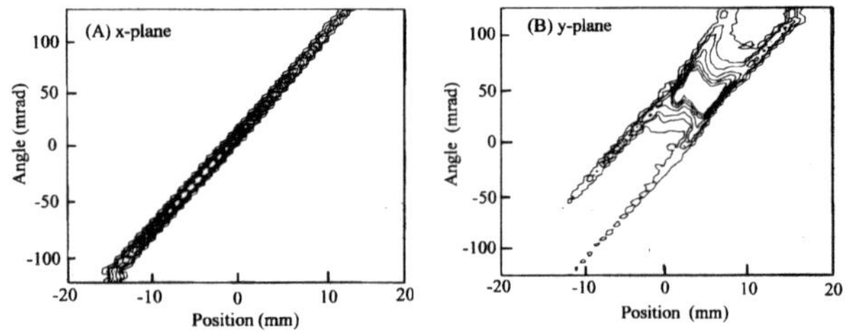

FIGURE 4. A set of x (A) and y (B) phase-space measurements on a 29-keV, 250-mA H^- beam extracted from a slit emitter.

be reduced with improved extractor design. The 250-mA H^- current and emittance measurements were made in two days in 1987, and no further characterization of the source was made as no apparent application was envisioned.

Table 3 summarizes the 1X to 4X scaling results[3] for the slit beam. Column 1 gives the ion source parameter, column 2 gives the 1X actual design and column 3 gives the 4X design based on the scaling laws of Table 1. Column 4 shows the actual result for the optimized 4X source, and column 5 shows results for an aggressively cooled 4X cathode operating in the discharge-only mode[18]. There is generally good agreement between column 3 (4X design) and column 4 (4X actual). The discharge depth W_{4X} was increased from 12 to 17 mm to obtain a quiescent discharge that produces a low-noise beam. To reach the j_{4X} design a cathode power density of only 2.2 kW/cm^2 was required, about a factor of two lower than the 4.2kW/cm^2 scaling law prediction. This reduction results in a factor 2 increase in the 4X H^- beam production efficiency (cf. Fig. 2) compared to that of the 1X sources. Operating at j_{4X} = 783 mA/cm^2 resulted in i_{4X}=250mA. One may expect a factor of 5.6 increase in the 4X source x-plane emittance[17] based on the increase of slit width from 0.5mm to 2.8mm, but only a factor of 2.5 emittance increase is observed (0.15 πmm-mrad vs. 0.06 πmm-mrad).

There is a 70% increase in the y-plane emittance, which may be decreased by optimized extractor design.

After completion of the high-current slit beam work on the 4X source in 1987, addition of 1-3% of N_2 gas to the Penning H_2 discharge was found to give beneficial effects for quieting the extracted beam current. This effect is shown in Fig. 5, where the quieting effect of N_2 gas in the Penning 4X discharge is demonstrated. All source parameters were the same in Figs. 5(A) and 5(B), except that approximately 1%

TABLE 3. Experimental results for the application of the scaling laws to the 4X H^- Penning SPS using slit emitters. No slit aspect ratio study has been made on the 4X source to optimize H^- beam brightness.

ION SOURCE PARAMETER	1X ACTU AL	4X DESI GN	4X ACTU AL	4X DISCH. ONLY
L (mm)	4.3–5	17	17	17
W (mm)	2 – 3	12	17	17
T (mm)	12	16	16	16
B-field (G)	2200	550	460	500
Duty factor (%)	0.5	2-3	0.5 meas	6%
Disch. voltage (V)	100	100	93	115
I (A)	180	240	197	125
i (mA)	160	160	250	Not meas
j_{H^-} (mA/cm^2)	3200	800	783	Not meas
F (kW/cm^2)	16.7	4.2	2.24	1.76
ξ (mA/kW)	192	192	349	Not meas
Emission slit (y,x) (mm)	10, 0.5	10, 2.0	11.4, 2.8	
Emittance, rms norm (πmm-mrad)	0.17y0 .06x		0.29y 0.15x	Not meas.

N_2 gas was introduced in the discharge of Fig. 5(B). The N_2 effects were mostly studied for circular aperture beams[3], and the effect on the beam noise

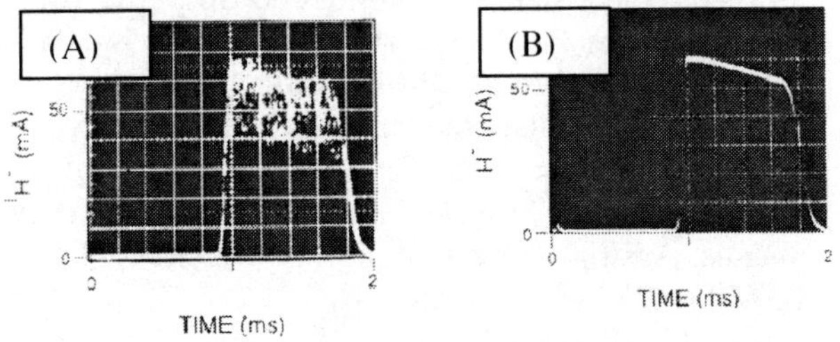

FIGURE 5. Photos of beam extracted from a circular aperture Penning 4X source. The discharge parameters are identical, except for the addition of 1% N_2 gas in (B).

and phase space of H^- beams extracted from slit emitters remains largely unknown. Emission spectroscopy measurements[19] showed that the Cs^+ and Mo^+ ion densities were increased when the 4X Penning discharge was operated with small (1%) N_2 concentrations.

This paper is concluded by a discussion of the application of the 4X technology to 100-mA, 5% df_{4X} H^- beams. First, from the fifth column of Table 3 the 4X source has already operated at 6% df_{4X} at I_{4X} = 125A of arc discharge current[18]. Thus, to reach nominal 6% df at 100-mA H^- from the measured 4X power efficiency in column 4, only I_{4X} = 80A of arc discharge current would be required, as $I_{4X}(i_{4X}$=100mA) = (100mA/250mA)197A = 80A. This scaling implies that the H^- beam production is a linear function of the Penning discharge current. This scaling has been demonstrated from 10^{-2} to 10^2 A arc current in the 1X source[13]. Another argument is the observation that a circular emitter beam on the 4X source operated at 180A discharge with 2.3% df_{4X}[3]. Assuming the product of i_{4X} and discharge df_{4X}(%) is constant (as observed in the LANL and RAL 1X sources – see discussion above), one can write $(i_{4X})(df_{4X})$ = 250mA(180A/197A)(2.3%) = 525(mA%), or at 5% df_{4X} a 105 mA H^- beam could be obtained. Using the linear H^- beam/arc discharge current scaling the 105 mA H^- would require I_{4X} = 83A of discharge current, as given by $I_{4X}(i_{4X}$=105mA) = (105mA/250mA)*197A = 83A. These two rather independent scalings indicate order 100 mA H^- beam may be available from the 4X slit source at 80A discharge current for df_{4X} = 5%. If such performance could be demonstrated, the Penning 4X technology could be considered for application to high-power linac drivers[2].

There is an approximate factor of 6 increase in $(i_{nX})(df_{nX})$ when the 4X ($i_{4X}df_{4X}$ = 525mA%) is compared to the 1X source ($i_{1X}df_{1X}$ = 80mA%). A factor of 4 increase may arise from the decrease in F_{4X} when compared to F_{1X}. The other unexpected factor of 2 may arise from the increased H^- production efficiency of the 4X compared to the 1X. The enhancement factors are apparently additive. One may conclude that this empirical observation of times 6 enhancement of $(i_{4X})(df_{4X})$ compared to $(i_{1X})(df_{1X})$ represents a confirmation of the original 4X technology hypothesis: reducing the pulsed cathode temperature rise would allow extended df operation of the H^- Penning SPS.

Assuming ion source development confirms the Penning H^- discharge scaling law assertions contained in this paper and in the earlier work[3], substantial issues for a 100-mA H^- source and injector remain. The source must be consistent with reliability and lifetime requirements of the total accelerator facility. Given the I_{4X} = 80A for 100-mA H^- is realized for the 4X source and assuming V_{4X} = 100V, the cathode power loading $F_{4X}(i_{4X}$=100mA) = (100V)(80A)/(3*CA_{4X}) = 0.99kW/cm^2, where CA_{4X} = W_{4X}*T_{4X} = 2.7cm^2 (cf Table 2.). This scaled F_{4X} is already lower than the cathode power loads shown in Fig. 1. The 1X Penning SPS H⁻ source is continuing to demonstrate reliable operation on the ISIS facility at RAL[20], with F_{1X} in the range of 4kW/cm^2, so there is hope the 4X would offer improved lifetime, at least from the viewpoint of cathode erosion. If a magnetic focusing, beam space-charge neutralized LEBT were chosen for RFQ matching of a 100-mA H^- beam, little emittance growth would be allowed to maintain an overall 0.3 (πmm-mrad) injector emittance specification. The extracted ion beam energy would have to be carefully chosen to minimize emittance growth in the transport elements, emittance growth from beam plasma effects[14,21], and then satisfying minimal emittance growth in the extraction process. The beam plasma effects can be controlled by operating at higher H^- beam energies in the LEBT, introduction of xenon gas in the LEBT, and keeping the LEBT length short[14].

REFERENCES

1. G. I. Dimov, "Use of Hydrogen Negative Ions in Particle Accelerators", Rev. Sci. Instrum. 67, (October, 1996), 3393.
2. R. Duperrier, et. al., "The ESS Front End Associated with the SC Linac", ESS TAC Meeting n° 1, FZ Juelich, January 7-9, 2002.
3. H. Vernon Smith, Jr., et. al., "H^- and D^- Scaling Laws for Penning Surface-Plasma Sources", Rev. Sci. Instrum. 65, (January, 1994), 123.
4. C. W. Planner, "Matching Unequal Transverse Emittances from an H^- Ion Source into a RFQ", Particle Accelerators 48, (1995), 243.
5. O. R. Sander, et. al., "Transverse Emittance of a 2.0-MeV RFQ Beam with High Brightness", in the Proc. of the 1985 Particle Accelerator Conference, Vancouver, BC, Canada, IEEE Trans. on Nuclear Sci., (Oct., 1985), 2588.
6. Yu I. Belchenko, et. al., "Physical Principles of the Surface Plasma Method for Producing Beams of Negative Ions", in the Proc. of the Symposium on the Production and Neutralization of Negative Hydrogen Ions and Beams, Brookhaven New York, BNL 50727, (1977), 79.
7. A. von Engel, *Ionized Gases*, Second Edition, Oxford University Press (1965), 288.
8. G. E. Derevyankin and Vadim Dudnikov, "Production of High-Brightness H^- Beams in Surface Plasma Sources", in the Proc. of the Third International Symposium on the Production and Neutralization of Negative Hydrogen Ions and Beams, Brookhaven New York, AIP Conference Proceedings No. 111, (1983), 376.
9. H. S. Carslaw and J. C. Jaeger, *Conduction of Heat in Solids*, Second Edition, Oxford University Press (1959), 401.
10. P. E. Gear and R Sidlow, Proc. of the Second International Conference on Low-Energy Ion Beams, Institute of Physics Conf. Series No 54(1980), 284.
11. R Sidlow and N D West, "Performance of the Penning H- Ion Source at ISIS", in the Proc. of the Third European Particle Accelerator Conference, Berlin, Editions Frontieres, (March, 1992), 1005.
12. R. Sidlow, et. al., "Operational Experience of Penning H^- Ion Sources at ISIS", in the Proc. of the Fifth European Particle Accelerator Conference, Sitges (Barcelona), Instituite of Physics Publishing, (June, 1996), 1525.
13. Paul W. Allison, "A Direct Extraction H^- Ion Source", IEEE Transactions on Nuclear Science, Vol. NS-24, No. 3 (June 1977), 1594.
14. Paul Allison and Joseph D. Sherman, "Operating Experience with a 100-keV, 100-mA H^- Injector", in the Proc. of the Third International Symposium on the Production and Neutralization of Negative Hydrogen Ions and Beams, Brookhaven New York, AIP Conference Proceedings No. 111, (1983), 511.
15. P. W. Allison, et. al., "H^- Ion Source Research at Los Alamos", in the Proc. of the Second International Symposium on the Production and Neutralization of Negative Hydrogen Ions and Beams, Brookhaven New York, BNL 51304, (Oct., 1980), 171.
16. Paul W. Allison, et. al., "An Emittance Scanner for Intense Low-Energy Ion Beams", IEEE Transactions on Nuclear Science, Vol. NS-30, No. 4, 2204 (1983).
17. Paul Allison, et al., "Comparison of Measured Emittance of an H^- Ion Beam with a Simple Theory", Los Alamos Technical Report, LA-8808-MS (1981).
18. H. Vernon Smith, Jr., et. al., "H^- Ion source with High Duty Factor", in Proc. of the 1987 IEEE Particle Accelerator Conf., (March, 1987), 301.
19. H. Vernon Smith, Jr., "Emission Spectroscopy of the 4X Source Discharge with and without N_2 Gas", AT-10 Technical Note: 89-07, (April, 1989).
20. J. W. G. Thomason and R. Sidlow, "ISIS Ion Source Operational Experience", in the Proc. of the 7th European Particle Accelerator Conference, Vienna, Austria, Austrian Academy of Sciences Press, (2000), 1625.
21. D. G. Dzahbarrov and A. P. Naida, "Motion of Positive Ions in the Plasma Produced by a Dense Beam of Negative Ions in a Gas", Sov. J. Plasma Phys. 6(3) (1980), 316.

The ISIS Penning Ion Source and the Volume Ion Sources from Frankfurt and DESY for H⁻ Production

J. W. G. Thomason*, H. Klein#, J. Peters@

*CLRC RAL, Didcot, Oxon, UK
#IAP, Germany
@DESY, Germany

Abstract. H⁻ ion source development work at Rutherford Appleton Laboratory (RAL), the Institut für Angewandte Physik (IAP) in Frankfurt and the Deutsches Elektronen-Synchrotron (DESY) will be described, along with future avenues of research. The most recent values of important parameters for each ion source will be presented in the context of next generation particle accelerators, particularly the proposed European Spallation Source (ESS). A brief progress report on the ISIS RFQ test stand, on which many of the RAL ion source measurements were taken, will be included.

INTRODUCTION

A key element in the design of the ESS is the ion source for producing pulsed currents at relatively high duty cycles. The minimum requirement for the ESS short pulse option is a source delivering a 60 mA H⁻ ion beam in 1.0 ms pulses at a repetition rate of 50 Hz (5% duty cycle) and with a low normalised rms emittance ~ 0.1 π mm mrad (although a normalised rms emittance of 0.2 – 0.3 π mm mrad at the entrance to the RFQ would be acceptable). If H⁻ ions are also to be used for the ESS long pulse option a similar source will be required, but delivering 2.0 ms pulses at a repetition rate of 50/3 Hz (3.3% duty cycle). In addition to meeting these requirements these sources must operate with high availability and reliability[1].

It has been recognised that no existing H⁻ ion source can immediately fulfil the needs of the ESS, but that a source to meet the minimum requirements could be developed from present technologies within the time scale of building the ESS[1, 2]. Two types, the volume source and the Penning source, have been confirmed as suitable candidates for development.

THE FRANKFURT VOLUME SOURCE

The low electron temperature and the high achievable current densities which characterise volume sources make them one of the candidates for the ESS injector. Several source prototypes have been used in experiments at IAP in Frankfurt to test different source geometries and methods of plasma confinement, and to optimise the cathode configuration.

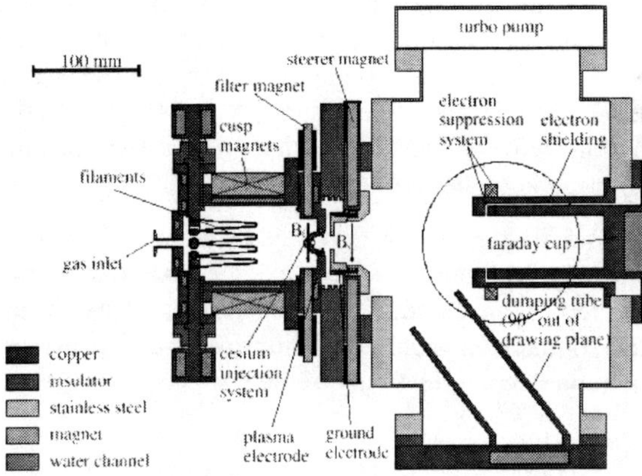

FIGURE 1. Schematic cross section of the Frankfurt H⁻ volume source.

As a result of these experiments, the ion source shown in figure 1 has been chosen for further study[3]. The plasma chamber of the ion source is made of a water-cooled copper cylinder 7.0 cm in diameter and 10.0 cm in depth. It is surrounded by ten CoSm magnets in cusp field arrangement. Near the chamber axis are mounted four tungsten filaments of 1.8 mm diameter. The front end of the chamber is enclosed by the plasma electrode. An electromagnet is installed in the flange of the plasma electrode. Its transverse magnetic field (B_f) acts as an electron filter. An external oven for introducing caesium is mounted on the flange of the plasma electrode.

The electron beam which is extracted along with the H⁻ ions is dumped behind the extractor at full energy into a water-cooled cup. The extension of the filter magnet field into the extraction gap deflects the electron beam out of the beam axis and an additional transverse magnetic field (B_s) is then used to steer the electron beam into the dumping tube.

CP642, *High Intensity and High Brightness Hadron Beams: 20th ICFA Advanced Beam Dynamics Workshop on High Intensity and High Brightness Hadron Beams,* edited by W. Chou, Y. Mori, D. Neuffer, and J.-F. Ostiguy

Experiments have shown that, in operation with caesium, the ion source is capable of producing 120 mA of H^- beam, measured using a Faraday cup, at an arc power of 47.5 kW (50 Hz, 1.2 ms). With caesium the H^- emission current is up to 4.5 times higher, and the e/H^- ratio is about 7 at 47.5 kW. The emittance of the source has yet to be measured, but other beam diagnostics have provided a plausible estimate that the normalised rms emittance should be < 0.1 π mm mrad.

With the present filament configuration the ion source has been run for 180 hours at an arc power of 40 kW, during which the diameter of the filaments was reduced from 1.8 mm to 1.5 mm. Since the filaments can be used down to a thickness of about 0.9 mm the ion source lifetime could be expected to be about 15 days.

Caesium is injected on demand only, so there is wear of the caesium layer, which has to be rebuilt every 10 to 18 hours, requiring a reduction in the arc power for about 15 minutes. A caesium injection system will be developed that will allow caesium injection at higher powers.

The first phase of this study has shown that the Frankfurt volume ion source has the potential to fulfil the requirements for ESS, but further research and development is required to improve the lifetime and demonstrate the reliability of the source.

THE DESY VOLUME SOURCE

The DESY rf-driven volume source was developed from a design from Lawrence Berkeley National Laboratory, but incorporating an antenna situated at the wall of the source, completely covered by an Al_2O_3 ceramic[4], as shown in figure 2. By removing the antenna from the discharge chamber it is protected from sputtering damage, and so antenna lifetime no longer presents a limit to source performance. The DESY source has now been run for more than 25,000 hours with no antenna failure. In normal operation, without the addition of caesium, the source produces 48 mA of H^- ions (10 Hz, 150μs) with a normalised rms emittance of 0.2 π mm mrad.

A collar has been introduced into the source for electron suppression, and considerable effort has gone into the optimisation of its geometry, bias and material. With this collar in place an H^- current of 80 mA has been reached.

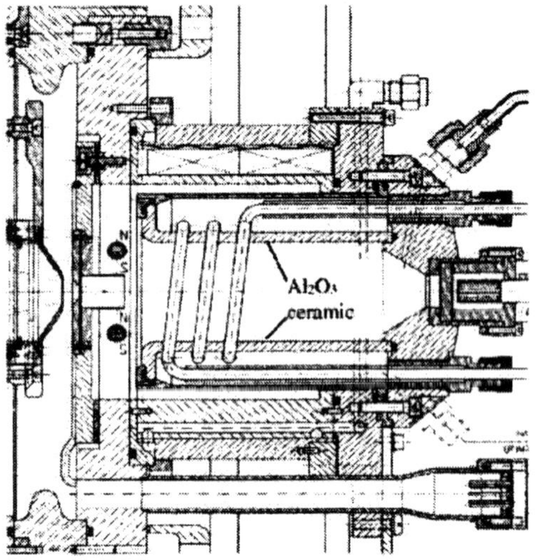

FIGURE 2. The DESY volume source with Al_2O_3 antenna cover.

An investigation of the production of an rf-driven version of the Frankfurt volume source using an external antenna based on the DESY design is planned as part of a European initiative on negative ion sources[5], and may provide a source for ESS with enhanced lifetimes.

THE ISIS PENNING SOURCE

The ISIS ion source in operation at RAL is shown in figure 3. It is a surface plasma ion source of the Penning type, and routinely produces 35 mA of H^- ions (50 Hz, 200μs for uninterrupted periods of up to 50 days[6, 7]. Caesium is used to enhance the production of H^- ions by lowering the work function of the cathode surface. In non-operational periods of ISIS the source has delivered 55 mA for sustained periods in its present form, which corresponds to a current density of ≈ 1 A cm^{-2}, which is considerably higher than is possible in volume sources. The e/H^- ratio is ≈ 1 and so a relatively low electron current is extracted simultaneously with the H^- ion beam. The ISIS source has slit extraction geometry, but the extracted beam is immediately passed through a 90° sector magnet with gradient n=1, which separates out any electrons and produces an approximately round beam profile.

A dedicated ion source development rig (ISDR) has been built at RAL that will allow further development of the source to meet the duty cycle and intensity requirements of the ESS[8]. The ISDR will allow extraction at energies greater than the present 18 keV and emittance measurements to be made close to the source.

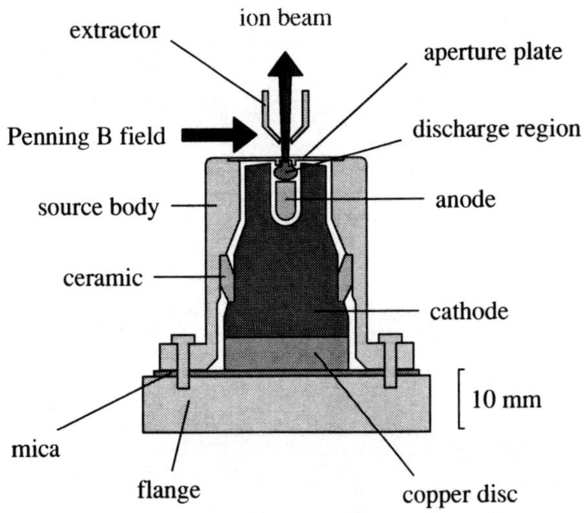

FIGURE 3. Schematic of the ISIS ion source.

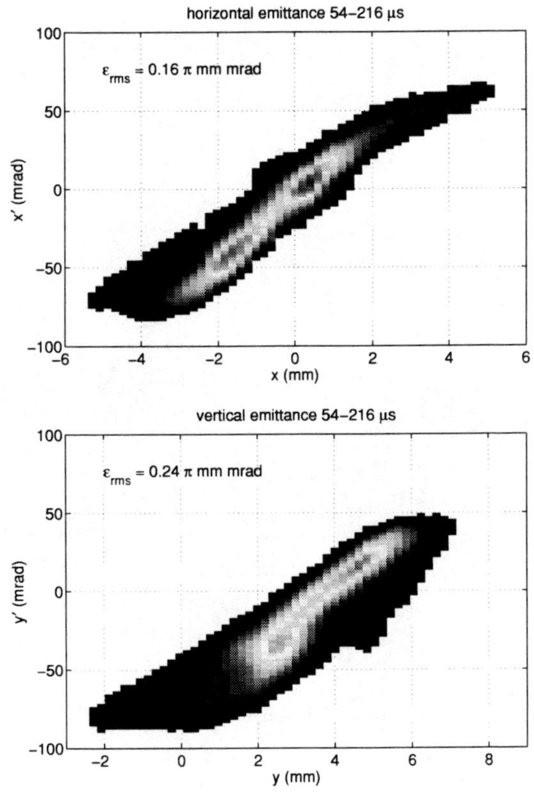

FIGURE 4. Ion source emittances from the ISIS RFQ test stand.

The source is very compact, and the primary problem to solve is that of stabilisation of the cathode and anode temperatures at the higher power levels required for the ESS short and long pulse options by means of controlled cooling. The thermal properties of the standard ISIS source will be modelled and this simulation will then be extended to longer duty cycles in order to establish how additional cooling can be introduced most efficiently. If this proves ineffective a scaling approach similar to that previously adopted on the 4X and 8X sources at Los Alamos[9] may be adopted, where increasing component dimensions has been shown to offset heating effects.

The most recent ion source results at RAL have been produced on the ISIS RFQ test stand[10] and have demonstrated normalised rms emittances of $\varepsilon_H = 0.16$ π mm mrad and $\varepsilon_V = 0.24$ π mm mrad for the standard ISIS source with a beam pulse of 200 μs at ≈ 35 mA (figure 4). In addition the solenoid matching system has been shown to provide excellent matching into the RFQ, allowing transmission factors through the RFQ of, typically, 91% for an input current of 36 mA.

Further research and development of the Penning source is needed to meet the ESS specifications, but this is not a large extrapolation from current performance and is considered feasible in the timescale of the ESS.

ACKNOWLEDGMENTS

The authors would like to thank K. Volk and R. Sidlow for invaluable contributions to ion source development at IAP and RAL, respectively, and A. P. Letchford and C. P. Bailey for data from the ISIS RFQ test stand.

REFERENCES

1. Ferdinand, R., Klein, H. and Thomason, J. W. G., ESS Workshop on Sources, 28 November 2001, Saclay, ESS report, (April 2002)
2. Schröder, L., Leung K-N., and Alonso, J. (Ed.), *Proc. of Workshop on Ion Source Issues Relevant to a Pulsed Spallation Neutron Source,* Lawrence Berkeley National Laboratory, USA, (1994).
3. Volk, K., Maaser, A. and Klein, H., LINAC 98, TH4057 (1998).
4. Peters, J., RSI 69(2), (1998).
5. High Performance Negative Ions Sources Network, HPRI-CT2001-50021, Supported by European community (2001).
6. Sidlow, R., *et al.,* EPAC 96, THP084L, (1996).
7. Thomason, J. W. G. and Sidlow, R., EPAC 2000, THP4A07, (2000).
8. Thomason, J. W. G., *et al.,* RSI 73(2), (2002).
9. Smith, H. V., *et al.,* RSI 65(1), (1994).
10. Bailey, C. P., *et al.,* EPAC 2000, THP4A03, (2000).

SESSION I

SPACE CHARGE EXPERIMENTS

Summary Report

S. Machida

KEK, Tsukuba, Japan

Experiment at the KEK PS Main Ring
S. Igarashi

Beam profiles were measured at different vertical bare tunes and clear broadening is observed when the bare tune is approaching integer 7. As a function of time, the profile is also measured 3 ms after injection and compared with the ACCSIM simulation. Both agree with each other quite nicely when the horizontal bare tune is 7.05. The ACCSIM simulation clearly identifies the octupole resonance that produces the broadening.

Experiment at the KEK PS Booster
I. Sakai

Since the bare tune of the booster is (2.17, 2.32), just above the coupling resonance of $2nx - 2ny = 0$, space charge forces may induce emittance exchange. Experimentally, that was observed when the beam intensity is increased. Emittance evolution in the first 1 ms is measured and compared with Simpsons. The result confirms that space charge is the source of the exchange, at least qualitatively.

Experiment at PSR
S. Cousineau

Beam broadening has been observed both experimentally and in ORBIT simulations. An analysis based on coherent resonance theory is presented. RMS emittance is observed as a function of several parameters such as longitudinal charge distribution and vertical bare tune. The results between theory and simulation are indeed consistent.

Measurement of the PSR Inductive Insert Ferrite Material Characteristics
K.-Y. Ng

Although the first attempt of using an inductive insert at PSR to cancel the longitudinal space charge force was successful, longitudinal microwave instabilities were observed when the bunch length is shortened. Microwave instabilities were subsequently suppressed by heating the ferrite to 130 deg C.The real and imaginary part of the ferrite impedance were measured and the impedance reduction at 130 deg C was confirmed.

Flattening of Bunch shape with Longitudinal Hollow Distribution at the CERN PSB
C. Carli

Initially, a scheme employing RF gymnastics with complicated control of phase and amplitude of the first and second harmonics was envisioned to create a hollow bunch. Simulation confirmed that a hollow bunch would indeed be created as expected; however the scheme was difficult to implement. Studies subsequently led to an improved procedure which keeps the first harmonic RF voltage constant and involves less RF gymnastics of the other parameters. Experimentally, this improved procedure was employed. The resulting distribution was measured using a tomographic technique and confirmed that the creation of hollow bunches by redistribution of phase space is possible.

Measurement of the PSI Beam
A. Adelmann

Transverse beam profiles were measured in the beam line and in the cyclotron. Simulation with space charge with MAD9P is generally in agreement, however, but at some locations, the measurement results slightly deviate from what the code predicts. A possible explanation is that while neutralization is expected to vary, the calculations assumed it to be constant.

Current Status of Beam Experiments at the University of Maryland, Including Electron Ring and 5-beamlet Experiment
R. Kishek

The University of Maryland Electron Ring (UMER) is currently under construction and should be completed in a year or so. The goal is to provide a research platform for high intensity, space-charge dominated beams. Recently, an energy analyzer with a 10^{-4} resolution has been developed.

CP642, *High Intensity and High Brightness Hadron Beams: 20th ICFA Advanced Beam Dynamics Workshop on High Intensity and High Brightness Hadron Beams*, edited by W. Chou, Y. Mori, D. Neuffer, and J.-F. Ostiguy

Space Charge Effects at the KEK PS Main Ring

S. Igarashi, K. Koseki, E. Nakamura, Y. Shimosaki, M. Shirakata, K. Takayama and T. Toyama

KEK 1-1 Oho, Tsukuba, Ibaraki 305-0801, Japan

Abstract. Space charge effects during the injection period of the 12 GeV main ring of the KEK proton synchrotron have been studied. Measurement of the transverse beam profiles using flying wires has revealed a characteristic temporal change of the beam profile within a few milliseconds after the injection. Horizontal emittance growth was observed when the horizontal tune was close to the integer. The effect was more enhanced for higher beam intensity. Resonance created by the space charge field was the cause of the emittance growth. A multiparticle tracking simulation program, ACCSIM, taking account of space charge effects has successfully reproduced the beam profiles.

INTRODUCTION

The beam intensity of the KEK PS 12 GeV main ring has significantly increased since the K2K neutrino oscillation experiment started. Efforts to minimize beam loss have been continuously made. One of the issues is to reduce the loss during the injection period. Nine bunches of protons with the kinetic energy of 500 MeV are injected with the interval of 50 ms. About 10 % of protons are lost during the injection period of 510 ms. The highest operating intensity of the main ring is 1.4×10^{12} protons per bunch at the injection. The nominal operational tune has been optimized to be 7.15 and 5.25 for the horizontal and vertical tune respectively. The main ring has a circumference of 340 m and four-fold symmetry. A super period consists of seven FODO cells.

The incoherent tune shift is estimated to be 0.5 for the highest operating intensity without considering the effect of the image field or dispersion. The large value of the space charge tune shift is partly due to a small emittance of the injection beam. Emittance dilution and particle loss would occur under these circumstances. It is empirically known that emittance dilution observed after the injection to the main ring depends on the beam intensity and tune.

PROFILE MEASUREMENTS

Flying wire transverse beam profile monitors have been operated at the main ring and have demonstrated a wide dynamic range of more than two orders of magnitude and a position accuracy of 0.4 mm [1]. It takes about 4 ms for the wire to scan a typical size of the injection beam. The wire scanning can be initiated once in an acceleration cycle of the main ring at an arbitrary timing using a delayed trigger with respect to the injection kicker timing signals or the acceleration cycle timing signals. A new stroboscopic procedure has recently been established to reconstruct the beam profile that quickly changes with a time scale of 1 ms or less. A series of profile data are acquired by changing the trigger setting with an increment of 0.2 ms. They are then rearranged to reconstruct the simultaneous profiles.

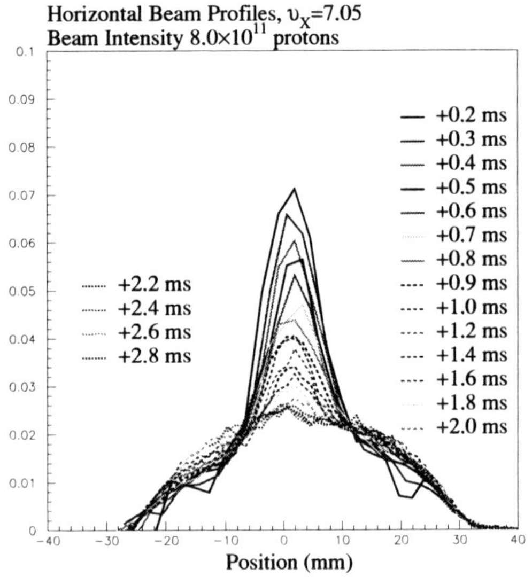

FIGURE 1. Horizontal beam profi les 0.2 ~ 2.8 ms after the injection when the horizontal tune was 7.05 and the injection beam intensity was 8.0×10^{11}.

When the injection beam intensity was set to 8.0×10^{11} protons, the beam profile 0.2 ms to 2.8 ms after the injection were reconstructed as in figure 1. The horizontal tune in this case was 7.05 which was not the nominal operational value. The vertical tune was 5.22. A significant beam loss was observed within 1 ms after the injection under the condition. The reconstructed profile shows a notable change of the distribution. The profile at 0.2 ms after the injection consists of a narrow peak and a broad

CP642, *High Intensity and High Brightness Hadron Beams: 20th ICFA Advanced Beam Dynamics Workshop on High Intensity and High Brightness Hadron Beams,* edited by W. Chou, Y. Mori, D. Neuffer, and J.-F. Ostiguy

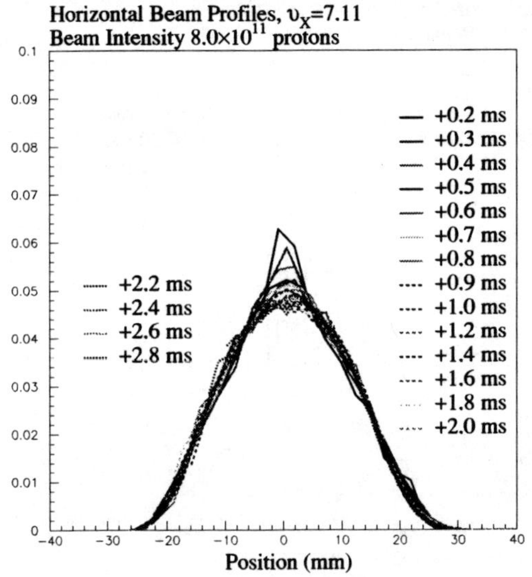

FIGURE 2. Horizontal beam profi les 0.2 $\sim$ 2.8 ms after the injection when the horizontal tune was 7.11 and the injection beam intensity was 8.0×10^{11}.

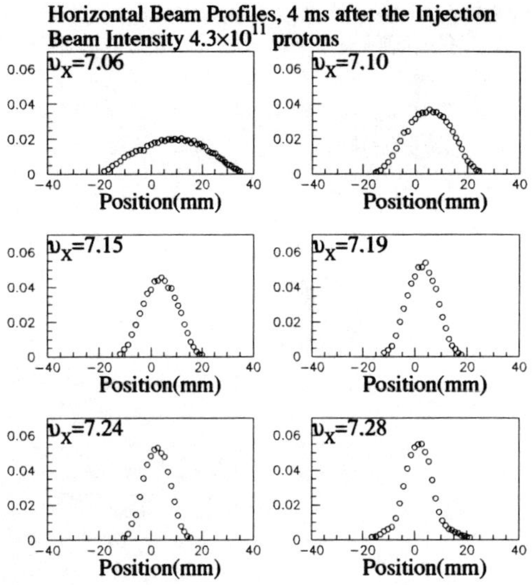

FIGURE 3. Horizontal beam profi les 4 ms after the injection when the injection beam intensity was 4.3×10^{11}.

distribution. The narrow peak diminishes in 2 ms and only the broad distribution remains.

The same procedure was applied for the horizontal tune of 7.11 and the vertical tune of 5.21 which was near the nominal operational value. The reconstructed profiles are shown in figure 2. The profile at 0.2 ms after the injection still consists of a narrow peak and a broad distribution. The narrow peak diminishes in 1 ms, and only the broad distribution remains as in the case of the tune of 7.05. The narrow peak of this case is, however, less significant than that of the previous tune, and the beam loss is not either significant in this case.

Horizontal beam profiles after the injection were measured for the injection beam intensity of 2.4, 4.3 and 8.3×10^{11} protons. The measurements were performed for a range of the horizontal tune from 7.05 to 7.3. The vertical tune was maintained to be between 5.2 and 5.25. The trigger was set to initiate the wire scanning to take beam profiles of about 4 ms after the injection when the rapid change of the profile was settled. The profiles for the intensity of 4.3×10^{11} protons are shown in figure 3. It was observed to be wide when the horizontal tune was 7.06, and became narrower as the tune was away from the integer. It is also noted that the profile of the tune of 7.28 has tails unlike any other profile.

A threshold was set to cut baseline noises and the full beam widths were measured from the profiles. Beam emittances were estimated with the full widths and plotted in figure 4. The injected beam 95 % emittance was measured at the beam transfer line to be 15.9±0.2, 17.8±0.1 and 18.6±0.2 πmmmrad for the intensity of 2.4, 4.3 and 8.3×10^{11} protons respectively. Emittance

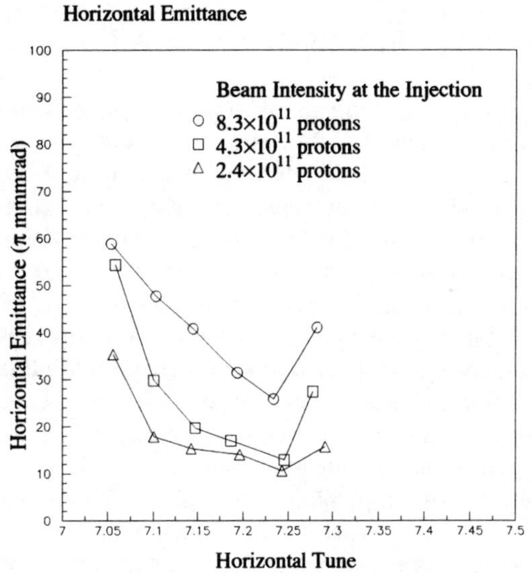

FIGURE 4. Horizontal emittance as a function of the horizontal tune. About 0.3 % of the tail is ignored in case of a Gaussian distribution.

growth was observed when the tune is close to the integer for all the measured intensity settings. A tune range that emittance growth occurs, however, depends on the intensity. The ranges are 0.1 and 0.15 for the intensity of 2.4 and 4.3×10^{11} protons respectively. Emittance growth was observed for all the measured tune range for the intensity of 8.3×10^{11} protons. The tune range of the emittance growth corresponds to naive estimate of the incoherent tune shift. It is inferred that the emittance growth is due to resonance created by the space charge.

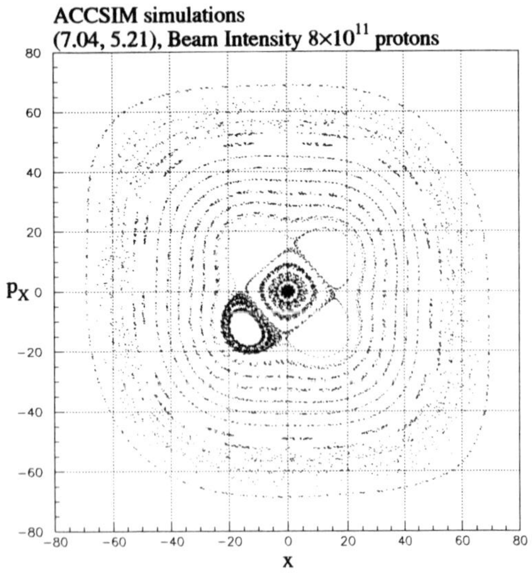

FIGURE 5. ACCSIM simulation of the x-p_x phase space plots of 20 test particles when the horizontal tune was 7.04 and the injection beam intensity was 8.0×10^{11}.

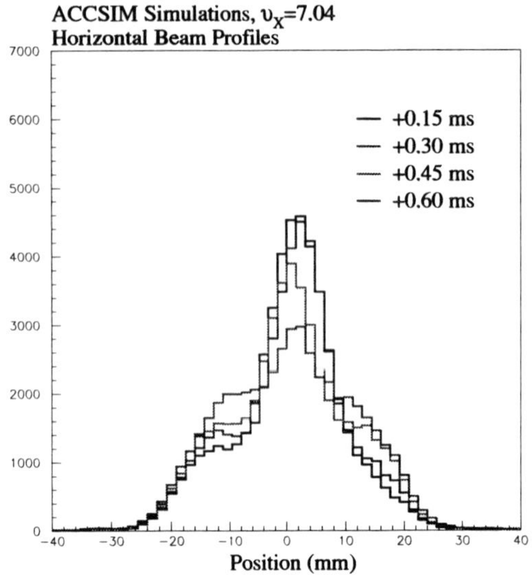

FIGURE 6. ACCSIM simulation of the Horizontal beam profi les 0.15 $\sim$ 0.6 ms after the injection when the horizontal tune was 7.04 and the injection beam intensity was 8.0×10^{11}.

ACCSIM SIMULATIONS

A multiparticle tracking simulation program, ACCSIM [2], taking account of space charge effects has been performed to understand the observed phenomena. Transverse space charge forces have been calculated for 10000 macro particles with a hybrid fast-multipole technique and grids of 1 mm $\times$ 1 mm every 0.76 m step. Thin lens kicks have been applied to simulate sextupole and octupole magnets. A fringing field from an injection septum magnets was suspected as one source of closed orbit distortion and included in simulations for some cases. Parameters for the injection beam emittance were based on the transfer line profile measurements.

Figure 5 is the x-p_x phase space plot of 20 test particles for 400 turns when the horizontal tune is 7.04. It shows patterns of resonance that was created by the space charge force. Octupole type space charge field would create the resonance. The resonant tune is 7/4 for a super period, because the main ring has four-fold symmetry.

Beam profiles at the tune of 7.28 has tails as in figure 3. The tails may also be due to resonance by the space charge field. The resonant tune is 7.25 in this case, assuming some imperfection in the four-fold symmetry.

Horizontal profiles from the ACCSIM simulation up to 400 turns which corresponds to 0.6 ms is plotted for the horizontal tune of 7.04 in figure 6. It shows a good agreement with the measurement. Another tracking simulation code, PATRASH [3], has also been applied and the results agreed with the ACCSIM results.

CONCLUSIONS

Measurement of the transverse beam profiles using flying wires has revealed a characteristic temporal change of the beam profile within a few milliseconds after the injection. Horizontal emittance growth was observed when the horizontal tune was close to the integer. The effect was more enhanced for higher beam intensity. Resonance created by the space charge field was the cause of the emittance growth. A multiparticle tracking simulation program, ACCSIM, taking account of space charge effects has successfully reproduced the beam profiles.

ACKNOWLEDGMENTS

We thank F. Jones for valuable advices and installation of ACCSIM in our computer. We also thank H. Sato and K. Sato for useful comments.

REFERENCES

1. Igarashi, S., Arakawa, D., Koba, K., Sato, H., Toyama, T., and Yoshii, M., *Nuclear Instruments and Methods in Physics Research A*, **482/1-2**, 32–41 (2002).
2. Jones, F., *Accsim Reference Guide Version 3.5s*, pp. 1–61 (1999).
3. Shimosaki, Y., and Takayama, K., *KEK Preprint*, **2001-111** (2001).

Intensity Dependent Emittance-Exchange in a High Intensity Proton Synchrotron

I. Sakai, S. Machida, T. Adachi, Y. Arakida, Y. Irie, K. Kitakawa, Y. Mori, Y. Shimosaki, H. Someya, M. Yoshimoto

KEK, 1-1 Oho, Tsukuba-shi, Ibaraki-ken, 305-0801, Japan

Abstract. The normalized acceptance of the KEK booster has large value in horizontal plane (80 π mm mrad in the horizontal plane and 18π mm mrad in the vertical plane). To increase the beam intensity we have been trying to achieve painting injection in the horizontal plane. The intensity-dependent emittance-exchanges in the horizontal and vertical plane were measured by a method using a scraper and bump magnets and extracted beam profiles measured by multi-wire profile monitor.

INTRODUCTION

The 500-MeV KEK booster has characteristics of rapid cycling (20Hz), being very compact and combined-function magnets. It supplies proton beams to both the 12-GeV main ring (MR) and the neutron and meson science laboratory (NML). The MR demands control of the extracted beam emittance of the booster for optimum injection, and the NML demands high-intensity beams that are as high as possible.

The features of the KEK booster are a wide horizontal aperture and a rather narrow vertical aperture, because of its initial design of septum-type multi-turn injection. The normalized value of the designed acceptance of the KEK booster is 80π mm mrad in the horizontal plane and 18π mm mrad in the vertical plane. The injection system was converted to one using a charge-exchanging method, and the injection energy has been upgraded to 40MeV to increase the beam intensity. The incoherent space-charge limit of the booster was calculated to be 3.2×10^{12} ppp assuming the full designed acceptance of the machine. Until now, a beam intensity of 2.2×10^{12} ppp was constantly obtained by normal H^- injection without painting. At this beam intensity, the emittance of the extracted beam is 40π mm mrad in the horizontal plane and 24π mm mrad in the vertical plane. An increase of the vertical beam size would seem to restrict the beam intensity. But the horizontal emittance of the extracted beams is half the value of the full acceptance.

To increase the beam intensity, painting injection in the horizontal plaine has been tried [1]. It had been expected that the beam intensity increases by 25%, if the horizontal emittance is extended to the full acceptance of the machine by painting injection. In the case of a low-intensity beam ($\leq1\times10^{12}$ ppp), the extracted beam emittance is well controlled by painting injection, and fills the full acceptance of the machine. However, regarding high-intensity mode operation ($\geq1.8\times10^{12}$ ppp), the horizontal emittance of the extracted beams dose not exceed the half value of the designed acceptance and only the vertical emittance grows; nevertheless, the injected beams of the booster were extended to the full acceptance by off-center injection in the horizontal plane.

This strange, but interesting, phenomenon has been carefully examined and studied from various angles. The beam envelopes were measured through an accelerating process, and it has been known that the shrinking of the horizontal emittance and the blow-up of the vertical emittance both occur in the bunching process just after injection, where the line-charge density increases in the RF bucket. In this paper we discuss the space-charge effects of flat beams in a high-intensity proton synchrotron based on the experimental results.

INJECTION SCHEME AND PAINTING DEVICE

The injection scheme of the KEK Booster is shown in Fig.1. The injection system consist of four closed-orbit bump magnets for charge-exchange injection and two orbit-shift bump magnets for phase-space painting. The position and angle of the injection point for charge-exchange injection are fixed by the closed-orbit bump magnets, and the center of the closed orbit

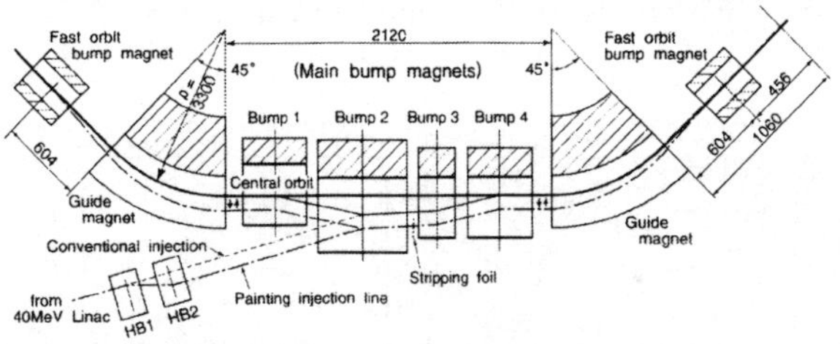

FIGURE 1. Charge-Exchange Injection System for Phase Space Painting

CP642, *High Intensity and High Brightness Hadron Beams: 20th ICFA Advanced Beam Dynamics Workshop on High Intensity and High Brightness Hadron Beams,* edited by W. Chou, Y. Mori, D. Neuffer, and J.-F. Ostiguy

at the injection point can be shifted by two orbit bump magnets which are placed at the upstream and the downstream positions. At the each position of orbit shift bump magnets, the phase of betatron oscillation leads and lags by $\pi/2$ from the injection point. Beams are injected gradually from the center of phase space to the outside by a shift of the closed orbit, which is controlled by the decay pattern of the magnetic field of these two bump magnets.

INTESITY DEPENDENCE OF THE EXTRACTED BEAM EMITTANCE

Observation of Extracted Beam Profiles under the Fixed Off-Center Injection

The projection of the extracted beam distribution is measured by multi-wire profile monitors at the dispersion-free points on the extracted beam line.

In low-intensity mode operation, lower than $1.0\text{x}10^{12}$ ppp, the profiles of the extracted beams are controllable by adjusting the injection devices [1]. A uniform distribution of charged particles is realized by the precise orbit shift painting, which is performed by two orbit shift bump magnets in the ring. The simple off-center injection, without two orbit shift bump magnets, forms a hollow distribution by adjusting the two steering magnets on the injection beam line. In the case of the hollow distribution, the center of the projection of charged-particle beams is 22% lower than the uniform distribution by precise orbit sift painting. The precise orbit shift painting is replaceable by simple off-center injection for the convenience in experiments.

Horizontal and vertical beam profiles under the 15mm off-center injection in the horizontal plane as parameters of the beam intensity are shown in Fig.2.

In the low-intensity mode, the extracted beam profile in the horizontal plane indicate the hollow distribution in the real space by the off-center injection. The vertical profile indicate the Gaussian distribution by the center injection without painting.

In the high-intensity mode operation, $2.0\text{x}10^{12}$ ppp, extracted beam profiles are not directly reflected by the situation of the off-center painting injection. In horizontal plane, the width of the extracted beam profile shrinks compared with low intensity mode operation. On the other hand, in vertical plane, the extracted beam profiles blow-up in the vertical plane.

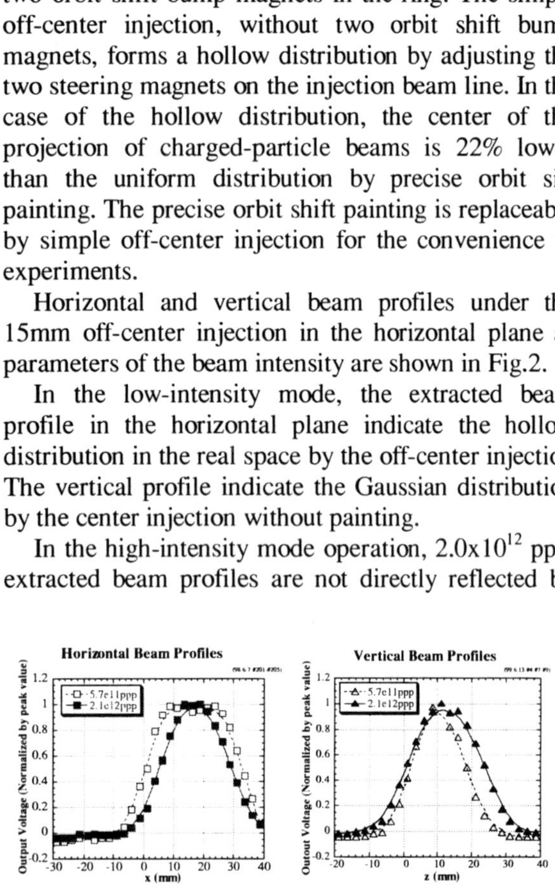

FIGURE 2. Projection of The Extracted Beam Profiles as Parameter of the Beam Intensity (Normalized by peak value)

The Change of Extracted Beam Emittance with the Width of Off-Center Injection

The extracted beam emittance with the width of off-center injection in the horizontal plane as the parameter of the beam intensity is shown in Fig.3. In the low-intensity mode, the increments of the horizontal emittance give good agreement with the calculated values by the width of the off-center injection, and the vertical emittance is only little affected by the horizontal expansion of the emittance. It can be considered that the emittnce is conserving through the accelerating process.

On the other hand, in the high-intensity mode, although the injected beams of the booster were expanded by off-center injection, the horizontal emittance dose not increase with the extended off-center injection. The distributions of injected beams are no longer conserved at the extraction.

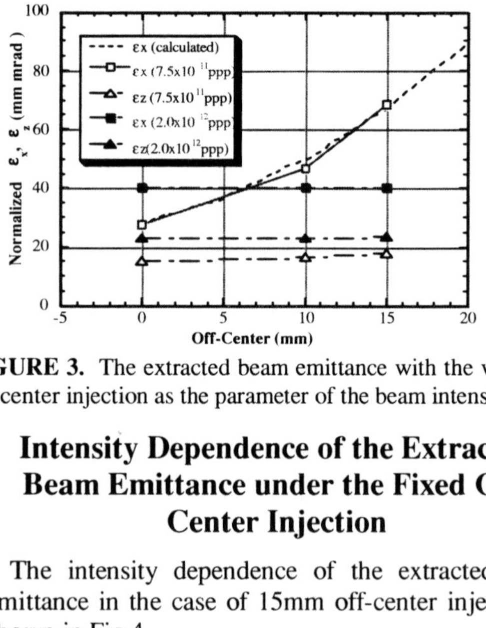

FIGURE 3. The extracted beam emittance with the width of off-center injection as the parameter of the beam intensity.

Intensity Dependence of the Extracted Beam Emittance under the Fixed Off-Center Injection

The intensity dependence of the extracted beam emittance in the case of 15mm off-center injection is shown in Fig.4.

The extracted beam emittance is measured by multi-wire profile monitors at the dispersion-free points on the extracted beam line. The off-center painting injection is tried only in the horizontal plane. In the vertical plane, the injection point and angle are adjusted to the center of the phase space to obtain the minimum emittance in the vertical plane.

The horizontal emittance decreases with the beam intensity nevertheless the injected beam is expanded to the full acceptance by the 15mm off-center injection. On the other hand, vertical emittance increases with the beam intensity. The results coincide with those of Fig. 2 and Figf.3.

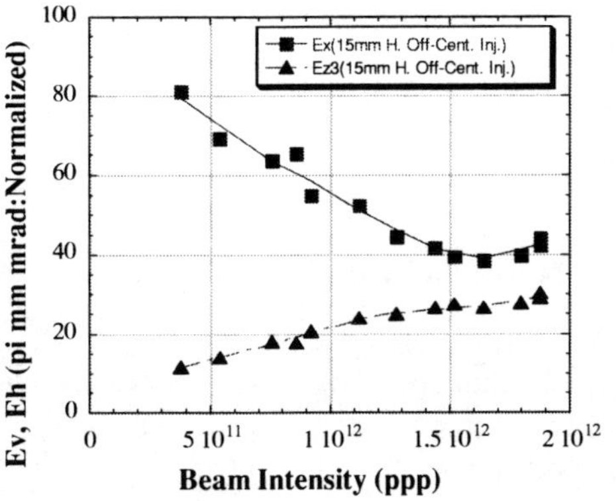

FIGURE 4. Intensity Dependence of the Extracted Beam Emittance under the 15mm Off center Injection in the Horizontal Plain.

EMITTANCE EXCHANGE THROUGH ACCELERATING PROCESS

To obtain information about shrinkage of the extracted beam emittance in the accelerating process, the beam size was measured using a system composing a scraper and bump magnets [3].

The beam sizes in both the horizontal and vertical planes were measured over a period of 0.06ms to 2.0 ms after injection, where the charge-exchange injection is completely finished and after which the injected coasting beams are gradually captured in a RF bucket. In this bunching process, the momentum spread and the line- charge density increases in the RF bucket. The bunching factor decreases from 1.0 to 0.4 during this period from 0.06ms to 1.5 ms after injection.

The experimental results are shown in Fig.5. Here, the values of the beam size at 95% beam intensity are converted to the value of the normalized emittance. In the horizontal plane, because the momentum dispersion of the KEK booster is not zero (1.4m) at the scraper position, the momentum spread is not constant throughout the accelerating process; we cannot directly calculate the horizontal emittance from the whole beam size. However, under the assumption that the momentum spread is constant at the same time after injection, irrespective of the beam intensity (neglecting the beam loading in RF acceleration and space charge effect on the RF bucket), the difference in the beam size is considered to be the difference in the emittance due to the beam intensity.

We can deduce the initial value of the injected beam emittance (at 0.06ms after injection) in the high-

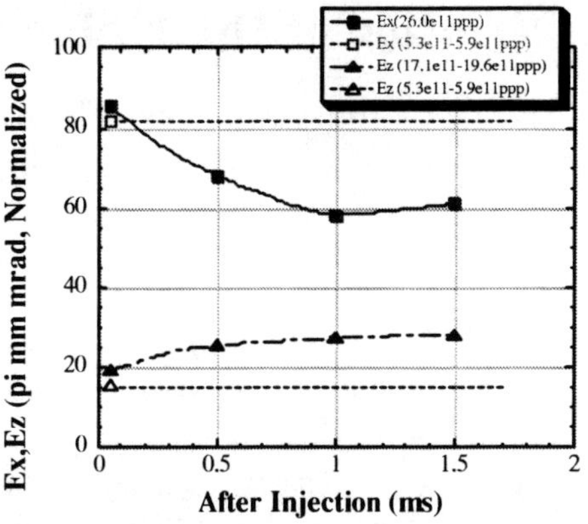

FIGURE 5. Normalized Ex, Ez Through Accelerating Process Just After Injection as a Parameter of High & Low Beam Intensity under the Horizontal Off Center Injection.

intensity mode operation from the absolute value of the extracted beam emittance in the low intensity mode operation.

As shown Fig.5, in high-intensity mode operation, a decrease in the horizontal emittance and the increase in the vertical emittance take place simultaneously during the period from 0.06ms (or faster) to 1.5ms after injection has been completed. The emittance changes saturate until 2.0ms and the values of the circulating beam emittance are close to the extracted beam emittance in Fig. 2. The vertical emittance growth reaches to the acceptance limit of the vertical plane. The experimental results suggest the existence of an X-Y coupling resonance by space-charge effects.

DISCUSSION AND SUMMARY

From the experimental results, the emittance exchange due to the coupling resonance induced by space charge effects is observed. The tricky measurement of beam envelopes reveals that the exchange occurs right after injection when the longitudinal bunching proceeds and correspondingly the line density increase. The simulation results well support the experimental results [3] [4].

REFERENCES

1. I. Sakai, et al, “Phase-space Painting of Charge-exchange Injection in the KEK Booster”, EPAC98, Stockholm, June 1998.
2. I. Sakai, et. al.,“Observation of Space-Charge Effects in the KEK Booster” EPAC2000, Vienna, June 2000
3. S. Machida, et. al., "Simulation of Space Charge Effects in a Synchrotron "AIP Conf. Proc. 448, p. 73, May 1998I.
4. Sakai, et. al., ”Intensity Dependent Emittance-Exchange in a High Intensity Proton Synchrotron” PAC 2001, Chicago, June 2001

Studies of the Coherent Half-Integer Resonance

Sarah Cousineau*, Jeff Holmes†, John Galambos†, Robert Macek**, Alexei Fedotov‡ and Jie Wei‡

*Indiana University, Bloomington, Indiana
†Oak Ridge National Laboratory, Oak Ridge, Tennessee
**Los Alamos National Laboratory, Los Alamos, New Mexico
‡Brookhaven National Laboratory, Upton, New York

Abstract. We present studies of space-charge-induced beam profile broadening at high intensities in the Proton Storage Ring (PSR) at Los Alamos National Laboratory. Previous work has associated the observed broadening in the vertical direction with the coherent half integer resonance [1]. Here, we study the effect of the space charge environment on this resonance; specifically, we investigate the strength of the resonance versus beam intensity, longitudinal bunching factor, transverse lattice tune, and two different beam injection scenarios. For each case, detailed particle-in-cell simulations are combined with experimental results to elucidate the behavior and sensitivity of the beam resonance response.

INTRODUCTION

The next generation of high intensity synchrotrons will require unprecedented minimization of beam loss in order to control radiation activation of the machine. Many of these accelerators will operate in an energy regime where space charge effects are a primary mechanism for halo development and beam loss. In striving towards successively higher beam intensities, an important task is to understand space charge from both theoretical and experimental standpoints. The Proton Storage Ring (PSR) at Los Alamos National Laboratory can operate with very high space charge effects and therefore provides an ideal setting for this type study. In this paper, we analyze experimentally observed beam broadening at high intensities in the PSR ring. Both particle core model and particle-in-cell (PIC) simulations are employed as tools to interpret the data. In the PIC simulations, we pay particular attention to lattice, injection, and RF settings, and in general find very good agreement between the simulated and experimentally measured beam profiles.

The quantity of merit in space charge studies has traditionally been the incoherent space charge tune shift limit. Recently, experimental and computational work has shown that the coherent tune bears more consequence on resonance behavior of the beam [2,3]. A complete analysis of the half-integer coherent resonance response was performed by Sacherer in his doctoral thesis [4], and we quote his results extensively here. He used the particle core model to demonstrate that the onset of the half-integer resonance occurs at space charge tune shifts in excess of the incoherent tune limit. He later extended his analysis to rms second moments of the beam distribution, thus making it applicable to realistic space charge scenarios. We have performed extensive work to tie Sacherer's core model together with an independently developed rms core model which includes dispersion [5,6] and also with PIC calculations and experimental data. The emphasis here is weighted heavily towards the experimental side.

BEAM BROADENING AT HIGH INTENSITY AT PSR

The PSR ring is 90 meters in length an can accumulate up to about 5×10^{13} protons. Beam profile broadening is observed at the highest operating intensities (above 3×10^{13}) and coincides with an escalation of beam losses. For the experiment presented here, $7\mu C$ of beam was accumulated over a period of 1.16ms, or about 3214 turns, injecting 1.36×10^{10} protons per turn. The lattice tunes were set to their nominal working values of $(\nu_x, \nu_y) = (3.19, 2.19)$, and optimal injection settings with vertical painting were used. Horizontal and vertical profiles were measured on a wire scanner in the extraction channel after the full accumulation of beam. The intensity of the beam was varied by chopping the linac beam pulses before injection into the ring. The profiles in Figure 1 show the vertical beam profile at one-fourth of the full intensity, one-half of the full intensity, and at the full intensity. No significant difference is observable between the one-fourth and one half intensity profiles, whereas the full intensity case exhibits a large amount of beam broadening.

In order to explore the area between the half and the full intensities, we invoke PIC model simulations of the full experiment. The simulations neglect magnetic errors, strip foil scattering, and chromatic effects, but include a complete set of transverse and longitudinal space charge

CP642, *High Intensity and High Brightness Hadron Beams: 20th ICFA Advanced Beam Dynamics Workshop on High Intensity and High Brightness Hadron Beams*, edited by W. Chou, Y. Mori, D. Neuffer, and J.-F. Ostiguy

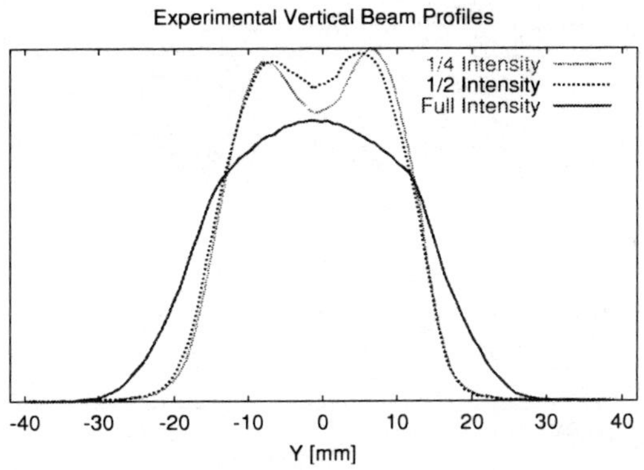

FIGURE 1. Vertical beam profiles measured on a wire scanner after extraction of the accumulated beam, shown for one-fourth, one-half, and full intensity beams.

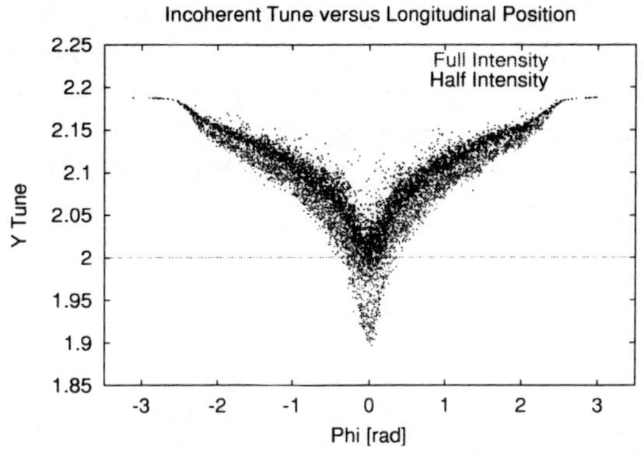

FIGURE 2. Incoherent tune shifts plotted as a function of longitudinal coordinate in the ring for full and half intensity beams. The dashed line corresponds to the half-integer stopband.

interactions; therefore, space charge provides the only nonlinearity. The details of the code are available in the ORBIT reference manual [7]. Although the benchmarks are not shown here, good agreement is reached for each of the profiles shown in Figure 1, as well as for the other experimental profiles discussed below.

Two quantities that are of importance in diagnosing space-charge-induced effects on a beam are the space-charge-depressed incoherent tunes, and the second moment of the beam, both available to us through the PIC simulations. Figure 2 below shows the former, plotted as a function of longitudinal distance in the ring, for both the half intensity and full intensity cases. The PSR longitudinal profile is sharply peaked in the center, leading to large tune depressions in this area. For the half size beam, the maximally depressed tune shifts slightly exceed the incoherent half-integer threshold, $\nu_y = 2.0$. Recall that no beam broadening occurs at this stage, implying that the incoherent threshold can be crossed without immediate consequence to the beam. At the full intensity, where the broadening is seen, the beam has crossed the incoherent limit by an additional 0.1. The limit for the onset of beam broadening relative to the tune shift points strongly towards a coherent response to the resonance condition.

Further evidence of the coherent resonance is presented in Figure 3, where the second order moments of the beam, normalized to the matched beam envelope, are plotted over one turn of the accelerator. Here, the rms beam envelope executes four oscillations per turn, a signature trait of the half-integer coherent resonance. The moments are plotted for particles in the middle of the longitudinal distribution, where the density is highest. An analysis of the moments outside of this range shows a sharp decrease in the moment amplitudes, and frequencies tending towards twice the bare tunes as we move away from the center. This is an indication that the observed beam profile broadening results mainly from particles located at the very center of the longitudinal bunch, a fact which makes for interesting experiments related to the longitudinal bunching factor.

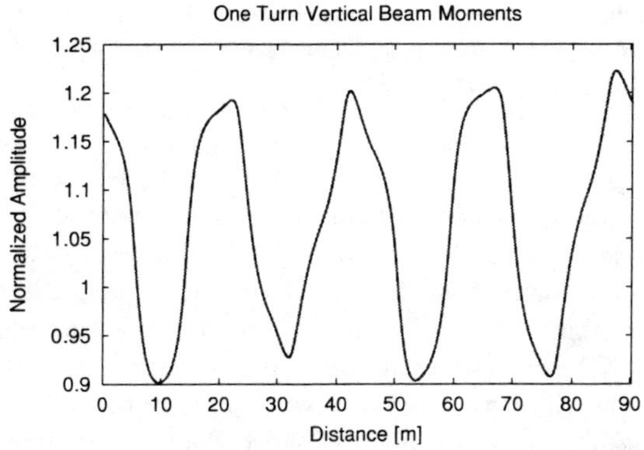

FIGURE 3. Normalized second moments of the full intensity vertical beam, plotted over one turn of the ring.

EXPLORATIVE EXPERIMENTS

This section is dedicated to three experiments performed to explore the onset of the resonance in detail. In the first experiment, the longitudinal profile shape was altered by placing a notch in the center of the longitudinal profile, i.e., no beam was injected within 15 degrees of the center of the bunch train; additional beam was injected outside of this range so that the final intensity of the beam was unaffected. Beam eventually diffused into the notched region, but the overall peaking was greatly reduced, and the bunching factor increased from the nominal value of about 0.3 to about 0.45. The effect on the transverse profiles is shown in Figure 4, where the vertical profiles after the full accumulation with and without the longitudinal notch are plotted. Note that the longitudinally notched beam is about 15% thinner. Full PIC simulations of the experiment show that the emittance growth for the highest intensity case is greatly reduced when the notch is in place, with the final rms emittance reaching about $7\pi mm \cdot mrad$ with the notch, compared to the $9.5\pi mm \cdot mrad$ reached without the notch. The notch was not observed to make any difference in the profiles in lower intensity cases, where the beam is not near the resonance.

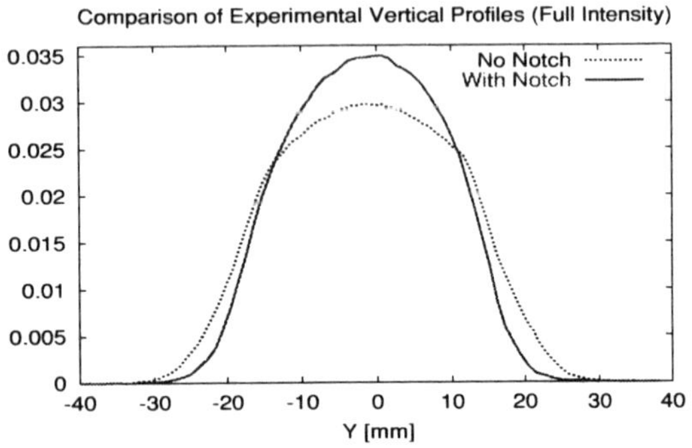

FIGURE 4. Vertical experimental beam profiles for a $7\mu C$ beam with and without a longitudinal notch. The red solid line is the beam without the longitudinal notch (bunching factor $\approx$ .3) and the blue dashed beam is the beam with the longitudinal notch (bunching factor $\approx$ .45).

In the second experiment, the horizontal tune was kept at the fixed value of $\nu_x = 3.19$, while the vertical tune was incrementally lowered from the nominal $\nu_y =$ 2.19 to $\nu_y = 2.09$ in steps of 0.02. The result was a dramatic increase in the rms beam width from 17mm at the highest tune value to 22.5mm at the lowest tune value. The corresponding experimental losses increased linearly with decreasing tune value up until the last data point, where losses jumped substantially, most likely caused by a violation of the limiting ring aperture.

Finally, beam broadening was measured as a function of painting scheme. A $5\mu C$ beam was painted vertically to two different sizes: one beam was painted with a 17mm maximum injection offset, and the other beam was painted with a 12mm maximum injection offset. Profiles recorded at half of the run intensity (2.5 μC) show that the smaller painted beam is smaller at the end of accumulation than the larger painted beam. However, with the full $5\mu C$ of beam current, the small and large painted vertical beam profiles at the end of accumulation are nearly identical (see Figure 5). The smaller painted beam, subject to a much more intense space charge environment, reacts more quickly and more severely to the coherent resonance than does the large painted beam.

PIC simulations of the experiment show that the small painted beam reacts to the resonance very early in the accumulation stage, and displays beam moment oscillations that are about 10% larger than those of the large painted beam. After this initial stage of growth, the small painted beam oscillates with moment amplitudes just slightly higher than those of the large painted beam, and the emittance of the small beam at the end of accumulation is about 8% greater than that of the large beam. This behavior is consistent with Sacherer's claims that the emittance of the beam grows in response to the half integer resonance and weakens the space charge environment. In an accumulation scenario, this is an ongoing process, and a comparison of Sacherer's analytic equations with the PIC simulations shows that the PSR beam is sitting continuously at the very edge of the coherent resonance threshold. The emittance of the beam is growing constantly in excess of pure painting effects in order to compensate for the resonance condition. Eventually the beam size exceeds the limiting ring aperture and losses escalate, ultimately limiting the attainable beam intensity.

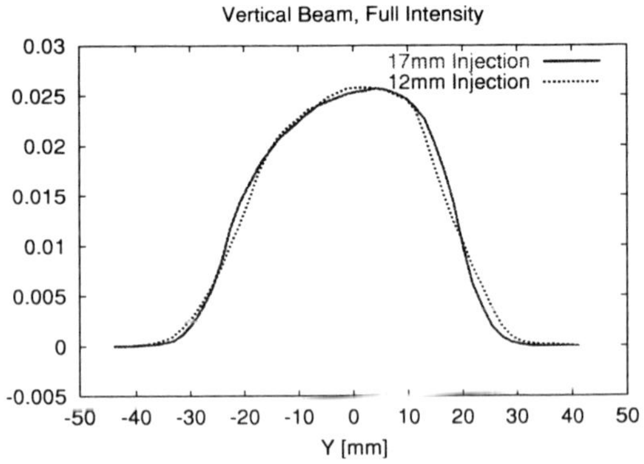

FIGURE 5. Vertical experimental profiles after accumulation of a $5\mu C$ beam painted to 17mm and a beam painted to 12mm. The solid red is the large painted beam and the dashed blue is the small painted beam.

CONCLUSIONS

Strong evidence supports the conclusion that beam broadening at high intensities in the PSR accumulator ring is a result of a coherent half-integer resonance response. PIC simulations, all well-benchmarked with experimental data, show that space-charge-depressed tunes are in excess of the incoherent stopband limit before the onset of broadening, and that the beam envelope executes four oscillations per turn, both strong indicators of a coherent resonance. Experiments performed to explore the resonance show that at high intensity, factors such as longitudinal profile shape, transverse painting scheme, and variation of bare tune all affect the response of the beam to the resonance; at low intensities, before the onset of resonance, no significant changes are observed from the modification of these parameters.

This work is supported by SNS through UT-Battelle, LLC, under contract DE-AC05-00OR22725 for the U.S. DOE, and by Indiana University through a DOE grant, DE-FG02-92ER40747. SNS is a partnership of six national laboratories: Argonne, Brookhaven, Jefferson,

Lawrence Berkeley, Los Alamos, and Oak Ridge. The Los Alamos National Laboratory is operated by the Univesity of California for the U.S. DOE under contract W-7405-ENG-36.

REFERENCES

1. J.A. Holmes, V. Danilov, J. Galambos, A. V. Fedotov, and R. L. Gluckstern, in *Proceedings of the Particle Accelerator Conference, Chicago, 2001*, p. 3188.
2. A. V. Fedotov and I. Hofmann, "Half-Integer Resonance Crossing in High-Intensity Rings," PRST-AB **5**, (2002) 024202.
3. A. Uesugi, S. Machida, and Yoshiharu Mori, "Experimental study of a half-integer resonance with space-charge effects in a synchotron," PRST-AB **5**, (2002) 044201.
4. F. J. Sacherer, "Transverse Space-Charge Effects in Circular Accelerators," PhD Thesis, University of California, Berkeley, (1968).
5. S.Y. Lee and H. Okamoto, Phy. Rev. Lett. 80, 124201 (1999).
6. J. Holmes, J. Galambos, D. Olsen, S.Y. Lee, in Workshop on Space Charge Physics in High Intensity Hadron Rings, Shelter Island, NY, 1998, p. 254.
7. J. Galambos et al., ORBIT User's Manual, SNS/ORNL/AP Technical Note 011, 1999.

Induction Inserts at the Los Alamos PSR

K.Y. Ng

Fermi National Accelerator Laboratory, P.O. Box 500, Batavia, IL 60540

Abstract. Ferrite-loaded induction tuners installed in the Los Alamos Proton Storage Ring have been successful in compensating space-charge effects. However, the resistive part of the ferrite introduces unacceptable microwave instability and severe bunch lengthening. An effective cure was found by heating the ferrite cores up to $\sim 130°$C. An understanding of the instability and cure is presented.

INTRODUCTION

Inductive tuners were installed into the PSR to cancel the space charge effects of the intense proton beam at 799 MeV. Each tuner consists of a stainless steel pill-box cavity closely packed with 30 Toshiba M_4C_{21A} ferrite cores, each 1.0 in thick, 5.0 in I.D. and 8.0 in O.D. When two tuners were installed in 1997 with the intention of two-third space-charge compensation, clear and consistent evidence was observed, including shorter bunch length, cleaner bunch gap, and smaller rf voltage required for stable operation. After an upgrade when 3 tuners were installed in 1999 and the beam intensity was raised, a longitudinal instability was observed. Figure 1 shows a chopped coasting beam accumulated for 125 μs and stored for 500 μs recorded at a wide-band wall current monitor. The ripples at the beam profile indicate a longitudinal microwave-like instability at 72.7 MHz, which is roughly the resonant frequency of the pill-box cavities housing the ferrite cores. The resonance also showed up as ripples at the rear half of a 250 ns (left) and 100 ns bunch (middle) in Fig. 2. Apparently, the instability is tolerable for the 250 ns bunch because the bunch shape distortion is small. However, the $\sim$100 ns bunch is totally disastrous because it was lengthened to 200 ns.

CAUSE OF INSTABILITY

To incorporate loss, the relative permeability of the ferrite can be made complex: $\mu \to \mu_s' + i\mu_s''$, where the subscripts denote "series". The impedance of the ferrite is therefore

$$Z_0^{\|} = -i(\mu_s' + i\mu_s'')\omega L\,, \tag{1}$$

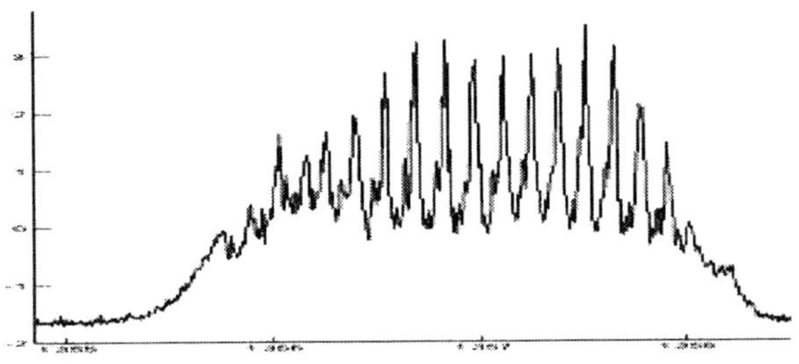

FIGURE 1. Longitudinal microwave-like instability recorded by wall-gap monitor of a coasting beam driven at 72.7 MHz.

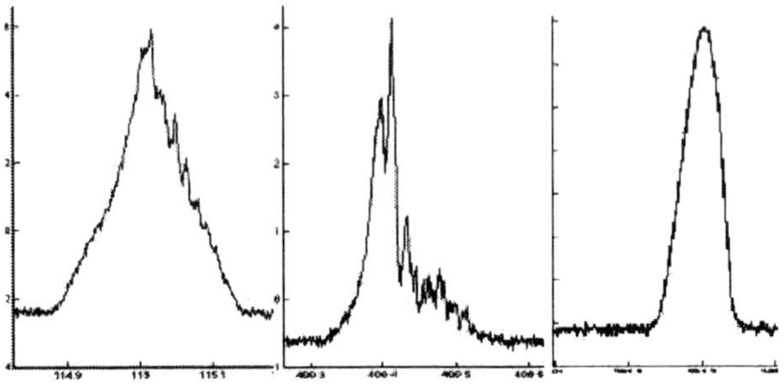

FIGURE 2. Instability perturbation on profiles of bunches with full width 250 ns (left) and 100 ns (middle). Right: Upon heating the ferrite to 130°C (see Sec. 4), the profile of the 100 ns bunch is no longer distorted.

where L denotes the inductance of the ferrite and ω the angular frequency. It is clear that μ_s' and μ_s'' must be frequency-dependent. The simplest circuit model is an inductor L_p in parallel with a resistor R_p. However, whenever ferrite is used, like the cores packed inside a pill-box cavity, there is an accompanying capacitor C_p in parallel. The impedance of the inductor tune can then be represented by

$$Z_0^{\|}(\omega) = \frac{R_p}{1 - iQ(\omega/\omega_r - \omega_r/\omega)}\,, \tag{2}$$

where ω_r is roughly where μ'' peaks. If we denote μ_L' as the value of μ' at low frequencies and μ_R'' the value of μ'' at resonant frequency $\omega_r/(2\pi)$, we readily obtain $\mu_R'' = Q\mu_L'$. Note that Q here is the quality factor describing the μ'' peak and is *not* the usual industrial-quoted Q. For a space-charge dominated beam, the actual area of beam stability in the complex $Z_0^{\|}/n$-plane (or the traditional U'-V' plane), where n is the revolution harmonic, is somewhat different from the commonly quoted Keil-Schnell estimation [2]. In Fig. 3, the heart-shape solid curve 1 is the threshold for parabolic distribution in momentum spread, where the momentum gradient is discontinuous at the ends of the spread. Instability develops and a smooth momentum gradient will result, changing the threshold curve to that of a distribution represented by 2. Further smoothing of the momentum gradient at the ends of the spread to a Gaussian distribution will change the threshold curve to 3. On the other hand, the commonly known Keil-Schnell criterion is denoted by the circle of unit radius in dots. This is why in many low-energy ma-

CP642, *High Intensity and High Brightness Hadron Beams: 20th ICFA Advanced Beam Dynamics Workshop on High Intensity and High Brightness Hadron Beams*, edited by W. Chou, Y. Mori, D. Neuffer, and J.-F. Ostiguy

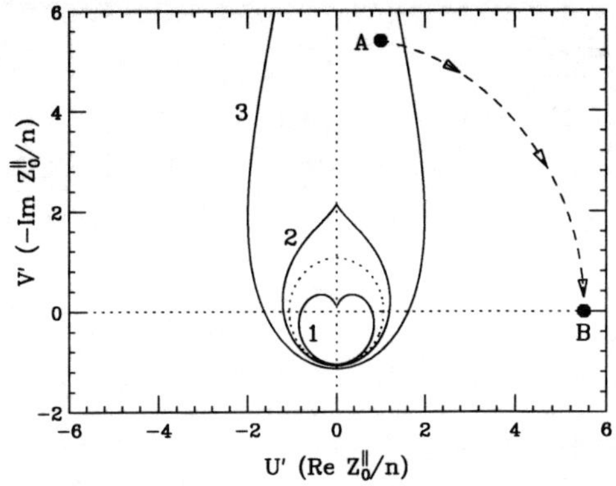

FIGURE 3. An intense space-charge beam may have impedance at A outside the Keil-Schnell circle (dots) but is stable inside the stability curve 3 for Gaussian distribution. Ferrite inserts compensating the space charge completely will lead to a resistive impedance roughly at B, which introduces instability.

chines the Keil-Schnell limit has been significantly overcome by a factor of about 5 to 10.

At the PSR, the space charge is almost the only source of the impedance, $\mathrm{Re}Z_0^{\|}/n$ typically orders of magnitude smaller. As an example, if the impedance of the Los Alamos PSR is at A, the beam is within the microwave stable region if the momentum spread is Gaussian like, although it exceeds the Keil-Schnell limit. Now, if we compensate the space-charge potential-well distortion by the ferrite inductance, the ferrite required will have an inductive impedance at low frequency equal to the negative value of the space charge impedance at A, for example, about -5.5 units according to Fig. 3. However, the ferrite also has a resistive impedance or $\mathrm{Re}Z_0^{\|}/n$ coming from μ''. Although $\mathrm{Re}Z_0^{\|}/n$ is negligible at low frequencies (for example, the rf frequency of 2.796 MHz of the PSR), it reaches a peak value near $\omega_r/(2\pi)$ (about 50 to 80 MHz for the Toshiba M_4C_{21A} inside the pill-box container) with the peak value the same order of magnitude as the low-frequency $\mathrm{Im}Z_0^{\|}/n$.

Thus the ferrite will contribute a resistive impedance denoted roughly by B (~ 5.5 units) when $Q \sim 1$. This resistive impedance of the ferrite will certainly exceed the threshold curve of any momentum distribution and we believe that the longitudinal instability observed at the PSR is a result of this consideration. In order to avoid this instability, we must search for a ferrite sample having a small $\mathrm{Re}Z_0^{\|}/n$ to $\mathrm{Im}Z_0^{\|}/n$ ratio.

HEATING THE FERRITE

Past experience indicates that when a piece of ferrite is heated up, μ_s' will increase and and loss at high frequencies will decrease, thus having exactly the same properties we are looking for.

A measurement has been made by sending a sinusoidal wave from one end of the ferrite tuner via an antenna and receiving it with an antenna at the other end, while the loss is recorded as a function of frequency of the wave [3]. The results are shown in Fig. 4 and reveal that the resonant peak drops by a factor of about 6 when the ferrite cores are heated from the room temperature of 23°C to 100°C. At the same time the resonant frequency moves to lower frequencies.

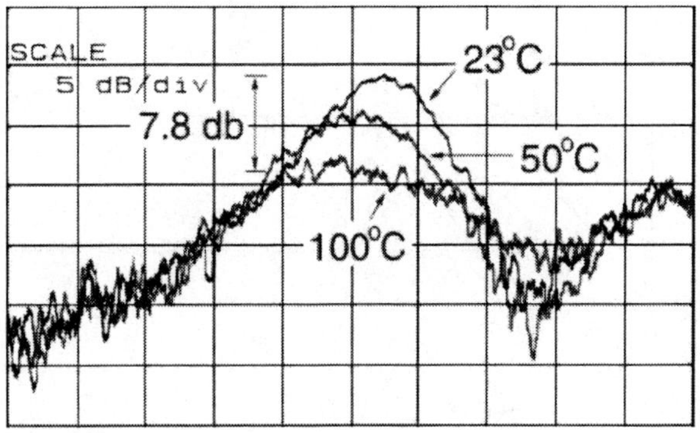

FIGURE 4. (color) As the ferrite cores are heated from room temperature to 100°C, the loss at the resonant peak reduces by almost 6 times. Abscissa: center 75.0MHz, span 149.4MHz. Ordinate: 5 db/div.

A measurement of the permeability of the ferrite has also been made on a single Toshiba M_4C_{21A} ferrite core as a function of core temperature [4]. To provide both a good electrical circuit path and a uniform core temperature, the core was encased in an aluminum test fixture before being placed on a hot plate. The top half of the test fixture consisted of a machined aluminum disk, 9 in in diameter and 1.25 in thick. The inner section of the disk was machined out 0.005 in undersize to accommodate the ferrite core. The disk was then heated and the core was slipped into the disk. Upon cooling, the aluminum disk contracted and made a good thermal contact with one side and the outer edge of the ferrite core. The aluminum fixture and core were then flipped over onto a flat aluminum plate so that only the inner edge of the core was exposed. A good electrical connection between the aluminum disk and flat plate was made using strips of adhesive backed copper tape. The test fixture was placed on a hot plate and covered with two fire bricks. The test fixture was then heated to 175°C and allowed to cool slowly. The impedance measurement was made by placing the probe of an HP4193A vector impedance meter directly across the inner edge of the ferrite core. Impedances were measured from 10 MHz to 110 MHz in 10 MHz steps from 150°C to 25°C. The temperature of the core was monitored by a Fluke 80T-150U temperature probe inserted into a small hole in the aluminum disk portion of the test fixture.

The capacitor C_p in parallel to the ferrite core was determined by adding additional fixed 100 pf capacitors

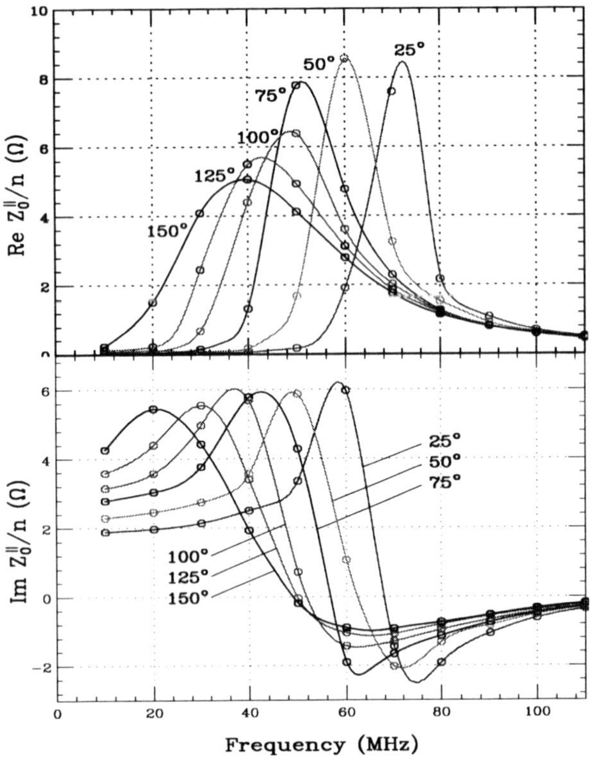

FIGURE 5. (color) Fermilab measurement of real and imaginary parts of impedance per revolution harmonic $Z_0^{\|}/n$ of a ferrite core inside a snugly fit pill box cavity.

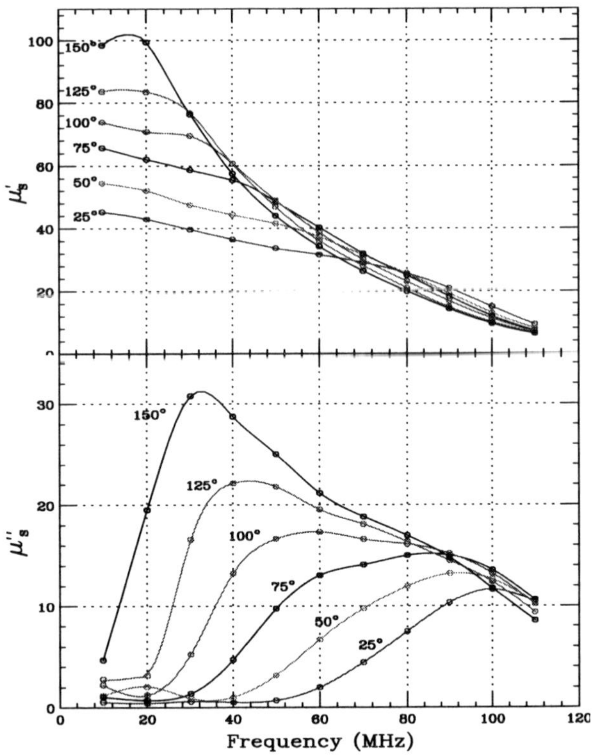

FIGURE 6. (color) Real and imaginary parts, μ_s' and μ_s'', of the permeability of the ferrite core as functions of frequency at different temperatures as derived from Fermilab measurement.

across the inner edge of the ferrite core and observing the change in the resonant frequency of the structure from 41 to 28 MHz, a frequency range in which the μ_s' of the ferrite is known to be relatively constant. In this manner, a capacitance of $C_p = 75$ pf was chosen to represent the equivalent parallel capacitance of the test circuit. There was also a residual resistance of $R_r = 0.55$ Ω in the probe in series with the *RLC* parallel equivalent circuit. This residual resistance introduces a large error at low frequencies (below $\sim$ 10 Hz) when the resistive part of the *RLC* circuit is small. From the measurements of the input impedance, R_p and L_p were computed. The real and imaginary parts of $Z_0^{\|}/n$ for one ferrite core encased in a cavity as seen by the beam are shown in Fig. 5. It is clear that the resonant peaks of $\mathrm{Re}Z_0^{\|}/n$ decreases twice as the temperature increases and the resonant frequency moves towards lower frequencies. At the same time, $\mathrm{Im}Z_0^{\|}/n$ increases by more than twice. If one plots $\mathrm{Re}Z_0^{\|}$ versus frequency instead, the decrease with temperature becomes more rapidly. For a coasting beam, the area under a $\mathrm{Re}Z_0^{\|}$ curves represents the energy loss to the ferrite assembly. We find that this loss is in fact temperature independent within 10%.

The real and imaginary parts of the series permeability of the ferrite core derived from the measurement are shown in Fig. 6. It is interesting to see that μ_s'' actually increases with temperature. However, an increase of the series resistance inside the ferrite core implies a decrease of the resistance R_p in the *RLC* parallel circuit, and therefore leads to a decrease in $\mathrm{Re}Z_0^{\|}/n$ of the resonant peak as depicted in Fig. 5.

APPLICATION TO THE PSR

When the induction tuners at the PSR is heated to 130°, the longitudinal microwave instability seen in Fig. 1 disappears. The profile of the 100 ns bunch now becomes as depicted in the right plot of Fig. 2, no longer distorted and lengthened. Further beam studies with the heated ferrites demonstrated other benefits of the inductor tuners without unmanageable operational impacts.

REFERENCES

1. M.A. Plum, et. al., *Phys. Rev. ST Accel. Beams*, **2**, 064201 (1999).
2. E. Keil and W. Schnell, CERN Report TH-RF/69-48 (1969); V.K. Neil and A.M. Sessler, Rev. Sci. Instr. **36**, 429 (1965). D. Boussard, CERN Report Lab II/RF/Int./75-2 (1975).
3. M. Popovic, private communication.
4. D. Wildman, private communication.

New Methods to Create Hollow Bunches

C. Carli and M. Chanel

CERN PS, CH-1211 Genève 23, Switzerland

Abstract. New methods to create hollow distributions in longitudinal phase space based on manipulations with a double harmonic RF system at high energy are presented with application to the PS Booster synchrotron (PSB). Whereas the first tentative to create hollow bunches at the PSB aimed to improve the performance of the PSB itself, these new methods are expected to reduce the limitations due to direct space charge forces in the receiving PS (where no double harmonic RF system is available) after transfer. One method aims to introduce empty phase space in the centre of the phase space by recombination of the beam in one bucket with another empty bucket. The second method is based on redistribution of phase space surfaces during the transfer of the beam from one second harmonic sub-bucket to another. During that process, phase space surfaces are exchanged and low density from the periphery ends up in the centre, whereas the high density surfaces from the centre are transferred to the periphery. Both methods have been simulated by particle tracking. The second method has been applied in practice at the PSB. The set-up turned out to be simple and fast, and to yield hollow distributions with good reproducibility.

INTRODUCTION

The performance of accelerators limited by transverse direct space charge forces, i.e. the "Laslett" tune shift, may be improved by reducing the peak current inside a bunch. For given bunch length and intensity, the peak current is decreased by creating "flat" bunches. One possibility to obtain flat bunches is to create hollow distributions in longitudinal phase space. The scheme used at the first attempts [1] at the PSB aimed to introduce empty phase space, before bunching, in the centre of the coasting beam. This was tested and brought close to operation recently [2].

In this paper, new methods to create hollow bunches in longitudinal phase space, based on RF gymnastics with a double harmonic RF system are presented. Those methods are in general not suitable to improve the performance of the machine in which they are applied. But they are intended to alleviate limitations due to space charge after transfer into a receiving machine without other means for bunch flattening.

PRINCIPLE

Both methods presented in this paper have been simulated by particle tracking in longitudinal phase space for a beam on a high energy flat-top (with increased length). Therefore, appropriate forms for the RF functions have been inserted into the equations of motion :

$$\frac{d}{dt}\Delta E = \frac{q}{T_0}\left[V_1(t)\sin\left(2\pi\frac{\tau}{T_0} - \phi_1(t)\right) + V_2(t)\sin\left(4\pi\frac{\tau}{T_0} - (2\phi_1(t) + \phi_{21}(t))\right)\right]$$

$$\frac{d}{dt}\tau = \frac{\eta}{\beta^2\gamma E_r}\Delta E \quad ,$$

where τ denotes the time (negative τ for the "head" of the bunch) and ΔE the energy offset of an individual particle w.r.t. a reference. T_0, η q and E_r are the revolution time, the momentum slip factor, the charge and the rest mass of the particle. V_1 and V_2 are the voltages of the first and second harmonic RF system. The phase ϕ_{21} between the two RF system is a crucial parameter, whereas the phase ϕ_1 of the principal RF system is not directly accessible (influenced by RF loops), but plays only a minor role and included for completeness only.

Recombination with an Empty Bucket

The method presented in this section has been triggered by tomographic reconstruction of bunch splitting and is similar to schemes in [3]. The basic idea is to merge the beam in one bucket with an empty bucket by appropriate gymnastics with a double harmonic RF system. Adjusting the RF parameters in an appropriate manner, one may insert a lot of empty phase space in the centre leading to a low density there, and insert very little empty phase space in the periphery.

A simulation of the whole process by particle tracking is shown in Fig. 1. RF parameters shown in the upper image are inserted into the equations of motion and yield phase space portraits below. During the first 14 ms, starting from a first harmonic bucket (with only a very small or no voltage of the second harmonic system), a double bucket structure has to be created such that the second bucket is outside the area occupied by the beam. This is achieved by increasing the voltage V_2 of the second harmonic RF system and shifting the phase

CP642, *High Intensity and High Brightness Hadron Beams: 20th ICFA Advanced Beam Dynamics Workshop on High Intensity and High Brightness Hadron Beams*, edited by W. Chou, Y. Mori, D. Neuffer, and J.-F. Ostiguy

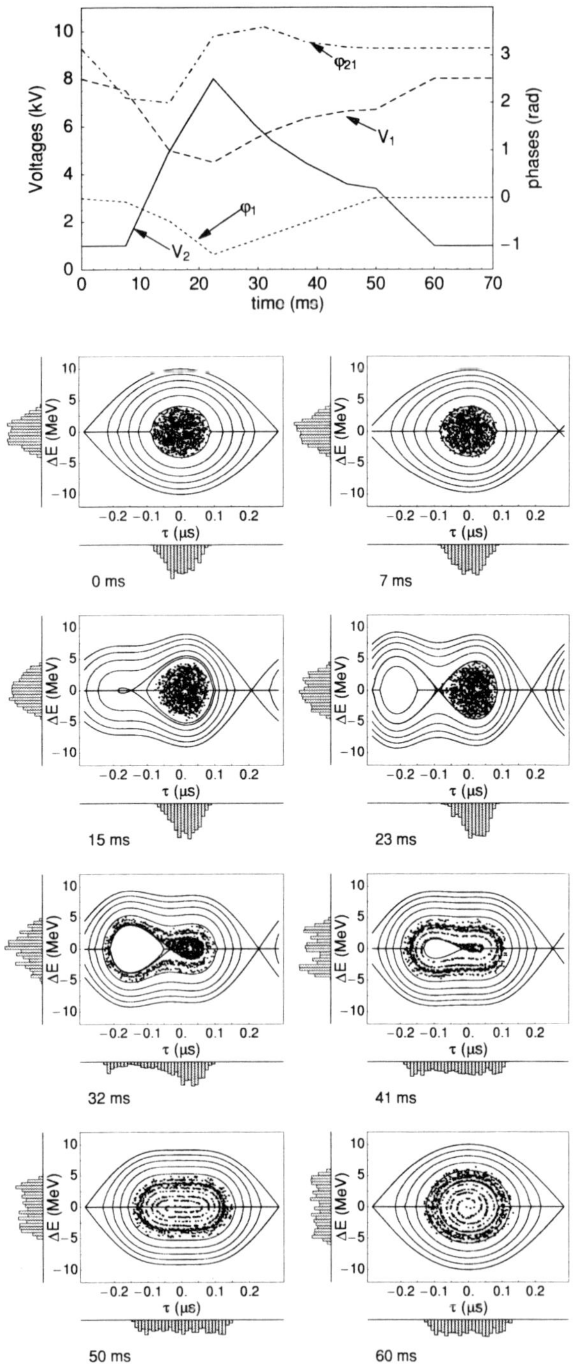

FIGURE 1. Creation of hollow bunches by recombination of the bunch with an empty bucket. RF parameters versus time (above) and phase space portraits.

ϕ_{21} to an appropriate value (sufficiently far from π). Then the empty bucket is increased to the desired size while keeping the full bucket just big enough to contain the beam. The situation after 23 ms is the starting point of the merging process. RF parameters are adjusted such that first mainly the full bucket collapses filling the periphery of the merged bunch and, towards the end more and more empty phase space is mixed in by collapsing the empty bucket. Care is taken that the buckets do not collapse too fast towards the end of the merging process, by changing the RF parameters slowly. At time 50 ms the merging is completed and during the last 10 ms, the RF system is brought back to the initial state.

Redistribution of surfaces in phase space

When applying asymmetric merging of a bunch with an empty bucket, one has to be careful to generate the double bucket structure such that the small bucket is outside the area occupied by the beam. For instance, if the emittance of the beam would be larger, the small bucket in Fig. 1 after 15 ms would not be empty. In the method outlined here, this situation is created on purpose (see situation after 10 ms in Fig. 2) and is the starting point of the redistribution process.

The process is best explained with the help of Fig. 2. Tracking with the RF functions shown in the upper image leads to phase space distributions plotted below. Particles have different shades (or colours if viewed with a device rendering colours) depending on their initial distance from the bunch centre in order to distinguish particles stemming from different initial locations. During the first 10 ms, a double-bucket structure is created with the small bucket inside the area occupied by the beam. Then, the RF parameters are adjusted such that the initially large bucket shrinks and releases phase space surfaces, which start to surround the two buckets. Simultaneously, the initially small bucket grows and captures phase space surfaces surrounding it. Thus, along the process, surfaces are transferred from one bucket to the other and high density surfaces from the centre are brought to the periphery. The total acceptance of the two second harmonic buckets together is slightly larger at the end of the process in order to mix some low density from the periphery with the dense core.

EXPERIMENTAL RESULTS

Redistribution of phase space surfaces necessitates relativeley simple RF gymnastics and does not lead to a significant blow-up of the total longitudinal emittance. This makes it particularly interesting for the PSB and, thus, this method has been tested experimentally. After a short set-up time, hollow bunches have been created with good reliability and reproducibility, even with the highest intensity possible in the PSB. Whereas the simulations had only been done for a long flat-top, the method has been tested experimentally on the high energy part of the ramp as well. The advantage is that the total cycle time is not increased and the process may take longer.

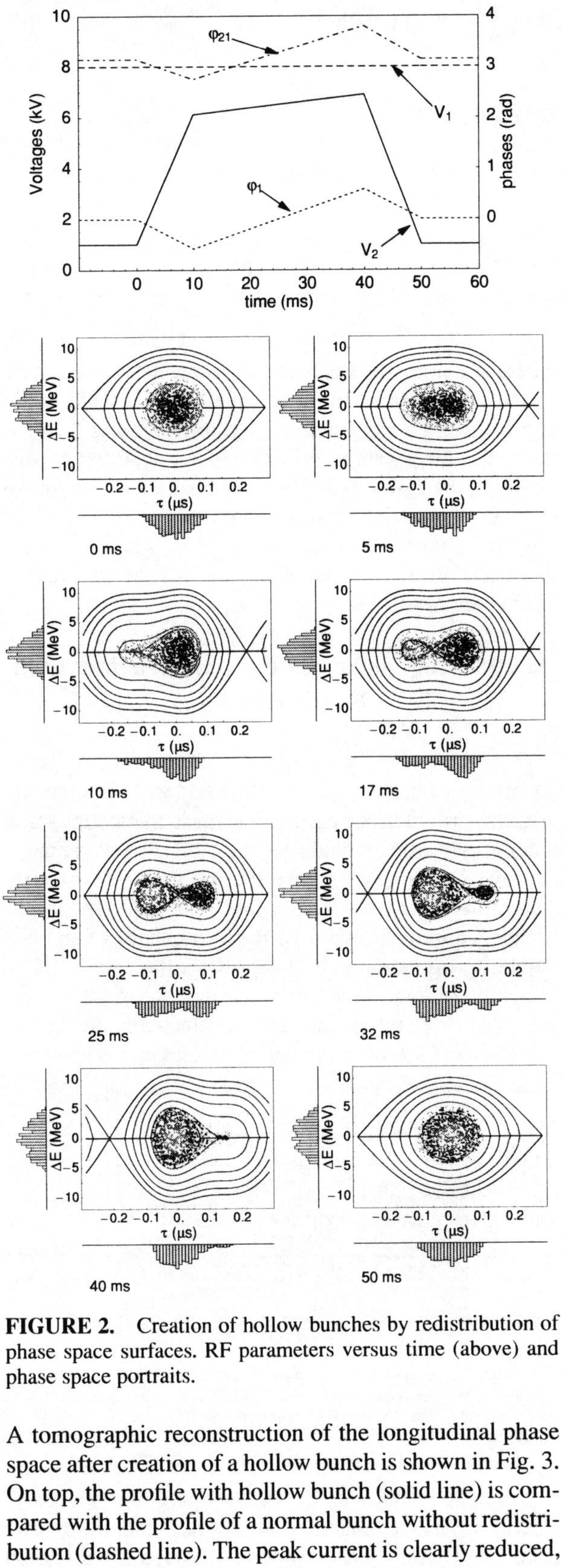

FIGURE 2. Creation of hollow bunches by redistribution of phase space surfaces. RF parameters versus time (above) and phase space portraits.

A tomographic reconstruction of the longitudinal phase space after creation of a hollow bunch is shown in Fig. 3. On top, the profile with hollow bunch (solid line) is compared with the profile of a normal bunch without redistribution (dashed line). The peak current is clearly reduced, with a negligible increase of the bunch length.

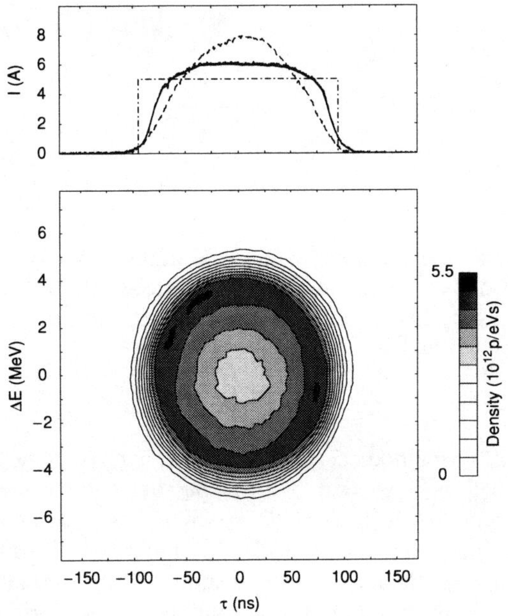

FIGURE 3. Tomographic reconstruction of the phase after redistribution of phase space surfaces.

SUMMARY AND OUTLOOK

New methods to create hollow bunches by gymnastics of a double harmonic RF system have been presented. One method based on redistribution of phase space surfaces has been tested successfully at the PSB. The aim is to alleviate limitations due to direct transverse space charge forces in a receiving synchrotron (in our case the PS).

The next step is to verify that the performance of space charge limited beams is indeed improved when the beam is transferred with hollow distribution in the longitudinal phase space.

ACKNOWLEDGEMENTS

We would like to Steve Hancock for stimulating discussions about the use of the "Tomoscope" and many aspects of longitudinal dynamics.

REFERENCES

1. J.P. Delahaye, G. Gelato, L. Magnani, G. Nassibian, F. Pedersen, K.H. Reich, K. Schindl and H. Schönauer, *Shaping of Proton Distribution for Raising the Space-charge of the PS Booster*, Proc. 11th Int. Conf. on High-Energy Accelerators, Geneva, 1980, p299.
2. A. Blas, S. Hancock, M. Lindroos, S. Koscielniak, *Hollow Bunch Distributions at High Intensity in the PS Booster*, Proc. 7th European Particle Accelerator Conference, Vienna, 2000.
3. S. Hancock, *(Unpublished) Simulations using ESME*
4. C. Carli, *Creation of Hollow Bunches Using a Double Harmonic RF System*, CERN/PS 2001-073 (AE).

Space Charge Studies at PSI

Andreas Adelmann

Paul Scherrer Institut & Lawrence Berkeley National Laboratory

Abstract. MAD9P (**m**ethodical **a**ccelerator **d**esign version **9** - **p**arallel) is a general purpose parallel particle tracking program including 3D space charge calculation. First simulation results on the B870 injection line and coasting beam studies of the PSI Injector 2 cyclotron are presented. Good agreement between measured beam profiles and calculations are reported.

INTRODUCTION

MAD9P (**m**ethodical **a**ccelerator **d**esign version **9** - **p**arallel) is a general purpose parallel particle tracking program including 3D space charge calculation. It is based on MAD9, and two frameworks:[1] CLASSIC (class library for accelerator system simulation and control) and POOMA (parallel object oriented methods and applications) [1]. A detailed description of MAD9P and more design studies can be found in [2]. The canonical MAD web page is located at: *www.cern.ch/mad/mad9.html*.

Problem Frame

PSI's current situation and the planned update is summarized in [3]. In order to achieve the challenging goals of reliability and high power operation we need a better qualitative understanding of the beam behavior in the transport lines and the cyclotrons.

PHYSICAL AND MATHEMATICAL MODEL

MAD9P is based on the Vlasov-Maxwell equations. In this model, particle motion is governed by external fields and a mean-field approach for the space-charge fields. Particle collisions and radiation are neglected. All physical elements are assumed to be perfectly aligned. The total Hamiltonian for a beam line element can be written as a sum of two parts, $\mathscr{H} = \mathscr{H}_1 + \mathscr{H}_2$, which correspond to the external and space charge contributions. A second-order integration algorithm (split operator) for a single step is then given by

$$\mathscr{M}(\tau) = \mathscr{M}_1(\tau/2)\,\mathscr{M}_2(\tau)\,\mathscr{M}_1(\tau/2) + \mathscr{O}(\tau^3) \qquad (1)$$

where τ denotes the step size, $\mathscr{M}_1$ is the map corresponding to $\mathscr{H}_1$ obtained by differential algebra (DA) methods from a general relativistic Hamiltonian and $\mathscr{M}_2$ is the map corresponding to $\mathscr{H}_2$. $\mathscr{M}_2$ is obtained by discretizing the resulting Poisson problem on a rectangular mesh using Fourier techniques to solve the time consuming cyclic convolution in $\mathscr{O}(M \log M)$, where M is the number of grid points. Open and periodic boundary conditions are available.

MAD9P a general 3D Particle Tracker with Space Charge

In order to ease the task of writing efficient parallel applications, we employ the POOMA framework. POOMA provides abstraction for mathematical/physical quantities (particles, fields, meshes, differential operators etc.) in an n-dimensional parallel fashion. For an architectural overview of MAD9P see Figure 1. The object-oriented approach manages the complexity of explicit parallel programming; it encapsulates the data distribution and communication between real or virtual processors. POOMA and all the other components are implemented as a set of templated C++ classes. When used

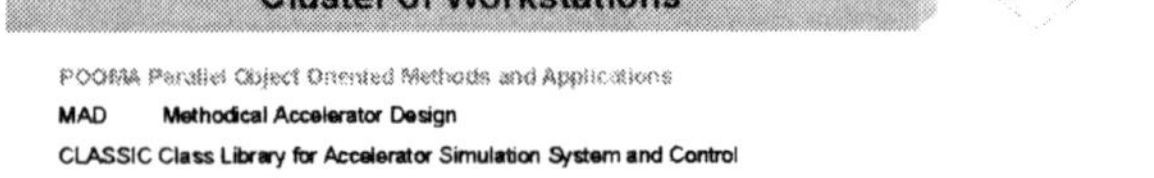

FIGURE 1. Architectural overview of MAD9P.

in a parallel environment, MAD9P partitions the particles in a particle container among the separate processors. The particle spatial layout will keep a particle on

[1] We use the notion of framework in the following sense: a framework is a set of co-operating classes in a given problem frame.

CP642, *High Intensity and High Brightness Hadron Beams: 20th ICFA Advanced Beam Dynamics Workshop on High Intensity and High Brightness Hadron Beams*, edited by W. Chou, Y. Mori, D. Neuffer, and J.-F. Ostiguy

the node which contains the section of the field in which the particle is located. If the particle moves to a new position, this layout will reassign it to a new node when necessary. This will maintain locality between the particles and any distributed field and it will help keep particles which are spatially close to each other local to the same processor as well. With this concept (inherited in the particle and field class) we do not need an explicit and complicated particle manager class. A 3D parallel particle-mesh solver is implemented on top of the POOMA framework and makes use of their efficient parallel Fourier transformation routines. Using up to 32 processors on a Beowulf cluster at PSI with flat communication structure, we obtain 87.5% of the optimal speedup. Using 128 processors, again on a Beowulf cluster, but with a non flat communication structure, we get still 37.5% of the optimal speedup (using ten million particles and a mesh with 128^3 points). On the other hand, it is well known that the numerical noise only decreases with $1/\sqrt{N}$ when the number of particles N is increased. The parallel approach allows us to use routinely configurations of $N = 10^7 \cdots 10^8$ and mesh sizes from 128^3 to $128^2 \times 2048$. MAD9P runs on a variety of UNIX platforms such as the SGI-ORIGIN 2000, IBM-SP2 and Linux clusters. Validation of the code is done by comparison with (semi) analytic models.

First Simulation Results with MAD9P

In the context of a feasibility study on how to accelerate a 3 mA proton beam with the PSI cyclotron facility, necessary upgrade steps for the different system components are under consideration. In particular, the qualitative **and** quantitative knowledge of phase space transformations of the proton beam at injection into the PSI Injector 2 cyclotron are essential for the successful production of high intensity beams with low losses. To obtain proper initial conditions for the Injector 2 cyclotron, we start with modeling the 870 keV injection line (B870) from Cockroft-Walton pre-injector to the Injector 2.

B870 Injection Line

The starting point for all B870 injection line calculations is a 4-dimensional transverse phase space distribution, which has been proven to be physically satisfactory in the daily operation of the beam line. The longitudinal dimensions are uniform in space and momenta. The initially DC beam is modeled by using a characteristic longitudinal beam length of $\beta\lambda$, where λ is the wave length of the rf and β the relativistic factor. The double gap buncher is modeled by (analytic) sinusoidal momenta modulation of the beam. Figure 2 shows the horizontal beam envelope (similar results are obtained in the vertical direction see Figure 3) after fitting the 4-dimensional transverse distribution and a global space-charge neutralisation factor f_e using a stochastic fit algorithm based on

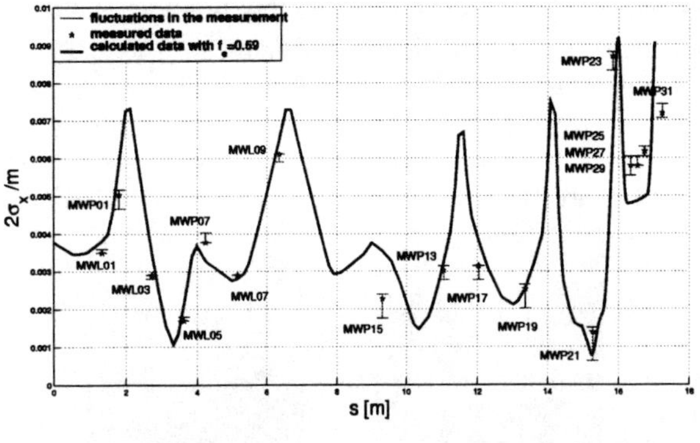

FIGURE 2. Best fit: horizontal beam profiles in the B870 line.

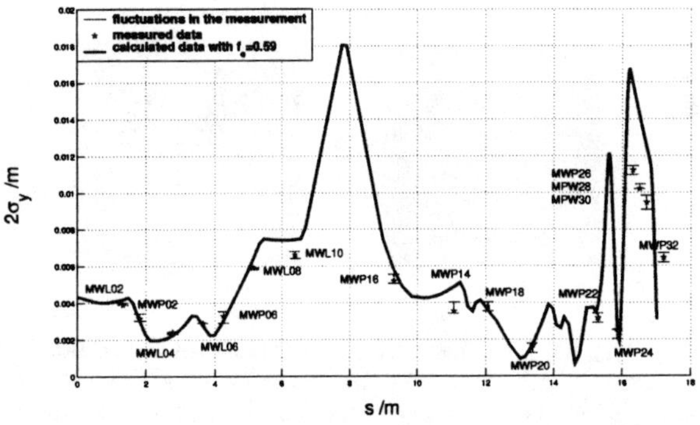

FIGURE 3. Best fit: vertical beam profiles in the B870 line.

Simulated Annealing. Define F as follows,

$$F = \sum_{n=1}^{\#monitors} (X_{mea}(s_n) - X_{sim}(s_n))^2. \quad (2)$$

This function is a measure of the degree of conformity between simulation and profile monitor measurements, where $X_{mea}(s_n)$ is a measured rms quantity at the position s_n along the beam line and $X_{sim}(s_n)$ is the corresponding calculated quantity obtained by MAD9P. The fitting procedure then minimizes F in (2). As shown in Figure 2 and 3 we obtain good agreement between measurement and simulation. The space-charge neutralisation factor $f_e = 0.59$ obtained is in the expected range [4]. The discrepancy in $MWP15$ is not fully understood. One possible reason is the large background which is not included by the simulation model at the moment. The deviations seen at $MWP25$ *to* $MWP31$ are related to the buncher and the high dispersive region in this part of the beam line. More detailed modeling is needed in order to minimize the gap between theory and observation.

Injector 2, Coasting Beam

A model of the Injector 2 lattice based on hard-edge elements is used for various coasting beam simulations. The early 2D results of Stefan Adam [5], which predict a stable round distribution in horizontal- longitudinal configuration space has been verified (see Figure 4 and 5) with the full 3D model. The data shown in the Figures 4 and 5 are for 5 *MeV* and 1 *mA*. The effect of the

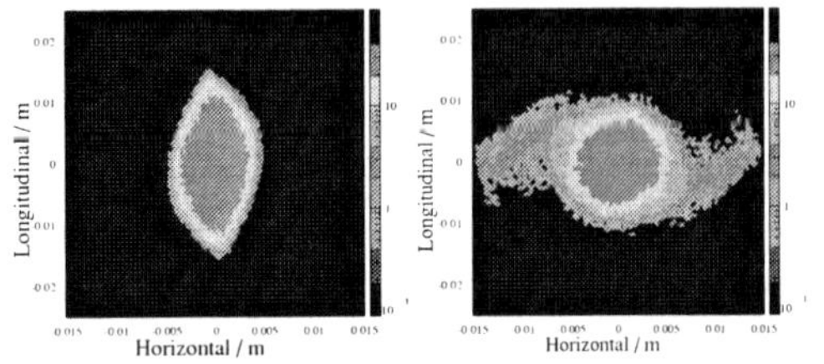

FIGURE 4. Charge density in a.u.: Turn 1 and 6.

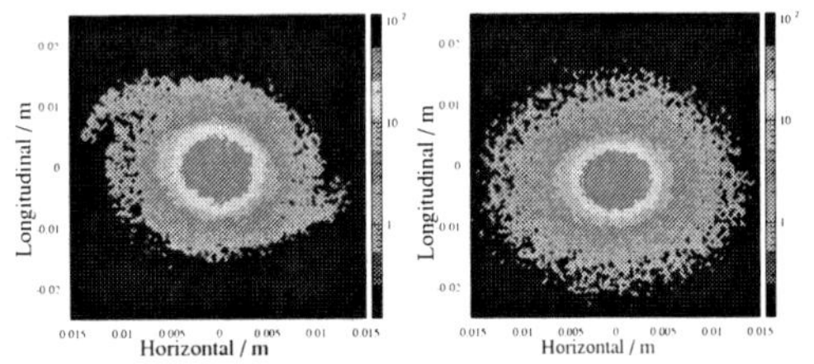

FIGURE 5. Charge density in a.u.: Turn 10 and 60.

beam intensity on the development of the rms beam sizes in the horizontal and longitudinal directions is shown in Figure 6 for 60 turns. The strong oscillations in the first few turns are due to an initial 'mismatch' of the beam. The fact that the rms beam size increases with increasing beam current strongly suggests that the matching of the incoming beam has to be adapted to the beam intensity and might be the key for a very fast development of the desired round and stable distribution. This first truly 3D simulation suggests again, that the concept of an isochronous cyclotron is well suited for high intensity operation. More research is however needed in order to finally make predictions for the redesign of the B870 line, to allow operation beyond the presently achieved 2 mA.

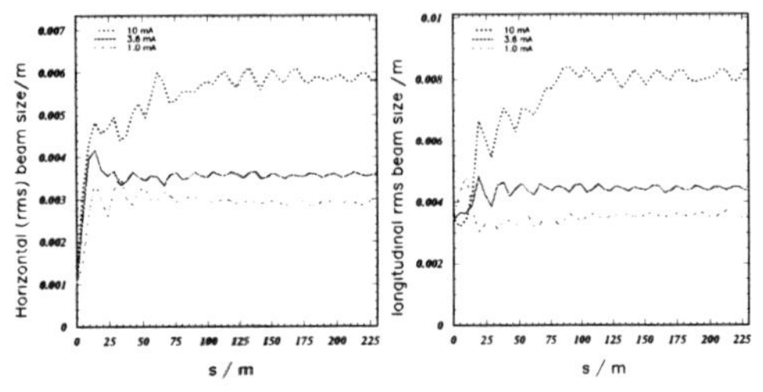

FIGURE 6. Horizontal and Longitudinal rms beam sizes at different intensities.

OUTLOOK

Adding acceleration to the hard-edge model of Injector 2 and using the initial conditions obtained from B870 injection line simulations will allow us to compare beam profile measured in the cyclotron [6, 7, 8] with calculations.

Missing physics such as a residual gas model, beam collimation including secondary effects, and neighboring turns in the cyclotron, will be added to MAD9P.

Different types of parallel field solver, such as tree-based, multigrid and hybrid versions are planned to be a part of MAD9P. Carefully benchmarking the field solvers against other codes and analytical models will help us to understand their different behavior.

ACKNOWLEDGMENTS

Many thanks to Martin Humbel and the PSI proton acceleration operation staff for providing the experimental data. I would also acknowledge the continuous support given by and the many fruitful discussion with Stefan Adam, Ralph Eichler, Rolf Jeltsch, Thomas Stammbach and Robert Ryne.

REFERENCES

1. Atlas, S., Banerjee, S., Cummings, J., Hinker, P., Srikant, M., Reynders, J., and Tholburn", M., ""POOMA: A high-performance distributed simulation environment for scientifi c applications"," in *In Supercomputing '95 Proceedings*, 1995.
2. Adelmann, A., *3D Simulations of Space Charge Effects in Particle Beams*, Ph.D. thesis, ETH, No 14545 (2002).
3. Schmelzbach, P. A., "PSI Cyclotron," in *20th ICFA Advanced Beam Dynamics Workshop High Intensity High Brightness Hadron Beams*, 2002.
4. Olivo, M., Mariani, E., and Sherman, J., *Rev. of Scientific Instrum.*, **63**, 2776 (1992).
5. Adam, S. R., *Methoden zur Berechnung der Longitudinalen Raumladungs Effekte in Isochronzyklotrons*, Ph.D. thesis, ETH (1985).
6. Dölling, R., "Measurement of the time structure of the 72 MeV Proton Beam in the PSI Injector 2 Cyclotron," in *DIPAC 2001 Proceedings*, 2001.
7. Rezzonico, L., and Olivo, M., "Profi le Measurement of high initensity-low energy ion beams," in *ICIS 95, Whistler Canada*, 1995.
8. Rezzonico, L., Adam, S. R., and Humbel, M., "Diagnostics for High Intensity Beams," in *6th Beam Instrumentation Workshop*, 1995.

Intense Beam Experiments at the University of Maryland Electron Ring (UMER)

R. A. Kishek, P. G. O'Shea, M. Reiser, B. Beaudoin, S. Bernal, Y. Cui, A. Diep, D. Feldman, M. Glanzer, T. F. Godlove, I. Haber, J. Harris, H. Li, J. Neumann, B. Quinn, M. Qurius, M. Snowel, A. Valfells, M. Virgo, M. Walter, R. Yun, Y. Zou

Institute for Research in Electronics and Applied Physics, Univ. of Maryland, College Park, MD 20742-3511, USA

Abstract. A detailed understanding of the physics of space-charge dominated beams is vital for many modern accelerators. In that regard, low-energy, high-intensity electron beams provide an excellent model system. The University of Maryland Electron Ring (UMER) has been designed to study the physics of space-charge dominated beams with extreme intensity in a strong focusing lattice with dispersion. At 10-keV, 100 mA, the UMER beam has a generalized perveance of 0.0015. Though compact (11-m in circumference), UMER is a very complex device. An update on construction and early experimental results is presented.

INTRODUCTION

There has been an increased interest in space charge effects in recent years. In linear accelerators, the demand for ever lower beam emittance and increased current has meant that space-charge driven emittance growth and halo formation can be a consideration at energies as high as 100 MeV in some electron linacs proposed for fourth generation light sources [1]. Circular machines, on the other hand, have been limited to much lower intensities than linacs due to destructive resonances which can severely limit the achievable number of turns. With space charge, the resonances are no longer caused by the interaction between the single particle orbits and the harmonics of the field errors. Instead, a resonance can occur when the frequency of a collective beam mode coincides with one of the harmonics of the error frequency spectrum.

At the University of Maryland, we have a tradition of using low-energy electron beams as model systems for studying space charge phenomena that are of general interest in intense beam systems [2]. Previously, all of our research had been done with straight-line systems. Our motivation to develop a ring system is twofold. There are many interesting phenomena in intense beam physics that evolve over longer distances than our 5-m linear solenoid channel. Furthermore, a ring with both strong focusing and dispersion would also allow us to study resonance issues in a realistic setting.

Currently we are constructing one major facility for conducting experiments in beam physics at extreme intensities in a circular lattice, namely, the University of Maryland Electron Ring (UMER) [3]. In addition, we have a number of side experiments, particularly, the Long Solenoid Experiment (LSE) [4], designed to study longitudinal beam physics in a straight geometry. These machines are designed to serve as a low-cost research tool for building higher intensity accelerators for a variety of applications, such as heavy-ion fusion drivers, the low energy end of high intensity electron linacs, ion booster synchrotrons, muon colliders, and spallation neutron sources.

The unknown territory in the extreme space-charge dominated regime is very challenging and should provide a wealth of new phenomena. The UMER facility will allow us to investigate emittance growth due to conversion of free energy, halo formation, and equipartitioning in a circular machine [5]. So far, these effects have only been studied in linear transport lines. In addition, UMER will permit experimental investigations of longitudinal-transverse coupling and beam profile changes resulting from dispersion; the behavior of bunch ends, resonance traversal; the longitudinal resistive wall instability; and other effects in the space-charge dominated regime that are currently inaccessible.

We are also investigating the possibility of using UMER and related systems for laboratory studies related to galactic dynamics. Many of the dynamical processes involved in galactic formation and evolution appear to have analogs in the physics of high intensity beams [6].

CP642, *High Intensity and High Brightness Hadron Beams: 20th ICFA Advanced Beam Dynamics Workshop on High Intensity and High Brightness Hadron Beams*, edited by W. Chou, Y. Mori, D. Neuffer, and J.-F. Ostiguy

KEY UMER FEATURES

Table 1 lists some relevant parameters of UMER. The beam is drifting at 10 keV, and 36 FODO cells around each turn provide transverse focusing, while 3 induction gaps provides longitudinal focusing. In a future stage, the induction gaps will also be used to accelerate the beam to 50 keV. At the present β of 0.2, the beam is nonrelativistic. The beam intensity can be quantified in a number of ways, for example using the dimensionless intensity parameter, χ (defined in ref. [3] as the ratio of the space-charge force to the external focusing force at the beam radius). The tune depression is expressed in terms of χ as $\frac{\nu}{\nu_0} = \frac{k}{k_0} = \sqrt{1-\chi}$, and the plasma wavenumber k_p as $\frac{k_p}{k_0} = \sqrt{2\chi}$. HIF drivers, for example, will likely operate with $0.89 < \chi < 0.98$. With a zero-current phase advance per period of $\sigma_0 = 76°$, a nominal emittance of $\varepsilon_n = 10$ μm and beam radius a = 1 cm, UMER can attain this region of extreme intensity. The intensity parameter, tune depression or beam current can be varied in UMER over a wide range by changing to different apertures sizes in the beam collimator at the exit of the gun, by changing the anode–cathode spacing, or by changing the beam energy.

TABLE 1. UMER design specifications. Data from [3]

Energy	10 keV
β (= v/c)	0.2
Current	<= 100 mA
Generalized perveance	0.0015
Emittance, 4x rms, norm	10 μm
Pulse Length	40-100 ns
Circumference	11.52 m
Lap time	197 ns
Pulse repetition rate	60 Hz
Mean beam radius	<= 1 cm
FODO period	0.32 m
0-current phase advance, σ_o	76 °
0-current Betatron tune, ν_o	7.6
Tune Depression	>= 0.2

The electron bunch is injected into the ring at a repetition rate of 60 Hz or less from the injector system [7] with the help of a DC Panofsky quadrupole and a pulsed dipole. The bunch can be extracted within the first turn or after any number of turns with a system that duplicates the features of the injector line except that the electron gun is replaced by a large diagnostic chamber with phosphor screen, emittance meter and energy analyzer [4, 8].

When complete, the ring magnetic lattice will consist of over 140 quadrupoles, dipoles and steering magnets. The typical focusing gradients and bending fields are on the order of 5 Gauss/cm and 10 Gauss, respectively. The use of iron-based magnets is impractical for such low fields. Therefore, the UMER magnets are based on an iron-free printed circuit (PC) design [9] which, while fabricated to a very high tolerance, considerably lowers the cost of the project. A new feature that we have introduced recently is a quadrupole with electronically adjustable skewness (roll). The feature allows us to both correct for residual skewness in our system, and also to deliberately introduce skewness at various location in the ring for beam dynamics studies [10].

The design of the electron gun and simulations of its performance are described elsewhere [11-12]. A grid near the cathode can be used to generate perturbations as means of investigating instabilities. In addition, we have been investigating the possibility of producing ultra-short (1 ns) current pulses on top of the main pulse. We have achieved this by using a nitrogen laser as a photocathode drive laser on the dispenser cathode [13] (Fig. 1). This will allow is to create perturbations on the beam that are well located in time and space. The evolution of such a perturbation will allow us to study dissipative processes, e.g. resistive wall phenomena.

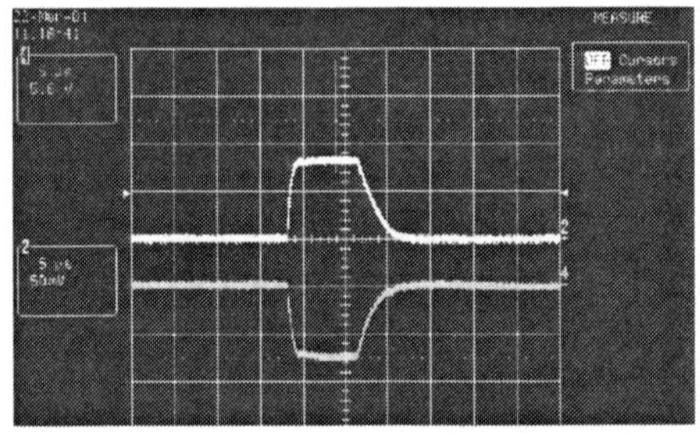

FIGURE 1. Combined photoelectron and thermionic current (upper trace). Note the sharp current spike corresponding to the photoelectron pulse. The lower trace is the gun voltage pulse.

To allow detailed comparison between theory and experiment, UMER will have a comprehensive set of beam diagnostics. Each of the 13 diagnostic stations around the ring will have a phosphor screen and capacitive beam position monitor [14]. In addition, fast current monitors and resistive beam position monitors will also be installed. A sophisticated diagnostic end-chamber has been fabricated by FM Technologies. The chamber houses emittance meters of the slit-wire and pepper-pot types; a retarding-field energy analyzer with eV resolution for energy and energy spread measurements; a movable phosphor screen with 1.5 meter travel for insertion into the complete transport line; and a Faraday cup for current measurement.

DEVELOPMENT OF ENERGY ANALYZER (EA)

To measure the energy spread of the beam, we have successively evolved the design of a parallel-plate electrostatic retarding voltage energy analyzer (EA) [4,8]. The first improvement consisted of adding a collimating cylinder attached to the retarding mesh. The cylinder collimates the particle trajectories such that they are normally incident on the mesh, hence reducing any errors from a transverse angle or spread in velocities. Simulations show the resolution of the device is about 8 eV (at 10 keV), and the measured values of the energy spread are just slightly greater than this value. This is a factor of 5 improvement over the previous design.

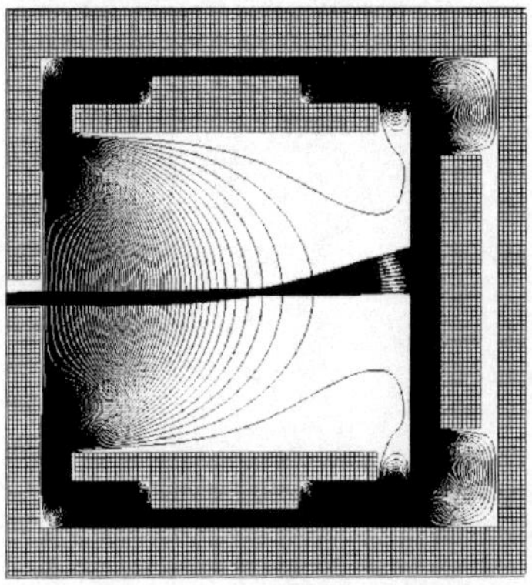

FIGURE 2. SimIon simulation of the latest EA Design, showing the structure of the device and the equipotentials, as well as some representative particle trajectories.

While the measured energy spread meets our specification of 20 eV, we are continuing to improve our diagnostic. By electrically separating the collimating cylinder from the retarding mesh [4], we gain additional control over the focusing of the trajectories, and simulations predict an extra factor of 10 in the resolution of the EA, i.e., down to 10^{-4}! This 3rd-generation EA (Fig. 2) is currently being tested.

PLANS

At present (May 2002), we are engaged in a phased installation of the ring. The first two our of 18 ring sections are currently being installed, and we are therefore about to complete 40 deg. shortly. The installation is done gradually to permit us to take measurements using the end diagnostic chamber around the entire perimeter of the ring during the first turn. This phased installation is scheduled to be completed by mid-2003.

After closure, we intend to attempt multi-turn operation, targeting 100 turns at low current (10 mA) and 10 turns at high current (100 mA), while maintaining beam emittance to within a factor of 4. At the end of that phase, we wish to upgrade UMER to a fast cycling synchrotron that accelerates the beam to 50 keV over 50-100 turns to study resonance crossing.

ACKNOWLEDGEMENTS

We wish to acknowledge Richard York and his colleagues at Michigan State University as well as FM Technologies, Inc. for their assistance with the design and construction of UMER components. This project is fully supported by the US Dept. of Energy grant numbers DEFG02-94ER40855 and DEFG02-92ER54178.

REFERENCES

Note: in the following, PAC01 means "Proceedings of the 2001 Particle Accelerator Conference, Chicago, IL, edited by P. Lucas and S. Weber, IEEE Cat. No. 01CH37268" (2001).

1. P.G. O'Shea and H.P. Freund, Science, **292**, 1854 (2001)
2. Martin Reiser, Theory and Design of Charged Particle Beams, (New York: John Wiley & Sons, Inc., 1994).
3. P.G. O'Shea, et. al., *NIM A* **464**, 646 (2001); also PAC01, p. 159; http://www.ireap.umd.edu/umer
4. Y. Cui *et al.*, PAC01, p. 2976.
5. R.A. Kishek, P.G. O'Shea, M. Reiser, Physical Review Letters **85**, 4514 (2000).
6. R. A. Kishek at al., PAC01, p. 151.
7. S. Bernal et al., PAC01, p. 2129.
8. A. Valfells *et al.*, PAC01, p. 3582.
9. W.W. Zhang, H. Li, S. Bernal, T. Godlove, R.A. Kishek, P.G. O'Shea, M. Reiser, V. Yun, and M. Venturini, Physical Review ST-AB **3**, 122401 (2000).
10. H. Li *et al.*, PAC01, p. 1802.
11. D. Kehne, *et al.*, NIM, **A464**, 605-609 (2001).
12. I. Haber *et al.*, PAC01, p. 2952; also these proceedings.
13. D. Feldman *et al.*, PAC01, p. 2132.
14. J. Harris *et al.*, PAC01, p. 1387.

SESSION J

HADRON BEAM COOLING

Limitations of Ion Beam Brightness with Electron Cooling – Theory and Experiment

V. Parkhomchuk

BINP, Novosibirsk, Russia

Abstract. Electron cooling is used for damping both single particle and coherent oscillations of ion beams. The extremely high phase space density of the cooled ion beam can become a source of stability problems ("electron heating"). For storage rings such as CELSIUS, COSY and the Indiana Cooler, this is a serious problem that limits the use of the electron cooling in physical experiments. Proper design of the electron cooler can improve the stability of the cooled ion beam.

SINGLE PARTICLE APPROACH

The cooling rate is determined by the cooling force – the friction force of an ion moving with velocity V with respect to an electron beam with density n_e. A longitudinal magnetic field is used to counteract the deflection caused by the space charge of electron beam. This field magnetizes the transverse thermal motion of the electrons. The cooling force can be calculated using the following equation [1,2]:

$$\vec{F}(\vec{V}) = \frac{4e^4 Z^2 n_e}{m} \frac{\vec{V}}{(\sqrt{(V^2+V_{eff}^2)})^3} ln\left(\frac{\rho_{max}+\rho_{min}+\rho_L}{\rho_{min}+\rho_L}\right) \quad (1)$$

where the impact parameters can be written in their simplest form as:

$$\rho_{max} = \frac{|V|}{\omega_e + 1/\tau}, \quad \omega_e = c\sqrt{4\pi n_e r_e},$$
$$\rho_{min} = \frac{Ze^2}{mV^2} = \frac{Zr_e}{(V/c)^2}, \quad r_e = \frac{e^2}{mc^2}, \quad (2)$$
$$\rho_L = \frac{mV_{e\perp}}{eB},$$

Here, Z is the ion charge, ω_e is the electron beam plasma frequency, τ is the time of flight in the cooling section (all quantities are evaluated in the beam reference frame), B is the longitudinal magnetic field, $V_{e\perp}$ is the thermal velocity of the electrons in the cooling section. V_{eff} is the effective velocity of the center of the electron Larmor orbits but is not the electron velocity. V_{eff} is determined by the electrical field from space charge of the electron beam and the transverse component of the magnetic field in the cooling section:

$$V_{eff} = c\frac{2\pi e n_e x}{B} + c\beta\gamma\theta_B, \quad (3)$$

where θ_B is the deflection of the direction of the magnetic line from the center of the ion beam orbit. For an ideal magnetic field in the cooler, $\theta_B = 0$. The ion beam emittance is ε_x. The cooling rate is maximized at a given electron current J_e with $n_e=(J_e/e)/(2\pi\varepsilon_x \beta_x)$, $<x^2>=\varepsilon_x \beta_x$, $V=c\beta\gamma(\varepsilon_x/\beta_x)^{1/2}$). The cooling rate can be written as:

$$\lambda cool = \frac{F}{MV} = \frac{2r_e r_i (Je/e)\eta_e}{\pi\gamma^5\beta^4} \times \frac{\sqrt{\beta_x}}{\varepsilon_x^{5/2}\left[1+\left(\frac{2J_e}{\varepsilon_x B\beta^2\gamma^2 c}\right)^2\right]^{3/2}} \ln\left(\frac{\rho_{max}+\rho_{min}+\rho_L}{\rho_{min}+\rho_L}\right) \quad (4)$$

It is easily seen that maximum cooling is achieved when the electron current is :

$$J_e = (0.5/\sqrt{2})\varepsilon_x B\beta^2\gamma^2 c = \frac{0.5}{\sqrt{2}}(\varepsilon_x / mm*mrad)(B/kGs)\beta^2\gamma^2(A) \quad (5)$$

The maximal cooling rate is:

$$\lambda_{max} = \frac{2r_e r_i (B/e)\eta_e c}{3^{3/2}\pi\gamma^{3/2}\beta^{1/2}} \times \frac{\sqrt{\beta_x}}{\varepsilon_{nx}^{3/2}} \ln\left(\frac{\rho_{max}+\rho_{min}+\rho_L}{\rho_{min}+\rho_L}\right) \quad (6)$$

For example, accumulation in the SIS synchrotron [3] for ε_x =150 mm*mrad, B=0.6kG, E = 11.4 MeV/n , J_{emax}=0.7A., and λ_{max}=0.5 1/s. Figure 1 shows the ion beam accumulation in multiturn injection, with injection at every 0.5s and simultaneous cooling. The time interval of 0.5 sec is not enough to cool all ions, a small fraction of the beam at the edge of the acceptance is lost at the moment of the next injection. (The first step is higher than the following ones.)

CP642, *High Intensity and High Brightness Hadron Beams: 20th ICFA Advanced Beam Dynamics Workshop on High Intensity and High Brightness Hadron Beams,* edited by W. Chou, Y. Mori, D. Neuffer, and J.-F. Ostiguy

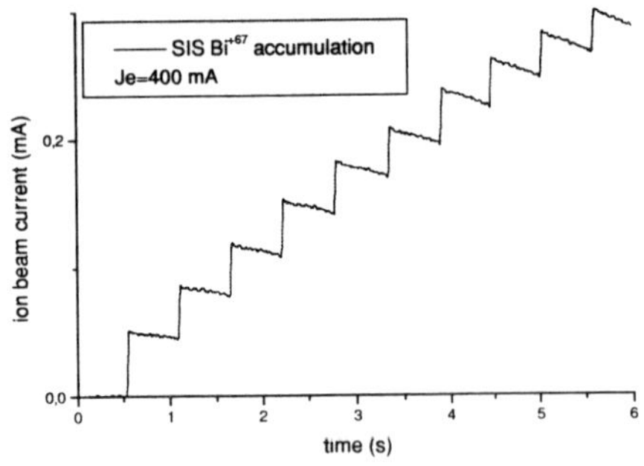

FIGURE 1. Ion beam accumulation in the SIS synchrotron GSI.

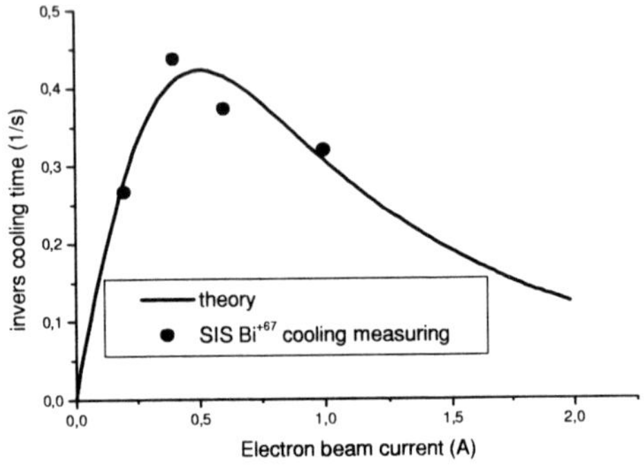

FIGURE 2. The cooling rate at an emittance of 150π mm-mrad versus the electron beam current, as measured in SIS (points) and calculated from equation (4).

The accumulation of beam is stopped when the losses of stored beam between new injections becomes equal to the injected increment. The lifetime of storage beam by recombination at the electron cooler is proportional to $\sqrt{(T_{e\perp})}$, where $T_{e\perp}$ is the temperature of the electron beam in the cooling section. In figure 3 can be seen the accumulation of Bi^{+67} in the SIS synchrotron and the loss of ion beam after end of injection. The lines show fitting formulas for accumulation and loss. (The losses are dominated by ion recombination.)

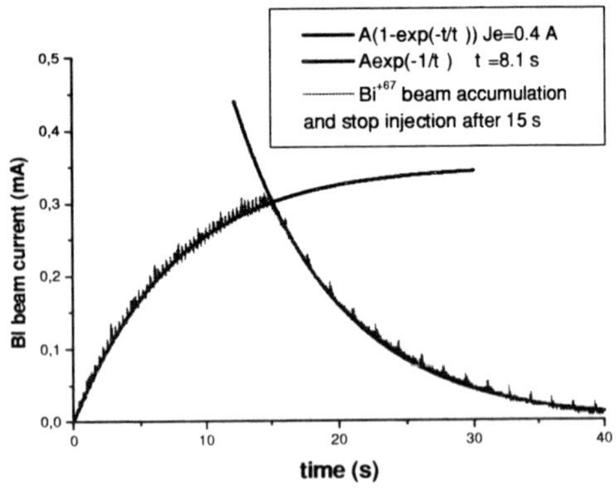

FIGURE 3. Accumulation and loss of Bi beam at SIS synchrotron.

Number of the storage cycles for saturation of storage current is estimated as:

$$N_{stor} = \lambda_{cool} \times \tau_{rec} \approx \frac{8500 \times \sqrt{T_{e\perp}}}{A(\gamma\beta\varepsilon_{nx} / \beta_x)^{3/2}} \frac{1}{(1 + (Je / Je_{\max})^2)^{3/2}} \frac{< n_e(r) >}{n_e(0)} \quad (7)$$

where $T_{e\perp}$ is the electron temperature measured in eV, $\varepsilon_{nx}=\beta\gamma\varepsilon_x$ is the normalized ion beam emittance, $<n_e(r)>$ is the average electron beam over tcross section of the electron beam and $n_e(0)$ is the electron beam density at the center of beam where storage ion beam. A cooling system with a large expansion of the electron beam from the gun to the cooling section obtains a low electron beam temperature ($T_{e\perp} = T_{cathode}$/expansion), but the accumulation potential is reduced by intensive recombination at the cooler. In the SIS cooler, direct experiments were made to compare the lifetime of the Bi beam with expansion factors of 1 and 6, with the same electron beam densities, and results are shown in figure 4.

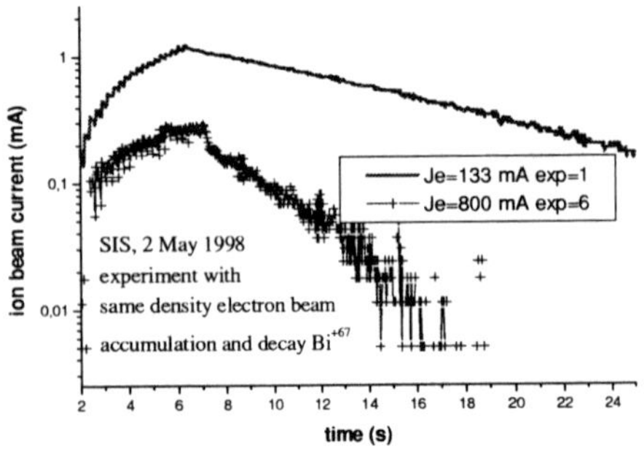

FIGURE 4. The accumulation and loss of Bi^{+67} ion beam with the same electron beam densities but with different expansion factors from the electron gun to the cooling section.

These results clearly demonstrate that for a high expansion factor the recombination rate increased without any increase in accumulation rate. A cooling system with high expansion factor is useful for experiments in atomic physics requiring high energy resolution, but it is at a disadvantage for systems requiring accumulation rate. Accumulation can be further increased by the use of an electron beam with lower density at the center ($n_e(0) << <n_e(r)>$). This can give an additional factor of about 10-20 in the stored ion beam current.

INSTABILITY FOR COOLED INTENSE ION BEAM

Experiments at the CELSIUS storage ring with electron cooling show that injection of a too intense proton beam results in high losses, and only a small fraction of beam is cooled[4]. This phenomenon was called "electron heating". As the injection intensities is higher for lighter ions and residual gas induced losses are not so high, the accumulation of these ions is limited, and usually develops this same instability. Figure 5 shows experiments with accumulation proton beam with an energy of 180 MeV in the CELSIUS ring.

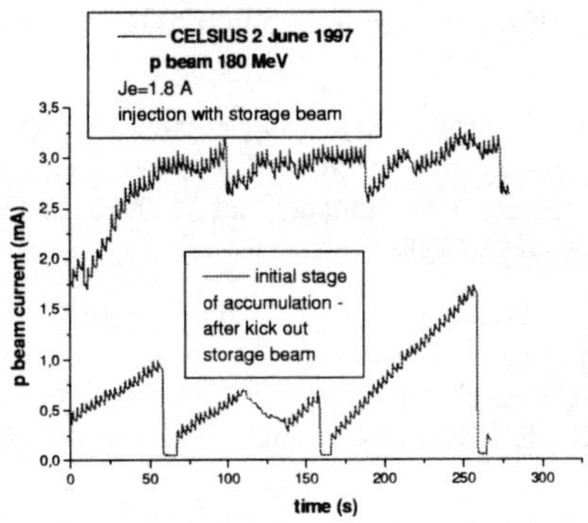

FIGURE 5. Accumulation of proton beam at CELSIUS multiturn injection from cyclotron.

The usual prescription for the suppression of instability is to increase the momentum spread of the ion beam. Direct application of an external rf noise field on the ion beam is not effective. Electron cooling at small amplitudes (in the cooled core) is extremely fast (a few ms) and an external noise applied to the ion beam can easily remove the tail of the beam without any useful effect on the beam core. Energy modulation of the electron beam appears to be a more reasonable solution. The ions cool toward the instantaneous electron energy and if the modulation has a high frequency (compared to the cooling rate) the ions will distribute their energies from the minimum to the maximum energies of the electron beam. Fig. 6 shows results of an experiment with modulation of the electron energy in the CELSIUS ring. Modulation of the electron beam energy at 218 keV ±300V produces an energy spread in the proton beam of $\pm1.5\times10^{-3}$ and stabilizes the cooling.

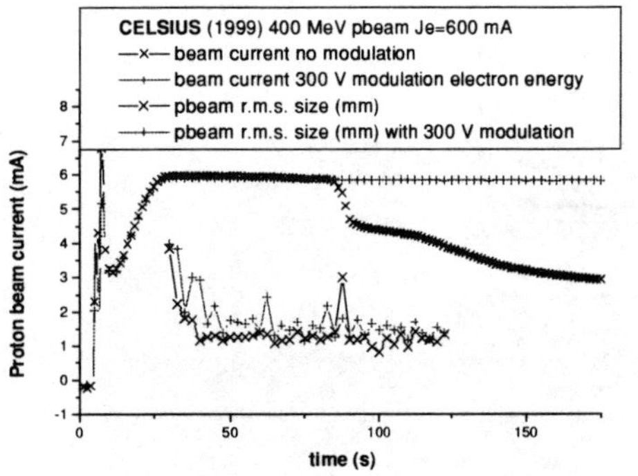

FIGURE 6. The proton beam current versus time. From 0 to 30 s the protons are injected and accelerated, from 30 to 170 s the prons are cooled.

NEW TOOLS TO CONTROL THE COOLING OF INTENSE ION BEAMS

Instability problems in intense ion beams stimulated the development of cooling systems that control the final ion-beam density. The simplest idea is using a hollow electron beam with empty space near thecenter where the core of the stored ion beam should be placed [5,6]. The cooling rate is inversely proportional to the cubic power of the amplitude of ion oscillation. It is possible to decrease the electron beam density in the center of beam without decreasing the cooling rate. These ideas were proposed for application in the CSR project (IMP, China). Two new coolers with variable electron beam profile were produced at BINP in collaboration with IMP. In August 2002, the first cooler will be commissioned at BINP. Figure 7 shows this cooler in March 2002.

FIGURE 7. The new cooler for IMP (China) built by BINP (Novosibirsk).

Figure 8 shows the measured profiles of the electron beam at a test bench for commissioning the electron gun and the electron collector. The electron beam diameter is 3 cm. By changing the voltage on the control electrode the electron beam is transformed from a solid to a hollow beam. An electron gun of this design can be used for optimization of the electron cooling process.

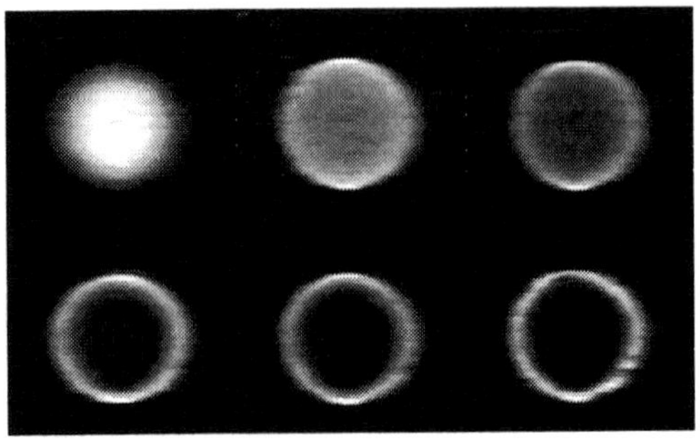

FIGURE 8. The electron beam distribution for different voltages on the control electrode - 0, 100, 200, 350, 400, 600 V. The measurement was made by scanning a tungsten wire across the electron beam.

Figure 9 shows a calculated example of cooling with hollow electron beams. The parameters of the beam are SIS injection parameters. The electron beam has a current of 0.5A and an empty central hole with a diameter of 1.5 cm. From fig.11 it is easily seen that the electron cooling formed a very specific ion beam profile. The center of the beam shows a Gaussian profile but with the tail cut off at 1.5 cm diameter. All ions in the initial tail are cooled into the central density region.

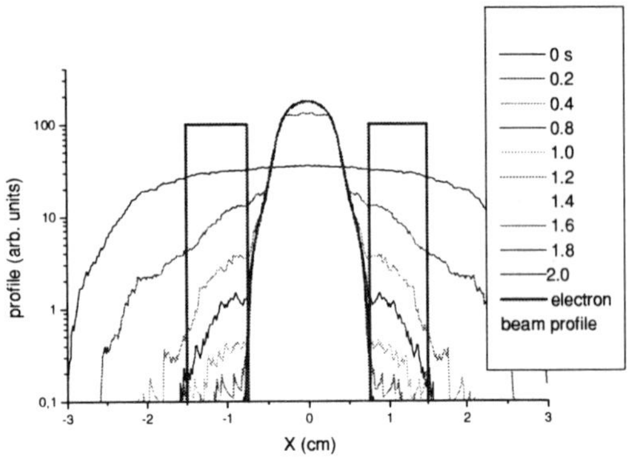

FIGURE 9. The development of the profile of a Bi ion beam with cooling by a hollow electron beam. Profiles at t=0, 0.2, 0.4, …, 2.0s are shown.

CONCLUSIONS

Electron cooling is very useful in the accumulation of rare ion beams. For example, the accumulation of secondary exotic nuclei prosduced at a target or of high-charge ions produced through stripping foils can be enhanced. Some new R.&.D projects in electron cooling for high luminosity colliders have begun (such as cooling for RHIC). A new application is the cooling of high intensity beams, where the system really cools only the tail of the main beam to reduce losses. Using cooling instead of only collimating can decrease the losses and the resulting activation of the vacuum chamber.

REFERENCES

1. V.V. Parkhomchuk, A.N.Skrinsky, Phusics-Uspekhi 43(5) 433-452(2000)
2. V.V. Parkhomchuk, "New Insights in the Theory of Electron Cooling", Nuclear Instruments and Methods in Physics Research A 441 (2000) 9-7
3. M. Steck, "Beam accumulation with the SIS Electron Cooler", Nuclear Instruments and Methods in Physics Research A 441 (200) 175-182.
4. D. Reistad, "Measurements of Electron Cooling and Electron Heating at CELSIUS", to appear in Proceedings of the workshop on Beam Cooling 2001, Bad Honeff, Germany.
5. E.I. Antohin. V.N.Bocharov, "Conceptual Project of an Electron Cooling System at an Energy of Electrons of 350 keV", Nuclear Instruments and Methods in Physics Research A 441 (2000) 87-91.
6. Ivanov A.V., Parkhomchuk V.V., Sukhina B.N., Tiunov M.A, "The Hollow Electron Beam Opportunities in Electron Cooling" to appear in Proceedings of the Workshop on Beam Cooling 2001, Bad Honeff, Germany.

Some Space-Charge Effects in Electron Cooling Devices

V.V. Parkhomchuk and V.B. Reva

The Budker Institute of Nuclear Physics, Novosibirsk, Russia

Abstract. The method of electron cooling uses an energy exchange between ions and electrons, both moving with the same average velocity in a special section of the ion ring [1]. The interaction between the ions and electrons leads to the various effects on the properties of the ion beam

INTERACTION OF THE SPACE-CHARGE FLUCTUATION IN THE ION AND THE ELECTRON BEAMS

It is convenient to represent the action of the Coulomb force on a single particle as a sum of two parts. The first part is a short-range interaction. This action is similar to the friction force of a fast particle moving in a cold gas. The second, long-range part of the interaction manifests itself as the action of a collective electric field. This interaction is more complicated and may lead to both decreasing ion energy and increasing ion energy. One of the reasons for this is related to the finite time of interaction between the ion and electron beams in the cooling section. In the case of a closed system the collective oscillation is reversible in time and does not lead to growth of the total energy of the ions. In our case, where the system is open and the electron beam is continuously renewed, the situation demands a special analysis.

A simple model illustrating the features of the finite time interaction is the the short-range longitudinal fluctuation model. Its dynamics can be described by the equations [2]:

$$\frac{d^2 s_i}{dt^2} + \omega_i^2 s_i = \omega_{ie}^2 s_i, \qquad (1)$$

$$\frac{d^2 s_e}{dt^2} + \omega_e^2 s_e = \omega_{ei}^2 s_e.$$

Here s_i and s_e are the longitudinal shift of the ion and the electron from an unperturbed location, ω_e and ω_i are the frequencies of longitudinal space-charge oscillations of the given wave vectors, and ω_{ei} and ω_{ie} are the oscillation frequencies of one type of particle in the collective field of the other particle type.

The determinant of the ion transfer matrix corresponding to these equations is

$$\text{Det}(M_{cool}) = 1 + \frac{2\omega_{ei}^2\omega_{ie}^2}{\omega_0^4}\cdot(\cos(\omega_0\tau) - 1) + \frac{\omega_{ei}^2\omega_{ie}^2}{\omega_0^4}\cdot\omega_0\tau\cdot\sin(\omega_0\tau) \qquad (2)$$

where $\omega_0 = \sqrt{\omega_e^2 + \omega_i^2}$, τ is the time of ion passage through the electron cooling section, M_{cool} is the transfer matrix from the state vector $[s_i, s_i']$ at the point before cooling section ($s = 0$) to the state vector $[s_i, s_i']$ at the point $s = L_{cool}$. The initial parameters of the electron beam are zero: $s_e = ds_e/dt = 0$. The determinant value is $\text{Det}(M_{cool}) < 1$ for small values of the interaction time $\omega_0\tau$, but the determinant $\text{Det}(M_{cool}) > 1$ for large $\omega_0\tau > 2\pi$. The parameter $\omega_{ei}\ \omega_{ie}/\omega_0^2$ defines the difference of the Det value from 1.

This model is applicable to high-frequency oscillations with a small wave vector. Such oscillations have small lifetime and fast mixing due to Landau damping. Thus this effect can be considered as an amplifier or a suppressor of a fluctuation. Its real role depends on the amount of energy contained in this short-wave fluctuation.

The equations of transverse motion of ion and electron beams in the magnetic field of the cooling section cooler are [3-5]

$$\frac{\partial^2\langle z_i\rangle}{\partial s^2} + \frac{i}{\lambda_{Hi}}\frac{\partial\langle z_i\rangle}{\partial s} + \frac{1}{\lambda_{ie}^2}\langle z_i\rangle = \frac{1}{\lambda_{ie}^2}\langle z_e\rangle,$$

$$\frac{\partial\langle z_e\rangle}{\partial s} + i\frac{\lambda_{He}}{\lambda_{ei}^2}\langle z_e\rangle = i\frac{\lambda_{He}}{\lambda_{ei}^2}\langle z_i\rangle. \qquad (3)$$

CP642, *High Intensity and High Brightness Hadron Beams: 20th ICFA Advanced Beam Dynamics Workshop on High Intensity and High Brightness Hadron Beams,* edited by W. Chou, Y. Mori, D. Neuffer, and J.-F. Ostiguy

Here $\langle z_i \rangle = \langle x_i \rangle + i\langle y_i \rangle$, $\langle z_e \rangle = \langle x_e \rangle + i\langle y_e \rangle$ is the centre-of-mass displacement of the ion and electron beams,

$$\lambda_{ie}^2 = \frac{m_i \gamma^3 (\beta c)^2}{2\pi e^2 Z_i n_e}, \quad \lambda_{ei}^2 = \frac{m_e \gamma^3 (\beta c)^2}{2\pi e^2 Z_i n_i},$$

$$\lambda_{Hi} = \frac{\gamma\beta m_i c^2}{Z_i eB}, \quad \lambda_{He} = \frac{\gamma\beta m_e c^2}{eB},$$

n_e and n_i are the densities of the ion and electron beams in the laboratory frame, B is the magnetic field, eZ_i is the ion charge, and m_i and m_e are ion and electron masses. We assume that the radii of the beams are equal, the disturbance of the beams is a long-wavelength transverse displacement, the radius of the vacuum chamber is large so that image forces are negligible, and the drift approximation is correct ($\lambda_{He} << \lambda_{Hi}, \lambda_{ei}, \lambda_{ie}$). The matrix of one revolution of the ion beam in the ring containing the cooling section is

$$M_{ring} = LRLM_{cool}. \qquad (4)$$

R is the Twiss matrix that is multiplied from the right and left sides by a matrix L of adrift section with a negative length of $-L_{cool}/2$.

The result of analysis in the case of small L_{cool} ($L_{cool} << \lambda_{ei}^2/\lambda_{He}, \lambda_{ei}, \lambda_{ie}$) is

$$\lambda^2 = 1 \pm \frac{1}{2}\beta_\perp L_{cool}^2 \frac{\lambda_{He}}{\lambda_{ie}^2 \lambda_{ei}^2}, \qquad (5)$$

where λ is the eigenvalue for matrix M_{ring}. Here $\beta_\perp$ is the beta-function, L_{cool} is the length of the cooling section.

The result is unusual. There is an eigenvalue greater than unity, which means that there is an unstable motion of the centre-of-mass of the ion beam. The decrease of the beta-function $\beta_\perp$ and the increase of the magnetic field in the cooling section leads to a decrease in the growth rate of this instability, but it does not suppress it completely. The effect is associated with the redistribution of energy between the various ion beam oscillation modes in the cooling section.

STOCHASTIC HEATING OF THE ION BEAM DUE TO ELECTRON BEAM NOISE FLUCTUATIONS

The electron beam can affect the ion beam as a lens with fluctuating strength. If the total space charge of the electron beam is changing at high frequency, then the life-time of the ion beam decreases. The noise in the electron beam can be induced by both the noise in the power supply and the intrinsic processes of the electron beam. The storage of secondary electrons and fast losses can also add space-charge fluctuations.

In a simple model the electron beam noise is defined by the current fluctuation, with the fluctuation independent for each ion turn. In this case we can write the increment of ion betatron amplitude a as

$$\frac{da^2}{dt} = f_0 \left[\frac{2\langle J_e \rangle r_i l_{cool} \beta_\perp}{q\beta^3\gamma^3 c}\right]^2 \begin{cases} a^2/a_e^4 & a < a_e \\ 1/a^2 & a > a_e \end{cases} \qquad (6)$$

The cooling time of ions with betatron amplitude equal a is [7]:

$$\tau^{-1}(a) = \frac{2\pi(J_e/q) r_e r_i Ln_c \eta}{\beta^4\gamma^5(\sqrt{a^2 + a_{eff}^2}/\beta_\perp)^3} \times \begin{cases} 1/(\pi a_e^2) & a < a_e \\ 1/(\pi a^2) & a > a_e \end{cases}$$

(7)

where J_e is the current of the electron beam, q is the electron charge, r_e, r_i are the classical electron and ion radii, Ln_c is the Coulomb logarithm, η is the fraction of the storage ring occupied by the cooling device, β,γ are beam kinetic parameters, a_{eff} is the effective amplitude defining the magnitude of the cooling force, a_e is the electron beam radius, $\beta_\perp$ is the betatron function at the cooling device, and f_0 is the revolution frequency. By comparing eqs. (6) and (7), one can see that the cooling time increases as a^3 if the ion moves inside the electron beam and as a^5 if ion is outside the electron beam. Thus it is possible that, for some amplitudes of ion betatron oscillation, the cooling process does not compensate the heating process. The ion amplitudes increase and the ions are lost from the storage ring. Fig.4 shows the heating and cooling power of a cooling device with parameters similar to that of ref. [6].

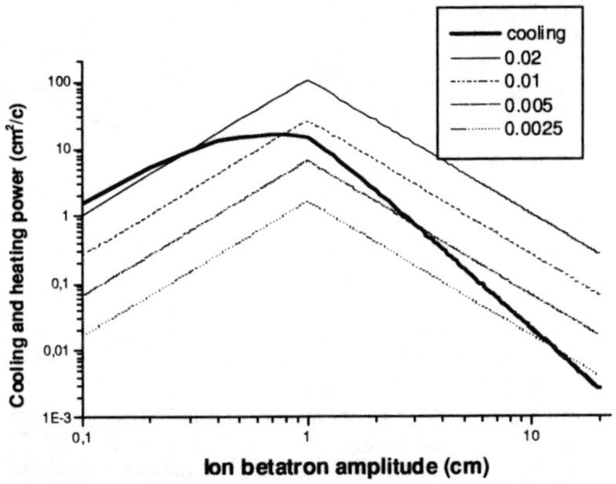

FIGURE 1. Cooling and heating power as a function of ion betatron amplitude.

One can see that a fluctuation of electron current by about 2% leads to the situation where ions with amplitude less than 3 mm can be cooled. The other ions will be lost. For normal operation of a cooling device the level of electron current fluctuation should be less than 0.1%. The threshold amplitude for ion loss is then more than 10 cm.

REFERENCES

[1] G. I. Budker, At.Energ. 22 (1967) 246-2582.

[2] V. V. Parkhomchuk, Nucl .Instr. Meth. A441 (2000) 9-17

[3] P. R. Zenkevich, A.E.Bolshakov, Nucl. Instr. Meth. A 441 (2000), p.36-39.

[4] A.Burov, Nucl. .Instr. .Meth , A 441 (2000), p.23-27.

[5] V. V. Parkhomchuk and V. B. Reva, Journal Exp. Theor. Phys., v.91 (2000), N 5, pp.975-982.

[6] E. I. Antokhin et al., "Conceptual project of an electron cooling system at an energy of electrons of 350 keV", Nucl. Instr. Meth. , A 441 (2000), p.87-91.

SESSION K

H^- STRIPPING

Summary Report

I. Yamane

KEK, Tsukuba, Japan

OUTLOOK

At the beginning of the session, the convener of the session announced that the motivation of this session was to find out a break through, or at least any hint leading it, to the present uneasy situation with the stripping media, and therefore, the session was concentrated upon stripping media. At present, solid foils don't seem enough durable against MW proton beams, but nevertheless we had not yet found out a realistic way of the laser stripping. The session was composed of two parts; Laser stripping and Foil stripping part, and four and three presentations were given in the respective part.

LASER STRIPPING

Session chaired by Y. Y. Lee

In the laser stripping part, three stripping schemes were proposed and an advanced technique on Fabry-Perot resonator was reported. Of these three laser stripping schemes, two were three-step scenarios:$H^- \to H^0$, $H^0_{gs} \to H^{0*}$, $H^{0*} \to (p+e)$. Another one was two-step scenario; $H^- \to H^0$, $H^0_{gs} \to H^{0*} \to (p+e)$. In both scenarios, the first step,$H^- \to H^0$, is commonly done by Lorentz stripping in a magnetic field. Difference is in following two processes; excitation of H^0 atom from the ground state to a higher excited state and decay of the excited H^0 atom to proton and electron. In the three-step scenario, these processes are done separately. In the two-step scenario, these are inseparable neither spatially nor temporally. In the three-step scenario, H^0 atoms are excited in the second step by collision with a laser beam in a free space or in a slight magnetic field. Then excited H^0 atoms are stripped by Lorentz stripping in the third step. In this case, following problems take place with the second step. 1) Doppler broadening of the transition frequency distribution occurs due to the momentum spread of H^0 beam. 2) Population of the upper state is saturated to one half by competition of pumping up and down by the laser beam. As a result, if any special measure like the Rabi-oscillation is not taken, stripping efficiency is limited to one half at most. The scheme proposed by U. Gastaldi is a three-step scenario, and as a measure to the problem 1), scanning of the transition frequency in a slight gradient of magnetic field is adopted. Because no special measure is taken for the problem 2), four or five systems are necessary to achieve a stripping efficiency near to 100%. But such a scheme is considered to be useful when H^0 beams must be shared by many users. This scheme is planned to adopt a Favbry-Perot resonator that contains a 10kW/cm2. In the other three-step scenario proposed by S. Danilov, the measure to the problem 1) is cancellation of Doppler shift by adjusting dispersion derivative and colliding angle between H^0 and laser beam. As a measure to the problem 2), a Froissart-Stora solution is adopted. In order to obtain necessary high power of the laserbeam, recycling of laser beam is considered at the first stage. The two-step scenario proposed by I. Yamane is a resonant photo-ionization via a broad Stark state. It utilizes a broad Stark state to cover the Doppler broadening due to the momentum spread of H^0 beams. Since the lifetime of the state is very short, population of the upper state cannot be increased and therefore pumping down of excited H^0 atoms to the ground state is negligible. Thus the ionization efficiency reaches to 100%. Necessary laser power density is supplied by a Fabry-Perot resonator which contains a several kW/cm2 light beam. In the report on the development of Fabry-Perot resonator for the PVLAS experiment, G. Cantatore showed that a very powerful Fabry-Perot resonator, which is 6m long and contains a 10kW/cm2 light beam with a diameter of 10mm, was available by existing technique. Introduction of such an advanced technique of another field of science is expected to bring an important step-up to the accelerator technology.

FOIL STRIPPING

Session chaired by I. Yamane.

M. Borden presented a review on the precious experience of stripping foils at LANL-PSR. I. Sugai talked on the present status of development of long-lived foils, and indicated that "cluster foil" is a promising candidate. He suggested that temperature control of foils to keep at about 750(C under bombardment may help to elongate the lifetime of foils. Y. Y. Lee showed the "Diamond foil"

CP642, *High Intensity and High Brightness Hadron Beams: 20th ICFA Advanced Beam Dynamics Workshop on High Intensity and High Brightness Hadron Beams,* edited by W. Chou, Y. Mori, D. Neuffer, and J.-F. Ostiguy

was the most long-lived foils. But it needs a frame of Si substrate and the frame may introduce some problems when it is used in a high intensity proton ring. In order to overcome the problems, he proposed a new type of stripping foil; "Diamond foil cylinder".

SUMMARY

It is impressive that patient and steady efforts are continued to develop long-lived foils. These efforts should be encouraged. Development of laser stripping showed a spread and three schemes are reported. Introduction of a powerful Fabry-Perot resonator is expected to encourage and accelerate the development of the laser stripping. A demonstration to show that the laser stripping works well is desired to be performed as soon as possible.

Lorentz Laser-Assisted Stripping (Lolas) for H^-/H^0 Injection into Proton Drivers

Ugo Gastaldi

INFN – Laboratori Nazionali di Legnaro
Via Romea 4, I-35020 Legnaro (Pd), Italy

Abstract. We discuss the main components of schemes for Lorentz laser-assisted stripping (abbreviated Lolas henceforth) proposed for injection into proton driver accumulators: $H^- \rightarrow H^0 + e^-$ Lorentz stripping, $H^0 \rightarrow H^0(n)$ laser excitation, $H^0(n) \rightarrow p^+ + e^-$ Lorentz stripping. We mention results obtained in practice of H^- beam transport and storage and of experiments addressing physics of the H^- ion, of the H^0 atom and of vacuum, which prove the feasibility of each Lolas component. For high enough injection energies, it is feasible to split without losses the H^0 beam sent towards the accumulator into a fraction stripped to p^+s and stored inside the accumulator and a complementary fraction of H^0s delivered to high duty-cycle users. The fraction of stored beam can exceed 50% with one single Fabry-Perot cavity used to enhance the laser power density. Aspects of Lolas integration and optimization are pointed out.

INTRODUCTION

Schemes for laser excitation and (simultaneous or subsequent) Lorentz stripping of H^0 atoms in the injection straight section of the accumulator of a proton driver have been worked out in Japan, Europe and USA [1-5]. The driving motivation is to design an injection without losses. This requirement is of paramount importance in order to enable improvements by factors exceeding 100 above present injection intensities with foil stripping.

All schemes have common basic ingredients, which include:
1) $H^- \rightarrow H^0 + e^-$ Lorentz stripping.
2) H^0 (1S) $\rightarrow H^0$ (n) laser induced resonant transitions from the H^0 ground state to a Stark sublevel of an $n \geq 3$ exited H^0 (n) state,
3) H^0 (n) $\rightarrow p^+ + e^-$ Lorentz stripping in the peak field region of the injection wiggler.

Critical points in all the schemes include:
a) angular dispersion of the H^0 beam,
b) tuning the laser frequency to the resonant transition,
c) tuning stability,
d) power density of the light that induces transitions.

Lolas schemes differ because:
1) H^- stripping is contemplated before or inside the injection straight section of the accumulator,
2) H^0 laser excitation is foreseen in a valley or in a peak of the injection wiggler,
3) The high power density necessary to induce H^0 excitations is achieved with pulsed lasers or with a Fabry-Perot cavity coupled to a CW laser [4].

Because of space limitations we restrict to comments on the main components and critical parts of Lolas schemes, for which refs [1-5] give extended sets of references and the essential formulas.

We mention experimental results [6-14] that prove the feasibility of Lolas components or give direct measurements of parameters of interest. These results are extracted from practice of H^- beam transport and storage and from experiments addressing basic and applied physics of the H^- ion and of the H^0 atom, and of the structure of the physical vacuum.

In most Lolas schemes emphasis is put in maximizing the probability of H^0 stripping at injection. A complementary point of view originates from the observation that proton drivers are necessary both for scientific programs which require high power and low duty-cycle beams (neutrino factory and spallation neutron source) and for programs that require high power and high duty-cycle proton beams (ISOL sources and accelerator driven systems for nuclear waste incineration). The new viewpoint is that it is convenient to foresee from the beginning sharing of the H^0 beam injected into the accumulator between low duty-cycle and high duty-cycle users. The H^0 beam can be split continuously and without losses into two fractions (typically 50/50). One fraction contains H^0s that are excited to one Stark sublevel of an $n \geq 3$ level of the H^0 atom and are subsequently fully stripped to p^+s and stored into the accumulator. The second fraction contains H^0s that have remained in the ground state, exit as a good quality beam from the accumulator injection straight section, and are utilized for the high duty-cycle programs. Under these circumstances the high duty-cycle programs run with a fraction of the total H^0 beam intensity that varies between 50% and 100%. Higher intensities for the accumulator users can be obtained by increasing the beam power delivered by the H^- injector linac.

CP642, *High Intensity and High Brightness Hadron Beams: 20th ICFA Advanced Beam Dynamics Workshop on High Intensity and High Brightness Hadron Beams,* edited by W. Chou, Y. Mori, D. Neuffer, and J.-F. Ostiguy

H^- STRIPPING

It is convenient to Lorentz strip H^- to H^0 outside the injection section of the accumulator because there one can install a stripper magnet with a small aperture, that enables a high field gradient. This is necessary to minimize the H^0 beam angular dispersion generated by the finite length of the segment of H^- trajectory where H^0s originate from H^- stripping (see ref. 6).

Since the motional electric field E seen in the H^- rest frame by a H^- moving through a magnetic field B is $E \approx 3\beta\gamma B$, the higher the H^- beam energy and the magnetic field intensity and gradient, the easier the stripping and the lower the angular dispersion of the resulting H^0 beam. For calculations and measurements of Lorentz stripping of H^-s see refs [6-8] and [1]. Further relevant references are given in ref [4].

$H^0(1S) \rightarrow H^0(N)$ EXCITATION

The $H^0(n)$ binding energy is $E(n) \approx 13.6\ n^{-2}$ eV. The energy of the photons necessary to induce transitions from the H^0 ground state to $n \geq 3$ levels is above 10 eV. The energy of photons emitted by a NeYAG laser working on the first harmonic is $E_I \approx 1.17$ eV. The enhancement of the photon energy in the H^0 c.m. system for H^0 atoms colliding head on at high energies with a NeYAG laser beam is $\gamma\ (1+\beta) \approx 2\gamma$. Therefore at $\gamma \approx 2$ (LAMPF and SNS energies) transitions to all excited n levels of H^0 require to use a NeYAG working on the 4th harmonic, at $\gamma \approx 3$ (CERN SPL conceptual design) most transitions can be excited with a NeYAG working on the 2nd harmonic, and at $\gamma \approx 8$ (Fermilab 8 GeV Linac Design Study) all transitions could be induced with a NeYAG laser working on the first harmonic

Transitions from the H^0 ground state to $H^0(n)$ have been studied and measured extensively at LAMPF [9-12]. Photo-excitation of Lyman series of atomic hydrogen for $7 < n < 13$ shows the expected $1/n^3$ amplitude dependence, with saturation onset at $n \leq 8$ [9]. Shifts and broadening of n=4 Stark sublevels of H^0 atoms experiencing motional electric field have been measured with 6 meV experimental width [11].

The techniques used to perform the measurements contain all Lolas basic ingredients. The power of the laser was sufficient to saturate transitions to $n<9$ levels. Since the laser was pulsed, the power was high and not an issue in these measurements. The effects were studied in coincidence with the laser pulse. Frequency tuning was obtained by changing the angle between particle and light beams. In Lolas scenarios the power is instead a central issue, since one needs to act on all the H^0s of the beam. Pulsed lasers offer high power, but all the difficulties linked to synchronization and matching of laser and beam pulses are present.

Use of a CW-laser beam is obviously simpler and comfortable, and if the necessary power could be obtained the laser light could appear as a passive element just as the foil in foil stripping. The author has suggested the adoption of a Fabry-Perot cavity locked to a CW-laser in order to increase dramatically the laser power density and still work with a CW source. The PVLAS collaboration has developed a Fabry-Perot cavity locked to a NeYAG laser working on the first harmonic [13] and recently also on the second harmonic [14]. PVLAS uses this technique in order to observe and measure directly vacuum polarization effects [13]. In the experiment the FP mirrors are 6 meter apart and light goes back and forth through a 1 m long 6 T superconducting dipole magnet positioned between the two mirrors. So far measurements have been made with a NeYAG working on the first harmonic. The FP finesse is in excess of 10^5. The power density is > 1 $KW \cdot cm^{-2}$. Operating conditions are not less difficult than in a Lolas injection region if one considers that the magnet (and the associated cryostat) rotates mechanically around the FP cavity axis! In spite of all that the cavity stays tuned and locked for hours. This demonstrates that a Lolas scheme would be viable from the power and operation point of view with an accumulator working at $\gamma \geq 6$. Further planned developments will lead to applicability at $\gamma \approx 3$ of the PVLAS technology with a NeYAG working at the 2nd harmonic.

Tuning stability and line broadening are dominated by the energy fluctuations and the momentum dispersion of the linac beam. In order that all H^0s of the beam cross resonant conditions there are two approaches. In the first one there is a constant gradient low magnetic field in the region of crossing of the H^0 and laser beams. A H^0 Stark sublevel has a shift (due to the Lorentz electric field experienced by the H^0 in its rest frame) that varies linearly in first approximation along the crossing region. By adapting conveniently the free parameters dB/ds and B(s) one can manage that all H^0s cross the resonance energy and have $\approx 50\%$ probability to exit the interaction region in a $H^0(n)$ exited level. In the second approach the magnetic field is high enough in the interaction region to broaden the chosen Stark sublevel so to match the Doppler broadening of the laser light due to the momentum dispersion of the H^0 beam. In the first approach the excited level is narrow and, for a given available light power density, the transition probability is high. However the time for the transition to occur is extremely short since the H^0 stays tuned for a small length of the interaction region. In the second approach each H^0 is always within the resonance width, but the transition probability is lower for the same light power density, because of the broad width of the resonance line. The second approach has the advantage that light induces transitions only one way, since the $H^0(n)$ level is immediately Lorentz stripped in the zero magnetic field gradient interaction region.

$H^0(N) \rightarrow P^+ + E^-$ STRIPPING

The binding energy of the electron of an excited $H^0(n)$ atom is $E_n \approx 13.6\ n^{-2}$ eV. For $n \geq 4$ the $H^0(n)$ electron is less bound then the extra electron of the H^- ion, and field ionization is expected to be easier. This is nicely demonstrated by experiments at LAMPF [7, 8] where three beams of H^-, $H^0(n \leq 3)$ and $H^0(n=4)$ move with equal speed through a magnet with large dispersion: at a proper setting of the magnet current the H^- beam is deflected but not dispersed, the $H^0(n \leq 3)$ is not deflected, $H^0(n=4)$ atoms are stripped and the emerging p^+ dispersed. At higher current settings also the H^- beam is stripped, but one observers that the $H^0(n=4)$ beam features less dispersion than the H^- beam.

Dispersion of p^+ from $H^0(n)$ stripping helps for painting in Lolas scenarios. In the scenario where $H^0(1S)$ atoms are excited in the valley of the injection wiggler, and $H^0(n)$ stripping occurs in the peak, the dispersion of the injected p^+ beam is controllable by choosing the wiggler peak field value. In the scenario where transitions are induced to a broadened Stark sublevel, the magnetic field must be uniformed and its value is constrained to be the central one of resonance conditions.

INTEGRATION

The basic guideline behind a Lolas H^0 sharing oriented scenario is that efforts to increase the average current in the proton accumulator are directed towards increasing the number of cycles of the H^- linac, by adding the necessary RF power, rather than trying to improve the H^0 stripping efficiency. This procedure would be advantageous for all users, both high and low duty-cycle, because the fraction of H^0 stripped to p^+ and captured into the accumulator stays constant (≈50%). All user communities would be motivated to support intensity upgrades of the linac by acquisition and implementation of additional RF equipment and power. Communities joining later on could contribute additional RF equipment, so improving the average intensity and shortening data taking time for all the other users. In the case of the SPL injection H^- linac proposed at CERN it would be technically feasible to move from 4 to 20 MW beam power at the linac output [15].

In view of the fact that the H^- linac could run all the time, while low duty-cycle programs need shutdown periods, it will be convenient for flexibility and reliability to plan from the beginning:

a) splitting of the H^- beam before the stripping magnet used for injection into the H^0 beam line

b) splitting of the H^0 beam emerging from the accumulator injection straight section, and adequate downstream floor space for high duty-cycle users,

c) a bypass of the injection region of the accumulator, to bring the H^- beam at the start of the H^0 beam line downstream of the accumulator injection straight section (this would permit maintenance work on the accumulator without disturbing high duty-cycle operations of H^0 users).

In order to convert to laser stripping an injection section designed for foil stripping, it might be explored the possibility of exciting resonantly H^- s by inducing transitions to the narrow H^- Feshbach resonance [10].

REFERENCES

1. Yamane I, *Phys. Rev.* ***ST-AB 1***, 053501(1998), and this workshop.
2. Suzuki Y., in 12th Symp. Accel. Sci. and Tech., Oct 27-29, 1999, Riken, Japan.
3. Drumm P., in C.R. Prior (Ed.), Proc. 6th ICFA Beam Dynamics Miniworkshop on Injection and Extraction in High Intensity Proton Machines, Feb. 1999, Abingdon, U.K.
4. Gastaldi U. and Placentino M., *Nucl. Instr. Methods* **A451**(2000)318.
5. Danilov S. et al., this workshop
6. Jason A.J. et al., IEEE Trans. Nucl. Sci., **NS-28**(1981)2709
7. Gulley M.S. et al., *Phys. Rev.* **A53**(1996)3201
8. Keating P.B. et al., *Phys. Rev.* **A58**(1998)4526
9. Donahue J.B. et al., *IEEE Trans. Nucl. Sgi.* **NS-28**(1981)1203
10. Bryant H.C. et al, *Phys. Rev.* **A27** (1983)2689
11. Bergeman T. et al., contributed paper to Atomic Physics 9, Seattle, 1984
12. Mohagheghi A.H. et al., *Phys. Rev.* **A43** (1991)1345
13. Bakalov D. et al., (PVLAS Collab.), *Quantum Semiclaa. Opt.* **10**(1998)239
14. Cantatore G. et al. (PVLAS Collab.), this workshop and refs therein
15. Autin B. et al. CERN 2000-012(2000)

Feasibility of Laser Stripping via a Broad Stark State

Isao Yamane

KEK, High Energy Accelerator Research Organization
Oho 1-1, Tsukuba-shi, Ibaraki-ken, 301-0801 Japan

Abstract. Laser wave length, magnetic field and laser power density necessary for the laser stripping via a broad Stark state are estimated for several energies of H^0 beam. Application of a Fabry-Perot resonator and emittance growth accompanying stripping were studied. The laser stripping via a broad Stark state turned out to be feasible.

INTRODUCTION

High-energy beams of excited H^0 atoms, which have a principal quantum number: n no less than 3, can be stripped in a same manner as high-energy H^- beams[1]. So, we have studied laser stripping schemes including laser pumping of H^0 atoms from the ground state to a higher excited state. But, it turned out there are two serious problems which are inherent in the laser pumping of high-energy H^0 beams in the free space. One problem is the Doppler broadening of the transition frequency distribution due to the momentum spread of the H^0 beam. The other is that the population of the upper state is saturated to 1/2 by competition between pumping up and down. Therefore, we need to take such a measure as the Rabi-oscillation.

In order to overcome these problems, we introduced a method to utilize a broad Stark state[2]. For the convenience sake, this method of laser stripping is called here as the BSS laser stripping.

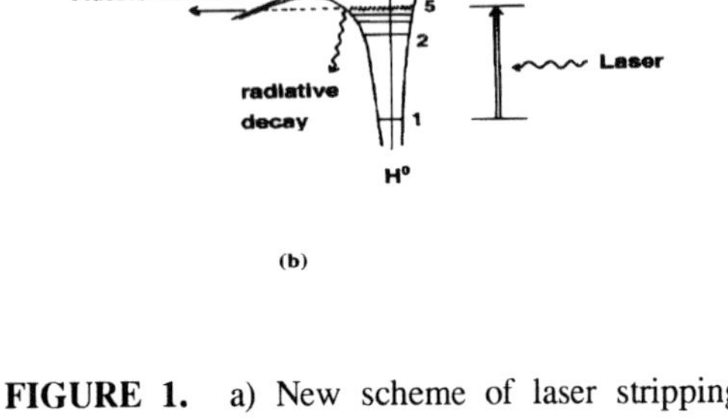

FIGURE 1. a) New scheme of laser stripping. (b) Conceptual picture of stripping process

BSS LASER STRIPPING

When we use Lorentz force to cause Stark effect, a laser stripping scheme like Fig.1 (a) may be possible.

H^0 beams, that have been injected into a straight section of a ring, are conducted into a magnetic field of an undulator, then pumped up to a broad Stark state through interaction with a laser beam. Excited H^0 atoms of the beam immediately decay to proton and electron and formed proton beam is captured by the ring.

The concept of strip.ping process is shown in Fig. 1 (b). Adjusting the magnetic field of the undulator so that the saddle point energy of the potential comes near to the aimed energy level of H^0 atoms, a broad Stark level as broad as the Doppler broadening of the transition frequency distribution is provided. Then, H^0 atoms are pumped up to the broad Stark level by a laser beam. Since the level width of the upper state is as broad as 10^{13}Hz, lifetime of excited H^0 atoms is very short. Excited H^0 atoms immediately decay to protons and electrons through autoionization. Therefore, the population of the excited state cannot be increased and pumping down of excited H^0 atoms to the ground state by stimulated emission of a photon is negligible. Thus, the reaction proceeds one-directionally from pumping up to autoionization. As a result, stripping efficiency can be made 100%.

CP642, *High Intensity and High Brightness Hadron Beams: 20th ICFA Advanced Beam Dynamics Workshop on High Intensity and High Brightness Hadron Beams,* edited by W. Chou, Y. Mori, D. Neuffer, and J.-F. Ostiguy

TABLE 1. Photo-ionization via Broad Stark State; (n=5,k=-4,m=0)

Interaction Length = 3m					
T (GeV)	0.800	1.000	1.300	2.000	H^0 kinetic energy
$\lambda_{1,5}^{LF}$ (nm)	327.9	372.2	437.4	586.2	Laser wavelength
I^{Laser}(W/cm^2)	39.1	41.5	43.9	46.9	Laser power density
B^{Stark} (T)	0.171	0.148	0.123	0.090	Magnetic field
Interaction Length = 0.3m					
I^{Laser}(kW/cm^2)	3.91	4.15	4.39	4.69	undulator/FP system

Since the physics of the Stark effect and the resonant photoionization is well established, numerical estimation of parameters, such as the magnetic field to cause necessary Stark effect and the necessary laser power density, is possible using formulae and data found in textbooks and data tables[3-5]. Table 1 is an example of numerical estimation made by a method described in the reference[2].

When we take the Stark state with n=5, k=-4 and m=0 and an interaction length of 3m, necessary laser power density is about 40W/cm^2. For 800MeV and 2GeV H^0 beam, necessary laser wavelengths are 328 nm and 586nm, and necessary magnetic fields are 0.17 and 0.09T, respectively. For the same Stark state and an interaction length of 30cm, necessary laser power density increases to about 4kW/cm^2.

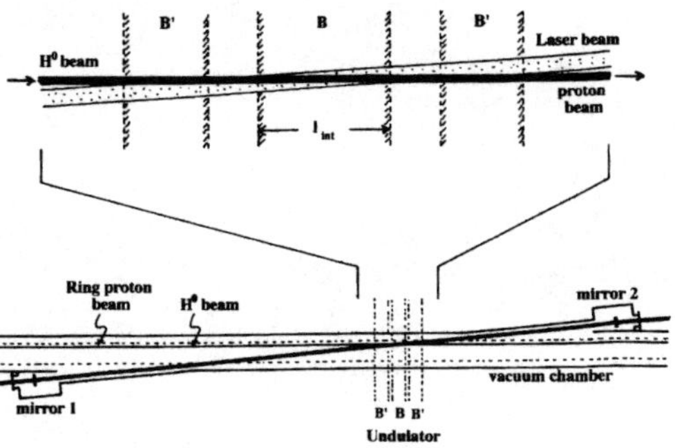

FIGURE 2. Laser stripping scheme using a Fabry-Perot resonator and an undulator.

APPLICATION OF FABRY-PEROT RESONATOR

Nowadays, such a Fabry-Perot resonator is available that has a finesse near to 10^5, is 6m or longer and stacks a 10kW/cm^2 light beam with a diameter of 10mm, as is reported by G. Cantatore [6] in this workshop. Using such a FP(Fabry-Perot) resonator, the interaction length can be reduced to about 30cm or shorter. Then, a laser stripping system like Fig. 2 may be possible. Here H^0 and laser beams collide with a very small angle. The interaction region, that is about 30cm long, is covered by a magnetic field to cause necessary Stark effect to H^0 atoms. Magnetic field of outer magnets which compose an undulator with the central one is set a little lower so that stripping does not occur in these magnets. When diameters of H^0 and laser beams are taken as 3mm and 10mm respectively, the colliding angle is 20mrad. Since the size of the beam duct is typically 10cm, the separation of two mirrors of the FP resonator becomes 6m or longer. However, this system is very simple and seems enough practical.

EMITTANCE GROWTH DUE TO STRIPPING

An important item to be checked is emittance growth due to stripping. Here we consider laser stripping in an undulator magnet as is shown in Fig. 3(a). For simplisity sake, magnetic field of outer magnets are taken same as that of the central magnet. Then the proton beam circulating in the ring takes an orbit shown by a thick line. Maximum deflection :θ and maximum displacement: d_0 are given by equations shown in the right side.

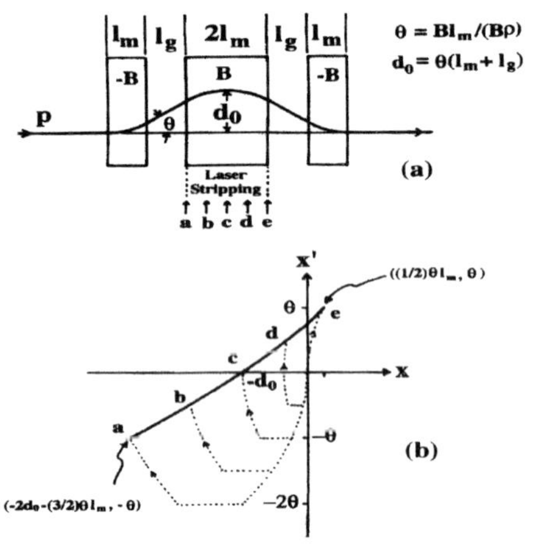

FIGURE 3. (a) Deflection and displacement in the undulator magnet. (b) Motion of formed proton in the phase space.

Laser stripping of H^0 beam occurs in the central magnet. Fig. 3(b) shows motion in the phase space of protons stripped from H^0 atoms at various points of the magnet. While protons formed at the point **a** in the central magnet go through and reach to the end of the undulator magnet, they walk along a dotted line and reach to the point **a** in the phase space. In the same manner, protons formed at the point **b** through **e** walk along respective dotted lines and reach the point **b** through **e** in the phase space. Therefore, protons formed by laser stripping receive a spread of deflection: 2θ, and a spread of displacement : $2(d_0+\theta l_m)$. For example, when 2GeV H^0 beams are stripped by a 532nm laser beam in an undulator magnet with B=0.09T, l_m=30cm and l_g=10cm, θ is 1.45mrad and $(d_0+\theta l_m)$ 0.58mm.

These spreads are considered to increase the emittance of the formed proton beam by several times from the original emittance of H^0 beam, that is typically 1πmmmrad. However, such an emittance growth is not considered to be so serious in the situation where we must form a ring beam with an emittance of several hundreds πmmmrad by a sophisticated phase space painting. Emittance growth accompanying laser stripping should be taken into account as a part of the phase space painting.

CONCLUSION

Laser stripping via a broad Stark state is feasible for GeV-energy H^0 beams, using a Fabry-Perot resonator and an undulator magnet.

REFERENCES

1. I. Yamane, PRST-AB Vol. 1, No 053501, 1998.
2. I. Yamane, KEK Report 2001-20, February 2002, A.
3. H. A. Bethe and E. E. Salpeter, *Quantum Mechanics of One and Two Electron Atoms* (Springer, Berlin, 1957).
4. V. S. Letokov, *Laser Photoionization Spectroscopy*, Academic Press, 1987.
5. K. Omidvar, Atom. and Nucl. Data Tables 28, 215-238(1983).
6. G. Cantatore et al., contribution to this workshop.

PVLAS Developments on Fabry-Perot Resonators Locked to CW Lasers and Suitable for Laser Assisted Lorentz Stripping of H^- Beams

G. Cantatore*, M. Bregant*, F. Della Valle*, G. Ruoso#, G. Zavattini@

**University and INFN – Trieste, Italy*
#INFN Legnaro National Laboratory, Legnaro, Padova, Italy
@University and INFN – Ferrara, Italy

Abstract. The PVLAS experiment, running at the Legnaro National Laboratory of INFN (Padova, Italy), is operating a 6.4 m long, 10^5 finesse Fabry-Perot (FP) cavity frequency locked to a 1064 nm, 100 mW output power Nd:YAG infrared laser. This cavity has a quality factor greater than 10^{12}, and is kept at resonance for periods of time in excess of several hours. A 100 mW, 532 nm, laser has been recently locked to an 87 cm long FP cavity with a finesse of 45000. Light power density at the center of this resonator was 1200 W/cm^2. Short term plans of the PVLAS experiment foresee the installation of a 6.4 m long, 10^5 finesse, Fabry-Perot resonator operating at 532 nm. Dedicated developments with a 532 nm, 5 W, input laser would provide a 10 kW/cm^2 power density, over a 1 cm^2 cross section, useful for laser assisted H^- full stripping.

INTRODUCTION

Fabry-Perot interferometers are becoming basic experimental tools in fields ranging from gravitational wave detection, to metrology, to the study of quantum vacuum[1]. The use of Fabry-Perot (FP) optical cavities has recently been suggested within a laser-assisted H^- double stripping scheme for the injection of high-intensity proton drivers[2]. The PVLAS experiment[3], running at the Legnaro National Laboratory of INFN, near Padova, Italy, is conducting an experimental study of the structure of quantum vacuum and is operating a high-finesse, high-quality factor, FP as a means to store photons in an interaction region, where an intense magnetic field is used to perturb the vacuum, and hence explore its structure. A similar cavity could, under certain conditions, be inserted without modifications in a laser-assisted stripping apparatus. The PVLAS collaboration has also successfully demonstrated the operation of a test FP optical cavity coupled to a 532 nm laser, based on the frequency-doubled emission from a Nd:YAG infrared laser[4]. Using existing technology, this device could be modified to accommodate the needs of the laser assisted stripping schemes being currently proposed. The first section below provides a short introduction to the PVLAS experiment, while the following section will present the current status of the optics in PVLAS. The third section will briefly illustrate the original techniques being used along with PVLAS driven developments. Finally, the last section will discuss the perspectives with respect to the H^- laser-assisted double-stripping technique.

THE PVLAS EXPERIMENT

The basic idea underlying the PVLAS experiment is to investigate the "material" properties of the Quantum Vacuum with a specific reference to its optical properties. It is well known that a gas-filled region of space where a magnetic field is present behaves as a birefringent crystal (Cotton-Mouton effect). This effect depends, among other things, on the gas pressure. In the zero pressure limit, contrary to the classical expectation, a residual effect must be present as evidence of the structure of the quantum vacuum: Quantum Electrodynamics gives an interpretation based on a photon-photon scattering process[1]. In the PVLAS experimental scheme, the vacuum is thought of as a target medium where an external magnetic field establishes an anisotropy. A linearly polarized light beam is then used as a probe to sample this target by measuring the polarization changes induced on the probe beam. In particular, one wishes to measure the ellipticity, that is the ratio of the semi-minor to the semi-major axis of the polarization ellipse. The main elements of the apparatus are an optical ellipsometer, a Fabry-Perot resonant optical cavity, and a superconducting dipole magnet housed in a rotating cryostat. The ellipsometer consists of two crossed polarizing prisms and of an ellipticity modulator, necessary to implement a heterodyne detection scheme capable of detecting small, time-varying, signals over a large background (index of refraction differences of the order of 10^{-22} must be detected). To provide the time variation of the effect, a 1 m long, 6 T dipole magnet, operating at 4.2 K, is

CP642, *High Intensity and High Brightness Hadron Beams: 20th ICFA Advanced Beam Dynamics Workshop on High Intensity and High Brightness Hadron Beams*, edited by W. Chou, Y. Mori, D. Neuffer, and J.-F. Ostiguy

rotated, along with its enclosing cryostat, around a vertical axis by means of a turntable. Finally, to increase the optical path in the magnetic field region up to 75 km, a 6.4 m long, 100000 finesse, FP cavity is continuously kept at the resonance condition. This last device will be illustrated in grater detail in the following section.

STATUS OF THE OPTICS

A Fabry-Perot optical resonator (or "cavity") consists of two facing mirrors set at a given distance L apart. In this device, light of wavelength λ can be accumulated by constructive interference when $L = n(\lambda/2)$, where n is an integer. For an ideal, lossless, FP the input light intensity equals the output light intensity, The performance of a FP is characterized by its finesse F: the larger F, the longer the optical path, the longer the average time τ a photon will spend inside the cavity. F is determined by mirror reflectivity: for identical mirrors $F = (\pi\sqrt{R})/(1-R)$, and $\tau = (FL)/(\pi c)$, where c is the speed of light. τ, and hence F, can be measured by observing the decay of the light intensity transmitted by the cavity once the input light is switched off. Such a decay curve, measured for the PVLAS 6.4 m long FP cavity at λ=1064 nm yields $\tau = 860\ \mu s$, that is $F = 1.4 \times 10^5$, for a total path length within the magnetic field region of $(1\ m)(2F/\pi) = 75$ km. During normal operation of the experiment, the PVLAS cavity stays in the resonance condition for a basically unlimited time while the magnet-cryostat assembly rotates around it.

ORIGINAL TECHNIQUES AND PVLAS DRIVEN DEVELOPMENTS

Locking Technique

To operate properly, a FP cavity must be at resonance. In practice, for mirror separations larger than a few cm, resonance can be kept only by a feedback scheme. One such scheme is the Pound-Drever-Hall (PDH) method[5], which exploits the fact that real FP cavities (with non-zero losses) reflect part of the incoming light even at resonance. This reflected light is used to sense the instantaneous frequency difference between input laser light and cavity: the laser wavelength is then continuously adjusted to follow changes in mirror separation. In the original PDH technique, the frequency difference between laser and cavity is converted into an error signal, which is then fed to a frequency modulator to close the feedback loop. An original variation of this method has been developed by the PVLAS collaboration[6]. By exploiting the fact that the Nd:YAG laser is based on a Non Planar Ring Oscillator (NPRO) crystal, which can be acted upon to change the beam wavelength, the external frequency modulator is eliminated from the feedback loop, providing enhanced noise and stability characteristics. The locking between laser and cavity can be maintained for extended periods of time and the frequency difference between laser and cavity can be kept at about 1 mHz/√Hz around 2 Hz. This translates, for instance, into the ability to sense changes in mirror separation down to a level of $\Delta L = 6\times10^{-18}$ m/√Hz.

PVLAS Driven Developments

The present wavelength of the PVLAS laser beam is 1064 nm. Operating at shorter wavelengths would be advantageous from several points of view, among which: the vacuum magnetic birefringence signal increases with photon energy, a visible beam is easier to align and operate, smaller beam radial dimensions result in lower noise from optical components. For these reasons a frequency doubled, continuous wave (CW), 532 nm laser has been tested and locked to a trial FP cavity, and is now ready to be installed on the main PVLAS apparatus[6]. The tests have been conducted on a Prometheus model laser, made by Innolight, Germany, having the following characteristics: infrared beam from a diode-pumped Nd:YAG NPRO crystal doubled to 532 nm by a single pass through a PPKTP non-linear crystal, visible 100 mW CW emission at 532 nm, secondary 1 W CW emission at 1064 nm, frequency tunable. An 89 cm long, 45000 finesse FP cavity was used for the locking tests following the modified PDH locking scheme outlined above. Stable locking was successfully achieved and overall performance proved similar to the performance of the laser-cavity system currently installed on the main PVLAS apparatus.

PERSPECTIVES

It has recently been suggested that the high power density which can be established inside a resonating FP could be exploited to achieve the excitation of H^0 atoms required by laser assisted stripping[2]. For a Fabry-Perot cavity at resonance, one can construct a power budget equation as follows:

$$W_L = W_{out} + W_R + W_{loss} + W_{H^0} \qquad (1)$$

where W_L is the light power output by the laser entering the FP cavity, W_{out} is the power in the transmitted beam, W_R is the reflected beam power, W_{loss} is the overall power loss due to diffraction and defects in the optical components, and W_{H^0} is a further loss due to photons "lost" to H^0 excitation. At resonance, light power within the cavity can be written as $W_C = (F/\pi)W_{out}$, where F is the cavity finesse, and the corresponding power density is $DW_C = (F/\pi)(W_{out}/S) = (\eta F W_L)/(\pi S)$, where S is the beam cross section and $\eta = W_{out}/W_L$ is the cavity transmission ratio. If one assumes that 10^{16} H^0/s must be stripped[2], then 10^{16} photons/s are lost from the cavity, leading (for λ = 532 nm), to a "stripping" loss given by:

$$W_{H^0} = 10^{16}\left[\frac{H^0}{s}\right]\cdot 2\left[\frac{eV}{ph}\right]\cdot 1.6\times 10^{-19}\left[\frac{J}{eV}\right] \approx 3\ \text{mW} \quad (2)$$

This loss is negligible compared to typical values of the laser light power within the FP cavities tested by PVLAS. The minimum power density required for laser assisted stripping is 10 Kw/cm^2, to be established over a 1 cm^2 beam cross section, that is a beam radius (waist) of 0.56 cm. Using Gaussian beam propagation equations, one can then estimate the FP cavity mirror radius of curvature required for proper matching of the beam to the cavity itself. Table 1 presents a summary of the relevant parameters for the current PVLAS 6.4 m long FP cavity, the shorter test FP cavity, the 6.4 m long cavity upgraded to the 532 nm wavelength, and for a projected cavity adjusted according to the requirements of laser assisted stripping. In this last case, only two of the parameters present technical challenges which have not already been met: light power input to the cavity and mirror radius of curvature. Mirror radius of curvature depends on the ability of manufacturers to prepare suitable mirror substrates, and it can safely be assumed that the only limiting factor would probably be cost. Assuming that, for ease of operation, a Nd:YAG frequency-doubled laser is chosen, laser power requirements could be met by both increasing pump power to the primary NPRO crystal and by increasing doubling efficiency. This could be achieved, for instance, by using a larger non-linear crystal and switching from a single pass doubling scheme to a multiple pass one. In conclusion, from the point of view of the construction of a Fabry-Perot cavity suitable for laser assisted stripping, the technology is either already available or it appears to be within reasonable reach.

TABLE 1. Summary of relevant Fabry-Perot parameters (see text)

	PVLAS current	PVLAS test	PVLAS upgrade	Projected
Laser power	100 mW	100 mW	100 mW	**5-10 W**
W_L	45 mW	20 mW	20 mW	**1.6 W**
λ	1064 nm	532 nm	532 nm	532 nm
η	0.13	0.07	0.13	0.2
F	100000	45000	100000	100000
Mirror radius of curv.	11 m	11 m	11 m	**110 m**
Beam waist	1.3 mm	0.7 mm	0.92 mm	5.6 mm
L	6.4 m	0.89 m	6.4 m	6.4 m
Optical losses	1.6x10^{-5}	1.6x10^{-5}	1.6x10^{-5}	1.6x10^{-5}
DW_C	3600 W/cm^2	700 W/cm^2	1200 W/cm^2	10 kW/cm^2
$W_H{}^0$	3 mW	3 mW	3 mW	3 mW
$W_H{}^0/W_C$	2x10^{-4}	2x10^{-4}	2x10^{-4}	3x10^{-7}

REFERENCES

1. Zavattini, E., et al., "The PVLAS Collaboration: Experimental Search For Anisotropy Of The Phase Velocity Of Light In Vacuum Due to a Static Magnetic Field" in *QED 2000*, edited by G. Cantatore, AIP Conference Proceedings 564, New York: American Institute of Physics, 2001, p. 77.

2. Gastaldi, U., and Placentino, M., *Nucl. Instr. Meth. A* **451**, 318 (2000).

3. http://www.ts.infn.it/experiments/pvlas

4. Bregant, M., Cantatore, G., Della Valle, F., Ruoso, G., and Zavattini, G., hep-ex/0202046, submitted to *Rev. Sci. Instr.*

5. Pound, R. V., *Rev. Sci. Instr.* **17**, 490 (1946); Pound, R. V., et al., *Appl. Phys. B* **31**, 97 (1983).

6. Cantatore, G., et al., *Rev. Sci. Instr.* **66**, 2785 (1995).

Present Status of Development of Thick Long-Lived Hybrid Carbon Stripper Foils for the 3 GeV Proton Booster Ring

I. Sugai[1*], Y. Takeda[1], M. Oyaizu[1], H. Kawakami[1], Y. Irie[1], Y. Arakida[1], I. Yamane[1], I. Sakai, M. Kinsho[2] and K. Kuramochi[2]

1) High Energy Accelerator Research Organization
2) Japan Atomic Energy Research Institute

Abstract. JAERI -KEK joint accelerator as a typified high power proton beam requires thick carbon stripper foils from 300 to 500 μg/cm^2 to stripe electrons from the H^- beam. In such high energy and high intensity beam bombardment, the foil lifetime is very short due to the radiation damage. Consequently, this short lifetime makes a great low operation efficiency of accelerator and highly hazardous work for the foil exchanger personal. Therefore, it is essentially important to make a breakthrough of the preparation method for long-lived carbon stripper foils. For this purpose, we have been performing systematic research, development and production with high reproducibility using a controlled AC□DC arc-discharge method.

INTRODUCTION

JKJ (JAERI-KEK joint) accelerator requires thick carbon stripper foils (300 to 500μg/cm2) to strip the electrons from the H- beam, supplied by the Linac, before injecting into the RCS (Rapid Cyclic Synchrotron). The 400 MeV H- beam from Linac has a pulse length of 0.5 ms with a repetition 25 Hz and an average current of 0.333 mA. Table 1 is the list of the injection beam properties of the joint accelerator. As mentioned below, long-lived thick carbon stripper foils are indispensable to the accelerator of the next generation.

Table 1. Injection beam properties of JKJ 3 GeV ring.

Kinetic energy	400 MeV
Beam pulse length	0.5 ms
Repetition rate	25 Hz
Average beam current	0. 333 mA
Beam size	10 mm x 10 mm

Lifetime of the stripper carbon foils depends very sensitive and strongly on the beam current intensity. Due to energy loss of H^- beam in the carbon foil and hitting by recirculated particles, the foil temperature will go up to high temperature. The foil rupture may be categorized mainly by the following two processes, depending on the foil temperature. Firstly, when the foil temperature is relatively low (<1000K), the foil ruptures will happen mainly due to the radiation damage on such slackening, thickening and shrinkage during irradiation. Secondly, when the foil temperature is high (>1500K), the ablating evaporation involving the sputtering effect becomes to dominate and the foil thickness will be rapidly reduced with increasing temperature, and the lifetime of the foil results in shortening.

The results of the short lifetime of stripper foils make following great problems. Namely, stripper foil replacement is one of the most dose-intensive activities and radiation hazardous work that maintenance personnel is concerned with, and the accelerator operational efficiency becomes very low.

As a matter of fact, in the carbon stripper foils of 300 and 500μg/cm^2 thickness cases, maximum temperatures of foil by beam injection, as listed in Table 1 are approximately 1200K to 2000K. The carbon stripper foils with temperature of 2000K will be rapidly evaporated by beam irradiation. For high temperature of 2000K, diamond foil is a very attractive candidate for a thick stripper foil, because of drastically high thermal conductivity. However, the diamond foil is not so easy to obtain as a self-supporting foil without Si substrate and also not yet so much creditable for practical use, according to test performance by high energy and high intensity ion beam accelerator so far.

Therefore, it is very difficult to obtain such stripper foils in the world.

Consequently, the development of a long-lived carbon stripper foil with high durability for 2000K is one of the key technologies for 3 GeV RCS, JKJ high power accelerator.

CP642, *High Intensity and High Brightness Hadron Beams: 20th ICFA Advanced Beam Dynamics Workshop on High Intensity and High Brightness Hadron Beams,* edited by W. Chou, Y. Mori, D. Neuffer, and J.-F. Ostiguy

FOIL PREPARATION AND DEVELOPMENT

The carbon foils prepared by conventional methods, for example, 1) electron beam evaporation-condensation, 2) glow discharge, 3) thermal cracking and 4) arc-discharge methods have been used [1~3].

The lifetime measurements of the foils made by different methods described above were performed for high energy 800 MeV proton beam of up to 85μA in PSR of Los Alamos National Laboratory [4], and also for low energy 3.2 MeV Ne^+ beam of 3-4μA from Van de Graaff accelerator of Tokyo Institute of Technology [3]. However, the both results showed short lifetime. We consider that these foils ruptured by the first process of the radiation damage. In order to overcome these short lifetimes, we have developed following mainly three carbon foil preparation methods of 1) Modified Controlled AC and DC arc-Discharge (mCADAD) method [5], 2) Heavy Ion Beam Sputtering (HIBS) and Mixed Ion Beam Sputtering (MIBS) methods and 3) Ion Beam Sputtering method with Reactive Nitrogen gas (IBSRN). Among them, the HIBIS and the MIBS, the IBSRN methods and the Laser plasma deposition (Lp-foil) method are available for thin foil of 5 to 30μg/cm^2 in thickness, but not for thicker than about 50μg/cm^2 due to its low sputtering rate of the carbon element, especially for the HIBS, MIBS and the IBSRN. On the Lp-foil method, a foil with thicker than ~10μg/cm^2 is not reported yet, but very attractive in the future development for its long lifetime. From these point of views, in practical foil preparation, the foils with thicker than 100μg/cm^2 can be produced only by the mCADAD method. This method, in principle, is unique and simple method for long-lived carbon foil preparation. Namely, the carbon foil formed by ablation from cathode electrode in DC arc-discharge is very strong against high intensity ion beam iraradiation, but very fragile for mechanical stress. On the other hand, the carbon foil by ablation from both electrodes in AC arc–discharge is very fragile for ion beam one, but very strong for mechanical stress. Therefore, by combining both positive merits, we can make long-lived carbon stripper foils. As stated in ref [6], the lifetime of the foil depends on the ratio R=Wc/(Wc+Wa)(%), where Wc and Wa are the carbon source weight losses due to ablation from the DC and AC arc discharge, respectively.

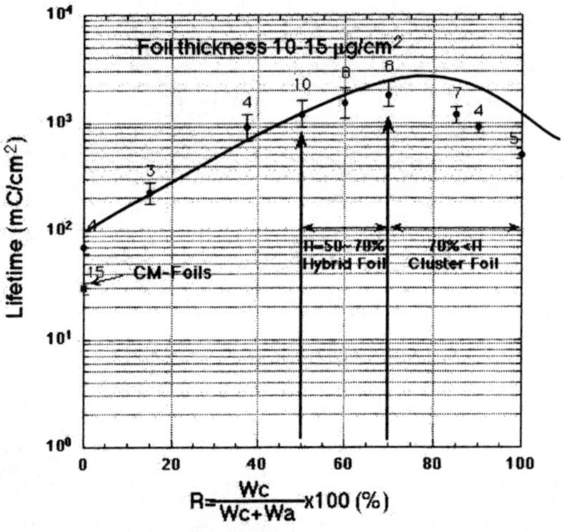

FIGURE 1. Relationship between the ratio R and the lifetime (Lifetime measurements were performed with a 3.2 MeV, Ne^+ at 3.0±0.5μA of 3.5mmϕ beam spot). Error bars represent mean square errors and the numbers attached to the points are measured sample numbers .

The lifetime is defined as integrated ion current (mC) per unit area (cm^2) until foil rupturing occurs. The foils are mounted on a stainless steel holder of 0.3 mm thick with a 10 mm diameter aperture. Figure 1 shows the relationship between the ratio R and the lifetime. The data of fig. 1 shows the very long lifetime corresponding 65 times, in maximum, of the best commercially available foil made by the evaporation-condensation method. We believe that this characteristic data of the lifetime versus R will be fruitful for lifetime estimations of the thick carbon stripper foil (>100μg/cm^2), although the tested foils shown in fig. 1 were thin, 10 to 15μg/cm^2. The foils made by the mCADAD method have been widely used not only for low energy heavy ion beams (thin ; 10-15μg/cm^2) , but also for high energies such as 800 MeV H^+ (thick; 100-300μg/cm^2)[5].

REFERENCES

1. P.Maier-Komor, *Nucl .Instr. and Meth.* **102**(1972)485-487

2. D.W.L. Tolfree, D.S. Whitmeel et al., *Nucl. Instr. and Meth.***158**(1979)333-336

3. I.Sugai, T. Fujino, K. Yamazaki and T. Hattori, *Nucl. Instr. and Meth.* **A236**(1985)576-589

4. M.J.Borden, G.E.Adamson, R.N.Johnson and W.F.Nicase, *Nucl. Instr. and Meth* .**A303**(1991)63-67

5. I.Sugai, M.Oyaizu, H.Kawakami, C.Ohmori, T.Hattori, K.Kawasaki, M.J.Borden and R.J.Macek, *Nucl. Instr. and Meth.* **A362**(1995)70-76

6. I . Sugai, T.Hattori, K.Yamazaki et al., *Nucl. Instr. and Meth.* **A265**(1988)376-382

SESSION L

ELECTRON CLOUD EFFECTS

Summary Report

Robert J. Macek

Los Alamos National Laboratory, Los Alamos, NM, 87544, USA

Abstract. This is a brief summary of the Electron Cloud Session L at the 20th ICFA Workshop, HB2002. Co-conveners were Robert Macek, LANL and Hitoshi Fukuma, KEK. The session was devoted to a review of the status, recent progress and impact of Electron Cloud Effects (ECE) on circular machines. Observations at both positron and proton machines were presented along with theory and simulations plus the impact of ECE on the joint KEK-JARI project.

INTRODUCTION

As noted in the SNOWMASS 2001 workshop report, electron cloud effects (ECEs) are now recognized as a serious problem for many existing and future high intensity accelerators and one that calls for a concerted R&D effort. ECEs have limited the performance and restricted the operating range of several modern storage rings around the world and are a major concern for future, high-intensity accelerators. Possible adverse effects include emittance dilution, single- and coupled- bunch instabilities, particle losses, vacuum degradation, excess power deposition on the chamber walls, and interference with certain diagnostics instrumentation. For these reasons, ECE is a very important and timely topic in the accelerator community and of particular importance to high intensity, high brightness rings, hence its inclusion in this workshop.

AGENDA

This session was organized to cover a broad range of topics in this field and review the considerable progress that has been made experimentally, on theory and simulations and in dealing with ECE in the operation of existing machines. The list of presentations included:

1. Kathy Harkay (ANL) - *ECE at e+/e- machines and electron cloud diagnostics.*
2. Hitoshi Fukuma (KEK) - *Observations of ECE at KEKB.*
3. Robert Macek (LANL) - *Observations of ECE at the Los Alamos PSR.*
4. Kazuhito Ohmi (KEK) - *Theory and Simulations of the Electron Cloud Instability.*
5. Mauro Pivi (LBNL) - *Simulations of electron cloud generation in the PSR and SNS.*
6. Hong Qin (PPPL) - *Nonlinear delta-f simulations of e-p two-stream instabilities in High-Intensity Beams.*
7. Takeshi Toyama (KEK) - *ECE in the KEK-PS and KEK-JAERI joint project.*

OBSERVATIONS OF ECE

A reasonably comprehensive list of electron cloud effects and the machines that are affected by ECE or, in the case of proposed machines, are at risk from the ECE includes:

- Beam induced multipacting
  - Resonant (APS, KEKB, PS, SPS), [LHC]
  - Trailing-edge multipactor (PSR), [SNS, JHF]
- Vacuum degradation i.e., electron-stimulated gas desorption, is the most common indication of beam induced multipactor
- Two-stream e-p instability (ISR, PSR, AGSB), [SNS, JHF]
- Transverse coupled bunch instability from EC wake (B factories, PS, SPS), [LHC]
- Single bunch (head-tail) instability; emittance blowup (B factories, PS, SPS)
- Tune shifts (KEKB, AGS Booster)
- Heat load on cryogenic wall [LHC]
- Cloud-induced noise or spurious signals in beam diagnostics (e.g., wire scanners, electrostatic pickups, ionization profile monitors) (PSR, PS, SPS)
- Electrons trapped in distributed ion pump leakage field (CESR)

Harkay reviewed various electron cloud diagnostics, in particular, the elegant retarding field analyzer (RFA) developed at ANL and the electron-sweeping detector developed at LANL. The RFA is a well-characterized device that measures the electron flux, energy spectra and time structure (when

CP642, *High Intensity and High Brightness Hadron Beams: 20th ICFA Advanced Beam Dynamics Workshop on High Intensity and High Brightness Hadron Beams,* edited by W. Chou, Y. Mori, D. Neuffer, and J.-F. Ostiguy

augmented with fast electronics) of electrons striking the wall. Since electrons enter the RFA through small slots in the wall and an outer, grounded grid, the collector and bias grid are shielded, and there is very little perturbation of the electron cloud or the beam/wall environment as compared with biased collection plates or buttons. The electron sweeper extends the RFA concept by using a pulsed electrode to sweep low energy electrons from the beam pipe into the RFA. These devices have been key elements for the observation of electron cloud characteristics at APS and PSR. RFA-like devices are now in use or planned at most machines that have encountered ECE. Harkay also discussed the information provided by the RFA to characterize the ECE buildup at the APS and comparisons with simulations by Furman and Pivi at LBNL.

Fukuma reviewed experience with ECE at the KEKB low energy ring (LER), which included observations of emittance growth, instability mode spectra and growth times, tune shifts and luminosity degradation both with and without solenoids. In the absence of mitigation, emittance growth and luminosity degradation greatly limited facility performance. Solenoids, which now cover 95% of the straight sections (~70% of the ring circumference), were most beneficial in suppressing the electron cloud effects and have resulted in a good improvement in the KEKB luminosity.

Beam blowup in the LER at KEKB was eliminated after the last (5th) installment of solenoids. The tune shift along the train (from the electron cloud), which seems to be a good measure of the cloud density, was reduced by at least 40% after the 4th installment of solenoids. Growth rate of the coupled bunch instability was reduced by a factor of two after the 4th installment of solenoids. Mode spectra from simulations of the electron cloud induced coupled bunch instability (solenoids off) are in general agreement with observations for the vertical plane but not for the horizontal plane if the photoelectrons are produced mainly at an illumination point of the synchrotron radiation. However, if the photoelectrons are produced uniformly over the surface of the vacuum chamber, then the simulated mode spectra are consistent with observations for both planes.

Macek discussed the extensive observations and experience with ECE at the Los Alamos PSR. It is now reasonably certain that trailing edge multipactor causes a good fraction of the prompt electron signal emerging at the end of the beam pulse and striking the chamber wall. The energy spectrum of the prompt component extends to ~400 eV and simulations of this spectrum by the LBNL code POSINST are in reasonable agreement with measurements. Measurements of the electrons surviving the beam-free gap between successive passages of the bunch were made with the electron-sweeping detector. These data provide a lower limit of 1-2% on the beam neutralization, which is sufficient to explain the e-p instability thresholds observed at PSR. Measurements with the electron sweeper also revealed a surprising slow decay (170 ns decay constant) of electrons after the beam was extracted. These data imply a high secondary emission coefficient for the low energy electrons lingering in the gap.

Tests and demonstrations of potential remedies for the e-p instability at PSR included Landau damping by increased rf buncher voltage, transverse coupling, multipoles and inductive inserts. These, along with a conditioning effect, have enabled stable storage of 6.3×10^{13} protons per pulse, which is 60% more than was previously possible without these measures.

Toyama presented data obtained from electrostatic pickups and exciter, which indicated ECE in the KEK-PS (12 GeV main ring). The indication was suppressed with bias voltage or solenoidal fields. Dedicated electrons detectors will soon be installed.

Toyama also presented data from KEK on measurements of SEY as a function of primary electron energy for a variety of chamber materials both as received and after sputter cleaning. Most interesting was the improvement in SEY with sputtering. TiN had a peak SEY of ~0.8 and stainless steel a peak SEY of ~ 1.2 after the sputtering treatment.

THEORY AND SIMULATION OF ECE

Ohmi discussed several approaches that he has used to model electron cloud buildup and the resulting instabilities for both positron and proton rings. The electron cloud buildup is calculated in a simulation model, which incorporates a rigid beam, and dynamical electrons that interact with the Coulomb field of the beam and the space charge of the other electrons. Primary electron yields are input to the simulations. In the case of positron rings, the primaries are the well understood photo electrons while for proton rings the source strength from losses and even residual gas ionization is much more uncertain. Instability dynamics were studied using both the wake field approach and direct tracking of both species (beam and cloud macro particles). Coupled bunch instabilities, which were observed at KEK-PF, BEPC and KEKB, are consistent with the simulations. The single bunch instability, which was observed at KEKB, is consistent with simulations for the threshold behavior and the effect of solenoids.

Toyama presented additional results on simulations of ECE for protons rings (JHF 3 GeV RCS, JHF 50 GeV MR, PSR, ISIS and AGS) using the codes and methods developed by Ohmi. He also discussed the implications of ECE for the KEK-JAERI joint project. The electron cloud buildup was presented as a source-independent amplification factor, which multiplies the primary electron yield to give the total electron yield along the beam bunch. PSR had the highest amplification followed by JHF rings and ISIS. The estimates of stability (coasting beam model) showed that PSR would be unstable while ISIS should be stable, which is in qualitative agreement with observations, and that the JHF rings are close to instability. A decision has been taken to use TiN coatings on the 3 GeV RCS while the use of coatings for 50 GeV main ring is still being evaluated.

Pivi described the main features of the LBNL code, POSINST, as modified and adapted to simulate electron cloud generation in PSR and SNS. The code employs a rigid proton beam using measured or calculated beam profiles. Electrons move in 3-dimensions subject to 2-dimensional forces from the proton beam and electron cloud space charge. Perfect conductor boundary conditions are imposed at the round or elliptical chamber walls. Primary electrons originate from proton beam losses ($4x10^{-6}$ lost protons/stored proton/turn for PSR) and/or residual gas ionization. For LHC, photoelectron emission is the dominant source of primary electrons. A very detail model of the secondary emission process at the wall is a hallmark of their code. Their simulations reproduce the general features of the RFA signals and cumulative energy spectra at PSR. For SNS they predict a large neutralization, which is reduced significantly by conditioned TiN coatings. As noted by both Pivi and Ohmi, primary electron source terms and SEY are key input parameters that must be determined by experiment or other analyses.

Hong Qin outlined the key features of the PPPL code (BEST) and discussed the main results to date. They employ self-consistent methods based on the Vlasov-Maxwell equations and use an assumed electron cloud distribution. The code has been parallelized but is still somewhat limited by available computing power. Simulation noise is reduced significantly by the delta-f method. This code does provide a self-consistent description of mode structures, instability thresholds and growth rates. To date, only coasting beams have been studied. Theory and simulations of the e-p instability in bunched beams are under development.

In general, the PPPL results confirm the main results of the centroid model for thresholds and mode structure. Landau damping emerges naturally. Mode structure, thresholds and growth times for the linear phase agree reasonably well with experimental observations. Both theory and simulations predict a saturation level that is probably below the experimental sensitivity and then a growing nonlinear phase. More computing time is needed to follow the nonlinear phase to large amplitude motion.

CONCLUSIONS

The study of ECE and their mitigation is a very active topic of investigation for both positron and proton machines for the simple reason that ECE is a limiting factor for a number of existing machines and a major concern for future high intensity accelerator facilities. This workshop provided a good overview and highlighted the impact of ECE on high intensity rings and high brightness beams. Significant progress has been made both to understand ECE at a fundamental level and to mitigate the adverse impact on accelerator design and operation. While there has been good progress, the problem is far from being resolved. The quantitative agreement between simulated results and measurements remains uneven, and the predictive power of the available tools is insufficient to extrapolate with high confidence the present results to future machines with higher beam intensity. A significant part of the problem for proton machines is the level of uncertainty and ambiguity on the input parameters associated with the primary. electrons and the SEY, which must be determined experimentally or by other analyses. More work is clearly needed

Electron Cloud Effects at Positron/Electron (e^+/e^-) Machines and Electron Cloud Diagnostics

K. C. Harkay[†], R. A. Rosenberg[†], R. J. Macek[‡], A. Browman[‡], T.-S. Wang[‡]

[†]*Argonne National Laboratory, Argonne, IL, 60439 USA*
[‡]*Los Alamos National Laboratory, Los Alamos, NM, 87544 USA*

Abstract. Background electrons are ubiquitous in high-intensity particle accelerators. Under certain operating conditions, amplification of the electron cloud can occur. The beam-cloud interaction can seriously degrade the accelerator performance with effects that range from vacuum degradation to collective beam instabilities. Although electron cloud effects (ECEs) were first observed 20 years ago in a proton ring, in recent years, they have been widely observed and intensely studied in e+/e- rings. This paper will focus on describing electron cloud diagnostics, which have led to an enhanced understanding of ECEs, especially details of beam-induced multipacting and saturation of the cloud. Such experimental results can be used to provide realistic limits on key input parameters for modeling efforts.

INTRODUCTION

Interactions between high-intensity particle beams and a background population of low-energy electrons, known as the electron cloud (EC), have been widely reported[1]. The mechanisms giving rise to the cloud vary[1,2], and in many cases, the cloud does not affect the beam. However, amplification of the cloud can occur under certain operating conditions, potentially giving rise to numerous effects that can seriously degrade accelerator performance. First observed in proton rings, electron cloud effects (ECEs) have in recent years been intensely studied in high-energy positron and electron rings as well. Mitigating ECEs remains an important concern in present and future high-intensity, high-energy accelerators.

An important tool in an improved understanding of ECEs has been the development of electron cloud diagnostics. Experimental data characterizing the EC properties can be used to provide realistic limits on key input parameters for modeling and analytical calculations. Combining EC and standard beam diagnostics allows EC effects to be separated from those arising from the geometric wake field. Although the nature of the beam-cloud interaction is rather different in positron vs. proton machines, there is a potential for synergy in the techniques applied to study the EC in both. This will be illustrated using results from the Advanced Photon Source (APS) at Argonne National Laboratory (ANL) and the Proton Storage Ring (PSR) at Los Alamos National Laboratory (LANL).

[†] Work supported by U.S. Department of Energy, Office of Basic Energy Sciences under Contract No. W-31-109-ENG-38.
harkay@aps.anl.gov, rar@aps.anl.gov
[‡] Work conducted at Los Alamos National Laboratory, operated by the University of California for the U.S. Department of Energy under Contract No. W-7405-ENG-36.
macek@lanl.gov, abrowman@lanl.gov, twang@lanl.gov

OVERVIEW OF EC EFFECTS

Many comprehensive review papers can be found in the proceedings of recent U.S. and European Particle Accelerator Conferences, giving a summary of experimental observations and modeling of electron cloud effects[1]. Only the briefest of discussions is given here. We will focus on two phenomena observed in both APS and PSR: beam-induced multipacting (BIM) and the build-up and saturation of the cloud. BIM is a resonance condition that can lead to large amplification of the EC if the secondary electron emission coefficient (δ) of the chamber surface is greater than unity. BIM can also be responsible for vacuum degradation due to electron-stimulated gas desorption, first observed at the CERN proton Intersecting Storage Ring (ISR)[3]. Depending on machine conditions, the cloud density can be observed to either grow or remain constant over a bunch train (or several turns). In either case, the cloud density appears to eventually reach saturation at a level related, in some cases nonlinearly, to the average beam charge density.

At APS[4], EC amplification and associated pressure rise due to BIM were clearly observed with positron beams at certain bunch currents and spacings (the Al vacuum chambers have a relatively high δ). A more modest effect is observed with electron beams in present APS operation. At BIM conditions, the cloud density saturates after 20-30 bunches. With positron beams only, a horizontal coupled-bunch instability was observed at conditions coinciding with the maximum saturated cloud density. At PSR[5], a long observed e-p instability occurs at a threshold bunch intensity and buncher voltage. BIM is observed on the trailing edge (tail) of the bunch, but these prompt electrons appear to have a less direct relationship to the instability. On the other hand, electrons that survive the gap are observed to saturate near the instability threshold.

CP642, *High Intensity and High Brightness Hadron Beams: 20th ICFA Advanced Beam Dynamics Workshop on High Intensity and High Brightness Hadron Beams,* edited by W. Chou, Y. Mori, D. Neuffer, and J.-F. Ostiguy

ELECTRON CLOUD DIAGNOSTICS

The planar retarding-field analyzer (RFA), first developed and implemented at APS[6], is a very simple device that measures electrons colliding with the chamber wall. The RFA consists of two grids and a graphite-coated collector (to lower its δ). The outer grid is grounded, while the inner grid can be biased for electron energy selection. The RFA is an integrating device; differentiating the collector signal gives the EC energy distribution. Figure 1 shows a schematic of two RFAs mounted on a standard APS chamber.

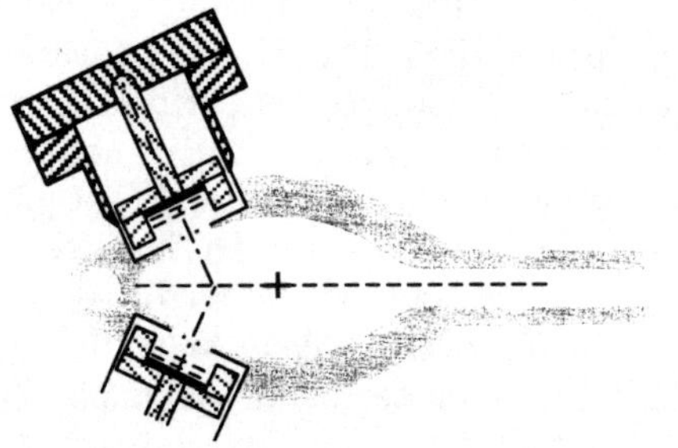

FIGURE 1: ANL schematic of two RFAs mounted on a standard APS chamber, cross-sectional view (42 × 21 mm chamber half-dimensions).

The benefit of shielding the collector and bias grid is that the measurement does not disturb the EC in the chamber. It is very difficult to measure the true wall flux or energy distribution with a biased beam position monitor (BPM), illustrated in Fig. 2. Secondary electron emission from the BPM surface affects the wall flux (the signal can even change sign). Also, the BPM bias changes the collection length (i.e., volume).

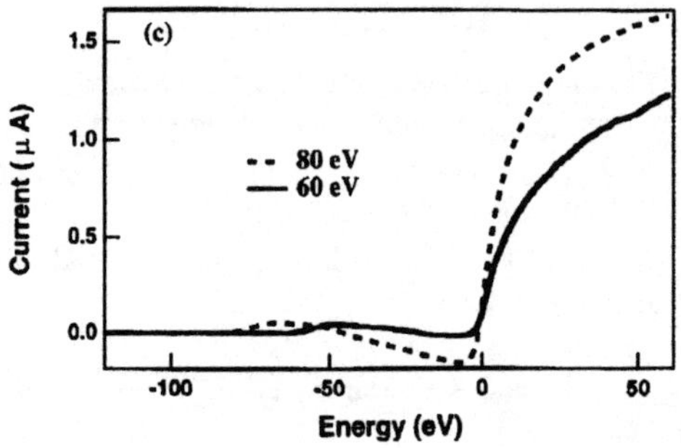

FIGURE 2: Signal produced from a BPM irradiated by 60-eV and 80-eV electrons as a function of bias voltage applied to the BPM[6].

ANL-type RFA devices have been widely implemented: Intense Pulsed Neutron Source (ANL), PSR (LANL), Beijing Electron-Positron Collider (IHEP), KEK B-factory (KEK), Super Proton Synchrotron (CERN), and Alternating Gradient Synchrotron booster (BNL). At PSR, the RFA was augmented with a broadband amplifier to allow time-resolved measurements. Also at PSR, the effect of chamber surface coatings with lower δ was measured *in situ*. At APS and PSR, RFAs were installed in regions free from external magnetic fields. At CERN SPS, an RFA-based detector was installed inside a specially-constructed dipole magnet, and multi-strip collector electrodes measure the transverse EC spatial density[7].

To detect electrons in the gap after the bunch passage, an electron sweeper was developed at PSR[8]. A curved electrode subtending a half-angle of 75º (~40 cm long) is placed opposite a large-aperture RFA (~8× higher sensitivity than the original design), shown in Fig. 3. A short (~20 ns), ~500 V pulse on the electrode selects the sampling time in the gap; the fractional acceptance area of electrons swept into the RFA is 0.3.

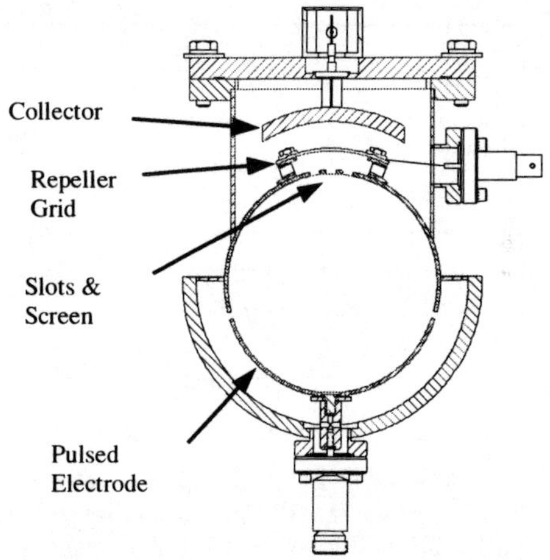

FIGURE 3: LANL electron sweeper (chamber diam. 10 cm).

A number of EC diagnostics concepts have been proposed or are under development, either for better energy resolution (Bessel Box at APS)[9], or less perturbative detection of the electrons in the vacuum chamber (e.g., by G. Lambertson (LBNL) and S. Heifets (SLAC)). At KEKB, the EC density is inferred indirectly by comparing the bunch-by-bunch space-charge tune shift with and without the EC.

EC MEASUREMENTS AT APS

In experiments[4] where the bunch spacing was varied, the primary (photoelectron) vs. secondary components of the EC production processes could be separated. While the RFA only measures the electron wall flux, analysis of the EC energy spectra gives an indirect measure of the electrons near the beam; a high-energy tail results from electrons near the beam that are accelerated to higher energy. The BIM resonance at 7 λ_{rf} and variation in EC distributions can be seen in Fig. 4. The collector current I_c is differentiated with respect to the bias voltage and normalized to the total beam current I_b. In Fig. 5, the measured EC build-up and saturation over positron bunch trains at the BIM bunch spacing is shown as a function of bunch current.

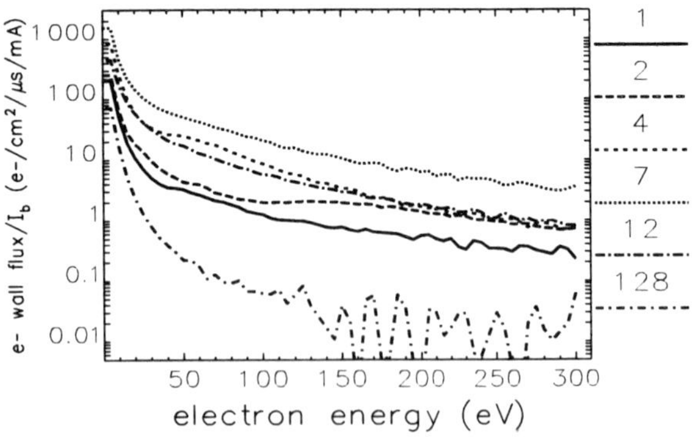

FIGURE 4: EC energy distribution vs. bunch spacing (units of rf wavelength, λ_{rf}); 2 mA/bunch, 10 positron bunches.

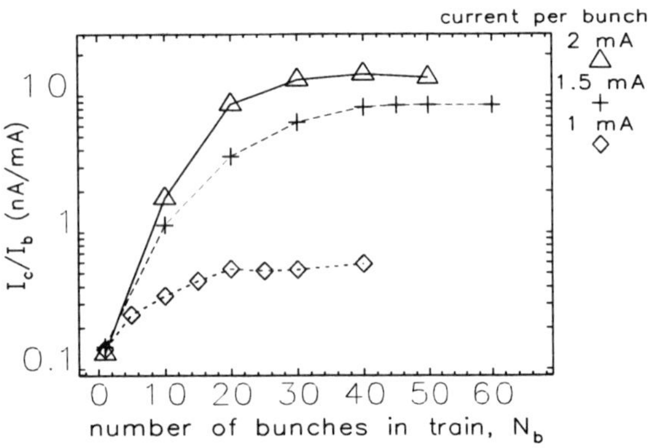

FIGURE 5: Measured EC build-up and saturation over positron bunch trains at the BIM bunch spacing. A horizontal coupled-bunch instability occurs for trains longer than ~20 bunches with a threshold current of about 2 mA/bunch.

EC MEASUREMENTS AT PSR

Extensive measurements have been carried out at PSR[5] to study the EC properties and correlation with the e-p instability. The electron sweeper data are particularly interesting; a typical result is shown in Fig. 6. The machine conditions are: 7.7 µC/pulse and 280-ns bunch length. BIM on the trailing edge of the beam accelerates and amplifies electrons that strike the wall

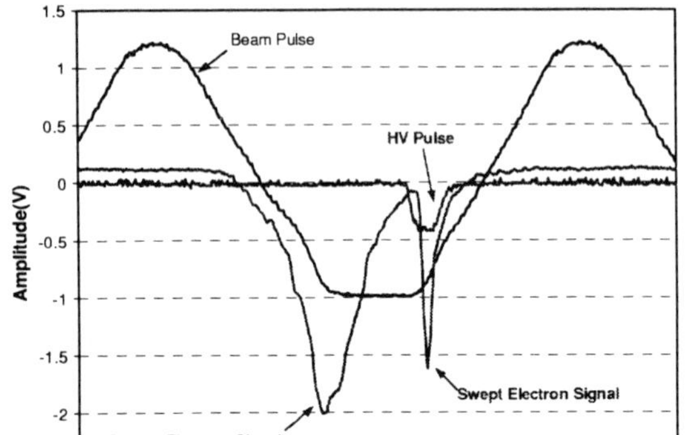

FIGURE 6: Typical PSR electron sweeper data (sweeper 500 V, RFA bias voltage –25 V, 32 macropulses averaged).

near the peak and in the tail of the bunch. These prompt electrons are believed responsible for populating the gap but the relationship is nonlinear. The electrons swept from the gap are more nearly linear with beam current for intensities in the region of greatest interest (> 4 µC/pulse) and saturate at 1-2% average beam neutralization near the instability onset.

SUMMARY

Electron cloud diagnostics based on the planar RFA have been successfully and broadly applied to study the EC properties and dynamics in positron, electron, and proton rings. The EC energy distribution (for flux at the wall) and estimate of density can be extracted from the RFA spectra – this is not possible with biased BPMs or collecting plates. EC diagnostics to date are mostly limited to electrons collected at the wall or swept to the wall. A number of concepts to measure the cloud nondestructively within the vacuum chamber have been proposed but not yet implemented.

Despite the obvious differences in the beam-cloud interaction, there is likely to be synergy between proton and positron rings in understanding the mechanisms of EC saturation and the correlation with cloud-induced instability thresholds. More analyses and experiments are planned.

ACKNOWLEDGMENTS

The authors would like to thank all who contributed to EC studies at APS and PSR.

REFERENCES

1. See links to workshop proceedings and review talks: http: //www.aps.anl.gov/asd/physics/ecloud/papers_gen.html
2. M. A. Furman and M. T. F. Pivi, Report No. LBNL-49711/CBP Note-415 (June 2002).
3. O. Gröbner, Proc. 10th Intl. Conf. on High Energy Accel., Protvino, Russia, 277 (1977).
4. K.C. Harkay and R.A. Rosenberg, Proc. of 1999 PAC, 1641 (1999) and K.C. Harkay et al., Report No. CERN-2002-001 (2002).
5. R.J. Macek et al., Proc. of 2001 PAC, 688 (2001) and references therein, and R. Macek et al., Report No. CERN-2002-001 (2002).
6. R.A. Rosenberg and K.C. Harkay, Nucl. Instrum. Methods **A453**, 507 (2000).
7. G. Arduini et al., "Measurement of the Electron Cloud Properties by Means of a Multi-strip Detector at the CERN SPS," to be published in Proc. of 2002 EPAC (2002).
8. T. Wang, D. Barlow, A. Browman, and R. Macek, "The Static Electric Field of a Curved Electrode in a Beam Pipe," PSR Technical Note 01-003 (May 2001).
9. R.A. Rosenberg and K.C. Harkay, Proc. of 2001 PAC, 2069 (2001).

Status of Solenoid System to Suppress the Electron Cloud Effects at the KEKB

H. Fukuma, J. Flanagan, K. Hosoyama, T. Ieiri, T. Kawamoto, T. Kubo
M. Suetake, S. Uno, S. S. Win and M. Yoshioka

High Energy Accelerator Research Organization (KEK)

Abstract. A large number of solenoids have been installed in the KEKB LER to remove the electron cloud. This paper describes recent solenoid winding work and the effect of it on electron cloud effects.

INTRODUCTION

A blow-up of the vertical beam size in single beam and multi-bunch operation has been observed in the KEKB LER [1]. Observations [1,2] have shown that the main characteristics of the blow-up are consistent with the model of the head-tail instability caused by the electron cloud which was proposed by F. Zimmermann and K. Ohm [3].

To remove the electron cloud a large number of solenoid magnets were installed in the KEKB LER several times since September 2000[2]. The number of the solenoids installed so far amount to about 8600. As the result the measurement by an interferometer in February 2002 showed no beam size blow-up up to 1300 mA in regular operation condition for the physics experiment.

This paper describes solenoid winding work in September 2001 and January 2002 and consequence of it on the electron cloud effects.

SOLENOID SYSTEMS

Parameters of solenoid system are shown in Tables 1 and 2. There are two kind of solenoids, one is a bobbin-type solenoid and the other a bobbinless-type solenoid. The length of the bobbin-type solenoid is from 150 to 650 mm depending on the length of available free space for winding. The bobbinless-type solenoid has the length of 40 mm and is located mainly on bellows and both sides of NEG pumps and ion pumps to cover regions where the bobbin-type solenoids can not be wound. The longitudinal magnetic field at the center of a solenoid is about 45 Gauss. A picture of the solenoids is shown in Fig. 1.

TABLE1. Parameters of solenoids.

Type	Length (mm)	Diameter (mm)	Turns	Bz (center) @ 5A (Gauss)
Bobbin	150 - 650	148	250(typ.)	45
Bobbinless	40	220	190, 200	48
Bobbinless	40	250	200	43
Bobbinless	40	300	200	37

TABLE2. Parameters of power supply for the solenoids.

	KEKB corrector P.S.	TRISTAN corrector P.S.
Current(A)	5	3

TABLE3. Installation history of the solenoids. Second and third columns show the number of the installed solenoids.

Date	Bobbinless	Bobbin	Location
2000. 9. to 2001. 8.	2204	2793	Arc section (Cu chamber, Bellows, NEG) Straight section (Cu chamber, Bellows)
2001. 9.	3411	43	Arc section (NEG, IP, Bellows) Straight section (Cu chamber, Bellows)
2002. 1.	119	0	Arc section (Between quad and sextupole magnet)
Total	5734	2836	
Units	616		40

CP642, *High Intensity and High Brightness Hadron Beams: 20th ICFA Advanced Beam Dynamics Workshop on High Intensity and High Brightness Hadron Beams,* edited by W. Chou, Y. Mori, D. Neuffer, and J.-F. Ostiguy

FIGURE 1. Solenoids in an ion pump (IP) section. Long solenoids are the bobbin-type solenoids and short solenoids arethe bobbinless-type solenoids.

Power supplies of the solenoids are those for correctors of the KEKB rings and the TRISTAN collider. The solenoids were installed in LER five times. Before September 2001 about 1280 m were covered by the solenoids. In September 2001 the bobbinless solenoids were mainly installed in NEG and ion pump sections and in January 2002 the bobbinless solenoids were installed between quadrupole and sextupole magnets. The length covered by the solenoid field larger than 20 Gauss is about 870 m in the installation in September 2001 and 28 m in the installation in January 2002. Now the solenoid field covers about 75% of the circumference. Table 3 shows installation history of the solenoids.

Beam blow-up

Fig. 2 shows the vertical beam size in a long train for physics experiment at several installation stages of the solenoids. The threshold current of the blow-up

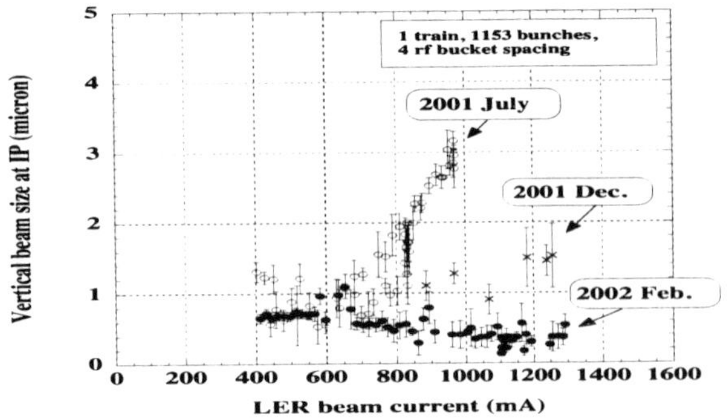

FIGURE 2. Effect of solenoids on the vertical beam size for a long train measured by the interferometer [2].

increased in each installation of the solenoids and finally the blow-up disappeared in the measurement in February 2002.

As expected by the measurement of the beam size the luminosity was improved after the installation of the solenoids. Fig. 3 shows the luminosity per bunch as a function of the bunch current product of HER and LER. An envelope curve of the luminosity taken for a

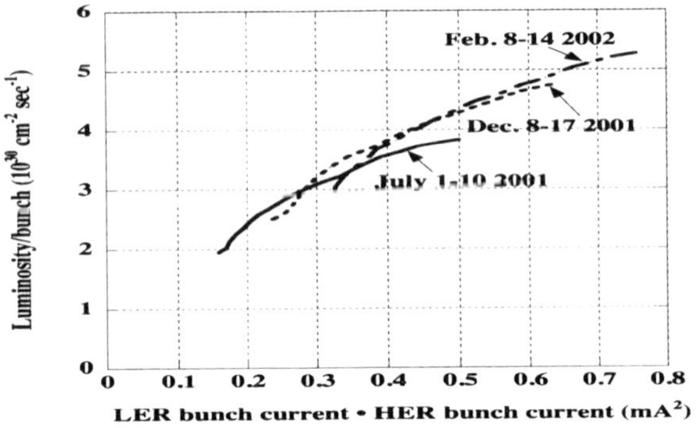

FIGURE 3. The luminosity per bunch at several installation stages of the solenoids.

week is plotted for each data set because the luminosity is affected by the beam tuning. As seen in Fig. 3 the luminosity was enhanced after the installation of the solenoids in September 2001. The luminosity may be slightly improved above 0.5 mA^2 after the installation in January 2002 though the improvement might be caused by the beam tuning.

According to a model of the single bunch instability due to the electron cloud the threshold bunch intensity of the blow-up $I_{b,\ th}$ is proportional to a bunch separation L_b for the regular and the strong head-tail instability and is proportional to the square root of L_b for the beam break-up instability [4]. The average beam size at several bunch separations was measured by the interferometer as a function of the beam current as shown in Fig. 4. In the measurements the number of bunches times the bunch separation was set to be equal. Comparing the blow-up for two fill patterns, if the blow-up starts at same beam current the scaling of $I_{b,\ th} \propto L_b$ holds. Fig. 4 shows that the scaling $I_{b,\ th} \propto L_b$ was observed and not changed after the installation of the solenoids in September 2001.

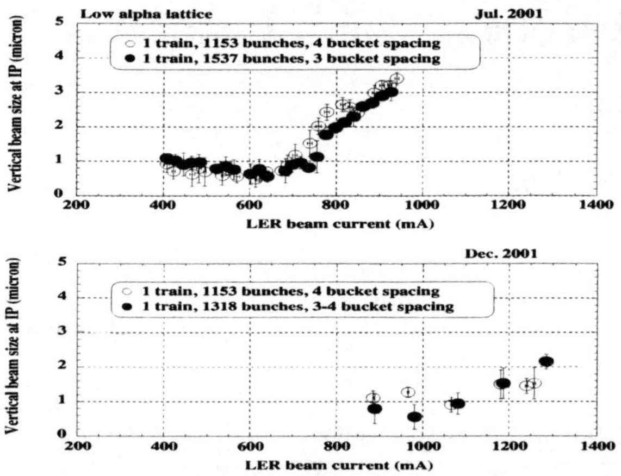

FIGURE 4. Vertical beam size in several rf bucket spacing measured by the interferometer. "3-4 bucket spacing means that the bunch spacing in the unit of rf bucket is 3,4,3,4, …

Tune shift and coupled bunch instability

The betatron tune of the bunch increases along the bunch train due to the electric force of the electron cloud [5]. Fig. 5 shows the tune shift before and after the installation of the solenoids in September 2001. The tune shift along the train decreased when the solenoids were added.

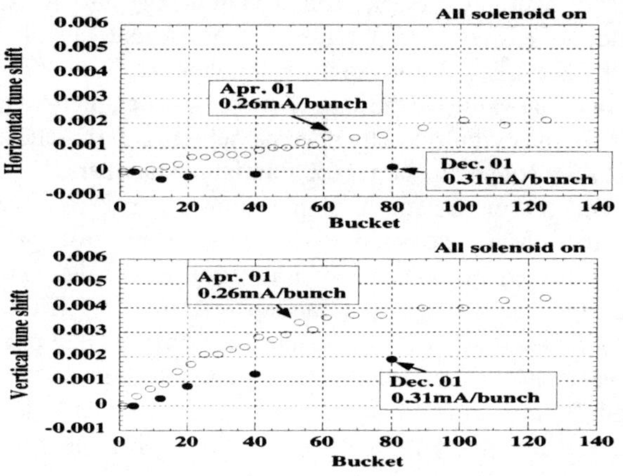

FIGURE 5. The effect of the solenoids on the horizontal and vertical tune shift along the train. Bunch current was 0.31 A.

The electron cloud can cause the transverse coupled bunch instability. In fact the mode spectrum of the coupled bunch instability observed in LER was changed with and without solenoid field [6]. The growth rate of the coupled bunch instability decreased by 74% and 66% in horizontal and vertical planes respectively after the installation of the solenoids in September 2001. The mode spectrum was not changed much, which can be expected because the addition of the solenoids would not change the transverse distribution of the electron cloud.

SUMMARY

The solenoids for suppressing the electron cloud were added in LER at September 2001 and January 2002 in the KEKB LER. After the addition of the solenoids the electron cloud effects were measured to see the effect of it. Results are summarized as

1) The blow-up of the beam size was disappeared up to 1300 mA with the fill pattern of 4 rf bucket spacing after the installation of the solenoids in January 2002.

2) The luminosity was improved after the installation of the solenoids in September 2001 and seems to be slightly improved after that in January 2002.

3) Tune shift along the train was decreased at least 40% after the installation in September 2001.

4) Growth rate of the coupled bunch instability decreased by 70 % after the installation in September 2001, while the mode spectrum was not changed much.

5) Threshold bunch current of the blow-up is proportional to the bunch separation.

ACKNOWLEDGMENTS

The authors would like to thank the members of KEKB group for their help for installing the solenoids and measuring the beam parameters. They also thank K. Oide for his support to winding the solenoids.

REFERENCES

1. K. Akai et al., "COMMISSIONING OF THE KEKB B-FACTORY", PAC'99, New York, Particle Accelerator Vol.1, 288-292(1999).

2. H. Fukuma, "ELECTRON CLOUD EFFECTS AT KEKB", ECLOUD02, CERN, April 2002, to be appeared in CERN-2002-001.

3. K. Ohmi and F. Zimmermann, "Head-Tail Instability Caused by Electron Clouds in Positron Storage Rings", Phys.Rev.Lett.85: 3821- 3824(2000).

4. F. Zimmermann, "THE ELECTRON CLOUD INSTABILITY: SUMMARY OF MEASUREMENTS AND UNDERSTANDING", 666, PAC2001, Chicago, June 2001.

5. K. Ohmi et al., "STUDY OF COHERENT TUNE SHIFT CAUSED BY ELECTRON CLOUD", APAC01, Beijing, September 2001.

6. S. S. Win et al., "Effect of Electron Cloud on the Bunch Oscillations in KEKB LER ", ECLOUD02, CERN, April 2002, to be appeared in CERN-2002-001.

Theory and Simulation of the Electron Cloud Instability

K. Ohmi

KEK, Tsukuba, Ibaraki, 305-0801 Japan

Abstract. We discuss two-stream instabilities of positron and proton beams caused by an electron cloud from the viewpoint of theory and simulation. The instabilities and their related phenomena are called the electron cloud effect. Build-up of an electron cloud and instabilities caused by it are discussed for various positron and proton rings.

INTRODUCTION

The instability caused by electron cloud has been discussed for a long time. The first work was done for CERN-ISR [1, 2], and the two-stream type of instability started to be discussed in the accelerator physics. After that the work was succeeded to ion trapping instability in electron storage rings [3, 4, 5], and then was succeeded to electron cloud instability in positron storage rings [6, 7, 8]. In the B factories of KEK (KEK-B) and SLAC (PEP-II), solenoid magnets which wipes off an electron cloud played an important role to realize higher luminosity [9, 10, 11]. Meanwhile, an instability has been observed for bunched proton beam in a proton synchrotron ring at Los Alamos National Laboratory (LANL-PSR). They reported that the instability was caused by an electron cloud [12].

We discuss the electron cloud instabilities of positron and proton beams. Electrons are produced by synchrotron light for the most part in positron ring. The number of photon and the conversion efficiency to electron are well understood. On the other hand there are some candidates of electron source in proton ring:i. e., ionization, proton loss, H^- injection etc. It is not clear which mechanism is important in proton rings. Anyway the initial yield is far less than that of positron ring, therefore Multipacting is essential to be built-up the electron cloud. We discuss the electron cloud build-up in Sec.2, and a single bunch instability driven by the electron cloud for positron and proton beam in Sec.3. The bunch structure is very different for positron and proton beams. Positron bunch has a typical size of $\sigma_x \times \sigma_y \times \sigma_z \sim 1 \times 0.1 \times 10mm^3$, while proton bunch has 1cm×1cm×100m in high intensity proton rings as are used for neutron sources.

ELECTRON CLOUD BUILD-UP

We discuss electron production mechanism at first. For positron beam, photo-emission at the chamber surface due to synchrotron radiation is dominant. The number of photon hitting the chamber wall is given by

$$n_\gamma = \frac{5\pi}{\sqrt{3}} \frac{\alpha\gamma}{L}. \quad (1)$$

for a positron in a meter, where α and γ are the fine structure constant 1/137 and the relativistic factor, respectively. The number of photo-electron produced by a positron at the chamber is given by

$$n_{e\gamma} = n_\gamma Y_\gamma. \quad (2)$$

The direct photo-emission rate was estimated to be $Y_\gamma = 0.1$ for an electron incidence of a shallow angle at the copper chamber. This value was consistent with an in situ measurement of electron current using button electrode in KEKB [13]. We installed a test ante-chamber in KEKB-LER to study density and yield of the electron cloud. The value for the ante-chamber was obtained to be 1/5 for the cylindrical chamber, namely, $Y_\gamma = 0.02$. These rates are somewhat larger for aluminum chambers.

For proton beam, it is not known clear electron sources. Many possibilities for primary electron production can be considered. Ionization of residual gas due to proton beam is a candidate. Proton beam ionizes residual gas with the result that electrons and positive ions are created. Ionization cross-section for CO and H_2 is estimated as $\sigma(CO) = 1.3 \times 10^{-22} m^{-2}$ and $\sigma(H_2) = 0.3 \times 10^{-22} m^{-2}$ using Bethe formula [14]. The molecular density d_m is related to the partial pressure in nPa by the relation at $20°C$, $d_m(m^{-3}) = 2.4 \times 10^{11} P_m(nPa)$. The electron production rate is $7.7 \times 10^{-9} e^-/(m \cdot p)$ at $2times10^{-7}$Pa. The production rate is 7 orders smaller than that of photoelectron in KEKB. The positive charged ions are repulsed by the proton beam, and hit the chamber wall. Electrons are created on the occasion of the ion absorption at the chamber surface [15, 16]. The number is known as $5 \sim 10$. The yield is $5 \times 10^{-8} e^-/(m \cdot p)$ at 2×10^{-7}Pa.

Electrons are created by proton absorption at the beam chamber surface. M. Furman et.al. [17] use an electron production rate $Y_1 = 4 \times 10^{-6} e^-/(m \cdot p)$ at the chamber surface. Proton loss of 4×10^{-6} has been observed in a

CP642, *High Intensity and High Brightness Hadron Beams: 20th ICFA Advanced Beam Dynamics Workshop on High Intensity and High Brightness Hadron Beams*, edited by W. Chou, Y. Mori, D. Neuffer, and J.-F. Ostiguy

revolution at PSR (L=90m). They assume that a proton creates 100 electrons at its loss.

H^- injection is a direct electron source. In JHF-RCS, since protons are injected during 500 turns, corresponding electron yield is estimated as $Y_1 = 2 \times 10^{-5} e^-/(m \cdot p)$. Most of the electrons are trapped in the magnetic field of the injection magnets, therefore the effective yield is smaller than the value.

Secondary electron production plays an important role for electron cloud build-up, especially quantity of primary electrons are much less than neutralization level and beam intensity is too high to accelerate the electrons over 100eV. Secondary yield [18], which is the number of electrons created by an electron incidence with an energy, is approximated by a formula as follows,

$$Y_2 = Y_s \times \frac{E}{E_{max}} \frac{1.44}{0.44 + (E/E_{max})^{1.44}}. \tag{3}$$

In recent studies, elastic reflection [19] of electrons is taken into account by $Y_2(E = 0) = 0.5 \sim 0.8$. In this paper, we use the original formula, Eq.(3).

We calculate electron density by the simulation code PEI [7]. Space charge force of electron cloud is taken into account. The beam chamber is assumed to be cylindrical and electrons are produced uniformly along azimuthal angle in the simulation. Electrons are produced at s_e and are tracked in two dimensional $x - y$ plane. The equation of motion of electrons is expressed by

$$\frac{d^2x(t)}{dt^2} = -\frac{2\lambda_p(s_e - vt) r_e c^2}{(\sigma_x + \sigma_y)\sigma_y} F(x(t)) - \frac{e}{m_e}\frac{\partial \phi}{\partial x}, \tag{4}$$

where the force $F_G(x)$ is expressed by the Bassetti-Erskine formula normalized so that $F_G \to x/|x|^2$ as $x \to \infty$.

Figure 1 shows variation of electron cloud density for the low energy ring (LER) of KEKB and the 3GeV rapid cycle synchrotron (RCS) of JHF as a function of bunch passage [1]. We note the bunch length at KEKB-LER is σ_z =5mm in the standard deviation while that at RCS is ℓ_p =90m in total length. The density increases and saturates at a certain density.

SINGLE BUNCH INSTABILITY CAUSED BY THE ELECTRON CLOUD

A single bunch instability is caused by a wake field induced by the electron cloud. We focus on the vertical instability in this paper. The wake field is represented by a resonator model. The resonator frequency (ω_e) corresponds to oscillation frequency of electrons in the beam field,

$$\omega_{e,y} = \sqrt{\frac{\lambda_+ r_e c^2}{\sigma_y(\sigma_x + \sigma_y)}}, \tag{5}$$

where λ_+ and $\sigma_{x(y)}$ are the beam line density in a bunch and transverse beam sizes, respectively, and r_e and c are the electron classical radius and the speed of light, respectively.

The impedance corresponding to the wake field is expressed by

$$Z(\omega)_\perp = \frac{cR_S}{Q\omega} \frac{Q}{1 + iQ\left(\frac{\omega_e}{\omega} - \frac{\omega}{\omega_e}\right)} \tag{6}$$

R_S/Q, which characterizes the coupling of beam and electron cloud, is given by

$$c\frac{R_S}{Q} = \frac{\lambda_e}{\lambda_+} \frac{L}{\sigma_y(\sigma_x + \sigma_y)} \frac{\omega_e}{c}. \tag{7}$$

Q which is called quality factor characterizes a correlation length of the wake field, and is ∞ in a simple linear theory.

λ_e is the line density of electron cloud which contributes to the interaction with the beam. The electrons distribute uniformly whole of the chamber at the start of the interaction with the beam. The positron beam gathers electrons in the cloud during the interaction. The positron beam size ($\sim$1mm) is much smaller than cloud (chamber) size ($\sim$ 5cm). We consider the line density as $\lambda_e = K \times 2\pi\rho_e\sigma_x\sigma_y$ [20], where K is an enhancement factor due to pinching of electrons and contributions of electrons near by the beam.

On the other hand, the proton beam size ($\sim$1cm) is ($\sim$5cm), and the bunch length is $\omega_e\ell/c$ is much larger than 1 ($\sim$100). Electrons are gathered at the position of the proton beam as soon as starting the interaction with the beam, therefore λ_e is considered to be the line density of electrons in the whole chamber [21].

The single bunch instability is estimated from the wake field. We use the coasting beam model to evaluate the instability, because of $\omega_e\sigma_z/c > 1$. The threshold of the instability is expressed by [22]

$$U \equiv \frac{\sqrt{3}\lambda_+ r_e \beta \omega_0}{\gamma \omega_e \eta \sigma_\delta} \frac{|Z_\perp(\omega_e)|}{Z_0} = \frac{\sqrt{3}\lambda_+ r_e \beta}{\gamma \nu_s \omega_e \sigma_z/c} \frac{|Z_\perp(\omega_e)|}{Z_0} = 1. \tag{8}$$

For $U > 1$, the beam is unstable.

We estimated the threshold value of electron cloud density for various positron and proton rings. The results are shown in Table 1 and 2.

The wake field approximated by the resonator model permits us to study the instability with simple analytic methods. However the estimation of the threshold includes somewhat ambiguous factors: i.e., for example,

[1] Space charge force has not been included in the calculation of 3GeV RCS yet

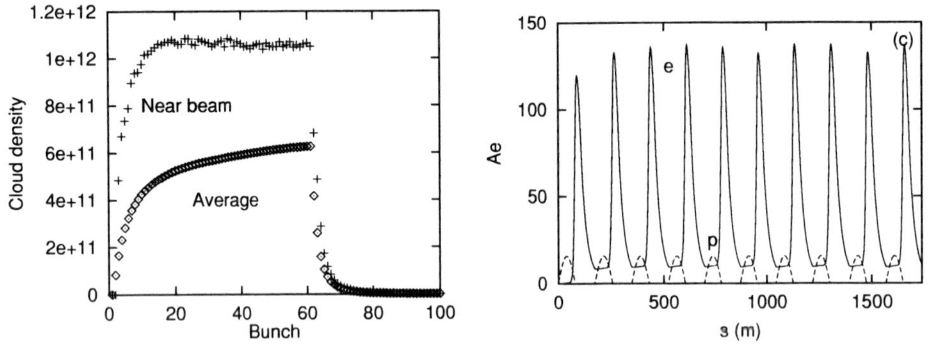

FIGURE 1. Electron cloud density as a function of bunch passage. The left and right pictures depict those of positron (KEKB) and proton (JHF-3GeV-RCS) beams, respectively.

TABLE 1. Single bunch electron cloud instability in positron storage rings. The enhancement factor is chosen to be $K = 3$. The impedance is evaluated at $\rho_e = 10^{12} m^{-3}$.

variable	KEKB	PEP-II	DAFNE	JLC-DR
E(GeV)	3.5	3.1	0.51	1.98
L(m)	3016	2200	97.7	398
$N_+(10^{10})$	3.3	6	4	0.75
ν_s	0.018	0.025	0.012	0.01
$\sigma_x(\mu$m$)$	420	700	2000	84
$\sigma_y(\mu$m$)$	60	120	63	7.1
σ_z(mm)	5	12	24	5
$\omega_c \sigma_z / c$	2.5	3.2	3.2	9.1
$Z/Z_0(m^{-1})$	2877	3363	511	2184
$\rho_{e,th}(10^{12} m^{-3})$	0.54	1.2	1.9	4.4

TABLE 2. Wake fi eld and stability for electron cloud instability.

variable	JHF				PSR	ISIS	AGS
	3GeV		50GeV				
	inj.	ext.	inj.	ext.			
L(m)	348.3	348.3	1567.5	1567.5	90	163	800
γ	1.4	4.2	4.2	54.	1.85	1.07	3.0
$N_p(\times 10^{13})$	4.15	4.15	4.15	4.15	3	1.25	1.2
σ_r(cm)	1.9	1.2	1.1	0.5	1.0	3.8	0.7
ℓ_p(m)	110	82	82	16	65	60	68
$\sigma_E/E(\%)$	0.6	0.7	0.7	0.25	0.25		0.28
η	-0.48	-0.047	-0.058	-0.0013	-0.187		-0.146
ν_s	0.0058	0.0005	0.0026	0.0001	0.0003	0.0003	0.0017
beam pipe radius $R(cm)$	12.5	12.5	6.5	6.5	5	8	5
$Z(\omega_e)_L/Q(M\Omega/m)$	0.29	0.24	0.68	0.019	0.46	0.0051	0.024
$Z(\omega_e)_H/Q(M\Omega/m)$	0.61	0.83	9.7	0.96	0.90	0.0085	0.37
$\omega_e \ell_p / c$	133	182	199	276	166	27	153
U_L	0.07	0.23	0.11	0.02	1.6	0.09	0.004
U_H	0.15	0.78	1.6	1.2	3.2	0.14	0.06

how to choose K and Q. To remove the ambiguity, we have to do tracking simulations [11, 23, 24, 25].

REFERENCES

1. H.G. Hereward, CERN 71-15 (1971).
2. E. Keil and B. Zotter, CERN-ISR-TH/71-58 (1971).
3. S. Sakanaka, M. Izawa, H. Kobayakawa and M. Kobayashi, Nucl. Instru. Methods, **A256**, 184 (1987).
4. D. Sagan and A. Temnykh, Nucl. Instru. Methods, **A344**, 459 (1994).
5. T.O. Raubenheimer and F. Zimmermann, Phys. Rev. **E52**, 5487 (1995); G. Stupakov et. al., Phys. Rev. **E52**, 5499 (1995).
6. M. Izawa, et. al., Phys. Rev. Lett., **74**,
7. K. Ohmi, Phys. Rev. Lett., **75**, 1526 (1995).
8. Z.Y. Guo, et. al., Proceedings of the PAC99, 633 (1999).
9. H. Fukuma, et. al., Proceedings of the HPAC2000, 1124 (2000).
10. A. Kurikov, et. al., Proceedings of the PAC2001, 1904 (2001).
11. K. Ohmi and F. Zimmermann, Phys. Rev. Lett., **85**, 3821 (2000).
12. R. J.Macek, et. al., Proceedings of the PAC2001, 688 (2001).
13. private communications.
14. Y. Baconnier, CERN report 85-19, pp.267 (1985).
15. R. J.Macek, private communications.
16. Y. Mori, private communications.
17. M. A.Furman, M. Pivi, Proceedings of the PAC2001, 707 (2001).
18. M. A.Furman and G. R.Lambertson, Proceedings of the PAC97, 1617 (1997).
19. V. Baglin et. al.,LHC-Project-Report-472.
20. K. Ohmi, F. Zimmermann, E. Perevedentsev, Phys. Rev. E **65**, 16502 (2002).
21. K. Ohmi, T. Toyama and C. Ohmori, Proceedings of the Mini-workshop on electron cloud simulations in proton and positron beams, CERN, 15-18, April 2002.
22. A.W. Chao, *Physics of Collective Beam Instabilities in High Energy Accelerators*, Wiley-Interscience Publication.
23. G. Rumolo et. al., Proceedings of PAC2001, 1889 (2001).
24. K.Ohmi, Proceedings of PAC2001, 1895 (2001).
25. Y. Cai, Proceedings of the Mini-workshop on electron cloud simulations in proton and positron beams, CERN, 15-18, April 2002.

Electron Cloud Effects in the KEK-PS and the KEK/JAERI Joint Project

T. Toyama*, Y. Irie*, S. Kato*, K. Ohmi*, C. Ohmori*, K. Satoh*, M. Uota*
N. Hayashi#

*KEK, Oho, Tsukuba, Ibaraki 305-0801, Japan
#JAERI, Tokai, Naka, Ibaraki 319-1195, Japan

Abstract. An e-p instability is potentially a serious problem for proton rings in JKJ, which forces us to study the e-p instability in several high intensity proton rings. This work informs JKJ whether we have to take measures to cure the instability. A TiN coating on the chamber surface is one of remedies. Results of SEY measurements performed at KEK are discussed. The observation of electron cloud candidates at the KEK 12 GeV PS Main Ring is also presented.

INTRODUCTION

A high-intensity proton accelerator facility has been proposed in Japan as a joint project of KEK and JAERI (JKJ). The facility would be equipped with two proton rings: a 3 GeV rapid cycling synchrotron and a 50 GeV proton synchrotron. The bunch population, which would be 4×10^{13}, compares with that of LANL PSR, RAL ISIS, BNL AGS. The e-p instability is potentially a serious problem for these two rings of JKJ. Not all high-intensity proton rings suffer from an electron cloud instability. It is worth comparing those proton rings from the viewpoint of the electron cloud instability.

It is important to know the secondary electron yield efficiency (SEY) not only as a candidate of remedies, but also as an input of a computer simulation. The results of ongoing measurements of SEY at KEK are discussed for several materials. The observation of electron cloud candidates at the KEK 12 GeV PS Main Ring is also discussed, although not yet confirmed.

ELECTRON CLOUD FORMATION AND INSTABILITY

Production rate of the primary electron generated due to the residual gas ionization by the proton beam ($Y_{1,i}$), due to proton beam loss ($Y_{1,\ell}$), at a stripping foil for H-minus charge exchange injection ($Y_{1,foil}$) are evaluated as $Y_{1,i} = 7.7\times10^{-9}$ $e^-/(m\cdot p)$, $Y_{1,\ell} = 4.4\times10^{-6}$ $e^-/(m\cdot p)$, $Y_{1,foil} = 1.9\times10^{-5}$ $e^-/(m\cdot p)$, respectively. Electrons at the stripping foil will be swept by the leakage field of bump magnets.

Electron cloud formation is estimated by tracking the transverse 2D motion of electrons produced by the primary and secondary electron emission. Primary electrons are produced at the chamber wall with energies of 10±5 eV. The secondary electron yield is approximated by true secondary electron yield of $Y_S(max)=2.1$ at $E_{max}=200$ eV. At position s along the ring, electrons move under an electric potential generated by a rigid "half-sin" proton beam. Space charge force between electrons is neglected because the average neutralization factor is less than 0.1 in the rings discussed here. The magnetic fields are also neglected in the whole ring for simplicity.

The amplification factor (A_e), the number of multiplied electrons divided by the number of primary electrons per one bunch passage, is calculated for several stages of the relevant rings. The results are shown in Table 1 and also in Figure 1. At every bunch passage a peak is formed by trailing-edge multipacting. Equilibrium is reached in 5 – 10 bunches passage. The neutralization factor strongly depends on several parameters: the beam size, chamber size, bunch length and bunch spacing.

The beam stability is evaluated based on a wake field approach and a coasting beam approximation and the instability may be fast enough regardless of the synchrotron oscillation. Both the proton beam and the electron cloud are assumed to have a rigid Gaussian distribution. By linearizing the coupled motion, the proton motion can be considered to be a forced oscillation with the wake field that is generated by the proton beam passing through the electron cloud. Including the damping effect due to electron oscillation frequency spread, the coupling impedance is reduced by Fourier transformation of the wake field. Making use of this coupling impedance, the dispersion relation, U=1, is obtained. For U>1, the beam is unstable. In Table 1, two values of U_H and U_L are listed. They are unstable criteria for the peak and bottom values of the neutralization factor, respectively. For ISIS, the slippage factor and the synchrotron tune are assumed to be the same as PSR. PSR is unstable. On the other hand, ISIS and AGS are stable. These results qualitatively agree with the observations. The rings of the joint project are in-between.

CP642, *High Intensity and High Brightness Hadron Beams: 20th ICFA Advanced Beam Dynamics Workshop on High Intensity and High Brightness Hadron Beams*, edited by W. Chou, Y. Mori, D. Neuffer, and J.-F. Ostiguy

SEY MEASUREMENTS

A series of measurements of the secondary electron yields (SEY) were made using an electron beam of ϕ0.5 mm with 100 - 5000 eV and of some tens nA, or using an argon ion beam with a raster-scanned size of a few mm^2, 5000 eV and some tens nA. Surfaces of sample materials were analysed with x-ray photoelectron spectroscopy or Auger electron spectroscopy.

Figure 2 shows the dependence of the SEY on the primary electron energy with normal incidence at the as-received surfaces and after sputtering with argon ions. Although a titanium as-received sample showed the highest yield, the yields of the others, except an isotropic graphite, were close to the peak of titanium. However, the yields of the carbon materials and the TiN film showed lower. The yields of the all materials were reduced after the argon ion sputtering. Since the carbon and oxygen impurities in the TiN film reached their saturation and remained even after sputtering of a thickness of 65nm, those impurities may have been included in the film, itself, during its preparation. Their reduction may reduce the yields of TiN further.

ELECTRON CLOUD IN THE KEK-PS

In the KEK-12 GeV-PS Main Ring, baseline drifts were observed around the transition energy and around the beginning of the flat top on newly installed electrostatic pick-ups and a previously installed exciter. Four electrodes were directly connected to the center control room on trial with a high impedance termination. The baseline drifts implies a charge-up of the pick-ups with a negative current of a few microamperes.

Although there seems to be no instability, the following experiments were performed to clarify the source of the baseline drift. Figure 3 shows the peak voltage of the baseline drift at the transition energy with seven, eight or nine successive bunches, with the same bunch population of 2.8×10^{11} protons. The baseline drift was saturated by applying more than ~40 V of positive bias. This saturation level decreased with increasing bunch gap. The baseline drift was not detectable if the bunch number was less than 6. At a fixed bunch number of 9, the peak voltage of the baseline drift was measured with solenoid currents of 0, 10, 20 and 30 A, as shown in Fig. 4. At 30 A no baseline drift was observed.

SUMMARY

Electron cloud build-up and beam stability were evaluated for high intensity proton rings: JKJ 3 GeV RCS, 50 GeV MR, PSR, ISIS and AGS. The number of primary electrons is amplified by trailing-edge multipacting. The amplification factor strongly depends on the secondary electron yield, beam shape, and chamber geometry. Then, using the neutralization factor obtained by the simulation, the single bunch stability was evaluated using a coasting beam model. The obtained stabilities agree qualitatively with observations in the existing machines.

Including both the elastic reflection in SEY estimation and the space charge in force evaluation is a subject for future study.

REFERENCES

1. References in T. Toyama, Y.Irie, S. Kato, K. Ohmi, C. Ohmori, K. Satoh and M. Uota, KEK Preprint 2002-53.

TABLE 1. Basic parameters of the proton rings.

Variable	symbol	Joint project 3GeV RCS inj.	50GeV MR ext.	inj.	ext.	PSR	ISIS inj.	AGS inj.
circumference	L(m)	348.3	348.3	1567.5	1567.5	90	163	800
Lorentz factor	γ	1.4	4.2	4.2	54	1.85	1.07	3.0
Bunch population	$N_p(x10^{13})$	4.15	4.15	4.15	4.15	3	1.25	1.2
Number of bunches	n_b	2	2	8	8	1	2	6
Harmonic number	h	2	2	9	9	1	2	6
Rms beam size	σ_r(cm)	1.9	1.2	1.1	0.35	1.0	3.8	0.7
Bunch length	ℓ_p(m)	110	82	82	16	65	60	68
Rms energy spread	σ_E/E(%)	0.6	0.7	0.7	0.25	0.25	0.025	0.28
Slippage factor	η	-0.48	-0.047	-0.058	-0.0013	-0.187		-0.146
Synchrotron tune	ν_s	0.0058	0.0005	0.0026	0.0001	0.0003		0.0017
Beam pipe radius	R(cm)	12.5	12.5	6.5	6.5	5	8	5
Amplification factor	Ae(bottom)	42.0	18.0	9.4	0.13	118	12.9	0.42
Amplification factor	Ae(peak)	87.6	62	136	6.9	236	17.5	5.18
Neutralization	η(bottom)	0.020	0.0067	0.0035	0.00001	0.034	0.003	0.0001
Neutralization	η(peak)	0.042	0.023	0.05	0.0005	0.067	0.005	0.0015
Stability (bottom)	U_L	0.07	0.23	0.11	0.02	1.6	0.09	0.004
Stability (peak)	U_H	0.23	0.78	1.6	1.2	3.2	0.14	0.06

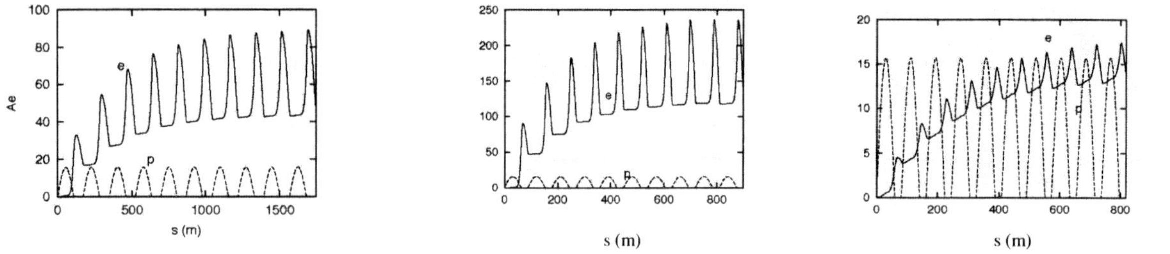

JKJ 3 GeV RCS injection LANL PSR RAL ISIS injection

FIGURE 1. Electron amplification factor and proton beam density for the JKJ 3 GeV RCS, PSR and ISIS. The dashed curves are the proton beam densities of a "half-sin" bunch (arbitrary unit).

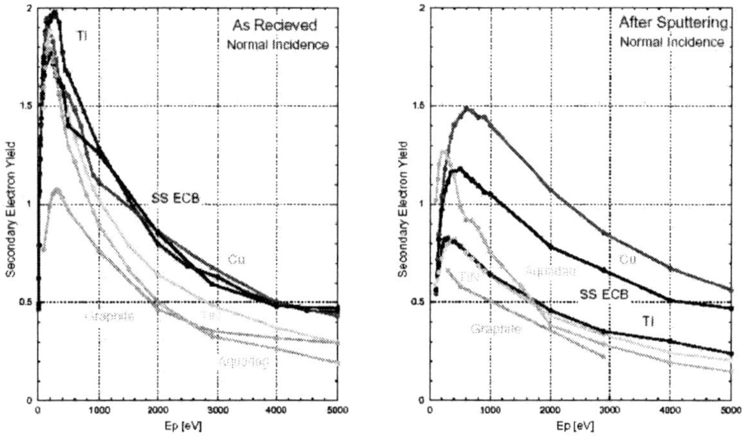

FIGURE 2. Dependence of the secondary electron yields on the primary electron energies at the surface as-received and after sputtering.

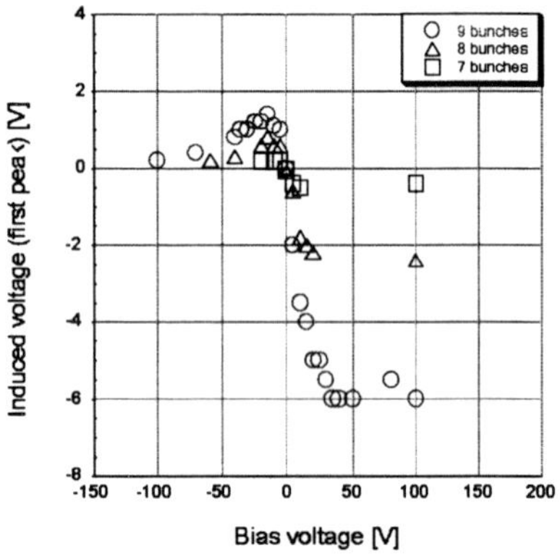

FIGURE 3. Peak voltage of the baseline drift at the transition energy with the number of bunches (7, 8 and 9).

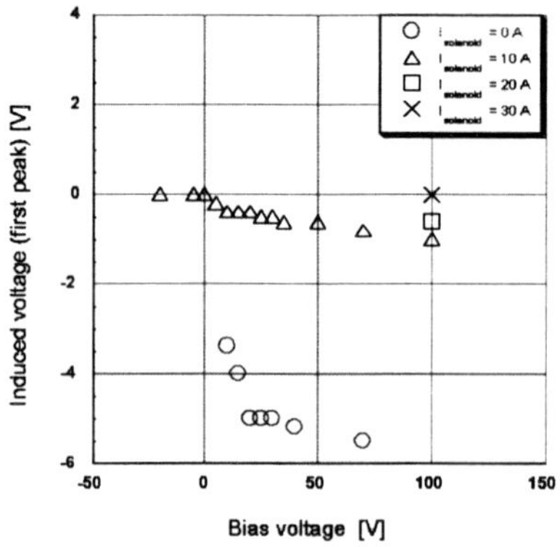

FIGURE 4. At a fixed bunch number 9, the peak voltage of the baseline drift measured with the solenoid current of 0, 10, 20 and 30 A.

Condition for Production of Circulating Proton Beam with Intensity Greater than Space Charge Limit

Vadim Dudnikov

Fermi National Accelerator Laboratory, Batavia, IL 60510, USA

Abstract. Transverse e-p instability in the proton rings could be damped by increase a beam density and the rate of secondary particles generation above a threshold level, corresponded a decrease of the unstable wavelength λ below a transverse beam size h(increase a beam density n_b and ion density n_i above a threshold level: $n_b + n_i > \beta^2 / r_e h^2$; $r_e = e^2/mc^2$).In high current proton rings is possible to reach this "Island of stability" by fast, concentrated charge exchange injection without painting and enhanced generation of secondary plasma as it was demonstrated in small scale PSR at BINP [1]. After damping of e-p instability has been accumulated a circulating proton beam with intensity, corresponding a tune shift $\Delta\nu$=0.85x6 in the ring with ν=0.85. Experimental observations of a transverse instability of proton beams in different accelerators and storage rings will be discussed. The methods of instability damping will be considered. These experimental dates could be useful for verification of computer simulation tools developed for investigation of space charge and instabilities in realistic conditions [4,5].

INTRODUCTION

Limitation of beam intensity and performance by repulsing space charge effects is accepted in general as a "natural" limit and a reaching of this limit is often a design goal of the projects in physics and technology. Some time this limit could be over passed with use of space charge neutralization (compensation) by particles with an opposite charge. The importance of the transverse beam instability driving by interaction with a plasma compensating particles has been considered in the first proposals of the high intense beam production as the Stabilized relativistic beam [6]. In the analyzes of stability of partly compensated electron beam by Chirikov [7] has been shown that threshold intensity and level of compensation for instability of coasting circulating beam can be low. In this model of electron-ion instability low energy compensating particles (electrons or ions) are trapped within the space- charge potential of the circulating beam. Coupled transverse oscillations of the beam and the trapped particles has develop with a transformation of the energy of beam to the oscillation energy, leading to the beam loss. For prevention of the beam degradation by compensating particles in electron-positron colliders from the first projects were used clearing electrodes for remove secondary particles along all orbits, and with a very deep vacuum this was enough for avoiding of this instability.

EXPERIMENTAL OBSERVATION OF e-p INSTABILITIES

A transverse instability of bunched proton beam driven by compensating particles has been observed first time at 1965 in small scale storage ring used for development of charge exchange injection [2,3]. In this small PSR with circumference L=2.5 m by charge exchange injection of 1 MeV H^- during 800 turns was reached linear proton density $\rho \sim 2.5\ 10^9$ p/cm, a volume density $n \sim 10^8\ cm^{-3}$ and corresponding beam potential relative the chamber wall U~1kV. This instability with a fast loss of bunched beam has been stabilized by simple negative feed back with pick up electrode, resonance amplifier and deflection electrode. At 1967 in other small PSR was observes an instability of coasting proton beam with a low threshold, connected with an accumulation of the compensating electrons [1,8]. The behavior of this instability was in good agreement with a Chirikov's analyses [7]. This instability has been damped by increase of the bounce frequency of electrons in the proton beam with increase of positive ion density [1,9]. These results were discussed in the reviews [10,11].

Observation of transverse instabilities in the AGS and in the ZGS has been presented in the Cambridge Accelerator Conference 1967 along with [8] but only recent was identified as e-p instability. Instability of the coasting proton beam with accumulation of electrons have been observed in the Bevatron and CERN ISR at 1971. In the Bevatron this instability was damped by feed back and by beam bunching, and in ISR by improving a vacuum from 10^{-10} Torr to 10^{-11} Torr and increase a number of clearing electrodes. A similar instability with accumulation of positive ions in negative charged electron and antiproton beams, has been observed in synchrotron radiation source ALADDIN and in the CERN and Fermilab antiproton accumulators [12]. Instability in AA was dumped by using a sophisticated methods if ion repulsing as beam shaking in supplement to vacuum improving and clearing by electric field. Further theoretical analyses of this instability of dipole and quadruple oscillation

CP642, *High Intensity and High Brightness Hadron Beams: 20th ICFA Advanced Beam Dynamics Workshop on High Intensity and High Brightness Hadron Beams*, edited by W. Chou, Y. Mori, D. Neuffer, and J.-F. Ostiguy

has been developed at 1972 by Koshkarev and Zenkevich [13] and has been extended in other publications.

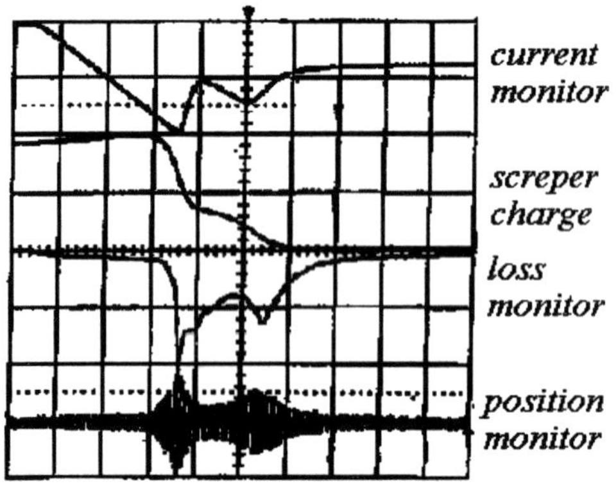

FIGURE 1. Oscilograms of coasting beam instability in the LANSCE PSR

New attention to the e-p instability was attracted after observation in the LANSE proton storage ring (PSR) a strong transverse instability with loss of bunched and unbunched beam [14]. This unpredicted and "mysterious" instability during 20 years limit the pulsed intensity of spallation neutron source in the level, below design goal. After many investigation was reached a good understanding of coasting beam instability and good agreement with theoretical models, but instability of the bunched beam have not an acceptable understanding. Typical development of an e-p instability of coasting beam in the LANSCE PSR shown in Fig.1. It is absolutely identical to the corresponding picture from the INP PSR in Fig.2. In several workshops have been discussed different aspects of the e-p instability in different accelerators and storage rings and further development for understanding and damping of the instability has been considered [15-17]. Transverse instability of coasting beam was observed at the BNL Booster and in Fermilab Booster with a DC field. With an increase of beam density the bounce frequency of secondary particles oscillations is increase and significant development of oscillation could be reached during one pass of bunch or trine of bunch through the portion of secondary particles accumulated during only one pass of the bunch or bunch trine. The attention to this possibility " Fast beam- ion instability" has been attracted in paper [18]. This type of instability has been observed in the ALS with an increased He gas density [19]. But this " Fast beam-ion instability" has been observed in low energy negative ion beam after ion source at 1975 [1] and after FNAL 0.75 MeV preaccelerator [20]. Space charge compensation by ions exhibit essential differences in comparison with compensation by electron one and corresponding instability have some features connected with mass difference and the ability to keep coherence.

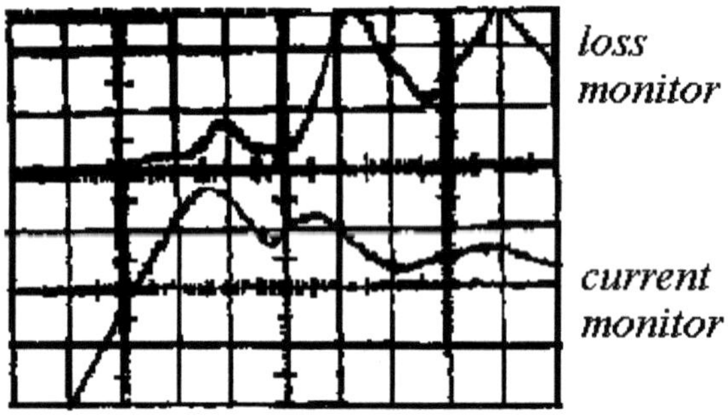

FIGURE 2. Coasting beam instability in the INP PSR.

DAMPING OF e-p INSTABILITY

For low beam density an electron bounce frequency $\omega^2 = 4c^2\ r_e\ n_i$ is comparable with a revolution frequency and bounce oscillation coupled with a low modes of betatron oscillations. For low modes, a magnitude of electron oscillations is more larger the beam oscillations and electrons removing from the beam by very small oscillations of protons. This mechanism used for remove ions from the antiproton beam. For a strong instability is need to have a high beam density and a strong enough source of a secondary particles. Suppression of a secondary particles generation and accumulation is a "traditional" method for e-p instability suppression: improve vacuum, use a gap for electron removing, use a cleaning electrodes, suppressing a secondary emission. A feed back system could be efficient in instability damping. Progress in suppressing of e-p instability in LA PSR was discussed in reports of R. Macek, presented in [15-17,21]. The need for higher beam intensity at PSR and for future high-intensity, proton drivers has motivated a multi-lab collaboration (LANL, ANL, FNAL, LBNL, BNL, ORNL, PPPL) to undertake research to better understand causes, dynamics and cures for e-p instability. Important characteristics of the electron cloud were recently measured with ANL electron analyzers and various collection electrodes [22]. Suppression of e generation by TiN coatings has confirmed the importance of secondary emission processes in generating the electron cloud. New tests of potential controls included dual harmonic RF, damping by higher order multipoles, damping by X,Y coupling and the use of inductive inserts to compensate longitudinal space charge forces. Use of a skew quadrupole, heated inductive inserts and higher RF voltage has enabled the PSR to accumulate stable beam intensity up to *9.7 mC (6 10^{13} p/p)*, which is a significant increase (60%) over the previous maximum of *6 mC*. This beam was stable with a high rate of secondary electron production

. Efficient damping of a e-p instability by increase a beam current density and increase a rate of secondary plasma generation has been demonstrated in small scale PSR at INP, Novosibirsk, at 1976 [1,9]. This process of the proton accumulation by charge-exchange injection is shown in Fig.3. With injected current below 2 mA and low gas density observed a fast excitation of dipole and quadrupole coherent oscillations and proton accumulation is saturated by loss on the level of N_p =1.2 10^{11} as shown by dotted lines. With increase an injected current and gas density above a threshold level the accumulation dynamic changed dramatically, as shown by solid lines. A number of circulating proton increase during the all injection time and limited only by injected current. Oscillations are damped after a short, fast growing.

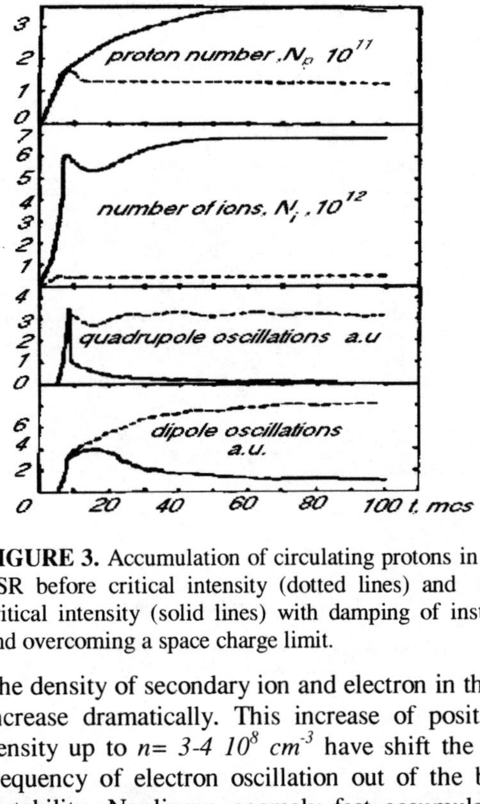

FIGURE 3. Accumulation of circulating protons in the INP PSR before critical intensity (dotted lines) and above a critical intensity (solid lines) with damping of instabilities and overcoming a space charge limit.

The density of secondary ion and electron in the beam increase dramatically. This increase of positive ion density up to $n= 3\text{-}4\ 10^8\ cm^{-3}$ have shift the bounce frequency of electron oscillation out of the bend of instability. Nonlinear, anomaly fast accumulation of secondary plasma is important for e-p stabilization. With injection current of 5 mA, 1 MeV H^- up to 1.8 10^{12} protons has been accumulated in PSR with a perimeter of 6 m. This intensity is corresponded a tune shift of Δv=0.85x6 in the PSR with v=0.85. This increase of beam and ion density above a second threshold for e-p instability could be used for production of extremely bright ion beams with increase a brightness by nonliowille charge exchange injection and for acceleration of high current ion beams in recirculators with an inductance linacs. It is looks as space charge compensation of the bunched beam is also possible.

REFERENCES

1. Budker G.,Dimov G.,Dudnikov V., et al., Proc.X Internat. Conf. Particle Accelerators, Protvino, 1977, v.2,p.287, 1978.
2. Budker G., Dimov G., Dudnikov V., Proc. Internat . Symp. Electron and Positron Storage Ring, France, Sakley,1966, rep. VIII, 6.1 (1966)
3. Budker G., Dimov G., and Dudnikov V., Soviet Atomic Energy, 22, 384 (1967).
4. Reiser M., The physics of High Brightness Beam, p.112, World Sci. 1999
5. Davidson R et al., in [17]
6. Budker G., Soviet Atomic Energy, 5, 9 (1956).
7. Chirikov B., Soviet Atomic Energy, 19, 239 (1965)
8. Budker G., Dimov G, Dudnikov V, Shamovsky V, Proc. sixth Internat. Conf. on High Energy Accelerators, 1967, MIT&HU,A-104, CEAL-2000, 1967, p.422.
9. Dimov G., Chupriyanov V., Particle accelerators, 14, 155-84 (1984).
10. Dudnikov V., Shabalin A., HIPAC'92, v.2,1025, Hamburg, 1993.
11. Dudnikov V., in [17].
12. Marriner J., Orlov Yu., et.al., Particle Accelerators, 30, 13 (1990).
13. Koshkarev D.and Zenkevich P., Part. Accel., 3, 1-9 (1972).
14. Colton E., et al., Particle Accelerators, 23,133 (1988).
15. Proc.of the Workshop on Accelerators for Future Spallation Neutron Sources, 1993, Santa Fe, NM.LA-UR-93-1356, Vol. II A,p.412-449(1993).
16. Proceedings of the Santa Fe Workshop on electron Effect, 1997, LA-UR-98-1601. (www.aps.anl.gov/conferences/icfa/two-stream).
17. ICFA Mini- Workshop on Two- Stream Instabilities in Particle Accelerators and Storage Rings, Santa Fe, 2000.
18. Stupakov G, Raubenheimer T.and Zimmermann F., Phys. Rev. E 52 (5), 5499 (1995).
19. Byrd J. et al., Phys. Rev. Let. 79(1),79 (1997). McCrory E.et al., Linac 88, CEBAF-89-001, p.182(1989).
20. Popovic M., Sullivan T., Linac2000.
21. Macek R., in [15-17] and this Proc.
22. Browman A., in [17].

List of Participants

Andreas Adelmann	Paul Scherrer Institute/LBNL	aaadelmann@lbl.gov
Masamitsu Aiba	KEK	masamitsu.aiba@kek.jp
James Alessi	Brookhaven National Laboratory	alessi@bnl.gov
John Anderson	Fermi National Accelerator Laboratory	janderson@fnal.gov
Terry Anderson	Fermi National Accelerator Laboratory	tander@fnal.gov
Charles Ankenbrandt	Fermi National Accelerator Laboratory	ankenbrandt@fnal.gov
Jeffrey A. Appel	Fermi National Accelerator Laboratory	appel@fnal.gov
Rick Baartman	TRIUMF	krab@triumf.ca
J. Scott Berg	Brookhaven National Laboratory	jsberg@bnl.gov
Chandrashekhara Bhat	Fermi National Accelerator Laboratory	cbhat@fnal.gov
Kip Bishofberger	UCLA/Fermilab	kpster@physics.ucla.edu
Michael Borden	Los Alamos National Laboratory	borden@lanl.gov
Alexey Burov	Fermi National Accelerator Laboratory	burov@fnal.gov
Giovanni Cantatore	INFN/University of Trieste	cantatore@ts.infn.it
Roberto Cappi	CERN	roberto.cappi@cern.ch
Christian Carli	CERN	christian.carli@cern.ch
Alex Chao	Stanford Linear Accelerator Center	achao@slac.stanford.edu
Weiren Chou	Fermi National Accelerator Laboratory	chou@fnal.gov
Brajesh Choudhary	California Institute of Technology	brajesh@hep.caltech.edu
Patrick Colestock	Los Alamos National Laboratory	colestock@lanl.gov
Sarah Cousineau	Indiana University	scousine@indiana.edu
Viatcheslav Danilov	Oak Ridge National Laboratory	danilovs@ornl.gov
Jeffrey Dooling	Argonne National Laboratory	jcdooling@anl.gov
Alexandr Drozhdin	Fermi National Accelerator Laboratory	drozhdin@fnal.gov
Robert J. Ducar	Fermi National Accelerator Laboratory	ducar@fnal.gov
Vadim Dudnikov	Fermi National Accelerator Laboratory	dudnikov@fnal.gov
Gerardo Dutto	TRIUMF	dutto@triumf.ca
Albert Rogers Ellingboe	Dublin City University	bert.ellingboe@dcu.ie
James Ellison	University of New Mexico	ellison@math.unm.edu
Béla Erdélyi	Fermi National Accelerator Laboratory	erdelyi@fnal.gov
Robin Ferdinand	CEA Saclay	robin.ferdinand@cea.fr
Dan Fitzgerald	Los Alamos National Laboratory	fitzg@lanl.gov
G. William Foster	Fermi National Accelerator Laboratory	gwf@fnal.gov
Giuliano Franchetti	GSI	g.franchetti@gsi.de
Hitoshi Fukuma	KEK	fukuma@post.kek.jp
Tony Gabriel	Oak Ridge National Laboratory	tag@ornl.gov

Roland Garoby	CERN	roland.garoby@cern.ch
Ugo Gastaldi	INFN Legnaro	gastaldi@lnl.infn.it
J. Douglas Gilpatrick	Los Alamos National Laboratory	gilpatrick@lanl.gov
Alain Girard	CEA Grenoble	agirard@cea.fr
Krishnaswamy Gounder	Fermi National Accelerator Laboratory	gounder@fnal.gov
James Griffin	Fermi National Accelerator Laboratory	jgriffin@fnal.gov.
Robert Hardekopf	Los Alamos National Laboratory	hardekopf@lanl.gov
Katherine Harkay	Argonne National Laboratory	harkay@aps.anl.gov
Gerd Heidenreich	Paul Scherrer Institute	gerd.heidenreich@psi.ch
Stuart Henderson	Oak Ridge National Laboratory	shenderson@sns.gov
Ingo Hofmann	GSI	i.hofmann@gsi.de
Stephen Holmes	Fermi National Accelerator Laboratory	holmes@fnal.gov
Martin Hu	Fermi National Accelerator Laboratory	martinhu@fnal.gov
Haixin Huang	Brookhaven National Laboratory	huanghai@bnl.gov
James Hylen	Fermi National Accelerator Laboratory	hylen@fnal.gov
Susumu Igarashi	KEK	susumu.igarashi@kek.jp
Masanori Ikegami	KEK	masanori.ikegami@kek.jp
Jun Imazato	KEK	jun.imazato@kek.jp
Cezary Jach	Fermi National Accelerator Laboratory	jach@fnal.gov
Jean-Bernard Jeanneret	CERN	bernard.jeanneret@cern.ch
David Johnson	LASL	itech@alum.calberkeley.org
Jeffrey Johnson	Oak Ridge National Laboratory	johnsonjo@ornl.gov
Rolland Johnson	IIT/Muons, Inc.	roljohn@aol.com
Carol Johnstone	Fermi National Accelerator Laboratory	cjj@fnal.gov
Yves Jongen	IBA s.a.	jongen@iba.be
Roderich Keller	Lawrence Berkeley National Laboratory	r_keller@lbl.gov
Rami Kishek	University of Maryland	ramiak@ebte.umd.edu
Junichi Kishiro	KEK	junichi.kishiro@kek.jp
Horst Klein	Frankfurt University	horst.klein@iap.uni-frankfurt.de
Thomas R. Kobilarcik	Fermi National Accelerator Laboratory	kobilarc@fnal.gov
Shane Koscielniak	TRIUMF	shane@triumf.ca
Mikhail Kostin	Fermi National Accelerator Laboratory	kostin@fnal.gov
Joe Kwan	Lawrence Berkeley National Laboratory	jwkwan@lbl.gov
Jean-Michel Lagniel	CEA Bruyeres-le-Chatel	jean-michel.lagniel@cea.fr
Valeri Lebedev	Fermi National Accelerator Laboratory	val@fnal.gov
Yong Lee	Brookhaven National Laboratory	yylee@bnl.gov
Trevor Linnecar	CERN	trevor.linnecar@cern.ch
Paul Lisowski	Los Alamos National Laboratory	lisowski@lanl.gov

Alfredo U. Luccio	Brookhaven National Laboratory	luccio@bnl.gov
Robert Macek	Los Alamos National Laboratory	macek@lanl.gov
Shinji Machida	KEK	shinji.machida@kek.jp
James MacLachlan	Fermi National Accelerator Laboratory	maclachlan@fnal.gov
Ernie Malamud	Fermi National Accelerator Laboratory	malamud@fnal.gov
Nikolay Malitsky	Brookhaven National Laboratory	malitsky@bnl.gov
Alberto Marchionni	Fermi National Accelerator Laboratory	alberto@fnal.gov
William Marciano	Brookhaven National Laboratory	marciano@bnl.gov
John Marriner	Fermi National Accelerator Laboratory	marriner@fnal.gov
Kirk McDonald	Princeton University	mcdonald@puphep.princeton.edu
Douglas Michael	California Institute of Technology	michael@hep.caltech.edu
Stephen Milton	Argonne National Laboratory	milton@aps.anl.gov
Nikolai Mokhov	Fermi National Accelerator Laboratory	mokhov@fnal.gov
Alexandre Molodojentsev	KEK	molodxx@yahoo.com
William Molzon	University of California, Irvine	wmolzon@uci.edu
Craig Moore	Fermi National Accelerator Laboratory	cmoore@fnal.gov
Yoshiharu Mori	KEK	yoshiharu.mori@kek.jp
Sergei Nagaitsev	Fermi National Accelerator Laboratory	nsergei@fnal.gov
Subrata Nath	Los Alamos National Laboratory	snath@lanl.gov
Filippo Neri	Los Alamos National Laboratory	fneri@lanl.gov
David Neuffer	Fermi National Accelerator Laboratory	neuffer@fnal.gov
George Neuschaefer	Los Alamos National Laboratory	neuschaefer@lanl.gov
King Y. Ng	Fermi National Accelerator Laboratory	ng@fnal.gov
Fumiaki Noda	JAERI	noda@linac.tokai.jaeri.go.jp
Kazuhito Ohmi	KEK	ohmi@post.kek.jp
Jean-Francois Ostiguy	Fermi National Accelerator Laboratory	ostiguy@fnal.gov
Carlo Pagani	INFN Milan	carlo.pagani@mi.infn.it
Richard Pardo	Argonne National Laboratory	pardo@phy.anl.gov
Vasily Parkhomchuk	Budker Institute for Nuclear Physics	parkhomchuk@inp.nsk.su
David Peaslee	University of Maryland	peaslee@umdhep.umd.edu
Nicolas Pichoff	CEA Saclay	nicolas.pichoff@cea.fr
Paolo Pierini	INFN Milan	paolo.pierini@mi.infn.it
Mauro Pivi	Lawrence Berkeley National Laboratory	mpivi@lbl.gov
Milorad Popovic	Fermi National Accelerator Laboratory	popovic@fnal.gov
Christopher Prior	Rutherford Appleton Laboratory	c.prior@rl.ac.uk
David Pushka	Fermi National Accelerator Laboratory	pushka@fnal.gov
Zubao Qian	Fermi National Accelerator Laboratory	zubao@fnal.gov
Ji Qiang	Lawrence Berkeley National Laboratory	jqiang@lbl.gov

Hong Qin	Princeton Plasma Physics Laboratory	hongqin@pppl.gov
Deepak Raparia	Brookhaven National Laboratory	raparia@bnl.gov
Grahame Rees	Rutherford Appleton Laboratory	ghrees@ukonline.co.uk
Validimir Reva	Budker Institute for Nuclear Physics	reva@csd.inp.nsk.su
Robert Ryne	Lawrence Berkeley National Laboratory	rdryne@lbl.gov
Izumi Sakai	KEK	izumi.sakai@kek.jp
Pierre Schmelzbach	Paul Scherrer Institute	pierre.schmelzbach@psi.ch
Martin Schulze	General Atomics	schulz@gat.com
Peter Schwandt	Indiana University	schwandt@iucf.indiana.edu
Andrew Sessler	Lawrence Berkeley National Laboratory	amsessler@lbl.gov
Elena Shaposhnikova	CERN	elena.chapochnikova@cern.ch
Joseph Sherman	Los Alamos National Laboratory	jsherman@lanl.gov
Jack Shi	University of Kansas	jshi@ku.edu
Kenta Shigaki	JAERI	shigaki@linac.tokai.jaeri.go.jp
Yoshito Shimosaki	KEK	shimo@www-accps.kek.jp
Ken Stanfield	Fermi National Accelerator Laboratory	stanfield@fnal.gov
James Steimel	Fermi National Accelerator Laboratory	steimel@fnal.gov
Gennady Stupakov	Stanford Linear Accelerator Center	stupakov@slac.stanford.edu
Isao Sugai	KEK	isao.sugai@kek.jp
Michael Syphers	Fermi National Accelerator Laboratory	syphers@fnal.gov
Akira Takagi	KEK	akira.takagi@kek.jp
Ken Takayama	KEK	takayama@post.kek.jp
Lee C. Teng	Argonne National Laboratory	teng@aps.anl.gov
H. Arch Thiessen	Los Alamos National Laboratory	hat@lanl.gov
John Thomason	Rutherford Appleton Laboratory	j.thomason@rl.ac.uk
Masahito Tomizawa	KEK	masahito.tomizawa@kek.jp
Takeshi Toyama	KEK	takeshi.toyama@kek.jp
Mayda Velasco	Northwestern University	mvelasco@lotus.phys.nwu.edu
Eyke Wagner	Paul Scherrer Institute	eyke.wagner@psi.ch
Werner Wagner	Paul Scherrer Institute	werner.wagner@psi.ch
Masanori Wakasugi	RIKEN	wakasugi@postman.riken.go.jp
Thomas Wangler	Los Alamos National Laboratory	twangler@lanl.gov
Rainer Wanzenberg	DESY Hamburg	rainer.wanzenberg@desy.de
Christopher Warsop	Rutherford Appleton Laboratory	c.m.warsop@rl.ac.uk
Robert Webber	Fermi National Accelerator Laboratory	webber@fnal.gov
Jie Wei	Brookhaven National Laboratory	wei1@bnl.gov
Wu-Tsung Weng	Brookhaven National Laboratory	weng@bnl.gov
Richard Werbeck	Los Alamos National Laboratory	rwerbeck@lanl.gov

David Wildman	Fermi National Accelerator Laboratory	wildman@fnal.gov
Vincent Wu	University of Cincinnati	vincentw@fnal.gov
Aimin Xiao	Fermi National Accelerator Laboratory	xiaoam@fnal.gov
Meiqin Xiao	Fermi National Accelerator Laboratory	meiqin@fnal.gov
Isao Yamane	KEK	isao.yamane@kek.jp
Masahito Yoshii	KEK	masahito.yoshii@kek.jp
Lloyd Young	Los Alamos National Laboratory	lyoung@lanl.gov
Xiaolong Zhang	Fermi National Accelerator Laboratory	zhangxl@fnal.gov

Workshop Program

Monday, April 8

Morning - Plenary Session, 1West

Chair: P. Lisowski (LANL)
Secretary: M. Kostin (Fermilab)

9:30 am	Opening address	K. Stanfield (Fermilab)
9:45 am	JHF physics	J. Imazato (KEK)
10:30 am	Coffee break	
11:00 am	Neutrino physics	D. Michael (CalTech)
11:45 am	Muon physics	W. Molzon (UC Irvine)
12:30 pm	Lunch break	

Afternoon - Plenary Session, 1West

Chair: A. Sessler (LBL)
Secretary: M. Xiao (Fermilab)

2:00 pm	Introductory remarks: Reflections on the reasons for building accelerators	A. Sessler (LBL)
2:15 pm	JHF machines	Y. Mori (KEK)
2:55 pm	Synchrotron based proton drivers	W. Chou (Fermilab)
3:35 pm	Coffee break	
4:00 pm	SNS Project	J. Wei (BNL/ORNL)
4:40 pm	Linac based proton drivers	T. Wangler (LANL)
5:20 pm	An 8 GeV SC proton linac	W. Foster (Fermilab)
6:00 pm	Reception at 15th Floor, Wilson Hall	

Tuesday, April 9

Morning - Parallel Sessions: 8:30 am – 12:30 pm, coffee break 10:15 am – 10:45 am

A. Lattice, Comitium (WH2SE)
Convener: A. Thiessen (LANL)
Secretary: B. Erdelyi (Fermilab)

1.	JHF lattice	K. Shigaki (JAERI)
2.	Tracking studies for JHF lattice	A. Molodojentsev (KEK)
3.	Proton driver lattice	C. Johnstone (Fermilab)
4.	Tracking of proton driver lattice	B. Erdelyi (Fermilab)
5.	AHF lattices	P. Schwandt (Indiana U.)
6.	Tracking AHF lattices	F. Neri (LANL)
7.	Magnets that meet tracking requirements for AHF	M. Schulze et al. (General Atomics)
8.	SNS Ring lattice	J. Wei et al. (BNL)
9.	Lattice design studies for CERN Proton Driver Accumulator and Compressor	C. Carli et al. (CERN)

B. Beam Loss, Collimation, and Shielding, Curia II (WH2SW)

Convener: N. Mokhov (Fermilab)
Secretary: M. Kostin (Fermilab)

1. Fermilab Booster beam loss and collimation — R. Webber (Fermilab)
2. SNS beam-in-gap cleaning and collimation — S. Cousineau et al. (ORNL/BNL)
3. Beam loss control on the ESS Accumulator Rings — C. Warsop (RAL)
4. Radiation handling in the slow extraction of the JHF 50 GEV ring — M. Tomizawa, T. Yokoi (KEK)
5. Beam collimation and shielding in the Fermilab Proton Driver — A. Drozhdin, N. Mokhov (Fermilab)
6. Shielding calculations for the SNS proton beam transport system — J. Johnson et al. (ORNL)
7. Development of high-power target systems for the SNS and the muon collider/neutrino factory — T. Gabriel et al. (ONAL)
8. Beam loss and collimation at the LHC — J.B. Jeanneret (CERN)

C. and G. Joint Session - FFAG and Cyclotrons, 1North
Conveners: Y. Mori (KEK), P. Schmelzbach (PSI), R. Baartman (TRIUMF)
Secretary: K. Gounder (Fermilab)

1. The PSI cyclotron and its extrapolation to a 10 MW driver — P. Schmelzbach (PSI)
2. High intensity cyclotron development at IBA — Y. Jongen (IBA)
3. FFAGs: scaling and isochronous — R. Baartman (TRIUMF)
4. Development of FFAG at KEK — Y. Mori (KEK)
5. Beam dynamics of FFAG accelerators — M. Aiba (KEK)
6. FFAG for neutrino factory — C. Johnstone (Fermilab)
7. RF acceleration in FFAG muon accelerator — S. Koscielniak (TRIUMF)
8. Stochastic cooling in FFAG — M. Wakasugi (RIKEN)
9. Longitudinal dynamics in FFAG — S. Berg (BNL)

D. New Ideas, 1East
Conveners: K. Takayama (KEK), C. Ankenbrandt (Fermilab)
Secretary: V. Wu (Fermilab)

1. Superbunch acceleration and induction devices — J. Kishiro et al. (KEK)
2. Beam echo in hadron accelerators — G. Stupakov (SLAC)
3. Beam echo experiment — P. Colestock (LANL)
4. Inductive inserts — J. Griffin (Fermilab)
5. Slip stacking — J. Steimel, K. Koba (Fermilab)
6. Barrier rf stacking — B. Ng (Fermilab)
7. Barrier bucket experiment in the Recycler — C. Bhat (Fermilab)
8. A linac Afterburner — M. Popovic (Fermilab)

Afternoon – Working Groups: 2:00 pm – 6:00 pm, coffee break 3:45 pm – 4:15 pm

WG I: Circular Accelertors, Curia II (WH2SW)
Conveners: L. Teng (ANL), R. Cappi (CERN)
Secretary: A. Burov (Fermilab)

1. 3-GeV ring at the JHF — F. Noda (JAERI)
2. AGS high power upgrade plan — W.T. Weng (BNL)
3. AHF project — A. Thiessen (LANL)
4. High density and high intensity beams at CERN-PS: present studies, achievements and future goals — R. Cappi (CERN)
5. Results from impedance reduction in the SPS — E. Shaposhnikova (CERN)
6. Emittance dilution in HERA-p: lessons and overview — R. Wanzenberg (DESY)

WG II: Linear Accelerators, 1North
Conveners: J-M. Lagniel (CEA), R. Garoby (CERN)
Secretary: X. Zhang (Fermilab)

1. Aspects of beam behavior in different sections of the SNS linac — S. Nath (LANL)
2. High-intensity proton linac for the KEK/JAERI joint project — M. Ikegami et al. (KEK)
3. ESS linac design overview — R. Ferdinand et al. (CEA)
4. Status and plans of the SPL study at CERN — R. Garoby (CERN)
5. The TRASCO project — C. Pagani (INFN)
6. IFMIF a challenging high intensity accelerator — R. Ferdinand et al. (CEA)

WG III: Beamlines and Targets, 1East
Conveners: C. Moore (Fermilab), K. McDonald (Princeton U.)
Secretary: A. Marchionni (Fermilab)

1. Target development for the SINQ high-power neutron spallation source — W. Wagner (PSI)
2. Carbon and beryllium targets and beam windows at PSI — G. Heidenreich (PSI)
3. NuMI target — J. Hylen (Fermilab)
4. Mercury target — K. McDonald (Princeton U.)
5. Radiation protection utilizing electronic berms — J. Anderson (Fermilab)

Wednesday, April 10

<u>Morning - Parallel Sessions</u>: 8:30 am – 12:30 pm, coffee break 10:15 am – 10:45 am

E. Space Charge Simulations, Curia II (WH2SW)
Conveners: I. Hofmann (GSI), R. Ryne (LBL)
Secretary: B. Erdelyi (Fermilab)

1. ORBIT: A code for collective beam dynamics in high intensity rings — S. Cousineau, J. Holmes (ORNL)
2. Behavior of intense beams simulated with Simpsons — S. Machida (KEK)
3. Coherent transverse resonance effects — I. Hofmann (GSI)
4. Space charge calculation in the code ORBIT — A.U. Luccio (BNL)
5. Simulation studies with PATRASH — Y. Shimosaki (KEK)
6. Space charge effects on beam loss due to resonance overlap — G. Franchetti (GSI)
7. Resonance tests of space charge — J. Qiang, R. Ryne(LBL)

F. Remote Handling, 1East
Convener: D. Pushka (Fermilab)
Secretary: M. Kostin (Fermilab)

1. Remote handling and shielding at PSI — E. Wagner (PSI)
2. Remote handling experience at LAMPF and LANSCE — R. Werbeck (LANL)
3. Dose calculations and remote handling plan for NuMI — J. Hylen (Fermilab)

G. (Joint with Session C on Tuesday)

H. Ion Sources, 1North
Conveners: K-N. Leung (LBL), H. Klein (U. of Frankfurt)
Secretary: V. Wu (Fermilab)

1. High brightness negative ion sources with high emission current density — V. Dudnikov (Fermilab)
2. Status of the SNS H^- source and LEBT — R. Keller et al. (LBL)
3. Performance of the magnetron H^- source on the BNL 200 MeV linac — J. Alessi (BNL)
4. Towards high current steady state H^- sources — B. Ellingboe (Dublin City U.)
5. ECR ion source for negative ions — A. Girard (CEA-Grenoble)
6. ECR ion sources and their roles in radioactive beam production — R. Pardo (ANL)
7. High current ECR source for protons and deuterons at Saclay — R. Ferdinand (CEA)
8. Review of a scaled Penning H^- surface plasma source — J.D. Sherman et al. (LANL)
9. The ISIS Penning source and the volume sources from Frankfurt and DESY for the production of H^- ions — J.W.G. Thomason (RAL)

Afternoon – Working Groups: 2:00 pm – 6:00 pm, coffee break 3:45 pm – 4:15 pm

WG I: Circular Accelertors (cont'd), Curia II (WH2SW)
Conveners: L. Teng (ANL), R. Cappi (CERN)
Secretary: A. Burov (Fermilab)

7. Multiparticle dynamics in the E-phi tracking code ESME : resources and results — J. MacLachlan (Fermilab)
8. Application of the UAL to high-intensity beam dynamics studies in the SNS ring — N. Malitsky et al. (BNL)
9. Longitudinal dynamics and rf hardware — M. Yoshii (KEK)
10. Diagnostic investigation of tune and tune shift in the IPNS RCS — J. Dooling et al. (ANL)
11. Hardware of beam injection and extraction — Y. Shirakabe (KEK)
12. Beam profile monitor of high intensity proton — Y. Hashimoto (KEK)
13. Magnets for high beam power synchrotrons — J.-F. Ostiguy (Fermilab)
14. Dual-harmonic resonant power supply — C. Jach (Fermilab)
15. Proton Driver vacuum system — T. Anderson (Fermilab)

WG II: Linear Accelerators(cont'd), 1North
Conveners: J-M. Lagniel (CEA), R. Garoby (CERN)
Secretary: X. Zhang (Fermilab)

7. An RFQ designed to accept beam from a weak focusing LEBT — L. Young (LANL)
8. Measurement of beam halo generation in an intense proton beam — P. Colestock et al. (LANL)
9. Wide dynamic-range beam-profile instrumentation for a beam-halo measurement: description and operation — J.D. Gilpatrick et al. (LANL)
10. Numerical code in use in CEA-Saclay — N. Pichoff et al. (CEA)
11. Modeling of mismatch induced emittance growth in linacs — I. Hofmann, G. Franchetti (GSI)
12. Halo and beam dynamics issues in high power linacs — N. Pichoff et al. (CEA)

WG III: Beamlines and Targets (cont'd), 1East
Conveners: C. Moore (Fermilab), K. McDonald (Princeton U.)
Secretary: A. Marchionni (Fermilab)

6. Automatic beam line tuning (AUTOTUNE) — T. Kobilarcik (Fermilab)
7. Sophisticated beam permit system — R. Ducar (Fermilab)

8. Tails of beam distribution — A. Marchionni (Fermilab)
9. Beam transfer lines for the SNS — D. Raparia (BNL)
10. Targetry issues for the Fermilab 2 MW neutrino superbeam — M. Kostin (Fermilab)

Thursday, April 11

Morning - Parallel Sessions: 8:30 am – 12:30 pm, coffee break 10:15 am – 10:45 am

I. Space Charge Experiments, Curia II (WH2SW)
Conveners: S. Machida (KEK), S.Y. Lee (Indiana U.)
Secretary: B. Erdelyi (Fermilab)

1. SC effects at the KEK PS Main Ring — S. Igarashi et al. (KEK)
2. Intensity dependent emittance-exchange in a proton synchrotron — I. Sakai (KEK)
3. Studies of the coherent half integer resonance at the PSR — S. Cousineau (Indiana U.)
4. Inductive inserts experiment at the PSR — K-Y. Ng (Fermilab)
5. New methods to create hollow bunches — C. Carli (CERN)
6. Space charge calculations at PSI — A. Adelmann (PSI)
7. Intense beam experiment at the U. of Maryland Electron Ring — R. Kishek (U. of Maryland)

J. Hadron Beam Cooling, Comitium (WH2SE)
Conveners: V. Parkhomchuk (INP), S. Nagaitsev (Fermilab)
Secretary: V. Wu (Fermilab)

1. Limitations of ion beam brightness under electron cooling – theory and experiments — V. Parkhomchuk (INP)
2. Plasma point of view on the electron cooling — V. Reva (INP)
3. Report from the Fermilab electron cooling group — S. Nagaitsev (Fermilab)

K. H^- Stripping, 1East
Conveners: I. Yamane (KEK), Y.Y. Lee (BNL)
Secretary: A. Xiao (Fermilab)

1. Motivation of this session — I. Yamane (KEK)
2. Review on laser assisted stripping — U. Gastaldi (INFN)
3. Feasibility of laser stripping via a broad Stark state for charge-exchange injection into a high intensity proton ring — I. Yamane (KEK)
4. A novel solution for H^- laser stripping — S. Danilov (ORNL)
5. PVLAS developments on Fabry-Perot resonators locked to CW lasers and suitable for laser assisted Lorentz stripping of H^- beams — G. Cantatore et al. (INFN)
6. Carbon stripping foil experience at PSR — M. Borden et al. (LANL)
7. Present status of development of carbon stripper foils at KEK — I. Sugai et al. (KEK)
8. The SNS strategy for H^- stripping — Y.Y. Lee (BNL)

L. Electron Cloud Effects, 1North
Conveners: R. Macek (LANL), H. Fukuma (KEK)
Secretary: M. Xiao (Fermilab)

1. Electron cloud effects (ECE) at e^+/e^- machines and electron cloud diagnostics — K. Harkay (ANL)
2. Observation of ECE at the KEKB — H. Fukuma (KEK)
3. Observation of ECE at the PSR — R. Macek et al. (LANL)
4. Theory and simulation of ECE — K. Ohmi (KEK)

5. Simulations of electron cloud generation in the PSR and SNS — M. Pivi (LBL)
6. Nonlinear delta-f simulation studies of e-p two-stream instabilities — H. Qin (PPPL)
7. ECE in high-intensity proton machines — J. Wei (BNL)
8. ECE in the KEK PS and the JHF project — T. Toyama et al. (KEK)
9. Condition for production of high intensity circulating proton beams beyond the sapce charge limit — V. Dudnikov (Fermilab)

Afternoon – Plenary Session, 1West

Chair: Y. Mori (KEK)
Secretary: M. Hu (Fermilab)

2:00 pm	Report from Fermilab physics study group	M. Velasco (Northwestern U)
2:30 pm	Report from BNL neutrino physics study group	W. Marciano (BNL)
3:00 pm	Progress in bright ion beams for medicine, industry and fusion	J. Kwan (LBL)
3:30 pm	Workshop photo session	
3:40 pm	Coffee break	
4:00 pm	Summary of Session A	A. Thiessen (LANL)
4:20 pm	Summary of Session B	R. Webber (Fermilab)
4:40 pm	Summary of Session I	S. Machida (KEK)
5:00 pm	Summary of Session D	C. Ankenbrandt (Fermilab)
5:20 pm	Summary of Session E	I. Hofmann (GSI)
5:40 pm	Summary of Session F	D. Pushka (Fermilab)
6:00 pm	Cash bar at User's Center	
7:00 pm	Dinner at Chez Leon	

Friday, April 12

Morning - Plenary Session, 1West

Chair: W. Chou (Fermilab)
Secretary: S. Seletskiy (Fermilab)

8:30 am	Summary of Session H	H. Klein (U. of Frankfurt)
8:50 am	Summary of Session C and G	R. Baartman (TRIUMF)
9:10 am	Summary of Session J	S. Nagaitsev (Fermilab)
9:30 am	Summary of Session K	I. Yamane (KEK)
9:50 am	Summary of Session L	R. Macek (LANL)
10:10 am	Coffee break	
10:40 am	Summary of Working Group I	L. Teng (ANL)
11:10 am	Summary of Working Group II	J-M. Lagniel (Saclay), R. Garoby (CERN)
11:40 am	Summary of Working Group III	C. Moore (Fermilab)
12:10 pm	Closing remarks	S. Holmes (Fermilab)
12:20 pm	Lunch break	

Afternoon

1:30 pm – 4:00 pm Fermilab Tour: MCR, A0, NLC, LHC magnets, Electron Cooling

Guides: K. Koba and Z. Qian

4:00 pm	Wine and cheese seminar, 1West From Snowmass to ICFA – Status of the Proton Driver Studies	W. Chou (Fermilab)
5:00 pm	Adjourn	

Andy Sessler (Lawrence Berkeley National Laboratory) is pondering the answer to a tough question.

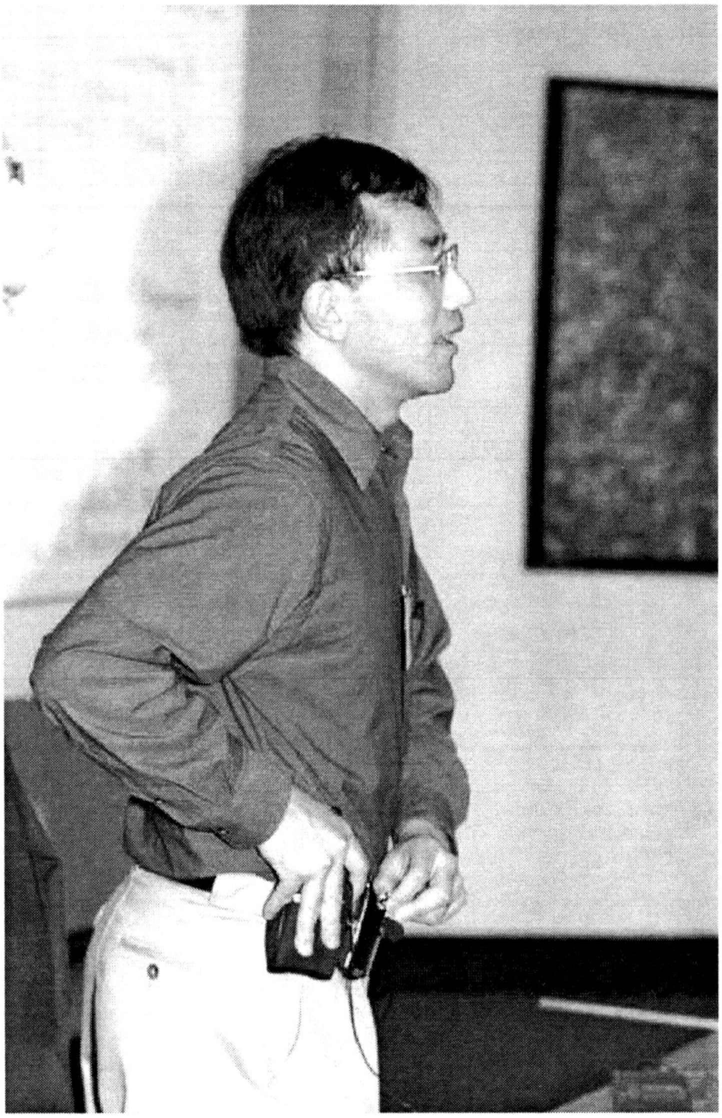

Jie Wei (Brookhaven National Laboratory) is turning on the microphone for a presentation on the status of the Spallation Neutron Source (SNS), the largest accelerator project under construction in the U.S.

Bill Foster (Fermi National Accelerator Laboratory) is promoting an 8-GeV Superconducting proton linac.

FERMILAB

Neutrino Physics
at the Intensity Frontier

A

B

C

D

E

F

G

H

Z